Medicinal Plants for Cosmetics, Health and Diseases

Medicinal Plants for Cosmetics, Health and Diseases

Edited by
Namrita Lall

CRC Press
Taylor & Francis Group
Boca Raton London

CRC Press is an imprint of the
Taylor & Francis Group, an **informa** business

First edition published 2023
by CRC Press
6000 Broken Sound Parkway NW, Suite 300, Boca Raton, FL 33487–2742

and by CRC Press
4 Park Square, Milton Park, Abingdon, Oxon, OX14 4RN

CRC Press is an imprint of Taylor & Francis Group, LLC

Library of Congress Cataloging-in-Publication Data

Names: Lall, Namrita, editor.
Title: Medicinal plants for cosmetics, health and diseases / edited by Namrita Lall.
Description: First edition. | Boca Raton : CRC Press, 2022. | Includes bibliographical references and index.
Summary: "This book reviews aspects of medicinal plants and their applications in treating various diseases. It provides comprehensive information on plant-derived alternative medical treatments for certain types of cancers, inflammatory conditions, tuberculosis, STDs, blood disorders, parasitic infections, and a variety of skin disorders"—Provided by publisher.
Identifiers: LCCN 2021060699 (print) | LCCN 2021060700 (ebook) | ISBN 9780367622114 (hardback) | ISBN 9780367622084 (paperback) | ISBN 9781003108375 (ebook)
Subjects: LCSH: Materia medica, Vegetable—Therapeutic use. | Medicinal plants. | Cosmetics. | Diseases—Alternative medicine.
Classification: LCC RS164 .M3544 2022 (print) | LCC RS164 (ebook) | DDC 615.3/21—dc23/eng/20211222
LC record available at https://lccn.loc.gov/2021060699
LC ebook record available at https://lccn.loc.gov/2021060700

ISBN: 978-0-367-62211-4 (hbk)
ISBN: 978-0-367-62208-4 (pbk)
ISBN: 978-1-003-10837-5 (ebk)

DOI: 10.1201/9781003108375

Typeset in Times LT Std
by Apex CoVantage, LLC

My humble effort I dedicate to my sweet and loving Father, Late Mr. Kamla Kant Sharan & Mother, Mrs Nira Sharan for always supporting me

Contents

Foreword

Long ago, traditional healers shared their knowledge on plants to ensure the next generation would uphold it. The knowledge of plants and the benefits they provided remained alive in their communities. Later, people started to document this knowledge to preserve it and ensure consistency within the community. Since this book is a comprehensive overview of medicinal plants found worldwide, it goes a considerable way to bring together what is known.

In this context, this book highlights the importance of traditional knowledge and how previously explored and unexplored plants may be useful to humankind. Though traditional knowledge plays a large role when selecting plants, this book explains how scientific analyses justify why plants are of importance. The information provided may lead to the discovery of new pharmacologically active compounds or a new natural products, it goes a considerable way to bring together what is known.

Concerning cosmetics, plant extracts and compounds which have been shown to be effective in the fight against aging, acne, eczema, wound healing and many other ailments have been reviewed and dealt with in detail. One unique aspect of the book is the broad range of diseases and conditions which have been documented. This includes prevalent diseases such as tuberculosis, cancer, fungal and parasitic ailments, the potential of plants to serve as an anti-coagulant, including neurodegenerative disease and the like in addition to areas which may affect the health wellness conditions, such as aging. The unique targets and treatment strategies of diseases have also been discussed. For example, if we consider cancer and plants, it is known that globally, cancer is the No. 1 killer. Thus, emphasis in this book is on a variety of cancers and the pharmacologically active plants and identified compounds that have been investigated within this field.

This book should inspire readers to examine the valuable use of potential plant-derived treatments for some of the most prevalent health conditions worldwide, as it highlights recent innovative advantages of phytomedicines. The book may further increase the demand for natural-based products over their commercialized synthetic counterparts. The history of the use of medicinal plants has been documented, which should highlight the exciting field of phyto-derived treatments. Molecular aspects of plants provide a clear understanding of how these compounds target the desired effects within the body.

Each chapter provides a clear perspective of each disease or condition and illustrates how plants can potentially benefit the pharmaceutical and cosmetic industries. In each chapter, the pathogenesis of the diseases is clearly and carefully presented in a way that ensures it will be easily understood by those new to the field. The book is written to lead the reader naturally through the subject, after giving an overview of particular ailments and relating those to the usage of medicinal plants and scientific investigations conducted to date. The author describes automated reasoning for furthering research on many candidate plants which may be developed further into fully fledged useful pharmaceutical and/or cosmeceutical products.

Though this book is informative about the diseases and conditions, it provides knowledge on different plants from all over the world that have shown potential for a particular ailment. This book provides insight into the plant world and how different industries and communities can use this knowledge to treat conditions and diseases that have been around for generations.

The target readers could be students, teachers and researchers engaged in universities, R&D institutions or the pharmaceutical or herbal industries, as well as health organizations. This hands-on guide is attractively illustrated, informative and easy to read, and so should prove an accessible and invaluable companion to professional and amateur naturalists. This book should also provide useful and meaningful information to phytochemists, combinatorial chemists, pharmacologists, scientists engaged in drug discovery and development, industrial research groups developing drugs from plants, departments of pharmaceutics, institutes for drug research, and pharmaceutical companies. Finally, this book will be deemed useful to manufacturers of herbal and various traditional medicines and cosmetic-based products.

The editor of the book, Prof. Namrita Lall, is a rising star in the discipline of plants with medicinal and cosmetic interests. She is passionate about evaluating medicinal plants and their potential as alternative pharmaceutical and cosmeceutical products. Over the years, she has received numerous recognitions, including "The Order of Mapungubwe," South Africa's highest honor from the Honorable South African President Jacob Zuma (April 2014) and Distinguished Young Women in Science Award by Naledi Pandor, Honorable Minister of the Science and Technology of South Africa (August 2011). She has recently published three books—*Medicinal Plants for Holistic Health and Well-Being*, *Underexplored Medicinal Plants from Sub-Saharan Africa* and *Aquatic Plants: Pharmaceutical and Cosmetic Applications*—which bear testimony to her enthusiasm for medicinal plants. This most recent book continues to display immense teamwork and collaboration and has helped compile something that can benefit global scientific knowledge.

Prof. Anastasia Karioti
Assistant Professor
School of Pharmacy
Aristotle University of Thessaloniki
Greece

Acknowledgments

First and foremost, praises and thanks to God for His showers of blessings throughout to complete this project successfully. You have given me the power to believe in my passion and pursue my dreams.

I am extending my heartfelt thanks to my committee, several colleagues, postgraduate students and collaborators for their continued support and encouragement. I would like to acknowledge the following contributors to this book, especially Ms. Marizé Nel and Mrs. Tenille Esmear, the editorial assistants and those who also served as project managers. I also wish to recognize the valuable help of the reviewers, namely Dr. Surjeet Verma, Dr. Vusani Mandiwana, Dr. Mohammad Isreb, Dr. Baby Chakrapani PS, Dr. Kumbukani K. Nyirenda, Prof. Asha Lakshmikuttyamma, Dr. Raghavendra Gowda, Prof. Wara, Sanusi Hassan, Dr. Velaphi C. Thipe, Dr. Analike Blom van Staden, Dr. Ashish Yadar, Dr. Ignatious Matimati, Dr. Ing. Zsófia Kutil, Dr. Ibukun Michael Famuyide, Dr. Rajendra Singh Bhakuni, Dr. Deborah M. Wendland and Dr. M. Narishma Reddy whose assistance was of high importance in accomplishing my end goal.

I would like to pay my regards to Prof. Anastasia Karioti for her kind and inspiring words in the Foreword. Lastly, but not least, I would like to thank all the scientists and knowledge-holders who have documented and published their wisdom and findings that form the basis of this work.

Editor

Professor Namrita Lall is truly passionate about evaluating the wonders of medicinal plants and not only concludes her findings on collating information about the ethnobotanical usage but takes it beyond, proving their efficacy and making all the efforts to obtain valuable pharmaceutical and cosmeceutical products from the medicinal plants. Prof. Lall is a distinguished scientist in medicinal plant science at the University of Pretoria in South Africa. She is internationally recognized as a leading scholar in the field of phytomedicine, which is reflected by her appointments as an adjunct professor at the School of Natural Resources at the University of Missouri, USA (one of the top 15 departments in the world), as a senior research fellow at the Bio-Tech R&D Institute, Jamaica and as an adjunct faculty position at JSS Academy of Higher Education and Research, India.

She has been placed in the Essential Science Indicators list of the top 1% of publication outputs (citations) in the discipline, pharmacology and toxicology. She has received international recognition for her research into the potential of medicinal plants for pharmaceutical and cosmeceutical purposes and has received numerous awards. These awards, which include "The Order of Mapungubwe," South Africa's highest honor from the Honorable South African President Jacob Zuma (April 2014) and Distinguished Young Women in Science Award by Naledi Pandor, Honorable Minister of the Science and Technology of South Africa (August 2011) bear testament to her immense contributions to the scientific community.

Dr. Lall has made significant contributions to the field of medicinal plant science. Several medicinal plants with valuable biological activities have been discovered, leading to many national and international patents. She has co-authored more than 170 research articles in peer-reviewed journals and 54 book chapters. In addition, she has edited three books: *Medicinal Plants for Holistic Health and Well-Being*, *Underexplored Medicinal Plants from Sub-Saharan Africa* and *Aquatic Plants: Pharmaceutical and Cosmetic Applications*. Prof. Lall has presented numerous keynotes and plenary talks at international conferences. She has demonstrated a commitment to community development by interacting positively with traditional health practitioners and engaging them in advancing traditional medicines toward conventional pharmaceutical products.

READERSHIP

Professors, academics; graduate students; researchers engaged in universities, R&D institutions, pharmaceutical or herbal industries; amateur botanists in the areas of medicinal plant science, pharmacognosy, phytochemistry and phytomedicine; pharmacologists; industrial research groups developing drugs from plants; manufacturers of herbal, complementary medicines and cosmetic products; individuals involved with herbal science and/or similar courses, as well as the scientists involved with drug discovery and engaged in natural product research in the pharmaceutical and cosmeceutical industries.

Contributors

Abdullahi A. AbdulRahaman
Department of Plant Biology
Faculty of Life Sciences
University of Ilorin
Kwara State, Nigeria

Mary Frances Vest
Department of Pharmacology, Toxicology and Neuroscience
School of Graduate Studies
Louisiana State University Health Sciences Centre
Shreveport, Louisiana

Mahwahwatse J. Bapela
Department of Plant and Soil Sciences
University of Pretoria
South Africa

Venugopal R. Bovilla
Centre of Excellence in Molecular Biology and Regenerative Medicine
Department of Biochemistry
JSS Medical College
JSS Academy of Higher Education & Research
Mysuru, Karnataka, India
And
Global Health Equity Scholars (GHES)—Post Doctoral Fellow (PDF)
Public Health Research Institute of India (PHRII)
Yadavagiri, Mysuru
Karnataka, India

Tafadzwa J. Chiome
Division of Nanoscience & Technology
Faculty of Life Sciences
JSS Academy of Higher Education & Research
Mysuru, Karnataka, India

Priya Darshani
National Institute of Pharmaceutical Education and Research Kolkata
Kolkata, India
And
CSIR-Indian Institute of Chemical Biology
Kolkata, India

Marco N. De Canha
Department of Plant and Soil Sciences
University of Pretoria
South Africa

Lilitha L. Denga
Department of Plant and Soil Sciences
University of Pretoria
South Africa

Marcela Dvorakova
The Czech Academy of Sciences
Institute of Experimental Botany
Prague, Czech Republic

Tenille Esmear
Department of Plant and Soil Sciences
University of Pretoria
South Africa

Jyoti Gaur
Medicinal Chemistry Department
Central Institute of Medicinal and Aromatic Plants
Lucknow, India

Rashmi Gaur
Natural Products Laboratory
International Joint Laboratory of Tea Chemistry and Health Effects
State Key Laboratory of Tea Plant Biology and Utilization
Anhui Agricultural University
Hefei, China

Lydia Gibango
Department of Plant and Soil Sciences
University of Pretoria
South Africa

Murthy V. Greeshma
Centre of Excellence in Molecular Biology and Regenerative Medicine
Department of Biochemistry
JSS Medical College
JSS Academy of Higher Education & Research
Mysuru, Karnataka, India

Lani Irvin
Agricultural Research Station
Fort Valley State University
Fort Valley, Georgia

Muhali O. Jimoh
Department of Horticultural Sciences
Cape Peninsula University of Technology
Bellville, South Africa

Nirmal Joshee
Agricultural Research Station
Fort Valley State University
Fort Valley, Georgia

Learnmore Kambizi
Department of Horticultural Sciences
Cape Peninsula University of Technology
Bellville, South Africa

Ramar Krishnamurthy
C. G. Bhakta Institute of Biotechnology
UkaTarsadia University
Mahuva Road, Bardoli, Surat (District)
Gujarat State, India

Quenton Kritzinger
Department of Plant and Soil Sciences
University of Pretoria
South Africa

Nikhilesh Kumar
Department of Burns and Plastic Surgery
AIIMS, Rishikesh
Uttarakhand, India

Mahadevaswamy G. Kuruburu
Center of Excellence in Molecular Biology and Regenerative Medicine
Department of Biochemistry
JSS Medical College
JSS Academy of Higher Education & Research
Mysuru, Karnataka, India

Namrita Lall
Department of Plant and Soil Sciences
University of Pretoria
South Africa
And
School of Natural Resources
University of Missouri
Columbia, Missouri
And
College of Pharmacy
JSS Academy of Higher Education and Research, India
And
Senior Research Fellow
Bio-Tech R&D Institute
University of West Indies
Kingston, Jamaica

Isa A. Lambrechts
Department of Plant and Soil Sciences
University of Pretoria
South Africa

Premysl Landa
The Czech Academy of Sciences
Institute of Experimental Botany
Prague, Czech Republic

Lenka Langhansova
The Czech Academy of Sciences
Institute of Experimental Botany
Prague, Czech Republic

Samantha R. Loggenberg
Department of Plant and Soil Sciences
University of Pretoria
South Africa

Mcebisi J. Mabuza
Department of Plant and Soil Sciences
University of Pretoria
South Africa

SubbaRao V. Madhunapantula
Centre of Excellence in Molecular Biology and Regenerative Medicine
Department of Biochemistry
JSS Medical College
JSS Academy of Higher Education & Research, Mysuru, Karnataka, India
And
Leader, Special Interest Group in Cancer Biology and Cancer Stem Cells (SIG-CBCSC)
JSS Academy of Higher Education & Research, India

Jacqueline Maphutha
Department of Plant and Soil Sciences
University of Pretoria
South Africa

Kevin S. Murnane
Department of Pharmacology
Toxicology & Neuroscience
School of Graduate Studies
Louisiana State University Health Sciences Centre
Shreveport, Louisiana
And
Department of Psychiatry
School of Medicine
Louisiana State University Health Sciences Centre
Shreveport, Louisiana
And
Louisiana Addiction Research Centre
Louisiana State University Health Sciences Centre
Shreveport, Louisiana

Sarah Murnane
Department of Physical Therapy
Louisiana State University Health
Shreveport, Louisiana

Marizé Nel
Department of Plant and Soil Sciences
University of Pretoria
South Africa

Prahlad Parajuli
Department of Oncology
Wayne State University & Karmanos Cancer Institute
Detroit, Michigan

Nikita Patel
C. G. Bhakta Institute of Biotechnology
UkaTarsadia University
Mahuva Road, Bardoli, Surat (District)
Gujarat State, India

Swetal Patel
C. G. Bhakta Institute of Biotechnology
UkaTarsadia University
Mahuva Road, Bardoli, Surat (District)
Gujarat State, India

Bianca D. Payne
Department of Plant and Soil Sciences
University of Pretoria
South Africa

V.S. Pragadheesh
CSIR-Central Institute of Medicinal and Aromatic Plants, Research Centre
Bengaluru, India

Tefo K. Pule
Department of Plant and Soil Sciences
University of Pretoria
South Africa

Sujatha Puttalingaiah
Centre of Excellence in Molecular Biology and Regenerative Medicine
Department of Biochemistry
JSS Medical College
JSS Academy of Higher Education & Research
Mysuru, Karnataka, India

Precious B. Ramontja
Department of Plant and Soil Sciences
University of Pretoria
South Africa

Anna-Mari Reid
Department of Plant and Soil Sciences
University of Pretoria
South Africa

Jonathan L. Seaman
Department of Plant and Soil Sciences
University of Pretoria
South Africa

Samantha H. Sherman
Agricultural Research Station
Fort Valley State University
Fort Valley, Georgia

Asha Srinivasan
Centre of Excellence in Molecular Biology & Regenerative Medicine
JSS Academy of Higher Education & Research
Mysuru, Karnataka, India

Md TanjimAlam
CSIR-Indian Institute of Chemical Biology
Kolkata, India

Prem P. Tripathi
CSIR-Indian Institute of Chemical Biology
Kolkata, India

Talita Turvey
Department of Plant and Soil Sciences
University of Pretoria
South Africa

Harish C. Upadhyay
Laboratory of Chemistry
Department of Applied Sciences
Rajkiya Engineering College
Churk, Sonbhadra, India

Surjeet Verma
Department of Plant and Soil Sciences
University of Pretoria
South Africa

Catherine Wilkinson
Department of Plant and Soil Sciences
University of Pretoria
Department of Psychiatry
South Africa

Abbreviations

2-AG	2-arachidonoylglycerol
3MG	Erythrocyte 3-O-metylglucose
3-NP	3-Nitropropionic acid
3TC	Lamivudine
5-ALA	5-aminolevulinic acid
5-HT$_{2A}$	Serotonin 2A receptor
5-LOX	5-Lipoxygenase
6-OHDA	6-hydroxydopamine
A2058	Human Caucasian metastatic melanoma cell line
A375	Malignant human melanoma cell line
A549 cell line	Adenocarcinomic human alveolar basal epithelial cells
ABPA	Allergic bronchopulmonary aspergillosis
ABTS	2, 2′-Azino-bis (3-ethylbenzothiazoline-6-sulfonic acid) diammonium salt
ACh	Acetylcholine
AChE	Acetylcholinesterase
AChEI	Acetylcholinesterase inhibitors
ACTs	Artemisinin-based combination therapies
Acyl-ACP	Acylated acyl carrier proteins
AD	Alzheimer's disease
ADA	Azodicarbonamide
ADHD	Attention deficit hyperactivity disorder
ADP	Adenosine diphosphate
Ag85B	Antigen 85B
AGEs	Advanced glycation end products
AgNP	Silver nanoparticle
AGS	Human Caucasian gastric adenocarcinoma
AHL	Acylated homoserine lactones
AI-2	Autoinducer 2
AIDS	Acquired immunodeficiency syndrome
AKBA	Acetyl-11-keto-beta-boswellic acid
AKT	Protein kinase B
AL	Artemether-lumefantrine
ALK	Anaplastic lymphoma kinase
ALM	Acral lentiginous melanoma
ALP	Alkaline phosphatase
ALS	Amyotrophic lateral sclerosis
AM	Allyl mercaptan
AM	Amelanotic melanoma
AMB cells	Atypical memory B cells
AmB	Amphotericin B
AMPK	5' adenosine monophosphate-activated protein kinase
AP-1	Activator protein 1
APC	Activated Protein C
APC	Antigen-presenting cells
APOE	Apolipoprotein
APP	Amyloid precursor proteins
APRIL	A proliferation-inducing related ligand

AROS	Active regulator of SIRT1
ARVs	Antiretroviral therapy
AS	Alpha-synuclein
ASAQ	Artesunate-amodiaquine
ASDs	Anti-seizure drugs
aSMase	Acidic sphingomyelinase
ASMQ	Artesunate-mefloquine
AS-SP	Artesunate-sulfadoxine-pyrimethamine
AT	Antithrombin
ATM	Ataxia telangiectasia mutated protein
ATP	Adenosine triphosphate
ATV	Atazanavir (atazanavir sulfate)
AUD	Alcohol abuse disorder
AuNP	Gold nanoparticle
AZT/ZDV	Aidovudine/azidothymidine
Aβ	Beta-amyloid
B16-F10	Murine melanoma cell line from C57BL/6J mice
Bap	Biofilm associated protein
Bax	Bcl-2-associated X protein
B-BA	Beta-boswellic acid
BBB	Blood-brain barrier
Bcap-37	Human breast cancer cell line
BCC	Basal cell carcinoma
BCC-1/KMC	Basal cell carcinoma cell line
BCG	Bacilli Calmette-Guerin
BCH	Bornyl cis-4-hydroxycinnamate
Bcl-2	B-cell lymphoma 2
Bcl-xL	B-cell lymphoma extra large
BCNU	1,3-bis (2-chloroethyl)-1-nitrosourea
BCP	Beta-caryophyllene
BCPO	Beta-caryophyllene oxide
BEAS-2B	Human normal bronchial epithelial cell line
B-GlcCer'ase	Beta glucocerebrosidase
Bhp	Bap-homologous protein
BRAF V600E	Serine/threonine-protein kinase B-Raf
BRAF	v-Raf murine sarcoma viral oncogene homolog B
BuOH	Butanol
C/EBPs	CCAAT element-binding proteins
CA	Carnosic acid
CAG	Cytosine, adenine and guanine
CAGR	Compound annual growth rate
CAM	Complementary and alternative medicine
cAMP	Cyclic adenosine monophosphate
CAT	Catalase
CB	Cell-based assay
CBP	CREB binding proteins
CC_{50}	Concentration of the sample that reduces the viability of cells by 50%
CCD-112CoN	Normal colon cells
CCD-966SK	Normal human skin fibroblasts
CCR	C-C chemokine receptor
CD	Circular dichroism

CD4	Cluster of differentiation 4
CD80	Cluster of differentiation 80
CD86	Cluster of differentiation 86
CDC	Centers for Disease Control and Prevention (United States)
CDK	Cyclin dependent kinase
CDT1	Chromatin licensing and DNA replication factor 1
CEM-SS	Human lymphoblastoid
CF	Cell-free assay
cGMP	Cyclic guanosine monophosphate
c-Myc	Cellular myelocytomatosis oncogene
CNS	Central nervous system
COBI	Cobicistat
COMT	Catechol-O-methyltransferase
CoREST	Co-repressor of RE1 silencing transcription
COSY	Correlation Spectroscopy
COVID-19	Coronavirus disease 2019
COX-1	Cyclooxygenase-1/Prostaglandin H synthase-1
COX-2	Cyclooxygenase-2/Prostaglandin H synthase-2
CPA	Chronic pulmonary aspergillosis
CPD	Cyclobutene pyrimidine dimers
CPE	Cytopathic effect
cPLA2	Cytosolic phospholipase A2
cPLA2α	Cytosolic phospholipase A2 alpha
CPP	Conditioned place preference
CREB	cAMP response element-binding protein
CRH	Corticotropin-releasing hormone
CS	Carnosol
CSF	Cytokine, granulocyte-macrophage colony-stimulating factor
CTLA-4	Cytotoxic T-lymphocyte associated protein 4
CTP's	Biological or cellular tissue products
CYP P450	Cytochrome P450
CYP	Cytochrome
CysD	Sulfate adenylyltransferase subunit 2
DA	Dopamine
DADS	Diallyl disulfide
DAT	Dopamine transporter
DATS	Diallyl trisulfide
DC	Dendritic cells
DCM	Dichloromethane
DC-SIGN	102-fold downregulation of CD209
DDA	Bis (4′-chlorophenyl) methane
DDD	1,1-dichloro-2,2-bis (p-chlorophenyl) ethane
DDE	2,2-bis (p-chlorophenyl)-1,1-dichlorethylene
DDM	Bis (4′-chlorophenyl) acetate
DDMU	1-chloro-2,2-bis (4′-chlorophenyl) ethylene
DDT	1,1,1-Trichloro-2,2-bis (p-chlorophenyl) ethane
DDT	Dichlorodiphenyltrichloroethane
DEPT	Distortionless Enhancement by Polarization Transfer
DFU	Diabetic foot ulcers
DHA-PQ	Dihydroartemisinin-piperaquine
DHF	Dihydrofolate

DHFR	Dihydrofolate reductase
DHPS	Dihydropteroate synthase
DiHETrEs	Dihydroxyeicosatrienoic acids
DL	Murine ascites Dalton's lymphoma
DLA	Dalton's lymphoma ascites
DM	Desmoplastic melanoma
DM	Diabetes mellitus
DMS114	Human lung carcinoma
DMT	N, N-Dimethyl tryptamine
DNA	Deoxyribonucleic acid
DOB	2,5-dimethoxy-4-bromoamphetamine
DOI	dimethyoxy-4-iodoamphetamine
DOPA	2-amino-3-(3',4'-dihydroxyphenyl) propionic acid
DOR	Doravirine
DPD	4,5-dihydroxy-2,3-pentanedione
DPPH	2, 2-diphenyl-1-picrylhydrazyl
DRV	Darunavir
DSLR	Digital single-lens reflex
DTG	Dolutegravir
DU-145	Human prostate cancer cell line
DVT	Deep vein thrombosis
EAC	Ehrlich's ascites carcinoma
EC	Euplotin C
ECM	Extracellular matrix
ECS	Endocannabinoid system
ED	Extensive-stage disease
ED_{50}	Concentration whereby 50% of the sample is effective
EETs	Epoxyeicosatrienoic acids
EFV	Efavirenz
EGCG	Epigallocatechin gallate
EGFR	Epidermal growth factor receptor
EIVs	Ether injection vesicles
EMT	Epithelial to mesenchymal transition
EOs	Essential oils
EPR	Enhanced permeation and retention
EPSFs	8-*C*-*N*-ethyl-2-pyrrolidinone-substituted flavan-3-ols
ER	Endoplasmic reticulum
ERK	Extracellular signal-regulated kinase
ERP	Enhanced permeation and retention
ESAT6	Early secretory antigenic target—6
Esp	*Enterococcus faecalis* surface proteins
EtOAc	Ethyl acetate
ETR	Etravirine
FADD	Fas-associated protein with death domain
FAME	Fatty acid modifying enzyme
Fas/APO1	Fas receptor/apoptosis antigen 1
FDA	Food and Drug Administration (United States)
Fe_2O_3	Iron (III) oxide
Fe_3O_4	Iron (II, III) oxide
FGF	Fibroblast growth factor
FII	Prothrombin

FIIa	Thrombin
FIX	Coagulation factor IX
FIXa	Activated coagulation factor IX
FLAP	5-lipoxygenase activating protein
FPV	Fosamprenavir
FPVs	French press vesicles
FRAP	Ferric reducing antioxidant power
FTC	Emtricitabine
FTIR	Fourier-transform infrared spectroscopy
FTR	Fostemsavir
FUS/TLS	Translocation liposarcoma/fused in sarcoma
FV	Coagulation factor V
FVa	Activated coagulation factor V
FVIIa	Activated coagulation factor VII
FVIII	Coagulation factor VIII
FVIIIa	Activated coagulation factor VIII
FX	Coagulation factor X
FXa	Activated coagulation factor X
FXIa	Activated coagulation factor XI
G0	G zero phase or resting phase
G1	Gap 1 phase
GABA	Gamma-aminobutyric acid
GABAergic	Action of GABA to function as a transmitter
GBD	Global Burden of Diseases, Injuries and Risk Factors
GBM	Glioblastoma multiforme
GC	Gas chromatography
GC-FID	Gas chromatography with flame ionization detection
GCLC	Gutamyl-cysteine ligase catalytic subunit
GC-MS	Gas chromatography with mass spectroscopy
GCN-5	General control nonderepressible-5
GDM	Gestational diabetes
GI_{50}	Whereby 50% of the cell's growth is inhibited
GLP-1	Glucagon-like peptide-1
GLUT 1	Glucose transporter 1
GLUT 2	Glucose transporter 2
GLUT 3	Glucose transporter 3
GLUT 4	Glucose transporter 4
GLUT	Glucose transporter
GM1	Monosialotetrahexosylganglioside
GNAT	GCN-5 related NAT
GP	Glutathione peroxidase
GP60	Glycoprotein
GPCRs	G protein-coupled receptors
GPDH	Glycerol-3-phosphate dehydrogenase
GPIIb/IIIa	Glycoprotein IIb/IIIa
GPX	Glutathione peroxidase
GR	Glutathione reductase
GSH	Glutathione
GSK3β	Glycogen synthase kinase 3 beta
GST	Glutathione S-transferase
H_index	Hirsch index

H_2O_2	Hydrogen peroxide
H441	Human distal lung epithelial cells
H-460	Large cell lung cancer
H520	Human lung squamous carcinoma
H69	SV40-transformed human biliary epithelial cells
HA	Hyaluronic acid
HaCaT	Immortal keratinocyte cells from adult human skin
HAT	Histone acetyltransferase
hBD2	Human beta-defensin 2
HCMM	Human cutaneous malignant melanoma
HCT116	Human colon cancer cell line
HD	Huntington's disease
HDACs	Histone deacetylases
HDACis	HDAC inhibitors
HDFs	Human dermal fibroblast cells
HDL	High-density lipoprotein
HeLa	Human cervix epithelioid carcinoma
HepG2	Human hepatocellular carcinoma
HER 2	Human epidermal growth factor 2
HETEs	Hydroxyeicosatetraenoic/monohydroxyeicosatetraenoic acids
HIV	Human immunodeficiency virus
HL-60	Human promyelocytic leukemia cells
HMBC	Heteronuclear multiple bond correlation
HNSCC	Head and neck squamous cell carcinoma
HO-1	Heme oxygenase 1
HONE-1	Human nasopharyngeal carcinoma
HPETEs	Hydroperoxyeicosatetraenoic acids
HPLC	High-performance liquid chromatography
HPTLC	High-performance thin-layer chromatography
HPV	Human papilloma virus
Hs68 cells	Human skin fibroblast cells
HSQC	Heteronuclear single quantum coherence
HS-SPME	Headspace solid-phase micro-extraction
HSV-1/2	Human herpes simplex virus type 1/2
HT-1080	Human fibrosarcoma cells
HT1299	Human non-small lung cancer cells
HT-29	Human colon adenocarcinoma cell line
HUVECs	Human umbilical vein endothelial cells
IA	Isoangustone A
IAP	Inhibitors of apoptosis proteins
IBA	Ibalizumb-uiyk
IBD	Irritable bowel disorder
IBS	Irritable bowel syndrome
IC_{50}	Whereby 50% of the targeted activity is inhibited
ICSN	Institut de Chimie des Substances Naturelles
IDF	International Diabetes Federation
IFN-α	Interferon-alpha
IFN-γ	Interferon gamma
IgE	Immunoglobulin E
IGF-1	Insulin-like growth factor 1
IgG	Immunoglobulin G

IIET	Institute of immunology and experimental therapy
IKS	Indigenous knowledge system
IL	Interleukin
iNOS	Inducible nitric oxide synthase
INR	International normalization ratio
IPA	Invasive pulmonary aspergillosis
IR	Infrared radiation
ISM	Indian traditional system of medicine
IX	Activated coagulation factor IX
Iκβ	Inhibitors of NF-κβ
JAK	Janus kinase
JNK	c-Jun N-terminal kinases
JWH	Jing Wan Hong
K562	Human erythroleukemic cell line
KATs	Lysine acetyltransferase
KBA	11-keto-beta-boswellic acid
KMSB	α-keto-γ-methylselenobutyrate
LAKs	Lymphokine activated killer cells
LC_{50}	Lethal concentration whereby 50% of the dosed population dies
LC_{90}	Lethal concentration whereby 90% of the dosed population dies
LCMS	Liquid chromatography-mass spectrometry
LD	Limited-stage disease
LDH	Lactate dehydrogenase
LDL	Low-density lipoprotein
L-DOPA	Levodopa
LipoLA	Liposomal lauric acid
LLC	Lewis lung carcinoma
LMM	Lentigo maligna melanoma
LMWH	Low molecular weight heparin
LNCaP	Androgen-sensitive human prostate adenocarcinoma cell
LOPS	Loss of protective sensation
LOX	Lipoxygenase
LPS	Lipopolysaccharide
LSD	Lysergic acid diethylamide
LT	Leukotrienes
LTA_4	Leukotriene A4
LTB_4	Leukotriene B4
LTC_4	Leukotriene C4
LTD_4	Leukotriene D4
LTE_4	Leukotriene E4
LUV	Large unilamellar vesicles
LX	Lipoxins
LXFL 529	Human non-small cell lung tumor cell line
MAO	Monoamine oxidase
MAPKs	Mitogen associated protein kinases
MASD	Moisture-associated dermatitis
MBC	Minimum bactericidal concentration
MBD2/3	Methyl-CpG binding domain
McA-RH7777	Rat hepatocellular carcinoma
MCF-7 TAX30	Human breast adenocarcinoma cell lines that are resistant to docetaxel
MCF-7	Human breast adenocarcinoma cell line

MDA	Malanodialdehyde
MDA-MB-231	Human mammary adenocarcinoma
MDBK	Madin Darby bovine kidney
MDR	Multidrug resistance
MDR-TB	Multidrug resistant TB
MeanTCperArt	Mean total citation per article
MeanTCperYear	Mean total citation per year
MEDS	Melanoma diagnosis system
MEF-2	Myocyte enhancer factor 2
MEK 1	Mitogen-activated protein kinase 1
MEK 2	Mitogen-activated protein kinase 2
MeO	Methoxy
MeOH	Methanol
MET	Tyrosine-protein kinase
MFC	Minimum fungal concentration
MGC-803	Human gastric cancer cell line
MHC-I/II	Major histocompatibility complex class I/II
mHTT	mutant Huntington protein
MIB	Minimum bactericidal concentration
MIC	Minimum inhibitory concentration
Mitf	Microphthalmia transcription factor
MLV	Multilamellar vesicles
MMPs	Matrix metalloproteinase
mPGES-1	Microsomal prostaglandin E2 synthase-1
mPGES-2	Microsomal prostaglandin E2 synthase-2
MPL	Monophosphoryl lipid A
MPLA	3-O-desacyl-4'-monophosphoryl lipid A
MPP^+	1-methyl-4-phenylpyridinium ion^-
MPTP	1-methyl-4-phenyl-1,2,3,6-tetrahydropyridine
MRSA	Methicillin-resistant *Staphylococcus aureus*
MSC	Se-methylselenocysteine
MSP	β-methylselenopyruvate
MTA2	Metastasis associated protein 2
mTOR	Mechanistic target of rapamycin
MTT assay	3-(4,5-Dimethylthiazol-2-yl)-2,5-diphenyltetrazolium bromide assay
MVC	Maraviroc
MYPT1	Myosin phosphatase targeting subunit 1
NAB	Nanoparticle Albumin bound technology
NAcc	Nucleus accumbens
nAChR	Nicotinic acetylcholine receptor
NAD^+	Nicotinamide adenine dinucleotide
NADH	Nicotinamide adenine dinucleotide hydrogen
NADPH	Nicotinamide adenine dinucleotide phosphate
NAFLD	Non-alcoholic fatty liver disease
NAMPT	Nicotinamide phosphoribosyltransferase
NAT	N-acetyltransferase
NCEs	New chemical entities
NCI	National Cancer Institute
NCI-H23	Human non-small cell lung carcinoma
N-CoR	Nuclear hormone receptor co-receptor
NDD	Neurodegenerative disorders

NDGA	Nordihdroguaiaretic acid
NFκB	Nuclear factor-kappa beta
NH3T3	Mouse fibroblasts
NIDA	National Institute on Drug Abuse
NIDDM	Non-insulin-dependent diabetes mellitus
NLRs	Nod-like receptors
NM	Nodular melanoma
NMDA	N-methyl-D-aspartate
NMR	Nuclear magnetic resonance spectroscopy
NO	Nitric oxide
NOESY	Nuclear Overhauser Effect Spectroscopy.
NOS	Nitric oxide synthase
NP	Nanoparticle
NP	Number of publications
NPK	Nitrogen, phosphorus and potassium
NQO1	NAD(P)H dehydrogenase quinone 1
NRAS	Neuroblastoma rat sarcoma viral oncogene homolog
NSAIDs	Non-steroidal anti-inflammatory drugs
NSCLC	Non-small cell lung cancer
NuRD	Nucleosome remodeling and deacetylase
NVP	Nevirapine
ORC	Powdered or oxidized regenerated cellulose
OTC	Over-the-counter
OXPHOS	Oxidative phosphorylation
p53	Tumour protein 53
PAD	Peripheral arterial disease
PAHs	Polycyclic aromatic hydrocarbons
PAR-2	Proteinase activated receptor-2
PARP	Poly (ADP-ribose) polymerase
PC12	Rat adrenal phaeochromocytoma
PC-3	Human prostate cancer cell line
PCA	Primary cutaneous aspergillosis
PCAF	P300/CBP-associating factor
PCL	Polycaprolactone
PCNA	Proliferating cell nuclear antigen
PD	Parkinson's disease
PD	Pharmacodynamic
PD-1	Programmed death 1
PDGF	Platelet-derived growth factor
PDGF-AA	Platelet-derived growth factor-AA
PDGF-BB	Platelet-derived growth factor-BB
PDL-1	Programmed death-ligand 1
PDL-2	Programmed death-ligand 2
PE	Pulmonary embolism
PEDF	Pigment epithelium-derived factor
PEG	Polyethylene glycol
PEGylation	Chemical modification to active PEG
PFC	Prefrontal cortex
Pfs	Methylthioadenosine (MTA)/SAH nucleosidase
PG	Prostaglandins
PGD_2	Prostaglandin D_2

PGE_2	Prostaglandin E_2
PGE_2	Prostaglandin E_2
$PGF_{2\alpha}$	Prostaglandin $F_2\alpha$
PGH_2	Prostaglandin H_2
PGI_2	Prostaglandin I_2
PI	Protease inhibitors
PI3K	Phosphoinositide-3-kinase
PI3K-Akt-mTOR	Phosphatidylinositol-3-kinase-AKT-mammalian target of rapamycin
PIA	Polysaccharide intracellular adhesion molecule
PIK3CA	Phosphatidylinositol-4,5-Bisphosphate 3-kinase catalytic subunit alpha
PK	Pharmacokinetic
PKA	Protein kinase cAMP-dependent
PLA-801	Human pulmonary giant cell carcinoma
PLC	Paclitaxel-loaded poly
pLDH	Plasmodium lactate dehydrogenase
PLGA	Poly(lactic-co-glycolic acid)
polyQ	Polyglutamine
PP1	Phosphate protein 1
PP1b	Protein phosphatase 1 beta
PP2A	Protein phosphatase 2
PP4	Serine/threonine-protein phosphatase 4
PPAR	Peroxisome proliferator-activated receptor
PS	Phosphatidylserine
PSB	Phosphate solubilizing bacteria
PSM	Phenol-soluble modulins
PTEN	Phosphatase tensin homologue
PTMs	Post-translation modifications
PVA	Polyvinyl acetate
PVC	Polyvinyl chloride
PXD101	Belinostat
PY_Start	Publication year
PYY	Peptide hormone
QS	Quorum sensing
QSM	Quorum sensing signaling molecules
R&D	Research and development
RAF	Rapidly accelerated fibrosarcoma
RAL	Raltegravir
RAS	Rat sarcoma
Rb	Retinoblastoma protein
RbAp	Retinoblastoma binding protein
RBC	Red blood cells
RCC	Renal cell carcinoma
RCTs	Randomized clinical trials
RES	Reticulo-endothelial system
RET	Transfection proto-oncogene
REVs	Reverse phase evaporation vesicles
RIP1	Receptor interacting serine/threonine kinase 1
RNS	Reactive nitrogen species
ROS	Reactive oxygen species
ROS1	Proto-oncogene tyrosine-protein kinase
RPV	Rilpivirine

rRNA	Ribosomal ribonucleic acid
RSA	Radical scavenging activity
RSV	Human respiratory syncytial virus
RT	Reverse transcriptase
RTI	Reverse transcriptase inhibitors
RTV	Ritonavir
SA	Self-administration
SADC	South African development community
SAH	S-adenosylhomocysteine
SAHA	Suberanilohydroxaxmic acid
SAM	S-adenosylmethionine
SAMHS	Substance abuse and mental health service administration
SAP18/30	Sin3-associated polypeptides
SAR	Structure–activity relationship
SBT-A	Scutebarbatine-A
SCA	Secondary cutaneous aspergillosis
SCC	Squamous cell carcinoma
SCC-25 KB	Squamous cell carcinoma
SCFAs	Short-chain fatty acids
SCLCs	Small cell lung cancers
SDE	Simultaneous distillation and extraction
SDH	Succinate dehydrogenase
sEH	Soluble epoxide hydrolase
SFN	Sulforaphane
SFN-Cys	SFN-cysteine
SFN-NAC	SFN-N-acetylcysteine
SGOT	Serum glutamate oxalate transaminase
SGPT	Serum glutamate pyruvate transaminase
SH-SY5Y	Thrice subclones derived from the neuroblastoma cell line
SI	Selectivity index
Sin3	Switch intensive 3
SiO_2	Silicon dioxide
SIR2	Silent information regulator 2
SIRTs	Sirtuins
SK-MEL-2/28	Human melanoma cell line
SKOV-3	Ovarian cancer cell line
SLN	Solid lipid nanoparticle
SM	Selenomethionine
SMC3	Structural maintenance of chromosomes protein 3
SMRT	Silencing mediator of retinoid and thyroid hormone receptors
SocL	*Scutellaria ocmulgee*
SOD	Superoxide dismutase
sPLA2	Phospholipase A2
SPRR3	Small proline-rich protein 3
SQV	Saquinavir
SRH	S-ribosylhomocysteine
SSF	Solid-state fermentation
SSM	Superficial spreading melanoma
SSRI	Selective serotonin reuptake inhibitors
ssRNA	Single-stranded RNA
STAT-3	Signal transducer and activator of transcription 3

SUD	Substance abuse disorder
SUV	Small unilamellar vesicles
T1DM	Type 1 diabetes mellitus
T-20	Enfuvirtide
T2DM	Type 2 diabetes mellitus
TACE	TNF-α converting enzyme
TAM	Tumor-associated macrophage
TAR	Trans-activation response element
TAT	Trans-activator of transcription
TB	Tuberculosis
TC	Total citations
TCM	Traditional Chinese medicine
TCN	Tricaproin
TCT	Tissue culture technology
TDB-43	TAR DNA-binding protein 43
TDF/DF	Tenofovir
TF	Tissue factor
TFC	Total flavonoid content
TF-FVIIa	Tissue factor and activated coagulation factor VII complex
TFPI	Tissue factor pathway inhibitor
TGF-β	Transforming growth factor-beta
Th	T-helper lymphocytes
THC	Tetrahydrocannabinol
THF	Tetrahydrofolic acid
TI	Therapeutic index whereby IC_{50} of a sample is divided by its EC_{50}
TICAM	Traditional Iranian and Central Asian medicine
TIMP	Tissue inhibitor of MMP
TIP60	TAT-interacting protein 60
TLC	Thin layer chromatography
TLR	Toll-like receptor
TM	Thrombomodulin
TNFR1/2	TNF-α receptor 1/2
TNF-α	Tumor necrosis factor-alpha
TNM	Breslow's depth; tumor, node and metastasis
TOSC	Total oxidant scavenging capacity
TPA	Tissue plasminogen activator
TPC	Total phenolic content
TPV	Tipranavir
TRADD	TNF-α receptor-associated death domain
TRAF2	TNFR associated factor 2
TRAIL	Tumor necrosis factor-related apoptosis-inducing ligand
TRAP1	Tumor necrosis factor receptor-associated protein
Treg	Regulatory T cells
TSA	Trichostatin A
TTO	Tea tree oil
TX	Thromboxane
TXA_2	Thromboxane X_2
U87-MG/U251	Human malignant glioma
UCSF	University of California, San Francisco
UFH	Unfractionated heparin
UNICEF	United Nations International Children's Emergency Fund

UPA	Urokinase plasminogen activator
UV	Ultraviolet
UVA	Ultraviolet radiation A
UVB	Ultraviolet radiation B
UVC	Ultraviolet radiation C
UVI	Ultraviolet radiation index
Va-Xa	Prothrombinase complex
VEGF	Vascular endothelial growth factor
VGP	Vertical growth phase
VHL	von Hippel-Lindau
VIIIa-IXa	Tenase complex with VIIIa and IXa
VLDL	Very low-density lipoprotein
VPA	Valproate/valproic acid
VTA	Ventral tegmental area
VTE	Venous thromboembolism
WBC	White blood cells
WHO	World Health Organization
WRL-68	Human liver cells
X	Activated coagulation factor X
XIAP	X-chromosome linked IAP
XO	Xanthine oxidase
ZOI	Zone of inhibition
ZPI	Protein Z dependant protease inhibitor
Δ9-THC	Deta-9 tetrahydrocannabinol

Introduction

The age-old tradition of using plants to treat a myriad of human conditions is without exception our greatest natural resource. However, over the years, synthetic compounds have been developed to replace or substitute plant extracts or plant-derived compounds with conventional drugs that are used commonly for various ailments. This has resulted in undermining the advantages of the medicinal properties that plants have to offer over the years and traditional knowledge to some extent, which is alarming. Drug resistance has become a real issue in addition to people developing allergies toward numerous synthetic compounds. This has resulted in demand or an opportunity of going 'back to our roots' and thus increasing the need for medicinal plant-based treatments.

In this book, numerous diseases and conditions are discussed, providing the readers with a clearer understanding of each ailment and also plant-based products being used for combating such problems. This will allow the reader to gain a different perspective and outlook on misunderstood and underestimated diseases and conditions. Pharmacologically, the results obtained from scientific investigations into plant extracts and compounds have been provided against cancer, tuberculosis, blood disorders, parasitic infections, skin disorders and wound healing, just to name a few. The use of plants for cosmeceutical purposes has also been demonstrated for efficacy in the fight against eczema, acne and the effects of aging.

Since plants are rarely considered as effective treatments, many medicinal plants do not undergo scientific analyses. This book will encourage scientists to consider the effectiveness of plants once again by promoting plant-based analyses and drug discoveries, either in conjunction with conventional drugs or as a novel drug. The aim is to provide readers with a clearer understanding of various diseases and conditions and to change their perspective on the use of plants. Overall, this book should encourage scientists across the world to analyze plants for their potential in treating other diseases, as well those that have not been mentioned, and to preserve the traditional knowledge of these medicinal plants worldwide.

Prof. Namrita Lall
Editor

Mrs. Tenille Esmear and Ms. Marizé Nel
Editorial Assistant and Project Manager

Disclaimer

This book provides a summary of diseases and ailments that occur worldwide and includes medicinal plants that have shown some potential for such ailments. The information presented in its current form has been extracted from scientific journals, book chapters and online resources. The chapters have been peer-reviewed by specialists in the fields of medicinal plant sciences, phytomedicine and ethnopharmacology. This book does not make claims concerning the pharmacology and possible toxicity of the plants, and care should be taken when considering the plants' application. Many of you have already experienced the positive effects of herbal medicines and or plant-derived compounds on your health. We hope that through the content of this book, you gain tools to make an even greater role in your well-being. The information provided in the book is not intended to be a substitute for medical treatment. If you have any specific questions about any medicinal matter, you should consult your doctor or another professional healthcare provider. The authors take no responsibility for any wrongful or misuse of any plant mentioned in this book.

1 Introduction

Talita Turvey, Samantha R. Loggenberg and Namrita Lall

CONTENTS

1.1 AN INTRODUCTION TO MEDICINAL PLANTS

Medicinal plants have been utilized since ancient times across the world and in various cultures. The use of plants in medicinal practices was instinctive at first as observed in animals, with some of the earliest documented evidence of the use of medicinal plants found in the form of the Egyptian Ebers Papyrus (1500 BC), found in Luxor, and the Sumerian clay slab, found in Nagpur (India) (700 BC) (Stojanoski, 1999; Qiu, 2007). Medicinal plants were primarily used in traditional herbal medicine in both developing and developed countries, and are known to be the oldest form of healthcare. In many cultures such as in Africa, Asia and Latin America, traditional medicine forms part of the cultural practices ingrained within these cultures for centuries and still forms part of modern medicinal practices (Jamshidi-Kia et al., 2018).

The traditional remedies for various ailments within these countries depended on the experience of healers and their predecessors. In most cases, the knowledge was passed on from ancestors to younger family members, but in other cases, it was only due to personal trial and error over time that healers could share their knowledge. This knowledge differed from one geographical location to another, and as a result, alternative treatments for specific ailments were developed from multiple plant species across the globe. Traditional medicine is defined by the World Health Organization (WHO) as the knowledge, skills and practices based on the theories, beliefs and experiences indigenous to different cultures, whether explicable or not, used in the maintenance of health, as well as in the prevention, diagnosis, improvement or treatment of physical and mental illnesses (WHO, 2000). It is also known as health practices and approaches that are based on the knowledge and beliefs that incorporate plants as medicine, spiritual therapies and physical therapy that can be applied singularly or in combination to diagnose, treat and prevent diseases or to maintain well-being. Occasionally, traditional medicine has been also referred to as complementary and alternative medicine (CAM) (Gunjan et al., 2015).

DOI: 10.1201/9781003108375-1

1.2 OVERVIEW OF THE HISTORY OF MEDICINAL PLANTS

Papyrus Ebers, an Egyptian medical document of the knowledge of herbal remedies used, was documented and used in 1550 BC (Ramawat and Mérillon, 2008). This document recorded medicinal formulations such as gargles, poultices, infusions, pills, ointments and snuffs that mostly used beer, milk, wine and honey as the vehicles of these remedies (Newman et al., 2000). Thereafter, *De Materia Medica* was written by Dioscorides between 60 and 78 AD (first century), and around 200 AD (second century), *Pen Ts'ao Ching Classic of Materia Medica* was written (Ramawat and Mérillon, 2008). In the ancient Western world, Greek civilization grew in abundance in the fields of both science and philosophy, thus making a significant contribution to the phytopharmaceutical sciences (Halberstein, 2005). Aristotle described 500 crude drugs that were used to treat various pathological conditions (Chatard, 1908; Khan, 2014).

Hippocrates (460–337 BC) was the first to formulate a scientific medical paradigm of treatment and proposed that the majority of pathological conditions were caused due to a disturbance in the physiology of the human body (Sykiotis et al., 2006). Claudius Galen Pergamum, another significant contributor to medicinal sciences, wrote over 30 books on botanicals in which he proposed various techniques involved in pharmaceutical formulations to create stable and effective drugs (Newman et al., 2000; Buerki and Higby, 2007). In the Dark Ages and Middle Ages (fifth to the 12th centuries), monasteries in England, Ireland, France and Germany preserved some of the knowledge acquired from the Greek civilization. The Arabian people were mostly responsible for the preservation of the knowledge of the ancient Western world of medicinal plants and later expanded to include plants extensively used in their own culture and the plants used by Chinese and Indian cultures that were unknown in the Greco-Roman world (Grabley and Thiericke, 1999; Newman et al., 2000).

In the eighth century, the Arabs were the first to establish privately owned pharmacies or drug stores. Avicenna was a Persian pharmacist, physician, philosopher and poet who contributed to the sciences of pharmacy and medicine by his work. His book *Canon Medicinae* was considered as the final compilation or codification of all documented Greco-Roman medicine. After this, *Corpus of Simples*, a comprehensive compilation of the data of Dioscorides and medicinal records from the Middle East and Far East, was written by Ibn al-Baitar. In 1618, the *London Pharmacopoeia* formally published these documents, and various other medicinal literature, in the UK. The first discovery of 'pure' compounds—or at least the idea thereof—can be traced back to the isolation of the active principles of the most commonly used plants. This included strychnine, morphine, atropine and colchicine in the early 1800s. Shortly after these isolations were successful, E. Merck produced the first commercial pure natural product, morphine, in 1826. Thereafter, the first semi-synthetic pure drug that was based on a natural product—namely aspirin—was conducted by Bayer in 1899 (Grabley and Thiericke, 1999).

This contributed to the commercialization of cocaine, derived from the coca plant (*Erythroxylum coca* Lam.), for medicinal purposes in the late 1800s. Strangely enough, it was used as a treatment for opiate addiction and to perform eye surgery, because it acted as a local anesthetic. During the late 1860s, it became popular to drink coca wines, of which the most famous was Vin Mariani. Two glasses of wine contained approximately 50 mg of cocaine, and it was even endorsed by the Pope at the time (Karch, 1999). Heroin or diacetylmorphine, which is derived from morphine that is isolated from the opium poppy plant (*Papaver somniferum* L.), is another example of a synthetic compound that had a variety of controversial uses in the past. It was first developed by Bayer Pharmaceuticals in 1898 as an anti-tussive to suppress coughing and was thought to act as a cure for alcoholism. During this time, it was possible to order heroin together with appropriate syringes for administration to be delivered to the individual's doorstep (Sneader, 1998).

During the mid-20th century, novel synthetic drugs were more relevant when compared to herbal products (Tyler, 1999). However, thanks to scientific advancements, the efficacy of herbal drugs could be established to some extent in most of the cases. In the 1960s, it was discovered that conventional medicine at the time had toxic and iatrogenic effects, which once again led to an increased demand for safer and economically cheaper herbal drugs. Additionally, the popularity of herbal medicine increased worldwide due to the WHO, which encouraged developed countries to incorporate traditional phytomedicine to accomplish the unmet needs of the modern system (Miller, 1998). It was reported by the WHO that approximately four billion people (80% of the world's population) use herbal medicine for some aspect of their primary healthcare, and during the last 20 years, the amount of research conducted on medicinal plants increases steadily, proving that this medicinal practice is here to stay (Fabricant and Farnsworth, 2001).

1.3 THE SIGNIFICANCE OF MEDICINAL PLANTS IN DRUG DISCOVERY

Due to rapid development in industrialized countries, there are at least 119 chemical substances derived from 90 plant species that can be considered as important drugs commonly used in one or more countries, of which 74% were discovered due to the research and isolation of plants that are used in traditional medicine (Arvigo et al., 1993). It was reported, after a study was conducted on US-based prescription drug data, that natural products play an important role in drug treatment, with over 50% of the most prescribed drugs in the United States containing either a natural product or plant-derived compound that functioned as the active ingredient (Grifo et al., 1997). The use of alternative medicine has increased in developed countries in modern times, due to an increase in reported cases of chronic diseases. Despite the accessibility of modern treatments, the use of CAM was deemed a comforting choice of treatment by patients dissatisfied with synthetic-based drugs (Thorne et al., 2002). Medicinal plants play an important role in modern drug development, with nearly 25% of modern drugs being derived from plants that are used in traditional medicine, according to WHO (Khan and Ahmad, 2019).

The pharmacological effects of plants are due to the presence of organic compounds produced by plants for various purposes, known as primary and secondary metabolites. Primary metabolites play a role in the growth, development, initiation of defense response and reproduction of plants. Secondary metabolites function to protect plants against various biotic and abiotic stress factors by acting as metal transporting agents, as agents of symbiosis between other organisms, as sexual hormones and finally as differentiation effectors (Demain and Fang, 2000).

Primary metabolites include compounds such as glucose, starch, proteins, lipids, polysaccharides and nucleic acids, while secondary metabolites include alkaloids, flavonoids, saponins, terpenoids, steroids, glycosides, volatile oils, tannins and many other compounds. Secondary metabolites, known as 'plant-derived compounds,' are present in all plants and due to several genetic and environmental factors, the ratios of these metabolites may be unique to each specific species. This leads to varying therapeutic properties of the different species of plants (Martinez et al., 2008).

Therapeutic properties, such as anti-malarial, analgesic, anti-spasmodic and diuretic activities have been linked to alkaloids. Terpenoids are known to exhibit anti-bacterial, anti-cancer, anti-malarial, anti-inflammatory, anti-viral and anti-helmintic properties, while glycosides are reported to have anti-fungal and anti-bacterial properties. Phenols and flavonoids are known to have anti-oxidant, anti-allergic, and anti-bacterial properties, and saponins have anti-inflammatory and anti-viral activity (Chopra and Doiphode, 2002; Maurya et al., 2008).

Between 1950 and 1970, it was reported that 100 new drugs were developed based on plants and introduced in the US pharmaceutical industry. These drugs consisted primarily of alkaloids and included vincristine, deserpidine, reseinnamine, reserpine and vinblastine. Vinblastine was derived from *Catharanthus roseus* (L.) G.Don, which is now used successfully for treating Hodgkin's disease, non-Hodgkin's lymphomas, choriocarcinoma, leukemia in children, neck and testicular cancer (Farnsworth and Bingel, 1977). After that, between 1971 and 1990, new drugs isolated from plants included etoposide, guggulsterone, teniposide, artemisinin, teniposide and many others. From 1991–1995, drugs such as paclitaxel, topotecan, gomishin, and irinotecan were developed (Vickers and Zollman, 1999). In recent decades, several other chemical compounds were derived from plant sources, which included quinine, aspirin, ephedrine, atropine, digoxin and colchicine (Moteriya et al., 2015; Ram et al., 2015).

A compound isolated from *Podophyllum emodi* Wall. ex Royle, namely podophyllotoxin, is used effectively against lung cancer, testicular cancer and lymphomas. A chemotherapeutic drug, taxol, was derived from *Taxus brevifolia* Nutt. and is used to treat metastatic ovarian cancer and lung cancer (Khan and Ahmad, 2019). A drug containing the compound, serpentine isolated from the root of *Rauwolfia serpentine* (L.) Benth. ex Kurz, is now used to treat hypertension and to reduce blood pressure (Hasan et al., 2009).

1.4 APPLICATION OF MEDICINAL PLANTS

As previously mentioned, a large percentage of the world's emerging population relies on traditional medicine and plant-derived products as a primary source of healthcare (Hamilton, 2004). It is estimated that between 70% and 80% of the emerging world's population relies on traditional medicine and plant-derived products as their primary source of healthcare (Hamilton, 2004). Furthermore, over 25% of modern Western pharmaceuticals contain ingredients from plant origins, including the top 20 best-selling prescription drugs in the United States (Mukeshwar et al., 2011). There are various potential therapeutic applications of plant-derived drugs in both traditional and modern medical practices, some of which are discussed in this book.

1.4.1 Anti-Cancer Therapy

Medicinal plants have been utilized in traditional anti-cancer regimes throughout the world to either combat the formation and progression of cancer or provide relief of symptoms caused by the disease. Despite the abundance of over 260,000 medicinal plants utilized in traditional cancer practices, only 5–15% of these medicinal plants have been screened for effective anti-cancer compounds, leading to a surplus of unexplored potential (Amin et al., 2009). Histone deacetylases (HDACs) inhibitors have been reported as a novel category of anti-cancer drugs. It has been reported that HDACs are used to treat hematologic cancers that have fewer adverse side effects (Yoon and Eom, 2016).

Naturally occurring HDAC inhibitors are discussed in this book, and the *in vitro* and *in vivo* characterization will be explored. Human malignant melanoma is the cancer of the skin caused by malignant melanoma cells and is the lead cause of deaths related to skin cancer (Matthews et al., 2017). Melanoma is highly prone to metastasis and as a result, conventional skin cancer therapy (such as surgery, radiation therapy and chemotherapy) are less effective. Therefore, modern-day melanoma treatments include the downregulation of the phosphatidylinositol-3-kinase/protein kinase B/mammalian target of rapamycin (PI3K/Akt/mTOR) and Mitogen-activated protein kinase (MAPK) pathways which are hyperactive in cases of melanoma (Lopes et al., 2018).

Due to the high metastatic potential of melanoma, there is a constant search for respective drugs that target more than one cellular pathway to minimize the development of drug resistance (Öztürk et al., 2020). Modern-day chemotherapy treatments against other fatal cancers, such as non-small cell lung cancers (NSCLC), may result in severe side effects. The search for novel medicinal agents is crucial against lung cancers which are mainly caused by chronic inflammatory conditions (McFarland, 2019). Thus, various anti-inflammatory and potential anti-cancer phyto-derived compounds have been discussed within this book.

1.4.2 Treatment and Prevention in Skin Conditions

An abundance of phyto-derived products has been incorporated into various skin cosmetics and ointments for cosmeceutical and anti-bacterial purposes (Aburjai and Natsheh, 2003). Medicinal plant products displaying both anti-inflammatory and hydrating properties are utilized for eczema treatment, as well as for anti-aging and anti-scarring aesthetics (Mabona and Vuuren, 2013). Plant-derived products that exhibit anti-microbial activity may harbor major potential in the treatment and prevention of pathogenic skin conditions, such as severe wound infection, acne and dermal infections caused by the Acne vulgaris bacterium and *Aspergillus* spp. fungi, respectively (Mummed et al., 2018). Wounds formed on the skin begin with the inflammatory response due to some form of damage. The process of wound healing begins by restoring the wound surface coating tissue, then fibroblasts migrate to form the required collagen to induce healing in the tissue and lastly, this leads to contortion and extraction of the wound. Various drugs can be used to heal wounds, but there are side effects and defects in these drugs; therefore, the development of a plant-based drug for wound healing is essential (Solati et al., 2021).

1.4.3 Therapeutics against Pathogenic Diseases

Multiple phyto-derived medicinal agents act as effective therapeutics against infectious diseases throughout the world. Medicinal plants from the *Artemisia*, *Cinchona*, *Cryptolepis* and *Tabebuia* genera have been utilized for millennia in the treatment and prevention of malaria (Mohammadi et al., 2020). In this book, different indigenous plants will be explored that exhibit anti-malarial or anti-plasmodial activity and the prospects for anti-malarial drug discovery. Over 30 plants spread, over 21 families are used traditionally to treat tuberculosis (TB), with the most frequent species being *Clausena anisate* (Willd.) Hook.f. ex Benth., *Haemanthus albiflos* Jacq. and *Artemisia afra* Jacq. ex Willd. (Lawal et al., 2014).

However, more medicinal plants should be screened for novel therapeutic agents that may be utilized for the control of TB or to act as an adjuvant to improve the efficacy of modern TB treatments (Oosthuizen et al., 2018). Different indigenous plants are highlighted that could potentially be candidates to be used as adjuvants or commercial products for TB. Various plant-derived compounds, such as alkaloids, display promising anti-viral activity which have subsequently lead to further developmental studies for potential medicinal agents against the human immunodeficiency virus (HIV) (Valadão et al., 2015).

1.4.4 Systemic Diseases

The promising bioactivity of natural products may contribute to the development of future treatments to prevent the progression of systemic diseases, such as neurodegenerative diseases, including diseases such as Alzheimer's disease, Parkinson's disease, etc. (Khalili and Hamzeh, 2010). Plant-derived compounds may further provide prevention of disease contraction by displaying activity in maintaining the homeostatic state within the body to prevent systemic diseases such as gut dysbiosis (resulting in various bowel ailments) and severe blood clotting leading to venous thrombosis (act as an anti-coagulant) (Cryan et al., 2019). Diabetes is known to be a chronic disease where either the pancreas doesn't make insulin, or the body cannot use the insulin properly. There are two types of diabetes, type 1 which is present in an individual from childhood, and type 2 which can be developed later on in life (Zimmet et al., 2014).

1.4.5 Nanotechnology

Nanotechnology became an important research endeavor early in the 21st century. Researchers developed the ability to manipulate and control the chemical, biological and physical properties of these particles that can now be used for various applications, which includes a more effective drug delivery system, for diagnostic purposes, as image contrast agents and more (McNeil, 2005). The synthesis of nanoparticles containing plant extract or isolated compounds introduced a new way to deliver drugs to treat various ailments; this book focuses on the use of nanotechnology to treat Acne vulgaris and cancer.

1.4.6 Bibliometric Analysis of Medicinal Plants

Bibliometric data is a statistical analysis of articles, books or other publications. The citation indexes and co-citation analyses are important factors to determine more sensitive measures of the research quality that was conducted; it is also vital to track the growth of certain fields in science and what the latest trends are within these research fields (Broadus, 1987). In this book, scientific contributions of medicinal applications for different plant species were reviewed. Furthermore, trends on the medicinal uses of these plant species were monitored using a bibliometric index in order to identify gaps that are present between researchers, the exchange of materials and human resources.

1.5 MEDICINAL PROPERTIES OF A FEW TRADITIONALLY USED PLANTS

1.5.1 *Echinacea* spp.

The *Echinacea* genus consists of herbaceous flowering plants consisting of ten coneflower species which are mainly found in eastern and central North America (Figure 1.1A). Plants of this species have been incorporated into customary treatments against upper respiratory tract infections and minor inflammatory conditions, and are said to improve immunity, high blood pressure, anxiety and skin conditions (Percival, 2000). *Echinacea* spp. has displayed anti-inflammatory and potential anti-cancer bioactivity (Rad et al., 2018). Although human-focused clinical testing is limited, consumption of *Echinacea* botanicals has been considered safe for short-term use (Rad et al., 2018; Percival, 2000).

1.5.2 *Salvia rosmarinus*

Salvia rosmarinus Spenn., also known as rosemary, is an aromatic evergreen shrub with needle-like leaves and flowers which may be white, pink, purple or blue (Figure 1.1B) (Macedo et al., 2020). This plant is native to the Mediterranean region and is said to harbor anti-oxidant and anti-inflammatory activity (Macedo et al., 2020). Rosemary is commonly administered to improve digestion, enhance cognitive function and alleviate symptoms associated with cancer (Brindisi et al., 2020). Bioactivity studies have indicated that the plant displays anti-bacterial, anti-diabetic, anti-inflammatory, anti-thrombotic and hepatoprotective effects (Macedo et al., 2020; Brindisi et al., 2020).

1.5.3 *Cirsium* spp.

Cirsium is a genus of perennial and biennial flowering plants known as plume thistles, belonging to the Asteraceae family and native to parts of Europe, northern Africa and Asia (Figure 1.1C)

FIGURE 1.1 Common traditionally utilized medicinal plants. A, *Echinacea* spp.; B, *Salvia rosmarinus*; C, *Cirsium* spp. (Delso, 2013; Montillot, 2020; Bergsma, 2017).

(Ali et al., 2011). Decoctions of the leaves are used to treat muscle rigor, seizures, fever, high blood pressure, hemorrhoids and nervous disorders, whereas steam infusions of the whole plant are used to alleviate rheumatic joints (Nazaruk, 2008). Species of this genus are traditionally used to treat peptic ulcers, epitasis, diabetes and leukemia (Ali et al., 2011; Nazaruk, 2008).

1.6 THE IMPORTANCE OF PRESERVATION OF MEDICINAL PLANTS

The conservation and sustainable use of medicinal plants is important to ensure the preservation of our biodiversity and provide prolonged access to these promising natural resources for generations to come. The unprecedented use of medicinal plants and nonregulated development of their materials into natural products threatens viable cultivation and conservation of these specimens, thereby hindering potential future utilization in pharmaceutical advancements (World Health Organization, 2019). The uncontrolled management of outsourcing potential therapeutic agents from medicinal plants has led to overexploitation—and in some cases, extinction—of rare plant biodiversity around the globe (Van Wyk, 2011).

The conservation of these plant species may be obtained by domesticated cultivation of these specimens rather than a collection of their material from the wild (Leonti and Casu, 2013). Sourcing medicinal plant material from articulated cultivations could ensure correct botanical and genetic identity, as well as provide stable quality and supply control. In addition, such agricultural practices should be guided under well-defined conditions concerning the climate requirements of the various species in order to obtain sustainable distribution and controlled use of these medicinal plants (Van Wyk, 2011; Leonti and Casu, 2013).

REFERENCES

Aburjai, T., and Natsheh, F. M. 2003. "Plants used in cosmetics." *Phytotherapy Research* 17: 987–1000. doi: 10.1002/ptr.1363.

Ali, F., Khan, S., and Ali, I. 2011. "Phytochemical study on the constituents from cirsium arvense." *Mediterranean Journal of Chemistry* 2. doi: 10.13171/mjc.1.2.2011.07.10.00.

Amin, A., Gali-muhtasib, H., Ocker, M., and Schneider-stock, R. 2009. "Overview of major classes of plant-derived anticancer drugs." *International Journal of Biomedical Science* 5, 1: 1–11.

Arvigo, R., Balick, M. J., and Evans, L. 1993. *Rainforest remedies: One hundred healing herbs of Belize*. Twin Lakes, WI: Lotus Press.

Bergsma, D. J. 2017. "*Cirsium* vulgare." Obtained from Wikimedia under the Creative commons Attribution-Share Alike 4.0 International license. https://commons.wikimedia.org/wiki/File:Mooie_bloeiwijze_van_een_Speerdistel_(Cirsium_vulgare)_03.jpg (Accessed 22 June 2021).

Brindisi, M., Bouzidi, C., Frattaruolo, L., Loizzo, M. R., Tundis, R., Dugay, A., Deguin, B., Cappello, A. R., and Cappello, M. S. 2020. "Chemical profile, antioxidant, anti-inflammatory, and anti-cancer effects of Italian *Salvia rosmarinus* Spenn. Methanol leaves extracts." *Antioxidants* 9, 826: 1–21.

Broadus, R. N. 1987. "Toward a definition of bibliometrics." *Scientometrics* 12, 5–6: 373–379.

Buerki, R., and Higby, G. 2007. *En: Encyclopedia of pharmaceutical technology: Dosage forms and basic preparations: History*. Edición, Editor James Swarbrick. New York: Informa Health Care, 948.

Chatard, J. A. 1908. "Avicenna and Arabian medicine." *Bulletin of the Johns Hopkins Hospital* 19: S157–S160.

Chopra, A., and Doiphode, V. V. 2002. "Ayurvedic medicine: Core concept, therapeutic principles, and current relevance." *Medical Clinics* 86: 75–89.

Cryan, J. F., O'riordan, K. J., Cowan, C. S. M., Sandhu, K. V., Bastiaanssen, T. F. S., Boehme, M., Codagnone, M. G., et al. 2019. "The microbiota-gut-brain axis." *Physiological Reviews* 99, 4: 1877–2013. doi: 10.1152/physrev.00018.2018.

Delso. 2013. "*Echinacea purpurea*." Obtained from Wikimedia under Creative commons Share-Alike Attribution license. https://commons.wikimedia.org/wiki/File:Echinacea_purpurea,_Jard%C3%ADn_Bot%C3%A1nico,_M%C3%BAnich,_Alemania,_2013-09-08,_DD_01.jpg (Accessed 22 June 2021).

Demain, A. L., and Fang, A. 2000. "The natural functions of secondary metabolites." *History of Modern Biotechnology I* 69: 1–39. doi: 10.1007/3-540-44964-7_1

Fabricant, D. S., and Farnsworth, N. R. 2001. "The value of plants used in traditional medicine for drug discovery." *Environmental Health Perspectives* 109: 69–75.

Farnsworth, N. R., and Bingel, A. S. 1977. "Problems and prospects of discovering new drugs from higher plants by pharmacological screening." Wagner, H., and Wolff, P. (eds). *New natural products and plant drugs with pharmacological, biological or therapeutical activity. Proceedings in life sciences.* Berlin, Heidelberg: Springer. doi: 10.1007/978-3-642-66682-7_1.

Grabley, S., and Thiericke, R. 1999. "Bioactive agents from natural sources: Trends in discovery and application." *Thermal Biosensors, Bioactivity, Bioaffinitty* 64: 101–154. doi: 10.1007/3-540-49811-7_4.

Grifo, F., Newman, D., Fairfield, A. S., Bhattacharya, B., and Grupenhoff, J. T. 1997. "The origins of prescription drugs." *Biodiversity and Human Health*: 131–163.

Gunjan, M., Naing, T. W., Saini, R. S., Ahmad, A., Naidu, J. R., and Kumar, I. 2015. "Marketing trends & future prospects of herbal medicine in the treatment of various disease." *World Journal of Pharmaceutical Research* 4: 132–155.

Halberstein, R. A. 2005. "Medicinal plants: Historical and cross-cultural usage patterns." *Annals of Epidemiology* 15: 686–699.

Hamilton, A. C. 2004. "Medicinal plants, conservation and livelihoods." *Biodiversity & Conservation* 13, 8: 1477–1517.

Hasan, S. Z., Misra, V., Singh, S., Arora, G., Sharma, S., and Sharma, S. 2009. "Current status of herbal drugs and their future perspectives." *Biological Forum—An International Journal* 1, 1: 12–17.

Jamshidi-Kia, F., Lorigooini, Z., and Amini-Khoei, H. 2018. "Medicinal plants: Past history and future perspective." *Journal of Herbmed Pharmacology* 7, 1.

Karch, S. B. 1999. "Cocaine: History, use, abuse." *Journal of the Royal Society of Medicine* 92, 8: 393–397.

Khalili, M., and Hamzeh, F. 2010. "Effects of active constituents of crocus sativus L, crocin on streptozocin-induced model of sporadic Alzheimer's disease in male rats." *Iranian Biomedical Journal* 14, 1–2: 59–65.

Khan, H. 2014. "Medicinal plants in light of history: Recognized therapeutic modality." *Journal of Evidence-based Complementary & Alternative Medicine* 19: 216–219.

Khan, M. S. A., and Ahmad, I. 2019. "Herbal medicine: Current trends and future prospects." *New Look to Phytomedicine*: 3–13.

Lawal, I. O., Grierson, D. S., and Afolayan, A. J. 2014. "Phytotherapeutic information on plants used for the treatment of tuberculosis in Eastern Cape Province, South Africa." *Evidence-Based Complementary and Alternative Medicine*: 1–11.

Leonti, M., and Casu, L. 2013. "Traditional medicines and globalization: Current and future perspectives in ethnopharmacology." *Frontiers in Pharmacology* 4: 1–13. doi: 10.3389/fphar.2013.00092.

Lopes, J. M., Soares, P., and Pópulo, H. 2018. "Melanoma treatment in review." *ImmunoTargets and Therapy* 7: 35–49.

Mabona, U., and S. F. Van Vuuren. 2013. "South African journal of botany Southern African medicinal plants used to treat skin diseases." *South African Journal of Botany* 87: 175–93. doi: 10.1016/j.sajb.2013.04.002.

Macedo, L. M. De, Mendes, É., Milit, L., Tundisi, L. L., Souto, E. B., and Mazzola, P. G. 2020. "Rosemary (*Rosmarinus officinalis* L., Syn *Salvia rosmarinus* Spenn.) and its topical applications: A review." *Plants* 9, 651: 1–12.

Martinez, M. J. A., Lazaro, R. M., Del Olmo, L. M. B., and Benito, P. B. 2008. "Anti-infectious activity in the anthemideae tribe." *Studies in Natural Products Chemistry* 35: 445–516. doi: 10.1016/S1572-5995(08)80012-9.

Matthews, N. H., Li, W. Q., Qureshi, A. A., Weinstock, M. A., and Cho, E. 2017. "Epidemiology of melanoma." *Cutaneous Melanoma: Etiology and Therapy* 3–22.

Maurya, R., Singh, G., and Yadav, P. P. 2008. "Antiosteoporotic agents from natural sources." *Studies in Natural Products Chemistry* 35: 517–548. doi: 10.1016/S1572-5995(08)80013-0.

McFarland, D. C. 2019. "New lung cancer treatments (immunotherapy and targeted therapies) and their associations with depression and other psychological side effects as compared to chemotherapy." *General Hospital Psychiatry* 60: 148–155. doi: 10.1016/j.genhosppsych.2019.04.001.

McNeil, S. E. 2005. "Nanotechnology for the biologist." *Journal of Leukocyte Biology* 78, 3: 585–594.

Miller, L. G. 1998. "Herbal medicinal: Selected clinical considerations focusing on known or potential drug-herb interactions." *Archives of Internal Medicine* 158: 2200–2211.

Mohammadi, S., Sharifi-rad, J., Asgharian, P., and Martorell, M. 2020. "Medicinal plants used in the treatment of malaria : A key emphasis to *Artemisia*, *Cinchona*, *Cryptolepis*, and *Tabebuia* genera." *Phytotherapy Research* 34: 1556–1569. doi: 10.1002/ptr.6628.

Montillot, M. 2020. "*Salvia rosmarinus*." Obtained from Wikimedia under the Creative commons Attribution-Share Alike 4.0 International license. https://commons.wikimedia.org/wiki/File:Salvia_rosmarinus_(10-04-2020),_vue_en_gros_plan.jpg (Accessed 22 June 2021).

Moteriya, P., Satasiya, R., and Chanda, S. 2015. "Screening of phytochemical constituents in some ornamental flowers of Saurashtra region." *Journal of Pharmacognosy and Phytochemistry* 3: 112–120.

Mukeshwar, P., Debnath, M., Gupta, S., and Chikara, S. K. 2011. "Phytomedicine: An ancient approach turning into future potential source of therapeutics." *Journal of Pharmacognosy and Phytotherapy* 3, 2: 27–37. doi: 10.5897/JPP.

Mummed, B., Abraha, A., Feyera, T., Nigusse, A., and Assefa, S. 2018. "In vitro antibacterial activity of selected medicinal plants in the traditional treatment of skin and wound infections in eastern Ethiopia." *BioMed Research International*. doi: 10.1155/2018/1862401.

Nazaruk, J. 2008. "Antioxidant activity and total phenolic content in *Cirsium* five species from North-East region of Poland." *Fitoterapia* 79: 194–196. doi: 10.1016/j.fitote.2007.11.008.

Newman, D. J., Cragg, G. M., and Snader, K. M. 2000. "The influence of natural products upon drug discovery." *Natural Product Reports* 17: 215–234.

Oosthuizen, C. B., Reid, A. M., and Lall, N. 2018. "Can medicinal plants provide an adjuvant for tuberculosis patients?" *Medicinal Plants for Holistic Health and Well-Being*: 213–253. doi: 10.1016/B978-0-12-812475-8.00007-X.

Öztürk, F. A., Jahan, H., Öztürk, M., and Günaydın, K. 2020. "Preclinical study of the medicinal plants for the treatment of malignant melanoma." *Molecular Biology Reports* 47, 8: 5975–5983. doi: 10.1007/s11033-020-05671-5.

Percival, S. S. 2000. "Use of echinacea in medicine." *Biochemical Pharmacology* 60, 99: 155–158.

Qiu, J. 2007. "Traditional medicine: A culture in the balance." *Nature Publishing Group* 448, 7150: 126–129.

Ram, J., Moteriya, P., and Chanda, S. 2015. "Phytochemical screening and reported biological activities of some medicinal plants of Gujarat region." *Journal of Pharmacognosy and Phytochemistry* 4, 2: 192–198.

Rad, M. S., Mnayer, D., Morais-Braga, M. F. B., Carneiro, J. N. P., Bezerra, C. F., Coutinho, H. D. M., Salehi, B., Martorell, M., del Mar Contreras, M., Soltani-Nejad, A., and Uribe, Y. A. H. 2018. "Echinacea plants as antioxidant and antibacterial agents: From traditional medicine to biotechnological applications." *Phytotherapy Research* 32, 9: 1653–1663. doi: 10.1002/ptr.6101.

Ramawat, K. G., and Mérillon, J. M. 2008. *Bioactive molecules and medicinal plants*. Berlin: Springer.

Sneader, W. 1998. "The discovery of heroin." *The Lancet* 352: 1697–1699.

Solati, K., Karimi, M., Rafieian-Kopaei, M., Abbasi, N., Abbaszadeh, S., and Bahmani, M. 2021. "Phytotherapy for wound healing: The most important herbal plants in wound healing based on Iranian ethnobotanical documents." *Mini Reviews in Medicinal Chemistry* 21, 4: 500–519.

Stojanoski, N. 1999. "Development of health culture in Veles and its region from the past to the end of the 20th century." *Veles: Society of Science and Art* 13: 34.

Sykiotis, G. P., Kalliolias, G. D., and Papavassiliou, A. G. 2006. "Hippocrates and genomic medicine." *Archives of Medical Research* 37: 181–183.

Thorne, S., Paterson, B., Russell, C., and Schultz, A. 2002. "Complementary/alternative medicine in chronic illness as informed self-care decision making." *International Journal of Nursing Studies* 39: 671–683.

Tyler, V. E. 1999. "Phytomedicines: Back to the future." *Journal of Natural Products* 62: 1589–1592.

Valadão, A. L. C., Abreu, C. M., Dias, J. Z., Arantes, P., Verli, H., Tanuri, A., and De Aguiar, R. S. 2015. "Natural plant alkaloid (emetine) inhibits HIV-1 replication by interfering with reverse transcriptase activity." *Molecules* 20, 6: 11474–11489. doi: 10.3390/molecules200611474.

Van Wyk, B. E. 2011. "The potential of South African plants in the development of new medicinal products." *South African Journal of Botany* 77, 4: 812–829. doi: 10.1016/j.sajb.2011.08.011.

Vickers, A., and Zollman, C. 1999. "Herbal medicine." *BMJ* 319: 1050–1053. doi: 10.1136/bmj.319.7216.1050

World Health Organization. 2000. "General guidelines for methodologies on research and evaluation of traditional medicine." https://apps.who.int/iris/bitstream/handle/10665/66783/WHO_EDM_TRM_2000.1.pdf

World Health Organization. 2019. "WHO global report on traditional and complementary medicine 2019." https://apps.who.int/iris/bitstream/handle/10665/312342/9789241515436-eng.pdf?ua=1.

Yoon, S., and Eom, G. H. 2016. "HDAC and HDAC inhibitor: From cancer to cardiovascular diseases." *Chonnam Medical Journal* 52, 1: 1.

Zimmet, P. Z., Magliano, D. J., Herman, W. H., and Shaw, J. E. 2014. "Diabetes: A 21st century challenge." *The Lancet Diabetes & Endocrinology* 2, 1: 56–64.

2 The Potential of Microbial Mediated Fermentation Products of Herbal Material in Anti-Aging Cosmetics

Bianca D. Payne and Namrita Lall

CONTENTS

2.1 WHY SKIN AGES: A LOOK AT CELLULAR SIGNALING

Aging affects all tissues of the body, most visibly the skin. It comprises complex molecular changes in each layer of the skin that ultimately manifest as the appearance of fine lines, wrinkles, abnormal pigmentation and dryness. The different types of aging, various theories and most popular current treatments have been previously reported (Fibrich and Lall, 2018; Fibrich, Lambrechts, and Lall, 2019). To better understand how and why the skin ages, one needs to consider the physical and chemical changes which occur in response to time, as well as aging accelerants such as ultraviolet (UV) exposure and pollution. Aging of the skin may be classified as intrinsic or extrinsic, depending on the causative agents. Intrinsic aging is the result of factors such as genetic predisposition, hormones and telomere shortening, while extrinsic aging is the result of external stressors such as pollution, smoking, alcohol consumption, nutrition and UV radiation (Sahu et al., 2013; Ganesan and Choi, 2016). Disruptions in the homeostasis of physiological levels of androgens have been revealed to contribute to dryness of the skin, loss of elasticity and degradation of collagen, and consequently the formation of wrinkles and epidermal atrophy (Khol et al., 2011). UV rays are the largest contributing factor to extrinsic skin aging. Therefore, this type of aging is often interchangeably called photoaging. In combination, these factors perturb physiological homeostasis within the skin that contributes to the formation of wrinkles (Sahu et al., 2013).

2.1.1 The Production of Collagen in the Dermis

As discussed, collagen and elastin are found in the dermis. The dermis is divided into the compact lower reticular and upper, more loosely organized, papillary dermis, which are distinguished by the

DOI: 10.1201/9781003108375-2

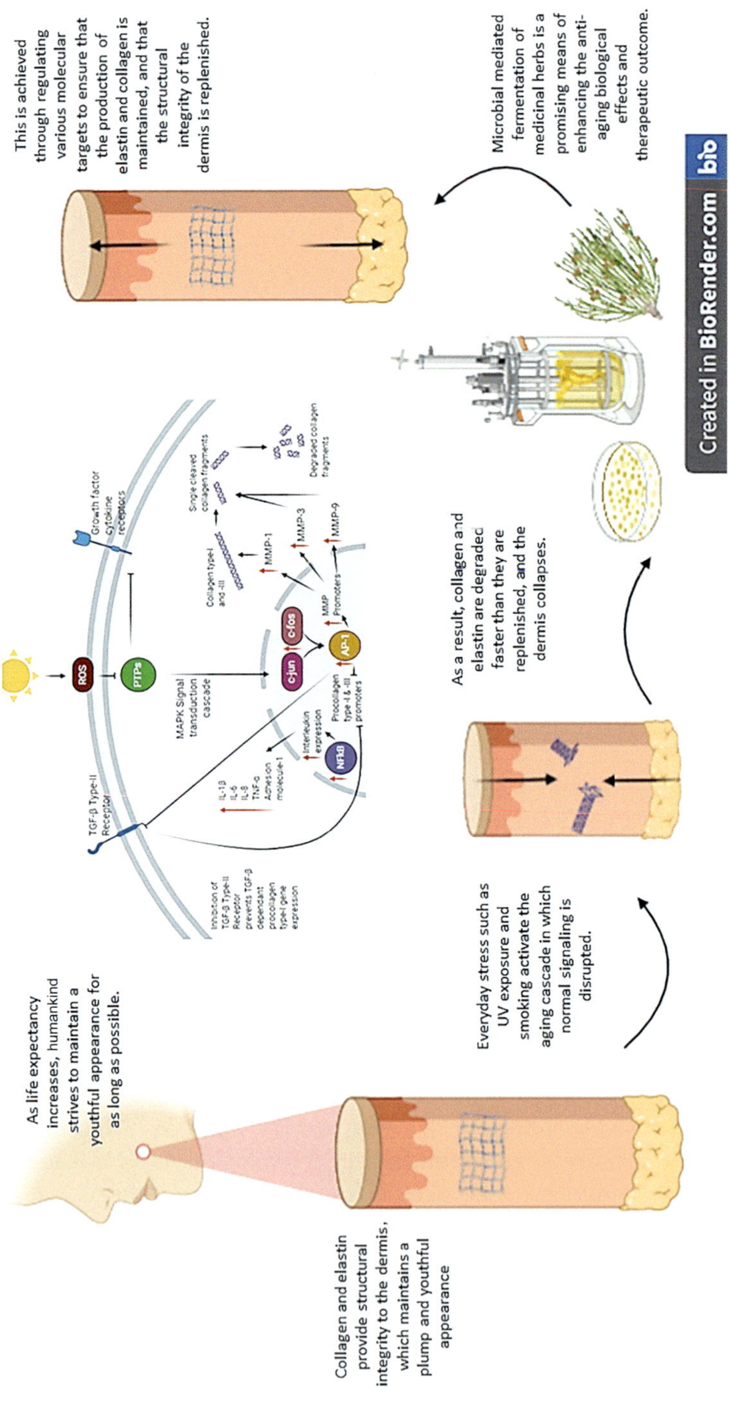

An overview of the aging cascade as discussed in this chapter.

composition and organization of elastin and collagen found in each layer (Cole et al., 2018). Mature collagen consists of three chains, two of which are non-collagenous chains called telopeptides, while the other is collagenous. The collagen chain is characterized by distinct amino acid repeats of glycine followed by two residues, normally proline or hydroxyproline. The presence of glycine at every third repeat establishes the signature triple helix structure associated with collagen, while the variation in amino acid sequences following glycine determine the type of collagen, of which 28 have been identified (Cole et al., 2018).

Type III collagen comprises approximately 20% of the total collagen in the skin, concentrated in the mesh-like papillary dermis, while the remaining 80% is comprised of type-I collagen, concentrated in the reticular dermis. Structurally, type III collagen is a homotrimer of α1(III) chains and type I, a heterotrimer of two α1(I) chains and one α2(I) chain arranged in an anti-parallel conformation to produce more dense fibers (Cole et al., 2018). These fibers are enzymatically cross linked following the secretion of procollagen and its maturation to confer resistance to proteolytic cleavage. The proteolytic removal of C and N terminal pro-peptides facilitates the maturation of procollagen into mature collagen fibers. For instance, in type-I collagen various intracellular lysine residues containing α1(I) and α2(I) chains are converted to hydroxylysine through the action of lysyl hydroxylase, while extracellular lysine and hydroxylysine residues are converted to aldehydes through the action of lysyl oxidase. This process enables spontaneous, non-reducible inter- and intra-peptide crosslinking. The action of lysyl oxidase is crucial in facilitating matrix deposition of elastin fibers and prevention of excessive elasticity (Cole et al., 2018).

The term "matrisome" refers to the extracellular matrix (ECM)–associated proteins, as well as the collective group of heterogeneous ECM components that comprise the core matrisome. The core matrisome includes collagens, proteoglycans and glycoproteins, while matrisome proteins include bound and secreted factors such as transforming growth factor beta (TGF-β), cytokines and matrix metalloproteinases (MMPs). Aging skin can hence be described as a disorder of the dermal matrisome (Cole et al., 2018).

2.1.2 Molecular Signaling That Causes Skin Aging

With time, as is the case with chronological aging, or following exposure to stress, as in extrinsic aging, the initiation of chemical and eventually physical changes are initiated by the generation of reactive oxygen species (ROS) and free radicals. Under normal physiological conditions, the action of these reactive species would swiftly be neutralized through the activity of endogenous antioxidants. With age, however, the efficacy of these systems becomes diminished, and the requirement for external supplementation becomes more pronounced (Xu et al., 2006).

Conditions of oxidative stress stimulate alternative signaling pathways through the modification of proteins and enzymes, two of which include the mitogen-associated protein kinases (MAPK) and nuclear factor kappa beta (NF-κB) pathways. Activation of MAPK signaling induces expression of the transcription factor activator protein-1 (AP-1), which significantly reduces the efficacy of the TGF-β pathway through increasing the expression and activity of existing MMPs. This simultaneously reduces the expression of procollagen and endogenous inhibitors of MMPs, which are tissue inhibitors of matrix metalloproteinases (TIMPs). MMPs are considered some of the most UV-inducible genes, as their expression has been shown to be up-regulated up to five-fold following UV exposures. They comprise a range of zinc-dependent proteinases—of which more than a dozen are known—responsible for the degradation of the structural components within the dermis (Pillai, Oresajo, and Hayward, 2005).

MMPs are classified into various groups, namely collagenases, gelatinases, stromelysins, and membrane-bound MMPs. They degrade proteins found in the ECM, such as collagen, and are overproduced in aging human keratinocytes and fibroblasts. This overexpression, however, has not been shown to be accompanied by elevated levels of TIMPs, thus contributing to the abnormal fibrosis of the ECM and the disrupted protease-antiprotease balance characterized in photoaged skin. MMP-1, belonging to the collagenase subgroup of MMPs, is considered the main contributor to the initial fragmentation of type-I and -III collagen (Pillai, Oresajo, and Hayward, 2005; Jang et al., 2020).

MMP-1 cleaves collagen type-I and -III at a single site in the helix, after which MMP-3 and -9 degrade these collagen fragments further. Collagen type-I that has only been cleaved by MMP-1 serves as a stimulus which negatively regulates the synthesis of procollagen type-1, and so complete fragmentation by the action of MMP-3 and -9 is crucial to mitigate the inhibition of new collagen synthesis. In aging skin, stabilization of fragmented collagen through the formation of cross links makes these dysfunctional collagen fibrils highly resistant to proteolytic cleavage, and subject to accumulation which is thought to be a major contributing factor to the manifestation of the aged phenotype (Cole et al., 2018).

The TGF-β pathway has been identified as a significant role player in aging skin, as many ECM-related genes that have been shown to be down-regulated in aging skin are regulated by this pathway. Additionally, many components of this pathway are down-regulated in aging skin (Cole et al., 2018). TGF-β controls the synthesis and degradation of collagen in the ECM. To do this, it binds to its cell surface receptor complex, which consists of TGF-β type I-and -II receptors, which in turn activates transcription factors Smad2 and Smad3. Once activated, these transcription factors form heteromeric complexes with Smad4 and translocate to the nucleus, where they interact with Smad binding elements situated in the promoter region of TGF-β target genes. This TGF-β/Smad regulatory pathway controls at least 265 known ECM and ECM-related genes (Cole et al., 2018).

TGF-β has been shown to down-regulate the activity of MMPs and up-regulate the activity of TIMPs and plasminogen activator inhibitor 1. This evidence has implicated the TGF-β/Smad pathway as a primary regulator of ECM homeostasis. This has severe implications in the pathogenesis of many conditions, including systemic sclerosis, chronic and progressive diseases characterized by fibrosis. Photo aged skin exhibits roughly a 60% reduction in the transcription of TGF-β type II receptors, while TGF-β type I remains approximately stable. The consequences are such that TGF-β binding to dermal fibroblasts occurs at drastically reduced rates and hence signal transduction is diminished. Overexpression of TGF-β type II restores the activity of TGF-β, prevents abnormal degradation of the ECM and stimulates the synthesis of procollagen type-I (Pillai, Oresajo, and Hayward, 2005; Xu et al., 2006; Cole et al., 2018).

Activation of NF-κB signaling induces the expression of inflammatory genes, including interleukins (IL)-1β, -6, -8 and tumor necrosis factor alpha (TNF-α), and recruits neutrophils to the affected site which proceed to release elastase. This contributes to the collapse of the ECM through degradation of collagen and elastin. Oxidative stress contributes to the oxidation of membrane lipids and subsequent release of arachidonic acid and ceramides that continue to recruit inflammatory mediators to the site, promoting a status associated with *inflamm-aging*. This process forms part of a negative feedback loop as ceramides and the accumulation of dysfunctional proteins serve as an activating stimulus for AP-1 (Figure 2.1) (Fisher et al., 1998; Fisher, Kang, and Varanti, 2002; Pillai, Oresajo, and Hayward, 2005; Yaar and Gilchrest, 2007; Xu et al., 2006).

Cysteine-rich proteins of the CCN protein family are expressed under conditions of oxidative stress, and six distinct cysteine rich proteins (CCN1–6) are distinguished. These are associated with the ECM and confer a myriad of functions, including regulation of ECM synthesis and degradation, inflammation and wound healing. Cysteine-rich protein 61, CCN1, is predominantly expressed in human dermal fibroblasts and expression in both chronologically and photoaged skin is elevated. The contribution of CCN1 to aberrant homeostasis within the dermis is three-fold (Quan and Fisher, 2015), as described following.

- Overexpressed CCN1 reduces the production of structural components within the skin such as type-I and -III collagen by inhibiting TGF-β signaling through down-regulation of TGF-β type II receptor, thus directly contributing to dermal thinning.

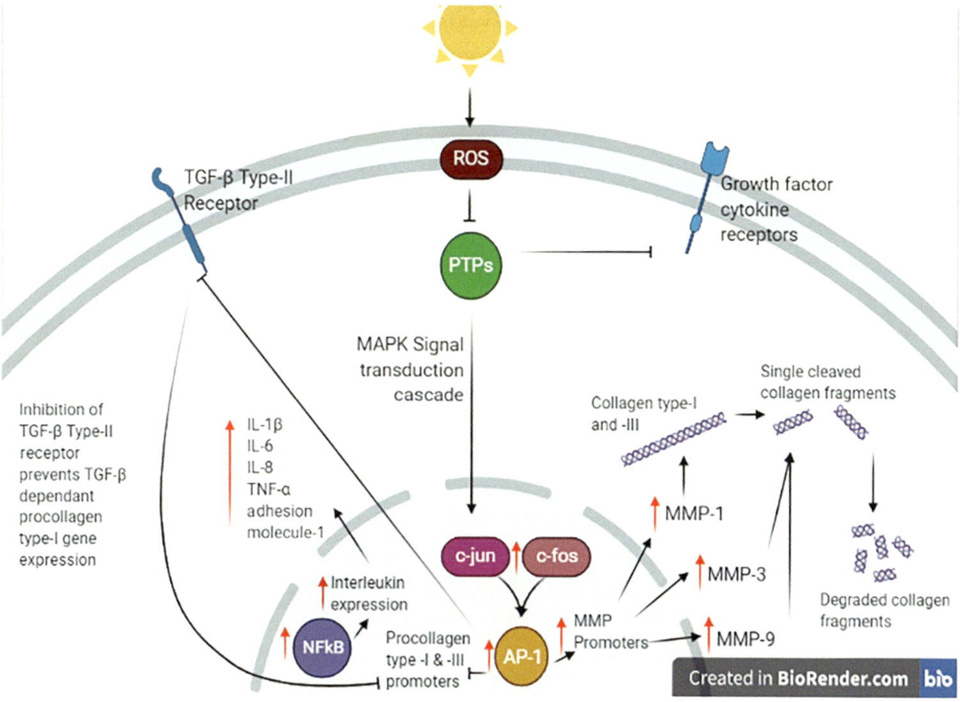

FIGURE 2.1 The molecular signaling in keratinocytes and fibroblasts, which leads to the formation of wrinkles following UV exposure (ceated in Biorender.com).

- The induction of MMP-1, -3, -9, -10 and -23 indirectly through the up-regulation of transcription factor AP-1, further contributing to thinning through the fragmentation of existing collagen.
- Mediation of states of chronic inflammation often associated with aging skin through the induction of many inflammatory cytokines including IL-1β, IL-6 and IL-8. These further act through a feedback loop. For example, IL-1β up-regulates MMPs (such as MMP-1) and down-regulates the synthesis of collagen type-I.

2.1.3 Cell-Based Mechanisms of Aging within the Skin

Fibroblasts are predominantly responsible for producing the ECM. In young skin, they attach to receptors on the surface of collagen fibrils, which facilitates the generation of contractile forces on the surrounding ECM. Mechanical resistance promotes polymerization of the actin cytoskeleton and assembly of intracellular scaffolds, which allow fibroblasts to spread. Fragmentation of the ECM during aging impairs the attachment of fibroblasts to collagen fibrils, drastically reducing their size and mechanical rigidity. This is because dermal fibroblasts achieve greater size on stiffer substrates. The abnormal degradation of the ECM associated with the aging dermis is accompanied by morphological changes of dermal fibroblasts. These changes are observed where flattened and spread-out cells in contact with collagen fibers become collapsed cells with reduced cytoplasm. Indirect association with fragmented collagen fibers is observed as an 80% reduction in network connectivity and 75% reduction in the surface area of fibroblasts, respectively. This in turn results in a reduction in the synthesis of ECM components. The molecular basis of the relationship between fibroblast morphology and ECM production has in part been attributed to the fact that reduced fibroblast size and mechanical force down-regulates the activity of the TGF-β type II receptor (Cole et al., 2018).

2.2 FERMENTATION AS A MEANS OF ENHANCING PHYTOCHEMICAL POTENCY

Research has generated a wealth of supporting data confirming the efficacy of medicinal plants in the treatment of a plethora of conditions; however, evidence confirms the necessity of transformation in some instances to facilitate biological activation. This may be easily facilitated through the action of microorganisms, which are naturally found throughout the human body. Fermentation is a microorganism-facilitated process in which carbohydrates are metabolized and holds the potential to yield high-value products from low-grade/raw substances. This is through the catabolism or conversion of undesirable substrates into compatible components by microbial enzymes. The value of fermentation in medicinal plant research is realized as a potential benefit, which among others, includes enhanced nutritional value of the extract and increased bioavailability of the product (Hussain et al., 2016). According to Hussain et al., (2016), the principles of microbial fermentation of herbal medicines are summarized as follows.

- A wide range of biologically active macro-molecules are produced by microorganisms in their exponential or growing phase. These include enzymes such as amylases, proteases and esterases, which are members of decomposing enzyme families that are responsible for reactions involving decomposition of herbal remedies and the formation of new products.
- The production of novel compounds through the alteration of normal microbial metabolism by the herbal constituents.
- The concentration of medicinal compounds through the increased consumption of non-medicinal constituents during microbial metabolism or through the perturbation of the nutritive and non-nutritive balance of the constituents.
- New and sometimes novel compounds can either be formed by the interaction of microbial secondary metabolites and plant constituents, or by the metabolism of existing herbal constituents.

Biotransformation of phytochemicals as a result of fermentation may result in various changes to the structure and function of these compounds. For example, this biotransformation may cause the conversion of an active compound to a less active metabolite, also referred to as inactivation or detoxification. 1,1,1-Trichloro-2,2-bis (p-chlorophenyl) ethane (DDT) has been used for the control of agricultural pests, and for protection against malaria and typhus. Due to its adverse environmental effects and potential risks to human health, it was subsequently banned in many parts of the world. DDT and its metabolites persist in soils for extended periods, and this is exasperated by the continued use of dicofol. Physical and chemical remediation offer disadvantages including that they are costly an often result in secondary pollution. Microbial mediated remediation offers a more economic and safer alternative as a few strains have been shown to utilize DDT as a carbon source, including *Alcaligenes eutrophus* A5, *Alcaligenes* sp. KK, *Alcaligenes* sp. DG-5, *Chryseobacterium* sp. PYR2, *Rhodococcus* sp. IITR03 *Aeromonas hydrophila* HS01, *Pseudomonas putida* T5, *Pseudomonas aeruginosa*, *Flavimonas oryzihabitans* and *Sphingobacterium* sp. D-6 (Barragán-Huerta et al., 2007; Hsu and Chiang, 2009; Gao et al., 2011; Xie et al., 2011; Cao et al., 2012; Rangachary et al., 2012; Bajaj et al., 2014; Qu et al., 2015; Erdem and Cutright, 2016).

Based on genomic and GC-MS analyses, Pan et al. (2017) propose the degradation of DDT by the isolate DDT-2 as follows: DDT could first be transformed into 1,1-dichloro-2,2-bis (p-chlorophenyl) ethane (DDD) and 2,2-bis (p-chlorophenyl)-1,1-dichlorethylene (DDE) via dechlorination at the trichloromethyl group, subsequently converted to the metabolite 1-chloro-2,2-bis (4′-chlorophenyl) ethylene (DDMU) via further dechlorination, and then form bis (4′-chlorophenyl) acetate (DDM) and bis (4′-chlorophenyl) methane (DDA) via carboxylation. Finally, DDA may be mineralized to carbon dioxide by cleavage of the aromatic ring and oxidation (Pan et al., 2017). Huang et al. (2018) propose a range of transformation pathways depending on the specific conditions (Figure 2.2).

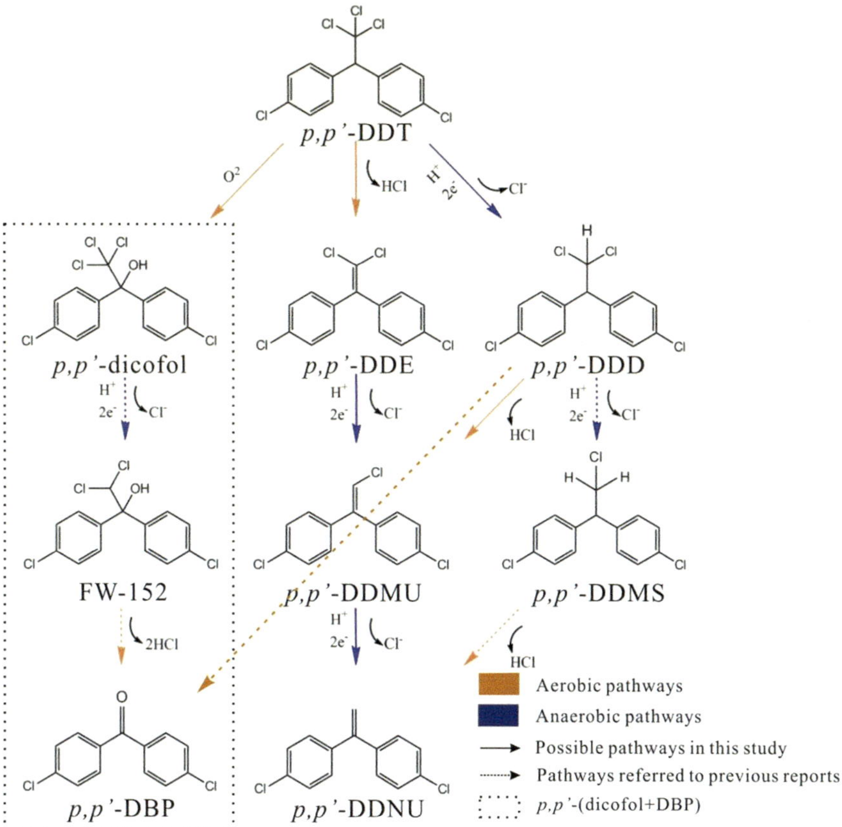

FIGURE 2.2 1, p,p'-DDT was transformed to p,p'-(dicofol þ DBP); 2, p,p'-DDT was transformed to p,p'-DDE under aerobic condition, but was reductively dechlorinated to p,p'-DDD in submerged soils where anaerobic conditions presumably prevail; 3, p,p'-DDD and p,p'-DDE was further converted to p,p'-DDMU (Huang et al., 2018).

Biotransformation may also constitute the conversion of an active metabolite to a more active metabolite, as in the instance of tectoridin to 6-hydroxygenistein and kakkalide to 6-hydroxybiochanin (Figure 2.3). This is referred to as bioactivation. Tsuchihashi et al. (2009) cultured *Peptostreptococcus productus* with tectoridin and kakkalide isoflavones from *Pueraria* flowers, which yielded 6-hydroxygenistein and 6-hydroxybiochanin A, respectively. While tectoridin and kakkalide did not show any hepatoprotective activity (EC_{50} > 200 μM), 6-hydroxygenistein and 6-hydroxybiochanin A showed significant activity, producing EC_{50} values of 18.3 μM and 14.5 μM, respectively (Tsuchihashi et al., 2009).

In activation, an inactive metabolite such as a pro-drug or precursor are activated, for example the activation of CB1954, known as tretazicar, a potent DNA alkylating and neoplastic agent by *E. coli*. Lehouritis et al. (2016) showed the *in vitro* biochemical activation of CB1954 following co-culture with various probiotic bacteria, however, *E. coli* showed the most promising result as it exhibited the least susceptibility to toxicity of the drug itself. HPLC analysis and spectrophotometric analysis of the absorbance at 420 nm of the co-culture supernatant confirmed the bioactivation of CB1954, which correlated with enhanced *in vitro* cytotoxicity against CT26 colon carcinoma cells. Experiments in mice with murine tumors validated the *in vitro* findings, with the results confirming a significant reduction in tumor growth and increase in survival of mice treated with probiotic bacteria and the pro-drug (Lehouritis et al., 2016).

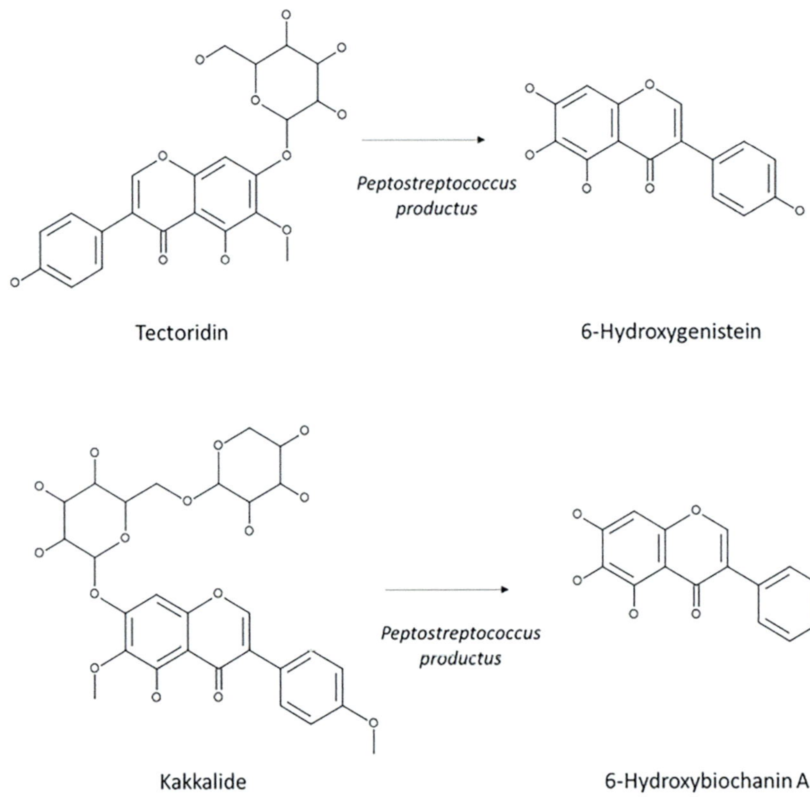

FIGURE 2.3 Bioactivation of the isoflavones from *Pueraria* flowers by *Peptostreptococcus productus*.

Biotransformation may also result in the conversion of an active compound to an equally active compound, as in the instance of the cardiac glycosides lanatoside C to digoxin by *Penicillium vermiculatum* and *Arthrobacter citreus* (Figure 2.4). In this two-step reaction, lanatoside C co-cultured with *P. vermiculatum* for 12–20 hours yields acetyldigoxin as the sole metabolite through deglycosylation, with a yield of 60%. Incubation of acetyldigoxin with *A. citrus* for a further 12 hours results in a deacetylation reaction which yields digoxin with approximately 60% efficiency. Alternatively, lanatoside C may be incubated with *P. vermiculatum* or *A. citreus* for 240 and 120 hours, respectively; however, in addition to the extended period required for the incubation, the reaction yields are low (Figure 2.4) (Fuska et al., 1987).

Finally, biotransformation may result in the conversion of an active metabolite to an active metabolite with an entirely different activity; for example, N-dealkylation at the piperazinyl nitrogen of nefazodone. An antagonist to the 5-HT2C serotonin receptor forms m-chlorophenylpiperazine, an agonist of this receptor (Figure 2.5) (Von Moltke et al., 1999; Kebamo and Tesema, 2015).

Thus, through fermentation of the plant material, the advantages to be gained include bioactivation and improved efficacy of the extracts, as well as diminished toxicity. Fermentation is thought to be one of the most feasible methods to achieve the discussed advantages, making the approaches involved more attractive (Rasor and Voss, 2001).

2.2.1 Types of Microbial Fermentation

Generally, four ubiquitous types of fermentation reactions are distinguished: lactic acid, acetic acid, alcohol/ethanol and alkaline fermentation. Lactic acid fermentation describes a process in which microorganisms such as yeast and bacteria convert carbohydrates into lactic acid, without requiring

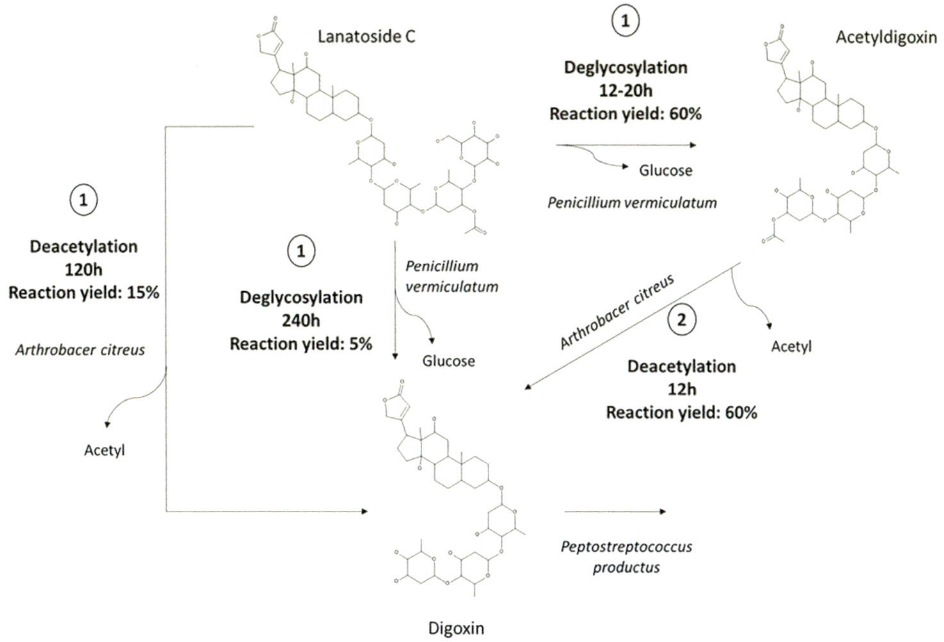

FIGURE 2.4 Biotransformation of lanatoside C by *Penicillium vermiculatum* and *Arthrobacter citreus.*

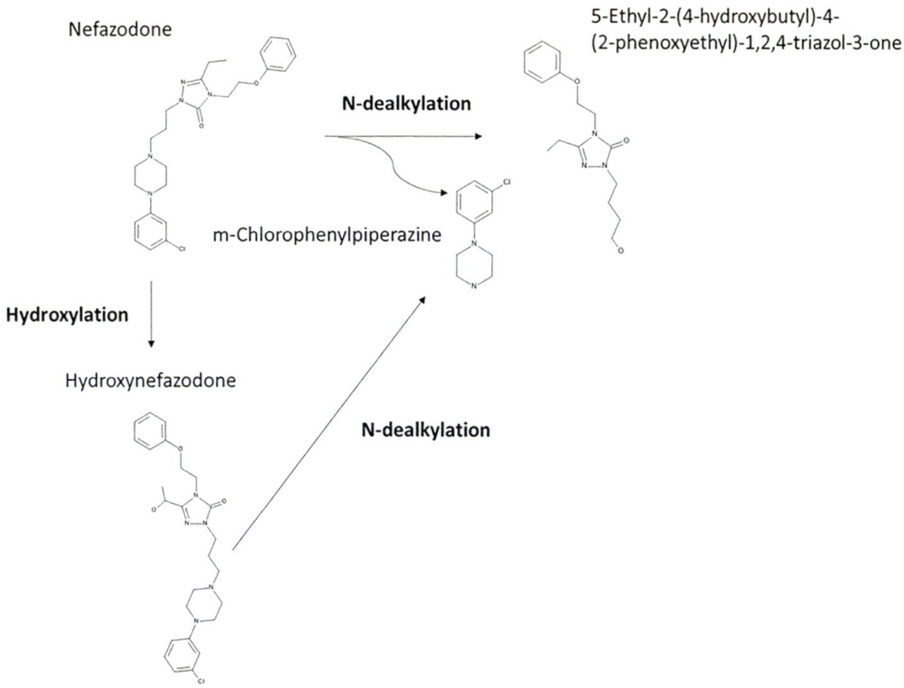

FIGURE 2.5 Biotransformation of nefazodone.

heat to initiate the reaction. Under anaerobic conditions, pyruvate generated from glycolysis is converted to pyruvic acid, which uses nicotinamide adenine dinucleotide hydrogen (NADH) to form lactic acid and nicotinamide adenine dinucleotide (NAD^+). This type of fermentation is inexpensive and frequently used to preserve healthy foods such as yogurt, sauerkraut, pickles, and sourdough bread. In addition to lactic acid, lactic acid fermentation produces (in smaller quantities), acetic acid, ethanol, aromatic compounds, polysaccharides and enzymes (Doelle, 1975; Anal, 2019). Some examples of heterofermentative pathways producing lactic acid, acetate and alcohol using glucose as an example for hexose starting substrates or xylose as an example of pentose starting substrates are presented in Figure 2.6 and Figure 2.7.

In alcohol fermentation, the glycolysis pathway produces pyruvate molecules, which are broken down into alcohol and carbon dioxide molecules by microorganisms such as yeast in the production of wine and beer. Alcoholic fermentation is divided into primary and secondary fermentation. In brief, for primary fermentation, microorganisms rapidly metabolize raw ingredients converting

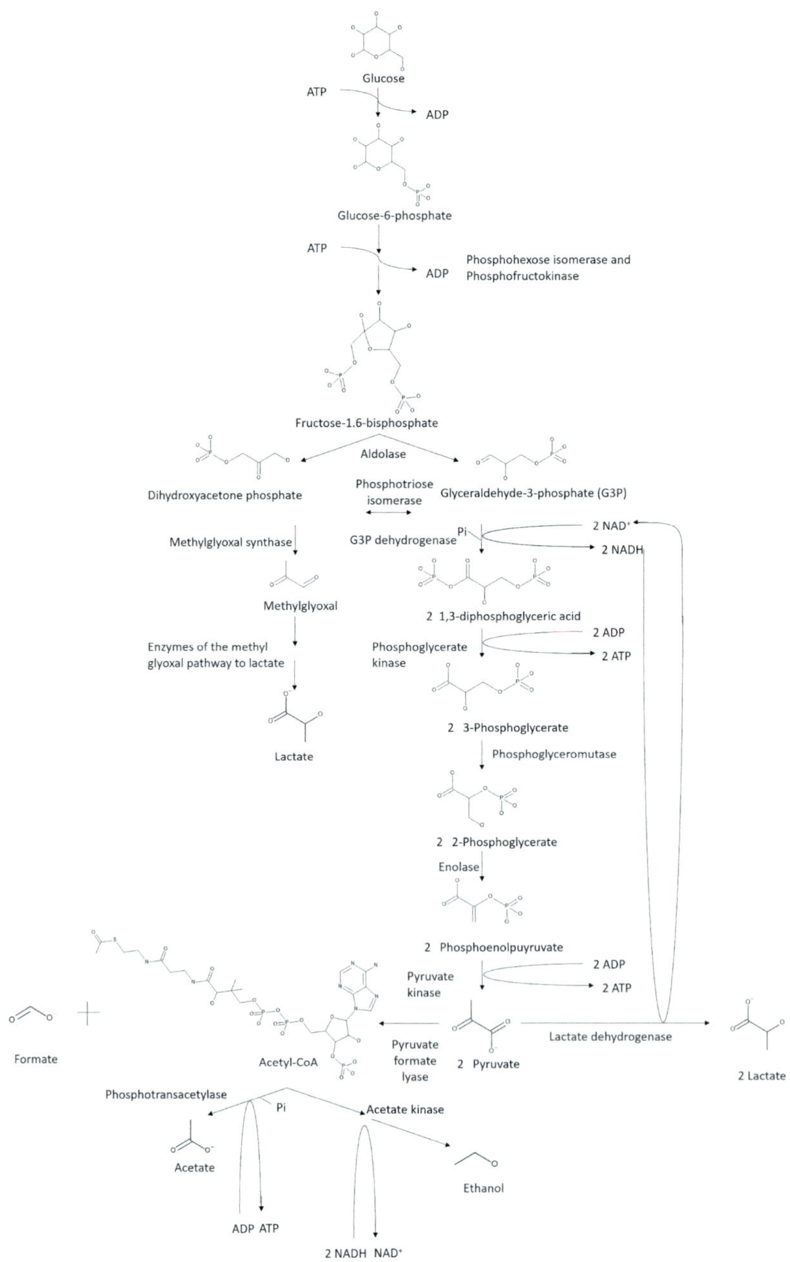

FIGURE 2.6 Heterofermentation using glucose as a substrate.

FIGURE 2.7 Heterofermentation using xylose as a substrate.

carbohydrates into alcohols and acids. Secondary fermentation, on the other hand, may last up to several days or weeks. In this time, the available carbohydrates become diminished and the products of fermentation, acids and alcohols accumulate. This results in microorganism death due to a significant drop in the pH and alcohol accumulation. In the instance of alcoholic beverage production, this fermentation product will be distilled to purify and concentrate the alcohol (Doelle, 1975; Anal, 2019).

Acetic acid fermentation is an aerobic process in which carbohydrates from grains and fruit are fermented into acetic acid. This is often used to produce condiments and beverages such as vinegar and kombucha by bacteria of the *Acetobacter* genus. Finally, in alkaline fermentation, proteins are broken down into their amino acid and peptide components, releasing ammonia in the process. *Bacillus* species are commonly employed with this type of fermentation and it is habitually associated with the fermentation of fish, soybeans and seeds. Therefore, although carbohydrates are the most ubiquitous substrate for fermentation, organic acids, peptides and molecules are fermentable substrates (Doelle, 1975; Anal, 2019).

2.2.2 Methods of Microbial Fermentation

When using microorganisms, there are typically two methods of fermentation: liquid-state and solid-state fermentation. Liquid-state, or submerged, fermentation involves the growth of microorganisms in an aqueous medium with free-flowing liquid substrates. This is best suited for fermentation employing bacteria as the driving mechanism, due to their necessity of high moisture content for survival. In this procedure, microorganisms are grown in a closed vessel containing the fermentation broth into which the bioactive compounds, extracts or fresh plant material is liberated. Rapid consumption of substrates by metabolizing microorganisms demands the replenishing of substrates recurrently for continued activity. Advantages of this technique include convenient concentration and purification of the desired product, especially secondary metabolites. Limitations include a strict necessity for sterile working conditions (Diaz-Godinez et al., 2017).

Solid-state fermentation (SSF) occurs on a solid matrix, which serves as a support or substrate on which the microorganism is added, either in the presence or absence of free water to sustain the growth of microorganisms. For this reason, fungi are typically the microorganism of choice. SSF is used in a variety of applications for producing enzymes commercially, antibodies, remediation and the production of bioethanol (Lizardi-Jiménez and Hernández-Martínez, 2017). In SSF, the plant material typically acts as a substrate and supports the microorganism. SSF generally takes longer than liquid-state fermentation, and as such, the substrate does not need to be replenished as frequently (Diaz-Godinez et al., 2017).

2.3 FERMENTATION OF PLANTS FOR ANTI-AGING COSMECEUTICALS

Anti-aging cosmetic creams increasingly muster popularity as the global life expectancy continues to rise, and with it the desire to look more youthful. Although the demand for natural anti-aging actives is growing, the use of fermentation of plant actives is generating a great deal of interest in this field. When appropriately prepared, fermented plant extracts offer rich sources of antioxidants, vitamins, minerals, polyphenols, proteins, fibers and prebiotics (Sivamaruthi, Chaiyasut, and Kesika, 2018).

However, improved phytochemical content is contingent on fermentation conditions, microorganism selection, and choice of plant species employed (Sivamaruthi, Chaiyasut, and Kesika, 2018). This increasingly popular tool has been used for the improvement of numerous herbs in various cosmetic applications such as anti-aging, oral hygiene, and hair loss. Table 2.1 summarizes the observed change in biological activity following the fermentation of several herbs against specific anti-aging targets for a range of medicinal plants (Figure 2.8 and 2.9). Unless indicated, these results are relative to the control sample (i.e. the plant species prepared identically in the absence of a fermenting agent).

TABLE 2.1
Effect of Microbial Fermentation on the Anti-Aging Activity of Several Medicinal Plants (Figure 2.8 and Figure 2.9)

Plant Species/Plant Parts/Extraction Solvents	Common Names	Microorganisms Used in Fermentation	Results of Investigation	References
Acanthopanax koreanum Nakai/ Root/Aqueous	NA	*Lactobacillus plantarum* KCCM 12116 and *Bifidobacterium bifidum* KCCM 12096	↑ ABTS scavenging ↑ DPPH scavenging ↓ p53 and p21Cip1/WAF1 induced by UVB or H_2O_2 treatment in HS68 cells ↓ Senescence-associated β-galactosidase ↓ UVB and H_2O_2 induced ROS ↓ UVB and H_2O_2 induced MMP-1, -3, and NF-κB activation	(Park and Bae, 2016)
Agastache rugosa Kuntze/Stems and leaves/70% Ethanol	Korean mint	*Lactobacillus rhamnosus* and *Lactobacillus paracasei*	↑ Flavonoid content ↓ Melanogenesis in B16F10 cells ↑ Reducing power ↑ TPC ↑ Tyrosinase inhibition	(Kim, Kwon, and Lee, 2017)
Agastache rugosa Kuntze/Leaves/ Aqueous	Korean mint	*Lactobacillus rhamnosus* HK-9	↑ SOD activity in UVB irradiated HaCaT cells ↑ Total glutathione levels in UVB irradiated HaCaT cells ↓ UVB-induced MMP-2 and -9 ↓ UVB-induced photooxidative stress in HaCaT cells No effect on ABTS radical scavenging	(Shin et al., 2018)
Aloe arborescens Mill./Aerial parts/ Aqueous	Candelabra Aloe, tree Aloe, mountain bush Aloe	*Lactobacillus plantarum*	↑ DPPH scavenging ↓ MMP-1 expression No significant improvement of collagen type 1 synthesis in human fibroblasts	(Ro et al., 2020)
Arctium lappa L/ Root, called *Fructus arctii*/Aqueous	Burdock, the seed is commonly referred to as the Great power seed	*Grifola frondosa* HB0071	↑ DPPH scavenging ↑ HDF proliferation ↑ 5-LOX inhibition ↓ MMP-1 expression in UVA irradiated HDF cells ↑ Superoxide scavenging	(Kim et al., 2010)
Asparagus cochinchinensis (Lour.) Merr./Root/ Aqueous, 95% ethanol, 50% ethanol, 100% ethyl acetate, and 50% ethyl acetate	Chinese asparagus	*Bifidobacterium bifidum* ATCC 29521	↑ DPPH scavenging ↑ Inhibition of mushroom tyrosinase ↑ Reducing power ↑ TPC	(Wang et al., 2016)

(*Continued*)

TABLE 2.1 (Continued)

Plant Species/Plant Parts/Extraction Solvents	Common Names	Microorganisms Used in Fermentation	Results of Investigation	References
Asparagus cochinchinensis (Lour.) Merr./Root/ Ethyl acetate	Asparagus	*Aspergillus oryzae*	↑ DPPH scavenging ↓ Melanin production in human epidermal melanocytes ↓ MITF ↓ Reducing power ↑ TPC ↓ TRP-1 ↑ Tyrosinase inhibition No significant effect on TRP-2	(Wang et al., 2019)
Astragalus membranaceus (Fisch.) Bge. var. *mongholicus* (Bge.)/ Not specified/ Aqueous	Mongolian milkvetch	*Bacillus subtilis* natto ATCC 7059	↑ Expression of hyaluronan synthase-2 and -3 in fibroblasts and HaCaT cells ↑ HA production ↓ Specific isoflavonoids	(Hsu and Chiang, 2009)
Avena sativa L./Not specified/Not specified	Oat	*Lactobacillus buchneri*	↑ Procollagen type-I synthesis in normal human dermal fibroblasts and epidermal keratinocytes No significant effect on elastase inhibition	(Kang et al., 2020)
Camellia sinensis var. *assamica*/Leaves/ Aqueous, 95% ethanol, 50% ethanol, 95% acetone and 50% acetone	Tea	*Trichoderma harzianum* CB-Pin-01	↓ ABTS scavenging ↓ DPPH scavenging ↓ Ferric reducing antioxidant power ↓ TFC ↓ TPC	(Kritsadaruangchai et al., 2019)
		Trichoderma sp. NTY 211	↓ ABTS scavenging ↓ DPPH scavenging ↓ Ferric reducing antioxidant power ↓ TFC ↓ TPC	
		Trichoderma sp. PSUT 001	↓ ABTS scavenging ↓ DPPH scavenging ↓ Ferric reducing antioxidant power ↓ TFC ↓ TPC	
		Trichoderma sp. PSUT 002	↓ ABTS scavenging ↓ DPPH scavenging ↓ Ferric reducing antioxidant power ↓ TFC ↓ TPC	
Citrus unshiu L./Peel/ Aqueous	Unshu mikan, cold hardy/satsuma mandarin, satsuma orange, naartjie, tangerine	*Schizophyllum commune* KACC 93060P	↑ Collagen biosynthesis ↓ MMP-1 expression in UVA irradiated HDF cells ↓ UVA induced cellular senescence	(Bae et al., 2012)

Plant Species/Plant Parts/Extraction Solvents	Common Names	Microorganisms Used in Fermentation	Results of Investigation	References
Combination of: *Atractylodes macrocephala* Koidz/*Angelica gigas* Nakai/*Broussonetia kazinoki* (L.) L'Herit ex Vent./*Glycyrrhiza glabra* L./*Lithospermum officinale* L./*Paeonia lactiflora* Pall/*Morus alba* L. (root bark)/ Fungi: *Poria cocos*/ Aqueous	Largehead atractylodes Korean angelica, giant angelica, purple parsnip, dangquai/Paper mulberry/ Licorice/ Common gromwell, European stone seed/Chinese peony/White mulberry	*Phellinus linteus*	↓ Melanin content in B16F10 mouse melanocyte cells ↓ Tyrosinase production in B16F10 cells No inhibitory effects on tyrosinase related protein-1 and tyrosinase related protein-2	(Cha et al., 2012)
Glycine max Merr./ Not specified/Not specified	Soybean	*Lactobacillus buchneri*	↑ Expression of superoxide and catalase mRNA in UVB irradiated human dermal fibroblasts and epidermal keratinocytes ↑ Expression of HA synthase-1 in keratinocytes ↓ Expression of MMP-1 and -9 mRNA in UVB irradiated human keratinocytes ↑ Procollagen type-I synthesis in normal human dermal fibroblasts and epidermal keratinocytes No significant effect on elastase inhibition	(Kang et al., 2020)
Helianthus tuberosus L/Not specified	Jerusalem artichoke	*Lactobacillus buchneri*	↑ Procollagen type-I synthesis in normal human dermal fibroblasts and epidermal keratinocytes No significant effect on elastase inhibition	(Kang et al., 2020)
Lavandula angustifolia Mill./ Flowers/70% Ethanol	Lavender	*Pediococcus pentosaceus* DK1	↑Collagen production in UVB irradiated fibroblasts ↓ ROS generation in UVB irradiated fibroblasts ↓ UVB-mediated MMP-1 expression in fibroblasts	(Ha et al., 2019)
Lespedeza cuneata G. Don/Aerial parts/70% Ethanol	Chinese lespedeza	*Lactobacillus pentosus*	↑ DPPH scavenging ↑ Elastase inhibition in Hs68 cells ↑ Ferric iron reducing activity ↑ Inhibition of UV-induced MMP-1 expression in Hs68 cells ↑ Procollagen type-I synthesis in Hs68 cells	(Seong et al., 2017)

(*Continued*)

TABLE 2.1 (Continued)

Plant Species/Plant Parts/Extraction Solvents	Common Names	Microorganisms Used in Fermentation	Results of Investigation	References
Lonicera japonica Thunb/Not specified/Ethanol	Japanese honeysuckle, Chinese honeysuckle	*Alcaligenes piechaudii* CC-ESB2	↑ DPPH scavenging ↑ Mushroom tyrosinase inhibition ↑ Superoxide scavenging activity ↑ TPC	(Chen, Liou, and Chan, 2013)
Oryza sativa L./Seed/Aqueous	Rice	*Aspergillus niger*	↑ DPPH scavenging ↑ Elastase inhibition ↑ FRAP ↑ TFC ↑ TPC ↑ Tyrosinase inhibition	(Razak et al., 2019)
		Rhizopus oligosporus	↑ DPPH scavenging ↑ Elastase inhibition ↑ FRAP ↑ TFC ↑ TPC ↑ Tyrosinase inhibition	
Paeonia suffruticosa Andr./Not specified/Aqueous, 95% ethanol, 50% ethanol, 100% ethyl acetate, and 50% ethyl acetate.	Moutan peony	*Bifidobacterium bifidum* ATCC 29521	↑ DPPH scavenging ↑ Inhibition of mushroom tyrosinase ↑ Reducing power ↑ TPC	(Wang et al., 2016)
**Panax ginseng* C.A. Mey./Root/Aqueous	Black Ginseng	*Saccharomyces cerevisiae*	↓ MMP-1, MMP-2, MMP-9 expression ↑ Procollagen type-I expression in HS68 cells ↑ TIMP-2 expression	(Pham, Jang, and Kim, 2017)
Panax ginseng C.A. Mey./Root/Aqueous	Red ginseng	*Lactobacillus brevis* M2 KCTC11390BP	↑ ABTS scavenging ↑ DPPH scavenging ↑ Elastase inhibition ↑ Ginsenoside metabolite content ↑ Mushroom tyrosinase inhibition ↑ TFC ↑ TPC ↑ Uronic acid content	(Lee et al., 2012)
Panax ginseng C.A. Mey./Berry/70% Ethanol	Ginseng	*Pleurotus ferulae*	↑ Collagen synthesis ↑ DPPH scavenging ↑ 5-LOX inhibition ↑ MMP-1 inhibition ↑ Tyrosinase inhibition	(Lee, 2019)
		Tricholoma matsutake	↑ Collagen synthesis ↑ DPPH scavenging ↑ Tyrosinase inhibition No significant inhibition of 5-LOX No significant inhibition of MMP-1	

Plant Species/Plant Parts/Extraction Solvents	Common Names	Microorganisms Used in Fermentation	Results of Investigation	References
		Sarcodon aspratus	↑ Collagen synthesis ↑ DPPH scavenging ↑ Tyrosinase inhibition No significant inhibitory effect on 5-LOX No significant inhibition of MMP-1	
		Phellinus linteus	↑ Collagen synthesis ↑ DPPH scavenging ↑ Tyrosinase inhibition No significant inhibition of 5-LOX No significant inhibition of MMP-1	
		Gonoderma lucidum	↑ Collagen synthesis ↑ DPPH scavenging ↑ 5-LOX inhibition ↑ Tyrosinase inhibition No significant inhibition of MMP-1	
Prunus persica (L.) Batsch/Not specified/Aqueous, 95% ethanol, 50% ethanol, 100% ethyl acetate, and 50% ethylacetate.	Peach	*Bifidobacterium bifidum* ATCC 29521	↑ DPPH scavenging ↑ Inhibition of mushroom tyrosinase ↑ Reducing power ↑ TPC	(Wang et al., 2016)
Punica granatum L./Not specified/Aqueous	Pomegranate	*Saccharomyces cerevisiae, Lactobacillus plantarum* and *Acetobacter aceti*	↑ Antiglycative activity ↑ COL4A4 Expression ↑ Collagen density in human skin fibroblasts ↓ Melanin production ↓ MMP-2 Expression ↑ Skin firmness ↑ TIMP-1 Expression ↑ TPC ↑ UVA protection in CCD-966SK No significant improvement in anti-inflammatory activity	(Lin and Hsiu, 2019)
Rhodiola rosea L./Not specified/Ethanol	Golden/Arctic root, rose root, Aaron's rod, king's crown, lignum rhodium, pin rose	*Alcaligenes piechaudii* CC-ESB2	↑ DPPH scavenging ↑ Mushroom tyrosinase inhibition ↑ Superoxide scavenging ↑ TPC	(Chen, Liou, and Chan, 2013)

(*Continued*)

TABLE 2.1 (Continued)

Plant Species/Plant Parts/Extraction Solvents	Common Names	Microorganisms Used in Fermentation	Results of Investigation	References
Scutellaria baicalensis Georgi/Not specified/70% Ethanol, aqueous	Baikal/Chinese skullcap	*Lactobacillus plantarum*	↑ Collagen type-I synthesis ↑ DPPH scavenging ↑ Involucrin expression in HaCaT cells ↑ Keratin10 expression in HaCaT cells ↑ Keratinocyte proliferation and differentiation ↑ TFC ↑ Transglutamase-1 expression in HaCaT cells ↑ TPC	(Lee et al., 2019)
Smallanthus sonchifolius (Poepp.) H. Rob./ Not specified/Not specified	Yacón	*Lactobacillus buchneri*	↑ Procollagen type-I synthesis in normal HDFs and epidermal keratinocytes No significant effect on elastase inhibition	(Kang et al., 2020)
**Trapa japonica* Flerow/Fruit/ Aqueous: Fermented extract was sequentially separated into hexane, chloroform, ethyl acetate, butanol, and water extracts	NA	*Bacillus methulotrophicus* and *Bacillus subtilis*	↑ Collagen synthesis through TGF-β1/GSK-3β/β-catenin pathway in HDF cells Regulates skin cell migration	(Nam et al., 2020)
		Bacillus subtilis (K007) and *Bacillus methylotrophicus* (S001)	↑ Collagen synthesis ↑ Inhibition of mRNA expression MMPs induced by TNF-α in HDFs ↑ Wound healing in HaCaT cells Clinical study of anti-aging activity was conducted	(Jang et al., 2020)
Triticum aestivum L./ Leaf/Not specified	Common wheat	*Lactobacillus buchneri*	↑ Procollagen type-I synthesis in normal HDFs and epidermal keratinocytes ↑ Expression of HA synthase-1 ↓ Expression of MMP-1 and -9 in UVB irradiated human keratinocytes ↑ Expression of superoxide and catalase mRNA in UVB irradiated human keratinocytes No significant effect on elastase inhibition	(Kang et al., 2020)
Zanthoxylum schinifolium Siebold. & Zucc./ Not specified/ Aqueous	Peppertree	*Lactobacillus rhamnosus* A6-5	↑ Antibacterial activity against *Propionibacterium acnes* KACC-11946 and *Staphylococcus epidermidis* KACC-13234 ↑ Anti-melanogenic activity ↑ DPPH scavenging ↓ IL-6 ↓ Nitric oxide production ↑ Tyrosinase inhibition No effect on TPC	(Lee et al., 2020)

*Activity not compared to control sample

FIGURE 2.8 Plant species represented in Table 2.1. a, *Agastache rugose*; b, *Aloe arborescens*; c, *Angelica gigas*; d, *Arctium lappa* aerial parts; e, *Arctium lappa* root; f, *Asparagus cochinchinensis*; g, *Astragalus membranaceus*; h, *Avena sativa*; i, *Broussonetia kazinoki*; j, *Camellia sinensis* var. *assamica*; k, *Citrus unshiu*; l, *Glycine max*; m, *Glycyrrhiza glabra*; n, *Helianthus tuberosus*; o, *Lavandula angustifolia*; p, *Lespedeza cuneate*; q, *Lithospermum officinale* (Данилевский, 2018; Bouba, 2007; Dalgial, 2010a; Dénes, 2019; Doronenko, 2007; Cliff, 2013; Giuşcă, 2011; Golik, 2018; JPLC, 2012a; Kenpei, 2007a; Kenraiz, 2018; Peganum, 2014; National Institute of Korean Language, 2017; Syrio, 2018; Qwert1234, 2009; Zell, 2010, 2011).

FIGURE 2.9 Plant species represented in Table 2.1. a, *Lonicera japonica*; b, *Morus alba*; c, *Oryza sativa*; d, *Paeonia lactiflora*; e, *Paeonia suffruticosa* flower; f, *Panax ginseng* aerial parts; g, *P. ginseng* root, red ginseng; h, *Prunus persica* flower; i, *Punica granatum*; j, *Rhodiola rosea*; k, *Scutellaria baicalensis*; l, *Smallanthus sonchifolius*; m, *Trapa japonica* floating on the surface of water; n, *T. japonica* fruit; o, *Triticum aestivum*; p, *Zanthoxylum schinifolium* (Σ64 2008; Dalgial 2010b, 2011; Dumat, 2011; Eggmoon, 2013; Garcia, 2015; Jaritz, 2014; JPLC, 2012b; Kenpei, 2005, 2007b; Melsi, 2020; Nguyen, 2013; Peka, 2019; Shizhao, 2005; Siebrand, 2005; Xaver, 2010).

Literature indicates that the most commonly observed mechanism by which fermentation transforms plant extracts is through the release of biologically active and more readily absorbed phenolics from their conjugated sugars. Glycosides are known to be hydrophilic due to the presence of a glycosyl group, which limits their application in topical cosmetics as they exhibit low skin permeability. The cleavage of glycosidic bonds to yield the aglycone form of several common phytochemicals is thought to primarily be caused by the action of enzymes such as β-glucosidases, which are particularly abundant in lactic acid bacteria. Aglycone components such as naringenin and caffeic acid tend to exhibit more hydrophobic properties and a lower molecular weight, which enables enhanced skin permeation. Alternative methods such as acidic or alkaline hydrolysis offer harsh conditions, which lead to the instability of active compounds, are not environmentally friendly and are cost intensive. This bids additional motivation for the adoption of techniques such as microbial fermentation. Similarly, enzymatic hydrolysis requires the implementation of cost-intensive purified enzymes, which is not economically feasible (Kim et al., 2010; Lee et al., 2012; Wang et al., 2016; Nam et al., 2020).

2.4 CONCLUSION

In understanding the molecular processing of aging skin, key focus areas and targets become apparent. Many scientific investigations have highlighted the importance of neutralizing oxidative stress in order to diminish the activation of the aging pathway. Additional measures all relate to the preservation of the structural components within the dermis, such as collagen and elastin. These measures may be directly exerted by identifying compounds capable of inhibiting proteolytic enzymes, which degrade these components directly or indirectly by mediating inflammatory markers, which sustain conditions of stress. Probiotics are beneficial microorganisms which—when administered in correct amounts—confer health benefits. Not only have they exhibited immune modulation through their beneficial effects on the gut microbiota, but they have similarly recorded effects on the skin.

Phytochemicals have long been known to have health-restorative effects in a range of life-threatening diseases such as cancer and tuberculosis, and can be used to rejuvenate the skin. Although biologically effective, the full healing effects of phytochemicals are often limited by their molecular weight, inability to reach the target site and bioavailability. By combining the therapeutic properties of plants with microbial fermentation, opportunities avail themselves for the development of novel phytochemicals with enhanced therapeutic efficacy. Although many studies have illustrated a promising future for fermentation technology in the development of natural cosmeceuticals, it should be noted that the process is highly dependent on several factors and combinations thereof, including the type of fermentation selected, microorganisms selected and plant species.

REFERENCES

Anal, Anil Kumar. 2019. "Quality Ingredients and Safety Concerns for Traditional Fermented Foods and Beverages from Asia: A Review." *Fermentation* 5, 1. doi: 10.3390/fermentation5010008.

Bae, Jun Tae, Hyun Ju Ko, Gyoung Bum Kim, Hyeong Bae Pyo, and Geun Soo Lee. 2012. "Protective Effects of Fermented Citrus Unshiu Peel Extract against Ultraviolet-a-Induced Photoageing in Human Dermal Fibrobolasts." *Phytotherapy Research* 26, 12: 1851–1856. doi: 10.1002/ptr.4670.

Bajaj, A., S. Mayilraj, M.K.R. Mudiam, D.K. Patel, and N. Manickam. 2014. "Isolation and Functional Analysis of a Glycolipid Producing *Rhodococcus* Sp. Strain IITR03 with Potential for Degradation of 1,1,1-Trichloro-2,2-Bis(4-Chlorophenyl)Ethane (DDT)." *Bioresource Technology* 167: 398–406.

Barragán-Huerta, B.E., C. Costa-Pérez, J. Peralta-Cruz, J. Barrera-Cortés, F. Esparza-García, and R. Rodríguez-Vázquez. 2007. "Biodegradation of Organochlorine Pesticides by Bacteria Grown in Microniches of the Porous Structure of Green Bean Coffee." *International Biodeterioration & Biodegradation* 59, 4: 239–244.

Bouba. 2007. "*Arctium Lappa* Aerial Parts." Obtained under the Creative commons Attribution-Share Alike 3.0 license. https://commons.wikimedia.org/w/index.php?curid=1644611 (Accessed 5 May 2021).

Cao, F., T.X. Liu, C.Y. Wu, F.B. Li, X.M. Li, H.Y. Yu, H. Tong, and M.J. Chen. 2012. "Enhanced Biotransformation of DDTs by an Iron- and Humic-Reducing Bacteria *Aeromonas hydrophila* HS01 upon Addition of Goethite and Anthraquinone-2,6-Disulphonic Disodium Salt (AQDS)." *Journal of Agricultural and Food Chemistry* 60, 45: 11238–11244.

Cha, Jae Young, Hyun Ju Yang, Hyung In Moon, and Young Su Cho. 2012. "Inhibitory Effect and Mechanism on Melanogenesis from Fermented Herbal Composition for Medical or Food Uses." *Food Research International* 45, 1. Elsevier Ltd: 225–231. doi: 10.1016/j.foodres.2011.11.002.

Chen, Yuh Shuen, Hua Chian Liou, and Chin Feng Chan. 2013. "Tyrosinase Inhibitory Effect and Antioxidative Activities of Fermented and Ethanol Extracts of *Rhodiola rosea* and *Lonicera japonica*." *The Scientific World Journal*. doi: 10.1155/2013/612739.

Cliff. 2013. "*Asparagus cochinchinensis*." Obtained under the Creative commons Attribution-Share Alike 2.0 license. https://commons.wikimedia.org/w/index.php?curid=29534806 (Accessed 5 May 2021).

Cole, Megan A., Taihao Quan, John J. Voorhees, and Gary J. Fisher. 2018. "Extracellular Matrix Regulation of Fibroblast Function: Redefining Our Perspective on Skin Aging." *Journal of Cell Communication and Signaling* 12, 1: 35–43. doi: 10.1007/s12079-018-0459-1.

Dalgial. 2010a. "*Lespedeza cuneata*." Obtained under the Creative commons Attribution-Share Alike 4.0 license. https://commons.wikimedia.org/w/index.php?curid=12343328 (Accessed 5 May 2021).

Dalgial. 2010b. "*Zanthoxylum schinifolium*." Obtained under the Creative commons Attribution-Share Alike 3.0 license. https://commons.wikimedia.org/w/index.php?curid=10748780%0A (Accessed 5 May 2021).

Dalgial. 2011. "*Scutellaria baicalensis*." Obtained under the Creative commons Attribution-Share Alike 3.0 license. https://commons.wikimedia.org/w/index.php?curid=15995530 (Accessed 5 May 2021).

Dénes, Emőke. 2019. "*Agastache rugosa*." Obtained under the Creative commons Attribution-Share Alike 4.0 license. https://commons.wikimedia.org/w/index.php?curid=81453366 (Accessed 5 May 2021).

Diaz-Godinez, G., M. Tellez-Tellez, C. Sanchez, and R. Diaz. 2017. "Characterization of the Solid State and Liquid Fermentation for the Production of Laccases of *Pleurotus ostreatus*." *Fermentation Processes*: 57–74. doi: 10.5772/64239.

Doelle, H.W. 1975. *Bacterial Metabolism*. 2nd ed. New York: Elsevier.

Doronenko, I. 2007. "*Astragalus membranaceus*." Obtained under the Creative commons Attribution-Share Alike 2.5 license. https://commons.wikimedia.org/w/index.php?curid=2293758%0A (Accessed 5 May 2021).

Dumat, Maja. 2011. "*Oryza sativa*." Obtained under the Creative commons Attribution-Share Alike 2.0 license. https://commons.wikimedia.org/w/index.php?curid=20840451%0A (Accessed 5 May 2021).

Eggmoon. 2013. "*Trapa japonica* Floating on the Surface of Water." Obtained under the Creative commons Attribution-Share Alike 3.0 license. https://commons.wikimedia.org/w/index.php?curid=29631015%0A (Accessed 5 May 2021).

Erdem, Z., and T.J. Cutright. 2016. "Biodegradation Potential of 1,1,1-Trichloro-2,2-Bis(Pchlorophenyl) Ethane (4,4′-DDT) on a Sandy-Loam Soil Using Aerobic Bacterium *Alcaligenes eutrophus* A5." *Environmental Engineering Science* 33, 3: 149–159.

Fibrich, Bianca Daphne, and Namrita Lall. 2018. "Fighting the Inevitable: Skin Aging and Plants." In *Medicinal Plants for Holistic Health and Well-Being*, edited by Namrita Lall, 77–114. Elsevier.

Fibrich, Bianca Daphne, Isa Anina Lambrechts, and Namrita Lall. 2019. "Nanotechnology and Anti-Ageing Skin Care." In *Natural Medicines: Clinical Efficacy, Safety and Quality*, edited by Dillip Ghosh and Pulok K. Mukherjee. Taylor and Francis.

Fisher, G.J., S. Kang, and J. Varanti. 2002. "Mechanisms of Photoageing and Chronological Skin Ageing." *Archives of Dermatology* 138: 1462–1477. doi: 10.1001/archderm.138.11.1462

Fisher, G.J., H.S. Talwar, J. Lin, P. Lin, P. McPhillips, Z. Wang, X. Li, Y. Wan, S. Kang, and J.J. Voorhees. 1998. "Retinoic Acid Inhibits Induction of C-Jun Protein by Ultraviolet Radiation That Occurs Subsequent to Activation of Mitogen-Activated Protein Kinase Pathways in Human Skin *in vivo*." *Journal of Clinical Investigation* 101: 1432–1440. doi: 10.1172/JCI2153.

Fuska, Ján, Bohumil Proksa, Alžbeta Khandlová, and Mária Šturdíková. 1987. "Microbial Transformations of Cardioglycosides." *Applied Microbiology and Biotechnology* 26, 4: 313–317. doi: 10.1007/BF00256660.

Ganesan, Palanivel, and Dong-Kug Choi. 2016. "Current Application of Phytocompound-Based Nanocosmeceuticals for Beauty and Skin Therapy." *International Journal of Nanomedicine* 11. Dove Press: 1987–2007. doi: 10.2147/IJN.S104701.

Gao, B., W.B. Liu, L.Y. Jia, L. Xu, and J. Xie. 2011. "Isolation and Characterization of an *Alcaligenes* Sp. Strain DG-5 Capable of Degrading DDTs under Aerobic Conditions." *Journal of Environmental Sciences Health B* 4, 3: 257–263.

Garcia. 2015. "*Morus alba*." Obtained under the Creative commons Attribution-Share Alike 2.1 Spain license. https://commons.wikimedia.org/w/index.php?curid=178195 (Accessed 5 May 2021).

Giuşcă, Bogdan. 2011. "*Lithospermum officinale*." Obtained under the Creative commons Attribution-Share Alike 1.0 license. https://commons.wikimedia.org/w/index.php?curid=15365025 (Accessed 5 May 2021).

Golik, Krzysztof. 2018. "*Camellia sinensis* var. *assamica*." Obtained under the Creative commons Attribution-Share Alike 4.0 license. https://commons.wikimedia.org/w/index.php?curid=67246344%0A (Accessed 5 May 2021).

Ha, Ji Hoon, A. Rang Kim, Keon Soo Lee, Song Hua Xuan, Hee Cheol Kang, Dong Hwan Lee, Mi Yeon Cha, Hye Jin Kim, Mi An, and Soo Nam Park. 2019. "Anti-Aging Activity of *Lavandula angustifolia* Extract Fermented with *Pediococcus pentosaceus* DK1 Isolated from Diospyros Kaki Fruit in UVB-Irradiated Human Skin Fibroblasts and Analysis of Principal Components." *Journal of Microbiology and Biotechnology* 29, 1: 21–29. doi: 10.4014/jmb.1809.09037.

Hsu, Mei Fang, and Been Huang Chiang. 2009. "Stimulating Effects of *Bacillus subtilis* Natto-Fermented *Radix astragali* on Hyaluronic Acid Production in Human Skin Cells." *Journal of Ethnopharmacology* 125, 3: 474–481. doi: 10.1016/j.jep.2009.07.011.

Huang, Huanfang, Yuan Zhang, Wei Chen, Wenwen Chen, Dave Yuen, Yang Ding, Yingjie Chen, Yao Mao, and Shihua Qi. 2018. "Sources and Transformation Pathways for Dichlorodiphenyltrichloroethane (DDT) and Metabolites in Soils from Northwest Fujian, China." *Environmental Pollution* 235: 560–570. doi: 10.1016/j.envpol.2017.12.071

Hussain, Ahtesham, Shambhunath Bose, Jing Hua Wang, Mukesh Kumar Yadav, Girish B. Mahajan, and Hojun Kim. 2016. "Fermentation, a Feasible Strategy for Enhancing Bioactivity of Herbal Medicines." *Food Research International* 81: 1–16. doi: 10.1016/j.foodres.2015.12.026.

Jang, Jin Dong, Minkyung Kim, Gun He Nam, Young Min Kim, Sang Moon Kang, Kee Young Lee, and Young Joon Park. 2020. "Antiaging Activity of Peptide Identified from Fermented *Trapa japonica* Fruit Extract in Human Dermal Fibroblasts." *Evidence-Based Complementary and Alternative Medicine.* doi: 10.1155/2020/5895029.

Jaritz, Johan. 2014. "*Paeonia lactiflora*." Obtained under the Creative commons Attribution-Share Alike 4.0 license. https://commons.wikimedia.org/wiki/File:Paeonia_lactiflora_24052014_966.jpg%0A (Accessed 5 May 2021).

JPLC. 2012a. "*Lavandula angustifolia*." Obtained under the Creative commons Attribution-Share Alike 3.0 license. https://commons.wikimedia.org/w/index.php?curid=19983389 (Accessed 5 May 2021).

JPLC. 2012b. "*Punica granatum*." Obtained under the Creative commons Attribution-Share Alike 3.0 license. https://commons.wikimedia.org/w/index.php?curid=20204581 (Accessed 5 May 2021).

Kang, Yun Mi, Chul Hee Hong, Sa Haeng Kang, Dong Seok Seo, Seong Oh Kim, Hoon Yeon Lee, Hyeon Jae Sim, and Hyo Jin An. 2020. "Anti-Photoaging Effect of Plant Extract Fermented with *Lactobacillus buchneri* on CCD-986sk Fibroblasts and HaCaT Keratinocytes." *Journal of Functional Biomaterials* 11, 1: 1–14. doi: 10.3390/jfb11010003.

Kebamo, Selamu, and Shibiru Tesema. 2015. "The Role of Biotransformation in Drug Discovery and Development." *Journal of Drug Metabolism & Toxicology* 06, 05. doi: 10.4172/2157-7609.1000196.

Kenpei. 2005. "*Prunus persica* Flower." Obtained under the Creative commons Attribution-Share Alike 3.0 license. https://commons.wikimedia.org/w/index.php?curid=1881170%0A (Accessed 5 May 2021).

Kenpei. 2007a. "*Citrus unshiu*." Obtained under the Creative commons Attribution-Share Alike 3.0 license. https://commons.wikimedia.org/w/index.php?curid=2144910 (Accessed 5 May 2021).

Kenpei. 2007b. "*Lonicera japonica*." Obtained under the Creative commons Attribution-Share Alike 3.0 license. https://commons.wikimedia.org/w/index.php?curid=2144947 (Accessed 5 May 2021).

Kenraiz, Krzysztof Ziarnek. 2018. "*Glycine max*." Obtained under the Creative commons Attribution-Share Alike 4.0 license. https://commons.wikimedia.org/w/index.php?curid=71029543 (Accessed 5 May 2021).

Khol, E., J. Steinbauer, M. Landthaler, and R.M. Szeimies. 2011. "Skin Ageing." *Journal of The European Academy of Dermatology and Venerology* 25: 873–884. doi: 10.1111/j.1468-3083.2010.03963.x.

Kim, Jin Hwa, Jun Tae Bae, Min Hyeon Song, Geun Su Lee, Soo Young Choe, and Hyeong Bae Pyo. 2010. "Biological Activities of *Fructus arctii* Fermented with the Basidiomycete *Grifola frondosa*." *Archives of Pharmacal Research* 33, 12: 1943–1951. doi: 10.1007/s12272-010-1209-y.

Kim, Nam Young, Hee Souk Kwon, and Hyeon Yong Lee. 2017. "Effect of Inhibition on Tyrosinase and Melanogenesis of *Agastache rugosa* Kuntze by Lactic Acid Bacteria Fermentation." *Journal of Cosmetic Dermatology* 16, 3: 407–415. doi: 10.1111/jocd.12264.

Kritsadaruangchai, Usa, Phanuphong Chaiwut, Putarak Chomnunti, Narit Thaochan, Angkana Saikeur, and Punyawatt Pintathong. 2019. "Effect of Solid State Fermentation with *Trichoderma* Spp. on Phenolic Content and Antioxidant Capacities of Mature Assam Tea Leaves." *Journal of Food Science and Agricultural Technology* 5: 106–113.

Lee, Hyeon Yong. 2019. "Improvement of Skin Barrier Dysfunction by *Scutellaria baicalensis* GEOGI Extracts through Lactic Acid Fermentation." *Journal of Cosmetic Dermatology* 18, 1: 183–191. doi: 10.1111/jocd.12521.

Lee, Hyun-sun, Mi-ryung Kim, Yooheon Park, Hyo Jung Park, and Un Jae Chang. 2012. "Fermenting Red Ginseng Enhances Its Safety and Efficacy as a Novel Skin Care Anti-Aging Ingredient: *In Vitro* and Animal Study." *Journal of Medicinal Food* 15, 11: 1015–1023. doi: 10.1089/jmf.2012.2187.

Lee, Kyung Rok, Il Hong, Do Gyeong Lee, Sung Min Park, Jung No Lee, and Nu Rim Lee. 2019. "Method for Preparing Cosmetic Composition Containing Fermented Ginseng Berry *Pleurotus ferulae* Product and Use Thereof." US 2016/0151273 A1, issued 2019.

Lee, Se Won, Jeong Muk Lim, Harshavardhan Mohan, Kamala Kannan Seralathan, Yool Jin Park, Jeong Ho Lee, and Byung Taek Oh. 2020. "Enhanced Bioactivity of *Zanthoxylum schinifolium* Fermented Extract: Anti-Inflammatory, Anti-Bacterial, and Anti-Melanogenic Activity." *Journal of Bioscience and Bioengineering* 129, 5: 638–645. doi: 10.1016/j.jbiosc.2019.12.003.

Lehouritis, Panos, Michael Stanton, Florence O. McCarthy, Matthieu Jeavons, and Mark Tangney. 2016. "Activation of Multiple Chemotherapeutic Prodrugs by the Natural Enzymolome of Tumour-Localised Probiotic Bacteria." *Journal of Controlled Release* 222: 9–17. doi: 10.1016/j.jconrel.2015.11.030.

Lin, Yung-Hsiang, and Wei Chuang Hsiu. 2019. "Fermentation Product of *Punica granatum* and Uses Thereof." US 2019/0224259 A1, issued 2019.

Lizardi-Jiménez, M.A., and R. Hernández-Martínez. 2017. "Solid State Fermentation (SSF): Diversity of Applications to Valorize Waste and Biomass." *3 Biotech* 7, 1. doi: 10.1007/s13205-017-0692-y.

Melsi. 2020. "*Panax ginseng* Aerial Parts." Obtained under the Creative commons Attribution-Share Alike 4.0 license. https://commons.wikimedia.org/w/index.php?curid=89577141 (Accessed 5 May 2021).

Nam, Gun He, Hye Won Kawk, Sang Yong Kim, and Young Min Kim. 2020. "Solvent Fractions of Fermented *Trapa japonica* Fruit Extract Stimulate Collagen Synthesis through TGF-B1/GSK-3β/β-Catenin Pathway in Human Dermal Fibroblasts." *Journal of Cosmetic Dermatology* 19, 1: 226–233. doi: 10.1111/jocd.13253.

National Institute of Korean Language. 2017. "*Arctium lappa* Root." Obtained under the Creative commons Attribution-Share Alike 2.0 Korea license. https://commons.wikimedia.org/w/index.php?curid=61569577 (Accessed 5 May 2021).

Nguyen, Marie-Lan. 2013. "*Paeonia suffruticosa* Flowers." Obtained under the Creative commons Attribution-Share Alike 2.5 license. https://commons.wikimedia.org/w/index.php?curid=25789312 (Accessed 5 May 2021).

Pan, Xiong, Tianheng Xu, Haoyu Xu, Hua Fang, and Yunlong Yu. 2017. "Characterization and Genome Functional Analysis of the DDT-Degrading Bacterium *Ochrobactrum* Sp. DDT-2." *Science of the Total Environment* 592: 593–599. doi: 10.1016/j.scitotenv.2017.03.052.

Park, Min Ja, and Young Seuk Bae. 2016. "Fermented *Acanthopanax koreanum* Root Extract Reduces UVB-and H_2O_2-Induced Senescence in Human Skin Fibroblast Cells." *Journal of Microbiology and Biotechnology* 26, 7: 1224–1233. doi: 10.4014/jmb.1602.02049.

Peganum. 2014. "*Angelica gigas*." Obtained under the Creative commons Attribution-Share Alike 2.0 license. https://commons.wikimedia.org/w/index.php?curid=37049814%0A (Accessed 5 May 2021).

Peka. 2019. "*Trapa japonica* Fruit." Obtained under the Creative commons Attribution-Share Alike 4.0 license. https://commons.wikimedia.org/w/index.php?curid=75812343 (Accessed 5 May 2021).

Pham, Quynh Lien, Hyun J.U.N. Jang, and K.Y.U. Bong Kim. 2017. "Anti - Wrinkle Effect of Fermented Black Ginseng on Human Fibroblasts." *International Journal of Molecular Medicine* 681–686. doi: 10.3892/ijmm.2017.2858.

Pillai, S., C. Oresajo, and J. Hayward. 2005. "Ultraviolet Radiation and Skin Aging: Roles of Reactive Oxygen Species, Inflammation and Protease Activation, and Strategies for Prevention of Inflammation-Induced Matrix Degradation—a Review." *International Journal of Cosmetic Science* 17–34. doi: 10.1111/j.1467-2494.2004.00241.x.

Qu, J., Y. Xu, G.M. Ai, Y. Liu, and Z.P. Liu. 2015. "Novel *Chryseobacterium* Sp. PYR2 Degrades Various Organochlorine Pesticides (OCPs) and Achieves Enhancing Removal and Complete Degradation of DDT in Highly Contaminated Soil." *Journal of Environmental Management* 161: 350–357.

Quan, Taihao, and Gary J. Fisher. 2015. "Role of Age-Associated Alterations of the Dermal Extracellular Matrix Microenvironment in Human Skin Aging: A Mini-Review." *Gerontology* 61, 5: 427–434. doi: 10.1159/000371708.

Qwert1234.2009. "*Broussonetia kazinoki*." Obtained under the Creative commons Attribution-Share Alike 3.0 license. https://commons.wikimedia.org/w/index.php?curid=6794487 (Accessed 5 May 2021).

Rangachary, L., R.P. Rajagopalan, T.M. Singh, and M.H. Krishnan. 2012. "Purification and Characterization of DDT-Dehydrohalogenase from *Pseudomonas putida* T5." *Preparative Biochemistry & Biotechnology* 42, 1: 60–76.

Rasor, J. Peter, and Edgar Voss. 2001. "Enzyme-Catalyzed Processes in Pharmaceutical Industry." *Applied Catalysis A: General* 221, 1–2: 145–158. doi: 10.1016/S0926-860X(01)00804-3.

Razak, Dang Lelamurni, Anisah Jamaluddin, Nur Yuhasliza Rashid, Amsal Ghani, and Musaalbakri Abdul Manan. 2019. "Assessment of Fermented Broken Rice Extracts for Their Potential as Functional Ingredients in Cosmeceutical Products." *Annals of Agricultural Sciences* 64, 2: 176–182. doi: 10.1016/j.aoas.2019.11.003.

Ro, Hyang Seon, Hyun Jun Jang, Gyu Rae Kim, Sang Jin Park, and Hyeon Yong Lee. 2020. "Enhancement of the Anti-Skin Wrinkling Effects of *Aloe arborescens* Miller Extracts Associated with Lactic Acid Fermentation." *Evidence-Based Complementary and Alternative Medicine*. doi: 10.1155/2020/2743594.

Sahu, R.K., A. Roy, M. Matlam, V.K. Deshmukh, J. Dwivedi, and A.K. Jha. 2013. "Review on Skin Ageing Compilation of Scientific Validated Medicinal Plants, Prominence to Flourish Better Research Reconnoiters in Herbal Cosmetics." *Research Journal Medicinal Plants* 7: 1–22. doi: 10.3923/rjmp.2013.1.22

Seong, Joon Seob, Song Hua Xuan, So Hyun Park, Keon Soo Lee, Young Min Park, and Soo Nam Park. 2017. "Antioxidative and Antiaging Activities and Component Analysis of *Lespedeza cuneata* g. Don Extracts Fermented with *Lactobacillus pentosus*." *Journal of Microbiology and Biotechnology* 27, 11: 1961–1970. doi: 10.4014/jmb.1706.06028.

Shin, Daehyun, Yoonjin Lee, Yu Hua Huang, Hye Won Lim, Kyounghee Jang, Dae Duk Kim, Chang Jin Lim, and Daehyun Shin. 2018. "Probiotic Fermentation Augments the Skin Anti-Photoaging Properties of *Agastache rugosa* through up-Regulating Antioxidant Components in UV-B-Irradiated HaCaT Keratinocytes." *BMC Complementary and Alternative Medicine* 18, 1: 1–10. doi: 10.1186/s12906-018-2194-9.

Shizhao. 2005. "*Panax ginseng* Root, Red Ginseng." Obtained under the Creative commons Attribution-Share Alike 2.0 license. https://commons.wikimedia.org/w/index.php?curid=250045 (Accessed 5 May 2021).

Siebrand. 2005. "*Triticum-aestivum*-Ear." Obtained under the Creative commons Attribution-Share Alike 3.0 license. https://commons.wikimedia.org/w/index.php?curid=253574 (Accessed 5 May 2021).

Sivamaruthi, Bhagavathi Sundaram, Chaiyavat Chaiyasut, and Periyanaina Kesika. 2018. "Cosmeceutical Importance of Fermented Plant Extracts: A Short Review." *International Journal of Applied Pharmaceutics* 10, 4: 31–34. doi: 10.22159/ijap.2018v10i4.26355.

Syrio. 2018. "*Helianthus tuberosus*." Obtained under the Creative commons Attribution-Share Alike 4.0 license. https://commons.wikimedia.org/w/index.php?curid=83181779 (Accessed 5 May 2021).

Tsuchihashi, Ryota, Mitsuru Kodera, Seiichi Sakamoto, Yoshijiro Nakajima, Takashi Yamazaki, Yujiro Niiho, Toshihiro Nohara, and Junei Kinjo. 2009. "Microbial Transformation and Bioactivation of Isoflavones from *Pueraria* Flowers by Human Intestinal Bacterial Strains." *Journal of Natural Medicines* 63, 3: 254–260. doi: 10.1007/s11418-009-0322-z.

Von Moltke, Lisa L., David J. Greenblatt, Brian W. Granda, Jeffrey M. Grassi, Jürgen Schmider, Jerold S. Harmatz, and Richard I. Shader. 1999. "Nefazodone, Meta-Chlorophenylpiperazine, and Their Metabolites *in Vitro*: Cytochromes Mediating Transformation, and P450–3A4 Inhibitory Actions." *Psychopharmacology* 145, 1: 113–122. doi: 10.1007/s002130051039.

Wang, Guey Horng, Chih Yu Chen, Chia Pei Lin, Chun Ling Huang, Chia hui Lin, Chiu Yu Cheng, and Ying Chien Chung. 2016. "Tyrosinase Inhibitory and Antioxidant Activities of Three *Bifidobacterium bifidum*-Fermented Herb Extracts." *Industrial Crops and Products* 89: 376–382. doi: 10.1016/j.indcrop.2016.05.037.

Wang, Guey Horng, Yi Min Lin, Jong Tar Kuo, Chia Pei Lin, Chin Feng Chang, Min Chi Hsieh, Chiu Yu Cheng, and Ying Chien Chung. 2019. "Comparison of Biofunctional Activity of *Asparagus cochinchinensis* (Lour.) Merr. Extract before and after Fermentation with *Aspergillus oryzae*." *Journal of Bioscience and Bioengineering* 127, 1: 59–65. doi: 10.1016/j.jbiosc.2018.06.015.

Xaver, Franz. 2010. "*Smallanthus sonchifolius*." Obtained under the Creative commons Attribution-Share Alike 3.0 license. https://commons.wikimedia.org/w/index.php?curid=10592914 (assessed).

Xie, H., L.S. Zhu, Q.F. Xu, J. Wang, W. Liu, J.H. Jiang, and Y. Meng. 2011. "Isolation and Degradation Ability of the DDT-Degrading Bacterial Strain KK." *Environmental Earth Sciences* 62, 1: 93–99.

Xu, Y., Y. Shao, J.J. Voorhees, and G.J. Fisher. 2006. "Oxidative Inhibition of Receptor-Type Epidermal Protein-Tyrosine Phosphatase Kappa by Ultraviolet Irradiation Activates Growth Factor Receptor in Human Keratinocytes." *Journal of Biological Chemistry*: 27389–27397. doi: 10.3923/rjmp.2013.1.22.

Yaar, M., and B.A. Gilchrest. 2007. "Photoageing: Mechanism, Prevention and Therapy." *British Journal of Dermatology* 157: 874–887.

Zell, H. 2010. "*Avena sativa*." Obtained under the Creative commons Attribution-Share Alike 3.0 license. https://commons.wikimedia.org/w/index.php?curid=9180071%0A (Accessed 5 May 2021).

Zell, H. 2011. "*Aloe aborescens*." Obtained under the Creative commons Attribution-Share Alike 3.0 license. https://commons.wikimedia.org/w/index.php?curid=17494434 (Accessed 5 May 2021).

Σ64.2008. "*Rhodiola rosea*." Obtained under the Creative commons Attribution-Share Alike 3.0 license. https://commons.wikimedia.org/w/index.php?curid=3386672 (Accessed 5 May 2021).

Данилевский, Юрий. 2018. "*Glycyrrhiza glabra*." Obtained under the Creative commons Attribution-Share Alike 4.0 license. https://commons.wikimedia.org/w/index.php?curid=69657365 (Accessed 5 May 2021).

3 Phytomedicines Targeting Antibiotic Resistance through Quorum Sensing and Biofilm Formation Associated with Acne Vulgaris

Isa A. Lambrechts and Namrita Lall

CONTENTS

3.1 INTRODUCTION

3.1.1 Traditional Medicine for Skin Diseases

Traditional medicine is used by around 80% of the world's population (Oyebode et al., 2016). Approximately 45% of South Africans rely on traditional medicines each year as their main healthcare source. This can be attributed to the different factors such as good accessibility to medicinal plants, affordability of traditional medicines compared to Western medicines, and extensive knowledge and expertise of ethnic groups and traditional healers with traditional medicines. Traditional medicine plays a significant role in managing skin disorders worldwide. In South Africa, rural communities extensively use plants to treat various skin maladies. Traditional healers in South Africa make use of indigenous plant species to treat various skin disorders. These skin ailments include burns, boils, rashes, wounds and acne. Treatment methods include herbal baths for treating rashes and itching skin. Traditional healers frequently use steam infused with medicinal plants to treat skin

DOI: 10.1201/9781003108375-3

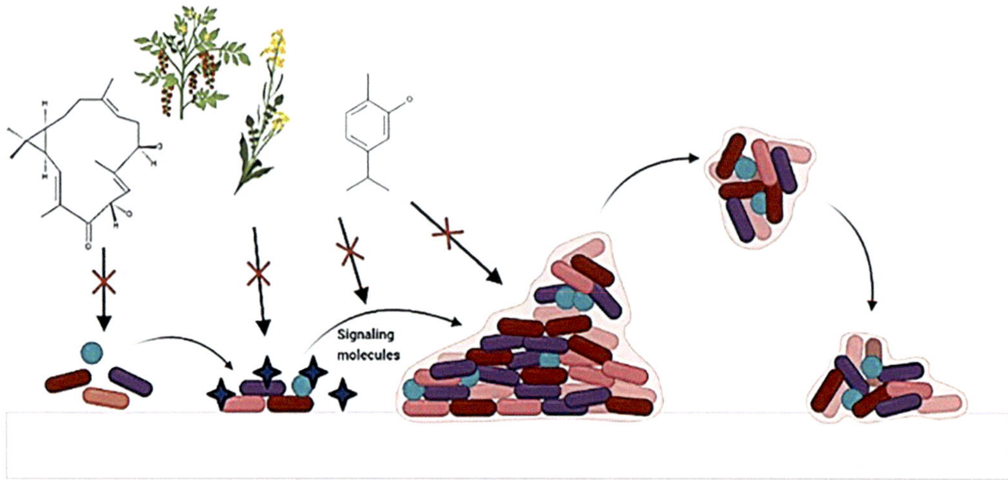

Graphical abstract representing the content of the chapter.

maladies such as acne vulgaris. There have been reported cases where traditional plant remedies have cleared acne after numerous months of unsuccessful treatment with Western medicine. In 2014, it was established that around 44.3% of the South African population lived with acne vulgaris. Of the 44.3% of South Africans having acne, Fitzpatrick skin-types V-VI was the most affected by the disease (Dlova, Mosam, and Tsoka-Gwegweni, 2018; Street, Stirk, and Van Staden, 2008; De Wet, Nciki, and van Vuuren, 2013; Williams, Dellavalle, and Garner, 2012). The pathogenesis of the disease, current treatments, side effects and potential alternative therapies are discussed in this chapter.

3.1.2 Acne Vulgaris a Global Problem

Acne vulgaris affects about 9.4% of the world's population. Therefore, it is the eighth-most prevalent disease worldwide and is moderate to severe in 15–20% of patients (Tan and Bhate, 2015). Acne vulgaris is commonly observed in the face, torso and back, which affects the patient's visual appearance. While acne vulgaris is a superficial disease that is not life-threatening, it can have seriously detrimental effects on the quality of life of a patient, affecting all aspects of their life, including emotional, social and physiological functions. Patients suffering from acne vulgaris are often self-conscious about their appearance, and the disease affects these patients' mental health, which snowballs into their daily lives and activities. Yazici et al. (2004) reported the psychological impact of the disease on a patient's mental health. Patients who had acne had an increase in depression and anxiety levels compared to healthy patients.

Furthermore, difficulties in academic and occupational fields have been observed with patients who have acne. They have shown a decreased career progression and an increase in unemployment since they are constantly self-conscious about their appearance, affecting their mood and productivity. Therefore, acne vulgaris should be acknowledged as a severe disorder (Gollnick, 2003; Tan and Bhate, 2015).

3.2 ACNE VULGARIS

Acne vulgaris is a disease related to the skin's pilosebaceous unit that includes the hair shaft, hair follicle and the sebaceous gland that produces sebum. Acne vulgaris predominantly occurs in the torso, back, neck and face. These areas are mostly affected due to the high amount of pilosebaceous

TABLE 3.1
Types of Non-Inflammatory and Inflammatory Acne Vulgaris (Singh, Hatwar, and Nayak, 2011)

Types	Description
Non-inflammatory acne	
Closed comedones (whiteheads)	Flesh or white-colored papules less than 1 cm in diameter and slightly raised. No apparent follicular opening or associated erythema.
Open comedones (blackheads)	The follicular opening is dilated, and a black or dark-colored comedone is visible.
Inflammatory acne	
Papule	Tender red blemishes can become severe forms of acne and cause scarring.
Pustule	Inflamed lesion with a puss-filled center. Red and can become large.
Nodule	Firm lesions that are deeper in the follicle compared to pustules. Nodules are firm, painful lumps filled with puss.
Cyst	Most severe form of inflammatory acne that is difficult to treat and can be present for long periods of time. Scarring can easily occur. Cystic acne mostly develops from trying to pop inflamed papules, pustules and nodules.

units in these sebum-rich areas. These sebum-rich areas are more specifically found in the area between the eyebrows known as the glabella, inside the ear, on the sides of the nostrils and the cheeks, behind the ear and the manubrium that is the upper chest and back. Acne vulgaris is divided into four categories of severity: mild, moderate, moderately severe and severe acne (Williams, Dellavalle, and Garner, 2012; Kaur et al., 2005; Singh, Hatwar, and Nayak, 2011). James (2005) described the severities of acne as follows.

- Mild: Non-inflammatory blackheads and whiteheads, although some papules and pustules might be present (fewer than ten).
- Moderate: Between 10–40 inflamed papules, pustules and comedones present. Lesions can be found on the torso.
- Moderately severe: Between 40–100 papules, pustules and comedones present on the back, torso and face. Up to five severe acne lesions such as nodules are present.
- Severe: Severe inflammation that includes cysts and nodules that are painful. Large pustules, papules and comedones present.

Microcomedones (earliest subclinical lesions) develop into inflamed or non-inflamed lesions and can be considered the primary acne lesion. Microcomedones can develop into full comedones that can either be open as in the case with blackheads or closed as for whiteheads. However, if the comedo ruptures into the dermis, it results in more severe forms of inflammatory acne such as pustules, papules, nodules and cysts (Table 3.1). This can lead to various forms of scarring and pigmentation (Williams, Dellavalle, and Garner, 2012; Kaur et al., 2005; Singh, Hatwar, and Nayak, 2011; Dessinioti and Katsambas, 2010).

3.2.1 Pathogenicity and Microbiology of Acne Vulgaris

The pathogenesis of acne vulgaris is divided into four major etiological factors involved in the development of the disease (Burkhart and Burkhart, 2003), as follows.

- Hyperkeratinization.
- Increased sebum production.

- The colonization and proliferation of *Cutibacterium acnes*.
- Inflammation due to the immunological consequences of the microbiological factor.

There is a range of theories about the sequence of these pathogenic factors and their contribution to the progression of acne vulgaris, which have not yet been fully answered. It is believed that follicular hyperkeratinization (the abnormal shedding of the skin) is the first step in the pathogenic pathway, blocking the pilosebaceous unit and leading to excess sebum accumulation. Comedone formation is a combined effect between the overproduction of keratinocytes and a reduction in the shedding of keratocytes due to the skin cells' inability to separate from one another. Comedones form when keratinocytes block the pilosebaceous unit that results in the accumulation of sebum in the unit. Several factors contribute to hyperkeratinization. These factors include the lack of linoleic acid in the sebum due to hyperseborrhea, resulting in abnormal keratinocyte differentiation. Cytokines such as cytokine interleukin-1 alpha (IL-1α) present in the follicle are believed to induce hyperkeratinization. Furthermore, it was recently discovered that the acne-inducing bacteria *Cutibacterium acnes* secrete a glycocalyx polymer that is a component of the bacterial biofilm. This secretion is incorporated in the sebum, which increases cohesion between the keratinocytes. This results in a blockage in the pilosebaceous unit and results in the formation of a comedone. It is therefore plausible that the initial step of acne pathogenesis could be the colonization and biofilm formation of *C. acnes* instead of hyperkeratinization (Dessinioti and Katsambas, 2010; James, Burkhart, and Morrell, 2009; Pawin et al., 2004).

Sebum production is hypothesized to be the second pathogenic factor in the acne pathogenic pathway. Human sebum comprises several compounds that include cholesteryl esters, cholesterol, fatty acids, triglycerides, squalene, diglycerides and wax esters. Androgens are a group of hormones present in males and females. These hormones can stimulate sebaceous secretion, and it is hypothesized that some acne patients can possess sebocytes that have an increased sensitivity to androgens. Sebum serves as a carbon and nutrient source for acne-causing bacteria such as *C. acnes*. The bacteria that proliferate in the pilosebaceous unit produce inflammatory and chemotactic mediators. For the bacteria to release these mediators, they attract neutrophils, which drive the inflammatory process (James, Burkhart, and Morrell, 2009; Bhate and Williams, 2013; Pawin et al., 2004).

3.2.2 Microbiology of the Skin and Its Role in the Pathogenesis of Acne Vulgaris

The skin microenvironment is affected by several extrinsic and intrinsic factors, such as temperature, pH, sweat, sebum production and moisture, affecting how these microorganisms thrive. These microorganisms are not generally harmful to the skin and play an essential role in protecting the host and serve as the first line of defense against pathogens. Failure to control these resident microbes is linked to the onset of various skin diseases such as acne vulgaris. It is believed that microorganisms associated with the microflora of the skin co-exist to protect the host. However, change in this microbial balance due to changes in extrinsic and intrinsic factors contributes to various dermatological conditions. With this in mind, microbial communities associated with the skin have become a subject of interest to understand cutaneous diseases better and develop novel therapies for these maladies (Sanford and Gallo, 2013).

Sixty percent (60%) of the skin's microbiome involves three co-existing bacterial genera: *Corynebacterium*, *Staphylococci* and the *Cutibacterium* (*Cutibacterium granulosum* and *Cutibacterium acnes*). The Gram-positive bacteria *Corynebacterium* and *Staphylococci* are aerobic bacteria that can be found in the epidermal (top) layer of the skin. The anaerobic diphtheroid *Cutibacterium* predominantly resides in the infra-infundibulum of the sebaceous follicle. Although all three genera have been isolated from the pilosebaceous unit, *Staphylococci* and *Cutibacterium* species are predominantly isolated from acne lesions (Sanford and Gallo, 2013; Kumar et al., 2016; Burkhart and Burkhart, 2003). Since *Staphylococcus epidermidis* and *C. acnes* are considered the primary cause of inflammatory acne and are predominantly isolated from acne lesions, the rest of the chapter will focus on these two bacteria.

As commensal bacteria, *S. epidermidis* and *C. acnes* play essential roles in protecting the skin from pathogenic microbes in two ways, indirectly and directly. These commensal bacteria compete with pathogens for resources such as space and nutrients, limiting the growth of these harmful microorganisms. Directly, commensal bacteria produce antimicrobial compounds called bacteriocins that directly restrict the growth of the pathogens. These proteinaceous factors inhibit the growth of closely related bacterial species, but with no effect on the organism that produces them. *Staphylococcus epidermidis* is a skin commensal bacteria, as well as an opportunistic pathogen, and it has been shown to cause serious infections in humans. The bacteria releases several bacteriocins that inhibit the growth of various pathogens, including one within the same genus, *Staphylococcus aureus* (Sanford and Gallo, 2013).

Staphylococcus epidermidis produce *Enterococcus faecalis* surface proteins (Esp) that inhibit proteins involved in *S. aureus* biofilm production and growth. This serine protease enhances the antimicrobial activity of human beta-defensin-2 (hBD2) produced by keratinocytes due to inflammation but does not have antibacterial activity against Gram-positive bacteria alone. Furthermore, *S. epidermidis* releases antimicrobial molecules such as phenol-soluble modulins (PSMs) that cause membrane leakage and targets mostly *Streptococcus pyogenes* and *S. aureus*. *Cutibacterium acnes*, another commensal skin bacterium, can protect the skin from pathogens such as methicillin-resistant *S. aureus* (MRSA). *Cutibacterium acnes* protects the skin from pathogens through fermenting glycerol, a product of sebum triglyceride hydrolysis, lowering the pH and inhibiting MRSA growth (Schröder and Harder, 1999; Sanford and Gallo, 2013).

Although these commensal bacteria protect the host and the skin, they cause infection under certain conditions. *Cutibacterium acnes* is an opportunistic pathogen that plays an essential role in the advancement of inflammatory acne. These Gram-positive, non-motile, rod-shaped bacteria are aerotolerant, meaning they can grow under different oxygen tensions, although anaerobic conditions are preferred. *C. acnes* play a crucial function in acne vulgaris pathogenesis since enzymes produced by the bacterium aids in rupturing the follicular wall. *C. acnes* release three enzymes called hyaluronidases, lipases and proteases that contribute to the pathogenesis of the bacteria (Falcocchio et al., 2006; Holland, Ingham, and Cunliffe, 1981; Lee et al., 2010). Their role in acne vulgaris are as follows.

- The hyaluronidase enzyme produced by *C. acnes* breaks down intracellular substances within the wall of the pilosebaceous unit. This causes an increase in the permeability of the follicular epithelium and results in the diffusion of irritants into the dermis of the skin. Hyaluronidase acts as a spreading factor and increases the inflammatory response (Holland, Ingham, and Cunliffe, 1981).
- *Cutibacterium acnes* release proteases that stimulate cellular responses through the proteinase-activated receptor-2 (PAR-2) that mediates and contributes to inflammation associated with inflammatory acne (Lee et al., 2010).
- The lipase enzyme, which hydrolyzes sebum triacylglycerides into glycerol and free fatty acids like oleic and palmitic acid, is released by the bacteria. These free fatty acids are chemotactic and irritate the sebaceous glands, contributing to the onset of inflammation (Falcocchio et al., 2006). Free fatty acids produced during sebum hydrolysis act as chemokines involved in producing pro-inflammatory cytokines. These cytokines include interleukin-1 beta (IL-1β), IL-6, IL-8, IL-12 and tumor necrosis factor-alpha (TNF-α), which subsequently elicits an inflammatory response that is associated with inflammatory acne (Burkhart and Burkhart, 2003; Kumar et al., 2016).

It has been hypothesized that *C. acnes* in other infections is not fully understood and has been significantly underestimated. *Cutibacterium acnes* is no longer only associated with acne vulgaris, and is also linked to various other diseases. For instance, *C. acnes* has been linked to biofilm-associated indwelling medical device infections, necro suppurative placentitis and abortion in a Holstein cow (Ramanuj, Bachani, and Kothari, 2012).

Staphylococcus epidermidis is a facultative anaerobic bacterium of the skin that is harbored in acne lesions. Usually, *S. epidermidis* is a non-pathogenic bacterium of the human skin. However, it has been found to change into a pathogenic bacterium during a change in intrinsic factors, like an impaired immune system. During the onset of acne vulgaris, *S. epidermidis* protects itself from fatty acids that are bactericidal for the bacteria by releasing certain virulence factors. The virulence factor released by the bacteria is the fatty acid modifying enzyme (FAME) that esterifies these lipids to cholesterol that does not affect bacterial growth. Other virulence factors produced by the organism include delta-hemolysis and the lipase enzyme contributing to the inflammation seen in acne vulgaris (Kumar et al., 2016).

When evaluating acne lesions, Nishijima et al. (2000) found that when isolated alone, *C. acnes* and *S. epidermidis* were only associated with pustule formation in 2 out of 30 and 1 out of 30 acne lesions, respectively. *Cutibacterium acnes* and *S. epidermidis* were isolated together in 50% of acne lesions (15 out of 30). Twelve of these lesions were pustules, two papules and one a comedone. Although it is widely recognized that *C. acnes* is the only pathogenic bacteria involved in the development of acne vulgaris, *in vivo* studies have confirmed that *S. epidermidis* play a central role in inflammatory acne. Nishijima et al. (2000) reported that the bacteriology of acne lesions is not complicated and that acne lesions consist of only two bacteria, *C. acnes* and *S. epidermidis*. Failure to detect either one of these bacteria reflects a flaw in the methodology (Nishijima et al., 2000).

Pathak et al. (2013) found that the population of *C. acnes* increased by 82%, while *S. epidermidis* increased by 70% in acne patients compared to the control, which consisted of non-acne patients. In the test samples obtained from patients suffering from acne vulgaris, it was found that the microbial load increased simultaneously for both *C. acnes* and *S. epidermidis* (Kumar et al., 2016; Packiavathy et al., 2012). This increase in the bacterial load of both bacteria in acne lesions could indicate that these two bacteria are closely linked and could work together to develop and regulate inflammatory acne vulgaris.

3.3 QUORUM SENSING

Quorum sensing is a concentration-dependent communication mechanism between bacteria that results in the expression of genes involved in the social activities of these bacteria, such as the release of virulence factors, biofilm formation and motility. Microorganisms release chemical signaling molecules to communicate either with the same species (intraspecies) or different species (interspecies). Only once a specific concentration of the bacterial population and signaling molecule is reached in the environment does it result in gene expression. Microorganisms use this type of communication to differentiate self from non-self and as a measure to determine the population density in a specific environment. Bacteria use quorum sensing to detect the bacterial population density since individual bacteria cannot elicit a strong enough reaction as compared to a large enough bacterial population to produce a more effective virulence response. These bacteria release signaling molecules called autoinducers to detect and respond to various environmental changes. Bacteria release molecules and autoinducers into the surrounding supernatant, where they can be detected. However, not all of these products can serve as a quorum-sensing signaling molecule (QSM). The QSM can be defined according to various criteria (Miller and Bassler, 2001; Waters and Bassler, 2005; Winzer, Hardie, and Williams, 2002):

- QSM production is stimulated because of physiological conditions, environmental factors or growth stages.
- QSMs accumulate in the biofilm's extracellular matrix, where bacteria receptors bind to them.
- A coordinated response is generated as a result of accumulating QSMs to a critical concentration threshold.

- The accumulation of the QSMs to a critical threshold concentration generates a combined response.
- The cellular response goes beyond the physiological changes needed for QSM to be metabolized or detoxified.

Bacteria have various receptors available to detect the accumulation of autoinducers, resulting in the alteration of gene expression. These activated genes encode either for symbiosis, biofilm formation, virulence factor release, competence, motility, sporulation or antibiotic production and conjugation (Miller and Bassler 2001; Waters and Bassler 2005).

Gram-negative and Gram-positive bacteria use quorum sensing as a way to control various physiological activities. Acylated homoserine lactones (AHL) are used as autoinducers by Gram-negative bacteria, whereas Gram-positive bacteria use processed oligopeptide (AL) autoinducers during intraspecies communication (Bassler, 1999).

3.3.1 Gram-Negative Quorum Sensing

N-Acyl homoserine lactones (AHLs) are quorum-sensing autoinducers produced by Gram-negative bacteria. AHLs are fatty acyl chains of between four and 18 carbon atoms long attached to an invariant homoserine lactone ring. Two regulatory proteins, LuxI and LuxR, are believed to control Gram-negative bacterial quorum sensing. LuxI AHL synthase enzymes biosynthesize AHL from S-adenosylmethionine (SAM) and an acylated acyl carrier protein (acyl-ACP). The membrane diffusible AHLs produced by LuxI bind to LuxR once a specific concentration is reached. The transcriptional activator protein LuxR promotes the transcription of the luxCDABE operon when bound to an autoinducer (Figure 3.1) (Bassler, 1999).

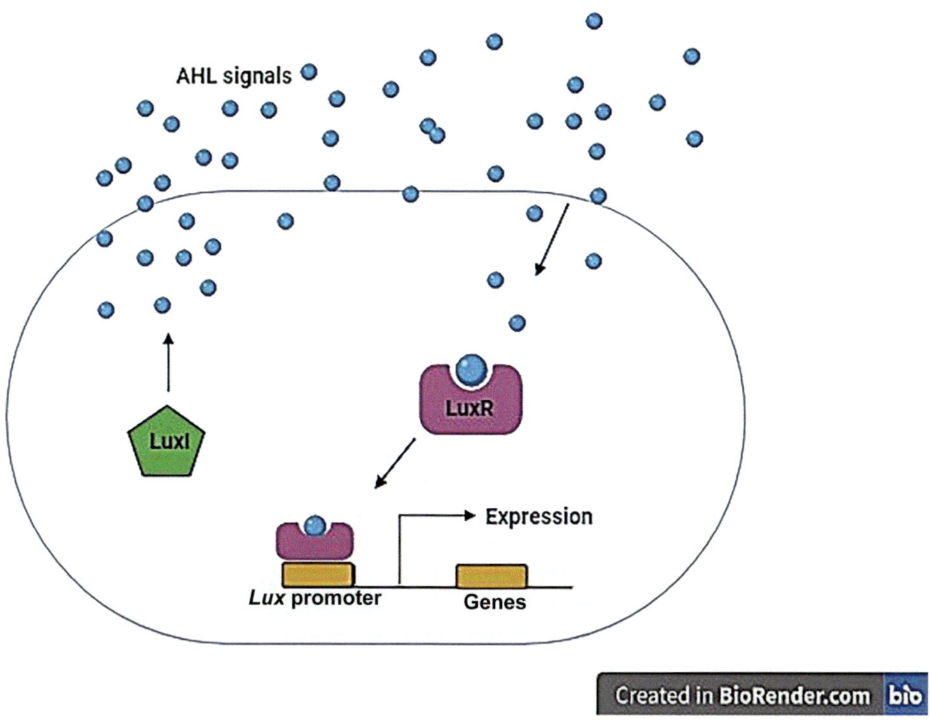

FIGURE 3.1 Quorum sensing in Gram-negative bacteria. AHL: N-Acyl homoserine lactones. Figure adapted from (Bassler, 1999).

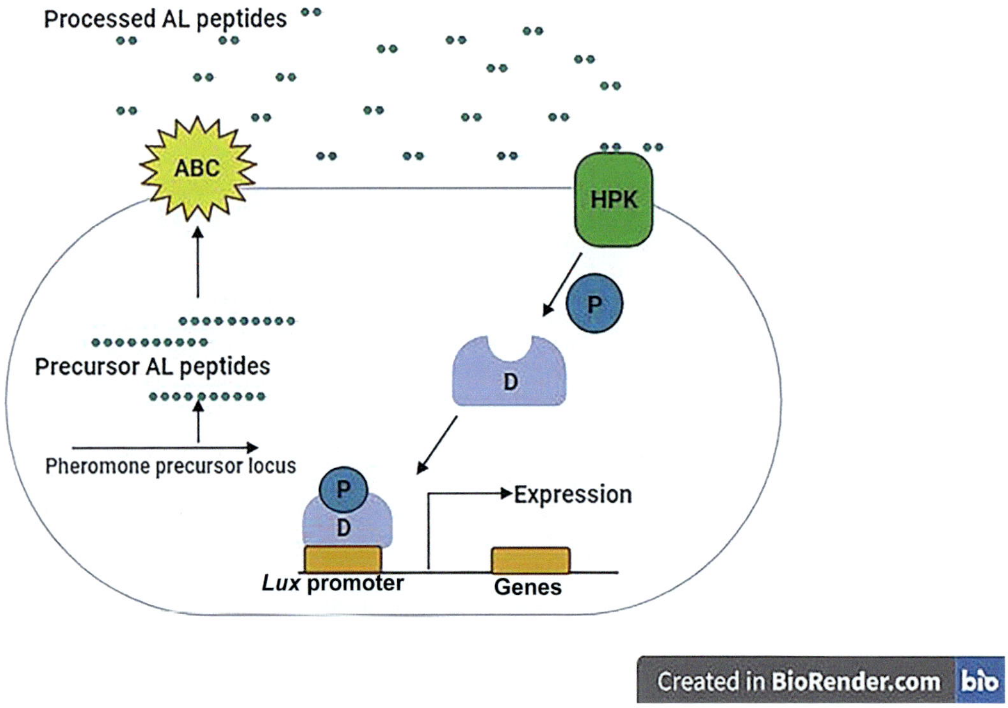

FIGURE 3.2 Quorum sensing in Gram-positive bacteria. ABC: ABC exporter protein, HPK: histidine kinase, D: response regulator, P: phosphorylation. Figure adapted from (Bassler, 1999).

3.3.2 Gram-Positive Quorum Sensing

Gram-positive bacteria use a two-component communication system. First, an adenosine triphosphate (ATP)-binding cassette known as ABC exporter protein transports processed AL molecules from the bacteria to the extracellular matrix. Once the peptide-signaling molecules reach a threshold concentration, the secreted peptide signals bind to a histidine kinase sensor protein on the bacterial membrane. Bassler (1999, pp. 582–587) reports that "autophosphorylation occurs on a conserved histidine residue, transferring the phosphoryl group to a cognate response regulator on a conserved aspartate residue." Following that, the phosphorylated response regulator binds to the *lux* promoters, suppressing or activating gene transcription (Figure 3.2) (Bassler, 1999).

3.3.3 Interspecies Quorum Sensing

Although many quorum-sensing signals are species-specific, interspecies communication has been intensively studied. The production and response to autoinducer-2 (AI-2) have been observed in various multispecies quorum sensing. During multispecies quorum sensing, one species of bacteria can produce AI-2 that affects the gene expression in another species of bacteria. This universal interspecies communication allows bacteria to change their behavior, and it affects social activities such as the release of virulent factors and biofilm formation between these different species. LuxS is a synthase enzyme widely conserved among Gram-positive and -negative bacteria and is required for AI-2 production. The AI-2 precursor, 5-dihydroxy-2,3-pentanedione (DPD), is produced from SAM through three enzymatic steps. S-adenosylmethionine serves

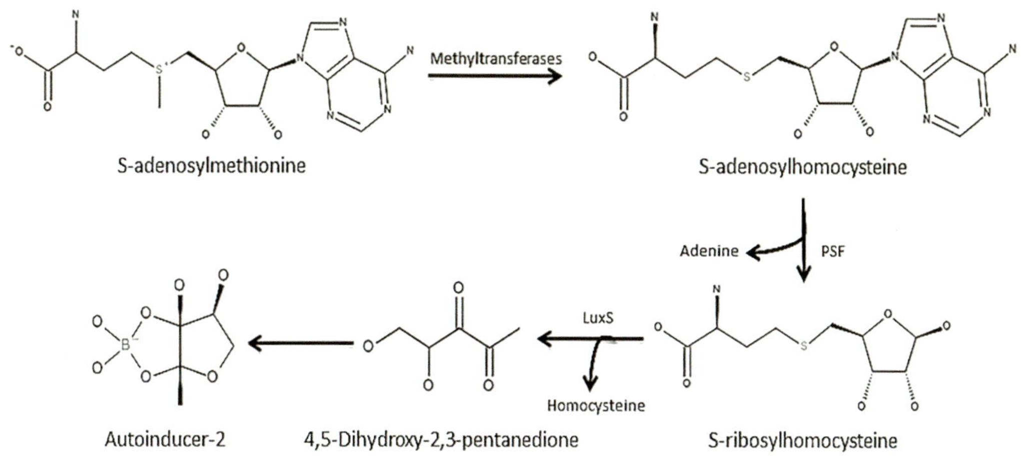

FIGURE 3.3 The synthesis of autoinducer-2 (AI-2). Figure adapted from (Chen et al., 2002).

as a methyl donor to form a toxic intermediate, S-adenosylhomocysteine (SAH). Subsequently, S-adenosylhomocysteine is hydrolyzed by 5'-methylthioadenosine/*S*-adenosylhomocysteine nucleosidase (MTA/SAHase), also known as Pfs, to yield S-ribosylhomocysteine (SRH) and adenine, reducing toxicity. The LuxS enzyme is responsible for catalyzing the reaction of SRH to DPD. DPD is thought to rearrange further to produce AI-2. (Figure 3.3). The AI-2 sensor protein LuxP forms a LuxP-AI-2 complex that interacts with LuxQ, a hybrid sensor kinase located in the inner bacterial membrane to transfer the autoinducer signal. It is hypothesized that during low bacterial cell densities and therefore low AI-2 availability in *Vibrio harveyi*, LuxQ autophosphorylate results in the phosphorylation of the hybrid sensor kinase proteins LuxU and LuxO. The phosphorylation of LuxO prevents transcription of the *lux* operon. When a specific concentration of AI-2 is reached, the kinase proteins change to phosphatases, resulting in dephosphorylation in the system and transcription of the *lux* operon commence. (Bassler and Miller, 2013; Chen et al., 2002; Pereira, Thompson, and Xavier, 2013; Winans and Bassler. 2002; Winzer, Hardie, and Williams, 2002).

Quorum sensing can be disrupted by targeting various processes involved during the communication process. These mechanisms include (Kalia, 2013) the following.

- Preventing the production of the signaling molecules.
- Degrading the signaling molecule.
- Interfering with the signal receptor.

To date, few studies have been conducted on the quorum-sensing mechanism of *C. acnes*. Coenye, Peeters, and Nelis (2007) confirmed that sessile and planktonic *C. acnes* bacteria produce AI-2. However, AI-2 production was three times higher in mature biofilms than planktonic cells. Considering these results, it was hypothesized that AI-2 could potentially contribute to the production of virulence factors as a result of virulence gene expression contributing to the pathogenesis of acne vulgaris (Coenye, Peeters, and Nelis, 2007).

The quorum-sensing mechanism in *S. epidermidis* has been intensively studied. The accessory gene regulator (agr) quorum-sensing mechanism controls modified peptides released as signals in *Staphylococci*. Previous genomic structures have confirmed the presence of the *luxS* homologs in the bacteria. Li et al. (2007) have confirmed that the LuxS/AI-2 system regulates key *S. epidermidis* social activities and virulence factor release (Li et al., 2007).

3.4 QUORUM SENSING, BIOFILM FORMATION AND ANTIBIOTIC RESISTANCE IN ACNE VULGARIS

3.4.1 Biofilm Formation

In nature, most microorganisms do not exist as planktonic bacteria floating in a suspension but rather as a biofilm. Biofilms form when populations or communities of bacteria arrange in layers, together with their decomposition and metabolic products, and adhere to abiotic and biotic surfaces such as the wall of the pilosebaceous unit. After the bacteria have adhered to surfaces, they start to secrete polysaccharides in which these bacteria are encased. The extracellular polysaccharide is around 70% of the biofilm mass and is essential for biofilm architecture. The biofilm is comprised of four components: water, extracellular DNA (eDNA), extracellular polysaccharides and cellular byproducts. Extracellular DNA is produced due to quorum sensing and plays a role in biofilm formation, gene transfer, biofilm stabilization and bacterial and biofilm attachment to a stratum. The extracellular polysaccharide is a glycocalyx polymer, which functions as a protective layer that prevents effective concentrations of antimicrobials from entering the microenvironment of the bacteria. Compared to planktonic bacteria, bacteria associated with a biofilm have altered growth and gene expression. It has been found that bacteria inside the biofilm is more resistant to antibiotics (50–500 times more) than planktonic bacteria not associated with a biofilm (Montanaro et al., 2011; Burkhart and Burkhart, 2003; Sandasi et al., 2011; Yarwood and Schlievert, 2003).

The formation of an established biofilm and distribution of the biofilm occurs in five stages (Figure 3.4) (Unosson, 2015; Center for Biofilm Engineering: Montana State University, 2020).

- Reversible adherence of the planktonic bacteria to a substrate or surface.
- Irreversible adherence of biofilm, quorum sensing and formation of monolayer microcolonies and the production of the extracellular polysaccharide matrix.
- Mature biofilm formation starts and multilayer cell formation.
- Further maturation of the biofilm into a three-dimensional structure forms channels for the movement of nutrients.
- Biofilm reaches a critical mass and releases infectious planktonic bacteria and biofilm fragments to infect and colonize other substratum.

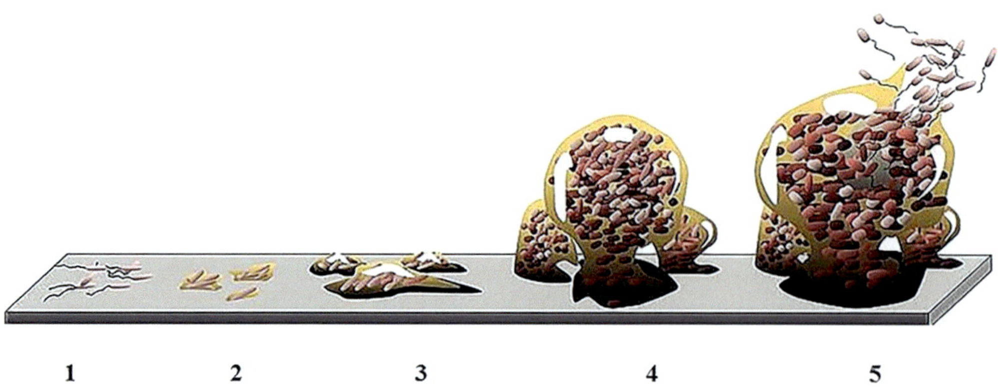

FIGURE 3.4 The stages of biofilm formation (Monroe 2007).

There are several advantages for a microorganism to form a biofilm that includes (Burkhart and Burkhart, 2003; Sandasi et al., 2011) the following.

- Gaining of new genetic material.
- Metabolic cooperation.
- Nutrient availability.
- Environmental protection whereby the biofilm community is protected and less susceptible to antimicrobials.

Antibiotic resistance can be defined as when the minimum inhibitory concentration against the targeted bacteria is higher than the concentration of the active that is available *in vivo*. Therefore, antibiotic resistance is when antibiotics cannot inhibit the microorganisms they previously were able to inhibit. The antibiotic resistance observed in biofilms could be due to two mechanisms of action: a reduction in the diffusion of antibiotics through the biofilm and a reduction in bacterial metabolism of the bacteria within the biofilm (Brandwein, Steinberg, and Meshner, 2016). Quorum sensing regulates genes involved in biofilm formation and dispersion; hence, biofilm formation and quorum sensing are linked.

3.4.2 *Cutibacterium acnes* Quorum Sensing and Biofilm Formation

Quorum sensing in *C. acnes* is still relatively unexplored; however, Coenye, Peeters, and Nelis (2007) reported the presence of the PPA0450 gene that encodes a LuxS homolog in *Vibrio harveyi*. The researchers demonstrated the release of lipase and AI-2 by *C. acnes* that were three times higher in mature biofilms of the bacteria than in planktonic cultures. This research suggests that AI-2 in *C. acnes* could contribute to the release of virulence factors and biofilm formation in the species (Coenye, Peeters, and Nelis, 2007).

Antibacterial agents play a significant role in treating acne vulgaris. Azelaic acid, benzoyl peroxide, clindamycin, erythromycin, tetracycline, triclosan and combinations of these drugs are examples of topical antibacterial agents used to treat acne vulgaris. However, recent studies have found an increase in antibiotic-resistant *C. acnes* strains in patients suffering from persistent acne vulgaris. The ability of *C. acnes* could contribute to antibiotic resistance reported in the disease. The sequenced genome of *C. acnes* revealed that the bacterial genome contains genes for biofilm formation. The genome of *C. acnes* has been found to contain genes coding to produce glucosyltransferases. The biofilm extracellular polysaccharide matrix is comprised of dextran (glucan), which is formed by glucosyltransferases. In addition, several genes that encode for adhesion proteins and a LuxS involved in interspecies communication through AI-2 have been identified in the genome of *C. acnes* (Coenye et al., 2008).

The establishment of the *C. acnes* biofilm in the pilosebaceous unit has been confirmed in recent *in vivo* investigations. Results have shown that *C. acnes* biofilm formation is not exclusively associated with the anaerobic portions of the skin. In some cases, *C. acnes* reside in the stratum corneum of the epidermis, between 80–1900 μm deep. More mature biofilms can be found in the deeper layer of the epidermis. Jahns and Alexeyev (2014) identified the following four patterns of how *C. acnes* biofilms occur in the pilosebaceous unit.

- *Cutibacterium acnes* attachment to the pilosebaceous wall.
- Attachment of the biofilm to the hair shaft.
- Spreading of the biofilm over the entire lumen.
- Biofilms are not attached to the pilosebaceous wall but occur in the center of the pilosebaceous unit.

3.4.3 *Staphylococcus epidermidis* Quorum Sensing and Biofilm Formation

The agr system is involved in quorum sensing in *Staphylococci* and increases bacterial pathogenicity and penetration into surrounding tissue. The agr system triggers the expression of accessory factors (Figure 3.5). There are two main components to the agr system. First, a two-part signal transduction pathway detects the number of autoinducer peptides (AIP) in the environment. The AgrC and AgrA complex controls the signal transduction pathway. Once a critical concentration of these AIP is reached outside of the bacterial cell wall, it binds to AgrC to elicit a response. Upon binding with AIP in the external environment, AgrC activates the DNA binding protein, AgrA, binds to the agr intergenic region (agr-IR) and upregulates the expression of the promoters P2 for the transcription of RNAII and P3 for the transcription of RNAIII. Second, the agr system has an AI production side that includes AgrD and AgrB. RNAII encodes the genes *agrB*, *agrD*, *agrC* and *agrA*. The production of AI is under the regulation of genes *agrB* and *agrD*. The *agrD* gene encodes a 7–9 amino acid–long AgrD autoinducer peptide (AIP precursor) trimmed by the enzyme AgrB, encoded by the *agrB* gene, into a ring structure. This structure contains thiolactone and is exported by the enzyme to the external environment (Le and Otto, 2015; Vuong et al., 2003; Kong, Vuong, and Otto, 2006).

RNAIII plays a role in the transcription of several virulent factors associated with the bacteria. RNAIII contains the *hld* gene that codes for the phenol-soluble modulin, a delta toxin. These delta toxins play a role in the detachment of bacteria from the biofilm. This detachment could spread the bacteria, establish biofilms and release virulence factors such as hemolysins, protease and lipase

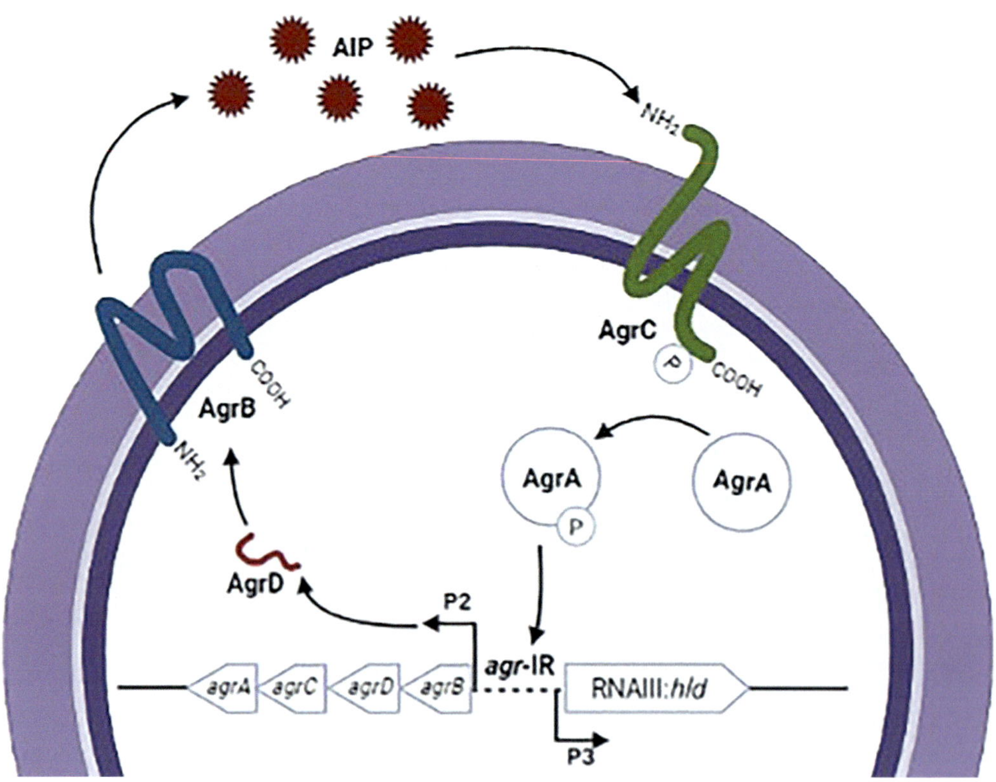

FIGURE 3.5 The accessory gene regulator system in *Staphylococcus* species (Painter et al., 2014).

enzymes associated with quorum sensing and various diseases. In addition, the RNAIII gene has been found to prevent the translation of the repressor of toxin protein (Rot) that results in the upregulation of enzymes such as lipase and protease involved in acne vulgaris (Le and Otto, 2015; Vuong et al., 2003; Kong, Vuong, and Otto, 2006).

Staphylococcus epidermidis virulence is under the LuxS/AI-2 control. Several studies have investigated the role of AI-2 in *S. epidermidis* virulence and have confirmed that AI-2 acts as a signaling molecule in the species. However, the role of AI-2 in biofilm formation seems to be strain-specific. In contrasting studies, Xu et al. (2006) found that *S. epidermidis* strain 1457 does not form a mature biofilm compared to the luxS mutant strain that formed a biofilm while confirming the formation of mature biofilm in the presence of AI-2 in *S. epidermidis* strain RP62A (ATCC 35984). Studies have demonstrated that *Staphylococci* species harboring the biofilm-associated protein (*bap*) gene can successfully attach to services and form a well-established mature biofilm. *Staphylococcus epidermidis* strain RP62A has been found to contain the Bap-homologous protein (*bhp*) gene that is upregulated in the presence of AI-2 to produce more biofilm compared to the wild type strain with no addition of AI-2 (Eiff, Peters, and Heilmann, 2002; Xue et al., 2015).

S. epidermidis uses biofilm formation as one of its main virulence mechanisms. The intracellular adhesion (icaA) gene encodes for the polysaccharide intracellular adhesion (PIA) molecule that regulates biofilm formation in *S. epidermidis*. PIA functions as an adhesion molecule for the bacteria to attach to the skin surface of the host. Several other adhesion factors are produced by *S. epidermidis*, which include surface fibrinogen binding proteins, anchored proteins, autolysin proteins and poly-N-succinyl-glucosamine. An *in vivo* study done by Jahns and Alexeyev (2014) observed biofilm-like colonization of *Staphylococci* in areas close to the skin surface. This observation could indicate that *Staphylococci* not only form biofilms, but could potentially contribute to antibiotic resistance in acne vulgaris (Jahns and Alexeyev, 2014).

3.5 PLANTS AND PURE COMPOUNDS TARGETING QUORUM SENSING AND BIOFILM FORMATION

Current research focuses on plants and their compounds to inhibit bacterial growth, biofilm growth and various other virulence factors associated with acne vulgaris. However, there is a lack of current research in using plants and plant-based compounds to target aspects of bacterial infections as a whole or against quorum sensing, which is hypothesized to be the root of numerous infections, including acne vulgaris. Furthermore, research on plants targeting acne-causing bacterial biofilms and quorum sensing is currently limited and more research in these areas is required (Table 3.2 and Table 3.3) (Figure 3.6 and 3.7) (Van Vuuren and Holl, 2017).

3.6 CONCLUSION

Acne vulgaris is a disease that has extensively been studied. However, research and our understanding of quorum sensing and multispecies biofilm formation related to the disease remains largely elusive. Furthermore, although there are drugs available on the market and plant-based therapy research for acne vulgaris, these therapies target the pathogenic factors associated with acne vulgaris. Currently, there is no acne therapy on the market that targets quorum sensing and multispecies biofilms associated with antibiotic-resistant acne vulgaris. Research on plants that target quorum sensing, antibiotic resistance, and virulence factor release only focuses on one aspect of quorum sensing. More research in the field is required to investigate and better understand multispecies acne vulgaris, the bacterial interactions, the natural environments of these bacteria and the plants that have the potential to treat this disease.

TABLE 3.2
Quorum Quenching and Biofilm Inhibitory Potential of Plants and Their Compounds against the Acne-Causing Bacteria *Cutibacterium acnes* and *Staphylococcus epidermidis*

Scientific Names and Plant Parts Used	Traditional Use	Active Compounds against *C. acnes*	Quorum Quenching Activity	Biofilm Inhibitory Activity	Other Targets	References
Epimedium brevicornum Plant	Impotence, numbness, rheumatic pain and seminal emission	Icariin	None reported	*E. brevicornum* extract inhibited *C. acnes* biofilm growth at sub-MIC values. Icariin inhibited 40–70% of *C. acnes* biofilm formation at a concentration between 0.01–0.08% (w/v).	*E. brevicornum* extract inhibited *C. acnes* planktonic growth at a MIC of 0.625% (w/v). Icariin displayed a MIC of 2.5% (w/v) against *C. acnes*.	(Coenye et al., 2012; C. Yu et al., 2011)
Malus pumila Fruit	Skincare and respiratory ailments such as coughs	Polyphenols	None reported	The extract effectively inhibited *C. acnes* biofilm at a concentration of 0.05% (w/v).	Inhibited planktonic *C. acnes* at a MIC of 0.04% (w/v).	(Coenye et al., 2012; Korkmaz and Karakuş, 2015)
Myrtus communis Leaves and stems	Bleeding, conjunctivitis, diarrhea, epistaxis, excessive perspiration, hemorrhoids, headaches, inflammation, leucorrhea, palpitation, peptic ulcers, pulmonary diseases, skin diseases and urethritis	5-acetoxy-4-hydroxy-4-isobutyl-2,2,6,6-tetramethylcyclohexan-1,3-dione and isomyrtucommulone-B	None reported	*M. communis* inhibited *C. acnes* biofilm formation at a concentration ranging between 0.0001% and 0.03% in a dose-dependent manner. At 0.001%, *M. communis* combined with 1000 mg/L erythromycin and or 500 mg/L clindamycin significantly inhibited *C. acnes* biofilm formation.	Leaf extract inhibited *C. acnes* growth with a MIC of 7.8 μg/mL. Noteworthy antioxidant activity was observed for *M. communism* acetone leaf extract against 2,2-diphenyl-1-picrylhydrazyl (DPPH) and ferric reducing antioxidant power (FRAP).	(Feuillolay et al., 2016; Alipour, Dashti, and Hosseinzadeh, 2014; Abood, Addai, and Abood, 2016; Hamdy et al., 2017)
Polygonum cuspidatum Rhizome	Traditionally used as an antipyretic, expectorant, analgesic and as a diuretic; it is also used to treat inflammation, skin burns and infection, jaundice and high cholesterol	Resveratrol	None reported	*P. cuspidatum* effectively inhibited *C. acnes* biofilm at 0.5% (w/v). Resveratrol inhibited 80% of *C. acnes* biofilm at 0.32% (w/v). Resveratrol enhanced *S. epidermidis* biofilms 1.5 fold at a concentration of 100 μg/mL.	*P. cuspidatum* extract inhibited *C. acnes* planktonic growth at a MIC of 0.625% (w/v). Resveratrol inhibited *C. acnes* planktonic bacteria at a MIC of 1.25% (w/v). Resveratol does not inhibit planktonic *S. epidermidis* at the highest concentration tested of 2 mg/mL.	(Coenye et al., 2012; Ta and Arnason, 2016; Morán et al., 2014; Wu et al., 2012; Peng et al., 2013)

Rhodiola crenulata Root	Traditionally used to treat mountain sickness and lung damage	Salidroside	None reported	*R. crenulat* effectively inhibited *C. acnes* biofilm at 0.5% (w/v). Salidroside inhibited *C. acnes* biofilms between 20 and 40% in a concentration range of 2.5–0.02% (w/v).	*R. crenulat* extract inhibited *C. acnes* planktonic growth at a MIC of 0.156% (w/v). No inhibition of *C. acnes* planktonic bacteria was observed for salidroside at the highest concentration tested of 2.5% (w/v).	(Coenye et al., 2012; Lee et al., 2013)
Origanum vulgare L. Essential oil	Traditionally used to treat cutaneous skin infections and respiratory ailments and help promote appetite and as an antispasmodic treatment	Carvacrol and thymol	None reported	At 1/2MIC *O. vulgare* essential oil inhibited 35% of *S. epidermidis* biofilm formation and displayed a biofilm inhibitory concentration at 0.125% (v/v).	*O. vulgare* essential oil inhibited *S. epidermidis* planktonic bacteria at a MIC of 0.125% (v/v).	(Nostro et al., 2007; Vale-Silva et al., 2012)
Rosmarinus officinalis	Traditionally used to relieve flatulence, indigestion, pain, inflammation, rheumatoid arthritis, relieve muscle spasms and is used as a choleretic treatment	Rosmarinic acid (phenolic compound)	None reported	Prevents *C. acnes* adhesion and has the potential for *C. acnes* biofilm disruption starting at 60 μg/mL.	Inhibits planktonic *C. acnes* at a MIC of 4 mg/mL.	(Tsai et al., 2013; Minaiyan et al., 2011)
Bauhinia acuruana Branches and fruit	Traditionally used to treat respiratory ailments such as tuberculosis, various skin disorders and urinary tract infections	Traditionally used as a tonic, as a depurative and to treat diabetes	None reported	At a concentration of 4 mg/mL, 18.3% *S. epidermidis* biofilm formation occurred of a branch extract and 22.2% in the presence of the fruit extract.	Bactericidal against *S. epidermidis* at 1 mg/mL.	(Trentin et al., 2011)

TABLE 3.3
Quorum Quenching and Biofilm Inhibitory Potential of Isolated Compounds against the Acne-Causing Bacteria *Cutibacterium acnes* and *Staphylococcus epidermidis*

Compounds	Plant Isolated from and Plant Part	Quorum Quenching Activity	Biofilm Inhibitory Activity	Other Targets	References
Allicin	*Allium sativum* L.	None reported	Inhibits more than 90% of *S. epidermidis* biofilm at 4 mg/mL. In combination with vancomycin at 4 mg/mL, inhibited *S. epidermidis* biofilms in prosthetic joints of a rabbit model.	Inhibits methicillin-susceptible and methicillin-resistant strains of *S. epidermidis* at a MIC of 8 µg/mL.	(Pérez-Giraldo et al., 2003; Ta and Arnason, 2016; Zhai et al., 2014)
Berberine	*Berberidaceae* and other species	None reported	Inhibits 50% of *S. epidermidis* biofilms at 30 µg/mL and 45 µg/mL, strain-dependent.	Berberine inhibited planktonic *S. epidermidis* at MIC values of 64 µg/mL and 128 µg/mL, depending on the strain.	(Wang et al., 2009; Ta and Arnason, 2016)
Carvacrol	*Origanum vulgare* L. Essential oil	None reported	At 1/2 MIC, carvacrol inhibited 29% of *S. epidermidis* biofilm formation and displayed a biofilm inhibitory concentration at 0.125% (v/v).	Carvacrol inhibited *S. epidermidis* planktonic bacteria at a MIC of 0.031% (v/v).	(Nostro et al., 2007)
Casbane diterpene	*Croton nepetaefolius*	None reported	*S. epidermidis* biofilm disruption at a concentration above 250 µg/mL.	Inhibits planktonic S. epidermidis at concentrations above 62.4 µg/mL.	(Carneiro et al., 2011)
Chelerythrine	Papaveraceae species	None reported	Inhibits 50% of *S. epidermidis* biofilms at 8.6 µM.	Chelerythrine inhibited planktonic *S. epidermidis* at a MIC range of 16.3–32.6 µM.	(Artini et al., 2012; Ta and Arnason, 2016)
Cinnamaldehyde	*Curcuma longa* Rhizome	Inhibit of 65% *Vibrio harveyi* AI-2-mediated bioluminescence at a concentration of 100 µM	In combination with curcumin, effective *S. epidermidis* biofilm formation inhibition was observed at a concentration of 15.62 µg/mL.	Inhibits *S. epidermidis* planktonic growth at 125 µg/mL. In combination with curcumin have a combined antibacterial effect against *S. epidermidis* at a MIC of 31.25 µg/mL.	(Sharma et al., 2014; Brackman et al., 2008; Ta and Arnason, 2016)
Curcumin	*Curcuma longa* L.	Inhibition of 88% *Vibrio harveyi* bioluminescence at 100 µg/mL	Inhibits *S. epidermidis* biofilm formation at a MIC of 25 µg/mL.	Curcumin significantly inhibited *C. acnes* bacteria after treatment with blue light at a concentration of 1.52 µM. Curcumin as a lipid vesicle significantly inhibited *C. acnes* growth at a concentration of 0.43 µg/mL.	(Sharma et al., 2014; Ta and Arnason, 2016; Yang et al., 2018; Yu et al., 2016; Liu and Huang, 2013)
Epigallocatechin gallate	*Camellia sinensis*	Inhibition of AI-2 at sub-MIC concentrations inhibit biofilm formation of *Eikenella corrodens*	None reported.	Targets sebum production through the AMPK-SREBP-1 pathway. Furthermore, it inhibits *C. acnes* bacteria *in vitro*.	(Matsunaga et al., 2010; Yoon et al., 2013)

Gallic acid	Various species	None reported	Increase *S. epidermidis* biofilm by threefold at a concentration of 188 µg/mL.	Does not inhibit planktonic *S. epidermidis* at the highest concentration tested of 2 mg/mL.	(Morán et al., 2014; Ta and Arnason, 2016)
Isolimonic acid	*Citrus aurantium* L. Seeds	Inhibited 99.23% *Vibrio harveyi* AI-2-mediated bioluminescence at a concentration of 100 µg/mL	None reported.	No activity reported against *C. acnes.*	(Vikram et al., 2011; Ta and Arnason, 2016)
Ichangin	*Citrus aurantium* L. Seeds	Inhibited 90% *Vibrio harveyi* AI-2-mediated bioluminescence at a concentration of 100 µg/mL	None reported.	No activity reported against *C. acnes.*	(Vikram et al., 2011; Ta and Arnason, 2016)
Proanthocyanidins A2-phosphatidylCholine	*Aesculus hippocastanum*	None reported	Inhibits 50% of *S. epidermidis* biofilms at 7.6 µM.	Does not inhibit planktonic bacteria of *S. epidermidis* at the highest concentration tested of 168.7 µM.	(Artini et al., 2012)
Quercetin	Various species	Inhibited 75% *Vibrio harveyi* AI-2-mediated bioluminescence at a concentration of 6.25 µg/mL	None reported.	Effectively inhibits *C. acnes* at a concentration less than or equal to 64 µg/mL.	(Vikram et al., 2010; Ta and Arnason, 2016; Lim, Kim, and Seo, 2007)
Rhodomyrtone	*Rhodomyrtus tomentosa* Leaves	None reported	Significant *C. acnes* biofilm inhibition at 1/16 MIC. At 4MIC and 8MIC, 40–80% of viable *C. acnes* in a mature biofilm is inhibited.	Inhibits planktonic *C. acnes* at MIC values ranging between 0.125–0.5 µg/mL. Significant *C. acnes* lipase and protease inhibition at 1/2 MIC.	(Wunnoo, Saising, and Voravuthikunchai, 2017)
Salvipisone	*Salvia sclarea* L. Roots	None reported	Reduce *S. epidermidis* biofilm between 85% and 90% at 37.5 µg/mL depending on the strain.	Inhibits *S. epidermidis* planktonic bacteria between 9.37 and 37.5 µg/mL, depending on the strain.	(Ta and Arnason, 2016; Kuźma et al., 2007; Walencka et al., 2007)
Sinensetin	*Orthosiphon stamineus*	Significant *Vibrio harveyi* AI-2-mediated bioluminescence observed at 100 µg/mL	None reported.	No activity reported against *C. acnes.*	(Vikram et al., 2010)
Sanguinarine	*Papaveraceae* species	None reported	Inhibits 50% of *S. epidermidis* biofilms at 5 µM.	Sanguinarine inhibited planktonic *S. epidermidis* at a MIC range of 8.5–17 µM.	(Artini et al., 2012; Ta and Arnason, 2016)
Thymol	*Origanum vulgare* L. Essential oil	None reported	At 1/2 MIC, thymol inhibited 28.3% of *S. epidermidis* biofilm formation and displayed a biofilm inhibitory concentration at 0.125% (v/v).	Thymol inhibited *S. epidermidis* planktonic bacteria at a MIC of 0.062% (v/v).	(Nostro et al., 2007)

FIGURE 3.6 Plants having quorum quenching and biofilm inhibitory activity against acne-causing bacteria: A, *Epimedium diphyllum*; B, *Malus pumila*; C, *Murtus communis*; D, *Polygonum cuspidatum*; E, *Rhodiola rosea*; F, *Origanum vulgare*; G, *Rosmarinus officinalis*; H, *Bauhinia kockiana* (Dessì, 2006; Qwert1234, 2016; Sehrg, 2011; Superior National Forest, 2010; Alpsdake, 2015; Friviere, 2019a, 2019b; Mokkie, 2020).

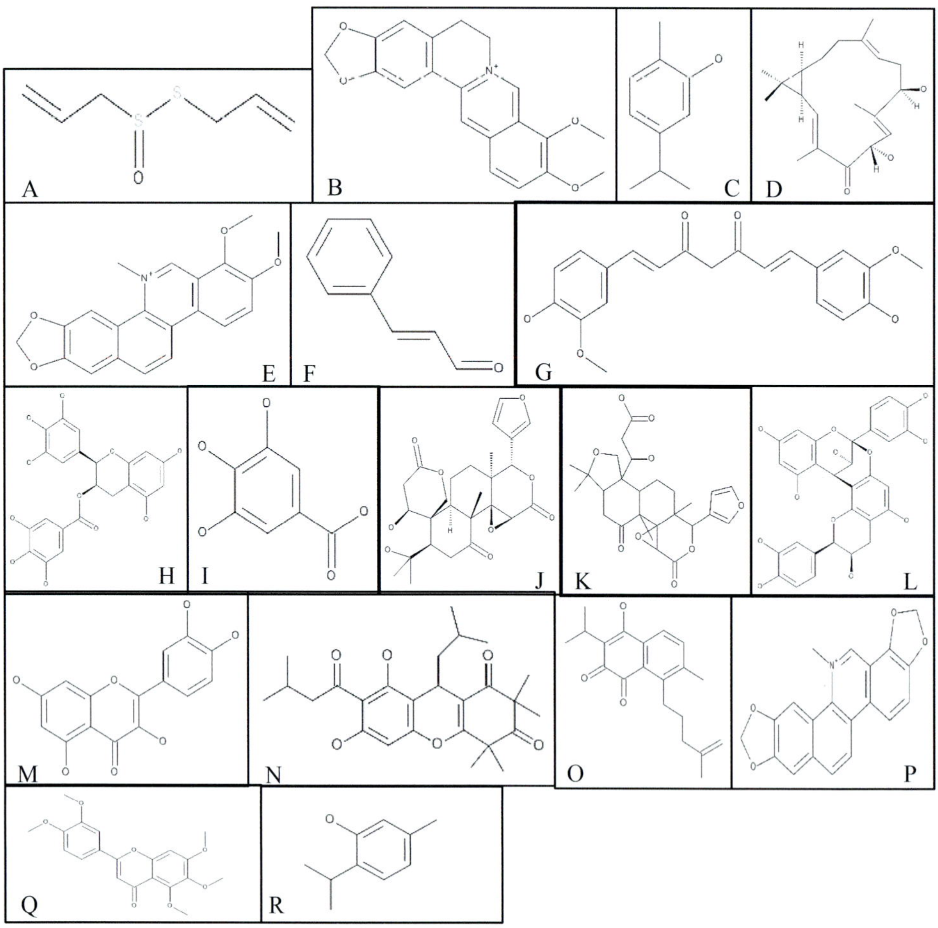

FIGURE 3.7 Compounds with quorum quenching and biofilm inhibition potential against acne-causing bacteria: A, Allicin; B, Berberine; C, Carvacrol; D, Casbane diterpene; E, Chelerythrine; F, Cinnamaldehyde; G, Curcumin; H, Epigallocatechin gallate; I, Gallic acid; J, Ichangin; K, Isolimonic acid; L, Proanthocyanidin-A2; M, Quercetin; N, Rhodomyrtone; O, Salvipison; P, Sanguinarine; Q, Sinensetin; R, Thymol.

REFERENCES

Abood, Methaq Satar, Zuhair Radhi Addai, and Methaq Sattar Abood. 2016. "Antioxidants, Total Phenolic Content and Antimicrobial Evaluation of *Myrtus Communis* Leaf and Stem Extract." *International Journal of ChemTech Research* 9, 9: 217–222.

Alipour, Ghazal, Saeedeh Dashti, and Hossein Hosseinzadeh. 2014. "Review of Pharmacological Effects of *Myrtus communis* L. and its Active Constituents." *Phytotherapy Research* 28, 8: 1125–1136. doi: 10.1002/ptr.5122.

Alpsdake. 2015. "*Rhodiola Rosea* (Female)." Obtained from Wikimedia Commons under the Creative Commons CC0 License. https://commons.wikimedia.org/wiki/File:Rhodiola_rosea_(female_s9).jpg (Accessed: July 2020)

Artini, M., R. Papa, G. Barbato, G.L. Scoarughi, A. Cellini, P. Morazzoni, E. Bombardelli, and L. Selan. 2012. "Bacterial Biofilm Formation Inhibitory Activity Revealed for Plant Derived Natural Compounds." *Bioorganic and Medicinal Chemistry* 20, 2: 920–926. doi: 10.1016/j.bmc.2011.11.052.

Bassler, Bonnie L. 1999. "How Bacteria Talk to Each Other: Regulation of Gene Expression by Quorum Sensing." *Current Opinion in Microbiology* 2, 6: 582–587. doi: 10.1016/S1369-5274(99)00025-9.

Bassler, Bonnie L., and Melissa B. Miller. 2013. "Quorum Sensing." In *The Prokaryotes: Prokaryotic Communities and Ecophysiology*, 9783642301230: 495–509. Springer-Verlag Berlin Heidelberg. doi: 10.1007/978-3-642-30123-0_60.

Bhate, K., and H.C. Williams. 2013. "Epidemiology of Acne Vulgaris." *British Journal of Dermatology* 168, 3: 474–485. doi: 10.1111/bjd.12149.

Brackman, Gilles, Tom Defoirdt, Carol Miyamoto, Peter Bossier, Serge Van Calenbergh, Hans Nelis, and Tom Coenye. 2008. "Cinnamaldehyde and Cinnamaldehyde Derivatives Reduce Virulence in *Vibrio* Spp. by Decreasing the DNA-Binding Activity of the Quorum Sensing Response Regulator LuxR." *BMC Microbiology* 8. doi: 10.1186/1471-2180-8-149.

Brandwein, Michael, Doron Steinberg, and Shiri Meshner. 2016. "Microbial Biofilms and the Human Skin Microbiome." *Npj Biofilms and Microbiomes* 2. doi: 10.1038/s41522-016-0004-z.

Burkhart, Craig N., and Craig G. Burkhart. 2003. "Microbiology's Principle of Biofilms as a Major Factor in the Pathogenesis of Acne Vulgaris." *International Journal of Dermatology* 42, 12: 925–927. doi: 10.1111/j.1365-4632.2003.01588.x.

Carneiro, Victor A., Hélcio S.D. Santos, Francisco V.S. Arruda, Paulo N. Bandeira, Maria R.J.R. Albuquerque, Maria O. Pereira, Mariana Henriques, Benildo S. Cavada, and Edson H. Teixeira. 2011. "Casbane Diterpene as a Promising Natural Antimicrobial Agent against Biofilm-Associated Infections." *Molecules* 16, 1: 90–201. doi: 10.3390/molecules16010190.

Center for Biofilm Engineering: Montana State University. 2020. "Biofilm Basics: Section 1." www.biofilm.montana.edu/biofilm-basics/what_are_biofilms.html (Accessed: July 2020).

Chen, Xin, Stephan Schauder, Noelle Potier, Alain Van Dorsselaer, Istva Â N Pelczer, Bonnie L. Bassler, and Frederick M. Hughson. 2002. "Structural Identification of a Bacterial Quorum-Sensing Signal Containing Boron." *Letters to Nature* 415, 31.

Coenye, Tom, Gilles Brackman, Petra Rigole, Evy De Witte, Kris Honraet, Bart Rossel, and Hans J. Nelis. 2012. "Eradication *of Propionibacterium Acnes* Biofilms by Plant Extracts and Putative Identification of Icariin, Resveratrol and Salidroside as Active Compounds." *Phytomedicine* 19: 409–412. doi: 10.1016/j.phymed.2011.10.005.

Coenye, Tom, Kris Honraet, Bart Rossel, and Hans J. Nelis. 2008. "Biofilms in Skin Infections : *Propionibacterium Acnes* and Acne Vulgaris." *Infectious Disorders-Drug Targets* 8: 156–159. doi: 10.2174/1871526510808030156.

Coenye, Tom, Elke Peeters, and Hans J. Nelis. 2007. "Biofilm Formation by *Propionibacterium Acnes* Is Associated with Increased Resistance to Antimicrobial Agents and Increased Production of Putative Virulence Factors." *Research in Microbiology* 158, 4: 386–392. doi: 10.1016/j.resmic.2007.02.001.

Dessì, Giancarlo. 2006. "*Myrtus Communis*." Obtained from Wikimedia Commons under the Creative Commons CC0 License. https://commons.wikimedia.org/wiki/File:Myrtus_communis4.jpg. (Accessed: July 2020)

Dessinioti, Clio, and Andreas D. Katsambas. 2010. "The Role of *Propionibacterium Acnes* in Acne Pathogenesis: Facts and Controversies." *Clinics in Dermatology* 28, 1: 2–7. doi: 10.1016/j.clindermatol.2009.03.012.

De Wet, H., S. Nciki, and S.F. van Vuuren. 2013. "Medicinal Plants Used for the Treatment of Various Skin Disorders by a Rural Community in Northern Maputaland, South Africa." *Journal of Ethnobiology and Ethnomedicine* 9, 1: 51. doi: 10.1186/1746-4269-9-51

Dlova, N.C., A. Mosam, and J. Tsoka-Gwegweni. 2018. "The Spectrum and Sequelae of Acne in Black South Africans Seen in Tertiary Institutions." *Skin Appendage Disorders* 4: 301–303. doi: 10.1159/000488689

Eiff, Christof Von, Georg Peters, and Christine Heilmann. 2002. "Pathogenesis of Infections due to Coagulase-Negative *Staphylococci*." *Lancet Infectious Diseases. Lancet Publishing Group* 2, 11: 677–685. doi: 10.1016/S1473-3099(02)00438-3.

Falcocchio, Serena, Cristian Ruiz, F.I. Javier Pastor, Luciano Saso, and Pilar Diaz. 2006. "*Propionibacterium acnes* GehA Lipase, an Enzyme Involved in Acne Development, Can Be Successfully Inhibited by Defined Natural Substances." *Journal of Molecular Catalysis B : Enzymatic* 40: 132–137. doi: 10.1016/j.molcatb.2006.02.011.

Feuillolay, Catherine, Sophie Pecastaings, Céline Le Gac, Christel Fiorini-Puybaret, Joëlle Luc, Philippe Joulia, and Christine Roques. 2016. "A *Myrtus communis* Extract Enriched in Myrtucummulones and Ursolic Acid Reduces Resistance of *Propionibacterium acnes* Biofilms to Antibiotics Used in Acne Vulgaris." *Phytomedicine* 23, 3: 307–315. doi: 10.1016/j.phymed.2015.11.016.

Friviere. 2019a. "*Origanum Vulgare*." Obtained from Wikimedia Commons under the Creative Commons CC0 License. https://commons.wikimedia.org/wiki/File:Orenga_OriganumVulgare.jpg (Accessed: July 2020).

Friviere. 2019b. "*Rosmarinus Officinalis*." Obtained from Wikimedia Commons under the Creative Commons CC0 License. https://commons.wikimedia.org/wiki/File:Romani_RosmarinusOfficinalis.jpg (Accessed: July 2020).

Gollnick, Harald. 2003. "Current Concepts of the Pathogenesis of Acne." *Drugs* 63, 15: 1579–1596. doi: 10.2165/00003495-200363150-00005.

Hamdy, A.A., H.A. Kassem, G.E.A. Awad, S.M. El-Kady, M.T. Benito, E.G. Doyagüez, M.L. Jimeno, N. Lall, and A.A. Hussein. 2017. "*In-Vitro* Evaluation of Certain Egyptian Traditional Medicinal Plants against *Propionibacterium Acnes*." *South African Journal of Botany* 109, March: 90–95. doi: 10.1016/j.sajb.2016.12.026.

Holland, K.T., Ingham, E. and W.J. Cunliffe. 1981. "A Review The Microbiology of Acne." *Journal of Applied Bacteriology* 51, 2: 195–215. doi: 10.1111/j.1365-2672.1981.tb01234.x.

Jahns, Anika C., and Oleg A. Alexeyev. 2014. "Three Dimensional Distribution of *Propionibacterium Acnes* Biofilms in Human Skin." *Experimental Dermatology* 23, 9: 687–689. doi: 10.1111/exd.12482.

James, Kirk A., Craig N. Burkhart, and Dean S. Morrell. 2009. "Emerging Drugs for Acne." *Expert Opinion on Emerging Drugs* 14, 4: 649–659. doi: 10.1517/14728210903251690.

James, William D. 2005. "Acne." *New England Journal of Medicine* 352, 14: 1463–1472. doi: 10.1056/NEJMcp033487.

Kalia, Vipin Chandra. 2013. "Quorum Sensing Inhibitors: An Overview." *Biotechnology Advances* 31, 2: 224–245. doi: 10.1016/j.biotechadv.2012.10.004.

Kaur, A., J. Singh, S. Kamboj, A.K. Sexana, and R.M. Pandita. 2005. "Isolation of an N-Acetyl-D-Glucosamine Specific Lectin from the Rhizomes of *Arundo Donax* with Antiproliferative Activity." *Phytochemistry* 66, 16: 933–940. doi: 10.1016/J.PHYTOCHEM.2005.06.026.

Kong, Kok Fai, Cuong Vuong, and Michael Otto. 2006. "*Staphylococcus* Quorum Sensing in Biofilm Formation and Infection." *International Journal of Medical Microbiology* 296, 2–3: 133–139. doi: 10.1016/j.ijmm.2006.01.042.

Korkmaz, Mustafa, and Sinem Karakuş. 2015. "Traditional Uses of Medicinal Plants of Üzümlü District, Erzincan, Turkey." *Pakistan Journal of Botany* 47, 1: 125–134.

Kumar, Bipul, Rajiv Pathak, P. Bertin Mary, Diksha Jha, Kabir Sardana, and Hemant K. Gautam. 2016. "New Insights into Acne Pathogenesis: Exploring the Role of Acne-Associated Microbial Populations." *Dermatologica Sinica* 34, 2: 67–73. doi: 10.1016/j.dsi.2015.12.004.

Kuźma, L., M. Różalski, E. Walencka, B. Różalska, and H. Wysokińska. 2007. "Antimicrobial Activity of Diterpenoids from Hairy Roots of *Salvia Sclarea* L.: Salvipisone as a Potential Anti-Biofilm Agent Active against Antibiotic Resistant *Staphylococci*." *Phytomedicine* 14, 1: 31–35. doi: 10.1016/j.phymed.2005.10.008.

Le, Katherine Y., and Michael Otto. 2015. "Quorum-Sensing Regulation in *Staphylococci*-an Overview." *Frontiers in Microbiology*. doi: 10.3389/fmicb.2015.01174.

Lee, Sang, Ji Min Kim, Se Kyoo Jeong, Jeong Eun Jeon, Hyun Ju Yoon, Min Kyung Jeong, and Seung Hun Lee. 2010. "Protease-Activated Receptor-2 Mediates the Expression of Inflammatory Cytokines, Antimicrobial Peptides, and Matrix Metalloproteinases in Keratinocytes in Response to *Propionibacterium Acnes*." *Archives of Dermatological Research* 302, 10: 745–756. doi: 10.1007/s00403-010-1074-z.

Lee, Shih-Yu, Min-Hui Li, Hsin Chu, Cheng-Wen Ho, and Tsu-Chung Chang. 2013. "*Rhodiola Crenulata* Extract Alleviates Hypoxic Pulmonary Edema in Rats." *Evidence Based Complimentary and Alternative Medicine* 718739. doi: 10.1155/2013/718739

Li, M., Y. Lai, A.E. Villaruz, D.J. Cha, D.E. Sturdevant, M. Otto, K.W. Bayles, et al. 2007. "AI-2-Dependent Gene Regulation in *Staphylococcus epidermidis*." *Proceedings of the National Academy of Sciences* 104, 22: 9469–9474. doi: 10.1073/pnas.0702159104.

Lim, Young Hee, In Hwan Kim, and Jung Ju Seo. 2007. "*In vitro* Activity of Kaempferol Isolated from the Impatiens Balsamina Alone and in Combination with Erythromycin or Clindamycin against *Propionibacterium acnes*." *Journal of Microbiology* 45, 5: 473–477.

Liu, Chi Hsien, and Hsin Ying Huang. 2013. "*In Vitro* Anti-*Propionibacterium* Activity by Curcumin Containing Vesicle System." *Chemical and Pharmaceutical Bulletin* 61, 4: 419–425. doi: 10.1248/cpb.c12-01043.

Matsunaga, Tetsuro, Aya Nakahara, Karim M. Minnatul, Yuichiro Noiri, Shigeyuki Ebisu, Akio Kato, and Hiroyuki Azakami. 2010. "The Inhibitory Effects of Catechins on Biofilm Formation by the Periodontopathogenic Bacterium, *Eikenella Corrodens*." *Bioscience, Biotechnology and Biochemistry* 74, 12: 2445–2450. doi: 10.1271/bbb.100499.

Miller, Melissa B., and Bonnie L. Bassler. 2001. "Quorum Sensing In Bacteria." *Annual Review of Microbiology* 55: 165–199. doi: 10.1146/annurev.micro.55.1.165

Minaiyan, Mohsen, A.R. Ghannadi, M. Afsharipour, and P. Mahzouni. 2011. "Effects of Extract and Essential Oil of *Rosmarinus officinalis* L. on TNBS-Induced Colitis in Rats." *Research in Pharmaceutical Sciences* 6, 1.

Mokkie. 2020. "*Bauhinia Kockiana*." Obtained from Wikimedia Commons under the Creative Commons CC0 License. https://commons.wikimedia.org/wiki/File:Bauhinia_kockiana_flowers.jpg (Accessed: July 2020).

Monroe, D. 2007. "Biofilm." Obtained from Wikimedia Commons under the Creative Commons CC0 License. https://commons.wikimedia.org/wiki/File:Biofilm.jpg (Accessed: July 2020).

Montanaro, Lucio, Alessandro Poggi, Livia Visai, Stefano Ravaioli, Davide Campoccia, Pietro Speziale, and Carla Renata Arciola. 2011. "Extracellular DNA in Biofilms." *International Journal of Artificial Organs*. doi: 10.5301/ijao.5000051.

Morán, A., S. Gutiérrez, H. Martínez-Blanco, M.A. Ferrero, A. Monteagudo-Mera, and L.B. Rodríguez-Aparicio. 2014. "Non-Toxic Plant Metabolites Regulate *Staphylococcus* Viability and Biofilm Formation: A Natural Therapeutic Strategy Useful in the Treatment and Prevention of Skin Infections." *Biofouling* 30, 10: 1175–1182. doi: 10.1080/08927014.2014.976207.

Nishijima, Setsuko, Ichiro Kurokawa, Naoki Katoh, and Kunitomo Watanabe. 2000. "The Bacteriology of Acne Vulgaris and Antimicrobial Susceptibility of *Propionibacterium Acnes* and *Staphylococcus Epidermidis* Isolated from Acne Lesions." *The Journal of Dermatology* 27, 5: 318–323. doi: 10.1111/j.1346-8138.2000.tb02174.x.

Nostro, Antonia, Andrea Sudano Roccaro, Giuseppe Bisignano, Andreana Marino, Maria A. Cannatelli, Francesco C. Pizzimenti, Pier Luigi Cioni, Francesca Procopio, and Anna Rita Blanco. 2007. "Effects of Oregano, Carvacrol and Thymol on *Staphylococcus Aureus* and *Staphylococcus Epidermidis* Biofilms." *Journal of Medical Microbiology* 56, 4: 519–523. doi: 10.1099/jmm.0.46804-0.

Oyebode, O., Kandalhilton, P.J., and Lilford, R.J. 2016. Use of traditional medicine in middle-income countries: A WHO-SAGE study. *Health Policy Plan* 31: 984–991. https://doi.org/10.1093/HEAPOL/CZW022

Packiavathy, Sybiya Vasantha, Issac Abraham, Palani Agilandeswari, Khadar Syed Musthafa, Shunmugiah Karutha Pandian, and Arumugam Veera Ravi. 2012. "Antibiofilm and Quorum Sensing Inhibitory Potential of *Cuminum Cyminum* and Its Secondary Metabolite Methyl Eugenol against Gram-negative Bacterial Pathogens." *Food Research International* 45, 1: 85–92. doi: 10.1016/j.foodres.2011.10.022.

Painter, K.L., A. Krishna, S. Wigneshweraraj, and A.M. Edwards. 2014. "What Role Does The Quorum-Sensing Accessory Gene Regulator System Play During *Staphylococcus Aureus* Bacteremia?" *Trends In Microbiology* 22, 12: 676–685.

Pathak, Rajiv, Nitesh Kasama, Raj Kumar, and Hemant K. Gautam. 2013. "*Staphylococcus Epidermidis* in Human Skin Microbiome Associated with Acne: A Cause of Disease or Defence?" *Research Journal of Biotechnology* 8, 12: 78–82. doi: 10.1007/s00253-013-5394-8

Pawin, Henry, Claire Beylot, Martine Chivot, Michel Faure, Florence Poli, Jean Revuz, and Brigitte Dréno. 2004. "Physiopathology of Acne Vulgaris: Recent Data, New Understanding of the Treatments." *European Journal of Dermatology* 14, 1: 4–12.

Peng, Wei, Rongxin Qin, Xiaoli Li, and Hong Zhou. 2013. "Botany, Phytochemistry, Pharmacology, and Potential Application of *Polygonum Cuspidatum* Sieb.et Zucc.: A Review." *Journal of Ethnopharmacology*. doi: 10.1016/j.jep.2013.05.007.

Pereira, Catarina S., Jessica A. Thompson, and Karina B. Xavier. 2013. "AI-2-Mediated Signaling in Bacteria." *FEMS Microbiology Reviews* 37, 2: 156–181. doi: 10.1111/j.1574-6976.2012.00345.x.

Pérez-Giraldo, C., G. Cruz-Villalón, R. Sánchez-Silos, R. Martínez-Rubio, M.T. Blanco, and Antonio C. Gómez-García. 2003. "In Vitro Activity of Allicin against *Staphylococcus Epidermidis* and Influence of Subinhibitory Concentrations on Biofilm Formation." *Journal of Applied Microbiology* 95, 4: 709–711. doi: 10.1046/j.1365-2672.2003.02030.x.

Qwert1234.2016. "*Epimedium Diphyllum*." Obtained from Wikimedia Commons under the Creative Commons CC0 License. https://commons.wikimedia.org/wiki/File:Epimedium_diphyllum_1.JPG?uselang=de (Accessed: July 2020).

Ramanuj, Krupali, Pooja Bachani, and Vijay Kothari. 2012. "*In Vitro* Antimicrobial Activity of Certain Plant Products/Seed Extracts against Multidrug Resistant *Propionibacterium Acnes*, *Malassezia Furfur*, and Aflatoxin Producing *Aspergillus Flavus*." *Research in Pharmacy* 2, 4: 1–9.

Sandasi, M., C.M. Leonard, S.F. Van Vuuren, and A.M. Viljoen. 2011. "Peppermint (*Mentha Piperita*) Inhibits Microbial Biofilms *In Vitro*." *South African Journal of Botany* 77, 1: 80–85. doi: 10.1016/j.sajb.2010.05.011.

Sanford, James A., and Richard L. Gallo. 2013. "Functions of the Skin Microbiota in Health and Disease." *Seminars in Immunology* 25, 5: 370–377. doi: 10.1016/j.smim.2013.09.005.

Schröder, Jens M., and Jürgen Harder. 1999. "Human Beta-Defensin-2." *International Journal of Biochemistry and Cell Biology* 31, 6: 645–651. doi: 10.1016/S1357-2725(99)00013-8.

Sehrg. 2011. "*Malus Pumila* 'Pendula Rubrifolia." Obtained from Wikimedia Commons under the Creative Commons CC0 License. https://commons.wikimedia.org/wiki/File:Malus_pumila_%27Pendula_rubrifolia%27.jpg (Accessed: July 2020)

Sharma, Garima, Kiran Raturi, Shweta Dang, Sanjay Gupta, and Reema Gabrani. 2014. "Combinatorial Antimicrobial Effect of Curcumin with Selected Phytochemicals on *Staphylococcus Epidermidis*." *Journal of Asian Natural Products Research* 16, 5: 535–541. doi: 10.1080/10286020.2014.911289.

Singh, Deepak, Bhushan Hatwar, and S. Nayak. 2011. "Herbal Plants and *Propionibacterium Acnes*: An Overview." *International Journal of Biomedical Research* 2, 9: 486–498. doi: 10.7439/ijbr.v2i9.164.

Street, R.A., W.A. Stirk, and J. Van Staden. 2008. "South African Traditional Medicinal Plant trade—Challenges in Regulating Quality, Safety and Efficacy." *Journal of Ethnopharmacology* 119, 3: 705–710. doi: 10.1016/j.jep.2008.06.019.

Superior National Forest. 2010. "*Polygonum Cuspidatum*." Obtained from Wikimedia Commons under the Creative Commons CC0 License. https://commons.wikimedia.org/wiki/File:Polygonum_cuspidatum_3-eheep_(5097370153).jpg (Accessed: July 2020)

Ta, Chieu Anh Kim, and John Thor Arnason. 2016. "Mini Review of Phytochemicals and Plant Taxa with Activity as Microbial Biofilm and Quorum Sensing Inhibitors." *Molecules* 21, 1: 29. doi: 10.3390/molecules21010029.

Tan, J.K.L., and K. Bhate. 2015. "A Global Perspective on the Epidemiology of Acne." *British Journal of Dermatology* 172, July: 3–12. doi: 10.1111/bjd.13462.

Trentin, Danielle da Silva, Raquel Brandt Giordani, Karine Rigon Zimmer, Alexandre Gomes da Silva, Márcia Vanusa da Silva, Maria Tereza dos Santos Correia, Israel Jacob Rabin Baumvol, and Alexandre José Macedo. 2011. "Potential of Medicinal Plants from the Brazilian Semi-Arid Region (Caatinga) against *Staphylococcus Epidermidis* Planktonic and Biofilm Lifestyles." *Journal of Ethnopharmacology* 137, 1: 327–335. doi: 10.1016/j.jep.2011.05.030.

Tsai, Tsung Hsien, Lu Te Chuang, Tsung Jung Lien, Yau Rong Liing, Wei Yu Chen, and Po Jung Tsai. 2013. "*Rosmarinus Officinalis* Extract Suppresses *Propionibacterium Acnes*-Induced Inflammatory Responses." *Journal of Medicinal Food* 16, 4: 324–333. doi: 10.1089/jmf.2012.2577.

Unosson, Erik. 2015. "Antibacterial Strategies for Titanium Biomaterials." *Doctoral dissertation, Acta Universitatis Upsaliensis.*

Vale-Silva, Luís, Maria-Joã Silva, Daniela Oliveira, Maria-José Gonç alves, Carlos Cavaleiro, Lígia Salgueiro, and Eugé nia Pinto. 2012. "Correlation of the Chemical Composition of Essential Oils from *Origanum Vulgare* Subsp. Virens with Their *In Vitro* Activity against Pathogenic Yeasts and Filamentous Fungi." *Journal of Medical Microbiology* 61: 252–260. doi: 10.1099/jmm.0.036988-0.

Van Vuuren, S., and D. Holl. 2017. "Antimicrobial Natural Product Research: A Review from a South African Perspective for the Years 2009–2016." *Journal of Ethnopharmacology* 208: 236–252. doi: 10.1016/j.jep.2017.07.011

Vikram, A., G.K. Jayaprakasha, P.R. Jesudhasan, S.D. Pillai, and B.S. Patil. 2010. "Suppression of Bacterial Cell-Cell Signaling, Biofilm Formation and Type III Secretion System by Citrus Flavonoids." *Journal of Applied Microbiology* 109, 2: 515–527. doi: 10.1111/j.1365-2672.2010.04677.x.

Vikram, A., Palmy R. Jesudhasan, G.K. Jayaprakasha, Suresh D. Pillai, and Bhimanagouda S. Patil. 2011. "*Citrus Limonoids* Interfere with *Vibrio Harveyi* Cell-Cell Signaling and Biofilm Formation by Modulating the Response Regulator LuxO." *Microbiology* 157, 1: 99–110. doi: 10.1099/mic.0.041228-0.

Vuong, Cuong, Christiane Gerke, Greg A. Somerville, Elizabeth R. Fischer, and Michael Otto. 2003. "Quorum Sensing Control of Biofilm Factors in *Staphylococcus Epidermidis*." *The Journal of Infectious Diseases* 188, 5: 706–718. doi: 10.1086/377239.

Walencka, Elzbieta, Sylwia Rozalska, Halina Wysokinska, Marek Rozalski, Lukasz Kuzma, and Barbara Rozalska. 2007. "Salvipisone and Aethiopinone from *Salvia Sclarea* Hairy Roots Modulate Staphylococcal Antibiotic Resistance and Express Anti-Biofilm Activity." *Planta Medica* 73, 6: 545–551. doi: 10.1055/s-2007-967179.

Wang, Xiaoqing, Xiao Yao, Zhen'an Zhu, Tingting Tang, Kerong Dai, Irina Sadovskaya, Sigrid Flahaut, and Said Jabbouri. 2009. "Effect of Berberine on *Staphylococcus Epidermidis* Biofilm Formation." *International Journal of Antimicrobial Agents* 34, 1: 60–66. doi: 10.1016/j.ijantimicag.2008.10.033.

Waters, Christopher M., and Bonnie L. Bassler. 2005. "Quorum Sensing: Cell-to-Cell Communication in Bacteria." *Annual Review of Cell and Developmental Biology* 21: 319–346. doi: 10.1146/.

Williams, Hywel C., Robert P. Dellavalle, and Sarah Garner. 2012. "Acne Vulgaris." *The Lancet* 379, 9813: 361–372. doi: 10.1016/S0140-6736(11)60321-8.

Winans, S.C., and B.L. Bassler. 2002. "Mob Psychology." *Journal of Bacteriology* 184, 4: 873–883. doi: 10.1128/JB.184.4.873-883.2002

Winzer, Klaus, Kim R. Hardie, and Paul Williams. 2002. "Bacterial Cell-to-Cell Communication: Sorry, Can't Talk Now-Gone to Lunch!" *Current Opinion in Microbiology*. doi: 10.1016/S1369-5274(02)00304-1.

Wu, Xiao Bo, Xian Qin Luo, Shu Ying Gu, and Jia Hong Xu. 2012. "The Effects of *Polygonum Cuspidatum* Extract on Wound Healing in Rats." *Journal of Ethnopharmacology* 141, 3: 934–937. doi: 10.1016/j.jep.2012.03.040.

Wunnoo, Suttiwan, Jongkon Saising, and Supayang Piyawan Voravuthikunchai. 2017. "Rhodomyrtone Inhibits Lipase Production, Biofilm Formation, and Disorganizes Established Biofilm in *Propionibacterium Acnes*." *Anaerobe* 43, February: 61–68. doi: 10.1016/j.anaerobe.2016.12.002.

Xu, Lin, Hualin Li, Cuong Vuong, Viveka Vadyvaloo, Jianping Wang, Yufeng Yao, Michael Otto, and Qian Gao. 2006. "Role of the luxS Quorum-Sensing System in Biofilm Formation and Virulence of *Staphylococcus Epidermidis*." *Infection and Immunity* 74, 1: 488–496. doi: 10.1128/IAI.74.1.488-496.2006.

Xue, Ting, Jingtian Ni, Fei Shang, Xiaolin Chen, and Ming Zhang. 2015. "Autoinducer-2 Increases Biofilm Formation via an Ica- and Bhp-Dependent Manner in *Staphylococcus Epidermidis* RP62A." *Microbes and Infection* 17, 5: 345–352. doi: 10.1016/j.micinf.2015.01.003.

Yang, Ming Yeh, Kai Chih Chang, Liang Yü Chen, and Anren Hu. 2018. "Low-Dose Blue Light Irradiation Enhances the Antimicrobial Activities of Curcumin against *Propionibacterium Acnes*." *Journal of Photochemistry and Photobiology B: Biology* 189, December: 21–28. doi: 10.1016/j.jphotobiol.2018.09.021.

Yarwood, Jeremy M., and Patrick M. Schlievert. 2003. "Quorum Sensing in Staphylococcus Infections." *Journal of Clinical Investigation* 112, 11: 1620–1625. doi: 10.1172/jci20442.

Yazici, K., K. Baz, A.E. Yazici, A. Kokturk, S. Tot, D. Demirseren, and V. Buturak. 2004. "Disease-Specific Quality of Life Is Associated with Anxiety and Depression in Patients with Acne." *Journal of the European Academy of Dermatology and Venereology* 18, 4: 435–439. doi: 10.1111/j.1468-3083.2004.00946.x.

Yoon, Ji Young, Hyuck Hoon Kwon, Seong Uk Min, Diane M. Thiboutot, and Dae Hun Suh. 2013. "Epigallocatechin-3-Gallate Improves Acne in Humans by Modulating Intracellular Molecular Targets and Inhibiting *P. Acnes*." *Journal of Investigative Dermatology* 133, 2: 429–440. doi: 10.1038/jid.2012.292.

Yu, Chunna, Xiaojuan Chai, Lushan Yu, Shuqing Chen, and Su Zeng. 2011. "Identification of Novel Pregnane X Receptor Activators from Traditional Chinese Medicines." *Journal of Ethnopharmacology* 136, 1: 137–143. doi: 10.1016/j.jep.2011.04.022.

Yu, Qilin, Jianrong Li, Yueqi Zhang, Yufan Wang, Lu Liu, and Mingchun Li. 2016. "Inhibition of Gold Nanoparticles (AuNPs) on Pathogenic Biofilm Formation and Invasion to Host Cells." *Scientific Reports* 6, 1: 26667. doi: 10.1038/srep26667.

Zhai, Haohan, Jianchao Pan, En Pang, and Bo Bai. 2014. "Lavage with Allicin in Combination with Vancomycin Inhibits Biofilm Formation by *Staphylococcus Epidermidis* in a Rabbit Model of Prosthetic Joint Infection." *PLoS ONE* 9, 7: e102760. doi: 10.1371/journal.pone.0102760.

4 Nanoparticle-Stabilized Liposomes as an Effective Bio-Active Drug Molecule Delivery for Acne Treatment

Catherine Wilkinson, Marco N. De Canha and Namrita Lall

CONTENTS

4.1 INTRODUCTION

Acne vulgaris is an inflammatory condition of the pilosebaceous unit affecting up to 650 million people worldwide, equivalent to approximately 9.4% of the world's population (Garg, 2016; Van De Kerkhof et al., 2006). It is a chronic condition that has a significant impact on quality of life, particularly on the psychological well-being of a patient's life (Mallon et al., 1999; Mosam et al., 2005). Up to 90% of adolescents suffer from acne, and up to 50% of teenage acne sufferers continue to experience acne symptoms as adults (Chernyshov et al., 2018; Garg, 2016). While this skin disease is not considered a life-threatening condition, one of the major psychological implications is its link to the onset of suicidal thoughts as a result of an appearance-related sense of shame and decrease in self-esteem (Fu, 2016; Simonart, 2013).

Acne treatment is largely dependent on the severity of the condition, as well as the morphology of the lesions. Different acne lesions (comedones, pustules, papules, nodules and cysts) respond differently to the variety of available treatment types, and therefore, no single treatment is considered suitable for all types of acne (Doshi et al., 1997; Jappe, 2003). Selection of a suitable treatment option is generally chosen by determining the type of acne which can be classified as comedolytic, papulo-pustular and/or nodular, and by observing any skin abnormalities (inherent abnormalities in skin barrier function or increased sebum secretion) and any previously attempted treatments

DOI: 10.1201/9781003108375-4

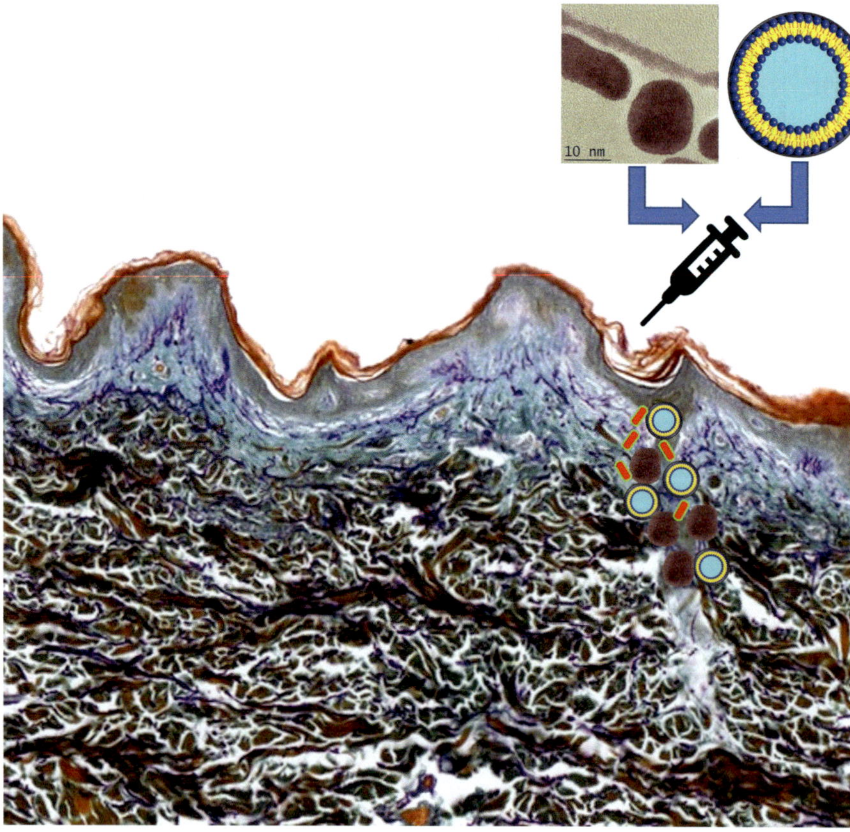

Graphical abstract representing the content of the chapter (Fionán, 2017; Kanasorhokeshawnidioescagozity, 2017; Mackenzie, 2021).

that were unsuccessful (Thiboutot and Del Rosso, 2013). A few treatment options include topical, systemic, hormonal, physical therapies and laser treatment (Van De Kerkhof et al., 2006). A targeted, multifactorial acne treatment is required in order to properly treat acne vulgaris, ideally with minimal side effects.

The use of nanoparticle (NP) technology may provide a more targeted drug-delivery system that could improve treatment efficacy by enhancing the localization of the treatment and allowing the controlled release of bio-active compounds or ingredients (Garg, 2016). The fundamental properties of nanoparticles are their miniscule size, the ability to manipulate their surface ligands and the biodegradable materials from which they are prepared (Sahoo and Labhasetwar, 2006). These properties confer many advantages, including the ability to minimize the toxicity of drugs/compounds, maintain the integrity of the molecules they are delivering, enhance penetration through capillaries and skin surfaces, and provide a targeted delivery of the drug/compound to an exact location, thereby reducing adverse effects on neighboring tissues and organs (Garg, 2016; Sahoo and Labhasetwar, 2006).

NPs have a variety of applications, and they are most commonly used to enhance the activity of certain drugs and therapies (Sahoo and Labhasetwar, 2006). There are several nanotechnological approaches that could potentially be applied to the formulation of an improved acne treatment. This chapter reviews different types of nanoparticle systems such as particulate, vesicular and colloidal, of which certain types may be better suited for the different types of acne based on factors such as the ability to use lower dosages with improved efficacy (Garg, 2016).

4.2 ACNE

4.2.1 Development and Causes

Acne is a relapsing condition that occurs primarily on the face, neck, chest and back, and is characterized by various physical manifestations appearing as comedones, nodules, pustules, papules and seborrhea (Garg, 2016; Holland et al., 1998; Simonart, 2012). The severity of acne is often determined by the presence of acne lesions of different morphologies, such as non-inflammatory lesions indicating a mild condition. This is often referred to as comedonal acne, characterized by open or closed comedones (commonly known as blackheads and/or whiteheads). Mild acne is often characterized by the presence of comedones with papules and pustules, and is referred to as papulopustular acne. In more severe cases, inflammatory lesions— including papules, pustules, nodules and cysts— may occur (Oon et al., 2019).

Hyperkeratinization, hypo-cornification and hypo-desquamation within the infundibulum are the primary causes of mild acne. In the most severe forms, nodules and cysts may occur, in which case it is referred to as nodulocystic acne (Oon et al., 2019). Severe acne is associated with inflammation caused by rupturing of the infundibulum, which allows intrafollicular substances, such as keratin, sebum, bacteria and hair, to leak into the dermis, exacerbating the inflammatory cascade (Figure 4.1). Constant damage and repair processes may result in the formation of scars (Decker and Graber, 2012; Jappe, 2003).

A multitude of factors cause acne, such as heredity, hormonal changes, anxiety and an imbalance in the skin microbiome resulting in opportunistic bacterial infection. However, there is still uncertainty regarding the exact order of events that occurs in acne development and progression (Garg, 2016; Holland et al., 1998; Khorvash et al., 2012). Environmental and other external factors—such as heat and humidity, smoke, industrial chemicals, hair sprays and comedogenic oil-containing products—can exacerbate the condition by reducing the breathability of the skin and blocking the follicles and pilosebaceous glands (Garg, 2016). One of the major underlying causes of acne progression is inflammation (Holland et al., 1998). However, the involvement of inflammation in the initiation of acne development is not yet clear. Inflammation may have an indirect effect of acne-causing factors, which includes an increase in sebum production and bacterial virulence factors, such as secretory proteins and biofilm formation (Achermann et al., 2014; Dréno et al., 2018; Holland et al., 1998; Jappe, 2003).

Since this skin disorder is observed predominantly in pubescent teenagers, androgenic hormones such as testosterone and dihydrotestosterone are said to play a major role in acne progression (Garg, 2016). Abnormalities in androgen levels, particularly during puberty and other hormonal changes, can cause inflammation and are therefore implicated in the onset of acne (Garg, 2016; Holland et al., 1998; Iftikhar and Choudhry, 2019; Simonart, 2012). A condition known as seborrhea (excessive

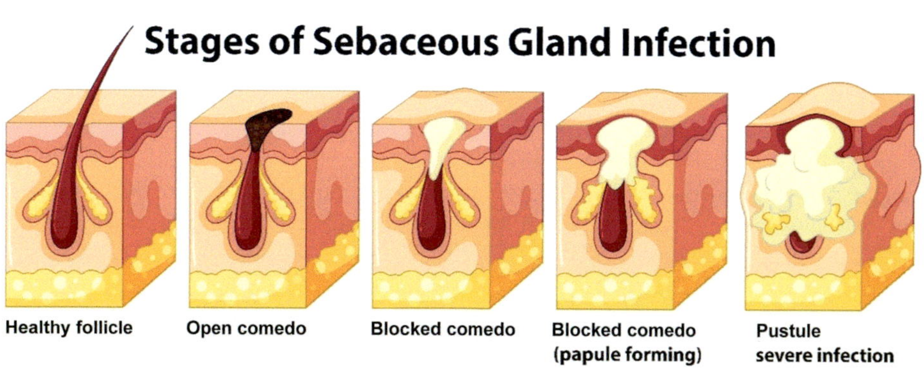

FIGURE 4.1 The stages of sebaceous gland infection (Graphics RF, 2020).

sebum production) is commonly observed in acne-afflicted individuals due to abnormal androgen levels. The abundance of androgen receptors present on the surface of sebaceous glands leads to the influence of androgenic hormones on the activity of these glands (Doshi et al., 1997; Jappe, 2003). Increased sebum production accompanied by abnormal keratinization of the follicles, which are both influenced by androgens, can lead to obstructed follicles and the eventual formation of non-inflammatory and inflammatory acne lesions depending on whether the bacterial infiltration and load is sufficient to cause an immune response (Garg, 2016).

Corticotropin-releasing hormone (CRH) regulates the body's behavioral and neuroendocrine stress responses. Sebocytes bind CRH via the presence of CRH receptors, thus linking the influence of stress to the development of acne. The essential fatty acid content in acne-afflicted patients is different when compared to patients without the skin disorder. Linoleic acid, in particular, regulates the secretion of Interleukin-8 (IL-8), a pro-inflammatory cytokine (Van De Kerkhof et al., 2006). Deficiencies of linoleic acid can be linked to comedone formation. This occurs when the infundibula and pilosebaceous cells are stimulated by the cytokine interleukin 1 alpha (IL-1α), which affects toll-like receptor (TLR) activity (Selway et al., 2013). Linoleic acid deficiencies have been correlated with increased sebaceous liquids in the pilosebaceous ducts, as well as with weakened integrity of the follicular epithelium barrier. An unstable follicular epithelium allows substances such as fatty acids to cross the barrier, thereby affecting the fatty acid composition of the follicle and leaving it vulnerable to the negative effects of reactive oxygen species (ROS). These ROS promote lysis and aid invading microorganisms (Garg, 2016; Ottaviani et al., 2010; Van De Kerkhof et al., 2006).

The role of microbial activity in acne progression has been a topic of debate in previous studies. Hypercornification of the follicular ducts results in complete blockage of the ducts, thereby creating an anaerobic environment and providing optimal conditions for microbial colonization (Jappe, 2003). *Cutibacterium acnes* is one of the major acne-causing microbes which form part of the human skin microbiome, generally existing as a commensal bacterium but capable of occasionally becoming pathogenic. *C. acnes* produces a secretory lipase with the ability to degrade sebum triglycerides into free fatty acids, which act directly by irritating the follicular epithelium, resulting in an inflammatory response. Studies have shown a correlation between an increased bacterial load of *C. acnes* and the production of sebum (Doshi et al., 1997; Jappe, 2003).

The role of other microbes such as *Staphylococcus aureus* and *Staphylococcus epidermidis* in the progression of acne are yet to be elucidated. It has been hypothesized that the commensal *S. epidermidis* inhibits the antibacterial activity of skin fatty acids through esterification reactions and excretes an exopolysaccharide that aids intracellular adhesion and biofilm formation. These biofilms create an anaerobic environment and allow resistant genes to be transferred between species (Adetutu et al., 2017; Jahns et al., 2012; Khorvash et al., 2012; Kumar et al., 2016).

C. acnes produces exocellular enzymes and other bio-active products, which enhance its virulence. It stimulates components of the immune system, resulting in downstream effects that aid pathogenicity. Cytokine production occurs as a result of the interaction of *C. acnes* with sebocytes and keratinocytes in the skin, and is initiated through TLRs which are present in these epithelial cells (Selway et al., 2013; Van De Kerkhof et al., 2006). Follicular hyperkeratinization caused by androgens can result in obstruction of the pilosebaceous canals, which may become colonized by *C. acnes*. Alternatively, or simultaneously, colonization of *C. acnes* may occur as a result of excessive sebum production initiated by androgens. Microbial colonization can cause the release of chemotactic factors, followed by neutrophilic phagocytosis and rupturing of the follicular wall due to the effect of hydrolytic enzymes, ultimately resulting in inflammation (Najafi-Taher and Amani, 2017).

4.2.2 Current Drugs and Treatment

First-line treatment for acne can be topical, systemic or a combination of both (Garg, 2016). Acne is treated using various drugs, many of which have specific forms of administration and mechanisms of action. Their use may be accompanied with adverse effects (Table 4.1). Antibiotics are still one

TABLE 4.1
Dosage Form, Mechanism of Action and Major Adverse Effects of Common Acne Treatments (sourced from Decker and Graber, 2012; Garg, 2016; Najafi-Taher and Amani, 2017; Simonart, 2012)

Treatments	Dosage Forms	Mechanisms of Action	Major Adverse Effects
Retinoids (Tretinoin, Tazarotene, Isotretinoin, Adapalene)	Topical	Anti-inflammatory, regulation of follicular cell hyperproliferation	Skin irritation, erythema, itching, stinging, dryness
Topical antibiotics (Erythromycin, Azithromycin, Clindamycin)	Topical	Antimicrobial (inhibition of *Cutibacterium acnes* growth)	Antibiotic resistance, disruption of normal skin microbiota
Herbal agents (*Aloe vera*, rose oil, tea tree oil, onion, azelaic acid)	Topical	Anti-inflammatory, antimicrobial, antioxidant, keratolytic	Skin irritation
Benzoyl peroxide	Topical	Keratolytic, antibacterial, moderate comedolytic	Erythema, skin irritation, dryness, dermatitis
Systemic antibiotics (Tetracycline, Minocycline, Doxycycline)	Oral	Antimicrobial, anti-inflammatory	Pigment deposition in skin and on teeth, gastrointestinal disorders, vaginal candidiasis
Hormonal therapy (Flutamide, Spironolactone, Cyproterone acetate)	Oral	Prevention of effects of androgen irregularities on sebaceous glands	Hyperkalemia, hormonal disruption (e.g. menstrual abnormalities)

of the most popular treatment choices for acne. Other treatment options, such as over-the-counter cleansers, vitamins, essential oils, mechanical treatments and leave-on products, are seldom used alone but rather in combination with prescribed treatments (Decker and Graber, 2012).

Topical retinoids act on several targets that contribute to their therapeutic effect. Their activities include the prevention of inflammation through the down-regulation of TLRs, nitric oxide production and cytokines secretion. In addition, they can act directly on mature non-inflammatory comedones, aid the skin in their expulsion and prevent the formation of micro-comedones by maintaining skin cell proliferation in homeostasis (Leyden et al., 2017). This class of compounds is therefore advantageous since acne is a multifactorial condition and requires versatile, multifaceted treatment. The treatment success of topical retinoids differs between patients and is largely dependent on the frequency of use, administrative form (topical or systemic), skin type and exposure to environmental factors. Retinoids, much like antibiotics, have a variety of side effects (Simonart, 2012). The combined administration of topical retinoids with topical antibiotics results in a higher rate of improvement in patients' symptoms when compared to the administration of a single treatment (Decker and Graber, 2012; Najafi-Taher and Amani, 2017).

Isotretinoin is one of the most effective acne treatments available, due to its ability to act on multiple acne-causing targets. Isotretinoin treatment is associated with the lowest rate of relapse and it is most often used for severe acne where nodules and cysts are present. However, in some cases, lower dosages have been prescribed for moderately severe acne. A major downfall of isotretinoin is its numerous, highly significant side effects, including increased serum lipids, photosensitivity, hyperlipidaemia, cheilitis, dry skin and mucous membranes, increased chance of cutaneous *S. aureus* infections, epistaxis, pseudotumor cerebri, teratogenicity, temporary worsening of lesions and myalgias (Simonart, 2012).

Antibiotic treatment prescribed at low doses over long periods—and often repeated with relapsing patients or patients with poor response—can lead to antibiotic resistance. This can cause selection of resistant strains and contribute to failure of treatment with other antibiotics of the same class. It is necessary to take into account the different microbial species present in the acne lesions before a course of antibiotics is administered, in order to prevent the development of resistant strains (Jappe, 2003).

Natural products, such as those derived from plants, offer a wide range of medicinally and cosmetically applicable activities as a result of their diverse chemistry. Several plant species have been shown to have anti-acne activity against a variety of acne-causing factors (Sinha et al., 2014; Soleymani et al., 2020). For example, polysaccharides of *Camellia sinensis* (L.) Kuntze have been shown to inhibit pathogen-mediated hemagglutination by *C. acnes*, *Heliobacter pylori* and *S. aureus*, while extracts of *Echinacea purpurea* (L.) Moench have been found to reverse the inflammatory reaction caused by *C. acnes* by preventing proliferation of the bacterium. Another mode of action of *E. purpurea* is the normalization of cytokine levels that may have become elevated as a result of bacterial colonization (Sinha et al., 2014). Curcumin, extracted from *Curcuma longa* (L.), has been found to have an inhibitory effect against *C. acnes*; it suppresses the inflammatory response and it elicits an antioxidant effect via the activation and reduction of certain pathways (Soleymani et al., 2020).

4.2.3 Future Treatments

Antibiotic-resistant strains of acne-causing bacteria, particularly *C. acnes*, are becoming more prevalent. The current conventional treatments also have adverse side effects, lengthy treatment plans, restricted use of certain pharmaceutical drugs (such as isotretinoin) and a lack of alternative treatment options. Given these negative aspects, it is necessary to develop novel, multifactorial therapies with more selective modes of action to minimize side effects and confer more efficacious treatment (Simonart, 2012, 2013).

Naturally derived antimicrobial peptides have the potential to be utilized as acne treatments and lower the probability of creating resistant bacterial strains. They are characterized by rapid activity at their target site and a degree of selectivity against the microbes and microbial structures they target (Simonart, 2012). Modern acne treatments focus on formulations designed for improved infiltration into the skin layers. These include novel delivery systems and carriers, such as liposomes, nano-emulsions, microsponges and aerosol foams (Garg, 2016; Simonart, 2012). These carriers increase the efficiency of treatments and the possibility of employing lower doses with the same activity as conventional treatment options (Simonart, 2012).

4.3 NANOPARTICLES

4.3.1 Different Types and Formulations

The term NP includes various types of colloidal particulate matter with dimensions of 100 nm or smaller (Khan et al., 2019; Sahoo and Labhasetwar, 2006). The unique properties of NPs—viz. the ability to manipulate their shape and size, their surface-to-volume ratio and their optical properties—suggest potential advantages for their use as therapeutic agents (Utreja et al., 2020). Metal NPs are commonly prepared from gold, silver, platinum and palladium, while metals such as copper, zinc, titanium and magnesium are less common. These substances vary in color according to the size and shape (nanorods, nanoshells and nanocages) of the synthesized NP (Hasan, 2014; Khan et al., 2019). Changes in the aspect ratio results in different colored solutions due to the absorption of the NPs being affected, which occurs when the concentration of the metal changes, or the thickness of the nanoshells increases (Bankar et al., 2010).

Nanoparticles consist of three layers: the surface, shell and core (Figure 4.2). The surface layer contains the metal ions, surfactants, polymers and several small molecules. The core is often referred to as the NP as it forms the central component of the particle and is chemically distinct from the

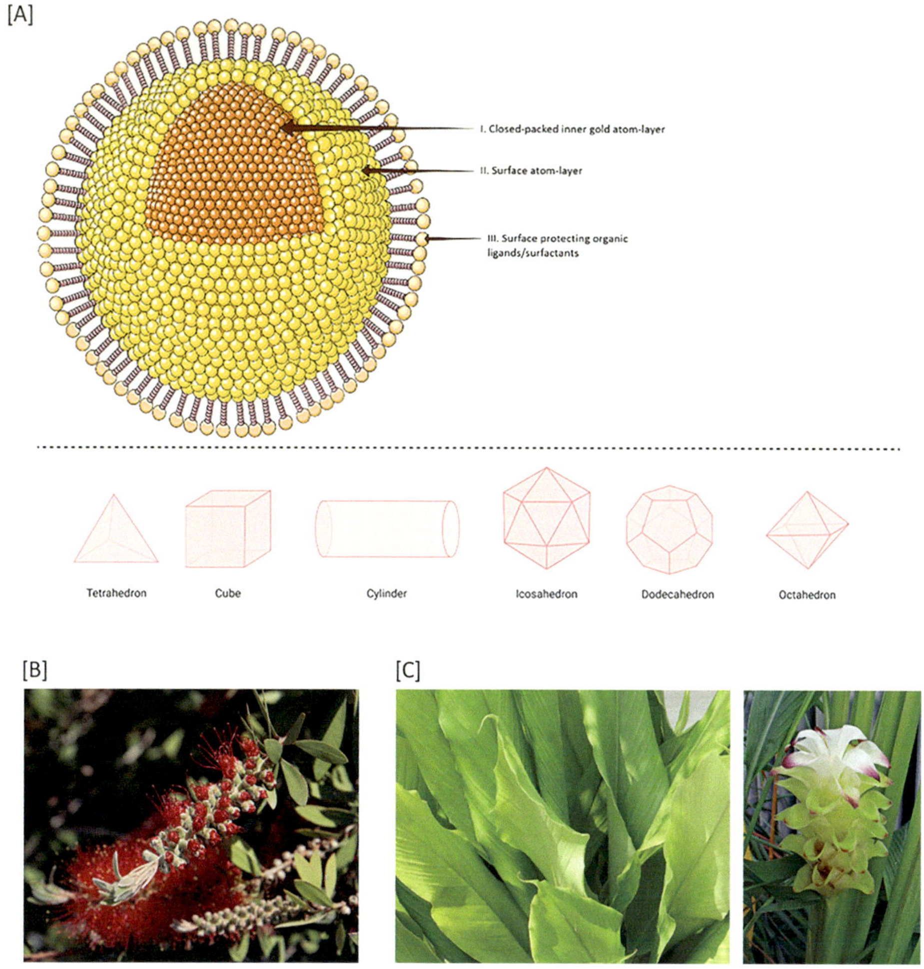

FIGURE 4.2 A, Schematic illustration of different shapes of a single crystalline gold nanoparticle; B, Inflorescence of *Callistemon citrinus*, from which certain isolated compounds have been used in the synthesis of silver nanoparticles; C, Leaves and inflorescence of *Curcuma longa*, from the rhizome of which the bioactive compound curcumin is extracted and can be delivered via a nano-emulsion formulation (Anichkova, 2009; Darweesh et al., 2019; Chan, 2014; Zell, 2009). (Reprinted from Darweesh et al., 2019).

shell (Khan et al., 2019). Nanoparticles are made from biodegradable polymers, which can be used to hold biologically active molecules via adsorption, entrapment or chemical coupling (Sahoo and Labhasetwar, 2006). Different methods of preparation can yield nanocapsules, nanospheres and a variety of other nanostructures, depending on the method used (Mohanraj and Chen, 2006).

Nanoparticles can be used for applications such as bio-imaging, medical treatments, energy storage in solar and oxide fuel batteries, and several others (Hasan, 2014; Khan et al., 2019). The chemical and physical characteristics of NPs are used to classify them into six main categories: metal, carbon-based, polymeric, ceramic, semiconductor and lipid-based NPs. Lipid-based NPs are used as drug carriers and targeted delivery mechanisms, among other biomedical applications. Top-down or bottom-up synthesis can be used to produce NPs (Khan et al., 2019). The top-down approach of nanostructure synthesis involves the breaking up of large molecular components, whereas bottom-up synthesis builds up nanostructures from single atoms and small molecules (Lee and Jun, 2019).

TABLE 4.2
Different Types of Inorganic Nanoparticles and Their Applications (sourced from Elbagory et al., 2017; Hasan, 2014; Paul and Sharma, 2020; Verma et al., 2019)

Nanoparticle Types	Advantages	Applications
Silver	Antimicrobial	Water treatment, cosmetics, pharmaceutical products, catalysis, bio-sensing, chemical sensing, photonics
Gold	Biostability, biocompatibility, antibacterial	DNA fingerprinting; identification of cancer cells, bacteria, heavy metals and aminoglycoside antibiotics; photothermal therapy; gene therapy; drug delivery
Alloy (e.g. Ag flakes)	Electrical conductivity	Metal fillers
Magnetic (e.g. Fe_3O_4[1] and Fe_2O_3[2])	Biocompatibility	Targeted cancer treatment, guided drug delivery, gene therapy, stem cell sorting and manipulation, magnetic resonance imaging, DNA analysis
Fluorescent (e.g. SiO_2[3] and quantum dots)	Biocompatibility	Biomedical imaging

Notes: **1** Iron (II, III) oxide; **2** Iron (III) oxide; **3** Silicon dioxide

The bottom-up approach can be used to synthesize NPs via biological systems using enzymes, fungi, plants, plant extracts and microorganisms (Khan et al., 2019).

Inorganic NPs (Table 4.2) such as gold and silver, magnetic and fluorescent NPs, have a wide variety of uses in medicine due to their chemical and physical nature. They are highly stable, biocompatible and hydrophilic, and have a high cellular uptake capacity. These characteristics allow them to enhance the therapeutic activity of drugs and reduce the occurrence of side effects (Paul and Sharma, 2020). Their rigidity prevents them from being easily deformed (Papakostas et al., 2011). Some inorganic NPs may have adverse effects on the environment or the individual exposed to them, as a result of their specific mechanism of action. An example of this is silver NPs (AgNPs) which generate ROS as a method of killing bacterial species. A build-up of ROS can cause cellular stress, thereby disrupting cellular transport processes (Yu et al., 2013). Organic NPs (e.g. proteins, polymers and lipids) may be less advantageous than inorganic NPs as they are not as easily modified or controlled. They are often softer and more easily affected by contact with other surfaces or by stress (Papakostas et al., 2011; Paul and Sharma, 2020).

4.3.2 Applications in Medicine and Dermatology

Nanoparticles have a wide range of applications in medicine, including therapies for treating cancer, diabetes, microbial infections, and allergies. They are used extensively for targeted drug delivery and the controlled release of bio-active compounds (Utreja et al., 2020). The term 'nanochemoprevention' has been created to describe the use of nanotechnology in the improvement of cancer prevention and treatment via nanotechnology (Siddiqui and Sanna, 2016). Nanoparticles can be designed for optimal drug delivery; several nanotechnology-based drug delivery systems are in clinical development and have either obtained U.S. Food and Drug Administration (FDA) approval or are already publicly available (Siddiqui and Sanna, 2016). The surface characteristics and particle size can be altered in order to allow parenterally administered drugs to have both passive and active targeting capabilities (Najafi-Taher and Amani, 2017).

Certain NPs can be used to detect microbial cell constituents, pre-cancerous cells and free viral fragments, in order to prevent the onset of disease (Utreja et al., 2020). Natural product delivery via NPs is associated with improved delivery and efficacy, and has been studied extensively. Examples of such products include the delivery of resveratrol, epigallocatechin-3-gallate (ECGG) isolated from green tea (*C. sinensis*), and curcumin from *C. longa*, for the treatment of cancer (Siddiqui and Sanna, 2016).

Nanoparticle treatments are usually administered intravenously, but other routes such as topical/transdermal, mucosal, oral and intraperitoneal can be used. Several barriers must be crossed before NPs can reach their target destination, which include mucous, skin, extracellular matrix and subcellular barriers. The main barriers to intravenous NP delivery are tissue penetration, the endolysosomal barrier and the reticuloendothelial system (Utreja et al., 2020). Nanoparticles enhance the permeation of drugs into the dermal and epidermal layers by increasing the aqueous solubility of the drugs. This is done via an interaction with the dermal and epidermal lipid and protein structures, which disrupts the structure of the skin layers and allows permeation of the drug molecules (Jijie et al., 2017).

The intravenous and intramuscular routes, respiratory tract, gastrointestinal tract and transdermal application are all possible administrative routes of NPs, from which they often travel via the bloodstream or lymphatic system to reach their target sites (Paul and Sharma, 2020). Methods such as magnetic guidance or attachment of ligands to particle surfaces, can be used to target specific sites in the body (Mohanraj and Chen, 2006). Transportation of drug molecules to their target site is enhanced by the small size of the NPs or nano-emulsion droplets (Banerjee et al., 2016).

Some nano-emulsions and surfactants can increase the permeation of the drug molecules as a result of their hydrophobic properties (Najafi-Taher and Amani, 2017). The action of NPs at their target sites includes even distribution of the drug to improve its therapeutic effect and removal of the drug once it has acted to prevent the occurrence of side effects. Different matrix constituents have different mechanisms of controlling the release of drugs and their subsequent degradation. No chemical reaction is required to load drug molecules into NPs; therefore, the activity of the drug molecules is preserved (Mohanraj and Chen, 2006).

Nanoparticles have a variety of dermatologically applicable properties, such as photoprotective, antiseptic, antioxidant, phototherapeutic, sebaceous gland targeting, topical dermatotherapy, hair disease treatment, and targeted and transdermal drug delivery. NP-based sunscreens convey protection against UV-induced oxidative stress and may offer added cosmetic benefits, such as anti-aging properties. Some NPs which are unable to cross the barrier into the epidermis due to their size are therefore replaced with structures such as liposomes or solid lipid NPs (SLN) that can be used in sunscreens in order to increase the barrier permeability, enhance the NP stability and improve the tolerance for the ingredients in the sunscreen. Daylong Actinica® is an example of a liposome-based sunscreen with greater compliance and enhanced activity as compared to non–liposome-based sunscreens (Papakostas et al., 2011).

Antioxidant-loaded NPs could potentially be used to prevent toxins excreted from the sweat glands from damaging the skin, via combination with barrier creams; this is an example of a barrier-forming particulate formation. Particulate-based barrier creams provide more protection than moisturizers or lotions in the prevention of water loss, thereby reducing possible irritation. In addition, barriers formed via surface coating or cross-linkage prevent rapid release of drug particles, thereby allowing sustained release over a longer time period and reducing possible toxicity of rapid drug release. Antioxidant-based treatments can be applied for anti-aging purposes, for example, Dermaviduals® and Regenerations Creme Intensive® NP-based barrier creams. The intrafollicular delivery of drugs via adapalene-loaded PLGA (poly glycolic-co-lactic acid) particles for the treatment of acne and other sebaceous gland conditions have been researched and applied. Particulate drug-delivery systems are being used increasingly in the treatment of hair conditions due to their enhanced permeation into hair follicles and sustained release of drug particles. (Papakostas et al., 2011).

4.3.3 Applications in Acne Treatment

In order for an acne treatment to be effective, it is required to be present at a high concentration within the follicles; this can be enhanced by the use of NPs (Najafi-Taher and Amani, 2017). There

TABLE 4.3
Different Nanoparticle Systems and Types Used as Forms of Delivery of Medical Treatments (sourced from Garg, 2016)

Systems	Delivery Types	Modes of Action	Examples
Particulate	Solid lipid nanoparticle	Protection of drug molecules from degradation; controlled drug release	Isotretinoin-loaded solid lipid nanoparticles
	Microspheres	Controlled drug release; reduction of irritation	Tretinoin microsphere gel; benzoyl peroxide
Vesicular	Liposomes	Controlled and optimized drug delivery; dual-release action	Liposomal lauric acids (LipoLA)
	Niosomes	Increase drug residence time; reduce systemic drug absorption; increase horny layer properties	Curcumin-loaded niosomes
Colloidal	Surfactant-free emulsion	Encapsulation of drug; emulsification and removal of epidermal lipids	Triclosan-loaded emulsion
	Micro-emulsion	Delivery of large quantity of drug; increased solubility of drug; increased transdermal drug permeation	Curcumin-loaded myristic acid micro-emulsions
	Nano-emulsion	Increase drug stability; effective transport of drug molecules	Tretinoin-loaded nano-emulsion
Miscellaneous	Aerosol foams	Increased drug bioavailability	Sodium sulfacetamide 10% sulfur, 5% emollient foam
	Microsponges	Delivery of large quantity of drug; protection of drug molecules from degradation; slow release of active molecules	Benzoyl peroxide microsponges
	Fullerenes	Entrapment of active compounds and release after application	Hydroxylated fullerene

are several different nanotechnological approaches for the delivery of acne treatments, each with their own advantages (Table 4.3) (Garg, 2016).

Inorganic NPs are highly stable and are therefore not degraded by microbial species (Paul and Sharma, 2020). AgNPs have antimicrobial activity due to their ability to produce ROS and free radicals, which destabilize microbial ribosomes and mitochondria and stress the endoplasmic reticulum of the microbe, thereby initiating its cell death. These properties render them useful for wound healing. The antibacterial strength of AgNPs is dependent on the type of bacteria and their cell wall composition, architecture and thickness (Lee and Jun, 2019).

Gold NPs (AuNPs) have a high biocompatibility and stability, and they therefore are the most commonly used NPs (Badeggi et al., 2020; Elbagory et al., 2017; Paul and Sharma, 2020). They are characterized by a low cytotoxicity, optical tunability, small size, solidity, easy synthesis and surface modification, among other properties (Badeggi et al., 2020; Rotimi et al., 2019). *Callistemon citrinus* (Curtis) Skeels (Figure 4.2B) is a South African plant whose bio-active compounds can be successfully synthesized into AuNPs to be used for targeted delivery of the compounds. The antibacterial activity of the compounds extracted from the seeds which were delivered as AuNPs was found to be stronger than those in crude form (Rotimi et al., 2019). While metallic NPs may have several advantages as carriers of drugs and bio-active molecules, there is limited information on their ability to penetrate the stratum corneum, epidermis and dermis (Jijie et al., 2017).

The non-uniform size distribution, low stability and aggregative nature of organic NPs, as well as their rapid clearance by the reticuloendothelial system (which inhibits the NP action), may reduce their suitability as drug-delivery systems (Paul and Sharma, 2020). However, organic NP systems can be stabilized by the addition of liposomes (Figure 4.2A) and other lipid-based substances, resulting in more stable drug-delivery systems with added therapeutic properties, as well as a greater tolerance for the drug constituents (Papakostas et al., 2011).

Microsphere carriers of tretinoin, a cream used for acne, can reduce inflammatory reactions, allow improved drug tolerability and increase the efficacy of a drug, as seen with microsphere-administered adapalene. Various anti-acne drugs are administered via nano-emulsions. Clindamycin phosphate and adapalene administered via soybean oil nano-emulsion have enhanced stability and efficacy when administered this way as compared to conventional delivery routes (Najafi-Taher and Amani, 2017). Inhibition of *C. acnes* is enhanced in drugs administered via oleic acid (e.g. nadifloxacin) and lemongrass oil (Najafi-Taher and Amani, 2017).

Isotretinoin administered in coconut oil has a more controlled release at the target site (Najafi-Taher and Amani, 2017). Solid lipid nanoparticles (SLNs) loaded with isotretinoin have a better therapeutic effect on all types and severities of acne as compared to conventionally administered isotretinoin. Other common acne treatments delivered by SLNs include tretinoin and retinoic acid, which offer enhanced efficiency of drug entrapment, reduced skin irritation and overall improved therapeutic efficacy (Garg, 2016). Studies have shown reduced irritation of retinoic acid when administered via SLNs as compared to delivery via the commercially sold retinoic acid cream (Papakostas et al., 2011).

Plant bio-active molecules are better candidates for NP synthesis than synthesized drug molecules, due to their lower toxicities and their natural abilities to cap NPs and prevent NP coagulation (Verma et al., 2019). The bio-active compound curcumin, extracted from the rhizomes of *C. longa* (Figure 4.2C), has several properties suitable for acne treatment i.e., its antimicrobial, anti-inflammatory, antioxidant and wound-healing properties, while reducing the hydrolytic and photochemical degradation of curcumin. The nano-emulsion of high lecithin content used as a carrier for curcumin increases penetration of the dermal layer, while improving the retention of curcumin in the epidermis (Caon et al., 2017).

Polycaprolactone (PCL)-based NPs can improve drug retention in the stratum corneum and epidermis, thereby preventing it from infiltrating other tissues and organs. PCL has applications in wound healing and acne treatment due to its high biocompatibility, flexible mechanical properties, microbial inhibition, inflammatory prevention and lack of immunogenicity (Caon et al., 2017). Liposomes are one of the best drug delivery forms for hair follicle–associated conditions. They are well-suited for the treatment of acne due to their additional therapeutic benefits such as moisturizing and wound healing, and are able to permeate the skin more effectively. This enhances the accumulation of drug molecules and improves the overall therapeutic activity (Carneiro et al., 2010; Rahman et al., 2016).

4.4 LIPOSOMES

Liposomes are highly biocompatible, biodegradable and versatile vesicular drug nanocarriers (Bruch et al., 2019; Paul and Sharma, 2020). They are made up of interior and exterior aqueous media separated by an amphiphilic phospholipid bilayer (Bruch et al., 2019; Spuch and Navarro, 2011). The amphiphilic nature enables the encapsulation of either lipophilic, hydrophilic or gaseous compounds. They are able to protect drugs from degradation, thus keeping them at a concentration at which they are active (Bruch et al., 2019).

Lipids used to synthesize liposomes can be obtained from either a natural (e.g., phosphatidylcholine) or synthetic source (Allen et al., 1991; Fu, 2016). Unstable liposomes may fuse together, which is undesirable as several fused liposomes are often too large to pass through the skin (Fu, 2016). Membrane stability and rigidity may be increased by the addition of saturated lipids such as cholesterol or sphingomyelin. This results in a more controlled release of bio-active molecules from the liposomes (Fu, 2016; Spuch and Navarro, 2011). Alternatively, liposomes can be stabilized by NPs such as AuNPs, which prevents the fusion of liposomes and improves drug encapsulation. AuNPs release from liposomes at about pH 4, allowing the liposome-encapsulated drug to freely interact with bacteria, i.e. *C. acnes* (Fu, 2016).

4.4.1 Different Types and Formulations

Liposomes can have multiple lipid bilayers and exist in varying vesicle sizes. Unilamellar vesicles consisting of a single lamella, range from small unilamellar vesicles (SUV) of a diameter < 100 nm to large unilamellar vesicles (LUV) of a diameter > 100 nm. Multilamellar vesicles (MLV) can reach sizes as large as micrometers and consist of two or more concentric phospholipid bilayers. Ether injection vesicles (EIV), French press vesicles (FPV) and reverse phase evaporation vesicles (REV) are types of liposomes classified according to the method used in their preparation (Bruch et al., 2019; Spuch and Navarro, 2011).

Liposomes can be structurally modified (Figure 4.3A) in order to optimize their activity for them to have the most effective therapeutic effect; according to this classification type, liposomes can be conventional, cationic, targeted, long circulating and pH-sensitive (Bruch et al., 2019; Spuch and Navarro, 2011). Targeted liposomes have been designed to have a suitable stability to allow the liposomes to cross specific barriers in the body (e.g., the blood-brain barrier) and be transported across membranes. Their characteristic surface features allow the targeting of precise locations where active compounds will be released, maintained at a certain concentration for a certain period of time and protected from the body's immune system reactions (Spuch and Navarro, 2011).

4.4.2 Applications in Medicine and Dermatology

Liposomes are able to increase drug bioavailability, allowing use of the drug at lower concentrations and, therefore, lower toxicity (Francolini et al., 2019). Examples of liposomal NP administered drugs include: Amphotericin B, Daunorubicin, Doxorubicin, Verteporfin and morphine sulfate, among others (Fu, 2016). Doxorubicin-loaded pH-sensitive liposomes were shown by Paliwal et al. (2012) to have enhanced intracellular accumulation and activity. These liposomes showed better activity than both free drug molecules and non-pH-sensitive targeted liposomes (Paliwal et al., 2012). Paromomycin-loaded conventional liposomes are able to permeate skin more effectively and have an improved therapeutic effect in the topical treatment of cutaneous leishmaniasis than free paromomycin molecules (Carneiro et al., 2010). Liposomes can be used for the delivery of a range of bio-active compounds for the treatment of a variety of diseases and conditions throughout various parts of the body; rendering that they are not a limited drug-carrier system (Lamichhane et al., 2018).

Liposomes are sensitive to degradation as a result of uptake by the reticuloendothelial system of macrophages and to environmental factors including light, pH changes, temperature, ultrasonic waves and magnetism. Liposome stability and protection can be achieved by the addition of more lipid bilayers, as well as a more controlled liposome release profile (Spuch and Navarro, 2011). Metallic NPs may have toxic effects, produce harmful by-products and need expensive apparatus to synthesize. Therefore, liposomal NPs may be better suited and more easily exploited for use in therapeutic applications (Elbagory et al., 2017).

4.4.3 Applications in Acne Treatment

Liposomes are more effective drug carriers for topical treatment than conventional topically administered treatment, and are particularly well-suited to the treatment of conditions associated with hair follicles (Rahman et al., 2016). Topically applied treatments can enter the skin via different variations of two pathways: the trans-epidermal and trans-appendageal pathways (Figure 4.3B). The trans-epidermal pathway involves the penetration of bio-active compounds across the intact stratum corneum and via the intercellular and transcellular pathways, into the epidermal and dermal layers. The trans-appendageal pathway involves the movement of bio-active compounds into the skin via hair follicles and sweat glands and is affected by follicular volume, number and size of openings (El Maghraby et al., 2008).

The multiple mechanisms whereby bio-active compounds can permeate the skin once applied topically include free drug permeation, liposome component-enhanced penetration, liposome

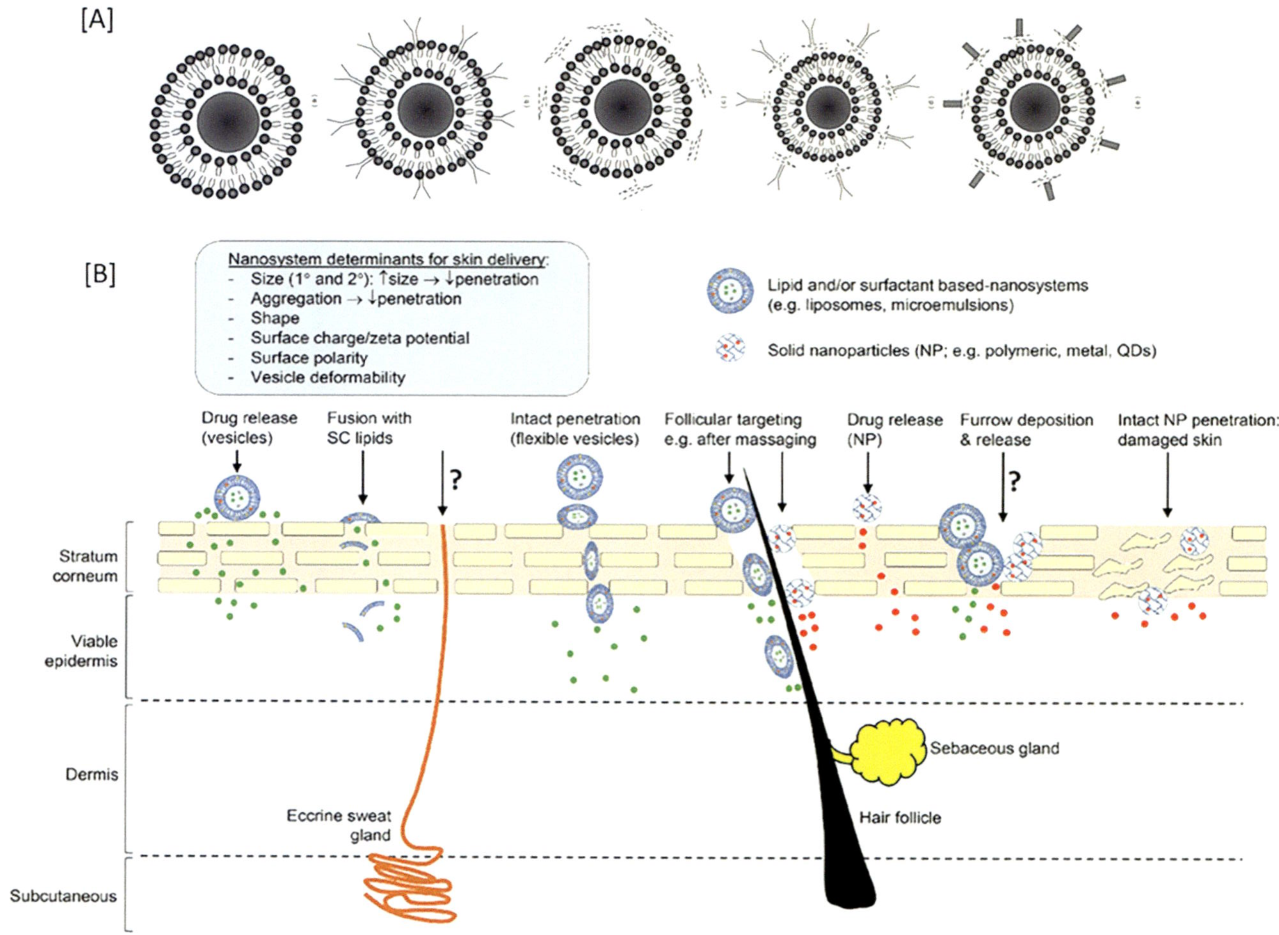

FIGURE 4.3 A, Schematic representation of liposome types: conventional; conventional directly tagged with antibodies; polymeric conjugate coated stealth liposome; polymeric conjugate coated stealth liposome tagged with antibodies; specific ligand coated liposome. Other liposomes include: immunoliposomes, gene-based liposomes, and virosomes; B, Representation of skin penetration via trans-epidermal and trans-appendageal pathways of liposome drug-carrier systems and other nanosystems, and the factors that determine the delivery path taken (Roberts et al., 2017; Spuch and Navarro, 2011). (Reprinted from Roberts et al., 2017, p. 90).

adsorption into the stratum corneum and intact liposome penetration through the skin (Figure 4.3B) (El Maghraby et al., 2008). Free drug permeation has been deemed an ineffective delivery mechanism by El Maghraby et al. (1999), as drug compounds are unable to penetrate healthy stratum corneum cells without assistance. Liposomes can enhance the delivery of drug molecules by interacting with the lipids in the stratum corneum and increasing skin permeability (El Maghraby et al., 1999). Liposome fusion with the lipid matrix of the stratum corneum greatly increases the permeation of the drug into the skin, while liposome adsorption by the stratum corneum allows release of drug molecules directly to the skin. These two mechanisms are depicted in Figure 4.3B as 'fusion with SC lipids' and 'intact penetration' (El Maghraby et al., 2008).

Incorporation of liposomes into the liposome phospholipid bilayer can enhance the stability and solubilization of lipophilic bio-active compounds. This will protect the compounds from degradation and removal. The structure of liposome membranes is similar to that of corneocytes in the stratum corneum as it allows liposomes to accumulate in the stratum corneum (Ternullo et al., 2019).

Tretinoin is a common topically administered acne treatment, but is the cause of several adverse effects such as photosensitivity, erythema, peeling, and burning. The activity of tretinoin is negatively affected by instability due to its sensitivity to external factors such as oxygen, low pH and light; this results in unpredictable effectivity due to the ease of removal. Liposome-delivered tretinoin is found to be present at higher concentrations in the dermal and epidermal layers due to its protection from degradation conferred by the liposomes (Rahman et al., 2016). Table 4.4 lists some of the commonly used acne treatments that are administered via liposomes.

There are several liposome-based drugs that have been approved by the FDA as a result of their non-toxic nature (Jijie et al., 2017). Laboratory and clinical studies performed on tretinoin-loaded liposomal formulations show that a 1% Carbopol gel formulated from tretinoin-loaded liposomes has a significantly enhanced effect on non-inflammatory acne lesions as compared to the commercially sold formulation (Rahman et al., 2016). The side effects of the conventional administrative form of tretinoin limit its usage, whereas the liposomal formulation shows better patient compliance, enhanced tolerability for the active ingredients and fewer to no side effects. This is due to the drug molecules' improved penetration of the stratum corneum and protection against photodegradation conferred by the liposomes (Malaekeh-Nikouei et al., 2011; Rahman et al., 2016). Liposomal tretinoin can be administered at a concentration 5–10 times lower than conventional tretinoin, since the liposomes enhance the concentration of the drug, thereby allowing it to reach its maximum comedolytic activity at a much lower concentration (Malaekeh-Nikouei et al., 2011).

A plant-derived treatment for acne could potentially be formulated from *Elephantorrhiza elephantina* (Burch.) Skeels, a South African plant traditionally used for a wide range of ailments including acne. The characteristics of *E. elephantina* that render it as a potential anti-acne treatment are antibacterial activity against *C. acnes*, high antioxidant capacity and anti-inflammatory activity. The saponin extracts of *E. elephantina* are able to form liposomes via the binding of cholesterol.

TABLE 4.4
Acne Treatments Delivered via Liposome (sourced from Garg, 2016; Fu, 2016)

Bio-Active Acne Treatment Molecules	Improvements/Results
Tretinoin	Improved treatment efficacy allowing a reduction of concentration of bio-active agent
Clindamycin	Improved treatment efficacy
Benzoyl peroxide	Improved antibacterial activity
5-aminolevulinic acid (5-ALA)	Reduction in inflammatory acne and reduced side effects such as photosensitivity
LipoLA	Antibacterial activity and reduction in toxicity

There is little research on this plant as an acne treatment, but it has shown potential in the few studies performed (Mpofu, 2012; Sinha et al., 2014).

Rhodomyrtone, a compound obtained from *Rhodomyrtus tomentosa* (Aiton) Hassk., has the potential to be used in an anti-acne formulation. This compound has antibacterial, antioxidant and anti-inflammatory properties. A study by Chorachoo et al. (2013), showed the improved therapeutic activity of rhodomyrtone against acne when encapsulated by liposomes, as compared to the free rhodomyrtone. The liposome-encapsulated compound showed better minimum inhibitory and minimum bactericidal concentrations (MIC and MBC) against *S. aureus*, *S. epidermidis* and *C. acnes*. Rhodomyrtone is released slowly from the liposomes, resulting in a prolonged therapeutic effect and a reduced incidence of side effects (Chorachoo et al., 2013).

4.5 CONCLUSION

Liposomes are an effective means of enhancing the therapeutic effect of bio-active molecules. They can be successfully utilized in the treatment of acne for the enhancement of drug delivery, drug molecule retention in the skin, protection of drug molecules from various types of degradation and the reduction of side effects (Bruch et al., 2019; Spuch and Navarro, 2011). Different nanostructures may be differently suited for the delivery of molecules to different layers of the skin (i.e. stratum corneum, epidermis, dermis) as a result of their respective penetrative abilities (Caon et al., 2017). The biochemical nature of liposomes determines their properties and characteristics that have high potential for the treatment of acne and its associated symptoms. Liposomes are efficient carriers for lipophilic substances, but can be carriers for a wide range of molecule types due to their highly biocompatible and versatile nature. They have potential value for the treatment of all types and severities of acne (Elbagory et al., 2017; Spuch and Navarro, 2011).

The lipophilic nature of liposomes has similarities with the stratum corneum and is therefore less invasive than the use of a completely foreign substance (e.g. metallic or ceramic nanoparticles), rendering liposomes as the most suitable nanoparticles for acne treatment (Ternullo et al., 2019). Liposomes provide therapeutic benefits such as moisturizing and wound healing, both of which are beneficial for acne treatment (Rahman et al., 2016). One of the major downfalls of liposomes as drug carriers is their low stability. Fortunately, this can be improved by the addition of stabilizing substances such as cholesterol and sphingomyelin (Spuch and Navarro, 2011).

The method used to administer a drug can affect its characteristics and activity, especially with regards to toxicity, stability and efficacy; therefore, it is necessary to choose the most suitable and effective form of drug delivery (Raza et al., 2013). Liposome-based methods used to deliver drug molecules for the treatment of acne can be used to enhance the delivery, efficacy and duration of treatment, therefore resulting in minimal or no side effects, as well as a reduced rate of relapse (Bruch et al., 2019). The advantages of liposomes combined with a plant-derived treatment, could potentially result in an improved acne treatment. Plant-based treatments can be more advantageous than synthesized drugs, due to their natural ability to cap NPs, prevent coagulation and reduce the occurrence of toxicity (Verma et al., 2019). Clinical trials are required for the progression of the development of a liposome-delivered acne treatment, especially regarding the liposomal encapsulation of plant-derived molecules, as little to no testing has been performed on this.

REFERENCES

Achermann, Y., E.J. Goldstein, T. Coenye and M.E. Shirtliff. 2014. "*Propionibacterium acnes*: From Commensal to Opportunistic Biofilm-associated Implant Pathogen." *Clinical Microbiology Reviews* 27, 3: 419–440.

Adetutu, A.A., B. Oritsewehinmi, O.M. Ikhiwili, A.O. Moradeke, A.S. Odochi and O.E. Adeola. 2017. "Studies on *Staphylococcus aureus* Isolated from Pimples." *Pakistan Journal of Biological Sciences* 20, 7: 350–354.

Allen, T.M., C. Hansen, F. Martin, C. Redemann and A. Yau-Young. 1991. "Liposomes Containing Synthetic Lipid Derivatives of Poly(Ethylene Glycol) Show Prolonged Circulation Half-Lives in Vivo." *BBA—Biomembranes* 1066, 1: 29–36.

Anichkova, A. 2009. "File:*Callistemon citrinus* 3588.Jpg." Obtained from wikimedia under the Creative common Share-Alike 3.0 license. https://commons.wikimedia.org/w/index.php?title=File:Callistemon_citrinus_3588.jpg&oldid=235047238%0A (Accessed 6 July 2020).

Badeggi, U.M., E. Ismail, A.O. Adeloye, S. Botha, J.A. Badmus, J.L. Marnewick, C.N. Cupido and A.A. Hussein. 2020. "Green Synthesis of Gold Nanoparticles Capped with Procyanidins from *Leucosidea sericea* as Potential Antidiabetic and Antioxidant Agents." *Biomolecules* 10, 3.

Banerjee, A., j. Qi, R. Gogoi, J. Wong and S. Mitragotri. 2016. "Role of Nanoparticle Size, Shape and Surface Chemistry in Oral Drug Delivery." *Journal of Controlled Release* 238: 176–185.

Bankar, A., B. Joshi, A.R. Kumar and S. Zinjarde. 2010. "Banana Peel Extract Mediated Novel Route for the Synthesis of Silver Nanoparticles." *Colloids and Surfaces A: Physiochemical and Engineering Aspects* 368, 1–3: 58–63.

Bruch, G.E., L.F. Fernandes, B.L.T. Bassi, M.T.R. Alves, I.O. Pereira, F. Frézard and A.R. Massensini. 2019. "Liposomes for Drug Delivery in Stroke." *Brain Research Bulletin* 152, June: 246–256.

Caon, T., L. Mazzarino, C.M.O. Simões, E.L. Senna and M.A.S. Silva. 2017. "Lipid- and Polymer-Based Nanostructures for Cutaneous Delivery of Curcumin." *AAPS PharmSciTech* 18, 3: 920–925.

Carneiro, G., D.C.M. Santos, M.C. Oliveira, A.P. Fernandes, L.S. Ferreira, G.A. Ramaldes, E.A. Nunan and L.A.M. Ferreira. 2010. "Topical Delivery and in Vivo Antileishmanial Activity of Paromomycin-Loaded Liposomes for Treatment of Cutaneous Leishmaniasis." *Journal of Liposome Research* 20, 1: 16–23.

Chan, W.C.E. 2014. "File:*Curcuma longa* (Leaves).Png." Obtained from wikimedia under the Creative common Share-Alike 3.0 license. https://commons.wikimedia.org/w/index.php?title=File:Curcuma_longa_(leaves).png&oldid=219509777 (Accessed 6 July 2020).

Chernyshov, P.V., C.C. Zouboulis, L. Tomas-Aragones, G.B. Jemec, L. Manolache, T. Tzellos and F. Sampogna. 2018. "Quality of Life Measurement in Acne. Position Paper of the European Academy of Dermatology and Venereology Task Forces on Quality of Life and Patient Oriented Outcomes and Acne, Rosacea and Hidradenitis Suppurativa." *Journal of the European Academy of Dermatology and Venereology* 32, 2: 194–208.

Chorachoo, J., T. Amnuaikit and S.P. Voravuthikunchai. 2013. "Liposomal Encapsulated Rhodomyrtone: A Novel Antiacne Drug." *Evidence-Based Complementary and Alternative Medicine* 2013.

Darweesh, R.S., N.M. Ayoub and S. Nazzal. 2019. "Gold Nanoparticles and Angiogenesis: Molecular Mechanisms and Biochemical Applications." *International Journal of Nanomedicine* 14: 7643–7663. Reprinted from Dove Medical Press Limited, with permission from Dove Medical Press Limited.

Decker, A. and E.M. Graber. 2012. "Over-the-Counter Acne Treatments." *Clinical Aesthetic Dermatology* 5, 5: 32–41.

Doshi, A., A. Zaheer and M.J. Stiller. 1997. "A Comparison of Current Acne Grading Systems and Proposal of a Novel System." *International Journal of Dermatology* 36: 416–418.

Dréno, B., S. Pécastaings, S. Corvec, S. Veraldi, A. Khammari and C. Roques. 2018. "*Cutibacterium acnes* (*Propionibacterium acnes*) and Acne Vulgaris: A Brief Look at the Latest Updates." *Journal of the European Academy of Dermatology and Venereology* 32, 2: 5–14.

Elbagory, A., M. Meyer, C. Cupido and A.A. Hussein. 2017. "Inhibition of Bacteria Associated with Wound Infection by Biocompatible Green Synthesized Gold Nanoparticles from South African Plant Extracts." *Nanomaterials* 7, 12.

El Maghraby, G.M., B.W. Barry and A.C. Williams. 2008. "Liposomes and Skin: From Drug Delivery to Model Membranes." *European Journal of Pharmaceutical Sciences* 34, 4–5: 203–222.

El Maghraby, G.M., A.C. Williams and B.W. Barry. 1999. "Skin Delivery of Oestradiol from Deformable and Traditiona Liposomes: Mechanistic Studies." *Journal of Pharmacy and Pharmacology* 51, 10: 1123–1134.

Fionán. 2017. "TEM Image of Gold Nanoparticles and Nanorods." Obtained from Wikimedia under the Creative Commons Attribution-Share Alike 4.0 International license. https://commons.wikimedia.org/w/index.php?curid=64480896 (Accessed 17 November 2021)

Francolini, I., L. Giansanti, A. Piozzi, B. Altieri, A. Mauceri and G. Mancini. 2019. "Glucosylated Liposomes as Drug Delivery Systems of Usnic Acid to Address Bacterial Infections." *Colloids and Surfaces B: Biointerfaces* 181, May: 632–638.

Fu, V. 2016. "Nanoparticle Stabilized Liposomes for Acne Therapy." *UC San Diego Electronic Theses and Dissertations.*

Garg, T. 2016. "Current Nanotechnological Approaches for an Effective Delivery of Bio-Active Drug Molecules in the Treatment of Acne." *Artificial Cells, Nanomedicine and Biotechnology* 44, 1: 98–105.

Graphics RF. 2020. "Stages of Sebaceous Gland Infection." Obtained from Vecteezy.Com under the Creative common Share-Alike 3.0 license. www.vecteezy.com/vector-art/419676-stages-of-sebaceous-gland-infection (Accessed 29 September 2020).

Hasan, S. 2014. "A Review on Nanoparticles : Their Synthesis and Types." *Research Journal of Recent Sciences. Uttar Pradesh (Lucknow Campus)* 4: 1–3.

Holland, K.T., O. Aldana, R.A. Bojar, W.J. Cunliffe, E.A. Eady, D.B. Holland, E. Ingham, C. McGeown, A. Till and C. Walters. 1998. "*Propionibacterium acnes* and Acne.Pdf." *Dermatology* 196: 67–68.

Iftikhar, U. and N. Choudhry. 2019. "Serum Levels of Androgens in Acne and Their Role in Acne Severity." *Pakistan Journal of Medical Sciences* 35, 1: 146–150.

Jahns, A.C., B. Lundskog, R. Ganceviciene, R.H. Palmer, I. Golovleva, C.C. Zouboulis, A. McDowell, S. Patrick and O.A. Alexeyev. 2012. "An Increased Incidence of Propionibacterium Acnes Biofilms in Acne Vulgaris: A Case-Control Study." *British Journal of Dermatology* 167, 1: 50–58.

Jappe, U. 2003. "Pathological Mechanisms of Acne with Special Emphasis on Propionibacterium Acnes and Related Therapy." *Acta Dermato-Venereologica* 83, 4: 241–248.

Jijie, R., A. Barras, R. Boukherroub and S. Szunerits. 2017. "Nanomaterials for Transdermal Drug Delivery: Beyond the State of the Art of Liposomal Structures." *Journal of Materials Chemistry B* 5, 44: 8653–8675.

Kanasorhokeshawnidioescagozity. 2017. "Liposome." Obtained from Flickr under the Public domain license. www.flickr.com/photos/152631188@N06/33355040604 (Accessed 17 November 2021)

Khan, I., K. Saeed and I. Khan. 2019. "Nanoparticles: Properties, Applications and Toxicities." *Arabian Journal of Chemistry* 12, 7: 908–931.

Khorvash, F., F. Abdi, H.H. Kashani, F.F. Naeini and T. Narimani. 2012. "*Staphylococcus aureus* in Acne Pathogenesis: A Case-Control Study." *North American Journal of Medical Sciences* 4, 11: 573.

Kumar, A., K. Dhar, S.S. Kanwar and P.K. Arora. 2016. "Lipase Catalysis in Organic Solvents: Advantages and Applications." *Biological Procedures Online* 18, 2.

Lamichhane, N., T.S. Udayakumar, W.D. D'Souza, C.B. Simone II, S.R. Raghavan, J. Polf and J. Mahmood. 2018. "Liposomes: Clinical Applications and Potential for Image-Guided Drug Delivery." *Molecules* 23, 2: 288.

Lee, S.H. and B.H. Jun. 2019. "Silver Nanoparticles: Synthesis and Application for Nanomedicine." *International Journal of Molecular Sciences* 20, 4.

Leyden, J., L. Stein-Gold and J. Weiss. 2017. "Why Topical Retinoids Are Mainstay of Therapy for Acne." *Dermatology and Therapy* 7: 293–304.

Mackenzie, K. 2021. "Human Skin Section, Histology." Obtained from Wellcome collection under the Creative Commons Attribution-Share Alike 4.0 International license. https://wellcomecollection.org/works/pbfv4uhw (Accessed 17 November 2021)

Malaekeh-Nikouei, B., S. Golmohammadzadeh, M. Hosseini and H. Nassirli. 2011. "Preparation and Characterization of Liposomes Encapsulated With Clindamycin and Tretinoin." *International Journal of Comprehensive Pharmacy* 6, 4.

Mallon, E., J.N. Newton, A. Klassen, S.L. Stewart-Brown, T.J Ryan and A.Y. Finlay. 1999. "The Quality of Life in Acne: A Comparison with General Medical Conditions Using Generic Questionnaires." *British Journal of Dermatology* 140: 672–676.

Mohanraj, V.J. and Y. Chen. 2006. "Nanoparticles—A Review." *Tropical Journal of Pharmaceutical Research* 5, 1: 561–573.

Mosam, A., N.B. Vawda, A.H. Gordhan, N. Nkwanyana and J. Aboobaker. 2005. "Quality of Life Issues for South Africans with Acne Vulgaris." *Clinical and Experimental Dermatology* 30, 1: 6–9.

Mpofu, S.J. 2012. "Elephantorrhiza elephantina and Pentanisia prunelloides; Antimicrobial Activities and Synergy." *University of Johannesburg (South Africa), ProQuest Dissertations Publishing.*

Najafi-Taher, R. and A. Amani. 2017. "Nanoemulsions: Colloidal Topical Delivery Systems for Antiacne Agents- A Mini-Review." *Nanomedicine Research Journal* 2, 1: 49–56.

Oon, H.H., S. Wong, D.C.W. Aw, W.K., Cheong, C.L., Goh and H.H. Tan. 2019. "Acne Management Guidelines by the Dermatological Society of Singapore." *The Journal of Clinical and Aesthetic Dermatology* 12, 7: 34–50.

Ottaviani, M., E. Camera and M. Picardo. 2010. "Lipid Mediators in Acne." *Hindawi Publishing Corporation*, Article ID 858176, 6 p.

Paliwal, S.R., R. Paliwal, H.C. Pal, A.K. Saxena, P.R. Sharma, P.N. Gupta, G.P. Agrawal and S.P. Vyas. 2012. "Estrogen-Anchored PH-Sensitive Liposomes as Nanomodule Designed for Site-Specific Delivery of Doxorubicin in Breast Cancer Therapy." *Molecular Pharmaceutics* 9, 1: 176–186.

Papakostas, D., F. Rancan, W. Sterry, U. Blume-Peytavi and A. Vogt. 2011. "Nanoparticles in Dermatology." *Archives of Dermatological Research* 303, 8: 533–550.

Paul, W. and C.P. Sharma. 2020. "Inorganic Nanoparticles for Targeted Drug Delivery." In *Biointegration of Medical Implant Materials*, edited by Chandra P. Sharma, 2nd ed., 333–373. Amsterdam: Elsevier.

Rahman, S.A., N.S. Abdelmalak, A. Badawi, T. Elbayoumy, N. Sabry and A. El Ramly. 2016. "Tretinoin-Loaded Liposomal Formulations: From Lab to Comparative Clinical Study in Acne Patients." *Drug Delivery* 23, 4: 1184–1193.

Raza, K., B. Singh, S. Lohan, G. Sharma, P. Negi, Y. Yachha and O.P. Katare. 2013. "Nano-Lipoidal Carriers of Tretinoin with Enhanced Percutaneous Absorption, Photostability, Biocompatibility and Anti-Psoriatic Activity." *International Journal of Pharmaceutics* 456, 1: 65–72.

Roberts, M.S., Y. Mohammed, M.N. Pastore, S. Namjoshi, S. Yousef, A. Alinaghi and I.N. Haridass. 2017. "Topical and Cutaneous Delivery Using Nanosystems." *Journal of Controlled Release* 247: 86–105. Reprinted from Journal of Controlled Release, with permission from Elsevier. doi: 10.1016/j.jconrel.2016.12.022.

Rotimi, L., M.O. Ojemaye, O.O. Okoh, A. Sadimenko and A.I. Okoh. 2019. "Synthesis, Characterization, Antimalarial, Antitrypanocidal and Antimicrobial Properties of Gold Nanoparticle." *Green Chemistry Letters and Reviews* 12, 1: 61–68.

Sahoo, S.K. and V. Labhasetwar. 2006. "Nanoparticle Interface: An Important Determinant in Nanoparticle-Mediated Drug/ Gene Delivery." In *Drugs and the Pharmaceutical Sciences: Nanoparticle Technology for Drug Delivery*, edited by Ram B. Gupta and Uday B. Kompella, 159th ed., 139–160. New York: Taylor and Francis Group.

Selway, J.L., T. Kurczab, T. Kealey and K. Langlands. 2013. "Toll-like Receptor 2 Activation and Comedogenesis: Implications for the Pathogenesis of Acne." *BMC Dermatology* 13.

Siddiqui, I.A. and V. Sanna. 2016. "Impact of Nanotechnology on the Delivery of Natural Products for Cancer Prevention and Therapy." *Molecular Nutrition & Food Research* 60, 6: 1330–1341.

Simonart, T. 2012. "Newer Approaches to the Treatment of Acne Vulgaris." *American Journal of Clinical Dermatology* 13, 6: 357–364.

Simonart, T. 2013. "Immunotherapy for Acne Vulgaris: Current Status and Future Directions." *American Journal of Clinical Dermatology* 14, 6: 429–435.

Sinha, P., S. Srivastava, N. Mishra and N.P. Yadav. 2014. "New Perspectives on Antiacne Plant Drugs: Contribution to Modern Therapeutics." *BioMed Research International* 2014.

Soleymani, S., M.H. Farzaei, A. Zargaran, S. Niknam and R. Rahimi. 2020. "Promising Plant-Derived Secondary Metabolites for Treatment of Acne Vulgaris: A Mechanistic Review." *Archives of Dermatological Research* 312, 1: 5–23.

Spuch, C. and C. Navarro. 2011. "Liposomes for Targeted Delivery of Active Agents against Neurodegenerative Diseases (Alzheimer's Disease and Parkinson's Disease)." *Journal of Drug Delivery* 2011, December: 1–12.

Ternullo, S., E. Gagnat, K. Julin, M. Johannessen, P. Basnet, Ž. Vanić and N. Škalko-Basnet. 2019. "Liposomes Augment Biological Benefits of Curcumin for Multitargeted Skin Therapy." *European Journal of Pharmaceutics and Biopharmaceutics* 144, September: 154–164.

Thiboutot, D. and J.Q. Del Rosso. 2013. "Acne Vulgaris and the Epidermal Barrier." *The Journal of Clinical and Aesthetic Dermatology* 6, 2: 18–24.

Utreja, P., S. Verma, M. Rahman and L. Kumar. 2020. "Use of Nanoparticles in Medicine." *Current Biochemical Engineering* 6, 1: 7–24.

Van De Kerkhof, P.C.M., M.M. Kleinpenning, E.M.G.J. De Jong, M.J.P. Gerritsen, R.J. Van Dooren-Greebe and H.A.C. Alkemade. 2006. "Current and Future Treatment Options for Acne." *Journal of Dermatological Treatment* 17, 4: 198–204.

Verma, V., M. Chaudhary and N. Srivastava. 2019. "Antioxidative Properties of Isolated Saponins of Verbesina Encelioides (Cav.) Benth. & Hook. Fil Ex Gray and SEM Studies of Synthesized Green Nanoparticles for Acne Management." *Plant Science Today* 6, 1: 575–582.

Yu, S.J., Y.G. Yin and J.F. Liu. 2013. "Silver Nanoparticles in the Environment." *Environmental Sciences: Processes and Impacts* 15, 1: 78–92.

Zell, H. 2009. "File:*Curcuma longa* 001.JPG." Obtained from wikimedia under the Creative common Share-Alike 3.0 license. https://commons.wikimedia.org/wiki/File:Curcuma_longa_001.JPG (Accessed 6 July 2020).

5 Medicinal Plants for Eczema

Marizé Nel and Namrita Lall

CONTENTS

5.1 INTRODUCTION

Atopic dermatitis, commonly known as eczema, is a prevalent chronic inflammatory skin condition that affects about 15–20% of preadolescent children and 1–3% of adults worldwide; moreover, 22.6% of children and 17.1% of adults have been diagnosed with AD annually (Gelmetti and Wollenberg, 2014; Chamlin et al., 2002; Bylund et al., 2020). In 2017, a study was conducted in the United States which indicated that approximately one-quarter to one-third of the population possess some form of eczema (Sullivan and Silverberg, 2017). These people experience eczematous lesions, epidermal barrier dysfunctions and immune dysregulations (Lee and Lee, 2014).

The most prevalent concern associated with eczema is that the condition can occur at any stage in one's life (Akdis et al., 2006). Though this condition is non-contagious, it is often hereditary and worsens over time (Palmer et al., 2006). The exact cause of eczema has not been fully identified; therefore, different hypotheses have been developed. The first hypothesis states that the condition could be caused by an immunological defect, which allows patients to become sensitive to immunoglobulin E (IgE). This hypothesis is regarded as the inside-out hypothesis (Brandt and Sivaprasad, 2011).

The second hypothesis states that the condition could be attributed to a disruption occurring in the skin barrier. This disruption can either be caused by genetic mutations which control the formation of the skin barrier, or environmental changes which lead to sensitization (Brandt and Sivaprasad, 2011). Medicinal plants have been and are currently used in numerous cultures around the world as a source of medicine. Though plants are primarily used as a food source within developing countries, developed and developing countries use plants as an alternative to conventional

DOI: 10.1201/9781003108375-5

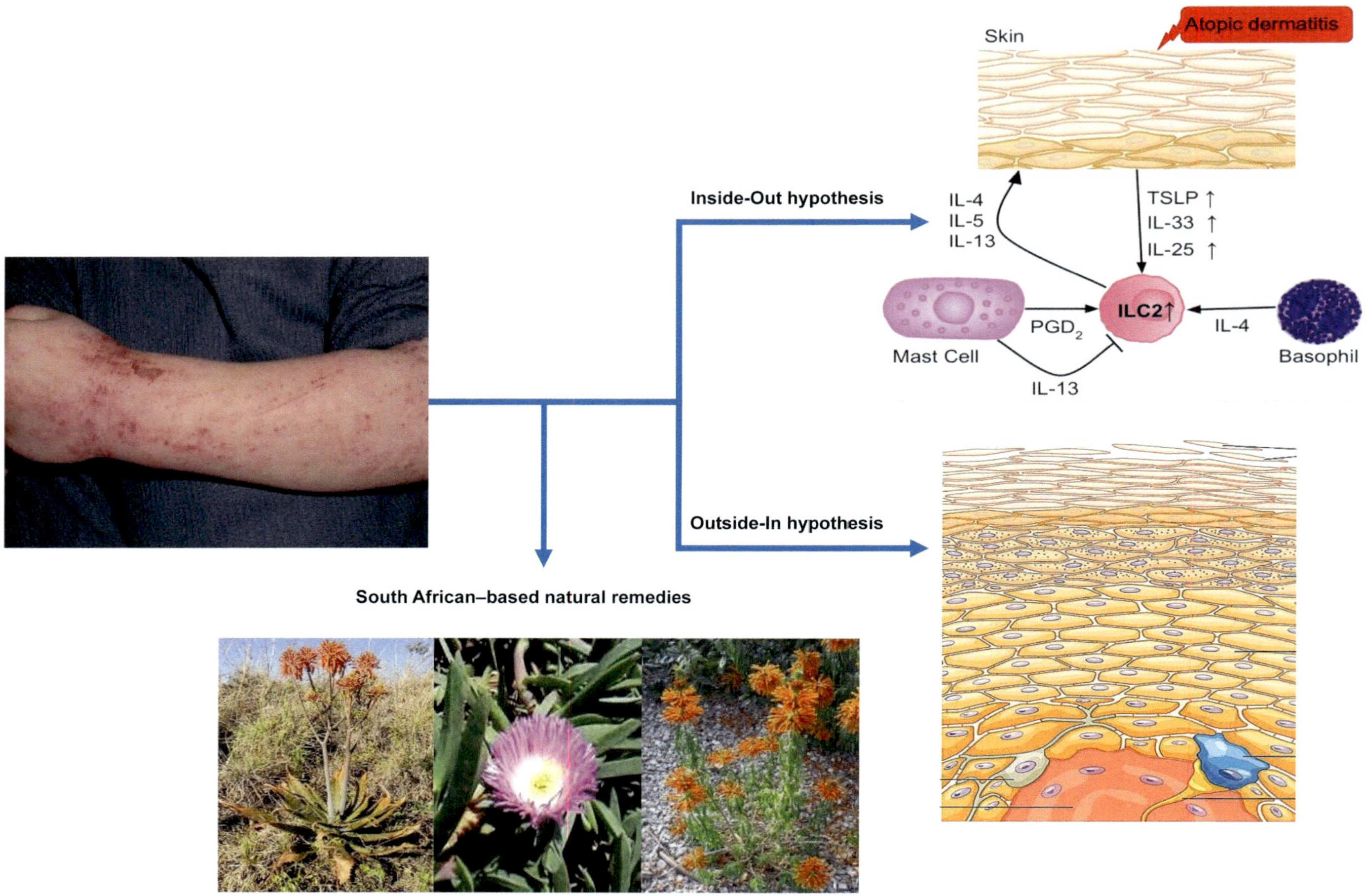

Graphical abstract representing the content of the chapter (Eisfelder, 2004; Fern, 2019; JoJan, 2005; Giraud, 2011; OpenStax College, 2013; Kelly, 2020).

medicines (Tadeg et al., 2005). This chapter elaborates on each hypothesis and indicates a few indigenous South African plants that have been used traditionally to reduce eczema, as well as plant bioactive compounds that have been shown to reduce aspects of the condition.

5.1.1 Inside-Out Hypothesis

The inside-out hypothesis states that atopic dermatitis is caused by an immunological defect which alters the permeability of the skin, allowing these patients to be sensitive to allergens (Brandt and Sivaprasad, 2011; Chamlin et al., 2002). The original model indicates that the condition is caused by an imbalance of T helper (Th) 1 and Th 2 lymphocytes, which stimulate cytokine production within the epidermal hyperplasia (Sullivan and Silverberg, 2017). Depending on the age and prevalence of the condition, there are several subtypes, which include intrinsic and extrinsic (Figure 5.1) (Mansouri and Guttman-Yassky, 2015).

Extrinsic subtype occurs in patients who experience acute atopic dermatitis commonly known as contact dermatitis (Mansouri and Guttman-Yassky, 2015). Within these patients, Th 2 and Th 22 lymphocytes are imbalanced, which promote the synthesis of interleukin 4 (IL-4), IL-5, tumor necrosis factor-alpha (TNF-α) and IL-13 and recruit eosinophils to the infected site (Sullivan

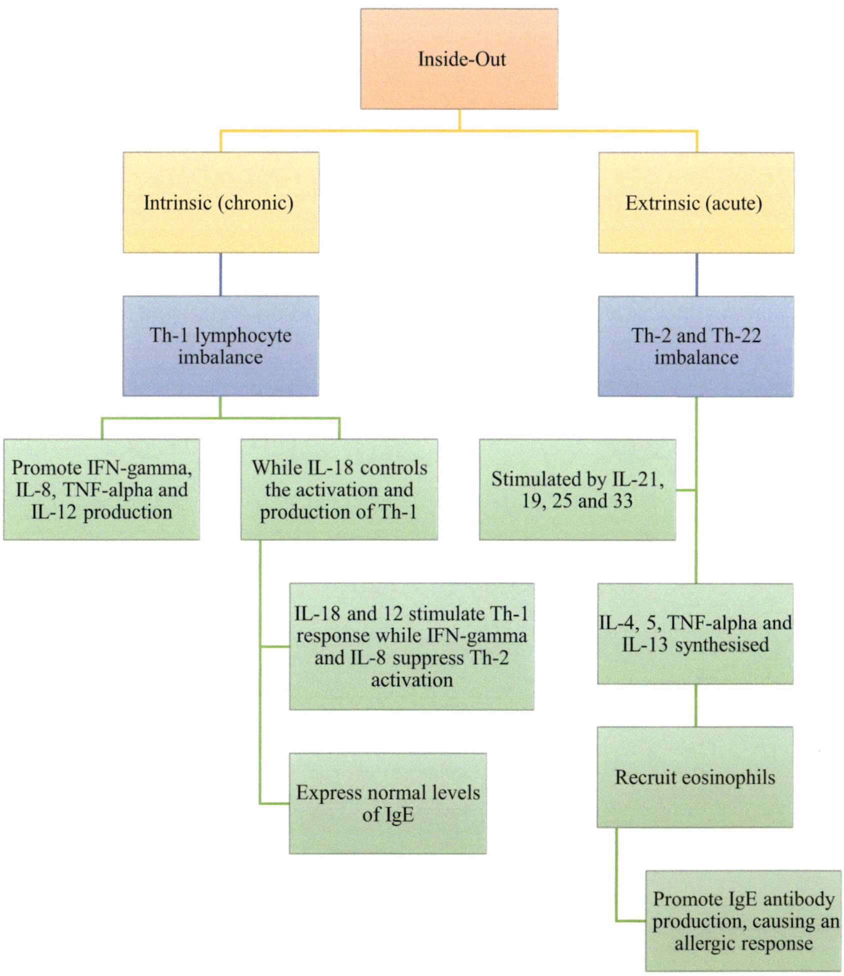

FIGURE 5.1 Intrinsic and extrinsic pathways associated with the inside-out hypothesis.

and Silverberg, 2017; Mansouri and Guttman-Yassky, 2015; Malajian and Guttman-Yassky, 2015; Nedoszytko et al., 2014). These cytokines promote immunoglobulin E (IgE) antibodies, which reduce extracellular pathogens and induce an allergic response (Sullivan and Silverberg, 2017).

In an earlier study, it was found that IL-4 and IL-13 inhibited the production of IL-17, which is prominent within psoriasis patients (Malajian and Guttman-Yassky, 2015). However, some studies indicated that IL-4, IL-31 and TNF-α compromised the epidermal barrier function (Gschwandtner et al., 2013). These cytokines reduced the expression of filaggrin and ceramide, which maintain the epidermal barrier (Cornelissen et al., 2012; Howell et al., 2009; Kim et al., 2011; Hatano et al., 2005).

Other cytokines that are stimulated by Th 2 and Th 22 include IL-21, IL-19, IL-25 and IL-33 (Brandt and Sivaprasad, 2011; Mansouri and Guttman-Yassky, 2015). These cytokines induced the expression of IL-5 and IL-13, modulate Th 2 responses and recruit dendritic cells, basophils and eosinophils to the infected site (Brandt and Sivaprasad, 2011). In some cases, bacteria such as *Staphylococcus aureus* trigger IgE antibody production. This is due to the abnormal bacterial colonization found in the infected sites. These bacteria present antigens or superantigens, which are detected by the immune system, which stimulates the immune system's proteolytic activity, resulting in a cytokine storm (hypercytokinemia) which damages the skin barrier (Gelmetti and Wollenberg, 2014).

The intrinsic subtype occurs in patients who experience chronic atopic dermatitis. Within these patients, Th 1 lymphocytes are imbalanced, which promote the synthesis of interferon-gamma (IFN-γ), IL-8, TNF-α and IL-12 (Mansouri and Guttman-Yassky, 2015; Nedoszytko et al., 2014). Pro-inflammatory cytokines, such as IL-18, are used to control the activation of Th 1 and Th 2 lymphocytes' response. During the acute stage, IL-18 increases the secretion of IL-4 and IL-13 promoting IgE synthesis. During the chronic stage, IL-18 and IL-12 stimulate the Th 1 response (Nedoszytko et al., 2014). Cytokines such as IFN-γ and IL-8 suppress Th 2 lymphocyte activation and reduce intracellular pathogens (Sullivan and Silverberg, 2017). These patients express normal levels of IgE antibodies and synthesize other cytokines such as CCL20, IFN-alpha (IFN-α) and IL-1 beta (IL-1β) (Mansouri and Guttman-Yassky, 2015). This subtype causes patients to experience the condition throughout their lives, or when exposed to stressful situations (Nedoszytko et al., 2014).

TNF-α is a pro-inflammatory cytokine produced by macrophages and monocytes that modulates inflammation within acute and chronic phases of eczema (Figure 5.2) (Song et al., 2008). When

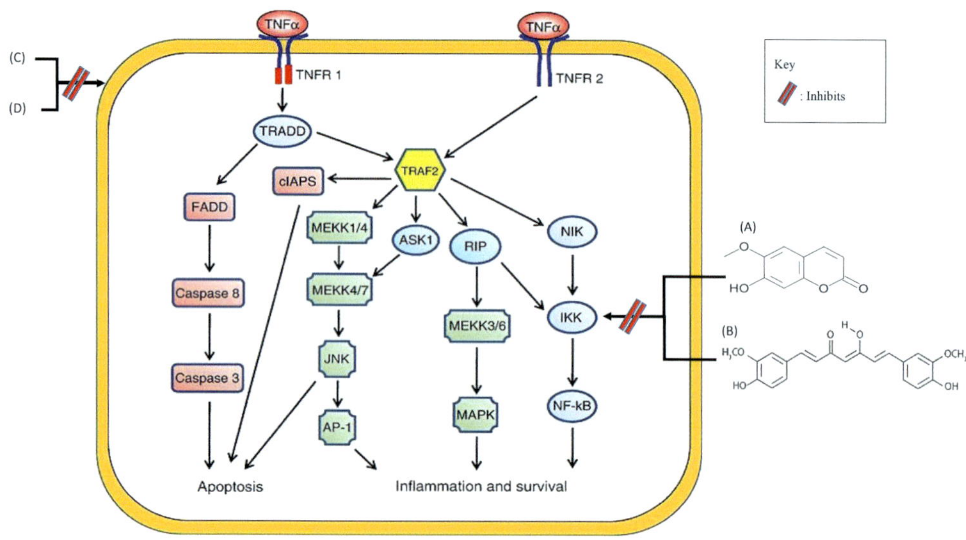

FIGURE 5.2 Tumor necrosis factor-alpha (TNF-α) pathway. A, Scopoletin; B, Curcumin; C, Magnosalin; D, Andamanicin. Adapted from (Wu and Zhou, 2010; Chrumps, 2010; Biophysik, 2008).

synthesized and released into the cytoplasm, TNF-α can bind to two receptors known as TNF-α receptor 1 (TNFR1) and TNFR2, which are found on cell membranes (Figure 5.2) (Palladino et al., 2003). During the synthesis of TNF-α, a pro-TNF trimer is cleaved from the cell membrane by the TNF-α converting enzyme (TACE) to form a mature TNF trimer that binds to the extracellular domain of TNFR1 (Palladino et al., 2003; Xu and Shi, 2007). This allows the release of the silencer of death inhibitory from the intracellular domain. Thereafter, adaptor proteins, known as TNF-α receptor-associated death domain (TRADD), recruits TNFR associated factor 2 (TRAF2) and receptor-interacting serine/threonine kinase 1 (RIP1), forming a plasma membrane-bound complex known as Complex I (Xu and Shi, 2007).

Once Complex I has formed, inhibitors of nuclear factor kappa B (IκB) become phosphorylated and dissociate from nuclear factor kappa B (NF-κB) transcription factors. This allows NF-κB to become active and translocate into the nucleus (Miller et al., 2010). Inside the nucleus, NF-κB transcribes sections of the DNA that are required to activate the synthesis of cytokines and chemokines such as TNF-α, IL-1β, IL-6, IL-8 and IFN-γ (Palladino et al., 2003). If IκB is unable to dissociate from NF-κB, the TNFR1 intracellular domain, known as the death domain, recruits a cytoplasmic complex, Complex II, which is comprised of RIP1 or TRADD and Fas-associated protein with death domain (FADD). Once formed, FADD recruits caspase 8 or 10 that are activated upon self-cleavage and cause a protease cascade which results in apoptosis (Xu and Shi, 2007). To prevent this, active NF-κB transcribes inhibitors of apoptosis proteins (IAP) such as X-chromosome-linked IAP (XIAP), which bind to and inhibit active caspases (Van Antwerp et al., 1998).

5.1.2 Outside-In Hypothesis

The outside-in hypothesis, commonly known as the hygiene hypothesis, is based on the condition occurring due to a disruption within the skin barrier (Brandt and Sivaprasad, 2011). The skin is an interface between the organism's body and the environment. This protects the organism from the environment and supports the internal environment. The primary mediator of the epidermal barrier is the stratum corneum (Figure 5.3) (Lee and Lee, 2014). The stratum corneum is comprised of corneocytes, which secrete intercellular or intracellular substances such as filaggrin (Sullivan and Silverberg, 2017).

These intercellular or intracellular substances degrade amino acids such as urocanic acid, pyrrolidone carboxylic acid, alanine, glutamine, arginine and histidine, and lipids such as ceramide 1, fatty acids and cholesterol responsible for creating the barrier layer and maintaining its integrity (Sullivan and Silverberg, 2017; Lee and Lee, 2014; Varothai, Nitayavardhana, and Kulthanan, 2013). Below the stratum corneum is keratinocyte membranes that contain tight junctions. These junctions form an additional barrier that protects the organism from the environment when the stratum corneum is dysfunctional. However, within eczema patients, the expression of claudins is downregulated, causing these tight junctions to become dysfunctional (Malajian and Guttman-Yassky, 2015).

The degradation of amino acids maintains hydration within the epidermal and reduces the pH of the skin between 4.7 and 5.75 (Gelmetti and Wollenberg, 2014; Lee and Lee, 2014). This low skin pH maintains the epidermal barrier homeostasis, ensures the stratum corneum is intact, activates ceramide metabolism and modulates serine protease cascade (Lee and Lee, 2014). Within eczema patients, the skin pH is slightly higher than normal skin at pH 6, which causes lipid processing enzymes such as beta-glucocerebrosidase (β-GlcCer'ase), acidic sphingomyelinase (aSMase) and phospholipase A2 (sPLA2) to be inhibited and the production of serine proteases to be increased (Gelmetti and Wollenberg, 2014; Lee and Lee, 2014). This increase in serine proteases downregulates the secretion of lamellar bodies, inhibits protective resident bacterial flora on the skin and causes barrier dysfunction (Gelmetti and Wollenberg, 2014).

The dysfunctional skin barrier is caused by several genetic mutations that impair the functions of genes required to maintain structural integrity (Elias and Wakefield, 2014). This includes the missense and nonsense mutations of TMEM79, which leads to skin lesions and inflammation (Voisin

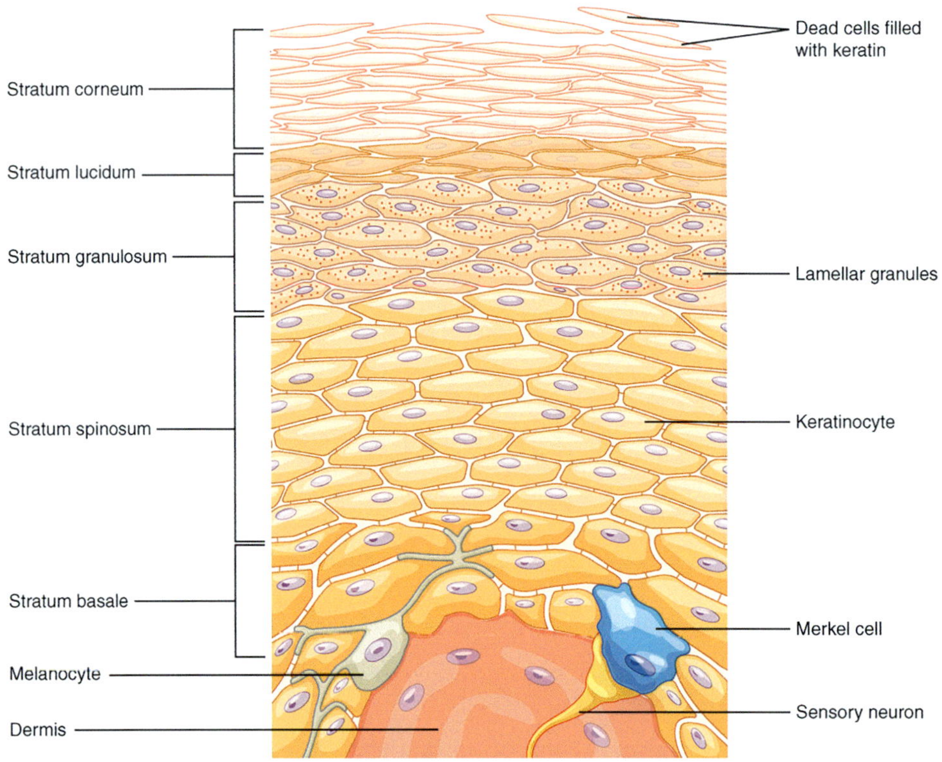

FIGURE 5.3 Different layers of the epidermis (OpenStax College, 2013).

and Chiu, 2018). Another mutation linked to atopic dermatitis is the overexpression of the small proline-rich protein 3 (SPRR3) gene due to a 24 base pair defect and an in-frame insertion and deletion (Elias and Wakefield, 2014).

The basis of this hypothesis is the loss-of-function mutation within the filaggrin (FLG) gene, which further impairs the barrier function (Sullivan and Silverberg, 2017). Studies have shown that patients with these mutations have a higher chance of developing atopic dermatitis than those who do not by one-fifth to one-third (Langan, Irvine, and Weidinger, 2020). The filaggrin gene is located on chromosome 1q21 and translates into a polyprotein known as profilaggrin, which is the main component of keratohyalin granules (Lee and Lee, 2014; Palmer et al., 2006). These profilaggrins are comprised of calcium-binding N-terminal domain and are dephosphorylated and cleaved into 10–12 filaggrin monomers that contribute to the strength and integrity of the epidermis (Lee and Lee, 2014; Brown and McLean, 2012).

When these monomers aggregate, the keratin cytoskeleton form a protein-lipid matrix, which maintains epidermal hydration (Voisin and Chiu, 2018). In eczema patients who have impaired barrier functions, there is an increase in transepidermal water loss and a decrease in water-binding capacity, which manifests into dry skin (Sullivan and Silverberg, 2017; Lee and Lee, 2014).

Within eczema patients, there is an increase in histamine production, which leads to an intolerable itching sensation (Buddenkotte, Maurer, and Steinhoff, 2010). This biogenic amine is synthesized during the decarboxylation of histidine when this amino acid is exposed to histidine decarboxylase (1-histidine decarboxylase EC 4.1.1.22) (Castells, 2006; Abe et al., 1993). Once synthesized, the histamine is stored in secretory granules within mature mast cells and basophils, along with other mediators such as proteoglycan heparin and chondroitin sulfate E (Castells, 2006; Abe et al., 1993; Hogan and Schwartz, 1997).

This causes a large number of mast cells to be present within the skin. These mast cells express NF-κB–related genes and Th-2 cytokines when exposed to an allergen (Jensen, Falkencrone, and Skov, 2014). However, these allergens release histamine into the extracellular space during a process known as degranulation (Abe et al., 1993; Hogan and Schwartz, 1997). Some of the histamines are metabolized via two alternative pathways (Abe et al., 1993). One of the pathways involves a catalyze known as histamine N-methyltransferase (EC 2.1.1.8), while the other involves a catalyze known as diamine oxidase (EC 1.4.3.6) (Castells, 2006; Abe et al., 1993).

Increased levels of histamine can affect the expression of genes associated with maintaining the epidermal barrier, which includes filaggrin, keratins and proteases. In response to this, filaggrin levels are reduced and the mRNA expression of some keratins is downregulated (Gschwandtner et al., 2013; Gutowska-Owsiak et al., 2014). However, the expression of proteases such as cathepsins and elastase are increased, which contributes to the inflammatory response and the dry effect that occurs around eczematous skin lesions (Agrawal and Woodfolk, 2014; Voegeli et al., 2009). These proteases cause the pH of the skin to increase, which enhances their effect on the inflammatory system and the epidermal barrier (Gelmetti and Wollenberg, 2014). Elastase is mainly used to remove foreign proteins during wound healing; however, when overexpressed, these proteases degrade elastin, which leads to wrinkle formation (Thring, Hili, and Naughton, 2009).

The intercellular or intracellular substances of the stratum corneum further increase transepidermal water loss and decrease water-binding capacity (Lee and Lee, 2014). This is an indication of an impaired barrier function, which leads to dry skin (Sullivan and Silverberg, 2017; Lee and Lee, 2014). This is due to the low ratio of intercellular lipids, an imbalance of protease and anti-protease present within the stratum corneum and an increase in cystatin A production, which leads to the degradation of the corneodesmosome (Gelmetti and Wollenberg, 2014).

5.2 SOUTH AFRICAN MEDICINAL PLANTS TO TREAT ECZEMA

Within South Africa, traditional healers use various indigenous plants to treat patients that suffer from eczema (Mabona and Van Vuuren, 2013). Of these, eight indigenous plants were selected and are described in Table 5.1. This selection was based on the amount of traditional-based information available specifically on how they were prepared and used to treat eczema. Though these plants are indigenous, some are found in other parts of Africa. The descriptions for each plant contain traditional knowledge on other conditions and diseases. This is to promote the medicinal potential of these plants and to provide a brief physical description of each plant.

TABLE 5.1
Traditional Methods Using South African Plants to Treat Eczema

South African Medicinal Plants	Plant Parts Used to Treat Eczema	Traditional Methods Used	References
***Aloe greatheadii* Schonland**	The sap of the leaves	Applied topically to the infected areas	(Fern, 2019; Lall and Kishore, 2014)
***Aspalathus linearis* (Burm f.) R. Dahlgren**	Shoots from the stems	Cut into smaller pieces, bruised and left to dry. After fermentation, the plants are brewed at low temperatures then applied topically to infected areas	(Joubert and de Beer, 2011; Joubert et al., 2008)
***Carpobrotus edulis* (L.) L. Bolus**	The juice and pulp from the leaves	Applied topically to the infected areas	(Deutschländer, Lall, and Van De Venter, 2009; Lall and Kishore, 2014)

(Continued)

TABLE 5.1 (Continued)

South African Medicinal Plants	Plant Parts Used to Treat Eczema	Traditional Methods Used	References
***Harpephyllum caffrum* Bernh. ex Krauss**	The bark of the tree	Applied topically to infected areas	(Lall and Kishore, 2014)
***Leonotis leonurus* (L.) R. Br.**	Leaves and stems	Infusions and decoctions are applied externally to infected areas	(El-Ansari et al., 2009)
***Trichilia emetica* Vahl.**	Roots, leaves and fruits	Leaves and roots are made into a poultice and applied externally; the roots are made into a powder and mixed with potassium hydroxide in water	(Lall and Kishore, 2014; Komane, Olivier, and Viljoen, 2011)
***Olea europaea* L. subsp. *africana* (Mill.) P. S. Green**	Bark, leaves and oils from the seeds	Bark decoctions and leaf infusions are used topically and internally to treat different types of dermatitis and inflammation; the oils are applied topically to treat hand, contact and atopic dermatitis specifically	(Lall and Kishore, 2014; Long, Tilney, and Van Wyk, 2010; Hashmi et al., 2015)
***Pelargonium graveolens* L Hér**	Oils from the seeds and leaves	The leaves are applied directly onto infected areas while the oils are used topically	(Lall and Kishore, 2014; Saraswathi et al., 2011)

5.2.1 *Aloe greatheadii* Schonland

Aloe greatheadii Schonland (Figure 5.4), commonly known as the spotted aloe, is part of the Aloacea family and is an important indigenous plant (Lall and Kishore, 2014; Human and Nicolson, 2006). This plant is widespread throughout South Africa, and is used in winter as a food source to increase the bee population within Pretoria (Human and Nicolson, 2006). Though these bees favor this plant, they become aggressive when exclusively bred on *A. greatheadii* (Grace et al., 2009). This evergreen succulent is harvested for food and medicinal properties (Fern, 2019).

The sap of the leaves is used as eye drops for chronic conjunction and applied topically for skin cancer, arthritis and skin irritation (Lall and Kishore, 2014; Fern, 2019). However, adverse side effects of the sap include congestion and irritation when consumed, and if used for a prolonged time may cause colorectal cancer. The leaves are made into infusions and decoctions to treat gonorrhea and purgatives to treat other diseases (Fern, 2019).

5.2.2 *Aspalathus linearis* (Burm f.) R. Dahlgren

Aspalathus linearis (Burm f.) R. Dahlgren (Figure 5.4), commonly known as rooibos tea, is part of the Fabaceae family and is natively found in the Cedarberg Mountains within the Western Cape Province (Lall and Kishore, 2014; McKay and Blumberg, 2007). This shrub-like leguminous bush is used worldwide as a beverage and for its medicinal properties (McKay and Blumberg, 2007). The shoots, which contain the leaves and stems, are used; however, the method used to prepare the plant material is different (Joubert and de Beer, 2011; Joubert et al., 2008).

In some cultures, the shoots are cut into smaller pieces that are bruised with a hammer to allow fermentation and pasteurized using steam. The steam prevents microbes from entering and contaminating the plant material. Afterward, the plants are left to dry and made into infusions (Joubert and de Beer, 2011). These infusions are red; however, if the plant material is dried before it is fermented, then the infusions are green in color (McKay and Blumberg, 2007). By drying the leaves before

FIGURE 5.4 Plants that are traditionally used against eczema. 1, *Aloe greatheadii*; 2, *Aspalathus linearis*; 3, *Carpobrotus edulis*; 4, *Harpephyllum caffrum*; 5, *Leonotis leonurus*; 6, *Trichilia emetic*; 7, *Olea europaea* subsp. *africana*; 8, *Pelargonium graveolens* (DerHexer, 2010; JMK, 2012; Bruenken, 2005; JoJan, 2005; Richfield, 2011; Giraud, 2011; Hectonichus, 2018; BotBln, 2010).

fermentation, the moisture content is reduced to prevent oxidation from occurring. This allows the leaves to retain their green coloration (Joubert and Schultz, 2012). Depending on the ailment, one of these infusions will be used in folk medicine to treat headaches, nausea, mild depression and colic disorders (McKay and Blumberg, 2007).

The Khoisans chop the shoots with an ax and crush the material with a mallet. The bruised sections are placed inside a hollow stone reef to allow fermentation to occur. Afterward, the plants are dried and boiled in hot water. Once brewed, the concentration of the infusion can be altered by adding water or shoots, depending on the condition (Joubert et al., 2008).

5.2.3 *Carpobrotus edulis* (L.) L. Bolus

Carpobrotus edulis (L.) L. Bolus (Figure 5.4), commonly known as sour fig, is part of the Aizoaceae family and is native to South Africa; however, it invades numerous coastal areas (Lall and Kishore, 2014; Mudimba and Nguta, 2019). These plants mainly inhabit sandy coastal areas, but have been found within sandy and marshy inland areas. The fruit, leaves and fruit are traditionally used for various ailments depending on the preparation method used (Mudimba and Nguta, 2019).

To treat bacterial and fungal infections, the leaves, fruit or flowers are chewed raw or boiled in water. The boiled leaves are used for tuberculosis and other respiratory diseases (Mudimba and Nguta, 2019). The juice from the leaves is gargled for throat infections; however, when taken orally, it can be used as a diuretic and a styptic. The juice from the fruit is often gargled to treat throat and mouth infections (Deutschländer, Lall, and Van De Venter, 2009). It is applied topically; the leave juice and leaf palp are used to treat burns, wounds and infections (Lall and Kishore, 2014; Deutschländer, Lall, and Van De Venter, 2009).

5.2.4 *Harpephyllum caffrum* Bernh. ex Krauss

Harpephyllum caffrum Bernh. ex Krauss (Figure 5.4), commonly known as the wild plum, is part of the Anacardiaceae family, which is the largest tree family in South Africa (Lall and Kishore, 2014; Chinyama, 2009). The bark of this evergreen tree is applied topically for acne and made into decoctions to be used as a blood purifier and emetic (Lall and Kishore, 2014; Chinyama, 2009; Maroyi, 2019).

The fruit is consumed and made into marketable products such as jams, jellies and non-alcoholic and alcoholic beverages (Chinyama, 2009; Maroyi, 2019). However, the fruit can be applied topically to treat wounds and sprains. Since *H. caffrum* is one of the smallest trees within the family, many people purchase this plant for ornamental purposes (Maroyi, 2019).

5.2.5 *Leonotis leonurus* (L.) R. Br.

Leonotis leonurus (L.) R. Br. (Figure 5.4), commonly known as wild dagga, is part of the Lamiaceae family and is found throughout South Africa (Lall and Kishore, 2014; Nsuala et al., 2017). These broadleaf evergreen woody shrubs are drought tolerant and are found near river banks and rocky hillsides (Nsuala et al., 2017; Nsuala, Enslin, and Viljoen, 2015). For many years, different sections of the plant have been used traditionally for numerous ailments. Leaf decoctions involve chopping leaves into small pieces and boiling them in water for an extended time; once cooled and strained, these decoctions are applied topically and taken orally (Nsuala, Enslin, and Viljoen, 2015).

Oral decoctions are used against coughs, colds, headaches, diarrhea and diabetes (Nsuala, Enslin, and Viljoen, 2015; El-Ansari et al., 2009). Topical decoctions are used to treat boils and muscular pains, and to reduce itching (El-Ansari et al., 2009). The leaves are sometimes dried and crushed into a powder then applied topically as an ointment to relieve pain. Stems and seeds are made into a powder and boiled with water to make decoctions before being used either orally to treat hemorrhoids or topically for sores. Finally, the flowers are made into a purgative or a tonic to treat tuberculosis, influenza and sores (Nsuala, Enslin, and Viljoen, 2015).

5.2.6 *Trichilia emetica* Vahl.

Trichilia emetica Vahl. (Figure 5.4), commonly known as Natal mahogany, is part of the Meliaceae family that is mainly located within tropical and subtropical regions; however, it has been seen within the savanna region (Lall and Kishore, 2014; Diallo et al., 2003). This widespread small tree is used throughout Africa for its medicinal uses (Diallo et al., 2003; Komane, Olivier, and Viljoen, 2011).

The leaves are made into a poultice to soothe bruises, and decoctions are taken orally for coughs and headaches (Lall and Kishore, 2014; Komane, Olivier, and Viljoen, 2011). However, in Nigeria, leaf poultices are used topically to treat syphilis (Komane, Olivier, and Viljoen, 2011). To treat fevers, coughs and syphilis, the bark of the tree is dried up and crushed into a powder (Diallo et al., 2003). The Xhosa tribe uses bark decoctions to treat kidney disorders and anemia, whereas the Zulu tribe uses bark decoctions to treat stomach ailments and backaches (Komane, Olivier, and Viljoen, 2011).

Powdered roots are used as a poison antidote when mixed with milk, and for asthma when mixed with honey (Diallo et al., 2003). When combined with potassium hydroxide and water, these powdered roots are used to reduce inflammation. Finally, roasted roots that have been crushed into a powder are used topically to heal wounds (Komane, Olivier, and Viljoen, 2011).

5.2.7 *Olea europaea* L. subsp. *africana* (Mill.) P. S. Green

Olea europaea L. subsp. *africana* (Mill.) P. S. Green (Figure 5.4), commonly known as the olive tree, is part of the Oleaceae family and is mainly found within the eastern regions of Africa and extends toward the southern tip of the continent (Lall and Kishore, 2014; Cuneo and Leishman, 2006). This small evergreen tree is found in woodland and mountain habitats and has a distinctive purple-black fruit color (Cuneo and Leishman, 2006). *Olea europaea* is one of the oldest cultivated trees, dating back to around 7,000 years ago. These plants were mainly cultivated for commercial purposes; however, the oils were known for their medicinal properties (Hashmi et al., 2015).

To treat diabetes, diarrhea, urinary tract infections and hypertension, the leaves are made into decoctions and are taken orally, while infusions are used to treat hypotension, eye irritation, sore throat, inflammation and hypoglycemia (Hashmi et al., 2015; Long, Tilney, and Van Wyk, 2010). The bark of the tree is used to treat colic when made into infusions and to reduce rashes and itching when made into decoctions (Long, Tilney, and Van Wyk, 2010). The oil obtained from the seeds is used as a laxative when taken orally, while oils obtained from the fruit are applied on fractured limbs (Hashmi et al., 2015).

5.2.8 *Pelargonium graveolens* L Hér

Pelargonium graveolens L Hér (Figure 5.4), commonly known as Rose-scented pelargonium, is part of the Geraniaceae family and is indigenous to South Africa, but was introduced to other parts of the world, including Europe (Lall and Kishore, 2014; Asgarpanah and Ramezanloo, 2015). This aromatic perennial shrub is found in moist habitats and its oils are used commercially within the cosmetic and aromatherapy industry and can be used as a substitute for rose perfume (Asgarpanah and Ramezanloo, 2015; Lawrence, 2002).

Traditionally, the oils are used to treat acne, while fresh leaves are used topically to reduce skin diseases, rashes and ulcers (Lall and Kishore, 2014; Saraswathi et al., 2011). The roots are made into decoctions and are used to treat malaria and urinary disorders. Infusions and decoctions of the leaves and stems are used topically for wound healing and orally for syphilis, duodenal ulcers, shingles and herpes (Asgarpanah and Ramezanloo, 2015; Saraswathi et al., 2011).

5.3 PLANT COMPOUNDS THAT REDUCE ECZEMA

Plants contain secondary metabolites that can be used to reduce specific aspects of the condition such as cytokine production, histamine release, wrinkle formation, hyperpigmentation or just to reduce the characteristic effects of eczema such as dryness, itching sensation or rash. Though there are clinical treatments that can be used to reduce these different aspects of the condition, such as those used to reduce TNF-α and NF-κB production, they are used for other conditions such as rheumatoid arthritis, cancer and psoriasis (Song et al., 2008; Miller et al., 2010; Parameswaran and Patial, 2010; Bradley, 2008; Serasanambati and Chilakapati, 2016). Following are some plant compounds that can be used to reduce either the internal aspects of the condition, such as the synthesis of cytokines and transcription factors, or the physical aspects, such as the rash, dryness or itching sensation.

5.3.1 TNF-α Production

Some plant-based compounds that are used to reduce TNF-α production include scopoletin, curcumin, magnosalin and andamanicin (Figure 5.2). Scopoletin is isolated from aqueous extracts of *Artemisia feddie* H. Lev. & Vaniot, and prevents the release of TNF-α, while curcumin, derived from the rhizomes of *Curcurma longa* L., inhibits the production of pro-inflammatory cytokines within lipopolysaccharide (LPS)-induced macrophage cell lines (Paul, Gohil, and Bhutani, 2006). Both of these compounds targeted the IκB cascade by inhibiting the activation of NF-κB (Figure 5.2A–B) (Moon et al., 2007; Kahkhaie et al., 2019). Magnosalin and andamanicin (Figure 5.2C–D) are isolated from the leaves of *Perilla frutescens* (L.) Britton. These compounds display a 26% cytotoxic effect toward active cells thereby reducing the production of TNF-α (Paul, Gohil, and Bhutani, 2006).

5.3.2 NF-κB Production

Plant-based compounds that are used to modulate NF-κB activity include kamebakaurin, acanthoic acid, isohelenin and parthenolide. Both kamebakaurin, isolated from *Isodon japonicus* (Burm. f.) H. Hara, and acanthoci acid, derived from *Acanthopanax koreanum* Nakai, have been shown to interfere with the NF-κB activation cascade. Isohelenin, isolated from *Arnica montana* L., prevents IκB degradation by alkylating the p69 subunit of NF-κB, and parthenolide, derived from *Tanacetum parthenium* (L.) Sch. Bip., binds to and inhibits I kappa B kinase beta (IKK-β) (Palladino et al., 2003).

5.3.3 Physical Aspects

Table 5.2 contains a list of plant compounds that are used to reduce common side effects associated with eczema. This includes the itching sensation, the rash and the dryness of the skin. Some of the compounds are used to reduce *Staphylococcus aureus* formation on eczematous lesions, which are thought to promote the formation of these lesions (Arima et al., 2003).

TABLE 5.2
Plant Compounds Used to Reduce Common Side Effects Associated with Eczema

Medicinal Plants	Family	Sections of the Plants	Secondary Metabolite	Location	References
Albizia adianthifolia (Schumach.) W. Wight	Fabaceae	The bark, leaves and stem roots	Anthraquinones, flavonoids, tannins, triterpenoids, sesquiterpene hydrocarbons, oxygenated sesquiterpenes and saponins	North Maputaland in South Africa and tropical areas of Africa	(Dlova and Ollengo, 2018; Maroyi, 2018)
Aloe ferox Mill.	Xanthorrhoeaceae	The leaves and the roots	Chrysophanol, aloe-emodin, aloin a, aloesin, aloenin and aloeresin a	Eastern Cape and cape coastal areas in South Africa	(Dlova and Ollengo, 2018; Chen et al., 2012)
Dysphania ambrosioides (L.) Mosyakin & Clements	Amaranthaceae	The leaves	Chenopodiumamine a, b and c	Eastern Cape in South Africa	(Dlova and Ollengo, 2018; Song et al., 2015)
Momordica balsamina L.	Cucubitaceae	The leaves	Anthraquinones, flavonoids, tannins and saponins	North Maputaland in South Africa, Namibia and Botswana	(Dlova and Ollengo, 2018; Thakur et al., 2009)
Portulacaria afra Jacq.	Portulacaceae	The leaves	Anthraquinones, flavonoids, tannins and saponins	North Maputaland, South Africa	(Dlova and Ollengo, 2018)
Chamaecyparis obtusa (Siebold & Zucc.) Endl.	Cupressaceae	The wood	B-thujaplicin, limonene, bornyl acetate and elemol	Native to Japan and southern parts of Korea	(Arima et al., 2003; Joo et al., 2010)

5.4 CONCLUSION

In conclusion, both the inside-out and outside-in hypotheses are considered as viable theories on the causation of eczema. Recently, more studies support the outside-in hypothesis due to the discovery of the loss-of-function mutation within the filaggrin gene in 2006. However, there are aspects of the inside-out hypothesis that have not been explored before. Thus, for future recommendations, more studies should be focused on the inside-out hypothesis and using plants to reduce aspects within this hypothesis. This chapter promotes the advantages of plant compounds that possess anti-eczematous properties. Furthermore, this chapter promotes the need to analyze indigenous plants for anti-eczematous properties using the same methods of extraction as traditional healers.

REFERENCES

Abe, Y., S. Ogino, M. Irifune, *et al.* 1993. "Histamine content, synthesis and degradation in human nasal mucosa." *Clinical & Experimental Allergy* 23, 2: 132–136. doi: 10.1111/j.1365-2222.1993.tb00308.x.

Agrawal, R., and J. A. Woodfolk. 2014. "Skin barrier defects in atopic dermatitis." *Current Allergy and Asthma Reports* 14, 5: 1–11. doi: 10.1007/s11882-014-0433-9.

Akdis, C. A., M. Akdis, T. Bieber, *et al.* 2006. "Diagnosis and treatment of atopic dermatitis in children and adults: European Academy of Allergology and Clinical Immunology/American Academy of Allergy, Asthma and Immunology/PRACTALL Consensus Report." *The Journal of Allergy and Clinical Immunology* 118, 1: 152–169. doi: 10.1111/j.1398-9995.2006.01153.x.

Arima, Y., Y. Nakai, R. Hayakawa, and T. Nishino. 2003. "Antibacterial effect of β-thujaplicin on *Staphylococci* isolated from atopic dermatitis: Relationship between changes in the number of viable bacterial cells and clinical improvement in an eczematous lesion of atopic dermatitis." *Journal of Antimicrobial Chemotherapy* 51, 1: 113–122. doi: 10.1093/jac/dkg037.

Asgarpanah, J., and F. Ramezanloo. 2015. "An overview on phytopharmacology of *Pelargonium graveolens* L." *Indian Journal of Traditional Knowledge* 14, 4: 558–563.

Biophysik. 2008. "Curcumin-enol.png." Obtained from Wikimedia under the Public Domain License. https://commons.wikimedia.org/wiki/File:Curcumin-enol.png (Accessed 20 March 2021).

BotBln. 2010. "*Olea europaea africana*." Obtained from Wikimedia under a Creative Common Attribution-Share Alike 3.0 Unported license https://commons.wikimedia.org/wiki/File:Olea_europaea_africana_KirstenboshBotGard09292010C.JPG (Accessed 8 June 2020).

Bradley, J. R. 2008. "TNF-mediated inflammatory disease." *The Journal of Pathology: A Journal of the Pathological Society of Great Britain and Ireland* 214, 2: 149–160. doi: 10.1002/path.2287.

Brandt, E. B., and U. Sivaprasad. 2011. "Th2 cytokines and atopic dermatitis." *Journal of Clinical & Cellular Immunology* 2, 3: 1–25. doi: 10.4172/2155-9899.1000110.

Brown, S. J., and W. H. I. McLean. 2012. "One remarkable molecule: Filaggrin." *Journal of Investigative Dermatology* 132, 3: 751–762. doi: 10.1038/jid.2011.393.

Bruenken, W. 2005. "Rooibos (*Aspalathus linearis*)." Obtained from Wikimedia under a Creative Common Attribution-Share Alike 2.5 Generic license https://commons.wikimedia.org/wiki/File:Rooibos_(Aspalathus_linearis).jpg (Accessed 8 June 2020).

Buddenkotte, J., M. Maurer, and M. Steinhoff. 2010. "Histamine and antihistamines in atopic dermatitis." In *Histamine in Inflammation*, 73–80. Boston: Springer doi: 10.1007/978-1-4419-8056-4_8.

Bylund, S., L. B. von Kobyletzki, M. Svalstedt, and Å. Svensson. 2020. "Prevalence and incidence of atopic dermatitis: A systematic review." *Acta dermato-venereologica* 100: 320–329. doi: 10.2340/00015555-3510.

Castells, M. 2006. "Mast cell mediators in allergic inflammation and mastocytosis." *Immunology and Allergy Clinics* 26, 3: 465–485. doi: 10.1016/j.iac.2006.05.005.

Chamlin, S. L., J. Kao, I. J. Frieden, *et al.* 2002. "Ceramide-dominant barrier repair lipids alleviate childhood atopic dermatitis: Changes in barrier function provide a sensitive indicator of disease activity." *Journal of the American Academy of Dermatology* 47, 2: 198–208. doi: 10.1067/mjd.2002.124617.

Chen, W., B.-E Van Wyk, I. Vermaak, and A. M. Viljoen. 2012. "Cape aloes—A review of the phytochemistry, pharmacology and commercialisation of *Aloe ferox*." *Phytochemistry Letters* 5, 1: 1–12. doi: 10.1016/j.phytol.2011.09.001.

Chinyama, R. F. 2009. "Biological activities of medicinal plants traditionally used to treat septicaemia in the Eastern Cape, South Africa." Masters, Biomedical Technology, Nelson Mandela Metropolitan University.

Chrumps. 2010. "Scopoletin.png." Obtained from Wikimedia under the Public Domain License. https://commons.wikimedia.org/wiki/File:Scopoletin.png (Accessed 20 March 2021).

Cornelissen, C., Y. Marquardt, K. Czaja, *et al.* 2012. "IL-31 regulates differentiation and filaggrin expression in human organotypic skin models." *Journal of Allergy and Clinical Immunology* 129, 2: 426–433.

Cuneo, P., and M. R. Leishman. 2006. "African Olive (*Olea europaea* subsp. *cuspidata*) as an environmental weed in eastern Australia: A review." *Cunninghamia* 9, 4: 545–577.

DerHexer. 2010. "*Pelargonium graveolens*." Obtained from Wikimedia under a Creative Common Attribution-Share Alike 3.0 Unported license. https://commons.wikimedia.org/wiki/File:Pelargonium_graveolens_in_Hel.jpg (Accessed 8 June 2020).

Deutschländer, M. S., N. Lall, and M. Van De Venter. 2009. "Plant species used in the treatment of diabetes by South African traditional healers: An inventory." *Pharmaceutical Biology* 47, 4: 348–365. doi: 10.1080/13880200902752959.

Diallo, D., B. S. Paulsen, T. H. A. Liljebäck, and T. E. Michaelsen. 2003. "The Malian medicinal plant *Trichilia emetica*; studies on polysaccharides with complement fixing ability." *Journal of Ethnopharmacology* 84, 2: 279–287. doi: 10.1016/S0378-8741(02)00330-6.

Dlova, N. C., and M. A. Ollengo. 2018. "Traditional and ethnobotanical dermatology practices in Africa." *Clinics in Dermatology* 36, 3: 353–362. doi: 10.1016/j.clindermatol.2018.03.009.

Eisfelder. 2004. "Atopic dermatitis child." Obtained from Wikimedia under the Creative Commons Attribution-Share Alike 3.0 unported license. https://commons.wikimedia.org/wiki/File:Atopic_dermatitis_child.JPG (Accessed 26 October 2021).

El-Ansari, M. A., E. A. Aboutabl, A. R. H. Farrag, *et al.* 2009. "Phytochemical and pharmacological studies on *Leonotis leonurus*." *Pharmaceutical Biology* 47, 9: 894–902. doi: 10.1080/13880200902942428.

Elias, P. M., and J. S. Wakefield. 2014. "Mechanisms of abnormal lamellar body secretion and the dysfunctional skin barrier in patients with atopic dermatitis." *Journal of Allergy and Clinical Immunology* 134, 4: 781–791. doi: 10.1016/j.jaci.2014.05.048.

Fern, K. 2019. "*Aloe greatheadii*." www.tropical.theferns.info/viewtropical.php?id=Aloe+greatheadii (Accessed 14 June 2020).

Gelmetti, C., and A. Wollenberg. 2014. "Atopic dermatitis—all you can do from the outside." *British Journal of Dermatology* 170, 1: 19–24. doi: 10.1111/bjd.12957.

Giraud, G. 2011. "*Leonotis leonurus*." Obtained from Wikimedia under a Creative Commons Universal Public Domain Dedication license. https://commons.wikimedia.org/wiki/File:Leonotis_leonurus_03.jpg (Accessed 8 June 2020).

Grace, O. M., M. S. J. Simmonds, G. F. Smith, and A. E. van Wyk. 2009. "Documented utility and biocultural value of *Aloe* L. (Asphodelaceae): A review." *Economic Botany* 63, 2: 167–178. doi: 10.1007/s12231-009-9082-7.

Gschwandtner, M., M. Mildner, V. Mlitz, *et al.* 2013. "Histamine suppresses epidermal keratinocyte differentiation and impairs skin barrier function in a human skin model." *Allergy* 68, 1: 37–47. doi: 10.1111/all.12051.

Gutowska-Owsiak, D., L. Greenwald, C. Watson, *et al.* 2014. "The histamine-synthesizing enzyme histidine decarboxylase is upregulated by keratinocytes in atopic skin." *British Journal of Dermatology* 171, 4: 771–778. doi: 10.1111/bjd.13199.

Hashmi, M. A., A. Khan, M. Hanif, U. Farooq, and S. Perveen. 2015. "Traditional uses, phytochemistry, and pharmacology of *Olea europaea* (olive)." *Evidence-Based Complementary and Alternative Medicine* 1–30. doi: 10.1155/2015/541591.

Hatano, Y., H. Terashi, S. Arakawa, and K. Katagiri. 2005. "Interleukin-4 suppresses the enhancement of ceramide synthesis and cutaneous permeability barrier functions induced by tumor necrosis factor-α and interferon-γ in human epidermis." *Journal of Investigative Dermatology* 124, 4: 786–792. doi: 10.1111/j.0022-202X.2005.23651.x.

Hectonichus. 2018. "Meliaceae -*Trichilia emetica*." Obtained from Wikimedia under a Creative Common Attribution-Share Alike 4.0 International license https://commons.wikimedia.org/wiki/File:Meliaceae-Trichilia_emetica_(Natal_Mahogany).JPG (Accessed 8 June 2020).

Hogan, A. D., and L. B. Schwartz. 1997. "Markers of mast cell degranulation." *Methods* 13, 1: 43–52. doi: 10.1006/meth.1997.0494.

Howell, M. D., B. E. Kim, P. Gao, *et al.* 2009. "Cytokine modulation of atopic dermatitis filaggrin skin expression." *Journal of Allergy and Clinical Immunology* 124, 3: R7–R12. doi: 10.1016/j.jaci.2009.07.012.

Human, H., and S. W. Nicolson. 2006. "Nutritional content of fresh, bee-collected and stored pollen of *Aloe greatheadii* var. *davyana* (Asphodelaceae)." *Phytochemistry* 67, 14: 1486–1492. doi: 10.1016/j.phytochem.2006.05.023.

Jensen, B. M., S. Falkencrone, and P. S. Skov. 2014. "Measuring histamine and cytokine release from basophils and mast cells." In *Methods in Molecular Biology*, edited by B. F. Gibbs and F. H. Falcone, 135–145. New York: Springer. doi: 10.1007/978-1-4939-1173-8_10.

JMK. 2012. "*Aloe greatheadii* var *davyana*." Obtained from Wikimedia under a Creative Common Attribution-Share Alike 3.0 Unported license. https://commons.wikimedia.org/wiki/File:Aloe_greatheadii_var_davyana,_habitus,_Louwsburg.jpg (Accessed 8 June 2020).

JoJan. 2005. "*Carpobrotus edulis*01.jpg." Obtained from Wikimedia under a public domain license. https://commons.wikimedia.org/wiki/File:Carpobrotus_edulis01.jpg (Accessed 8 June 2020).

Joo, S. S., Y-M. Yoo, S-H. Ko, *et al.* 2010. "Effects of essential oil from Chamaecypris obtusa on the development of atopic dermatitis-like skin lesions and the suppression of Th cytokines." *Journal of Dermatological Science* 60, 2: 122–125. doi: 10.1016/j.jdermsci.2010.08.008.

Joubert, E., and D. de Beer. 2011. "Rooibos (*Aspalathus linearis*) beyond the farm gate: From herbal tea to potential phytopharmaceutical." *South African Journal of Botany* 77, 4: 869–886. doi: 10.1016/j.sajb.2011.07.004.

Joubert, E., W. C. A. Gelderblom, A. Louw, and D. de Beer. 2008. "South African herbal teas: *Aspalathus linearis*, *Cyclopia* spp. and *Athrixia phylicoides*—A review." *Journal of Ethnopharmacology* 119, 3: 376–412. doi: 10.1016/j.jep.2008.06.014.

Joubert, E., and H. Schultz. 2012. "Production and quality aspects of rooibos tea and related products. A review." *Journal of Applied Botany and Food Quality* 80, 2: 138–144.

Kahkhaie, K. R., A. Mirhosseini, A. Aliabadi, *et al.* 2019. "Curcumin: A modulator of inflammatory signaling pathways in the immune system." *Inflammopharmacology* 27, 5: 885–900. doi: 10.1007/s10787-019-00607-3.

Kelly, M. 2020. "ILCs in atopic dermatitis." Obtained from Wikimedia under the Creative Commons Attribution-Share Alike 4.0 International license. https://commons.wikimedia.org/wiki/File:ILCs_in_atopic_dermatitis_9_PNG_FINAL.png (Accessed 26 October 2021).

Kim, B. E., M. D. Howell, E. Guttman, *et al.* 2011. "TNF-α downregulates filaggrin and loricrin through c-Jun N-terminal kinase: Role for TNF-α antagonists to improve skin barrier." *Journal of Investigative Dermatology* 131, 6: 1272–1279. doi: 10.1038/jid.2011.24.

Komane, B. M., E. I. Olivier, and A. M. Viljoen. 2011. "*Trichilia emetica* (Meliaceae)—A review of traditional uses, biological activities and phytochemistry." *Phytochemistry Letters* 4, 1: 1–9. doi: 10.1016/j.phytol.2010.11.002.

Lall, N., and N. Kishore. 2014. "Are plants used for skin care in South Africa fully explored?" *Journal of Ethnopharmacology* 153, 1: 61–84. doi: 10.1016/j.jep.2014.02.021.

Langan, S. M., A. D. Irvine, and S. Weidinger. 2020. "Atopic dermatitis." *The Lancet* 396, 10247: 345–360. doi: 10.1016/S0140-6736(20)31286-1.

Lawrence, E. 2002. "*Pelargonium graveolens*." http://pza.sanbi.org/pelargonium-graveolens (Accessed 15 June 2020).

Lee, H., and S. Lee. 2014. "Epidermal permeability barrier defects and barrier repair therapy in atopic dermatitis." *Allergy, Asthma & Immunology Research* 6, 4: 276–287. doi: 10.4168/aair.2014.6.4.276.

Long, H. S., P. M. Tilney, and B. E. Van Wyk. 2010. "The ethnobotany and pharmacognosy of *Olea europaea* subsp. *Africana* (Oleaceae)." *South African Journal of Botany* 76, 2: 324–331. doi: 10.1016/j.sajb.2009.12.005.

Mabona, U., and S. F. Van Vuuren. 2013. "Southern African medicinal plants used to treat skin diseases." *South African Journal of Botany* 87: 175–193. doi: 10.1016/j.sajb.2013.04.002.

Malajian, D., and E. Guttman-Yassky. 2015. "New pathogenic and therapeutic paradigms in atopic dermatitis." *Cytokine* 73, 2: 311–318. doi: 10.1016/j.cyto.2014.11.023.

Mansouri, Y., and E. Guttman-Yassky. 2015. "Immune pathways in atopic dermatitis, and definition of biomarkers through broad and targeted therapeutics." *Journal of Clinical Medicine* 4, 5: 858–873. doi: 10.3390/jcm4050858.

Maroyi, A. 2018. "Albizia adianthifolia: Botany, medicinal uses, phytochemistry, and pharmacological properties." *The Scientific World Journal*. doi: 10.1155/2018/7463584.

Maroyi, A. 2019. "Medicinal uses, biological and chemical properties of wild plum (*Harpephyllum caffrum*): An Indigenous fruit plant of Southern Africa." *Journal of Pharmacy and Nutrition Sciences* 9, 5: 258–268. doi: 10.29169/1927-5951.2019.09.05.4.

McKay, D. L., and J. B. Blumberg. 2007. "A review of the bioactivity of South African herbal teas: Rooibos (*Aspalathus linearis*) and honeybush (*Cyclopia intermedia*)." *Phytotherapy Research* 21, 1: 1–16. doi: 10.1002/ptr.1992.

Miller, S. C., R. Huang, S. Sakamuru, *et al.* 2010. "Identification of known drugs that act as inhibitors of NF-κB signaling and their mechanism of action." *Biochemical Pharmacology* 79, 9: 1272–1280. doi: 10.1016/j.bcp.2009.12.021.

Moon, P-D., B-H. Lee, H-J. Jeong, *et al.* 2007. "Use of scopoletin to inhibit the production of inflammatory cytokines through inhibition of the IκB/NF-κB signal cascade in the human mast cell line HMC-1." *European Journal of Pharmacology* 555, 2–3: 218–225. doi: 10.1016/j.ejphar.2006.10.021.

Mudimba, T. N., and J. M. Nguta. 2019. "Traditional uses, phytochemistry and pharmacological activity of *Carpobrotus edulis*: A global perspective." *The Journal of Phytopharmacology* 8, 3: 111–116. doi: 10.31254/phyto.2019.8305.

Nedoszytko, B., M. Sokołowska-Wojdyło, K. Ruckemann-Dziurdzińska, J. Roszkiewicz, and R. J. Nowicki. 2014. "Chemokines and cytokines network in the pathogenesis of the inflammatory skin diseases: Atopic dermatitis, psoriasis and skin mastocytosis." *Advances in Dermatology and Allergology* 31, 2: 84–91. doi: 10.5114/pdia.2014.40920.

Nsuala, B. N., G. Enslin, and A. Viljoen. 2015. " 'Wild cannabis': A review of the traditional use and phytochemistry of *Leonotis leonurus*." *Journal of Ethnopharmacology* 174: 520–539. doi: 10.1016/j.jep.2015.08.013.

Nsuala, B. N., G. P. Kamatou, M. Sandasi, G. Enslin, and A. Viljoen. 2017. "Variation in essential oil composition of *Leonotis leonurus*, an important medicinal plant in South Africa." *Biochemical Systematics and Ecology* 70: 155–161. doi: 10.1016/j.bse.2016.11.009.

OpenStax College. 2013. "Layers of epidermis." Obtained from Wikimedia under a Creative Common Attribution-Share Alike 3.0 Unported license https://commons.wikimedia.org/wiki/File:502_Layers_of_epidermis.jpg (Accessed 12 June 2020).

Palladino, M. A., F. R. Bahjat, E. A. Theodorakis, and L. L. Moldawer. 2003. "Anti-TNF-α therapies: The next generation." *Nature Reviews Drug Discovery* 2, 9: 736–746. doi: 10.1038/nrd1175.

Palmer, C. N. A., A. D. Irvine, A. Terron-Kwiatkowski, *et al.* 2006. "Common loss-of-function variants of the epidermal barrier protein filaggrin are a major predisposing factor for atopic dermatitis." *Nature Genetics* 38: 441–446. doi: 10.1038/ng1767.

Parameswaran, N., and S. Patial. 2010. "Tumor necrosis factor-α signaling in macrophages." *Critical Reviews™ in Eukaryotic Gene Expression* 20, 2: 87–103. doi: 10.1615/CritRevEukarGeneExpr.v20.i2.10.

Paul, A. T., V. M. Gohil, and K. K. Bhutani. 2006. "Modulating TNF-α signaling with natural products." *Drug Discovery Today* 11, 15–16: 725–732. doi: 10.1016/j.drudis.2006.06.002.

Richfield, J. 2011. "*Harpephyllum caffrum* Anacardiaceae." Obtained from Wikimedia under a Creative Common Attribution-Share Alike 3.0 Unported license https://commons.wikimedia.org/wiki/File:Harpephyllum_caffrum_Anacardiaceae_6397s.jpg (Accessed 8 June 2020).

Saraswathi, J., K. Venkatesh, N. Baburao, M. H. Hilal, and A. R. Rani. 2011. "Phytopharmacological importance of *Pelargonium* species." *Journal of Medicinal Plants Research* 5, 13: 2587–2598.

Serasanambati, M., and S. R. Chilakapati. 2016. "Function of nuclear factor kappa B (NF-kB) in human diseases-a review." *South Indian Journal of Biological Sciences* 2, 4: 368–387. doi: 10.22205/sijbs/2016/v2/i4/103443.

Song, K., J. Zhang, P. Zhang, *et al.* 2015. "Five new bioactive compounds from *Chenopodium ambrosioides*." *Journal of Asian Natural Products Research* 17, 5: 482–490. doi: 10.1080/10286020.2015.1042872.

Song, M-Y., S-K. Park, C-S. Kim, *et al.* 2008. "Characterization of a novel anti-human TNF-α murine monoclonal antibody with high binding affinity and neutralizing activity." *Experimental & Molecular Medicine* 40, 1: 35–42. doi: 10.3858/emm.2008.40.1.35.

Sullivan, M., and N. B. Silverberg. 2017. "Current and emerging concepts in atopic dermatitis pathogenesis." *Clinics in Dermatology* 35, 4: 349–353. doi: 10.1016/j.clindermatol.2017.03.006.

Tadeg, H., E. Mohammed, K. Asres, and T. Gebre-Mariam. 2005. "Antimicrobial activities of some selected traditional Ethiopian medicinal plants used in the treatment of skin disorders." *Journal of Ethnopharmacology* 100, 1–2: 168–175. doi: 10.1016/j.jep.2005.02.031.

Thakur, G. S., M. Bag, B. S. Sanodiya, *et al.* 2009. "*Momordica balsamina*: A medicinal and neutraceutical plant for health care management." *Current Pharmaceutical Biotechnology* 10, 7: 667–682. doi: 10.2174/138920109789542066.

Thring, T. S. A., P. Hili, and D. P. Naughton. 2009. "Anti-collagenase, anti-elastase and anti-oxidant activities of extracts from 21 plants." *BMC Complementary and Alternative Medicine* 9, 1: 27. doi: 10.1186/1472-6882-9-27.

Van Antwerp, D. J., S. J. Martin, I. M. Verma, and D. R. Green. 1998. "Inhibition of TNF-induced apoptosis by NF-κB." *Trends in Cell Biology* 8, 3: 107–111. doi: 10.1016/S0962-8924(97)01215-4.

Varothai, S., S. Nitayavardhana, and K. Kulthanan. 2013. "Moisturizers for patients with atopic dermatitis." *Asian Pacific Journal of Allergy and Immunology* 31, 2: 91–98.

Voegeli, R., A. V. Rawlings, M. Breternitz, *et al.* 2009. "Increased stratum corneum serine protease activity in acute eczematous atopic skin." *British Journal of Dermatology* 161, 1: 70–77. doi: 10.1111/j.1365-2133.2009.09142.x.

Voisin, T., and I. M. Chiu. 2018. "Molecular link between itch and atopic dermatitis." *Proceedings of the National Academy of Sciences* 115, 51: 12851–12853. doi: 10.1073/pnas.1818879115.

Wu, Ya-di, and B. P. Zhou. 2010. "TNF-α/NF-κ B/Snail pathway in cancer cell migration and invasion." *British Journal of Cancer* 102, 4: 639–644. doi: 10.1038/sj.bjc.6605530. Obtained under the Creative Commons CC-BY license.

Xu, G., and Y. Shi. 2007. "Apoptosis signaling pathways and lymphocyte homeostasis." *Cell Research* 17, 9: 759–771. doi: 10.1038/cr.2007.52.

6 Natural Products in the Treatment of Unremitting Wounds Secondary to Diabetes or Peripheral Vascular Disease

Sarah Murnane

CONTENTS

Graphical abstract representing the content of the chapter (Brooks, 2013; Dormiraubois, 2016; Douglas, 2011; Margenauer, 2005; Shebs, 2006; Zaynabazam7, 2018).

DOI: 10.1201/9781003108375-6

6.1 INTRODUCTION

According to the World Health Organization (WHO), diabetes is a global health issue affecting more than 422 million people as of 2014 (World Health Organization, 2020). Both the WHO and International Diabetes Federation (IDF) have named diabetes an epidemic. Worldwide, 1 in 11 people have been diagnosed with diabetes, while half of those remain undiagnosed (International Diabetes Federation, 2015). The number of people living with diabetes only continues to grow, with an estimate of 700 million people to be diagnosed by 2045 (International Diabetes Federation, 2015). Furthermore, there is an estimate of 374 million people with an impaired glucose tolerance currently "at-risk" for developing type 2 diabetes (International Diabetes Federation, 2015).

There remains a high economic cost of diabetes, which accounts for approximately 10% of all healthcare expenditure (International Diabetes Federation, 2015). People with diabetes will spend 2–5 times more on healthcare in their lifetime than people without diabetes (Tabish, 2007). One study projected the global cost of diabetes to be 1.31 trillion US dollars in 2015 alone (Bommer et al., 2017). In addition to the economic costs, there is a significant social cost to diabetes. Diabetes can lead to an extensive risk to a person's overall health. Heart disease, stroke, blindness, kidney damage and nerve damage are common secondary complications of diabetes. Microvascular and macrovascular disease are common, with cardiovascular complications as a leading cause of mortality among people with diabetes (Zheng, Ley, and Hu, 2018). Diabetes and peripheral arterial disease (PAD) are closely linked since diabetes is a known risk factor for developing arteriosclerosis and PAD (Murabito, D'Agostino, Silbershatz, and Wilson, 1997).

Another vascular complication of diabetes is the development of diabetic foot ulcers (DFU). It is estimated that approximately 50% of people with diabetes will develop a DFU within their lifetime (Hicks and Selvin, 2019). People with PAD who have diabetes are at a higher risk for developing an ischemic ulcer versus people with PAD without associated diabetes (Ikem, Ikem, Adebayo, and Soyoye, 2010). In people with diabetes with a foot ulcer, PAD is found in more than 50% of patients (Bandyk, 2018). The global prevalence of diabetic foot ulcers has been found to be 6.3%, with North America having the highest prevalence at 13.0% (Zhang et al., 2017).

There is a significantly higher mortality rate in people with diabetes due to the high rate of non-healing DFU leading to lower extremity amputations (Everett and Mathioudakis, 2018; Pecoraro, Reiber, and Burgess, 1990). In individuals who develop DFU, the mortality rate is 5% in the first 12 months and increases to 42% at five years (Everett and Mathioudakis, 2018). The rising incidence of diabetes, as well as its cutaneous complications, make finding viable alternative treatments of vital importance.

6.2 CATEGORIES OF DIABETES MELLITUS

Diabetes mellitus (DM) is classified as a metabolic disorder of persistent hyperglycemia caused by insufficient insulin secretion, ineffective insulin action or both. It was reported in the literature as early as 1788 in the *London Medical Journal* (Cawley, 1788). Understanding the etiology and pathophysiology of a disease is often helpful in categorizing its subtypes. Unfortunately, scientists and clinicians do not yet have a complete understanding of the etiology and pathophysiology of diabetes. However, the three most common forms are type 1 diabetes mellitus (T1DM), type 2 diabetes mellitus (T2DM) and gestational diabetes (GDM).

T1DM, previously known as juvenile dependent or insulin-dependent diabetes mellitus, accounts for 5–10% of people with diabetes. Traditionally diagnosed in children and teens, T1DM can occur in any age group. In T1DM, immunologic, genetic and environmental factors lead to a loss of beta cells within the pancreatic islets. The beta cells normally produce insulin, so when they are destroyed, there is a decrease in insulin production. Insulin normally functions to lower blood glucose, so the destruction of beta cells leads to hyperglycemia within the bloodstream. This prolonged state of hyperglycemia leads to the damage of other organs and vessels. Genetic factors, viral infections and toxins have all been found to be triggers for the loss of insulin production from pancreatic beta cells (Lucier and Weinstock, 2021).

T2DM, previously known as adult-onset diabetes or non-insulin-dependent diabetes mellitus (NIDDM), accounts for 90–95% of people with diabetes. In T2DM, insulin is produced by the pancreatic cells but becomes progressively ineffective, possibly from insulin resistance within the muscle and adipose tissue. Initially, the pancreatic cells still produce insulin to maintain blood glucose levels; however, over time with ongoing resistance, insulin production decreases leading to hyperglycemia. The decrease in glucagon secretion and other hormonal impairments are associated with this type of diabetes (Goyal and Jialal, 2021).

The third most common form, GDM, is a hyperglycemic state during pregnancy. The prevalence of GDM accounts for 1.7–11.7% of pregnancies in various countries (Schneider, Bock, Wetzel, Maul, and Loerbroks, 2012). Children of women with GDM are at an increased risk of developing diabetes, as well as obesity (Dabelea, 2007). Further complications can occur within the baby, while women with GDM are at an increased risk of developing diabetes later in life (Metzger, 2007). GDM is thought to occur due to insulin resistance primarily from the hormone human placental lactogen, as well as a delayed response of pancreatic beta cells to glycemic levels (Quintanilla Rodriguez and Mahdy, 2021).

6.3 DIABETIC FOOT ULCERS

Diabetic foot ulcers are a leading cutaneous secondary complication of diabetes. The development of a DFU is a complex process and often results from a combination of multiple factors including sensory neuropathy, motor neuropathy, and autonomic neuropathy compounded by impaired vascular status, infection and structural changes in the foot. Sensory neuropathy causes a loss of protective sensation (LOPS) in a distal to proximal pattern typically starting in the feet (Zakin, Abrams, and Simpson, 2019). If type A myelin fibers are affected, this may cause a loss of proprioception, pressure and vibration sensation. Neuropathic changes to type C sensory fibers, on the other hand, may cause a loss of ability to detect a painful stimulus (Bandyk, 2018). It is estimated that 66% of patients with diabetes will experience sensory neuropathy and LOPS in the lower extremities, placing these individuals at a much higher risk of developing a DFU (Zubair, Malik, and Ahmad, 2011).

Motor neuropathy involves damage to motor nerves leading to atrophy within the intrinsic muscles of the foot, loss of strength, changes in the range of motion and structural deformities of the foot and toes. More extrinsic muscle activation causes hammer-toe contractures, ankle equinus deformities and depression of the metatarsal heads (Bandyk, 2018). Significant damage can result in the complete collapse of the midfoot, called a Charcot's foot. These changes can affect a patient's gait pattern and place abnormal pressure on the plantar aspect of the foot and toes, leading to ulceration (Figure 6.1). A loss of thermoregulation within the autonomic system can lead to anhidrosis, or reduced sweating, and reduced sebaceous gland production. The skin's lubrication becomes diminished, leading to dry, cracked skin and fissures, increasing the risk of ulceration and infection (Schmidt and Holmes, 2018).

6.3.1 Complications of Peripheral Arterial Disease

The clinical complications of diabetes such as a DFU are compounded by peripheral arterial disease. While not completely understood, metabolic changes seen in diabetes such as oxidative stress, LDL (low-density lipoprotein) cholesterol and advanced glycation end products may worsen peripheral arterial disease (Yang et al., 2017). Both large and small vessels are affected by diabetes-induced peripheral arterial disease (Jude, Oyibo, Chalmers, and Boulton, 2001). Microvascular changes occur such as thickening of the capillary basement membrane and increased vascular permeability in a hyperglycemic state (Akbari and LoGerfo, 1999).

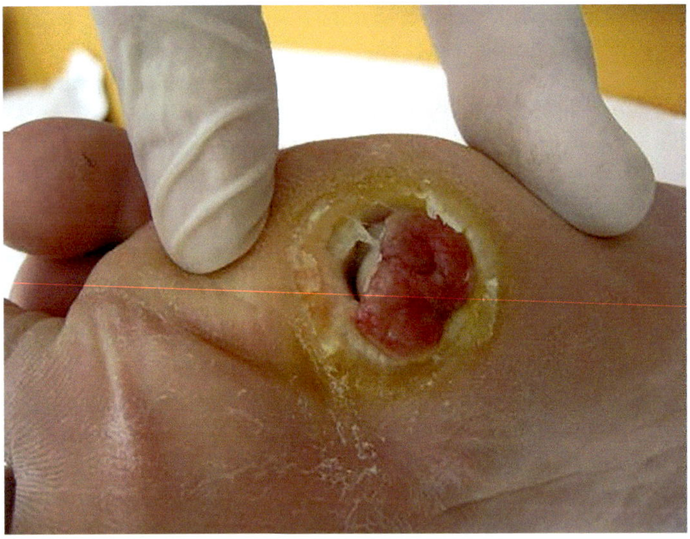

FIGURE 6.1 Diabetic foot ulcer (Intermedichbo at Serbian Wikipedia, 2012).

Arteriosclerosis and decreased blood flow contribute to ischemic conditions. Within similar levels of ischemia, one study found that in hyperglycemic conditions, there was significantly more tissue necrosis; thus, once a wound develops, this is one factor that can impair wound healing (Alizadeh et al., 2007; Levigne, Tobalem, Modarressi, and Pittet-Cuenod, 2013). Delayed myofibroblast development and delayed wound closure are seen in ischemic limb models (Alizadeh et al., 2007). In a hyperglycemic vs normoglycemic state, there was a five-fold delay in wound healing within an ischemic rat model as well as impaired myofibroblast development (Tobalem et al., 2015). In people with a DFU, the presence of peripheral arterial disease is a major risk factor for amputation and accounts for 70% of patient deaths in people with T2DM (Costa et al., 2017; Janka, Standl, and Mehnert, 1980).

6.3.2 Diabetes and Wound Healing

Normal wound healing progresses through four primary stages of healing. This includes hemostasis, inflammation, proliferation and tissue remodeling. Hemostasis involves vasoconstriction and the release of platelets for clot and fibrin plug formation. Diabetes can affect the formation of the fibrin plug, as well as delay the transition into the inflammatory phase (Scarborough, 2010).

In a normal wound healing process, inflammation begins quickly and lasts approximately 3–7 days, during which time neutrophils, macrophages and lymphocytes work toward phagocytosis, releasing growth factors and cytokines, and initiating pro-inflammatory conditions. As seen in people with poorly controlled diabetes, dysfunction arises in this phase, as well, initially with a decreased number of neutrophils limiting phagocytosis (Baltzis, Eleftheriadou, and Veves, 2014). In hyperglycemic conditions, studies have shown a decrease in leukocyte adherence, chemotaxis and the production of growth factors (Oguntibeju, 2019). Once inflammatory cells are present, an increase in pro-inflammatory cytokines can contribute to a prolonged inflammatory response and chronic inflammation (Patel, Srivastava, Singh, and Singh, 2019).

During the proliferation phase, fibroblasts are essential for the production of collagen to repair damaged tissue. In people with diabetes, there is a decrease in fibroblast proliferation and migration, thus delaying wound healing (Baltzis et al., 2014). Angiogenesis, or the influx of new vascular growth, is seen in the proliferative phase of normal wound healing. In people with diabetes, there is inhibition of angiogenesis in wound healing, possibly due to several factors including altered macrophage function, the production of anti-angiogenic factors, differential expressions of microRNAs and decreased platelet-derived growth factor with a decrease in total vascularity, as well as a decrease in capillary density (Okonkwo and DiPietro, 2017). With a decrease in angiogenesis, wounds are under more anaerobic conditions with less oxygen and nutrients to effectively heal (Blakytny and Jude, 2009).

6.3.2.1 Current Treatments

Current treatment strategies to manage DFUs include interventions such as offloading, pharmacological management of blood glucose levels, revascularization of the ischemic limb, a wide variety of pharmaceutical topical products and naturally occurring products. Topical products may be applied for various reasons, including wound debridement, antimicrobial coverage or to manage bioburden levels. Topical products are also used to support the phases of wound healing through exudate management and/or to promote angiogenesis with collagen, human and non-human acellular and cellular tissue-based products (Thomas, 2013).

Topical debriding agents may be used in the treatment of DFUs if removal of necrotic tissue is necessary to advance healing. Collagenase clostridium ointment is currently the only available prescription enzymatic debriding ointment available in the United States. It works to selectively target and cleave peptide bonds in collagen present in necrotic wounds (Shi and Carson, 2009). Wound debridement has been shown to decrease the overall bioburden within a wound, as well as harmful matrix metalloproteinases and cytokines to allow the proliferation of keratinocytes and healthy tissue (Ramundo and Gray, 2008). Studies have shown that compared to other debridement methods, collagenase clostridium ointment is cost-effective in treating DFUs (Waycaster et al., 2018).

There is a wide variety of antimicrobial dressings commercially available. Antimicrobial topical wound dressings may contain silver, iodine or a variety of synthetic compounds. Antimicrobial dressings are used when wounds are thought to be critically colonized or infected, inhibiting normal wound healing progression. However, additional randomized controlled trials are needed to determine the true effectiveness of antimicrobial dressings in patients with DFUs (Dumville et al., 2017).

Hundreds of commercially available wound products exist for topical use. Maintaining a moist wound bed environment to prevent tissue desiccation became a core concept of wound healing with research in the 1960s (Winter, 1962). The primary role of a topical wound dressing is to manage wound exudate by maintaining the balance of a moist wound bed. Dressings range from absorptive dressings such as hydrofibers, calcium alginates or foams to dressings that add moisture to a wound bed such as hydrogels. Other dressings go beyond maintaining a moist wound balance and enhance the proliferation of new tissue growth. Collagen dressings form a scaffold for new fibroblast activity, as well as inactivate matrix metalloproteinases (MMPs) that can contribute to a non-healing chronic wound (Chattopadhyay and Raines, 2014).

Biologics or cellular tissue products (CTPs) have experienced a massive growth within the wound market since 2000 when Apligraf was first approved for use for DFUs. Currently, over 70 CTPs are approved by the U.S. Food and Drug Administration (Abdo and Ortman, 2020). These dressings can be acellular or cellular, single or bilayered, and made of a variety of materials including naturally sourced from humans or animals, synthetic or a combination. Acellular products provide a scaffold on which cells can migrate into a wound bed, while cellular products contain live cells for which to enhance proliferation (Abdo and Ortman, 2020). Products may come from bovine, porcine or ovine sources; human amniotic tissue; human placental tissue; and/or synthetic substances. Common commercially available products are summarized in Table 6.1.

6.3.2.2 Natural Products in Wound Healing

Wound management has employed natural topical agents for centuries. Honey for medicinal purposes has been referenced on Sumerian clay tablets dating from 1250–1900 BC, in ancient Egyptian and Greek times, and by Hippocrates to clean wounds (Chepulis, 2008). Medicinal honey remains a widespread natural topical agent for wound management today. The bioactivity of honey to promote

TABLE 6.1
Current Commercially Available Products Utilized in Treating Diabetic Foot Ulcers

Treatment Types	Descriptions	Product Examples
Debriding agents	Enzymatic debriding ointment	Collagenase clostridium histolyticum/Santyl®
Collagen	Powdered or oxidized regenerated cellulose (ORC)/collagen sheet	Stimulen™, Puracol™ Plus, Promogran Prisma™ Matrix
Biologics/cellular tissue products	Cellular combination of human cultured neonatal skin and bovine dermal layer	Apligraf®
		Dermagraft®
	Cellular, a combination of human and synthetic material	Epifix®
	Acellular, derived from human placenta	Graftjacket™
	Acellular, autograft dermal matrix	

healing within cutaneous wounds is multifocal, including having antibacterial, antioxidant and anti-inflammatory effects. Honey thus acts as a debriding agent and stimulates tissue growth (Oryan, Alemzadeh, and Moshiri, 2016). Improved wound healing may be due to an increased number of keratinocytes with the application of topical honey (Majtan, Kumar, Majtan, Walls, and Klaudiny, 2010). Increased epithelialization within wound beds and a decrease in scar formation have been found with the use of topical honey (Al-Waili, Salom, and Al-Ghamdi, 2011). Manuka honey was shown to assist wound healing through the stimulation of monocytes that differentiate into macrophages and thus release TNF-α (Tonks, Cooper, Price, Molan, and Jones, 2001).

Various forms of honey have different potency levels. Manuka honey is widely used and commercially sold by numerous manufacturers for medicinal purposes. It possesses antibacterial properties and can withstand dilution from wound exudate. This prevents the honey from becoming inactive, as is the case with other types of honey (Molan and Rhodes, 2015). Microbial resistance is often a problem with topical wound dressings; however, no bacterial resistance has been found with honey-based dressings (Dixon, 2003). Manuka honey is antimicrobial against numerous bacteria, including *Staphylococcus aureus*, *Escherichia coli* and *Enterobacter aerogenes* (French, Cooper, and Molan, 2005; Lusby, Coombes, and Wilkinson, 2005). Honey's naturally acidic pH level assists in its antimicrobial activity. It may be due to the osmotic effect of the high sugar content of honey and drawing water out of bacterial cells within the lower water content wound bed (Mandal and Mandal, 2011).

Sugar, derived from sugarcane or *Saccharum officinarum* L., is another natural product used for cutaneous wound healing of DFUs and other types of wounds. Sugar dressings have been used for more than two centuries; however, they started being more widely used in contemporary medicine in 1976 by Herszage and Montenegro in Buenos Aires. It has been demonstrated to be non-toxic to wounds in people with diabetes (Chirife, Herszage, Joseph, and Kohn, 1983). Caster or granulated sugar is a suitable choice for topical application, as they are pure sucrose and do not contain additional additives. Sugar has been shown to have numerous beneficial effects within a cutaneous wound. Wound odor may decrease as sugar breaks down amino acids within the wound bed. Angiogenesis and epithelialization to promote healing may be stimulated by sugar-lowering the oxygen pressure in a wound bed. Like honey, sugar is thought to be antimicrobial due to its high osmolality (Wu, Figenshau, Rajpara, Aires, and Elston, 2020). Topical cutaneous sugar has been found to possess antibacterial activity against *E. coli*, *Pseudomonas aeruginosa*, *Klebsiella pneumoniae* and *S. aureus* (Chirife et al., 1983).

As a topical cutaneous dressing, olive oil derived from olive trees or *Olea europaea* L. has been shown to have antioxidant, antimicrobial and anti-inflammatory effects (El and Karakaya, 2009). There are limited *in vivo* studies using olive oil, but one study did find a decrease in ulcer area and increase incomplete wound closure with the topical application (Nasiri, Fayazi, Jahani, Yazdanpanah, and Haghighizadeh, 2015). It is thought the essential fatty acids linoleic and linolenic acids within the olive oil stimulate cell healing. The high amounts of polyphenols found in olive oil may work to decrease the inflammatory process, given their natural antioxidant properties (Cicerale, Lucas, and Keast, 2012; Rafehi, Ververis, and Karagiannis, 2012).

In addition to these more common topical dressings for DFUs, several other medicinal plants have shown promising effects within *in vivo* studies using diabetic wound models. They have demonstrated anti-inflammatory, antibacterial, antioxidant or tissue regeneration properties. These medicinal plants are summarized in Table 6.2 (Figure 6.2).

TABLE 6.2
Effect of Medicinal Plants in Healing Diabetic Wounds (sourced from Oguntibeju, 2019)

Botanical Names	Bioactive Compounds	Extract Types	Assessment Methods	Bioactivities in Wound Healing	Mechanisms of Action
Rosmarinus officinalis L.	Terpenoids and polyphenols such as carnosol, carnosic acid and rosmarinic acid	Topical essential oil from the aerial part of the plant	Full-thickness skin incision on diabetic mice (Abu-Al-Basal, 2010)	Antioxidant, antibacterial and anti-inflammatory	Increase in the amount of granulation tissue, increase in percentage of wound bed contraction
Carica papaya L.	Triterpenes, sterols, alkaloids, flavonoids and glycosides	Aqueous topical extract from seeds	Excision and dead space models diabetic rats (Nayak, Pinto Pereira, and Maharaj, 2007)	Antimicrobial, antioxidant and anti-inflammatory	Decrease in the epithelialization period, increase in the percentage of wound contraction, increase in a dry and wet weight of granulation tissue
Radix Rehmanniae		Herbal extract from the root of Rehmannia glutinosa	Patient-derived primary fibroblast cultures from foot ulcers of diabetic patients (Lau et al., 2009)	Tissue regeneration, angiogenesis and inflammation control	Decrease in ulcer area increased epithelialization
Annona squamosal L.	High phenolic content	Ethanolic leaf extract	Full-thickness excision in diabetic Wistar albino rats (Ponrasu and Suguna, 2014)	Collagen synthesis and wound contraction	Increase contraction, decrease wound size and increased tensile strength
Catharanthus roseus L.	Alkaloids, triterpenoids and tannins	Ethanolic leaf extract	Full-thickness wound in diabetic rats (Nayak, 2006)	Antimicrobial, tissue regeneration	An increase in granulation tissue increased collagen production
Acalypha langiana Müll.arg		Aqueous leaf extract	Excision and incision wounds in diabetic rats (Perez Gutierrez and Vargas, 2006)	Tissue regeneration	Increase tensile strength, increased granulation tissue
Hylocereus undatus Haw.		Aqueous extracts of leaves, rind, fruit pulp and flowers	Excision and incision wounds in diabetic rats (Perez Gutierrez and Vargas, 2006)	Collagen synthesis	Increase tensile strength, increased collagen content

(*Continued*)

TABLE 6.2 (Continued)

Botanical Names	Bioactive Compounds	Extract Types	Assessment Methods	Bioactivities in Wound Healing	Mechanisms of Action
Punica granatum L.		Extract of powdered flowers	Diabetic induced male rats (Pirbalouti, Azizi, Koohpayeh, and Hamedi, 2010)	Antibacterial, antifungal and antiviral	Increase in tissue regeneration
Aloe Vera L.		Extract of leaves	Excision and incision wounds in diabetic male rats (Chithra, Sajithlal, and Chandrakasan, 1998)	Anti-inflammatory and antibacterial	Increased fibroplasia, collagen formation and contraction

FIGURE 6.2 A, *Rosmarinus officinalis* L.; B, *Carica papaya* L.; C, *Annona squamosal* L.; D, *Catharanthus roseus* L.; E, *Acalypha langiana* Müll.arg.; F, *Hylocereus undatus* Haw.; G, *Punica granatum* L.; H, *Aloe vera* L. (Culbert, 2013; Fpalli, 2003; Rulkens, 2010; Silversmith, 2005; Vincentz, 2011; Zell, 2009a, 2009b, 2010).

6.4 CONCLUSION

The global health issue of diabetes-related cutaneous wounds is well documented. Complicating factors such as peripheral arterial disease can lead to further complications, wound healing issues and impaired quality of life. Multiple natural products have been used to treat diabetic cutaneous wounds. Numerous other compounds have been studied and show promising potential. Studies of these substances within diabetic models are limited, and most have not been studied *in vivo* on human dermal tissue. Further clinical studies are needed to isolate and identify active ingredients, the bioactivity of the ingredients and the physiological effects on the body.

REFERENCES

Abdo, J., and H. Ortman. 2020. "Biologic and synthetic cellular and/or tissue-based products and smart wound dressings/coverings." *Surgical Clinics of North America* 100, 4: 741–756. doi: 10.1016/j.suc.2020.05.006

Abu-Al-Basal, M. A. 2010. "Healing potential of Rosmarinus officinalis L. on full-thickness excision cutaneous wounds in alloxan-induced-diabetic BALB/c mice." *Journal of Ethnopharmacology* 131, 2: 443–450. doi: 10.1016/j.jep.2010.07.007.

Akbari, C. M., and F. W. LoGerfo. 1999. "Diabetes and peripheral vascular disease." *Journal of Vascular Surgery* 30, 2: 373–384. doi: 10.1016/S0741-5214(99)70154-0.

Alizadeh, N., M. S. Pepper, A. Modarressi, K. Alfo, K. Schlaudraff, D. Montandon, G. Gabbiani, M. L. Bochaton-Piallat, and B. Pittet. 2007. "Persistent ischemia impairs myofibroblast development in wound granulation tissue: A new model of delayed wound healing." *Wound Repair and Regeneration* 15. 6: 809–816. doi: 10.1111/j.1524-475X.2007.00312.x.

Al-Waili, N., K. Salom, and A. A. Al-Ghamdi. 2011. "Honey for wound healing, ulcers, and burns; data supporting its use in clinical practice." *Scientific World Journal* 11: 766–787. doi: 10.1100/tsw.2011.78.

Baltzis, D., I. Eleftheriadou, and A. Veves. 2014. "Pathogenesis and treatment of impaired wound healing in diabetes mellitus: New insights." *Advances in Therapy* 31, 8: 817–836. doi: 10.1007/s12325-014-0140-x.

Bandyk, Dennis F. 2018. "The diabetic foot: Pathophysiology, evaluation, and treatment." *Seminars in Vascular Surgery* 31, 2: 43–48. doi: 10.1053/j.semvascsurg.2019.02.001.

Blakytny, R., and E. B. Jude. 2009. "Altered molecular mechanisms of diabetic foot ulcers." *International Journal of Lower Extremity Wounds* 8, 2: 95–104. doi: 10.1177/1534734609337151.

Bommer, C., E. Heesemann, V. Sagalova, J. Manne-Goehler, R. Atun, T. Barnighausen, and S. Vollmer. 2017. "The global economic burden of diabetes in adults aged 20–79 years: A cost-of-illness study." *Lancet Diabetes and Endocrinology* 5, 6: 423–430. doi: 10.1016/S2213-8587(17)30097-9.

Brooks, L. 2013. "Child foot." Obtained from Wikimedia under the Public Domain License. https://commons.wikimedia.org/w/index.php?title=File:Child_foot.jpg&oldid=456896999 (Accessed 21 October 2021).

Cawley, T. 1788. "A singular case of diabetes, consisting entirely in the quality of the urine; with an inquiry into the different theories of that disease." *London Medical Journal* 9, 3: 286–308.

Chattopadhyay, S., and R. T. Raines. 2014. "Review collagen-based biomaterials for wound healing." *Biopolymers* 101, 8:821–833. doi: 10.1002/bip.22486.

Chepulis, L. M. 2008. *An Investigation of the Health Benefits of Honey as a Replacement For Sugar In the Diet.* The University of Waikato, Hamilton, New Zealand.

Chirife, J., L. Herszage, A. Joseph, and E. S. Kohn. 1983. "In vitro study of bacterial growth inhibition in concentrated sugar solutions: Microbiological basis for the use of sugar in treating infected wounds." *Antimicrobial Agents and Chemotherapy* 23, 5: 766–773. doi: 10.1128/aac.23.5.766.

Chithra, P., G. B. Sajithlal, and G. Chandrakasan. 1998. "Influence of aloe vera on the healing of dermal wounds in diabetic rats." *Journal of Ethnopharmacology* 59, 3: 195–201. doi: 10.1016/s0378-8741(97)00124-4.

Cicerale, S., L. J. Lucas, and R. S. Keast. 2012. "Antimicrobial, antioxidant and anti-inflammatory phenolic activities in extra virgin olive oil." *Current Opinion in Biotechnology* 23, 2: 129–135. doi: 10.1016/j.copbio.2011.09.006.

Costa, Rafael Henrique Rodrigues, Natália Anício Cardoso, Ricardo Jayme Procópio, Túlio Pinho Navarro, Alan Dardik, and Ligia de Loiola Cisneros. 2017. "Diabetic foot ulcer carries high amputation and mortality rates, particularly in the presence of advanced age, peripheral artery disease and anemia." *Diabetes & Metabolic Syndrome: Clinical Research & Reviews* 11: S583-S587. doi: 10.1016/j.dsx.2017.04.008.

Culbert, D. 2013. "*Acalypha langiana*." Obtained from Wikimedia under the Creative Commons Attribution-Share Alike 2.0 Generic license. https://commons.wikimedia.org/w/index.php?title=File:Acalypha_langiana,_a_Hierba_de_Cancer_(11240005233).jpg&oldid=497395319 (Accessed 23 November 2020).

Dabelea, D. 2007. "The predisposition to obesity and diabetes in offspring of diabetic mothers." *Diabetes Care* 30, Suppl 2: S169–S174. doi: 10.2337/dc07-s211.

Dixon, B. 2003. "Bacteria can't resist honey." *Lancet Infectious Diseases* 3, 2: 116. doi: 10.1016/s1473-3099(03)00524-3.

Dormiraubois. 2016. "Buddy-taped toes." Obtained from Wikimedia under the Creative Commons Attribution-Share Alike 4.0 International License. https://commons.wikimedia.org/w/index.php?title=File:Buddy-taped_toes.jpg&oldid=476234213 (Accessed 21 October 2021).

Douglas, P. P. 2011. "Bowl of white sugar." Obtained from Wikimedia under the Creative Commons Attribution 3.0 Unported License. https://commons.wikimedia.org/w/index.php?title=File:Bowl_of_white_sugar.jpg&oldid=486041996 (Accessed 21 October 2021).

Dumville, J. C., B. A. Lipsky, C. Hoey, M. Cruciani, M. Fiscon, and J. Xia. 2017. "Topical antimicrobial agents for treating foot ulcers in people with diabetes." *Cochrane Database of Systematic Reviews* 6: CD011038. doi: 10.1002/14651858.CD011038.pub2.

El, S. N., and S. Karakaya. 2009. "Olive tree (*Olea europaea*) leaves: Potential beneficial effects on human health." *Nutrition Review* 67, 11: 632–638. doi: 10.1111/j.1753-4887.2009.00248.x

Everett, E., and N. Mathioudakis. 2018. "Update on management of diabetic foot ulcers." *Annuals of the New York Academy of Sciences* 1411, 1: 153–165. doi: 10.1111/nyas.13569.

Fpalli. 2003. "*Annona squamosal* (Martinique)." Obtained from Wikimedia under the Creative Commons Attribution-Share Alike 3.0 Unported license. https://commons.wikimedia.org/w/index.php?title=File:Annona_squamosa_(Martinique).jpg&oldid=464401111 (Accessed 23 November 2020).

French, V. M., R. A. Cooper, and P. C. Molan. 2005. "The antibacterial activity of honey against coagulase-negative staphylococci." *Journal of Antimicrobial Chemotherapy* 56, 1: 228–231. doi: 10.1093/jac/dki193.

Goyal, R., and I. Jialal. 2021. "Diabetes Mellitus Type 2." *StatPearls Publishing*, Treasure Island. PMID: 30020625.

Hicks, C. W., and E. Selvin. 2019. "Epidemiology of Peripheral Neuropathy and Lower Extremity Disease in Diabetes." *Current Diabetes Reports* 19, 10: 86. doi: 10.1007/s11892-019-1212-8.

Ikem, R., I. Ikem, O. Adebayo, and D. Soyoye. 2010. "An assessment of peripheral vascular disease in patients with diabetic foot ulcer." *The Foot* 20, 4: 114–117. doi: 10.1016/j.foot.2010.09.002.

Intermedichbo at Serbian Wikipedia. 2012. "Diabetic Wound." Obtained from Wikimedia under the Creative Commons Attribution-Share Alike 3.0 Unported license. https://commons.wikimedia.org/w/index.php?title=File:Diabetic_Wound_121.jpg&oldid=503242066 (Accessed 23 November 2020).

International Diabetes Federation. 2015. "Diabetes Atlas." 7th Edition, International Diabetes Federation, Brussels, Belgium. http://www.diabetesatlas.org (Accessed 15 May 2020).

Janka, H. U., E. Standl, and H. Mehnert. 1980. "Peripheral vascular disease in diabetes mellitus and its relation to cardiovascular risk factors: Screening with the doppler ultrasonic technique." *Diabetes Care* 3, 2: 207–213. doi: 10.2337/diacare.3.2.207.

Jude, E. B., S. O. Oyibo, N. Chalmers, and A. Boulton. 2001." Peripheral arterial disease in diabetic and non-diabetic patients: A comparison of severity and outcome." *Diabetes Care* 24, 8: 1433–1437. doi: 10.2337/diacare.24.8.1433.

Lau, T. W., Y. W. Chan, C. P. Lau, K. M. Lau, C. B. Lau, K. P. Fung, P. C. Leung, and Y. Y. Ho. 2009. "Radix Astragali and Radix Rehmanniae, the principal components of two antidiabetic foot ulcer herbal formulae, elicit viability-promoting effects on primary fibroblasts cultured from diabetic foot ulcer tissues." *Phytotherapy Research* 23, 6: 809–815. doi: 10.1002/ptr.2649.

Levigne, D., M. Tobalem, A. Modarressi, and B. Pittet-Cuenod. 2013. "Hyperglycemia increases susceptibility to ischemic necrosis." *Biomed Research International* 490964. doi: 10.1155/2013/490964.

Lucier, J., and R. S. Weinstock. 2021. "Diabetes Mellitus Type 1." *StatPearls Publishing*, Treasure Island. PMID: 29939535.

Lusby, P. E., A. L. Coombes, and J. M. Wilkinson. 2005. "Bactericidal activity of different honeys against pathogenic bacteria." *Archieves of Medical Research* 36, 5: 464–467. doi: 10.1016/j.arcmed.2005.03.038.

Majtan, J., P. Kumar, T. Majtan, A. F. Walls, and J. Klaudiny. 2010. "Effect of honey and its major royal jelly protein 1 on cytokine and MMP-9 mRNA transcripts in human keratinocytes." *Experimental Dermatology* 19, 8: e73–79. doi: 10.1111/j.1600-0625.2009.00994.x.

Mandal, M. D., and S. Mandal. 2011. "Honey: Its medicinal property and antibacterial activity." *Asian Pacific Journal of Tropical Biomedicine* 1, 2: 154–160. doi: 10.1016/S2221-1691(11)60016-6.

Margenauer. 2005. "Bottle of olive oil." Obtained from Wikimedia under the Creative Commons CC0 1.0 Universal Public Domain Dedication. https://commons.wikimedia.org/w/index.php?title=File:Bottle_of_olive_oil.jpg&oldid=487783329 (Accessed 21 October 2021).

Metzger, B. E. 2007. "Long-term outcomes in mothers diagnosed with gestational diabetes mellitus and their offspring." *Clinical Obstetrics and Gynecology* 50, 4: 972–979. doi: 10.1097/GRF.0b013e31815a61d6.

Molan, P., and T. Rhodes. 2015. "Honey: A biologic wound dressing." *Wounds* 27, 6: 141–151.

Murabito, J. M., R. B. D'Agostino, H. Silbershatz, and W. F. Wilson. 1997. "Intermittent claudication. A risk profile from The Framingham Heart Study." *Circulation* 96, 1: 44–49. doi: 10.1161/01.cir.96.1.44.

Nasiri, M., S. Fayazi, S. Jahani, L. Yazdanpanah, and M. H. Haghighizadeh. 2015. "The effect of topical olive oil on the healing of foot ulcer in patients with type 2 diabetes: A double-blind randomized clinical trial study in Iran." *Journal of Diabetes and Metabolic Disorders* 14: 38. doi: 10.1186/s40200-015-0167-9.

Nayak, S. B. 2006. "Influence of ethanol extract of *Vinca rosea* on wound healing in diabetic rats." *Journal of Biological Sciences* 6, 51–55.

Nayak, S. B., L. Pinto Pereira, and D. Maharaj. 2007. "Wound healing activity of Carica papaya L. in experimentally induced diabetic rats." *Indian Journal of Experimental Biology* 45, 8: 739–743.

Oguntibeju, O. O. 2019. "Medicinal plants and their effects on diabetic wound healing." *Veterinary World* 12, 5: 653–663. doi: 10.14202/vetworld.2019.653-663.

Okonkwo, U. A., and L. A. DiPietro. 2017. "Diabetes and Wound Angiogenesis." *International Journal of Molecular Sciences* 18, 7. doi: 10.3390/ijms18071419.

Oryan, A., E. Alemzadeh, and A. Moshiri. 2016. "Biological properties and therapeutic activities of honey in wound healing: A narrative review and meta-analysis." *Journal of Tissue Viability* 25, 2: 98–118. doi: 10.1016/j.jtv.2015.12.002.

Patel, S., S. Srivastava, M. R. Singh, and D. Singh. 2019. "Mechanistic insight into diabetic wounds: Pathogenesis, molecular targets and treatment strategies to pace wound healing." *Biomedicine and Pharmacotherapy* 112: 108615. doi: 10.1016/j.biopha.2019.108615.

Pecoraro, R. E., G. E. Reiber, and E. M. Burgess. 1990. "Pathways to diabetic limb amputation. Basis for prevention." *Diabetes Care* 13, 5: 513–521. doi: 10.2337/diacare.13.5.513.

Perez Gutierrez, R. M., and S. R. Vargas. 2006. "Evaluation of the wound healing properties of *Acalypha langiana* in diabetic rats." *Fitoterapia* 77, 4: 286–289. doi: 10.1016/j.fitote.2006.03.011.

Pirbalouti, A. G., S. Azizi, A. Koohpayeh, and B. Hamedi. 2010. "Wound healing activity of *Malva sylvestris* and *Punica granatum* in alloxan-induced diabetic rats." *Acta Poloniae Pharmaceutica* 67, 5: 511–516.

Ponrasu, T., and L. Suguna. 2014. "Efficacy of *Annona squamosa* L in the synthesis of glycosaminoglycans and collagen during wound repair in streptozotocin induced diabetic rats." *Biomed Research International* 124352. doi: 10.1155/2014/124352.

Quintanilla Rodriguez, B. S., and H. Mahdy. 2021. "Gestational diabetes." *StatPearls Publishing*, Treasure Island. PMID: 31424780.

Rafehi, H., K. Ververis, and T. C. Karagiannis. 2012. "Mechanisms of action of phenolic compounds in olive." *Journal of Dietary Supplements* 9, 2: 96–109. doi: 10.3109/19390211.2012.682644.

Ramundo, J., and M. Gray. 2008. "Enzymatic wound debridement." *Journal of Wound Ostomy and Continence Nursing* 35, 3: 273–280. doi: 10.1097/01.WON.0000319125.21854.78.

Rulkens, T. 2010. "*Catharanthus roseus*." Obtained from Wikimedia under the Creative Commons Attribution-Share Alike 2.0 Generic license. https://commons.wikimedia.org/w/index.php?title=File:Annona_squamosa_(Martinique).jpg&oldid=464401111 (Accessed 23 November 2020).

Scarborough, P. 2010. "Diabetes and wound care." McCulloch J.M., & Kloth L.C. (Eds.), *Wound Healing Evidence-Based Management*, 4e. McGraw Hill, Philadelphia.

Schmidt, B. M., and C. M. Holmes. 2018. "Updates on diabetic foot and Charcot osteopathic arthropathy." *Current Diabetes Reports* 18, 10: 74. doi: 10.1007/S11892-018-1047-8.

Schneider, S., C. Bock, M. Wetzel, H. Maul, and A. Loerbroks. 2012. "The prevalence of gestational diabetes in advanced economies." *Journal of Perinatal Medicine* 40, 5: 511–520. doi: 10.1515/jpm-2012-0015.

Shebs, S. 2006. "*Aloe ferox*." Obtained from Wikimedia under the Creative Commons Attribution-Share Alike 3.0 Unported License. https://commons.wikimedia.org/w/index.php?title=File:Aloe_ferox_1.jpg&oldid=485544157 (Accessed 21 October 2021).

Shi, L., and D. Carson. 2009. "Collagenase Santyl ointment: A selective agent for wound debridement." *Journal of Wound Ostomy and Continence Nursing* 36, 6 Suppl: S12–16. doi: 10.1097/WON.0b013e3181bfdd1a.

Silversmith, E. 2005. "*Aloe vera*." Obtained from Wikimedia under a GNU Free Documentation License. https://commons.wikimedia.org/w/index.php?title=File:Aloe_Vera.jpg&oldid=475898291 (Accessed 23 November 2020).

Tabish, S. A. 2007. "Is diabetes becoming the biggest epidemic of the twenty-first century?" *International Journal of Health Sciences (Qassim)* 1, 2: V-VIII.

Thomas, D. R. 2013. "Clinical management of diabetic ulcers." *Clinics in Geriatric Medicine* 29, 2: 433–441. doi: 10.1016/j.cger.2013.01.007

Tobalem, M., D. Levigne, A. Modarressi, F. Atashi, F. Villard, B. Hinz, and B. Pittet-Cuenod. 2015. "Hyperglycemia interacts with ischemia in a synergistic way on wound repair and myofibroblast differentiation." *Plastics and Reconstructive Surgery Global Open* 3, 7: e471. doi: 10.1097/GOX.0000000000000443.

Tonks, A., R. A. Cooper, A. J. Price, P. C. Molan, and K. P. Jones. 2001. "Stimulation of TNF-alpha release in monocytes by honey." *Cytokine* 14, 4: 240–242. doi: 10.1006/cyto.2001.0868.

Vincentz, F. 2011. "*Hylocereus undatus*." Obtained from Wikimedia under the Creative Commons Attribution-Share Alike 3.0 Unported license. https://commons.wikimedia.org/wiki/File:Teguise_Guatiza_-_Jardin_-_Hylocereus_undatus_02_ies.jpg (Accessed 23 November 2020).

Waycaster, C., M. J. Carter, A. M. Gilligan, E. S. Mearns, C. E. Fife, and C. T. Milne. 2018. "Comparative cost and clinical effectiveness of clostridial collagenase ointment for chronic dermal ulcers." *Journal of Comparative Effectiveness Research* 7, 2: 149–165. doi: 10.2217/cer-2017-0066.

Winter, G. D. 1962. "Formation of the scab and the rate of epithelization of superficial wounds in the skin of the young domestic pig." *Nature* 193: 293–294. doi: 10.1038/193293a0.

World Health Organization. 2020. "Diabetes." https://www.who.int/news-room/facts-in-pictures/detail/diabetes (Accessed 15 May 2020).

Wu, D. J., K. Figenshau, A. Rajpara, D. J. Aires, and D. Elston. 2020. "Sugar-based dressings for chronic wound care." *Journal of the American Academy of Dermatology* 82, 3: e73. doi: 10.1016/j.jaad.2019.09.049.

Yang, S. L., L. Y. Zhu, R. Han, L. L. Sun, J, X, Li, and J. T. Dou. 2017. "Pathophysiology of peripheral arterial disease in diabetes mellitus." *Journal of Diabetes* 9, 2: 133–140. doi: 10.1111/1753-0407.12474.

Zakin, E., R. Abrams, and D. M. Simpson. 2019. "Diabetic Neuropathy." *Seminars in Neurology* 39, 5: 560–569. doi: 10.1055/s-0039-1688978.

Zaynabazam7. 2018. "Honey." Obtained from Wikimedia under the Creative Commons Attribution-Share Alike 4.0 International License. https://commons.wikimedia.org/w/index.php?title=File:Honey-PNG-Image-83961.png&oldid=587982757 (Accessed 21 October 2021).

Zell, H. 2009a. "*Carica papaya*." Obtained from Wikimedia under the Creative Commons Attribution-Share Alike 3.0 Unported license. https://commons.wikimedia.org/w/index.php?title=File:Carica_papaya_005.JPG&oldid=505156023 (Accessed 23 November 2020).

Zell, H. 2009b. "*Punica granatum*." Obtained from Wikimedia under the Creative Commons Attribution-Share Alike 3.0 Unported license. https://commons.wikimedia.org/w/index.php?title=File:Punica_granatum_004.JPG&oldid=442209329 (Accessed 23 November 2020).

Zell, H. 2010. "*Rosmarinus officinalis*." Obtained from Wikimedia under the Creative Commons Attribution-Share Alike 3.0 Unported license. https://commons.wikimedia.org/w/index.php?title=File:Rosmarinus_officinalis_0001.JPG&oldid=506537845 (Accessed 23 November 2020).

Zhang, P., J. Lu, Y. Jing, S. Tang, D. Zhu, and Y. Bi. 2017. "Global epidemiology of diabetic foot ulceration: A systematic review and meta-analysis (dagger)." *Annuals of Medicine* 49, 2: 106–116. doi: 10.1080/07853890.2016.1231932.

Zheng, Y., S. H. Ley, and F. B. Hu. 2018. "Global aetiology and epidemiology of type 2 diabetes mellitus and its complications." *Nature Reviews Endocrinology* 14, 2: 88–98. doi: 10.1038/nrendo.2017.151.

Zubair, M., A. Malik, and J. Ahmad. 2011. "Clinico-microbiological study and antimicrobial drug resistance profile of diabetic foot infections in North India." *Foot (Edinburgh, Scotland)* 21, 1: 6–14. doi: 10.1016/j.foot.2010.10.003.

7 Global Medicinal Plants and Phyto-Derived Compounds in Wound Healing

Samantha R. Loggenberg and Namrita Lall

CONTENTS

7.1 INTRODUCTION

The skin acts as the first line of defense in the body by protecting the internal tissues and organs against pathogens and microbes from the external environment. Lesions and tears in the epidermal layer of the skin may expose the internal tissues, rendering them susceptible to pathogenic infection. Severe or non-healing wounds in the skin may lead to prolonged wound infection, which may cause adverse health conditions and may result in death (Aldridge, 2015).

Despite modern advances in medicine, chronic wounds are difficult to treat, due to the variation in severity and physiology between individual patients. Severe wounds adversely affect millions of individuals by causing impaired mobility or mental trauma, thus reducing the quality of life (Thangapazham, Sharad, and Maheshwari, 2016). Thus, appropriate wound care and treatment that facilitates healthy wound healing are important in order to maintain normal bodily functions and improve patient health (Aldridge, 2015).

Traditional wound care and treatment practices have been utilized for millennia globally and across various cultures, affording a vast range of potential wound healing treatments (Thangapazham, Sharad, and Maheshwari, 2016). However, these practices are limited by what resources are immediately available or what can be acquired locally, such as plant and animal products. In some underdeveloped parts of the world, the use of traditional healthcare regimes and the use of naturally derived products may be the only source readily available and affordable for wound care (Pereira and Bártolo, 2016). This further indicates the importance and opportunity for

DOI: 10.1201/9781003108375-7

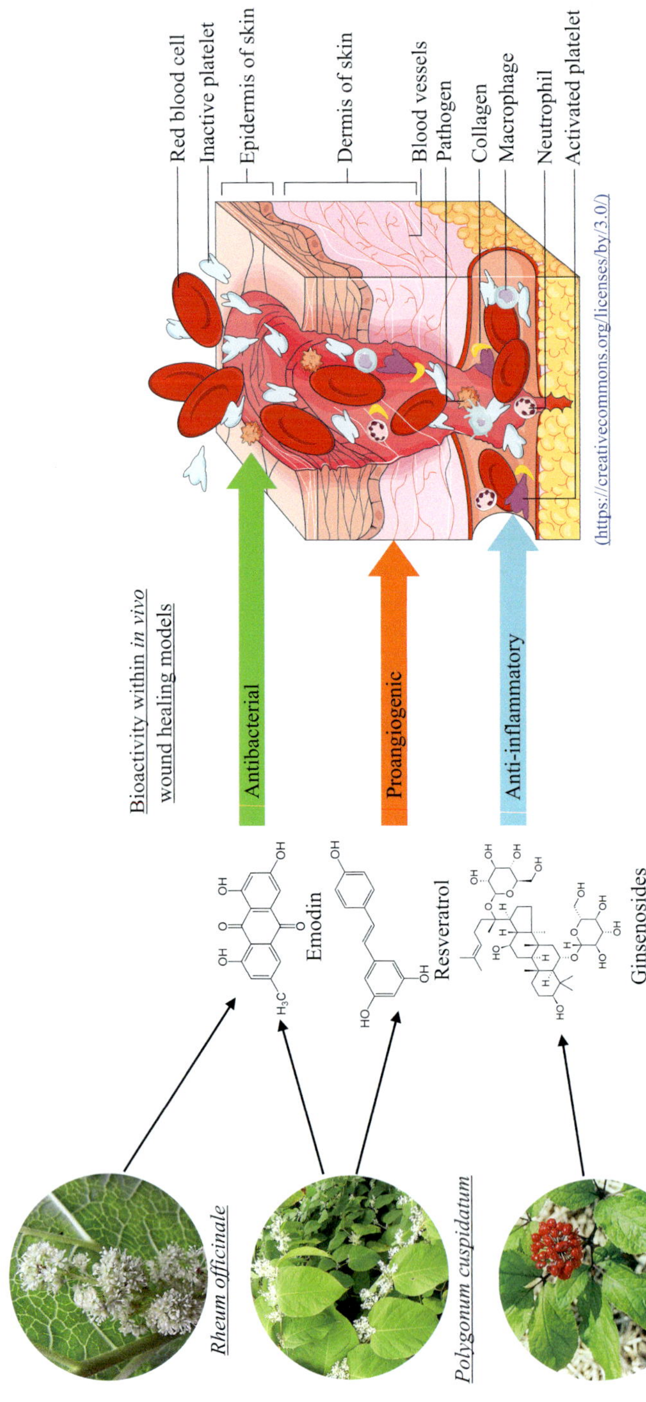

Graphical abstract representing the content of the chapter.

identifying potential medicinal resources for wound healing based on traditional knowledge and use (Shedoeva et al., 2019).

Medicinal plants and naturally derived products are abundantly utilized in traditional wound care practices that have been reported to facilitate and even accelerate healthy wound healing. The effectiveness of some of these medicinal plants—and compounds phyto-derived from them—in wound healing have been well established, and some have even been incorporated into modern commercial wound care products (Shedoeva et al., 2019; Pereira and Bártolo, 2016). However, to fully understand the effective use of these medicinal plants and phyto-derived compounds in wound healing products, it is important to understand the fundamentals of wound healing to identify molecular target processes for potentially effective wound care treatments (Aldridge, 2015).

7.2 WOUND HEALING

7.2.1 The Stages of Wound Healing

Wound healing is a dynamic process that involves numerous cellular factors with overlapping phases and molecular cascades that all interact in an orderly sequence to repair damaged tissue. The wound healing process may be categorized into four distinct phases: hemostasis, inflammation, cell proliferation and tissue remodeling (Figure 7.1) (Singer and Clark, 1999).

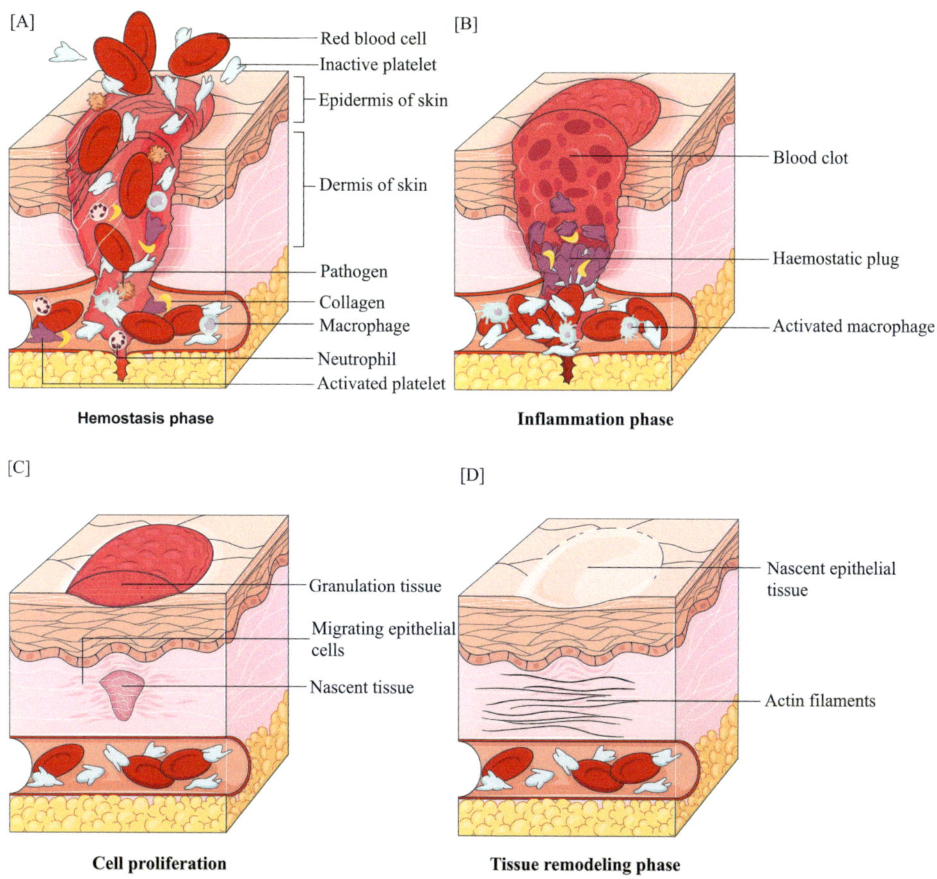

FIGURE 7.1 The stages of wound healing: A, Hemostasis phase; B, Inflammation phase; C, Cell proliferation phase; D, Tissue remodeling phase. Derived from (Servier, 2019).

The hemostasis phase takes place immediately after injury to the dermal layer has occurred (Figure 7.1A). Blood and leaked cellular contents mix, which results in the coagulation of blood and other cellular factors to prevent further bleeding from the wound. Inactive blood platelets come into contact with exposed collagen that has been released upon damage to the dermal cells (Aldridge, 2015). This interaction activates the blood platelets, causing them to secrete platelet-derived growth factor (PDGF), which coagulates with actin filaments and red blood cells to form a hemostatic plug in the wound site. This acts as a temporary barrier and prevents excess fluid loss (Singer and Clark, 1999).

This aggregation also acts as an extra-cellular matrix (ECM) through which blood constituents and other cellular components may continue to travel, thus allowing for the progression to successive phases of wound healing. Simultaneously, damaged endothelial cells secrete fibroblast growth factor (FGF), which stimulates the proliferation and migration of fibroblast cells; these are the connective tissue cells that facilitate the production of collagen and the ECM (Barrientos et al., 2008). Any invading pathogens that have entered the wound site are detected by proximate white blood cells (WBCs), known as leukocytes, and neutrophils, or natural killer cells. These WBCs destroy pathogens and simultaneously clear away cell debris. The detection of pathogenic components, such as lipopolysaccharides (LPS) within bacterial cell walls, activate the WBCs to secrete fibrogenic cytokines, such as transforming growth factor ß1 (TGF-ß1) and insulin-like growth factor 1 (IGF-1), and pro-inflammatory cytokines, such as tumor necrosis factor-α (TNF-α) and interleukins (IL) (such as IL-1 and IL-12), which initiate the inflammatory response (Lynch, Colvin, and Antoniades, 1989).

The detection of LPS by macrophages results in the activation of nitric oxide synthase (NOS), which upregulates the production of nitric oxide (NO), an inflammatory cytokine. Other secreted cytokines, such as IGF-1, initiate various cellular cascades that initiate the proliferation and migration of fibroblasts and epithelial cells, as observed in the phosphatidylinositol-3-kinase-AKT-mammalian target of rapamycin (PI3K-Akt-mTOR) signaling pathway (Barrientos et al., 2008).

Activated leukocytes further initiate the inflammatory response by inducing secretion of nuclear factor kappa-light-chain-enhancer of activated B cells (NF-κB), a transcription factor that upregulates the production of inflammatory cytokines. Inflammation is the second stage of wound healing, which results in swelling around the wound site as various cellular components migrate toward the damaged tissue (Figure 7.1B). Activated leukocytes also facilitate the production of anti-inflammatory cytokines, such as IL-10, to regulate the inflammation process. The inflammatory response aims to recruit macrophages and other immune cells to the wound site which secrete various cytokines to initiate the formation of granulation tissue to facilitate tissue repair, thus leading to the third stage of wound healing: cell proliferation and migration (Figure 7.1C) (Singer and Clark, 1999).

The formation of granulation tissue is a vital step in wound healing, as it acts as an ECM in which cell proliferation and migration may occur. The granulation tissue is filled with collagen and actin filaments, which actively contract, along with the proliferation and migration of epithelial cells to facilitate wound closure (Lynch, Colvin, and Antoniades, 1989). Once the wound is completely closed, the newly formed tissues may undergo maturation and increase in tensile strength. This is known as the final stage of wound healing, the tissue remodeling phase, in which collagen and other connective tissues re-organize to form stronger associations between epithelial cells (Figure 7.1D). The maturation of other dermal structures such as hair follicles, transmembrane proteins and associated cytoplasmic proteins also occurs during the remodeling phase of wound healing. The remaining aggregated-extracellular tissue is degraded until only the matured epithelial tissue is left. The recovery process after a dermal injury occurs is highly complex, and the duration may vary based on the physiology and pre-existing health conditions of the individual (Singer and Clark, 1999).

7.2.2 Angiogenesis in Wound Healing

Angiogenesis is a process that involves the regeneration and revascularization of blood vessels within newly formed tissue following tissue repair (Morgan and Nigam, 2013). The reformation of blood vessels plays a pivotal role in the wound recovery process, as well as the maturation of nascent tissue for normal cellular functions to occur. Angiogenesis is an intricate process that overlaps with the degradation of granulation tissue and cell proliferative stages of wound healing. Angiogenesis is regulated by various factors secreted by numerous cells involved in wound healing, such as PDGF-AA, PDGF-BB and vascular endothelial growth factor (VEGF), which are secreted by activated leukocytes (Lynch, Colvin, and Antoniades, 1989). These growth factors act as potent endothelial and fibroblast cell mitogens which ultimately facilitate the maturation of blood capillaries and surrounding connective tissue to re-establish the blood vessels and restore the blood supply to cells in nascent tissues (Aldridge, 2015).

Other cellular pathways, such as the PI3K-Akt-mTOR signaling pathway, ultimately play a role in angiogenesis by the upregulation of VEGF, which subsequently induces the secretion of proangiogenic hormones such as Angiotensin-II (Krikun et al., 2000). The process of angiogenesis is simultaneously downregulated by numerous cellular factors, such as pigment epithelium-derived factor (PEDF), to regulate capillary density and permeability in the maturing nascent tissues during the remodeling phase of wound healing. The regulation of angiogenesis plays an important role in facilitating healthy wound healing, and thus serves as a potential target process for medicinal wound healing treatments. Analyzing medicinal resources that may regulate the process of angiogenesis may allow for the identification of potential medicinal products that accelerate the wound recovery process (Morgan and Nigam, 2013).

7.3 COMPLICATIONS IN WOUND HEALING

The complex and dynamic process of dermal wound healing is a sensitive system that is prone to constant changes based on external factors, or stimulants, and patient physiology (Aldridge, 2015). Systemic factors that may result in impaired wound healing include nutritional health, diabetes, hormone treatments, irradiation, hypoxia, jaundice, renal failure and other pre-existing health conditions. Another continuous issue that causes a decline in wound healing efficacy is the deterioration of cellular function with old age (Halloran and Slavin, 2002). Dysregulation within the various wound healing cascades may delay the wound healing recovery period and result in various dermal-related pathologies, such as non-healing wounds, or pathogenic infection (Figure 7.2) (Lanas and Chan, 2017; Aldridge, 2015).

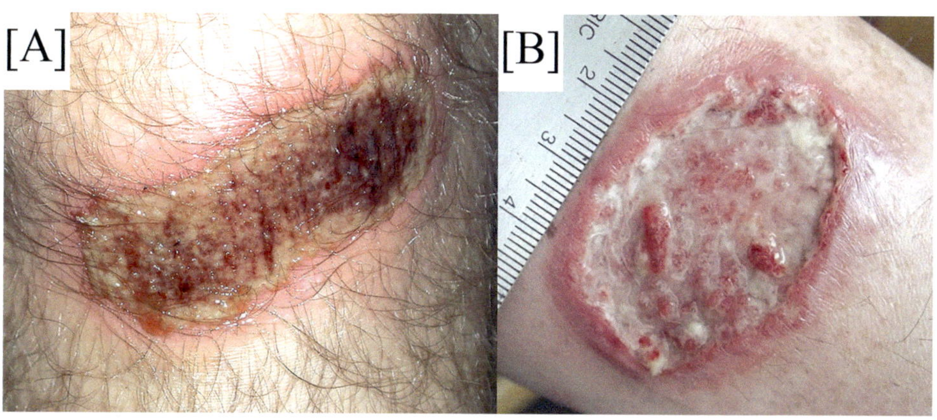

FIGURE 7.2 Complications in wound healing. A, Infected dermal wound; B, Dermal ulcer (Everson, 2004; Harris, 2008).

7.3.1 Wound Infection

The duration of the wound healing period may vary based on the severity of tissue damage leaving the exposed tissue susceptible to pathogenic infection as the wound provides a moist environment that may confer conditions for optimal bacterial growth (Aldridge, 2015). The main causes of pathogenic wound infection are due to bacterial colonization, originating either from the normal flora of the skin, normal microflora of the gut or pathogens from the external environment (Bessa et al., 2013). The most common causative pathogen of wound infection is the Gram-positive bacteria *Staphylococcus aureus* and other *Staphylococcus* spp. (Cohen and Kurzrock, 2004). Physiological imbalances in the wound healing process, such as a change in pH, may result in the facilitation of bacterial growth which may ultimately lead to pathogenic infection of the wound if left untreated (Halloran and Slavin, 2002).

Chronic wound infection may spread to deeper tissues of the body, ultimately resulting in adverse health conditions, such as septicemia or osteomyelitis (pathogenic infection of the blood or bone, respectively). Despite advances in modern medicine, more than 5.3 million people die annually from septicemia (de Janerio, 2019). Pathogenic infection of the blood and other vital organs may lead to an abrupt inflammatory response that spreads throughout the body, known as sepsis. This may result in excess tissue swelling which may decrease normal bodily functions or blood flow to body tissues, ultimately decreasing patient health by causing organ failure and tissue necrosis (Aldridge, 2015).

Wound infection is characterized by excessive and prolonged swelling, redness and pain around the wound site (Figure 7.2A). The pathogenic infection of wounds is treated by the administration of antibiotics and consistent sterilization of the wound (Halloran and Slavin, 2002). However, chronic infections are difficult to treat due to the increasing antibiotic resistance of certain bacterial strains. The administration of antibiotics may also not be applicable in certain medication regimes for patients with pre-existing health conditions (Aldridge, 2015).

7.3.2 Periwound Moisture-Associated Dermatitis

Periwound moisture–associated dermatitis (MASD) is the severe ulceration and inflammation that occurs in dermal wounds after excessive exposure to moisture, such as urine, stool, perspiration, wound drainage, saliva, or mucus (Figure 7.2B). Overexposure to moisture may compromise the integrity of the epidermal layer by disrupting the intricate intercellular lipid arrangement in the stratum corneum, or intercellular connections between epithelial tissues (such as corneocytes), increased inflammation and skin erosion (Lanas and Chan, 2017). The skin plays a crucial role in maintaining homeostasis within the body, including regulation of net trans-epidermal water loss, and thus a decrease in dermal layer integrity renders the skin susceptible to damage, ultimately hindering normal bodily functions (Gray et al., 2011).

However, MASD requires more than moisture damage in order to occur. Severe skin ulceration may be attributable to multiple factors, such as chemical irritants within the moisture source, changes in pH, mechanical factors such as friction and pathogenic wound-associated bacteria (Gray and Weir, 2007; Gray et al., 2011). Ulceration may result from chronic wound hypoxia and impaired micronutrient delivery caused by MASD, thus decreasing the process of angiogenesis which ultimately delays the wound healing process. Severe ulceration in wounds may be prevented with the use of wound dressings to limit moisture exposure. However, chronic skin ulcers may be susceptible to excessive inflammation and prolonged wound healing, which may ultimately hinder patient health and mobility (Gray and Weir, 2007).

7.4 CURRENT DRUGS AND TREATMENTS

There is a range of products commercially available that aid wound healing by providing antibacterial or anti-inflammatory effects (Öhnstedt et al., 2019). However, few drugs have been developed

TABLE 7.1
Summary of Available Current Treatments for Wound Healing (sourced from Öhnstedt et al., 2019; Pereira and Bártolo, 2016; Sritharadol et al., 2017; Linden, Emmans, and Safranek, 2014; Sood, Granick, and Tomaselli, 2014)

Treatment Types	Descriptions	Product Examples
Topical creams	Topical treatments in the form of creams, ointments, gels and serums that may contain growth stimulatory factors, antibacterial/antiseptic factors or anti-inflammatory factors	Neosporin® antibacterial ointment Bactroban® antibacterial ointment REGRANEX Gel® recombinant-PDGF therapy
Transdermal drug delivery systems	Dosage forms, such as adhesive patches filled with drug-encapsulating nanocarriers, that are placed directly onto the skin and are designed to deliver an effective dose of a drug across the dermal layers of the skin to allow for systemic circulation of the drug	Hydrosorb® Hydrogel Dressing Cutimed Sorbact® Antimicrobial Wound Dressing
Ingestible drugs	Orally administered medications that contain active compounds that provide antibacterial and anti-inflammatory properties	Cataflam® Diclofenac 50 mg Antibiotic tablets

that provide a growth stimulatory effect that aims to accelerate wound healing (Wicke et al., 2000). Current products available for the treatment of wounds, which includes topically or orally administered medications, as well as transdermal drug delivery systems or wound dressings, have been summarized in Table 7.1 (Pereira and Bártolo, 2016).

7.4.1 Topical Treatments

The topical application of antibacterial creams and ointments allows for the diffusion of the active compounds into the superficial layers of the skin, which limits the antibacterial action of the products to the application site (Ashok, Lakshmana, and Gopal, 2010). Antiseptic ointments generally contain antibacterial compounds, such as antibiotics, to prevent or treat wound infection. Commonly incorporated antibiotics include neomycin (kills pathogens such as *Escherichia coli*, *Haemophilus influenzae*, *Proteus species*, *S. aureus* and *Serratia* species) and mupirocin (kills pathogens such as *S. aureus* and *Streptococcus pyogenes*). These antibiotics inhibit bacterial protein synthesis and bacterial RNA synthesis, respectively (Rahal, 2006). Although the use of antibiotic treatments is effective, the exploitation of these products may perpetuate bacterial antibiotic resistance (Sood, Granick, and Tomaselli, 2014).

Other topical products providing antiseptic properties may contain alcohols or hydrogen peroxide, which tend to be harsh on damaged tissue and may ultimately increase discomfort and inflammation within the wound. Excess wound swelling may even cause further tissue tearing, which may delay the wound healing process (Loo et al., 2012). Aloe extracts, produced from *Aloe barbadensis* Miller and *Aloe ferox* Miller, are commonly utilized in many wound care products due to their wound healing properties and antibacterial effects (Hashemi, Madani, and Abediankenari, 2015).

Topical treatments containing growth factors aim to accelerate the recovery time of wounds via the stimulation of cellular cascades necessary for cell proliferation and migration involved in wound healing (Sood, Granick, and Tomaselli, 2014). However, growth factors appear to have an extremely reduced half-life upon topical application, so consistent re-application of these products must occur throughout the healing process (Kwon et al., 2006).

7.4.2 Transdermal Drug Delivery Systems

Transdermal drug delivery systems maximize the therapeutic effects of the drug by offering systemic circulation of the drug while avoiding first-pass metabolism (Ashok, Lakshmana, and Gopal, 2010). The steady and continuous release of the drug dosage allows for consistent blood plasma levels of the drug regardless of the half-life of the drug (Prausnitz and Langer, 2008). Transdermal patches are generally made up of a network of polymeric chains which form a network that retains water and active compounds, which ensure steady release into the wound for extended periods (Girard, Teferra, and Awika, 2019).

Studies conducted on the use of plant-derived polymers in transdermal patches have exhibited improved controlled release and permeation of drugs as opposed to synthetic polymers (Saidin, Anuar, and Affandi, 2018). Studies conducted on polymeric films consisting of polysaccharides isolated from *Aloe vera* exhibited a higher accumulation of the drug in deeper layers of the dermal skin (Pereira et al., 2014).

Polymeric matrix films may even be loaded with drug-encapsulated nanocarriers, such as liposomes, transfersomes, ethosomes, niosomes, dendrimers, lipid-based nanoparticles, polymer nanoparticles and nanoemulsions (Escobar-Chávez et al., 2012). Nanocarriers are colloidal systems, with a particle size of less than 500 nm, designed to transport a drug to a specific target organ within the body guided by the composition and structure of the nanocarrier (Tedesca, Primo, and Moura, 2017).

The use of nanotechnology in modern medicine aims to optimize the adsorption and delivery of administered drugs. Nanocarriers in transdermal drug delivery allow for an alternative administration of lipophilic and hydrophilic drugs. Nanocarriers may be utilized for the treatment of dermal wounds via encapsulating active compounds which provide either antibacterial, anti-inflammatory or immunostimulant properties (Yadav et al., 2016; Uchechi, Ogbonna, and Attama, 2014).

7.4.3 Ingestible Drugs

Ingestible drugs may contain steroids, such as glucocorticoids and retinoids, intended to accelerate wound healing via regulation of the inflammatory response to prevent excess swelling, which may negatively affect the wound healing process (Wicke et al., 2000). These anti-inflammatory drugs target general cellular mechanisms, or pro-inflammatory cytokines, involved in the initiation of the inflammatory cascade. The limitations thereof are that they may not apply to every individual patient due to physiological differences (Adebayo et al., 2015). The ingestion of antibiotics may also provide an effective treatment for severe wound infection. However, exploitation and over-use of these drugs may confer bacterial antibiotic resistance (Shajari, Khorshidi, and Moosavi, 2017).

7.5 MEDICINAL PLANTS FOR WOUND HEALING

Important aspects of wound healing are angiogenesis, inflammation regulation and prevention of bacterial infection for healthy wound healing to occur (Singer and Clark, 1999). Therefore, it is important to identify plant candidates that facilitate healthy wound healing by promotion of wound closure, prevention of pathogenic infection and anti-inflammatory activity (Shedoeva et al., 2019).

Numerous medicinal plants have been utilized in various cultures for the treatment of wounds and general wound care, some of which have been developed into commercial products for wound healing. The trend in choice of medicinal plant candidates for traditional wound healing practices is generally based on the antibacterial, anti-inflammatory or soothing effects of the plant preparation (Pereira and Bártolo, 2016). However, only a few of these medicinal plants have been validated through scientific *in vitro* and/or *in vivo* studies to support traditional wound healing claims. Although these medicinal plants may show exceptional wound healing activity by *in vitro* studies, their effective bioactivities may not always be supported by *in vivo* investigations (Shedoeva et al., 2019). Medicinal plants that exhibit either proangiogenic, anti-inflammatory/antioxidant or antibacterial activity using *in vivo* and/or *in vitro* wound healing-based methods are summarized in Table 7.2 (Figure 7.3).

TABLE 7.2
Traditionally Used Medicinal Plants Showing Wound Healing Activity in *in vivo* and/or *in vitro* Tests (sourced from Shedoeva et al., 2019).

Botanical Names (Plant Families)	Extract Types	Assessment Methods	Bioactivities in Wound Healing	Commercial Products
Aloe barbadensis Miller (Asphodelaceae)	Aqueous Gel extract	Punch biopsy wounds in Sprague-Dawley rats	Anti-inflammatory Immunomodulatory	Aloe vera gel
Arctium lappa (Asteraceae)	Aqueous whole leaves and aqueous whole root extracts	Human burn wounds Canine dermal fibroblast adhesion assay	Anti-inflammatory Antimicrobial	N/A
Andropraphis paniculata Nees (Acanthaceae)	Aqueous leaf extract	Excision model in albinos Wistar rats	Anti-inflammatory Antimicrobial Antioxidant Immunostimulatory	Chuan Xin Lian Nei Zhi Di Wan
Calendula officinalis Linnaeus (Asteraceae)	Hexane and EtOH[1] flower extracts	Scratch assay Excision wound model in BALB/c mice Punch wound model in rats	Antibacterial Anti-inflammatory Antioxidant	Calendula Herbal-Extract Toner Plenusdermax®
Centella asiatica Linnaeus (Apiaceae)	Hexane, EtOAc[2], MeOH[3] and aqueous extract of aerial parts	Incision and partial-thickness Burn Wound models in rats	Anti-inflammatory Antioxidant Proangiogenic	Madecassol® Centellase® Blastoestimulina® Collaven®
Daphne genkwa (Thymelaeceae)	Aqueous flower and root extract	Human wounds from anal fistula therapy	Antibacterial Anti-inflammatory Immunomodulatory	N/A
Panax ginseng (Araliaceae)	EtOH, DCM[4], EtOAc, BuOH[5] and MeOH extracts of leaves, roots and whole plant	Laser burn and excision wound models in mice cell migration and wound healing assays	Anti-inflammatory Antimicrobial Antioxidant Immunomodulatory	Ginseng Strong 200 mg Ginsemax® Ginseng Vita-Complex
Panax notoginseng (Araliaceae)	Essential oil of leaves, flower, roots and rhizome	*In vitro* anticoagulation and antioxidant test *In vivo* hemostasis and anti-inflammatory test	Anti-inflammatory Antimicrobial Antioxidant Pro-angiogenesis	San Qi Fen Yunnan Baiyao
Polygonum cuspidatum (Polygonaceae)	EtOH root extract	Full-thickness excision wounds in rats	Antibacterial Anti-inflammatory Antioxidant	Resveratrol supplement
Rheum officinale Baill (Polygonaceae)	EtOH root extract	Rat excisional wound model	Antibacterial Anti-inflammatory Antioxidant Immunomodulatory	Dahuang Zhechong Wan
Sanguisorba officinalis Linnaeus (Rosaceae)	EtOH root extract	Burn wound model in mice	Anti-inflammatory Antioxidant Immunomodulatory	*Sanguisorba officinalis* Mother Tincture

Notes: **1** Ethanol; **2** Ethyl acetate; **3** Methanol; **4** Dichloromethane; **5** Butanol

FIGURE 7.3 Medicinal plants traditionally used for wound healing. A, *Aloe vera*, B, *Arctium lappa*; C, *Andrographis paniculata*; D, *Candula officinalis*; E, *Centella asiatica*; F, *Daphne genkwa*; G, *Panax* spp.; H, *Polygonum cuspidatum*; I, *Rheum officinale*; J, *Sanguisorba officinalis* (Ā qiáo HQ, 2012, 2017; Grossir, 2011; Herring, 2010; HPN, 2020; Lohrie, 2006; Monkelbaan, 2020; Nijaki, 2013; Pethan, 2005; Zell, 2009).

7.6 PHYTO-DERIVED COMPOUNDS FOR WOUND HEALING

The medicinal plants discussed in Table 7.1 exhibit bioactivity in wound healing studies by displaying either anti-inflammatory, proangiogenic, promotion of cell proliferation or antibacterial properties. Isolation studies were conducted on some of these species to identify the main active compounds present within the previously mentioned medicinal plants. Compounds were further subjected to successive bioactivity assays associated with wound healing to determine their mechanisms of action. The modes of bioactivity in wound healing of the isolated active compounds have been summarized in Table 7.3.

TABLE 7.3
Phyto-Derived Compounds Showing Bioactivity in Wound Healing (sourced from Shedoeva et al., 2019; Morgan and Nigam, 2013; Thangapazham, Sharad, and Maheshwari, 2016)

Phyto-Derived Compounds	Plant Sources	Bioactivities in Wound Healing	Mechanisms of Action
Acemannan	*Aloe barbadensis* Miller inner leaf gel	Immunomodulatory	Promotes cell proliferation by inducing CDK-1[1] protein translation promotes fibroblasts proliferation by activation of PI3K-Akt-mTOR[2] signaling
Arctigenin	*Arctium lappa* roots and leaves	Anti-inflammatory	Promotes degradation of nitric oxide synthase
Arctiin	Seeds of *Arctium lappa*	Stimulates collagen production	Upregulates the expression of collagen mRNA in human dermal fibroblasts
Andrographolide	*Andropraphis paniculata* Nees stem and leaves	Anti-inflammatory	Inhibits NF-κß[3] binding to DNA

Phyto-Derived Compounds	Plant Sources	Bioactivities in Wound Healing	Mechanisms of Action
Esculetin	*Calendula officinalis* Linnaeus flowers	Anti-inflammatory	Antioxidant activity Radical scavenging
Quercetin-3-O-glucoside	*Calendula officinalis* Linnaeus flowers	Antibacterial	Peroxidation of bacterial membrane lipids Permeabilization of inner bacterial membranes
Asiaticoside	Aerial plant parts of *Centella asiatica* Linnaeus	Anti-inflammatory Proangiogenesis	Stimulates secretion of IL-1[4] stimulates the secretion of VEGF[5] and fibroblast production by induction of PI3K-Akt-mTOR signaling pathway
Daphnodorin derivatives	*Daphne genkwa* roots	Immunomodulatory	Regulation of angiogenesis by inhibition of Angiotensin-II production
Genkwanin derivatives	*Daphne genkwa* flowers	Antibacterial Anti-inflammatory	Mechanism not understood Upregulates the production of IL-10[6] and downregulates the production of IL-12[7] in macrophages
Ginsenosides	*Panax ginseng* roots	Anti-inflammatory Immunomodulatory	Inhibits the production of reactive oxygen species Regulation of PEDF[8] secretion
Notoginsenoside Ft1	*Panax notoginseng* roots	Proangeogenesis	Stimulates the proliferation of fibroblast by induction of the PI3K-Akt-mTOR signaling pathway
Resveratrol	*Polygonum cuspidatum* roots	Anti-inflammatory Immunomodulatory	Inhibit LPS-induced[9] production of NO[10] and TNF-α[11] Regulates angiogenesis by inhibiting the production of VEGF
Emodin	*Polygonum cuspidatum* and *Rheum officinale* Baill roots	Antibacterial Anti-inflammatory Immunomodulatory	Permeabilization of the inner bacterial cell membrane by lipid destabilization Suppression of cellular cascades such as PI3K-Akt-mTOR and NF-κß signaling pathways

Notes: **1** Cyclin dependent kinase 1; **2** Phosphatidylinositol-3-kinase-AKT-mammalian target of rapamycin; **3** Nuclear factor kappa-light-chain-enhancer of activated B cells; **4** Interleukin-1; **5** Vascular endothelial growth factor; **6** Interleukin-10; **7** Interleukin-12; **8** Pigment epithelium-derived factor; **9** Lipopolysaccharide-induced; **10** Nitric oxide; **11** Tumor necrosis factor-α

7.7 CONCLUSION

There are various medicinal plants and phyto-derived compounds that have been utilized in wound healing practices due to their antioxidant, anti-inflammatory, antibacterial and immunomodulatory activity. Various plant species have shown incredible bioactivity associated with wound healing using *in vitro* study methods. This has encouraged the incorporation of these medicinal extracts into the design of bioactive wound dressings, anti-inflammatory ingestible drugs or topical formulations. The efficacy of the wound healing potential of these medicinal plants discussed within this chapter has been made evident by the respective *in vivo* wound healing models (Shedoeva et al., 2019). However, there are inconsistencies among the different studies, and only a few of these medicinal plant species have been tested under *in vivo* conditions such as in human dermal wound models. Furthermore, there is a lack of clinical testing and reports on the potential adverse effects such as dermatitis associated with the long-term use of these medicinal plants.

Therefore, to better understand the potential use and development of the pharmacologically active extracts and compounds in wound healing products, standardized methodology and clinical trials are necessary to obtain validated evidence supporting the use of these traditional medicines in wound care treatments (Pereira and Bártolo, 2016). Based on this, medicinal plants throughout the world have been screened for their bioactivity in wound healing. However, there is still a large reservoir of potential medicinal plant candidates that may be investigated for development as effective novel wound healing agents.

There are some challenges with regards to the use of phytochemicals as therapeutic agents, such as poor solubility, bioavailability and lack of clinical research of the long-term effects of these medicinal agents (Thangapazham, Sharad, and Maheshwari, 2016). Thus, further testing should be conducted on potential wound healing agents in order to gain a better understanding of their specific mechanisms of action and subsequent physiological effects. This will assist in developing effective formulations of these phytochemicals in nascent wound healing technology, such as liposomes, micelles, nanoparticles and phospholipid complexes (Uchechi, Ogbonna, and Attama, 2014).

REFERENCES

Adebayo, Salmon A., Jean P. Dzoyem, Leshweni J. Shai, and Jacobus N. Eloff. 2015. "The Anti-Inflammatory and Antioxidant Activity of 25 Plant Species Used Traditionally to Treat Pain in Southern African." *BMC Complementary and Alternative Medicine* 15, 159: 1–10. doi: 10.1186/s12906-015-0669-5.

Aldridgc, Paul. 2015. "Complications of Wound Healing: Causes and Prevention." *Companion Animal* 20, 8: 453–459. doi: 10.12968/coan.2015.20.8.453.

Ā qiáo, HQ. 2012. "Rhubarb *Rheum officinale*." Obtained from Wikimedia under the Creative Commons Attribution-Share Alike 2.0 Generic license. 2012. https://commons.wikimedia.org/wiki/File:大黃_Rheum_officinale_-比利時_Ghent_University_Botanical_Garden,_Belgium-_(9227077897).jpg (Accessed 1 September 2020).

Ā qiáo, HQ. 2017. "*Daphne Genkwa*." Obtained from Wikimedia under the Creative Commons Attribution-Share Alike 2.0 Generic license. https://commons.wikimedia.org/wiki/File:芫花_Daphne_genkwa_-武漢清涼寨_Wuhan,_China-_(34282321295).jpg (Accessed 1 September 2020).

Ashok, J., S. Lakshmana, and V. Gopal. 2010. "Transdermal Drug Delivery System : An Overview." *International Journal of Pharmaceutical Sciences Review and Research* 3, 2: 49–54.

Barrientos, Stephan, Olivera Stojadinovic, Michael S. Golinko, Harold Brem, and Marjana Tomic-Canic. 2008. "Growth Factors and Cytokines in Wound Healing." *Wound Repair and Regeneration* 16, 5: 585–601. doi: 10.1111/j.1524-475X.2008.00410.x.

Bessa, Lucinda J., Paolo Fazii, Mara Di Giulio, and Luigina Cellini. 2013. "Bacterial Isolates from Infected Wounds and Their Antibiotic Susceptibility Pattern : Some Remarks about Wound Infection." *International Wound Journal* 12, 1: 47–52. doi: 10.1111/iwj.12049.

Cohen, Philip R., and Razelle Kurzrock. 2004. "Community-Acquired Methicillin-Resistant Staphylococcus Aureus Skin Infection: An Emerging Clinical Problem." *Journal of the American Academy of Dermatology* 50, 2: 277–280. doi: 10.1016/j.jaad.2003.06.005.

de Janerio, Rio. 2019. "Sepsis 2019." *Intensive Care Medicine Experimental* 7, S2: 1–19. doi: 10.1186/s40635-019-0254-1.

Escobar-Chávez, José Juan, Roberto Díaz-Torres, Isabela Marlen Rodríguez-Cruz, Clara Luisa Domínguez-Delgado, Rafael Sampere Morales, Enrique Ángeles-Anguiano and Luz Maria Melgoza-Contreras. 2012. "Nanocarriers for Transdermal Drug Delivery." *Research and Reports in Transdermal Drug Delivery* 1: 3–17. doi: 10.2147/RRTD.S32621.

Everson, B. 2004. "Infected Dermal Wound." Obtained from Flickr Creative Commons under the Attribution 2.0 Generic (CC BY 2.0). www.flickr.com/photos/editor/2722675201/in/album-72157606486639110/ (Accessed 20 August 2020).

Girard, Audrey L., Tadesse Teferra, and Joseph M. Awika. 2019. "Effects of Condensed vs Hydrolysable Tannins on Gluten Film Strength and Stability." *Food Hydrocolloids* 89, July 2018: 36–43. doi: 10.1016/j.foodhyd.2018.10.018.

Gray, Mikel, Joyce M. Black, Mona M. Baharestani, Donna Z. Bliss, Janice C. Colwell, Margaret Goldberg, Karen L. Kennedy-Evans, Susan Logan, and Catherine R. Ratliff. 2011. "Moisture-Associated Skin Damage: Overview and Pathophysiology." *Journal of Wound, Ostomy and Continence Nursing* 38, 3: 233–241. doi: 10.1097/WON.0b013e318215f798.

Gray, Mikel, and Dorothy Weir. 2007. "Prevention and Treatment of Moisture-Associated Skin Damage (Maceration) in the Periwound Skin." *Journal of Wound, Ostomy and Continence Nursing* 34, April: 153–157.

Grossir, Florian. 2011. "*Sanguisorba officinalis*." Obtained from Wikimedia Commons under the Creative Commons Attribution-Share Alike 3.0 Unported license. https://commons.wikimedia.org/wiki/File:Sanguisorba_officinalis_(Kastelberg).jpg (Accessed 20 August 2020).

Halloran, Christopher M., and John P. Slavin. 2002. "Pathophysiology of Wound Healing." *Surgery (Oxford)* 20, 5: i–v. doi: 10.1383/surg.20.5.0.14629.

Harris, Layne. 2008. "Dermal Ulcer." 2008. Obtained from Wikimedia under Public Domain. https://commons.wikimedia.org/wiki/File:Leishmaniasis_ulcer.jpg (Accessed 18 August 2020).

Hashemi, Seyyed Abbas, Seyyed Abdollah Madani, and Saied Abediankenari. 2015. "The Review on Properties of *Aloe vera* in Healing of Cutaneous Wounds." *BioMed Research International*. doi: 10.1155/2015/714216.

Herring, Albert. 2010. "*Polygonum cuspidatum*." Obtained from Wikimedia under the the Creative Commons Attribution 2.0 Generic license. https://commons.wikimedia.org/wiki/File:Polygonum_cuspidatum_3-eheep_(5097370153).jpg (Accessed 20 August 2020).

HPN. 2020. "*Centella Asiatica*." Obtained from Wikimedia under the Creative Commons Attribution-Share Alike 4.0 International license. https://commons.wikimedia.org/wiki/File:Centella_asiatica_-_large.jpg (Accessed 20 August 2020).

Krikun, Graciela, Frederick Schatz, Tom Finlay, Susan Kadner, Augusto Mesia, Rene Gerrets, and Charles J. Lockwood. 2000. "Expression of Angiopoietin-2 by Human Endometrial Endothelial Cells: Regulation by Hypoxia and Inflammation." *Biochemical and Biophysical Research Communications* 275, 1: 159–163. doi: 10.1006/bbrc.2000.3277.

Kwon, Young Bae, Hyun Woo Kim, Dae Hyun Roh, Seo Yeon Yoon, Rong Min Baek, Jeum Yong Kim, Hae Yong Kweon, Kwang Gill Lee, Young Hwan Park, and Jang Hern Lee. 2006. "Topical Application of Epidermal Growth Factor Accelerates Wound Healing by Myofibroblast Proliferation and Collagen Synthesis in Rat." *Journal of Veterinary Science* 7, 2: 105–109. doi: 10.4142/jvs.2006.7.2.105.

Lanas, Angel, and Francis K. L. Chan. 2017. "Peptic Ulcer Disease." *The Lancet* 390, 10094: 613–624. doi: 10.1016/S0140-6736(16)32404-7.

Linden, Leslie, Paul Emmans, and Sarah Safranek. 2014. "Topical Preparations for Wound Healing." *American Family Physician* 89, 12: 978–979.

Lohrie, Katharina. 2006. "Ginsengpflanze.Jpg." Obtained from Wikimedia under the Creative Commons Attribution 3.0 Unported license. https://commons.wikimedia.org/wiki/File:Ginsengpflanze.jpg (Accessed 20 August 2020).

Loo, Alvin Eng Kiat, Yee Ting Wong, Rongjian Ho, Martin Wasser, Tiehua Du, Wee Thong Ng, and Barry Halliwell. 2012. "Effects of Hydrogen Peroxide on Wound Healing in Mice in Relation to Oxidative Damage." *PLoS ONE* 7, 11. doi: 10.1371/journal.pone.0049215.

Lynch, S. E., R. B. Colvin, and H. N. Antoniades. 1989. "Growth Factors in Wound Healing. Single and Synergistic Effects on Partial Thickness Porcine Skin Wounds." *Journal of Clinical Investigation* 84, 2: 640–646. doi: 10.1172/JCI114210.

Monkelbaan, Agnes. 2020. "Goudsbloem (*Calendula officinalis*)." Obtained from Wikimedia under the Creative Commons Attribution-Share Alike 4.0 International license. https://commons.wikimedia.org/wiki/File:Goudsbloem_(Calendula_officinalis)_13-07-2020._(actm.).jpg (Accessed 20 August 2020).

Morgan, Claire, and Yamni Nigam. 2013. "Naturally Derived Factors and Their Role in the Promotion of Angiogenesis for the Healing of Chronic Wounds." *Angiogenesis* 16, 3: 493–502. doi: 10.1007/s10456-013-9341-1.

Nijaki, Nikodem. 2013. "*Aloe Vera* Plantation." Obtained from Wikimedia under the Creative Commons Attribution-Share Alike 3.0 Unported license. https://commons.wikimedia.org/wiki/File:Aloe_vera_plantation_IMGP0148.jpg (Accessed 20 August 2020).

Öhnstedt, E., H. Lofton Tomenius, E. Vågesjö, and M. Phillipson. 2019. "The Discovery and Development of Topical Medicines for Wound Healing." *Expert Opinion on Drug Discovery* 14, 5: 485–497. doi: 10.1080/17460441.2019.1588879.

Pereira, Gabriela Garrastazu, Sílvia Stanisçuaki Guterres, Anna Giulia Balducci, Paolo Colombo, and Fabio Sonvico. 2014. "Polymeric Films Loaded with Vitamin E and Aloe Vera for Topical Application in the Treatment of Burn Wounds." *BioMed Research International* 2, 1: 1–9. doi: 10.1155/2014/641590.

Pereira, R., and Paulo J. Bártolo. 2016. "Traditional Therapies for Skin Wound Healing." *Advances in Wound Care* 5, 5: 208–229. doi: 10.1089/wound.2013.0506.

Pethan. 2005. "*Arctium Lappa*." Obtained from Wikimedia under the Creative Commons Attribution-Share Alike 3.0 Unported license. https://commons.wikimedia.org/wiki/File:Arctium_lappa01.jpg (Accessed 20 August 2020).

Prausnitz, Mark R., and Robert Langer. 2008. "Transdermal Drug Delivery." *Nature Biotechnology* 26, 11: 1261–1268. doi: 10.1038/nbt.1504.

Rahal, James J. 2006. "Novel Antibiotic Combinations against Infections with Almost Completely Resistant Pseudomonas Aeruginosa and Acinetobacter Species." *Clinical Infectious Diseases* 43, Supplement_2: S95–99. doi: 10.1086/504486.

Saidin, Nor Maziah, Nor Khaizan Anuar, and M M R Meor Mohd Affandi. 2018. "Roles of Polysaccharides in Transdermal Drug Delivery System and Future Roles of Polysaccharides in Transdermal Drug Delivery System and Future Prospects." *Journal of Applied Pharmaceutical Science* 8, 3: 141–157. doi: 10.7324/JAPS.2018.8320.

Servier. 2019. "The Stages of Wound Healing." Obtained from Servier under the Creative Commons Attribution 3.0 Unported License. https://smart.servier.com/category/anatomy-and-the-human-body/cardiovascular-system/blood/. (Accsessed 20 August 2020).

Shajari, G., A. Khorshidi, and G. Moosavi. 2017. "Vancomycin Resistance in Staphylococcus Aureus Strains." *Archives of Razi Institute* 90, 54: 107–110.

Shedoeva, Aleksandra, David Leavesley, Zee Upton, and Chen Fan. 2019. "Wound Healing and the Use of Medicinal Plants." *Evidence-Based Complementary and Alternative Medicine* 2019, Figure 1. doi: 10.1155/2019/2684108.

Singer, A., and R. Clark. 1999. "Cutaneous Wound Healing." *New England Journal of Medicine* 341, 10: 738–746.

Sood, Aditya, Mark S. Granick, and Nancy L. Tomaselli. 2014. "Wound Dressings and Comparative Effectiveness Data." *Advances in Wound Care* 3, 8: 511–529. doi: 10.1089/wound.2012.0401.

Sritharadol, Rutthapol, Titpawan Nakpheng, Paul Wan Sia Heng, and Teerapol Srichana. 2017. "Development of a Topical Mupirocin Spray for Antibacterial and Wound-Healing Applications." *Drug Development and Industrial Pharmacy* 43, 10: 1715–1728. doi: 10.1080/03639045.2017.1339077.

Tedesca, Antonio C., Fernando L. Primo, and Marigilson P. de Siqueira Moura. 2017. "Nanodrug Delivery Systems for Dermal and Transdermal Photosensitizer Drugs." In *Nano- and Microscale Drug Delivery Systems*, edited by Alexandru Mihai Grumezescu, 267–280. Elsevier Ltd.

Thangapazham, Rajesh L., Shashwat Sharad, and Radha K. Maheshwari. 2016. "Phytochemicals in Wound Healing." *Advances in Wound Care* 5, 5: 230–241. doi: 10.1089/wound.2013.0505.

Uchechi, Okoro, John D. N. Ogbonna, and Anthony A. Attama. 2014. *Nanoparticles for Dermal and Transdermal Drug Delivery*. Edited by Ali Demir Sezer. 1st ed. Enugu State. doi: 10.5772/58672.

Wicke, Corinna, Betty Halliday, Daniel Allen, Nanette S. Roche, Heinz Scheuenestuhl, Martin M. Spencer, Anita B. Roberts, and Thomas K. Hunt. 2000. "Effects of Steroids and Retinoids on Wound Healing." *Archives of Surgery* 135, 11: 1265–1270. doi: 10.1001/archsurg.135.11.1265.

Yadav, J. P., S. Kumar, L. Budhwar, A. Yadav, and M. Yadav. 2016. "Nanomedicine & Nanotechnology Characterization and Antibacterial Activity of Synthesized Silver and Iron Nanoparticles Using Aloe Vera." *Journal of Nanomedicine & Nanotechnology* 7, 3: 384–391. doi: 10.4172/2157-7439.100038.

Zell, H. 2009. "*Andrographis paniculata*." Obtained from Wikimedia under the Creative Commons Attribution-Share Alike 3.0 Unported license. https://commons.wikimedia.org/wiki/File:Andrographis_paniculata_002.JPG (Accessed 20 August 2020).

8 Naturally Occurring Histone Deacetylase (HDAC) Inhibitors in the Treatment of Cancers

Sujatha Puttalingaiah, Murthy V. Greeshma, Mahadevaswamy G. Kuruburu, Venugopal R. Bovilla and SubbaRao V. Madhunapantula

CONTENTS

8.1 INTRODUCTION

In this section of the chapter, we cover recent developments demonstrating the: (a) role of histone deacetylases (HDACs) in the pathogenesis of cancer; (b) ability of pharmacological agents inhibiting HDACs for treating cancers; (c) HDAC-mediated molecular mechanisms involved in drug resistance;

DOI: 10.1201/9781003108375-8

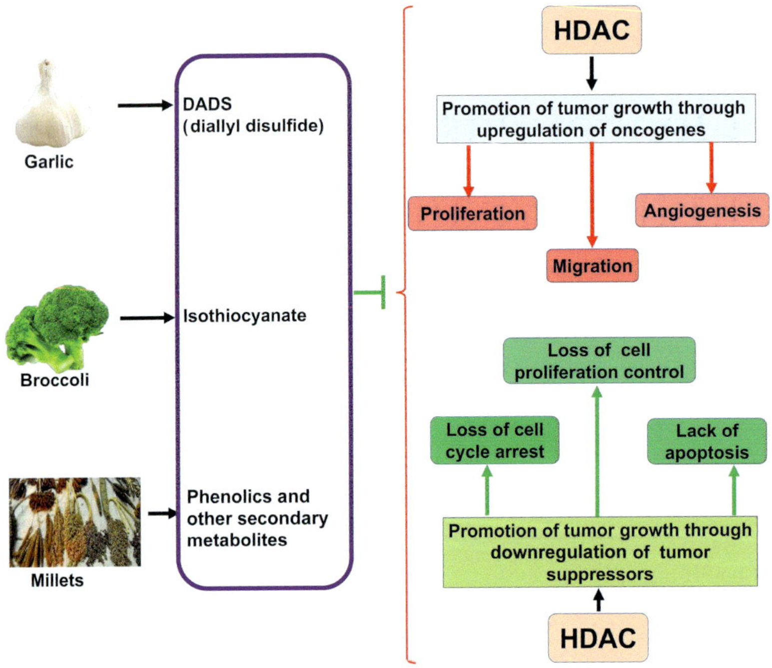

Graphical abstract representing the content in the chapter (Barneda-Zahonero and Parra, 2012; Eckschlager et al., 2017; Li and Seto, 2016; Ropero and Esteller, 2007).

and (d) the potential of knocking down HDACs in promoting sensitivity of cancer cells to chemo- and radiation therapeutics. Furthermore, a brief report on ongoing and recently completed clinical trials using HDAC inhibitors (HDACis) for the treatment of cancer is provided at the end of this chapter.

8.1.1 The Structure of Nucleosome

Epigenetic regulation of gene expression is one of the key mechanisms by which the expression of genes is controlled through the modification of histones (Stephens et al., 2013). Histones are a family of basic proteins (11,300–15,400 Daltons) found as a part of the nucleosome, which is the fundamental structural and functional unit of the chromatin (Kujirai et al., 2016). Nucleosomes are disc-shaped particles of 11 nm in diameter with a 5.7 nm height and are made up of 146 base pairs of DNA consisting of eight histone proteins viz., H2A, H2B, H3, H4 (2 units of each), and appear as beads on a string (McGinty and Tan, 2015).

Each nucleosome bead is separated from the next bead by a 54-base pair linker DNA. The linker DNA contains a single H1 histone protein, which seals off the two complete DNA turns (200 base pairs long) (Hergeth and Schneider, 2015). The addition of one H1 wraps another 20 base pairs, resulting in two full turns around the octamer, and forms a structure called 'chromatosome.' Long eukaryotic DNA wrapped around the histone proteins leads to the formation of a compact chromosome (Maeshima and Eltsov, 2008).

8.2 ACETYLATION AND DEACETYLATION OF HISTONES CONTROL THE EXPRESSION OF GENES

Since histones play an important role in the compactness of DNA, modifying these proteins through acetylation in turn modulates the expression of genes (Dong and Weng, 2013). Acetylation of histone proteins facilitates and enhances the gene transcription, whereas deacetylation suppresses the transcription by enhancing the compact packing of DNA (Park and Kim, 2020). Acetylation of histones is carried out by histone acetyltransferase (HAT), while deacetylation is performed by histone deacetylase (HDACs) enzymes (Legube and Trouche, 2003).

8.2.1 Histone Acetyl Transferases (HATs)

Acetylation of proteins occurs through a class of enzymes known as histone acetyltransferases (HATs, known as lysine acetyltransferases or KATs), which transfer acetyl group (CH_3CO-) to ε-amino group of lysine residues to form ε-*N*-acetyl lysines in histone proteins (Seto and Yoshida, 2014). HATs are classified into two major types (Type-A and Type-B) based on their cellular localization (Sun et al., 2015). Type-A HATs are located in the nucleus and control the gene expression through the acetylation of nucleosomal histones (Sterner and Berger, 2000). The bromodomain present in Type-A HATs recognizes acetylated lysine residues in histones (Hassan et al., 2007). Type-A HATs are further divided into three major families based on their primary structural homology (Bannister and Kouzarides, 2011).

The first family of HATs are known as general control nonderepressible-5 (GCN-5) related *N*-acetyltransferase (GNATs), represented by GCN-5 and p300/CREB binding protein (CBP)-associated factor (PCAF). The second family of HATs are p300/CBP, represented by p300 and CBP, while the third family is composed of MYST histone acetyltransferases, named after its four founding members MOZ, Ybf2 (Sas3), Sas2 and Tip60. The MYST family includes trans-activator of transcription (TAT) interacting protein—60 (TIP60) and monocytic leukemia zinc finger protein (Figure 8.1). Unlike Type-A HATs, the Type-B HATs are cytoplasmic and catalyze the acetylation of newly synthesized histones before they form nucleosomal complexes (Figure 8.1A) (Roth et al., 2001). Type-B HATs lack bromodomain, as they are primarily involved in recognizing non-acetylated lysine residues in the histones (Roth et al., 2001).

HATs use acetyl-CoA as a cofactor to transfer acetyl groups to the ε-amino group of lysine residues on histone tails. Acetylation masks the positive charge of the lysine residue, which reduces the interaction between histones and the negatively charged DNA backbone, and consequently increases gene transcription (Bannister and Kouzarides, 2011). Recent studies have shown that HATs can acetylate many proteins other than histones, and play an important role in the regulation of gene expression (Sterner and Berger, 2000). Dysregulation of HATs expression and activity is reported in several diseases, including cancer and a variety of neurodegenerative disorders (Saha and Pahan, 2006). Therefore, targeted modulation of HATs has gained importance in controlling diseases. Since the aim of this chapter is to provide an overview on HDAC enzymes, more emphasis is laid down on HDACs (Parthun, 2012; Audia and Campbell, 2016).

8.2.2 Histone Deacetylases (HDACs)

Modification of histones is one of the key epigenetic mechanisms implicated in the regulation of gene expression (Stephens et al., 2013). Acetylation and deacetylation reactions catalyzed by HATs and HDACs play an important role in mediating the unpacking and packing of DNA, thereby improving gene function (Shahbazian and Grunstein, 2007). HDACs remove the acetyl groups from histone and non-histone proteins, thereby controlling the expression of genes (Parbin et al., 2014). To date, 18 HDACs have been identified and characterized for their structure and function (Figure 8.1B) (Ropero and Esteller, 2007).

8.2.2.1 Classification of HDAC Enzymes

HDACs are classified into four major classes based on their homology to yeast HDAC proteins (Chen et al., 2015). HDACs belong to the α and β proteins class in the structural classification of proteins hierarchy (Seto and Yoshida, 2014). HDACs that deacetylate acetylated histones are classified in to Class I, Class II and Class IV enzymes, while those that deacetylate non-histone proteins

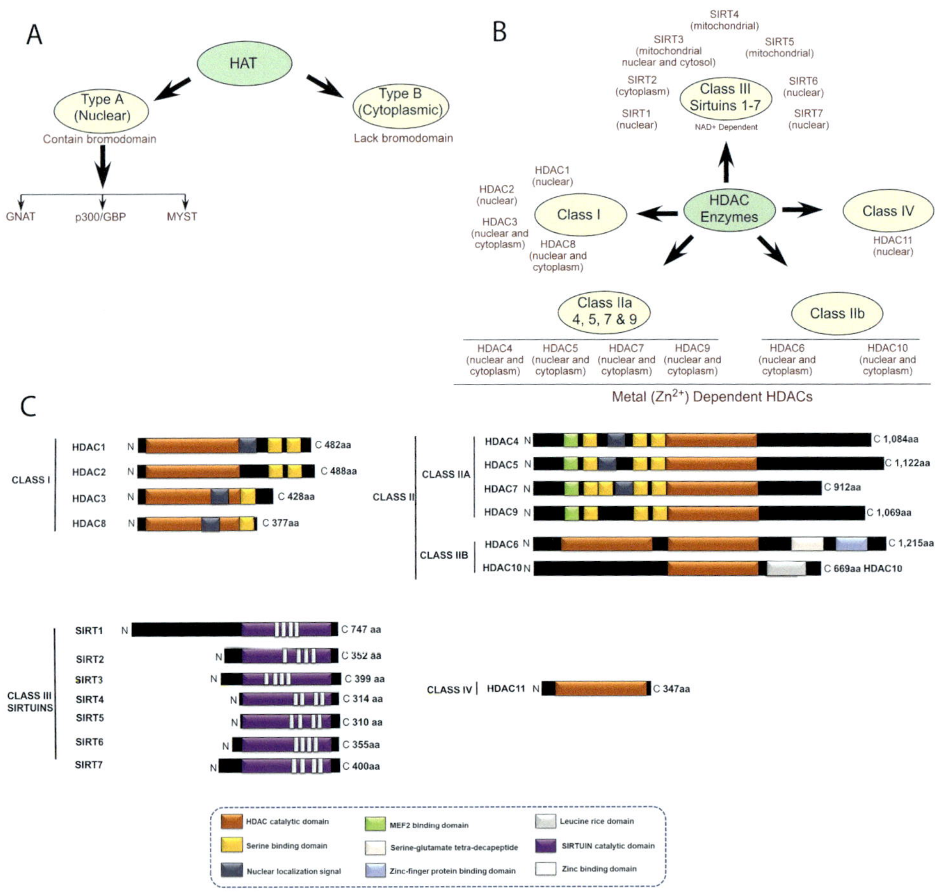

FIGURE 8.1 A, Classification of histone acetyl transferases (HATs); B, Histone deacetylases (HDACs); C, The domain structure of HDAC proteins.

such as tumor protein P53 (p53) and α-tubulin are grouped in to Class III known as sirtuins (SIRTs) (Peng and Seto, 2011). The Class I, II and IV HDACs are numbered according to their chronological order of discovery. Class I, II and IV HDACs are dependent on zinc for their catalytic activity, while SIRTs require nicotinamide adenine dinucleotide (NAD^+) for their function (Figure 8.1B) (Seto and Yoshida, 2014).

HDACs in each class are descended from a common ancestor and usually have similar three-dimensional structures, functions and sequence homology (Grozinger et al., 1999). Class I, II and IV HDACs belong to the arginase/deacetylase superfamily of proteins, which contain arginase-like amidino hydrolases, as well as the histone deacetylases (Seto and Yoshida, 2014). Interestingly, all Class I, II and IV HDACs precede the evolution of histone proteins, suggesting that the primary substrates for this HDAC family may be non-histones (Seto and Yoshida, 2014).

8.2.2.1.1 CLASS I Histone Deacetylase Enzymes

This class consists of HDAC1, HDAC2, HDAC3 and HDAC8 (Figure 8.1B) (Morris and Monteggia, 2013). Class I HDACs have sequence similarity to the yeast transcriptional regulator RDP3 and are primarily localized in the nucleus. Class I HDACs are the most active enzymes involved in the control of cell proliferation and survival (Bayat et al., 2018). This class contains deacetylated non-histone proteins such as AMP-activated protein kinase (AMPK), structural maintenance of chromosomes protein 3 (SMC3) and various transcription regulators to control cell growth and proliferation (Mihaylova and Shaw, 2013).

Overexpression of Class I HDACs is associated with neurological, cardiovascular and immune disorders, inflammation and carcinomas of breast, colon, lung and prostate (Yoon and Eom, 2016). Although the U.S. Food and Drug Administration (FDA) approved Class I HDAC inhibitors, which includes Vorinostat, Romidepsin, Belinostat, Chidamide and Panobinostat for inhibiting cancers, lack of isoform selectivity of these inhibitors caused systemic toxicity and poor clinical outcomes (Li and Seto, 2016). Therefore, future studies should focus more on developing isoform-selective HDAC inhibitors for further development in clinics.

8.2.2.1.2 CLASS II Histone Deacetylase Enzymes

This class comprises of HDAC4, HDAC5, HDAC6, HDAC7, HDAC9 and HDAC10 (Figure 8.2B) (Seto and Yoshida, 2014). Class II proteins exhibit similarity in amino acids sequence to the yeast Hda1 protein (Yang and Gregoire, 2005). Class II enzymes are further subdivided into class IIa—consisting of HDAC4, 5, 7 and 9—and Class IIb, consisting of HDAC6 and 10 (Park and Kim, 2020). Class IIa enzymes shuttle between nucleus and cytoplasm, and form large complex structures by interacting with other proteins, while Class IIb HDACs are located mainly in the cytoplasm. HDAC6 is a well-characterized cytoplasmic deacetylase (Figure 8.1C). The catalytic domains of Class II HDACs contain a structural zinc ion-binding subdomain and exhibit at least 1,000-fold lower enzymatic activity when compared with that of class I HDACs (Park and Kim, 2020).

Class IIa HDACs (HDAC 4, 5, 7 and 9) harbor a unique adapter domain in the N-terminus to facilitate the attachment of DNA-binding transcription factor MEF2 (Kee and Kook, 2011). Furthermore, the phosphorylation sites present in the C-terminus of Class Ia HDACs act as regulatory signals for the association of 14–3–3 proteins (Martin et al., 2007; Seto and Yoshida, 2014). In addition to the binding of 14–3–3- proteins, Class IIa HDACs form multimeric protein complexes with silencing mediator of retinoid and thyroid hormone receptors (SMRT)/nuclear hormone receptor co-repressor (N-CoR)-HDAC3 complex, thereby regulate chromatin modifications (Park et al., 2018). This class plays a key role in deacetylation reactions involving low enzymatic activity (Fischle et al., 2002).

Class IIb HDACs consisting of HDAC6 and HDAC10 are generally present in the cytoplasm and harbor a long tail domain at the C-terminus (Seto and Yoshida, 2014). HDAC6 contains two deacetylase domains and a C-terminal zinc finger ubiquitin-binding domain, whereas HDAC10 has only one deacetylase domain and a leucine-rich repeat at C-terminus (Zhang et al., 2006). Studies have shown that HDAC6 deacetylates α-tubulin, cortactin, chaperones and IFNαR, thereby regulating autophagy (Ran et al., 2015).

8.2.2.1.3 CLASS III Histone Deacetylase Enzymes

These enzymes are known as Sir2-like proteins due to their similarity to yeast silent information regulator-2s (Sir-2) (Seto and Yoshida, 2014). This class comprises of SIRT1, SIRT2, SIRT3, SIRT4, SIRT5, SIRT6 and SIRT7 (Figure 8.2B) (Poulose and Raju, 2015). Class III HDAC enzymes require nicotinamide adenine dinucleotide (NAD^+) to deacetylate lysine residues (North et al., 2005). These proteins are known to be involved in the process of aging. Among various SIRT families of proteins, the majority of research was conducted on SIRT1 and SIRT2 (Zhao et al., 2019). A unique characteristic of these sirtuins is that they exhibit mono-ADP-ribosyltransferase activity and histone deacetylase activity, whereby SIRT1 possesses the most robust histone deacetylase activity (Michan and Sinclair, 2007). Interestingly, SIRT5 exhibits lysine desuccinylase and demalonylase activity (Fischer et al., 2012; Kupis et al., 2016).

8.2.2.1.4 CLASS IV Histone Deacetylase Enzymes

This class is comprised of one HDAC enzyme, i.e. HDAC11 (Figure 8.1B) (Seto and Yoshida, 2014). The HDAC11 uniquely shares sequence homology with the catalytic domains of both Class I and

II HDACs. HDAC11 regulates the protein stability of DNA replication factor CDT1 (chromatin licensing and DNA replication factor 1) and the expression of interleukin 10 (IL-10) (Glozak and Seto, 2009). Further studies are warranted to delineate the structural and functional properties of HDAC11.

8.3 HDACS IN THE PATHOGENESIS OF CANCER

Accumulating evidences from *in vitro* and *in vivo* studies and recent epidemiological data suggested an important role for histone deacetylase (HDAC) proteins in the pathogenesis of various cancers (Ropero and Esteller, 2007). Deregulated HDACs promote the transformation of normal cells into cancer cells, which in turn make them less susceptible for various treatments (Barneda-Zahonero and Parra, 2012). Therefore, strategies targeting HDACs play vital roles in the treatment of cancers.

8.3.1 Role of HDACs in the Initiation and Progression of Cancers

Modification of histones by acetylation is one of the most common post-translational modifications, and was first defined in 1960. Inoue and Fujimoto (1969) observed that acetyl groups are removed from the histones in a calf thymus extract. Subsequently, during the 1970s, several studies identified HDAC activity in tissues; and in 1996, the first bona fide histone deacetylase, i.e. HDAC1, was isolated and cloned (Taunton et al., 1996). HDACs remove acetyl groups from histones, resulting in a more condensed, transcriptionally inactive chromatin state. Inactive chromatin causes the downregulation of tumor suppressor genes such as p21, p27, p53 and (retinoblastoma protein) Rb, etc., leading to uncontrolled cell proliferation, survival, migration and differentiation (Figure 8.2) (Parbin et al., 2014). Therefore, hyper-active HDACs are reported to induce cancer (Li and Seto, 2016). A number of studies further dissected the biochemical characterization of the histone deacetylase and elucidated structure–activity relationships.

HDACs play a key role in the progression of cancers by promoting the expression and activity of oncogenes while down regulating the tumor suppressors (Figure 8.2) (Ropero and Esteller, 2007). Deregulated expression of HDACs induces the levels of reactive oxygen species (ROS) and reactive nitrogen species (RNS), while promoting the secretion of various pro-inflammatory cytokines and growth factors in to the microenvironment, which subsequently promotes the growth of tumors (West and Johnstone, 2014). In addition, prior studies have reported mutations in HDACs (Kupis et al., 2016). For example, a truncating mutation of HDAC2 induced resistance to HDACis in cancer cells, suggesting the requirement of the assessment of HDAC2 mutational status before deciding on a treatment agent. Few other studies have shown that HDAC3 is associated with DNA damage control response, and inactivation of HDAC3 causes genomic instability (Bhaskara et al., 2010).

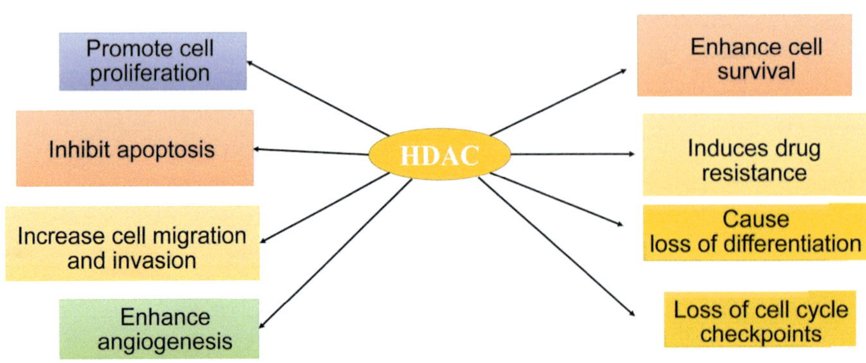

FIGURE 8.2 Cellular processes regulated by oncogenic HDAC in cancer cells.

Recently, mutations in HDAC4 have been identified at significant frequency in breast and colorectal cancers (Stark and Hayward, 2007). Similarly, mutations in SIRT2 (which acts as a tumor suppressor) contribute to genomic instability and tumorigenesis (Head et al., 2017). Therefore, HDACs are the key proteins involved in tumor initiation, progression and metastatic spread; hence, targeted inhibition of HDACs is likely to reduce tumor burden in cancer patients.

Variations in the levels of HAT and HDAC influence the compaction of chromatin, thereby causing improper expression of specific genes and culminating in genomic instability and epigenetic diseases (Park and Kim, 2020). Mechanistically, the amino groups of lysine residues in a protein undergoing different post-translational modifications such as acetylation, methylation, ubiquitination, sumoylation, propionylation, butyrylation, crotonylation, etc., thereby influencing the expression of genes (Seto and Yoshida, 2014). Class I HDACs play a prominent role in cell survival and proliferation, whereas Class II HDACs exhibit tissue-specific functions (Morris and Monteggia, 2013). For instance, HDAC1 knockout cells have a general proliferation and survival defect, despite increased levels of HDAC2 and HDAC3 activity. HDAC2 modulates transcriptional activity by regulating p53 binding. Abnormal HDACs play a key role in many human diseases including cancer, neurological and metabolic disorders, and inflammatory, cardiac and pulmonary diseases.

8.3.2 Regulation of the Activity of HDAC Enzymes

In general, the activity of HDAC enzymes is regulated at the transcription, post-transcription, translation and post-translational levels by a direct interaction with other biomolecules or through indirect effects mediated by other proteins or small molecule metabolites (Seto and Yoshida, 2014). Among these different regulatory mechanisms, the most well-defined mechanisms of HDAC regulation are protein–protein interactions and post-translational modifications (PTMs).

8.3.2.1 Regulation of HDAC Activity through Protein–Protein Interactions

Several studies have demonstrated that protein–protein interactions control the expression of HDACs through: (a) alternative RNA splicing; (b) the modulation of the availability of cofactors; (c) varied subcellular localization; and (d) different degrees of proteolytic processing (Gallinari et al., 2007; Seto and Yoshida, 2014). Individual HDAC proteins, especially HDAC1 and HDAC2, are generally low in enzyme activity; however, when associated with protein complexes such as Sin3, nucleosome remodeling and deacetylase (NuRD), and co-repressor for element-1-silencing transcription factor (CoREST), they exhibit enhanced function, indicating the importance of protein–protein interactions in controlling their biological activity (Banks et al., 2018).

For example, the nucleosome remodeling and histone deacetylase NuRD complex contain seven proteins that consists of HDAC1, HDAC2, RbAp46, RbAp48, Mi2, metastasis-associated protein 2 (MTA2) and methyl-CpG-binding domain protein 3 (MBD3) (Basta and Rauchman, 2015). Whereas MTA2 modulates the enzymatic activity of the histone deacetylase core complex, the MBD3 mediates the association of MTA2 with the core histone deacetylase complex. However, MBD3, although closely related to methylated DNA-binding MBD2 (methyl-CpG-binding domain protein 2), does not directly bind methylated DNA. MBD2 interacts with the NuRD complex and directs the complex to methylated DNA (Kupis et al., 2016). NuRD protein is known to silence the expression of genes through DNA methylation (Kupis et al., 2016). A separate study showed that the interaction between HDAC3 and silencing mediator for retinoid and thyroid hormone receptors/nuclear receptor co-repressor (SMRT/NCoR) stimulates HDAC3 enzyme, thereby reducing the expression of target genes (Guenther et al., 2001).

Similarly, Class III HDACs have been reported to interact with several proteins (Dai and Faller, 2008). For instance, interaction of SIRT1 with deleted in breast cancer 1 (DBC1) results in the downregulation of SIRT1 activity, which leads to an increase in p53 acetylation and up-regulation of p53-mediated function (Kupis et al., 2016). Another nuclear protein called active regulator of

SIRT1 (AROS), enhances SIRT1-mediated deacetylation of p53 and inhibits p53-mediated transcriptional activity (Kim et al., 2007). Similarly, Sin3 complex also contains HDAC1, HDAC2, RbAp46 and RbAp48, along with unique polypeptides Sin3, SAP18 and SAP30 (Kuzmichev et al., 2002). In conclusion, protein—protein interactions play a crucial role in the regulation of HDAC activity in cells.

8.3.2.2 Regulation of HDAC Activity through Post-Translational Modifications

HDAC1 is phosphorylated by cyclic adenosine monophosphate (cAMP)-dependent kinase PKA and protein kinase CK2 (Pflum et al., 2001). The two phosphate group acceptor sites on the carboxyl-terminal of HDAC1 are S421 and S423, which are essential for enzymatic activity and when mutated in to alanine causes a significant reduction in enzymatic activity and disrupts the complex formation. Phosphorylation of HDAC1 and HDAC2 is reversibly regulated by the protein phosphatase PP1 (Galasinski et al., 2002). Phosphorylation of HDAC3 by CK2 and DNA-PKcs significantly enhances HDAC3 activity (Zhang et al., 2005). HDAC3 can also be phosphorylated by Glycogen synthase kinase 3 (GSK-3b), and inhibition of GSK-3b protects against HDAC3-induced neurotoxicity (Bardai and D'Mello, 2011).

HDAC3's phospho-acceptor site, S424, which is a non-conserved residue among the Class I HDACs, when mutated to alanine severely compromises enzymatic activity of HDAC1 and HDAC2. Unlike HDAC1 and HDAC2, the HDAC3 associates with the catalytic and regulatory subunits of the protein serine/threonine phosphatase 4 complex (PP4c/PP4R1), and dephosphorylation of HDAC3 by PP4 down-regulates HDAC3 enzymatic activity (Zhang et al., 2005). Phosphorylation of HDAC8 at S39 leads to the disruption in the surface structure, which ultimately leads to the negative effect of HDAC8. Phosphorylation of Class IIa HDACs may lead to their ubiquitination and proteasomal degradation. Two phosphatases have been implicated in the regulation of Class IIa HDACs activities and functions: protein phosphatase 1b (PP1b), including myosin phosphatase targeting subunit 1 (MYPT1, a regulatory subunit of PP1), and protein phosphatase 2A (PP2A). Class IIb HDACs, global proteomic profiling of phosphopeptides, revealed that HDAC6 is phosphorylated at S22 and T30 (Beausoleil et al., 2004).

Class III HDACs, which contain SIRT1, are known phosphoproteins. The cell cycle–dependent kinase cyclin B/Cdk1 phosphorylates SIRT1 (Beausoleil et al., 2004). AMPK has been shown to increase intracellular NAD^+ levels, which in turn enhances SIRT1 deacetylation activity (Canto et al., 2009). Activation of the cAMP signaling pathway induces rapid deacetylation of SIRT1 substrates, which is independent of changes in NAD^+ levels (Gerhart-Hines et al., 2011). Activation of the cAMP signaling pathway induces rapid deacetylation of SIRT1 substrates independent of changes in NAD^+ levels (Gerhart-Hines et al., 2011). HDAC activities can be modulated by the protein–protein interaction and the post-translational modification by the phosphorylation of different complexes associated with the HDAC.

8.4 TARGETING HDACs FOR INHIBITING CANCERS

Due to their pivotal role in tumorigenesis, metastatic spread and drug resistance, several strategies have been developed by various investigators to inhibit HDACs in cancer cells (Li and Seto, 2016). Mechanistically, inhibition of HDAC enzymes using pharmacological agents arrests cell growth, promotes cell differentiation and induces apoptosis (Eckschlager et al., 2017). Reduction of HDAC activity promotes the expression of apoptotic proteins such as p53, Bax, Caspases and poly (ADP-ribose) polymerase (PARP) and the cell cycle inhibitors p21 and p27, while down-modulating cyclin D1 and cyclin B2 (Uo et al., 2009). Furthermore, targeted inhibition of HDACs promotes RAD-52 and ATM (serine/threonine kinase), thereby increasing cellular oxidative stress (Zhang et al., 2020). Realizing the importance of targeting HDAC proteins, several pharmacological industries have produced synthetic and naturally occurring inhibitors to mitigate HDACs (Singh et al., 2018).

8.4.1 Classification of HDAC Inhibitors (HDACis)

HDACis are broadly classified into six main classes according to their structure (Gottesfeld and Pandolfo, 2009). They are: (a) hydroxamic acids (trichostatin-A [TSA], suberoylanilide hydroxamic acid [SAHA], PXD101); (b) benzamides; (c) cyclic peptides and depsipeptides; (d) short-chain fatty acids (propionic acid, butyric acid, valeric acid, caproic acid and valproic acid); (e) sulfur-containing compounds (diallyl disulfide [DADS], diallyl trisulfide [DATS], sulforaphane [SFN]); (f) phenolic compounds (benzoic acid and cinnamic acid derivatives); and (g) hybrid molecules (Figure 8.3 and Table 8.1).

In addition, many naturally occurring inhibitors of HDAC have been reported in the literature. For example, sulfur-containing compounds from Cruciferous vegetables, phenolic and flavonoid compounds from millets and short-chain and medium-chain fatty acids from plants have also been shown to exhibit HDAC inhibitory activity in many *in vitro* and *in vivo* studies (Evans and Ferguson, 2018). In fact, clinically approved HDACis are either derived from naturally occurring compounds are having structural scaffolds similar to the ones identified from microorganisms (Bassett and Barnett, 2014).

Compared to synthetic HDAC inhibitors, compounds isolated from natural sources are in general less toxic and cause minimal adverse reactions; hence, a hunt for naturally occurring potent

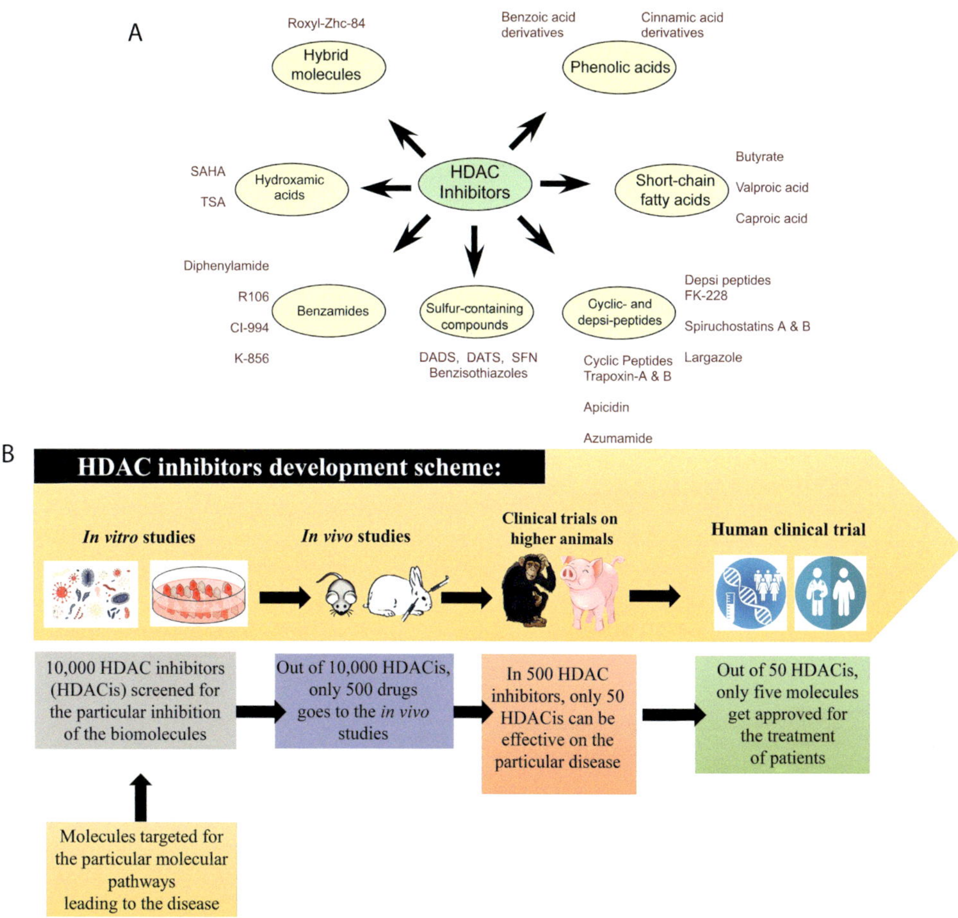

FIGURE 8.3 A, Classification of HDAC inhibitors; B, Discovery of HDAC inhibitors (Rafehi et al., 2017).

TABLE 8.1
List of Histone Deacetylase (HDAC) Inhibitors

Sl. Nos.	Name of Drug	Inhibits HDAC Classes	Natural/ Synthetic/ Semi-Synthetic	Classifications	Structural Formulas	*In vivo* Models Used	Clinical Trial Phases
1.	Trichostatin-A (Vigushin et al., 2001; De Souza and Chatterji, 2015)	Class I and II HDAC	**Natural** *Streptomyces hygroscopicus* Antifungal antibiotic	Organic oxygen compound	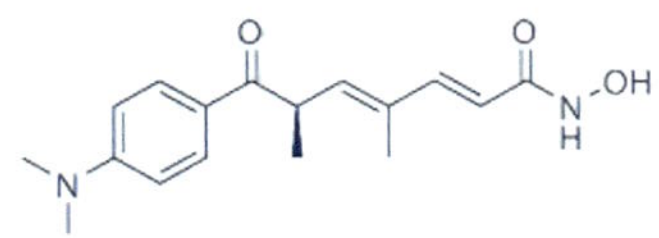	Inbred virgin female (Ludwig/ Wistar/Olac), Nu/nu athymic female mice MHC Beclin transgenic mice (carcinogen-induced rat mammary carcinoma model)	I-relapsed or refractory hematologic malignancies
2.	Romidepsin (Bertino and Otterson, 2011; Moodad et al., 2018)	HDAC1 and HDAC2 Class I and IV Classes: HDAC1, HDAC2, HDAC3, HDAC8 Location: Nuclear Substrates: Histones, Stat3, Bcl-6, P53, E2F-1	**Natural** *Chromobacterium violaceum* Antibiotic	Bicyclic depsipeptide	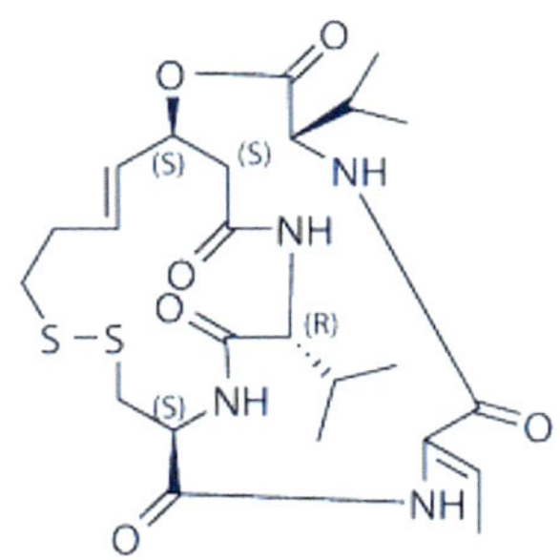	Male SCID mice	III-T-cell lymphoma II-sarcoma and leukemia
3.	Sodium butyrate (Jaworska et al., 2019)	Class 1 and 2	**Natural**	Short-chain fatty acid	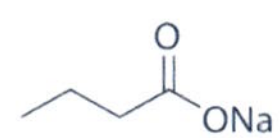	Mouse model of Huntington's disease	IV-alcohol withdrawal syndrome, atopic dermatitis, epilepsies II-Type 1 diabetes mellitus, shigellosis, ulcerative colitis and autism

4.	Curcumin Diferuloylmethane (Wang et al., 2015)	HDAC inhibitor	**Natural** *Curcuma longa* L.	Beta-diketone	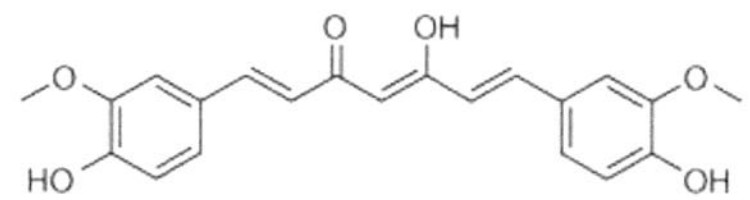	Female B6D2F1 mice	IV-autosomal dominant polycystic kidney disease, chronic schizophrenia, cervical squamous intraepithelial neoplasia, prostate cancer, gastric cancer
5.	Sinapinic acid (Senawong et al., 2013)	HDAC inhibitor	**Natural** Fruits, vegetables, wine and vinegar	Hydroxamic acid (Phenolic acid) phenylpropanoid family	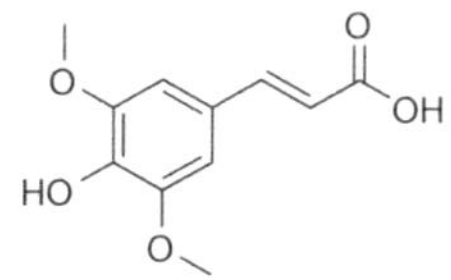	Amyloid β (Aβ)1–42 protein-induced Alzheimer's disease (Male ICR mice)	Stage I-prostate cancer
6.	Isoguanosine (Kim et al., 1994)	Inhibits HDAC3/5	**Natural** Seeds of *Croton tiglium* L.	Isomer of Guanosine	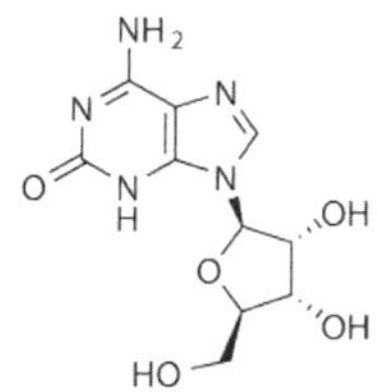	-	-
7.	(-)-Parthenolide (Sztiller-Sikorska and Czyz, 2020)	Inhibitor of the Nuclear Factor-κB Pathway, specifically depletes HDAC1 protein without affecting other class I/II HDACs	**Natural** Sesquiterpene lactone derived from *Tanacetum parthenium* (L.) Sch.Bip.	-	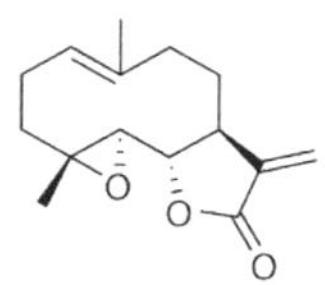	Mice model	I-leukemia, breast cancer, osteosarcoma, melanoma, mesenchymal tumors, prostatic carcinoma II-allergic contact dermatitis

(*Continued*)

TABLE 8.1 (Continued)

Sl. Nos.	Name of Drug	Inhibits HDAC Classes	Natural/ Synthetic/ Semi-Synthetic	Classifications	Structural Formulas	*In vivo* Models Used	Clinical Trial Phases
8.	Raddeanin A (Qian et al., 2014)	Moderate inhibitory activity against histone deacetylases	**Natural** From *Anemone raddeana* Regel	Triterpenoid saponin compound	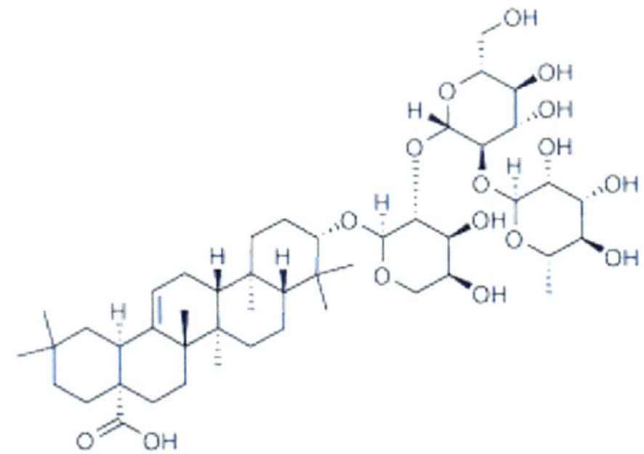	-	
9.	Dacinostat LAQ824 NVP-LAQ824 (Ganai, 2015)	HDAC inhibitor	**Natural** derived from the psammaplin	Hydroxamate	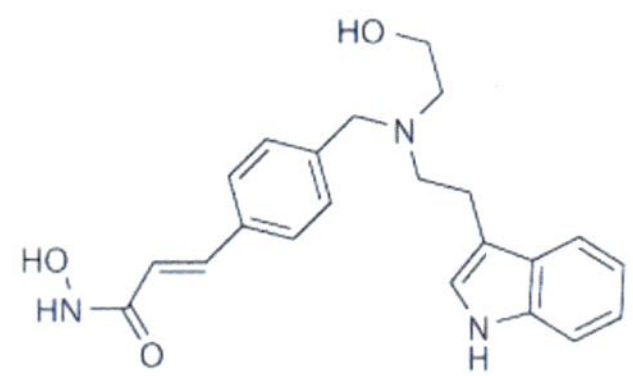	Athymic (nu/nu) female mice	Stage I-non–small cell lung cancer
10.	Valproic acid (Gottlicher et al., 2001)	Class 1 and IIa	**Synthetic** Derivative of propylpentanoic acid	Short-chain fatty acid, organic weak acid	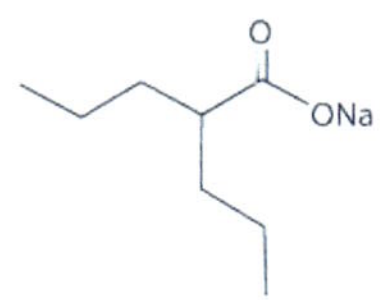	MT-450 rat breast cancer model	Approved for epilepsies, bipolar disorders and migraines I & II-non–small cell lung cancer

11.	Givinostat (Chifotides et al., 2020)	Clinical trials of givinostat as a salvage therapy for advanced Hodgkin lymphoma.	**Synthetic**	Hydroxamic acid, benzenoides	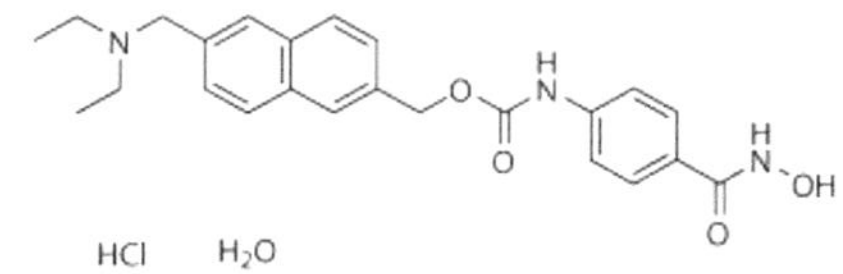	BALB/c mice	Stage II-relapsed leukemia and multiple myeloma III-Duchenne muscular dystrophy
12.	Panobinostat (Li et al., 2014)	Class I, II and IV Class I and IV Classes: HDAC1, HDAC2, HDAC3, HDAC8 Location: nuclear substrates: bistones, Stat3, Bcl-6, P53, E2F-1	**Synthetic**	Cinnamic hydroxamic acid analogue	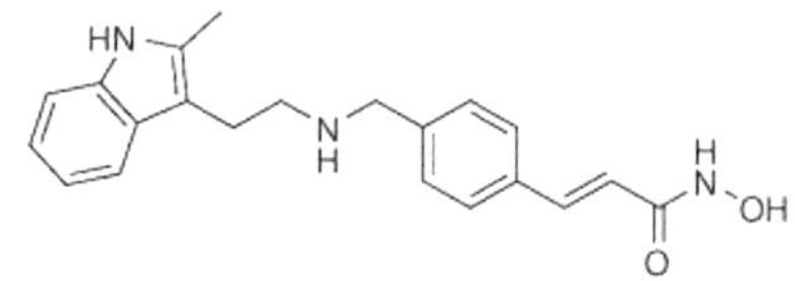	Severe combined immunodeficiency (SCID) mice	IV-chronic idiopathic myelofibrosis III-lymphoma, multiple myeloma, leukemia
13.	Sodium Phenylbutyrate (Ma et al., 2016)	Class I and IIa Urea cycle disorders	**Synthetic**	Benzenoids	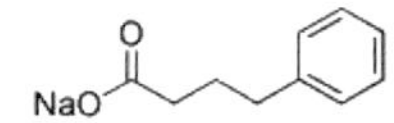	ALS mice	IV-diabetes insulin resistance, deficiencies in enzymes of the urea cycle III-Alzheimer's disease, inborn errors of amino acid metabolism and leukemia
14.	Mocetinostat or HBI-8000 (Nural-Guvener et al., 2015)	Class I and IV Classes: HDAC1, HDAC2, HDAC3, HDAC8 Location: Nuclear Substrates: Histones, Stat3, Bcl-6, P53, E2F-1	**Synthetic**	Benzenoids (aminobenzamide)	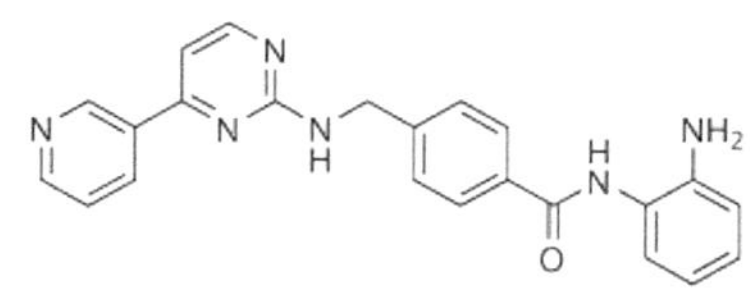	Tumor xenografts in nude mice	II-Hodgkin's lymphoma, leukemia, non–small cell lung cancer

(Continued)

TABLE 8.1 (Continued)

Sl. Nos.	Name of Drug	Inhibits HDAC Classes	Natural/ Synthetic/ Semi-Synthetic	Classifications	Structural Formulas	*In vivo* Models Used	Clinical Trial Phases
15.	Tucidinostat or Chidamide (Kong et al., 2020)	HDAC1, 2, 3, and 10	**Synthetic** analogue of MS-275	Benzamide	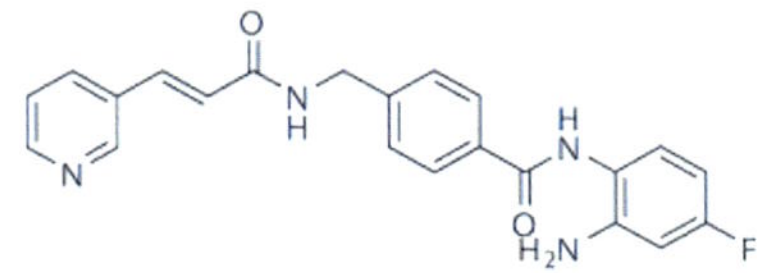	Athymic nude mice (BALB/c-nu)	IV-T-cell lymphoma III-breast cancer, B-cell lymphoma, adult T-cell lymphoma
16.	Tinostamustine (EDO-S101) (Besse et al., 2017)	Class I and II	**Synthetic**	Fusion molecule alkylator Bendamustine and Vorinostat	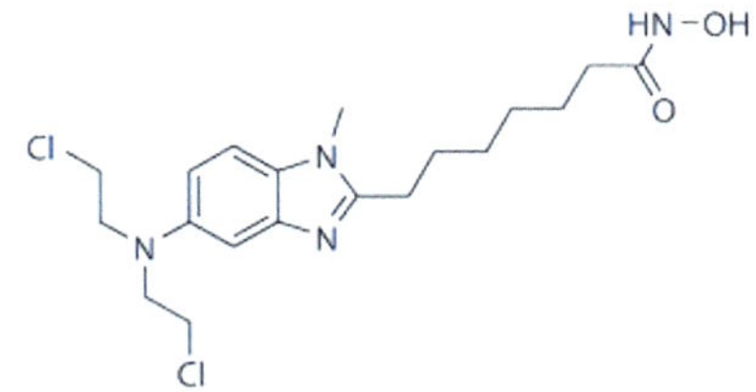	Burkitt's lymphoma xenograft model	II-multiple myeloma, endometrial cancer, small cell lung cancer I-glioblastoma, T-cell lymphoma
17.	Belinostat PXD101 NSC726630 PX-105684 (Chien et al., 2014)	Thymic cancer cells Classes: HDAC1, HDAC2, HDAC3, HDAC8 Location: nuclear substrates: bistones, Stat3, Bcl-6, P53, E2F-1	**Synthetic**	Benzamide	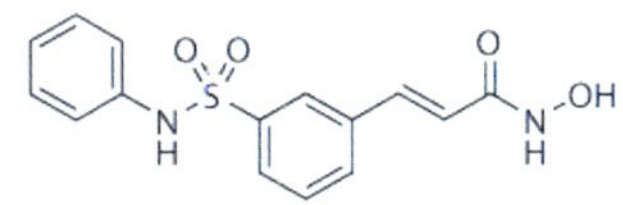	athymic mice	Status: approved for peripheral T-cell lymphoma II-glioblastoma, acute myeloid leukemia, thymic carcinoma, ovarian cancer, multiple myeloma
18.	MC1568 (Venza et al., 2013)	Selective class IIA	**Synthetic**	Derivative of (Aryloxopropenyl) pyrrolyl hydroxyamide	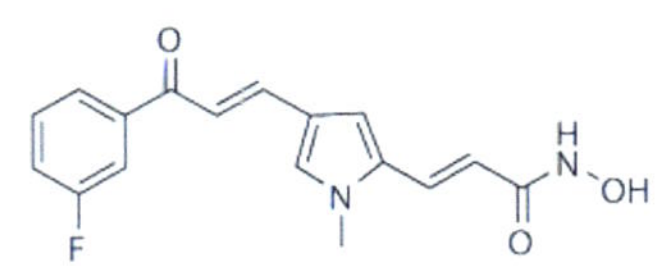	PPRE-Luc transgenic mouse (C57BL/6)	II-chronic lymphocytic leukemia

19.	Vorinostat or Suberoylanilide hydroxamic acid (SAHA) (Munshi et al., 2006)	HDAC1, HDAC2 and HDAC3 (Class I) and HDAC6 (Class II) Class I and IV Classes: HDAC1, HDAC2, HDAC3, HDAC8 Location: nuclear substrates: histones, Stat3, Bcl-6, P53, E2F-1	**Synthetic**	Hydroxamate	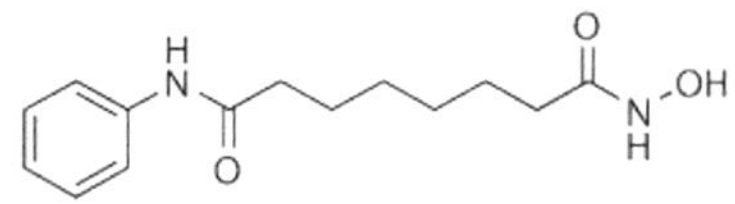	BALB/c nude mice, C57BL/6	Approved for Stage II and III-small cell lung cancer Cutaneous T-cell lymphoma
20.	CUDC-101 (Ji et al., 2018)	Potent multi-targeted inhibitor	**Synthetic**	Multi-targeted small molecule	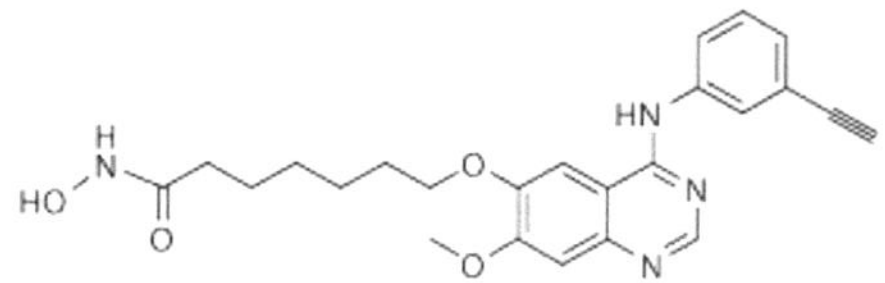	Female athymic mice (nude nu/nu CD-1)	Stage I-breast cancer, head and neck carcinoma, non–small cell lung carcinoma and malignancies
21.	Quisinostat (Orr and Doust, 1988)	HDAC1	**Synthetic**	Hydroxamic acid	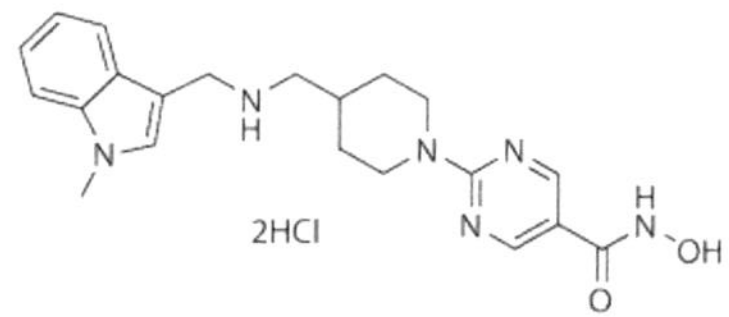	Male MFI nude mice and athymic male mice	Stage II T-cell lymphoma, ovarian cancer I-malignant lymphoma, non–small cell lung carcinoma, ovarian epithelial cancer
22.	Pracinostat SB939 (Novotny-Diermayr et al., 2010)	Potent pan-HDAC inhibitor	**Synthetic**	Hydroxamic acid	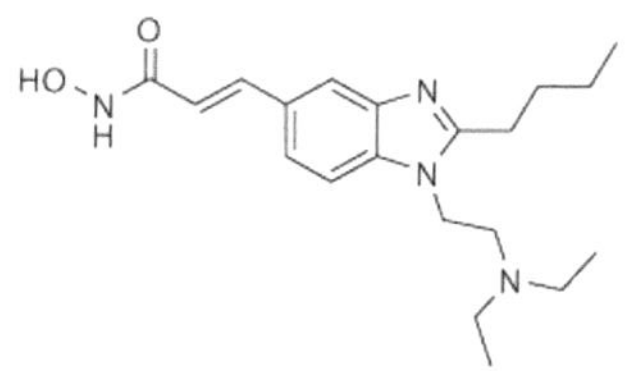	BALB/c nude, Male and APCm in mice	III-acute myeloid leukemia II-myelodysplastic syndrome, metastatic sarcoma, prostate cancer I-leukemia

(Continued)

TABLE 8.1 (Continued)

Sl. Nos.	Name of Drug	Inhibits HDAC Classes	Natural/ Synthetic/ Semi-Synthetic	Classifications	Structural Formulas	*In vivo* Models Used	Clinical Trial Phases
23.	Abexinostat CRA-024781 (Buggy et al., 2006)	Pan-HDAC inhibitor, Classes: HDAC1, HDAC2, HDAC3, HDAC8 Location: nucleus substrates: histones, Stat3, Bcl-6, E2F-1	**Synthetic**	Hydroxamic acid	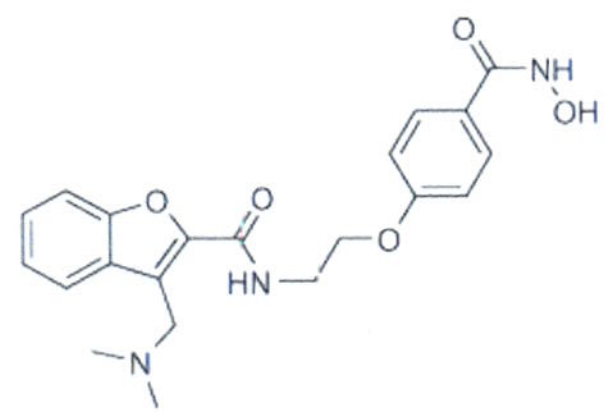	Female BALB/c nu/nu mice	Stage III-renal cell adenocarcinoma II-follicular lymphoma, B-cell lymphoma I-advanced melanoma, head and neck squamous cell carcinoma and lung cancer
24.	AR-42 HDAC-42 (Lu et al., 2007)	HDAC inhibitor	**Synthetic**	Short-chain fatty acid	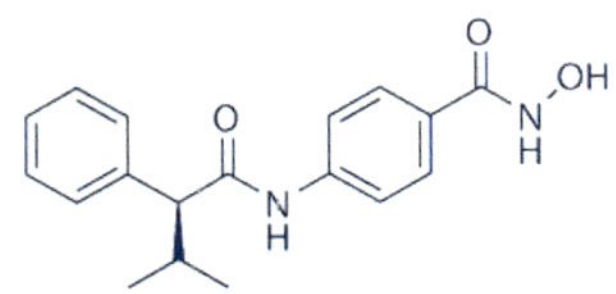	Male NCr athymic nude mice	I-acute myeloid leukemia, multiple myeloma, NK/T-cell lymphoma
25.	Rocilinostat (Cosenza et al., 2017)	Selective HDAC6 inhibitor	**Synthetic**	Benzamides	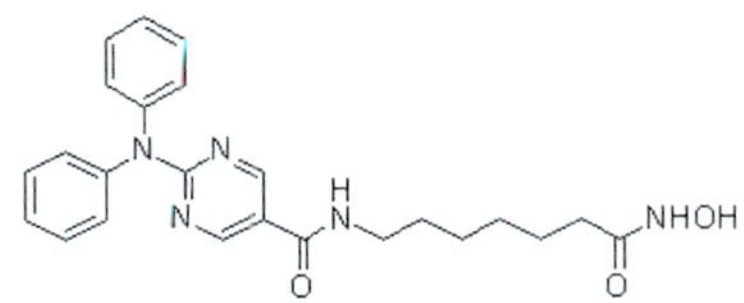	SCID mouse model	II-diabetic neuropathic pain I-metastatic breast cancer 3.3 m
26.	RG2833 (RGFP109) (Johnston et al., 2013)	Brain-penetrant HDAC inhibitor, histone deacetylase 3 (HDAC-3) and HDAC1	**Synthetic**	Benzamides	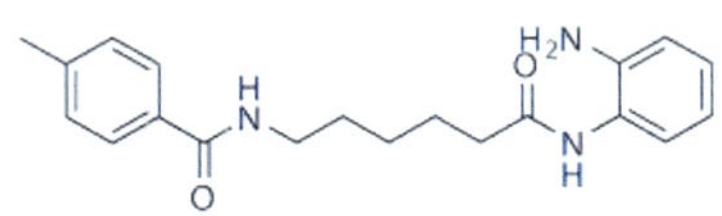	KIKI mice	Stage I-Friedreich's ataxia (FA)

27.	Resminostat (Mandl-Weber et al., 2010)	Selectively inhibits HDAC1/3/6, oral pan-HDAC inhibitor specifically targeting class I HDAC	**Synthetic**	Benzenesulphoneamides	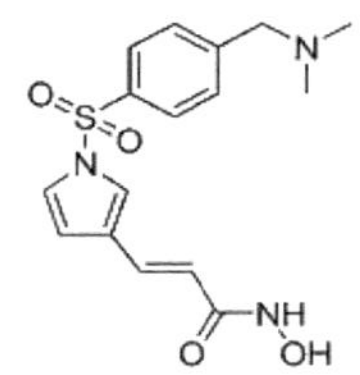	Human xenograft mice tumor models	Stage I and II-hepatocellular carcinoma, advanced colorectal carcinoma
28.	Scriptaid (Janaki Ramaiah et al., 2017)	Inhibits HDAC III and IV	**Synthetic**	Hydroxamic acid	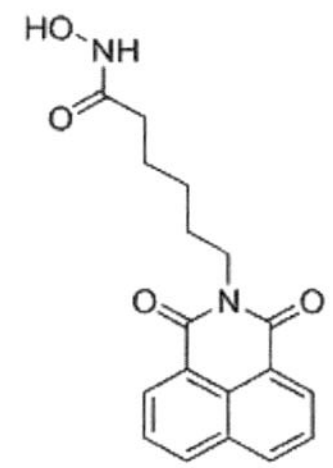	Human breast cancer cells MDA-MB-231	II-advanced breast cancer
29.	TMP269 (Kikuchi et al., 2015)	Selective class IIa HDAC inhibitor	**Synthetic**	Benzide	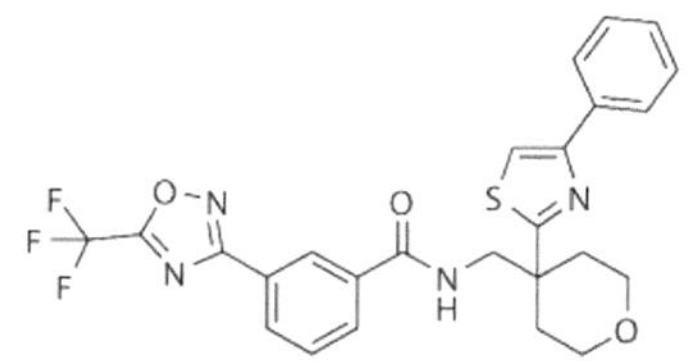	MMTV-PyMT transgenic mice	I-breast cancer
30.	Tasquinimod (Isaacs et al., 2006)	Allosterically inhibits HDAC4 signaling prostate cancer	**Synthetic**	Quinoline-3-carboxamide linomide analogue	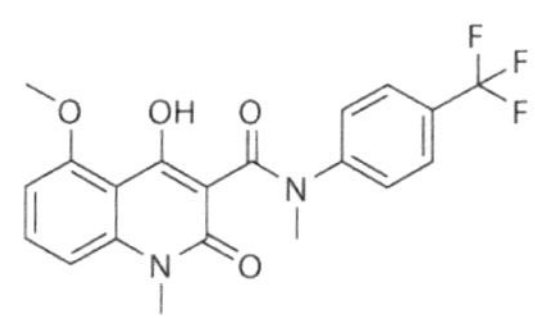	Mice bearing PC-82, CWR-22Rv1, LAPC-4, or LNCaP prostate cancer xenografts	III-prostate cancer I-multiple myeloma

HDAC inhibitors continues to take the priority (Kim and Hong, 2015). For example, the secondary metabolites of plants such as polyphenols/phenolics, alkaloids, N-containing compounds, organosulfur compounds, phytoesterols and carotenoids have been demonstrated to inhibit HDAC in cultured cells (Evans and Ferguson, 2018). Therefore, in the subsequent sections, current knowledge on widely studied naturally occurring HDAC inhibitors is discussed.

8.4.1.1 Hydroxamic Acids

Suberanilohydroxaxmic acid (SAHA; also known as Vorinostat) is a broad-spectrum HDAC inhibitor approved by the FDA in October 2006 for the treatment of cutaneous manifestations in patients with cutaneous T-cell lymphoma who had progressive, persistent or recurrent disease on or following two systemic therapies (Duvic and Vu, 2007). Vorinostat is a linear hydroxamate compound, developed by Merck & Co. Inc. In addition to cutaneous T-cell lymphomas, SAHA was also tested against solid tumors such as the carcinomas of breast, head and neck, and Hodgkin's lymphoma (Mottamal et al., 2015). Oral administration of SAHA (400 mg every day) produced an objective response of ~31% in a Phase II trial compared to an intermittent dosing, which yielded only 9% response (Pourrajab et al., 2020). However, severe side effects that include anemia and thrombocytopenia were reported in patients who had received Vorinostat by intravenous administration (Bubna, 2015). Vorinostat is currently being studied in 39 ongoing clinical trials as a monotherapeutic agent against leukemia, non–small cell lung cancer, multiple myelomas and breast, pelvic and prostate cancers (Siegel et al., 2009).

Mechanistically, treatment with SAHA reduced phosphorylated Akt/FOXO3a signaling thereby promoted apoptosis in prostate cancer cell lines DU145 and PC-3 (Shi et al., 2017). Additional studies have shown that SAHA can prevent the early-stage melanocytic lesion development in skin; hence, several derivatives of SAHA containing selenium were prepared and tested for safety and efficacy in cultured cells and in xenografted animals (Gowda et al., 2012). In small-cell lung cancer lines NCI-H520, NCI-H460, NCI-H522, NCI-H1299 AND SK-MES-1, addition of SAHA blocked cell growth at G0/G1 and G2/M phases (Asgar et al., 2016). Future studies should focus on combining Vorinostat—along with other chemotherapeutic drugs—to identify a synergistically acting combination therapy for treating cancers. Initial studies in this direction are currently under progress (Suraweera et al., 2018).

8.4.1.2 Benzamides

Benzamides are another class of HDAC inhibitors synthesized and tested against several carcinomas (Eckschlager et al., 2017). Although benzamides are primarily produced by chemical synthesis, a brief note on this group is provided as a part of this chapter to highlight the importance of this class of HDAC inhibitors in developing potent anti-cancer agents. Earlier studies have demonstrated that benzamide type HDAC inhibitors are relatively non-toxic (compared to hydroxamates) and more specifically target a particular class of HDAC enzymes (Chou et al., 2008). For example, a study showed that pimelic diphenylamide specifically inhibits Class I HDAC, with no demonstrated activity against Class II HDACs (Chou et al., 2008). Among Class I HDACs, benzamide inhibitor R106 inhibited HDAC3 with 15 times lower inhibitor constant compared to HDAC1 (Chou et al., 2008). Similarly, many other benzamide type inhibitors were synthesized and tested against multiple cancers (Abdizadeh et al., 2017).

CI-994 is another benzamide-based HDAC inhibitor, which was produced from the original compound Dinaline (PD 104 208) (Zhang et al., 2018). CI-994 is an oral cytostatic drug with potent activity against leukemic and normal stem cells. Further, CI-994 is evaluated in a combination therapy against non–small cell lung, pancreatic, breast and colorectal cancers (Li and Seto, 2016). CI-994 blocks cells in the G1-S phase of the cell cycle by inhibiting Class I HDACs. Very recently, a new 5-aryl-substituted 2-amino benzamide type HDAC inhibitor K-856 was synthesized by incorporating diketopiperazine group and tested against ischemia-induced neuronal cell death. K-856 was shown to exhibit a potent neuronal survival activity (Hirata et al., 2018). However, its efficacy against preventing the transformation of normal cells in to cancer cells yet to be tested (Hirata et al., 2018).

8.4.1.3 Cyclic Peptides and Depsipeptides

Cyclic peptides, also referred as 'macrocyclic' compounds, inhibit HDAC by interacting with the outer rim of the enzyme (Mwakwari et al., 2010). These cyclic compounds usually contain complex cap-groups which bind to the HDAC's outer rim regions, thereby inhibiting the enzyme activity with more potency and isoform selectivity (Maolanon et al., 2016). Based on their macrocyclic moieties, these HDACi are subdivided into: (a) cyclic peptides; and (b) depsipeptides. The first class contains a mixture of L- and D-amino acids and cyclic amino acids such as proline or pipecolic acid (Mwakwari et al., 2010). Examples of cyclic peptides are trapoxin-A and trapoxin-B, apicidin (a fungal metabolite from fusarium species), azumamide (isolated from Mycale izuensis) and HC-toxin (Mwakwari et al., 2010).

The second class comprises of bicyclic depsipeptides isolated from bacterial fermentation extracts (Mwakwari et al., 2010). These bicyclic depsipeptides are less stereochemically constrained, due to larger macrocycles compared to the cyclic tetrapeptides (Batiste and Johnston, 2016). They are prodrugs and require intracellular reactions to unmask their thiol groups. Once unmasked, the thiol groups bind to enzyme hydrophobic residues present in the outer rings (Mwakwari et al., 2010). As the enzyme rim residues are less conserved, these depsipeptides discriminate all 11 HDAC isoforms in classes I, II and IV (Nielsen et al., 2005). FK-228 is the most widely studied depsipeptide isolated from *Chromobacterium violaceum*. Other examples of depsipeptides are Spiruchostatin-A and -B (isolated from *Pseudomonas* species), and Largazole (isolated from cynobacterium *Symploca* species) (Furumai et al., 2002; Narita et al., 2009; Potharla et al., 2014).

8.4.1.4 Short-Chain Fatty Acids (SCFAs)

SCFAs are another class of HDAC inhibitors known to impede cancer cell growth (Anantharaju et al., 2017a). SCFAs are produced in humans as a result of gut-microbial activity on dietary fiber components. The most predominantly produced SCFAs are acetate, propionate and butyrate (den Besten et al., 2013). Several studies have shown that butyrate, but not propionate, inhibit HDAC activity more effectively (Silva et al., 2018). Currently, butyrate is being clinically evaluated as an antineoplastic agent not only due to its ability to induce cell cycle arrest, differentiation, and/or apoptosis in cancer cells, but also due to its favorable safety profile in humans (Chen et al., 2003). Mechanistically, gut microbiota–derived butyrate inhibits Class I HDACs, thereby promoting crotonylation of histone-3 (H3) at lysine 18 residue, which subsequently regulates the expression of several genes involved in cells' proliferation and survival (Fellows et al., 2018). Structure-inhibition relationship (SIR) studies further confirm that butyrate more specifically inhibits Class I HDACs, in particular HDAC3 compared to Class II HDACs such as HDAC7 (Steliou et al., 2012).

Another widely studied SCFA is Valproic acid (VPA), which is an eight-carbon short-chain fatty acid reported to mitigate epilepsy (Licciardi et al., 2011). Although the molecular mechanisms leading to the reduction of complications of epilepsy are not fully understood, it has been well documented that the level of inhibitory neurotransmitter i.e., GABA (γ-amino butyric acid) increases in the brain upon treatment with VPA (Nagaki et al., 1990). Recent studies have shown that VPA can induce differentiation of human leukemia cells (Rucker et al., 2016). However, further studies evaluating VPA in clinical trials are required to recommend this eight-carbon fatty acid for further use as a treatment agent for cancers.

Recent studies have reported the isolation and characterization of tricaproin (TCN) from *Simarouba glauca* DC. leaves and demonstrated its ability to inhibit Class I HDACs (Jose et al., 2018). TCN is made up of glycerol and hexanoic acid (known as caproic acid, with a molecular formula of $C_6H_{12}O_2$) (da Rocha Ataide et al., 2007). Hexanoic acid, a six-carbon SCFA, is released as a result of hydrolysis of TCN and inhibits Class I HDACs, thereby reducing the proliferation and migration of cancer cells without affecting the normal ones (Yuille et al., 2018). Additional studies have further shown that TCN impedes the growth of Ehrlich ascites carcinoma (EAC) cells growth in Swiss albino mice (Jose et al., 2018).

In summary, SCFAs—which are either produced in the gut by the microbial action or acquired from dietary sources—can inhibit HDAC enzymes, thereby impeding the proliferation and migration of cancer cells. However, to date, no single fatty acid is in clinical use for treating cancers. Therefore, additional studies are immediately warranted to test fatty acids as anti-cancer agents.

8.4.1.5 Hybrid Molecules

In order to overcome various drawbacks associated with HDAC inhibitors that include a lack of efficacy as monotherapeutic agents, systemic toxicity and poor stability, several groups have synthesized hybrid molecules that can inhibit other oncogenic targets, along with HDAC (Li and Seto, 2016). For instance, Roxyl-zhc-84, which inhibits HDAC along with CDK, has been shown to be more effective in impeding solid tumors (Heijkants et al., 2018). Mechanistically, Roxyl-zhc-84 induced apoptosis and G1-phase arrest in breast cancer and ovarian cancer cell lines (Huang et al., 2018). Further, in mice, oral administration of Roxyl-zhc-84 led to significant tumor regression and improved the limited response of traditional HDAC inhibitors in solid tumors by overcoming JAK1-STAT3-BCL2–mediated drug resistance (Huang et al., 2018). Many such hybrid molecules were synthesized and tested against carcinomas of breast, colon and rectum, and prostate and pancreatic cancers (Nagy et al., 1998). A hybrid molecule targeting topoisomerase-1 and HDAC was synthesized and shown to be more effective against cancer cells *in vitro* and *in vivo* (Cincinelli et al., 2018). Likewise, another hybrid drug 2–75 was synthesized and demonstrated to produce better anti-cancer effects against carcinomas of prostate (Hu et al., 2019).

8.5 OTHER HDAC INHIBITORS

8.5.1 Phenolic Compounds

In a recent study, Anantharaju et al. (2017b) showed that phenolic compounds derived from benzoic acid and cinnamic acid backbones could inhibit the growth of colorectal and cervical cancer cells *in vitro* by interacting with the trichostatin-A binding regions of HDAC2 (Anantharaju et al., 2017b). Among various phenolic acids tested, protocatechuic acid and caffeic acid were reported to inhibit HDAC2 compared to their methoxy-group counterparts Veratric acid (3,4-dimethoxy benzoic acid) and caffeic acid dimethyl ether (3,4-dimethoxy cinnamic acid) (Anantharaju et al., 2016). Mechanistically, inhibiting HDAC-elevated cellular reactive oxygen species (ROS) and induced cell cycle arrest at G2/M phase while promoting caspase-3 mediated apoptosis in cancer cells. A separate study showed HDAC inhibitory potential of gallic acid in prostate cancer cells (Jang et al., 2020). Treating prostate cancer cell lines LNCaP and PC-3 with gallic acid reduced HDAC1 and HDAC2, thereby mitigating mitochondrial membrane potential, induced apoptosis by augmenting the expression of cleaved caspase-3 and poly (ADP-ribose) polymerase 1 (PARP-1) (Jang et al., 2020).

Additionally, inhibition of HDAC upregulated the expression of acetyl-p53 and downregulated Cyclin D1 and E1, proliferating cell nuclear antigen (PCNA) and Bcl-2 (Jang et al., 2020). In animals, administration of gallic acid reduced the xenografted tumors' growth (Kaur et al., 2009). Ethanolic extract of *Hydnophytum formicarum* Jack, which is rich in sinapinic acid, inhibited cervical cancer cell line HeLa and colorectal cancer cell line HT29 by downregulating HDAC activity (Senawong et al., 2013). In summary, phenolic acids isolated from plants inhibit HDAC activity thereby reduce tumor growth. Future studies should focus on identifying the most potent phenolic acid, which can inhibit HDAC and develop a nutraceutical formulation to reduce the complications associated with deregulated HDAC activity.

8.5.2 Sulforaphane (SFN)

Sulforaphane is a sulfur-containing aliphatic isothiocyanate present in abundant quantities in various cruciferous vegetables such as *Brassica cretica* Lam. (broccoli) (Kiani et al., 2018). Several studies have shown that SFN and its metabolites inhibit HDACs, thereby increasing acetylated histones and cell cycle inhibitors p21 and p27 (Ho et al., 2009). SFN is metabolized via the mercapturic acid pathway, beginning with the conjugation of glutathione by glutathione-S-transferase (GST), followed by generation of SFN-cysteine (SFN-Cys) and SFN-*N*-acetylcysteine (SFN-NAC). SFN-NAC or SFN-Cys are the active inhibitors of HDAC (Myzak et al., 2006).

In addition to inhibiting HDAC, SFN is reported to promote Phase II enzymes, thereby downregulating the activation of carcinogens (Jiang et al., 2018). For instance, in a recent study, it has been observed that SFN promotes the expression of HO-1 protein, NADPH quinone oxido-reductase 1 (NQO1), glutamate-cysteine ligase catalytic subunit (GCLC), epoxide hydrolase, etc., thereby preventing the induction of prostate cancers (Lan et al., 2016). The effects of SFN are specific to cancer cells. A study demonstrated that about 15 μM SFN inhibited HDAC activity and arrested prostate cancer cells in G2/M phase without causing any effects in normal prostate cells (Lan et al., 2016). Based on these encouraging *in vitro* studies and supporting data from preclinical animal models, the bioavailability and distribution of SFN have been tested in humans (Atwell et al., 2015).

Results of these studies have demonstrated that SFN is efficiently absorbed and conjugated to GSH. Moreover, the pharmacokinetic studies showed that SFN is distributed in the body and reach micromolar concentrations in the blood (Ho et al., 2009). For example, a single dose of 200 μmol broccoli sprouts ITC preparation showed a peak concentration between 0.943 and 2.27 μmol/L one hour after feeding (Ullah, 2015). Furthermore, oral administration of broccoli sprouts containing 200 μmol SFN showed a mean accumulation of SFN in both the breasts (Ho et al., 2009). Accumulated SFN induced the detoxification genes NQO1 and heme oxygenase-1 indicating that the tissue incorporated SFN is functionally active and assists in reducing the oxidative stress (Guerrero-Beltran et al., 2012). However, to date, only minimal success for preventing/treating cancers is observed with SFN, hence warranting additional studies evaluating SFN in combination with other established chemotherapeutic agents.

8.5.3 Allyl Sulfide (AS)

Allyl sulphides, which are derived from *Allium sativum* L. (garlic), are known to inhibit HDAC activity in cancer cells (Nian et al., 2009b). In general, the metabolites of organo-sulphur allyl mercaptan (AM), organo-selenium compound β-methylselenopyruvate (MSP) and α-keto-γ-methylselenobutyrate (KMSB) have been shown to exhibit better HDAC inhibitory activity than the native compounds (Lee et al., 2009). MSP and KMSB are α-keto acid metabolites of Se-methylselenocysteine (MSC) and selenomethionine (SM), respectively. Computational studies suggested that these metabolites fit into the active site of HDAC enzymes and chelate catalytic Zn^{2+} via sulfhydryl group (AM) or keto acid group (MSP and KMSB). Further, these compounds induce cell cycle arrest and apoptosis via p21WAF1-mediated mechanism in human colon cancer cells (Nian et al., 2009a). A study by Nian et al. (2008) reported that allyl mercaptan inhibits the HDAC activity in HT29 human colon adenocarcinoma cell lines (Nian et al., 2008).

8.5.4 Diallyl Disulfide (DADS)

DADS is another sulfur-containing compound isolated from *A. sativum*. Several *in vitro* and preclinical studies have demonstrated the ability of DADS for inhibiting HDAC (Nian et al., 2008). Inhibition of HDAC by DADS increased acetylated histones and promoted cell cycle arrest through the elevation of p21 in many cancer cell lines (Druesne et al., 2004). In animals, intraperitoneal or oral administration of DADS impeded xenografted tumors through induction of apoptosis and reduction of proliferation rates (Yin et al., 2014). Molecularly, administered DADS resulted in the hyperacetylation of histones, which subsequently triggered apoptosis and autophagic cell death in tumors (Kim et al., 2016). However, it is currently unknown whether DADS induced cell death is due to the direct action of this compound on HDAC, and if so, which HDAC is inhibited by DADS. Addressing these questions, along with measuring the biodistribution and pharmacokinetics of DADS, will provide necessary information for suggesting DADS as a chemotherapeutic agent for treating cancers.

8.5.5 Benzisothiazol

Benzisothiazol is another sulfur-containing compound with more selectivity towards HDAC6 (Wang et al., 2017). It has a core structure of ebsulfur. Ebsulfur is a known analog of Ebselen, which is a clinically safe and a weak inhibitor of HDAC enzymes (Wang et al., 2017). Derivatives of Benzothiazole having sulfur exhibited potent anti-proliferative activity (GI_{50} = 2.01 μM) against SH-SY-5Y cancer cell line in a dose and time-dependent manner, which is slightly better than the reference drug SAHA (GI_{50} = 2.90 μM). This compound efficiently inhibits HDAC1 and HDAC6 isoforms with IC_{50} values in nanomolar concentrations. Furthermore, Benzothiazole inhibited colony formation capacity of SH-SY-5Y cells (Choi et al., 2019). Further studies on structure-inhibition relationships lead to the development of another compound called RBC-2008. RBC-2008 is synthesized based on the structure of RBC-2004 by adding a hydroxamic acid to the benzoyl moiety (Wang et al., 2017). RBC-2008, although inhibited all 11 HDAC enzymes, showed more selectivity towards HDAC6 (Wang et al., 2017).

8.6 IMMUNOMODULATORY EFFECTS OF HDAC INHIBITORS

Recent studies have shown that HDACis have the potential of modulating immune functions as they were involved in controlling the NF-κB and STAT3 pathways (Dickinson et al., 2010; Schotterl et al., 2015). NF-κB plays a critical role in the innate and adaptive immune responses, cell proliferation, inhibition of apoptosis, promotion of cell migration, invasion and stimulation of angiogenesis and metastasis, which lead to the acceleration of cancer progression (Grivennikov et al., 2010).

Mechanistically, the activation of NF-κB depends on the degradation of the specific inhibitor of NF-κB, i.e., IκB, followed by the phosphorylation by IκB kinase (IKK) complex (Karin, 1999). HDAC1 and HDAC2 interacts with NF-κB family proteins by binding with the co-repressor protein p65 (RELA) and p50 (NFκB1) and downregulate NF-κB-mediated gene transcription. HDAC3-mediated deacetylation of p65 results in shuttling of the NF-κB complex from the nucleus to the cytoplasm, thus controlling the pro-inflammatory response. NF-κB activation is also dependent on HDAC3 in the regulation of target genes IκB-α, IL-2, IL-6 promoter hyperacetylation (Leus et al., 2016).

HDAC enzymes could help control Th1 and Th2 differentiation of naive CD4+ T cells by reversing the hyperacetylation of histones three and four at the IFN-γ promoter (Licciardi and Karagiannis, 2012). Moreover, HDACs are also involved in the balance of inflammatory responses mediated by IL-6, IL-8, IL-1β, and GM-CSF by regulating the histone acetylation status of NF-κB and AP-1 (Licciardi and Karagiannis, 2012; Gatla et al., 2019). HDAC inhibitors entinostat and valproic acid (VPA) have been shown to impair poly-I:C-stimulated DC activation *in vitro*, thereby decreasing the nuclear RelB, IRF-3 and IRF-8 (McCaw et al., 2017). The pan-HDAC inhibitor Panobinostat reduces antigen uptake, expression of costimulatory molecules and production of Th1 polarizing inflammatory cytokines by poly-I:C-stimulated DCs, which in turn impairs the differentiation of IFN-γ–producing T cells (McCaw et al., 2017).

Modulation of CD8 T-cell responses with HDAC inhibitors is another important aspect in controlling tumor growth. Tumor-oriented cytolytic CD8 effector T cells, which produce IFN-γ and TNF-α, cytokines that activate APCs can make the tumor cells more immunogenic by enhancing MHC expression, activating the immunoproteasome and promoting tumor cytotoxicity (de Charette et al., 2016; Jorgovanovic et al., 2020). For example, a combination of entinostat and IL-2 elicits dramatic increases in IFN-γ production and CD8 T-cell cytotoxicity in a mouse model of renal cancer (Shen et al., 2012).

Treatment with HDAC inhibitors also increases the susceptibility of tumor cells to other mechanisms of T-cell–mediated killing (Li and Seto, 2016). For example, studies have shown that the treatment of primary melanoma samples with the inhibitor Vorinostat leads to DR5-dependent killing of tumor cells by CD8 T cells and the treatment of human colon carcinoma cells with Vorinostat elevated the expression of death receptor Fas and Fas-dependent cytotoxicity by T cells (Walker et al., 2009). Many studies have shown that the combination of HDACi and the immunotherapy could yield better treatment outcomes (Bridle et al., 2013; Li and Seto, 2016). The potential use of HDACis in the treatment of cancer and immune disorders or as adjuvants that direct the generation of specific immune phenotypes has significant implications in the design of new pharmacological drugs. Increased knowledge about the biological properties of HDACis will help in understanding the molecular mechanisms of immune system-mediated tumor therapies.

8.7 TOXICITY ASSOCIATED WITH HDAC INHIBITORS

Even though several naturally occurring and synthetic HDAC inhibitors have been developed, only very few agents have reached the clinical trials stage. Even among those which have reached clinical trials stage, many have failed to hit the market due to various toxicities and poor efficacy in inhibiting the disease symptoms at the dose tested. For example, SAHA (Vorinostat) is an FDA-approved drug for treating T-cell lymphoma. However, when administered at a dose of 300 mg/day, the patients have reported diarrhea, thrombocytopenia, nausea and pyrexia. Adverse events such as cardio-toxicity were reported when tested at a dose of 400 mg/day (Duvic and Vu, 2007). Similarly, systemic toxic effects have been reported for other FDA-approved HDAC inhibitors, which include Belinostat, Panobinostat, etc. (Allen and Lechowicz, 2018). Therefore, care must be executed when recommending HDAC inhibitors for treatment. Recent studies have focused on reducing the toxicity of HDAC inhibitors by: (a) combining with other pharmacological agents or other treatment methods such as radiation therapy, surgery etc.; (b) developing targeted nanodrug formulations; and (c) antibody conjugates. However, further studies are currently required to determine the safety and efficacy of these methods or formulations.

8.8 SUMMARY AND CONCLUSIONS

In summary, this chapter highlighted the role of HDAC proteins in the tumor development and chemoresistance of cancers. Furthermore, a detailed analysis of HDAC inhibitors, their safety and their efficacy for inhibiting tumor cell growth was discussed. In conclusion, HDAC is a potent therapeutic target to treat cancers. Future studies should focus more on establishing clinically safer approaches for delivering these potent HDAC inhibitors into tumor cells.

REFERENCES

Abdizadeh, T., Kalani, M.R., Abnous, K., Tayarani-Najaran, Z., Khashyarmanesh, B.Z., Abdizadeh, R., et al. 2017. "Design, synthesis and biological evaluation of novel coumarin-based benzamides as potent histone deacetylase inhibitors and anticancer agents." *European Journal of Medicinal Chemistry* 132: 42–62. doi: 10.1016/j.ejmech.2017.03.024.

Allen, P.B., and Lechowicz, M.J. 2018. "Hematologic toxicity is rare in relapsed patients treated with belinostat: A systematic review of belinostat toxicity and safety in peripheral T-cell lymphomas." *Cancer Management and Research* 10: 6731–6742. doi: 10.2147/CMAR.S149241.

Anantharaju, P.G., Gowda, P.C., Vimalambike, M.G., and Madhunapantula, S.V. 2016. "An overview on the role of dietary phenolics for the treatment of cancers." *Nutrition Journal* 15, 1: 99. doi: 10.1186/s12937-016-0217-2.

Anantharaju, P.G., Reddy, B.D., Padukudru, M.A., Chitturi, C.M.K., Vimalambike, M.G., and Madhunapantula, S.V. 2017a. "Naturally occurring benzoic acid derivatives retard cancer cell growth by inhibiting histone deacetylases (HDAC)." *Cancer Biology & Therapy* 18, 7: 492–504. doi: 10.1080/15384047.2017.1324374.

Anantharaju, P.G., Reddy, D.B., Padukudru, M.A., Chitturi, C.M.K., Vimalambike, M.G., and Madhunapantula, S.V. 2017b. "Induction of colon and cervical cancer cell death by cinnamic acid derivatives is mediated through the inhibition of Histone Deacetylases (HDAC)." *PLoS One* 12, 11: e0186208. doi: 10.1371/journal.pone.0186208.

Asgar, M.A., Senawong, G., Sripa, B., and Senawong, T. 2016. "Synergistic anticancer effects of cisplatin and histone deacetylase inhibitors (SAHA and TSA) on cholangiocarcinoma cell lines." *International Journal of Oncology* 48, 1: 409–420. doi: 10.3892/ijo.2015.3240.

Atwell, L.L., Hsu, A., Wong, C.P., Stevens, J.F., Bella, D., Yu, T.W., et al. 2015. "Absorption and chemopreventive targets of sulforaphane in humans following consumption of broccoli sprouts or a myrosinase-treated broccoli sprout extract." *Molecular Nutrition & Food Research* 59, 3: 424–433. doi: 10.1002/mnfr.201400674.

Audia, J.E., and Campbell, R.M. 2016. "Histone Modifications and Cancer." *Cold Spring Harbor Perspectives in Biology* 8, 4: a019521. doi: 10.1101/cshperspect.a019521.

Banks, C.A.S., Miah, S., Adams, M.K., Eubanks, C.G., Thornton, J.L., Florens, L., et al. 2018. "Differential HDAC1/2 network analysis reveals a role for prefoldin/CCT in HDAC1/2 complex assembly." *Scientific Reports* 8, 1: 13712. doi: 10.1038/s41598-018-32009-w.

Bannister, A.J., and Kouzarides, T. 2011. "Regulation of chromatin by histone modifications." *Cell Research* 21, 3: 381–395. doi: 10.1038/cr.2011.22.

Bardai, F.H., and D'Mello, S.R. 2011. "Selective toxicity by HDAC3 in neurons: Regulation by Akt and GSK3beta." *Journal of Neuroscience* 31, 5: 1746–1751. doi: 10.1523/JNEUROSCI.5704-10.2011.

Barneda-Zahonero, B., and Parra, M. 2012. "Histone deacetylases and cancer." *Molecular Oncology* 6, 6: 579–589. doi: 10.1016/j.molonc.2012.07.003.

Bassett, S.A., and Barnett, M.P. 2014. "The role of dietary histone deacetylases (HDACs) inhibitors in health and disease." *Nutrients* 6, 10: 4273–4301. doi: 10.3390/nu6104273.

Basta, J., and Rauchman, M. 2015. "The nucleosome remodeling and deacetylase complex in development and disease." *Translational Research* 165, 1: 36–47. doi: 10.1016/j.trsl.2014.05.003.

Batiste, S.M., and Johnston, J.N. 2016. "Rapid synthesis of cyclic oligomeric depsipeptides with positional, stereochemical, and macrocycle size distribution control." *Proceedings of the National Academy of Sciences U S A* 113, 52: 14893–14897. doi: 10.1073/pnas.1616462114.

Bayat, S., Shekari Khaniani, M., Choupani, J., Alivand, M.R., and Mansoori Derakhshan, S. 2018. "HDACis (class I), cancer stem cell, and phytochemicals: Cancer therapy and prevention implications." *Biomedicine and Pharmacotherapy* 97: 1445–1453. doi: 10.1016/j.biopha.2017.11.065.

Beausoleil, S.A., Jedrychowski, M., Schwartz, D., Elias, J.E., Villen, J., Li, J., et al. 2004. "Large-scale characterization of HeLa cell nuclear phosphoproteins." *Proceedings of the National Academy of Sciences U S A* 101, 33: 12130–12135. doi: 10.1073/pnas.0404720101.

Bertino, E.M., and Otterson, G.A. 2011. "Romidepsin: A novel histone deacetylase inhibitor for cancer." *Expert Opinion on Investigational Drugs* 20, 8: 1151–1158. doi: 10.1517/13543784.2011.594437.

Besse, L., Kraus, M., Besse, A., Bader, J., Silzle, T., Mehrling, T., et al. 2017. "The first-in-class alkylating HDAC inhibitor EDO-S101 is highly synergistic with proteasome inhibition against multiple myeloma through activation of multiple pathways." *Blood Cancer Journal* 7, 7: e589. doi: 10.1038/bcj.2017.69.

Bhaskara, S., Knutson, S.K., Jiang, G., Chandrasekharan, M.B., Wilson, A.J., Zheng, S., et al. 2010. "Hdac3 is essential for the maintenance of chromatin structure and genome stability." *Cancer Cell* 18, 5: 436–447. doi: 10.1016/j.ccr.2010.10.022.

Bridle, B.W., Chen, L., Lemay, C.G., Diallo, J.S., Pol, J., Nguyen, A., et al. 2013. "HDAC inhibition suppresses primary immune responses, enhances secondary immune responses, and abrogates autoimmunity during tumor immunotherapy." *Molecular Therapy* 21, 4: 887–894. doi: 10.1038/mt.2012.265.

Bubna, A.K. 2015. "Vorinostat-an overview." *Indian Journal of Dermatology* 60, 4: 419. doi: 10.4103/0019-5154.160511.

Buggy, J.J., Cao, Z.A., Bass, K.E., Verner, E., Balasubramanian, S., Liu, L., et al. 2006. "CRA-024781: A novel synthetic inhibitor of histone deacetylase enzymes with antitumor activity *in vitro* and *in vivo*." *Molecular Cancer Therapeutics* 5, 5: 1309–1317. doi: 10.1158/1535-7163.MCT-05-0442.

Canto, C., Gerhart-Hines, Z., Feige, J.N., Lagouge, M., Noriega, L., Milne, J.C., et al. 2009. "AMPK regulates energy expenditure by modulating NAD+ metabolism and SIRT1 activity." *Nature* 458, 7241: 1056–1060. doi: 10.1038/nature07813.

Chen, H.P., Zhao, Y.T., and Zhao, T.C. 2015. "Histone deacetylases and mechanisms of regulation of gene expression." *Critical Reviews in Oncogenesis* 20, 1–2: 35–47. doi: 10.1615/critrevoncog.2015012997.

Chen, J.S., Faller, D.V., and Spanjaard, R.A. 2003. "Short-chain fatty acid inhibitors of histone deacetylases: Promising anticancer therapeutics?" *Current Cancer Drug Targets* 3, 3: 219–236. doi: 10.2174/1568009033481994.

Chien, W., Lee, D.H., Zheng, Y., Wuensche, P., Alvarez, R., Wen, D.L., et al. 2014. "Growth inhibition of pancreatic cancer cells by histone deacetylase inhibitor belinostat through suppression of multiple pathways including HIF, NFkB, and mTOR signaling *in vitro* and *in vivo*." *Molecular Carcinogenesis* 53, 9: 722–735. doi: 10.1002/mc.22024.

Chifotides, H.T., Bose, P., and Verstovsek, S. 2020. "Givinostat: An emerging treatment for polycythemia vera." *Expert Opinion on Investigational Drugs* 29, 6: 525–536. doi: 10.1080/13543784.2020.1761323.

Choi, M.A., Park, S.Y., Chae, H.Y., Song, Y., Sharma, C., and Seo, Y.H. 2019. "Design, synthesis and biological evaluation of a series of CNS penetrant HDAC inhibitors structurally derived from amyloid-beta probes." *Scientific Reports* 9, 1: 13187. doi: 10.1038/s41598-019-49784-9.

Chou, C.J., Herman, D., and Gottesfeld, J.M. 2008. "Pimelic diphenylamide 106 is a slow, tight-binding inhibitor of class I histone deacetylases." *Journal of Biological Chemistry* 283, 51: 35402–35409. doi: 10.1074/jbc.M807045200.

Cincinelli, R., Musso, L., Artali, R., Guglielmi, M.B., La Porta, I., Melito, C., et al. 2018. "Hybrid topoisomerase I and HDAC inhibitors as dual action anticancer agents." *PLoS One* 13, 10: e0205018. doi: 10.1371/journal.pone.0205018.

Cosenza, M., Civallero, M., Marcheselli, L., Sacchi, S., and Pozzi, S. 2017. "Ricolinostat, a selective HDAC6 inhibitor, shows anti-lymphoma cell activity alone and in combination with bendamustine." *Apoptosis* 22, 6: 827–840. doi: 10.1007/s10495-017-1364-4.

Dai, Y., and Faller, D.V. 2008. "Transcription regulation by class III histone deacetylases (HDACs)-sirtuins." *Translational Oncogenomics* 3: 53–65. doi: 10.4137/tog.s483.

da Rocha Ataide, T., d.L.M., Valentim, I.B, Pinheiro, D.M., and Sant'Ana, A.E.G. 2007. "Chemical synthesis of tricaproin, trienantin and tricaprylin." *International Journal of Food Science & Technology* 42, 12: 1504–1508. doi: 10.1111/j.1365-2621.2006.01376.x.

de Charette, M., Marabelle, A., and Houot, R. 2016. "Turning tumour cells into antigen presenting cells: The next step to improve cancer immunotherapy?" *European Journal of Cancer* 68: 134–147. doi: 10.1016/j.ejca.2016.09.010.

den Besten, G., van Eunen, K., Groen, A.K., Venema, K., Reijngoud, D.J., and Bakker, B.M. 2013. "The role of short-chain fatty acids in the interplay between diet, gut microbiota, and host energy metabolism." *Journal of Lipid Research* 54, 9: 2325–2340. doi: 10.1194/jlr.R036012.

De Souza, C., and Chatterji, B.P. 2015. "HDAC inhibitors as novel anti-cancer therapeutics." *Recent Patents on Anti-Cancer Drug Discovery* 10, 2: 145–162. doi: 10.2174/1574892810666150317144511.

Dickinson, M., Johnstone, R.W., and Prince, H.M. 2010. "Histone deacetylase inhibitors: Potential targets responsible for their anti-cancer effect." *Investigational New Drugs* 28, Suppl 1: S3–20. doi: 10.1007/s10637-010-9596-y.

Dong, X., and Weng, Z. 2013. "The correlation between histone modifications and gene expression." *Epigenomics* 5, 2: 113–116. doi: 10.2217/epi.13.13.

Druesne, N., Pagniez, A., Mayeur, C., Thomas, M., Cherbuy, C., Duee, P.H., et al. 2004. "Diallyl disulfide (DADS) increases histone acetylation and p21(waf1/cip1) expression in human colon tumor cell lines." *Carcinogenesis* 25, 7: 1227–1236. doi: 10.1093/carcin/bgh123.

Duvic, M., and Vu, J. 2007. "Update on the treatment of cutaneous T-cell lymphoma (CTCL): Focus on vorinostat." *Biologics* 1, 4: 377–392.

Eckschlager, T., Plch, J., Stiborova, M., and Hrabeta, J. 2017. "Histone Deacetylase Inhibitors as Anticancer Drugs." *International Journal of Molecular Sciences* 18, 7. doi: 10.3390/ijms18071414.

Evans, L.W., and Ferguson, B.S. 2018. "Food bioactive HDAC inhibitors in the epigenetic regulation of heart failure." *Nutrients* 10, 8. doi: 10.3390/nu10081120.

Fellows, R., Denizot, J., Stellato, C., Cuomo, A., Jain, P., Stoyanova, E., et al. 2018. "Microbiota derived short chain fatty acids promote histone crotonylation in the colon through histone deacetylases." *Nature Communications* 9, 1: 105. doi: 10.1038/s41467-017-02651-5.

Fischer, F., Gertz, M., Suenkel, B., Lakshminarasimhan, M., Schutkowski, M., and Steegborn, C. 2012. "Sirt5 deacylation activities show differential sensitivities to nicotinamide inhibition." *PLoS One* 7, 9: e45098. doi: 10.1371/journal.pone.0045098.

Fischle, W., Dequiedt, F., Hendzel, M.J., Guenther, M.G., Lazar, M.A., Voelter, W., et al. 2002. "Enzymatic activity associated with class II HDACs is dependent on a multiprotein complex containing HDAC3 and SMRT/N-CoR." *Molecular Cell* 9, 1: 45–57. doi: 10.1016/s1097-2765(01)00429-4.

Furumai, R., Matsuyama, A., Kobashi, N., Lee, K.H., Nishiyama, M., Nakajima, H., et al. 2002. "FK228 (depsipeptide) as a natural prodrug that inhibits class I histone deacetylases." *Cancer Research* 62, 17: 4916–4921.

Galasinski, S.C., Resing, K.A., Goodrich, J.A., and Ahn, N.G. 2002. "Phosphatase inhibition leads to histone deacetylases 1 and 2 phosphorylation and disruption of corepressor interactions." *Journal of Biological Chemistry* 277, 22: 19618–19626. doi: 10.1074/jbc.M201174200.

Gallinari, P., Di Marco, S., Jones, P., Pallaoro, M., and Steinkuhler, C. 2007. "HDACs, histone deacetylation and gene transcription: From molecular biology to cancer therapeutics." *Cell Research* 17, 3: 195–211. doi: 10.1038/sj.cr.7310149.

Ganai, S.A. 2015. "Strategy for enhancing the therapeutic efficacy of histone deacetylase inhibitor dacinostat: The novel paradigm to tackle monotonous cancer chemoresistance." *Archives of Pharmacal Research*. doi: 10.1007/s12272-015-0673-9.

Gatla, H.R., Muniraj, N., Thevkar, P., Yavvari, S., Sukhavasi, S., and Makena, M.R. 2019. "Regulation of chemokines and cytokines by histone deacetylases and an update on histone decetylase inhibitors in human diseases." *International Journal of Molecular Sciences* 20, 5. doi: 10.3390/ijms20051110.

Gerhart-Hines, Z., Dominy, J.E., Jr., Blattler, S.M., Jedrychowski, M.P., Banks, A.S., Lim, J.H., et al. 2011. "The cAMP/PKA pathway rapidly activates SIRT1 to promote fatty acid oxidation independently of changes in NAD(+)." *Molecular Cell* 44, 6: 851–863. doi: 10.1016/j.molcel.2011.12.005.

Glozak, M.A., and Seto, E. 2009. "Acetylation/deacetylation modulates the stability of DNA replication licensing factor Cdt1." *Journal of Biological Chemistry* 284, 17: 11446–11453. doi: 10.1074/jbc.M809394200.

Gottesfeld, J.M., and Pandolfo, M. 2009. "Development of histone deacetylase inhibitors as therapeutics for neurological disease." *Future Neurology* 4, 6: 775–784. doi: 10.2217/fnl.09.55.

Gottlicher, M., Minucci, S., Zhu, P., Kramer, O.H., Schimpf, A., Giavara, S., et al. 2001. "Valproic acid defines a novel class of HDAC inhibitors inducing differentiation of transformed cells." *EMBO Journal* 20, 24: 6969–6978. doi: 10.1093/emboj/20.24.6969.

Gowda, R., Madhunapantula, S.V., Desai, D., Amin, S., and Robertson, G.P. 2012. "Selenium-containing histone deacetylase inhibitors for melanoma management." *Cancer Biology & Therapy* 13, 9: 756–765. doi: 10.4161/cbt.20558.

Grivennikov, S.I., Greten, F.R., and Karin, M. 2010. "Immunity, inflammation, and cancer." *Cell* 140, 6: 883–899. doi: 10.1016/j.cell.2010.01.025.

Grozinger, C.M., Hassig, C.A., and Schreiber, S.L. 1999. "Three proteins define a class of human histone deacetylases related to yeast Hda1p." *Proceedings of the National Academy of Sciences U S A* 96, 9: 4868–4873. doi: 10.1073/pnas.9.6.9.4868.

Guenther, M.G., Barak, O., and Lazar, M.A. 2001. "The SMRT and N-CoR corepressors are activating cofactors for histone deacetylase 3." *Mol C Molecular and Cellular Biology ell Biol* 21, 18: 6091–6101. doi: 10.1128/mcb.21.18.6091-6101.2001.

Guerrero-Beltran, C.E., Calderon-Oliver, M., Pedraza-Chaverri, J., and Chirino, Y.I. 2012. "Protective effect of sulforaphane against oxidative stress: Recent advances." *Experimental and Toxicologic Pathology* 64, 5: 503–508. doi: 10.1016/j.etp.2010.11.005.

Hassan, A.H., Awad, S., Al-Natour, Z., Othman, S., Mustafa, F., and Rizvi, T.A. 2007. "Selective recognition of acetylated histones by bromodomains in transcriptional co-activators." *Biochemical Journal* 402, 1: 125–133. doi: 10.1042/BJ20060907.

Head, P.E., Zhang, H., Bastien, A.J., Koyen, A.E., Withers, A.E., Daddacha, W.B., et al. 2017. "Sirtuin 2 mutations in human cancers impair its function in genome maintenance." *Journal of Biological Chemistry* 292, 24: 9919–9931. doi: 10.1074/jbc.M116.772566.

Heijkants, R., Willekens, K., Schoonderwoerd, M., Teunisse, A., Nieveen, M., Radaelli, E., et al. 2018. "Combined inhibition of CDK and HDAC as a promising therapeutic strategy for both cutaneous and uveal metastatic melanoma." *Oncotarget* 9, 5: 6174–6187. doi: 10.18632/oncotarget.23485.

Hergeth, S.P., and Schneider, R. 2015. "The H1 linker histones: Multifunctional proteins beyond the nucleosomal core particle." *EMBO Reports* 16, 11: 1439–1453. doi: 10.15252/embr.201540749.

Hirata, Y., Sasaki, T., Kanki, H., Choong, C.J., Nishiyama, K., Kubo, G., et al. 2018. "New 5-aryl-substituted 2-aminobenzamide-type HDAC inhibitors with a diketopiperazine group and their ameliorating effects on ischemia-induced neuronal cell death." *Scientific Reports* 8, 1: 1400. doi: 10.1038/s41598-018-19664-9.

Ho, E., Clarke, J.D., and Dashwood, R.H. 2009. "Dietary sulforaphane, a histone deacetylase inhibitor for cancer prevention." *Journal of Nutrition* 139, 12: 2393–2396. doi: 10.3945/jn.109.113332.

Hu, W.Y., Xu, L., Chen, B., Ou, S., Muzzarelli, K.M., Hu, D.P., et al. 2019. "Targeting prostate cancer cells with enzalutamide-HDAC inhibitor hybrid drug 2–75." *Prostate* 79, 10: 1166–1179. doi: 10.1002/pros.23832.

Huang, Z., Zhou, W., Li, Y., Cao, M., Wang, T., Ma, Y., et al. 2018. "Novel hybrid molecule overcomes the limited response of solid tumours to HDAC inhibitors via suppressing JAK1-STAT3-BCL2 signalling." *Theranostics* 8, 18: 4995–5011. doi: 10.7150/thno.26627.

Inoue, A., and Fujimoto, D. 1969. "Enzymatic deacetylation of histone." *Biochemical and Biophysical Research Communications* 36, 1: 146–150. doi: 10.1016/0006-291x(69)90661-5.

Isaacs, J.T., Pili, R., Qian, D.Z., Dalrymple, S.L., Garrison, J.B., Kyprianou, N., et al. 2006. "Identification of ABR-215050 as lead second generation quinoline-3-carboxamide anti-angiogenic agent for the treatment of prostate cancer." *Prostate* 66, 16: 1768–1778. doi: 10.1002/pros.20509.

Janaki Ramaiah, M., Naushad, S.M., Lavanya, A., Srinivas, C., Anjana Devi, T., Sampathkumar, S., et al. 2017. "Scriptaid cause histone deacetylase inhibition and cell cycle arrest in HeLa cancer cells: A study on structural and functional aspects." *Gene* 627: 379–386. doi: 10.1016/j.gene.2017.06.031.

Jang, Y.G., Ko, E.B., and Choi, K.C. 2020. "Gallic acid, a phenolic acid, hinders the progression of prostate cancer by inhibition of histone deacetylase 1 and 2 expression." *Journal of Nutritional Biochemistry* 84: 108444. doi: 10.1016/j.jnutbio.2020.108444.

Jaworska, J., Zalewska, T., Sypecka, J., and Ziemka-Nalecz, M. 2019. "Effect of the HDAC inhibitor, sodium butyrate, on neurogenesis in a rat model of neonatal hypoxia-ischemia: Potential mechanism of action." *Molecular Neurobiology* 56, 9: 6341–6370. doi: 10.1007/s12035-019-1518-1.

Ji, M., Li, Z., Lin, Z., and Chen, L. 2018. "Antitumor activity of the novel HDAC inhibitor CUDC-101 combined with gemcitabine in pancreatic cancer." *American Journal of Cancer Research* 8, 12: 2402–2418.

Jiang, X., Liu, Y., Ma, L., Ji, R., Qu, Y., Xin, Y., et al. 2018. "Chemopreventive activity of sulforaphane." *Drug Design, Development and Therapy* 12: 2905–2913. doi: 10.2147/DDDT.S100534.

Johnston, T.H., Huot, P., Damude, S., Fox, S.H., Jones, S.W., Rusche, J.R., et al. 2013. "RGFP109, a histone deacetylase inhibitor attenuates L-DOPA-induced dyskinesia in the MPTP-lesioned marmoset: A proof-of-concept study." *Parkinsonism & Related Disorders* 19, 2: 260–264. doi: 10.1016/j.parkreldis.2012.07.001.

Jorgovanovic, D., Song, M., Wang, L., and Zhang, Y. 2020. "Roles of IFN-gamma in tumor progression and regression: A review." *Biomarker Research* 8: 49. doi: 10.1186/s40364-020-00228-x.

Jose, A., Chaitanya, M., Kannan, E., and Madhunapantula, S.V. 2018. "Tricaproin Isolated From Simarouba glauca inhibits the growth of human colorectal carcinoma cell lines by targeting class-1 histone deacetylases." *Frontiers in Pharmacology* 9: 127. doi: 10.3389/fphar.2018.00127.

Karin, M. 1999. "How NF-kappaB is activated: The role of the IkappaB kinase (IKK) complex." *Oncogene* 18, 49: 6867–6874. doi: 10.1038/sj.onc.1203219.

Kaur, M., Velmurugan, B., Rajamanickam, S., Agarwal, R., and Agarwal, C. 2009. "Gallic acid, an active constituent of grape seed extract, exhibits anti-proliferative, pro-apoptotic and anti-tumorigenic effects against prostate carcinoma xenograft growth in nude mice." *Pharmaceutical Research* 26, 9: 2133–2140. doi: 10.1007/s11095-009-9926-y.

Kee, H.J., and Kook, H. 2011. "Roles and targets of class I and IIa histone deacetylases in cardiac hypertrophy." *Journal of Biomedicine and Biotechnology* 928326. doi: 10.1155/2011/928326.

Kiani, S., Akhavan-Niaki, H., Fattahi, S., Kavoosian, S., Babaian Jelodar, N., Bagheri, N., et al. 2018. "Purified sulforaphane from broccoli (Brassica oleracea var. italica) leads to alterations of CDX1 and CDX2 expression and changes in miR-9 and miR-326 levels in human gastric cancer cells." *Gene* 678: 115–123. doi: 10.1016/j.gene.2018.08.026.

Kikuchi, S., Suzuki, R., Ohguchi, H., Yoshida, Y., Lu, D., Cottini, F., et al. 2015. "Class IIa HDAC inhibition enhances ER stress-mediated cell death in multiple myeloma." *Leukemia* 29, 9: 1918–1927. doi: 10.1038/leu.2015.83.

Kim, B., and Hong, J. 2015. "An overview of naturally occurring histone deacetylase inhibitors." *Current Topics in Medicinal Chemistry* 14, 24: 2759–2782. doi: 10.2174/1568026615666141208105614.

Kim, E., Bisson, W.H., Lohr, C.V., Williams, D.E., Ho, E., Dashwood, R.H., et al. 2016. "Histone and non-histone targets of dietary deacetylase inhibitors." *Current Topics in Medicinal Chemistry* 16, 7: 714–731. doi: 10.2174/1568026615666150825125857.

Kim, E.J., Kho, J.H., Kang, M.R., and Um, S.J. 2007. "Active regulator of SIRT1 cooperates with SIRT1 and facilitates suppression of p53 activity." *Molecular Cell* 28, 2: 277–290. doi: 10.1016/j.molcel.2007.08.030.

Kim, J.H., Lee, S.J., Han, Y.B., Moon, J.J., and Kim, J.B. 1994. "Isolation of isoguanosine from Croton tiglium and its antitumor activity." *Archives of Pharmacal Research* 17, 2: 115–118. doi: 10.1007/BF02974234.

Kong, Y.L., Pan, B.H., Liang, J.H., Zhu, H.Y., Wang, L., Xia, Y., et al. 2020. "Chidamide, a histone deacetylase inhibitor, inhibits autophagy and exhibits therapeutic implication in chronic lymphocytic leukemia." *Aging (Albany NY)* 12, 16: 16083–16098. doi: 10.18632/aging.103536.

Kujirai, T., Horikoshi, N., Sato, K., Maehara, K., Machida, S., Osakabe, A., et al. 2016. "Structure and function of human histone H3.Y nucleosome." *Nucleic Acids Research* 44, 13: 6127–6141. doi: 10.1093/nar/gkw202.

Kupis, W., Palyga, J., Tomal, E., and Niewiadomska, E. 2016. "The role of sirtuins in cellular homeostasis." *Journal of Physiology and Biochemistry* 72, 3: 371–380. doi: 10.1007/s13105-016-0492-6.

Kuzmichev, A., Zhang, Y., Erdjument-Bromage, H., Tempst, P., and Reinberg, D. 2002. "Role of the Sin3-histone deacetylase complex in growth regulation by the candidate tumor suppressor p33(ING1)." *Molecular and Cellular Biology* 22, 3: 835–848. doi: 10.1128/mcb.22.3.835-848.2002.

Lan, A., Li, W., Liu, Y., Xiong, Z., Zhang, X., Zhou, S., et al. 2016. "Chemoprevention of oxidative stress-associated oral carcinogenesis by sulforaphane depends on NRF2 and the isothiocyanate moiety." *Oncotarget* 7, 33: 53502–53514. doi: 10.18632/oncotarget.10609.

Lee, J.I., Nian, H., Cooper, A.J., Sinha, R., Dai, J., Bisson, W.H., et al. 2009. "Alpha-keto acid metabolites of naturally occurring organoselenium compounds as inhibitors of histone deacetylase in human prostate cancer cells." *Cancer Prevention Research (Phila)* 2, 7: 683–693. doi: 10.1158/1940-6207.CAPR-09-0047.

Legube, G., and Trouche, D. 2003. "Regulating histone acetyltransferases and deacetylases." *EMBO Reports* 4, 10: 944–947. doi: 10.1038/sj.embor.embor941.

Leus, N.G., Zwinderman, M.R., and Dekker, F.J. 2016. "Histone deacetylase 3 (HDAC 3) as emerging drug target in NF-kappaB-mediated inflammation." *Current Opinion in Chemical Biology* 33: 160–168. doi: 10.1016/j.cbpa.2016.06.019.

Li, X., Zhang, J., Xie, Y., Jiang, Y., Yingjie, Z., and Xu, W. 2014. "Progress of HDAC inhibitor panobinostat in the treatment of cancer." *Current Drug Targets* 15, 6: 622–634. doi: 10.2174/1389450115666140306152642.

Li, Y., and Seto, E. 2016. "HDACs and HDAC inhibitors in cancer development and therapy." *Cold Spring Harbor Perspectives in Medicine* 6, 10. doi: 10.1101/cshperspect.a026831.

Licciardi, P.V., and Karagiannis, T.C. 2012. "Regulation of immune responses by histone deacetylase inhibitors." *ISRN Hematol* 2012: 690901. doi: 10.5402/2012/690901.

Licciardi, P.V., Ververis, K., and Karagiannis, T.C. 2011. "Histone deacetylase inhibition and dietary short-chain Fatty acids." *ISRN Allergy* 2011: 869647. doi: 10.5402/2011/869647.

Lu, Y.S., Kashida, Y., Kulp, S.K., Wang, Y.C., Wang, D., Hung, J.H., et al. 2007. "Efficacy of a novel histone deacetylase inhibitor in murine models of hepatocellular carcinoma." *Hepatology* 46, 4: 1119–1130. doi: 10.1002/hep.21804.

Ma, J., Luo, T., Zeng, Z., Fu, H., Asano, Y., Liao, Y., et al. 2016. "Histone deacetylase inhibitor phenylbutyrate exaggerates heart failure in pressure overloaded mice independently of HDAC inhibition." *Scientific Reports* 6: 34036. doi: 10.1038/srep34036.

Maeshima, K., and Eltsov, M. 2008. "Packaging the genome: The structure of mitotic chromosomes." *Journal of Biochemistry* 143, 2: 145–153. doi: 10.1093/jb/mvm214.

Mandl-Weber, S., Meinel, F.G., Jankowsky, R., Oduncu, F., Schmidmaier, R., and Baumann, P. 2010. "The novel inhibitor of histone deacetylase resminostat (RAS2410) inhibits proliferation and induces apoptosis in multiple myeloma (MM) cells." *British Journal of Haematology* 149, 4: 518–528. doi: 10.1111/j.1365-2141.2010.08124.x.

Maolanon, A.R., Madsen, A.S., and Olsen, C.A. 2016. "Innovative strategies for selective inhibition of histone deacetylases." *Cell Chemical Biology* 23, 7: 759–768. doi: 10.1016/j.chembiol.2016.06.011.

Martin, M., Kettmann, R., and Dequiedt, F. 2007. "Class IIa histone deacetylases: Regulating the regulators." *Oncogene* 26, 37: 5450–5467. doi: 10.1038/sj.onc.1210613.

McCaw, T.R., Randall, T.D., Forero, A., and Buchsbaum, D.J. 2017. "Modulation of antitumor immunity with histone deacetylase inhibitors." *Immunotherapy* 9, 16: 1359–1372. doi: 10.2217/imt-2017-0134.

McGinty, R.K., and Tan, S. 2015. "Nucleosome structure and function." *Chemical Reviews* 115, 6: 2255–2273. doi: 10.1021/cr500373h.

Michan, S., and Sinclair, D. 2007. "Sirtuins in mammals: Insights into their biological function." *Biochemical Journal* 404, 1: 1–13. doi: 10.1042/BJ20070140.

Mihaylova, M.M., and Shaw, R.J. 2013. "Metabolic reprogramming by class I and II histone deacetylases." *Trends in Endocrinology and Metabolism* 24, 1: 48–57. doi: 10.1016/j.tem.2012.09.003.

Moodad, S., Akkouche, A., Hleihel, R., Darwiche, N., El-Sabban, M., Bazarbachi, A., et al. 2018. "Mouse models that enhanced our understanding of adult T cell leukemia." *Frontiers in Microbiology* 9, 558. doi: 10.3389/fmicb.2018.00558.

Morris, M.J., and Monteggia, L.M. 2013. "Unique functional roles for class I and class II histone deacetylases in central nervous system development and function." *International Journal of Developmental Neuroscience* 31, 6: 370–381. doi: 10.1016/j.ijdevneu.2013.02.005.

Mottamal, M., Zheng, S., Huang, T.L., and Wang, G. 2015. "Histone deacetylase inhibitors in clinical studies as templates for new anticancer agents." *Molecules* 20, 3: 3898–3941. doi: 10.3390/molecules20033898.

Munshi, A., Tanaka, T., Hobbs, M.L., Tucker, S.L., Richon, V.M., and Meyn, R.E. 2006. "Vorinostat, a histone deacetylase inhibitor, enhances the response of human tumor cells to ionizing radiation through prolongation of gamma-H2AX foci." *Molecular Cancer Therapeutics* 5, 8: 1967–1974. doi: 10.1158/1535-7163.MCT-06-0022.

Mwakwari, S.C., Patil, V., Guerrant, W., and Oyelere, A.K. 2010. "Macrocyclic histone deacetylase inhibitors." *Current Topics in Medicinal Chemistry* 10, 14: 1423–1440. doi: 10.2174/156802610792232079.

Myzak, M.C., Ho, E., and Dashwood, R.H. 2006. "Dietary agents as histone deacetylase inhibitors." *Molecular Carcinogenesis* 45, 6: 443–446. doi: 10.1002/mc.20224.

Nagaki, S., Kato, N., Minatogawa, Y., and Higuchi, T. 1990. "Effects of anticonvulsants and gamma-aminobutyric acid (GABA)-mimetic drugs on immunoreactive somatostatin and GABA contents in the rat brain." *Life Sciences* 46, 22: 1587–1595. doi: 10.1016/0024-3205(90)90396-9.

Nagy, A., Schally, A.V., Halmos, G., Armatis, P., Cai, R.Z., Csernus, V., et al. 1998. "Synthesis and biological evaluation of cytotoxic analogs of somatostatin containing doxorubicin or its intensely potent derivative, 2-pyrrolinodoxorubicin." *Proceedings of the National Academy of Sciences U S A* 95, 4: 1794–1799. doi: 10.1073/pnas.95.4.1794.

Narita, K., Kikuchi, T., Watanabe, K., Takizawa, T., Oguchi, T., Kudo, K., et al. 2009. "Total synthesis of the bicyclic depsipeptide HDAC inhibitors spiruchostatins A and B, 5"-epi-spiruchostatin B, FK228 (FR901228) and preliminary evaluation of their biological activity." *Chemistry* 15, 42: 11174–11186. doi: 10.1002/chem.200901552.

Nian, H., Bisson, W.H., Dashwood, W.M., Pinto, J.T., and Dashwood, R.H. 2009a. "Alpha-keto acid metabolites of organoselenium compounds inhibit histone deacetylase activity in human colon cancer cells." *Carcinogenesis* 30, 8: 1416–1423. doi: 10.1093/carcin/bgp147.

Nian, H., Delage, B., Ho, E., and Dashwood, R.H. 2009b. "Modulation of histone deacetylase activity by dietary isothiocyanates and allyl sulfides: Studies with sulforaphane and garlic organosulfur compounds." *Environmental and Molecular Mutagenesis* 50, 3: 213–221. doi: 10.1002/em.20454.

Nian, H., Delage, B., Pinto, J.T., and Dashwood, R.H. 2008. "Allyl mercaptan, a garlic-derived organosulfur compound, inhibits histone deacetylase and enhances Sp3 binding on the P21WAF1 promoter." *Carcinogenesis* 29, 9: 1816–1824. doi: 10.1093/carcin/bgn165.

Nielsen, T.K., Hildmann, C., Dickmanns, A., Schwienhorst, A., and Ficner, R. 2005. "Crystal structure of a bacterial class 2 histone deacetylase homologue." *Journal of Molecular Biology* 354, 1: 107–120. doi: 10.1016/j.jmb.2005.09.065.

North, B.J., Schwer, B., Ahuja, N., Marshall, B., and Verdin, E. 2005. "Preparation of enzymatically active recombinant class III protein deacetylases." *Methods* 36, 4: 338–345. doi: 10.1016/j.ymeth.2005.03.004.

Novotny-Diermayr, V., Sangthongpitag, K., Hu, C.Y., Wu, X., Sausgruber, N., Yeo, P., et al. 2010. "SB939, a novel potent and orally active histone deacetylase inhibitor with high tumor exposure and efficacy in mouse models of colorectal cancer." *Molecular Cancer Therapeutics* 9, 3: 642–652. doi: 10.1158/1535-7163.MCT-09-0689.

Nural-Guvener, H., Zakharova, L., Feehery, L., Sljukic, S., and Gaballa, M. 2015. "Anti-Fibrotic Effects of Class I HDAC Inhibitor, Mocetinostat Is Associated with IL-6/Stat3 Signaling in Ischemic Heart Failure." *International Journal of Molecular Sciences* 16, 5: 11482–11499. doi: 10.3390/ijms160511482.

Orr, N.M., and Doust, B.D. 1988. "The impact of intravenous digital subtraction angiography on the evaluation of carotid artery disease: Review of the first 600 cases." *Australasian Radiology* 32, 1: 21–23. doi: 10.1111/j.1440-1673.1988.tb02687.x.

Parbin, S., Kar, S., Shilpi, A., Sengupta, D., Deb, M., Rath, S.K., et al. 2014. "Histone deacetylases: A saga of perturbed acetylation homeostasis in cancer." *Journal of Histochemistry and Cytochemistry* 62, 1: 11–33. doi: 10.1369/0022155413506582.

Park, S.Y., Kim, G.S., Hwang, H.J., Nam, T.H., Park, H.S., Song, J., et al. 2018. "Structural basis of the specific interaction of SMRT corepressor with histone deacetylase 4." *Nucleic Acids Research* 46, 22: 11776–11788. doi: 10.1093/nar/gky926.

Park, S.Y., and Kim, J.S. 2020. "A short guide to histone deacetylases including recent progress on class II enzymes." *Experimental & Molecular Medicine* 52, 2: 204–212. doi: 10.1038/s12276-020-0382-4.

Parthun, M.R. 2012. "Histone acetyltransferase 1: More than just an enzyme?" *Biochimica et Biophysica Acta* 1819, 3–4: 256–263. doi: 10.1016/j.bbagrm.2011.07.006.

Peng, L., and Seto, E. 2011. "Deacetylation of nonhistone proteins by HDACs and the implications in cancer." *Handbook of Experimental Pharmacology* 206: 39–56. doi: 10.1007/978-3-642-21631-2_3.

Pflum, M.K., Tong, J.K., Lane, W.S., and Schreiber, S.L. 2001. "Histone deacetylase 1 phosphorylation promotes enzymatic activity and complex formation." *Journal of Biological Chemistry* 276, 50: 47733–47741. doi: 10.1074/jbc.M105590200.

Potharla, V.Y., Wang, C., and Cheng, Y.Q. 2014. "Identification and characterization of the spiruchostatin biosynthetic gene cluster enable yield improvement by overexpressing a transcriptional activator." *Journal of Industrial Microbiology & Biotechnology* 41, 9: 1457–1465. doi: 10.1007/s10295-014-1474-8.

Poulose, N., and Raju, R. 2015. "Sirtuin regulation in aging and injury." *Biochimica et Biophysica Acta* 1852, 11: 2442–2455. doi: 10.1016/j.bbadis.2015.08.017.

Pourrajab, F., Zare-Khormizi, M.R., Hekmatimoghaddam, S., and Hashemi, A.S. 2020. "Molecular targeting and rational chemotherapy in acute myeloid leukemia." *Journal of Experimental Pharmacology* 12, 107–128. doi: 10.2147/JEP.S254334.

Qian, S., Chen, Q.L., Guan, J.L., Wu, Y., and Wang, Z.Y. 2014. "Synthesis and biological evaluation of Raddeanin A, a triterpene saponin isolated from Anemone raddeana." *Chemical and Pharmaceutical Bulletin (Tokyo)* 62, 8: 779–785. doi: 10.1248/cpb.c14-00138.

Rafehi, H., Kaspi, A., Ziemann, M., Okabe, J., Karagiannis, T.C., and El-Osta, A. 2017. "Systems approach to the pharmacological actions of HDAC inhibitors reveals EP300 activities and convergent mechanisms of regulation in diabetes." *Epigenetics* 12, 11: 991–1003. doi: 10.1080/15592294.2017.1371892.

Ran, J., Yang, Y., Li, D., Liu, M., and Zhou, J. 2015. "Deacetylation of alpha-tubulin and cortactin is required for HDAC6 to trigger ciliary disassembly." *Scientific Reports* 5, 12917. doi: 10.1038/srep12917.

Ropero, S., and Esteller, M. 2007. "The role of histone deacetylases (HDACs) in human cancer." *Molecular Oncology* 1, 1: 19–25. doi: 10.1016/j.molonc.2007.01.001.

Roth, S.Y., Denu, J.M., and Allis, C.D. 2001. "Histone acetyltransferases." *Annual Review of Biochemistry* 70, 81–120. doi: 10.1146/annurev.biochem.70.1.81.

Rucker, F.G., Lang, K.M., Futterer, M., Komarica, V., Schmid, M., Dohner, H., et al. 2016. "Molecular dissection of valproic acid effects in acute myeloid leukemia identifies predictive networks." *Epigenetics* 11, 7: 517–525. doi: 10.1080/15592294.2016.1187350.

Saha, R.N., and Pahan, K. 2006. "HATs and HDACs in neurodegeneration: A tale of disconcerted acetylation homeostasis." *Cell Death and Differentiation* 13, 4: 539–550. doi: 10.1038/sj.cdd.4401769.

Schotterl, S., Brennenstuhl, H., and Naumann, U. 2015. "Modulation of immune responses by histone deacetylase inhibitors." *Critical Reviews in Oncogenesis* 20, 1–2: 139–154. doi: 10.1615/critrevoncog.2014012393.

Senawong, T., Misuna, S., Khaopha, S., Nuchadomrong, S., Sawatsitang, P., Phaosiri, C., et al. 2013. "Histone deacetylase (HDAC) inhibitory and antiproliferative activities of phenolic-rich extracts derived from the rhizome of Hydnophytum formicarum Jack.: Sinapinic acid acts as HDAC inhibitor." *BMC Complementary and Alternative Medicine* 13, 232. doi: 10.1186/1472-6882-13-232.

Seto, E., and Yoshida, M. 2014. "Erasers of histone acetylation: The histone deacetylase enzymes." *Cold Spring Harbor Perspectives in Biology* 6, 4: a018713. doi: 10.1101/cshperspect.a018713.

Shahbazian, M.D., and Grunstein, M. 2007. "Functions of site-specific histone acetylation and deacetylation." *Annual Review of Biochemistry* 76, 75–100. doi: 10.1146/annurev.biochem.76.052705.162114.

Shen, L., Ciesielski, M., Ramakrishnan, S., Miles, K.M., Ellis, L., Sotomayor, P., et al. 2012. "Class I histone deacetylase inhibitor entinostat suppresses regulatory T cells and enhances immunotherapies in renal and prostate cancer models." *PLoS One* 7, 1: e30815. doi: 10.1371/journal.pone.0030815.

Shi, X.Y., Ding, W., Li, T.Q., Zhang, Y.X., and Zhao, S.C. 2017. "Histone deacetylase (HDAC) inhibitor, suberoylanilide hydroxamic acid (SAHA), induces apoptosis in prostate cancer cell lines via the Akt/FOXO3a signaling pathway." *Medical Science Monitor* 23, 5793–5802. doi: 10.12659/msm.904597.

Siegel, D., Hussein, M., Belani, C., Robert, F., Galanis, E., Richon, V.M., et al. 2009. "Vorinostat in solid and hematologic malignancies." *Journal of Hematology & Oncology* 2, 31. doi: 10.1186/1756-8722-2-31.

Silva, L.G., Ferguson, B.S., Avila, A.S., and Faciola, A.P. 2018. "Sodium propionate and sodium butyrate effects on histone deacetylase (HDAC) activity, histone acetylation, and inflammatory gene expression in bovine mammary epithelial cells." *Journal of Animal Science* 96, 12: 5244–5252. doi: 10.1093/jas/sky373.

Singh, A.K., Bishayee, A., and Pandey, A.K. 2018. "Targeting histone deacetylases with natural and synthetic agents: An emerging anticancer strategy." *Nutrients* 10, 6. doi: 10.3390/nu10060731.

Stark, M., and Hayward, N. 2007. "Genome-wide loss of heterozygosity and copy number analysis in melanoma using high-density single-nucleotide polymorphism arrays." *Cancer Research* 67, 6: 2632–2642. doi: 10.1158/0008-5472.CAN-06-4152.

Steliou, K., Boosalis, M.S., Perrine, S.P., Sangerman, J., and Faller, D.V. 2012. "Butyrate histone deacetylase inhibitors." *BioResearch Open Access* 1, 4: 192–198. doi: 10.1089/biores.2012.0223.

Stephens, K.E., Miaskowski, C.A., Levine, J.D., Pullinger, C.R., and Aouizerat, B.E. 2013. "Epigenetic regulation and measurement of epigenetic changes." *Biological Research for Nursing* 15, 4: 373–381. doi: 10.1177/1099800412444785.

Sterner, D.E., and Berger, S.L. 2000. "Acetylation of histones and transcription-related factors." *Microbiology and Molecular Biology Reviews* 64, 2: 435–459. doi: 10.1128/mmbr.64.2.435-459.2000.

Sun, X.J., Man, N., Tan, Y., Nimer, S.D., and Wang, L. 2015. "The role of histone acetyltransferases in normal and malignant hematopoiesis." *Frontiers in Oncology* 5, 108. doi: 10.3389/fonc.2015.00108.

Suraweera, A., O'Byrne, K.J., and Richard, D.J. 2018. "Combination therapy with histone deacetylase inhibitors (HDACi) for the treatment of cancer: Achieving the full therapeutic potential of HDACi." *Frontiers in Oncology* 8, 92. doi: 10.3389/fonc.2018.00092.

Sztiller-Sikorska, M., and Czyz, M. 2020. "Parthenolide as cooperating agent for anti-cancer treatment of various malignancies." *Pharmaceuticals (Basel)* 13, 8. doi: 10.3390/ph13080194.

Taunton, J., Hassig, C.A., and Schreiber, S.L. 1996. "A mammalian histone deacetylase related to the yeast transcriptional regulator Rpd3p." *Science* 272, 5260: 408–411. doi: 10.1126/science.272.5260.408.

Ullah, M.F. 2015. "Sulforaphane (SFN): An isothiocyanate in a cancer chemoprevention paradigm." *Medicines (Basel)* 2, 3: 141–156. doi: 10.3390/medicines2030141.

Uo, T., Veenstra, T.D., and Morrison, R.S. 2009. "Histone deacetylase inhibitors prevent p53-dependent and p53-independent Bax-mediated neuronal apoptosis through two distinct mechanisms." *Journal of Neuroscience* 29, 9: 2824–2832. doi: 10.1523/JNEUROSCI.6186-08.2009.

Venza, I., Visalli, M., Oteri, R., Cucinotta, M., Teti, D., and Venza, M. 2013. "Class II-specific histone deacetylase inhibitors MC1568 and MC1575 suppress IL-8 expression in human melanoma cells." *Pigment Cell & Melanoma Research* 26, 2: 193–204. doi: 10.1111/pcmr.12049.

Vigushin, D.M., Ali, S., Pace, P.E., Mirsaidi, N., Ito, K., Adcock, I., et al. 2001. "Trichostatin A is a histone deacetylase inhibitor with potent antitumor activity against breast cancer *in vivo*." *Clinical Cancer Research* 7, 4: 971–976.

Walker, T., Mitchell, C., Park, M.A., Yacoub, A., Graf, M., Rahmani, M., et al. 2009. "Sorafenib and vorinostat kill colon cancer cells by CD95-dependent and -independent mechanisms." *Molecular Pharmacology* 76, 2: 342–355. doi: 10.1124/mol.109.056523.

Wang, S.H., Lin, P.Y., Chiu, Y.C., Huang, J.S., Kuo, Y.T., Wu, J.C., et al. 2015. "Curcumin-mediated HDAC inhibition suppresses the DNA damage response and contributes to increased DNA damage sensitivity." *PLoS One* 10, 7: e0134110. doi: 10.1371/journal.pone.0134110

Wang, Y., Wallach, J., Duane, S., Wang, Y., Wu, J., Wang, J., et al. 2017. "Developing selective histone deacetylases (HDACs) inhibitors through ebselen and analogs." *Drug Design, Development and Therapy* 11, 1369–1382. doi: 10.2147/DDDT.S124977.

West, A.C., and Johnstone, R.W. 2014. "New and emerging HDAC inhibitors for cancer treatment." *Journal of Clinical Investigation* 124, 1: 30–39. doi: 10.1172/JCI69738.

Yang, X.J., and Gregoire, S. 2005. "Class II histone deacetylases: From sequence to function, regulation, and clinical implication." *Molecular and Cellular Biology* 25, 8: 2873–2884. doi: 10.1128/MCB.25.8.2873-2884.2005.

Yin, X., Zhang, J., Li, X., Liu, D., Feng, C., Liang, R., et al. 2014. "DADS suppresses human esophageal xenograft tumors through RAF/MEK/ERK and mitochondria-dependent pathways." *International Journal of Molecular Sciences* 15, 7: 12422–12441. doi: 10.3390/ijms150712422.

Yoon, S., and Eom, G.H. 2016. "HDAC and HDAC inhibitor: From cancer to cardiovascular diseases." *Chonnam Medical Journal* 52, 1: 1–11. doi: 10.4068/cmj.2016.52.1.1.

Yuille, S., Reichardt, N., Panda, S., Dunbar, H., and Mulder, I.E. 2018. "Human gut bacteria as potent class I histone deacetylase inhibitors in vitro through production of butyric acid and valeric acid." *PLoS One* 13, 7: e0201073. doi: 10.1371/journal.pone.0201073.

Zhang, B., Lyu, J., Yang, E.J., Liu, Y., Wu, C., Pardeshi, L., et al. 2020. "Class I histone deacetylase inhibition is synthetic lethal with BRCA1 deficiency in breast cancer cells." *Acta Pharmaceutica Sinica B* 10, 4: 615–627. doi: 10.1016/j.apsb.2019.08.008.

Zhang, L., Zhang, J., Jiang, Q., Zhang, L., and Song, W. 2018. "Zinc binding groups for histone deacetylase inhibitors." *Journal of Enzyme Inhibition and Medicinal Chemistry* 33, 1: 714–721. doi: 10.1080/14756366.2017.1417274.

Zhang, X., Ozawa, Y., Lee, H., Wen, Y.D., Tan, T.H., Wadzinski, B.E., et al. 2005. "Histone deacetylase 3 (HDAC3) activity is regulated by interaction with protein serine/threonine phosphatase 4." *Genes & Development* 19, 7: 827–839. doi: 10.1101/gad.1286005.

Zhang, Y., Gilquin, B., Khochbin, S., and Matthias, P. 2006. "Two catalytic domains are required for protein deacetylation." *Journal of Biological Chemistry* 281, 5: 2401–2404. doi: 10.1074/jbc.C500241200.

Zhao, E., Hou, J., Ke, X., Abbas, M.N., Kausar, S., Zhang, L., et al. 2019. "The roles of sirtuin family proteins in cancer progression." *Cancers (Basel)* 11, 12. doi: 10.3390/cancers11121949.

9 Synthetic Compounds vs. Phytochemicals for the Treatment of Human Cutaneous Malignant Melanoma

A Review on Human Malignant Melanoma, the PI3K/Akt/mTOR Pathway and the Potential of Phytochemicals Isolated from Medicinal Plants and Various Dietary Sources

Jacqueline Maphutha and Namrita Lall

CONTENTS

DOI: 10.1201/9781003108375-9

9.1 INTRODUCTION

9.1.1 Carcinogenesis

Carcinogenesis refers to the initiation of a tumor due to cancer-causing agents referred to as carcinogens. Carcinogens can be characterized as chemical carcinogens, biological carcinogens and physical carcinogens. Chemical carcinogens such as polycyclic aromatic hydrocarbons (PAHs), released into the environment from the combustion of materials containing carbon, elicit mutations in the tumor suppressor gene TP53 (Soliman, 2018). The protein (p53) enables uncontrolled cellular proliferation, which is evident in breast cancer. Furthermore, human papillomaviruses (HPV) such as HPV16 and HPV18 (biological carcinogens) produce an oncoprotein E6, which elicits mutations in the tumor suppressor gene TP53, resulting in uncontrolled proliferation. Human papillomaviruses predominantly lead to the initiation and progression of cervical cancer (Burd, 2003; Pal and Kundu, 2020; Scheffner et al., 1990).

Finally, physical carcinogens such as ultraviolet (UV) radiation, particularly ultraviolet B (UVB) radiation (280–315 nm), elicit loss of function mutations in the phosphatase tensin homologue (PTEN) tumor suppressor gene. This tumor suppressor gene is a negative regulator of the phosphatidylinositol-3-kinase (PI3K)/protein kinase B (Akt)/mechanistic target of rapamycin (mTOR) pathway that controls cell growth, survival and metabolism. The deactivation of PTEN, which is prevalent in melanoma, results in the auto phosphorylation of PI3K that subsequently activates a downstream kinase known as protein kinase B (Akt) (Ming and He, 2009; Stahl et al., 2003). The activation of Akt results in uncontrolled cell division and cell growth (Ming and He, 2009).

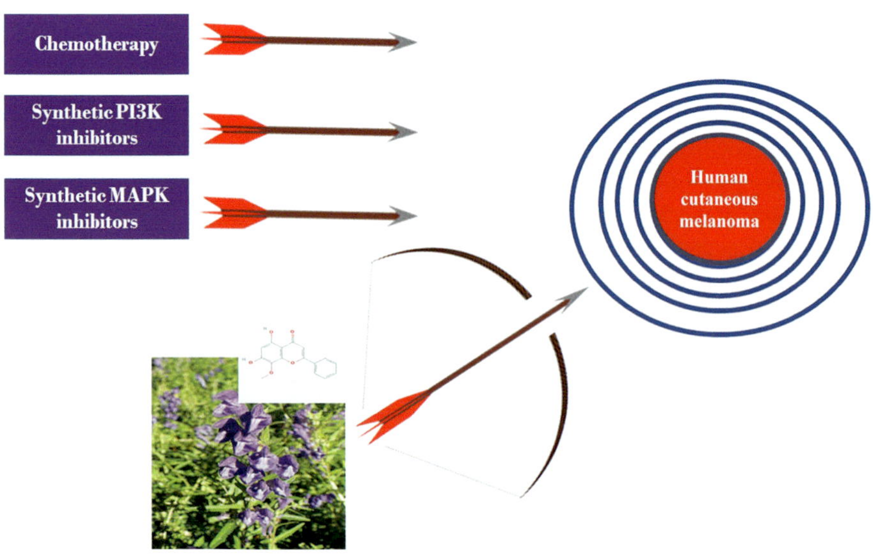

Graphical abstract representing the content of the chapter. Drawn using BIOVIA Draw (Kenraiz, 2016).

9.1.2 Prevalence of Cancer and Skin Cancer Worldwide

According to Bray et al. (2018), 18.1 million new cancer cases and 9.6 million deaths have been estimated. The most prevalent cancers for males were lung cancer, prostate cancer and colorectal cancer, while liver and stomach cancer accounted for the majority of deaths. For females, breast cancer, colorectal cancer, lung cancer and cervical cancer were prevalent for both incidence and mortality (Bray et al., 2018). The World Health Organization (WHO) reported that 2–3 million non-melanoma skin cancers and 132,000 melanoma skin cancers occur globally each year. Furthermore, one in three cancers diagnosed is a skin cancer (WHO, 2017b).

9.2 THE STRUCTURE OF THE SKIN AND SKIN CANCER

The skin is composed of three main layers: the epidermis, dermis and hypodermis. The epidermis is the outer layer of the skin that serves as a physical and biological barrier to the external environment, e.g. protection from ultraviolet radiation and allergens. The epidermis is made up of five layers, namely the stratum corneum, stratum lucidum (found where the skin is the thickest i.e. palms and soles of feet), stratum granulosum, stratum spinosum and stratum basale. The dermis, which is thicker than the epidermis, sustains and supports the epidermis, while the hypodermis consists mainly of fat that supports the skin and insulates the body from cold and shock (Lawton, 2019). Melanocytes are distributed in the stratum basale layer of the epidermis and melanin (produced by melanocytes) displays photoprotective effects through the absorption of UV radiation. However, the exposure of the skin to excessive UV radiation, especially UVB radiation (280–315 nm), which cannot be absorbed by melanin, results in mutagenesis in melanocytes leading to HCMM (Lawton, 2019).

9.2.1 Skin Cancer

Skin cancer is comprised of non-melanoma and melanoma skin cancer. Basal cell carcinoma (BCC) and squamous cell carcinoma (SCC) are classified as non-melanoma skin cancers. Basal cell carcinoma arises from the stratum basale layer in the epidermis, wherein metastasis is rare (if treated early), whereas the potential for metastasis in squamous cell carcinoma is significant if treatment is not administered early. Melanoma arises from melanocytes in the stratum basale layer of the epidermis and is the most lethal, due to signifi-cant metastatic potential, and accounts for the majority of skin cancer deaths that have been reported (Lawton, 2019; Matthews et al., 2017).

9.2.2 Types of UV Radiation and the UV Index

Ultraviolet radiation is part of the electromagnetic spectrum between visible light and X-rays, and has a wavelength of 100–400 nm; however, three subtypes of UV radiation exist, namely ultraviolet radiation A (UVA), ultraviolet radiation B (UVB) and ultraviolet radiation C (UVC). UVA has the longest wavelength (315–400 nm) that is not absorbed by the ozone layer or the atmosphere and penetrates the stratum basale layer where melanocytes are present. UVB has a shorter wavelength (280–315 nm), and most UVB radiation is absorbed by the ozone layer. Extended periods outdoors often leads to the accumulation of UVB radiation, which penetrates the skin leading to swelling and pain characteristic of sunburn. Finally, UVC radiation (100–280 nm) has the shortest wavelength and is absorbed by the ozone layer and atmosphere. UVA and UVB are the main risk factors for skin cancer (Watson, Holman, and Maguire-Eisen, 2016).

The UV index (UVI) was developed by the WHO, United Nations Environment Program and the World Meteorological Organization. The UVI serves as a tool to raise public awareness on the level of UV radiation emitted throughout the day such that protective measures can be adopted e.g.

application of sunscreen and decreasing time spent outdoors. The range of the UVI is 0–11; the higher the value, the greater the extent of damage to the skin and eyes (WHO, 2017a).

9.3 MELANOMA

9.3.1 Diagnosis and Staging of Melanoma

For the diagnosis of melanoma, a skin lesion is often assessed using the ABCDE rule. The ABCDE rule indicates the following: (A) asymmetry (two halves are not equal across a plane); (B) irregular border (no defined edges on the lesion); (C) color variation (black, brown, tan etc.); (D) a diameter less than 6 mm; and (E) elevated surface. The ABCDE rule has led to several false negatives, and thus, advanced digital systems such as the melanoma diagnosis system (MEDS) improve the accuracy of melanoma diagnosis through the use of several algorithms that analyzes the skin lesion (Mar and Soyer, 2018; Nachbar et al., 1994). Furthermore, Microphthalmia transcription factor (Mitf) is highly expressed in melanoma melanocytes and serves as a biomarker for the diagnosis of melanoma. However, Mitf cannot be used to distinguish between benign and malignant tumors (King et al., 1999). For the distinction between benign and malignant tumors, four staging systems are used, viz. the Clark scale; Breslow's depth; tumor, node and metastasis (TNM); and the number staging system (Liu and Sheikh, 2014).

The Clark scale evaluates the spread of melanoma from the epidermis, papillary dermis, reticular dermis and finally, the subcutaneous fat layer. The characteristics of each level are as follows (Davis, Shalin, and Tackett, 2019).

- Level one: The melanoma cells are confined to the epidermis (melanoma *in situ*).
- Level two: Invasion of the papillary dermis by a subset of melanoma cells.
- Level three: Infiltration of the papillary dermis by a large subset of melanoma cells.
- Level four: Infiltration into the reticular dermis.
- Level five: Invasion of the subcutaneous fat layer.

Levels three to five characterize a more aggressive form of malignant melanoma, wherein the dissection of lymph nodes and adjuvants would be accompanied with chemotherapy, radiation therapy, targeted therapies and immunotherapies. Breslow's depth is similar to the Clark scale, as they both describe how deeply the melanoma has penetrated the skin; however, the prognosis of melanoma in this system is based on the depth of the tumor in the skin, which is measured in millimeters (mm). The characteristics of each stage are as follows (Davis, Shalin, and Tackett, 2019).

- Stage I: Less than or equal to 0.75 mm.
- Stage II: 0.76–1.5 mm.
- Stage III: 1.51–2.25 mm.
- Stage IV: 2.26–3.0 mm.
- Stage V: Greater than 3.0 mm.

According to Breslow's depth, patients with tumors less than 1.5 mm require the dissection of lymph nodes and adjuvants for a higher survival rate and to decrease the likelihood of the recurrence of melanoma. The TNM staging system and number system were introduced due to variations in the prognosis of melanoma in conjunction with the Clark scale and Breslow's depth (Colloby, West, and Fletcher, 1991). The TNM staging system describes tumor thickness (T), the presence of melanoma in nearby lymph nodes (N) and whether the melanoma has spread to different sites in the body (M). The rate of mitosis, another factor in the TNM staging system, was used to determine the spread of melanoma to different sites in the body; however, this factor is no longer considered (Cancer Research UK, 2020b; Liu and Sheik, 2014). Finally, the number staging system (similar to the TNM staging system) describes the metastasis of melanoma according to the stages as follow (Cancer Research UK, 2020a).

- Stage zero (Melanoma *in situ*): The cancer cells are present in the stratum basale layer.
- Stage one: Cancer cells are in their early stage and are only located in the skin
- Stage two: Cancer cells are evident in the skin and there is no sign of spread to the lymph nodes or any other parts of the body.
- Stage three: Cancer cells have spread to nearby lymph nodes or to an intermediary site between the primary tumor and lymph nodes.
- Stage four: Cancer cells have spread to other parts of the body beyond the lymph nodes, e.g. liver, brain and bones.

9.3.2 Growth Phases of Melanoma

Human cutaneous melanoma is characterized by the radial growth phase and the vertical growth phase. The radial growth phase is characterized by the dispersal of melanoma cells in the epidermis that form a papule and appear planar. Similarly, the vertical growth phase melanoma cells are dispersed in the epidermis, as well as the dermis. However, a nodule forms in the vertical growth phase and the predominant v-Raf murine sarcoma viral oncogene homologue B (BRAF) and neuroblastoma rat sarcoma viral oncogene homologue (NRAS) mutations in melanoma increase significantly from the radial growth phase to the vertical growth phase (Dong et al., 2003, Greene et al., 2009).

9.3.3 Subtypes of Melanoma

Six subtypes of melanoma exist, viz. superficial spreading melanoma, nodular melanoma, lentigo maligna melanoma, acral lentiginous melanoma, amelanotic melanoma and desmoplastic melanoma. Superficial spreading melanoma and nodular melanoma are the most frequently diagnosed subtypes of melanoma. Table 9.1 outlines the subtypes of melanoma.

TABLE 9.1
Subtypes of Melanoma

Subtypes	Descriptions	References
Superficial spreading melanoma (SSM)	SSM arises from a small, asymmetrical lesion with variable coloration. The radial growth phase is evident for long periods of time, followed by the vertical growth phase. Approximately 70% of melanomas are SSM.	(Aquilato et al., 2014)
Nodular melanoma (NM)	NM accounts for 15% of all melanomas diagnosed, and is common in males above the age of 60. The growth of nodular melanomas occurs spontaneously as they do not arise from pre-existing lesions. These lesions are characterized by variable coloration (darkly pigmented, multi-colored or colorless). The shift from the radial growth phase to the vertical growth phase occurs in a shorter time period compared to SSMs.	(Aquilato et al., 2014)
Lentigo maligna melanoma (LMM)	LMM accounts for 10% of all melanomas diagnosed, and occurs in middle aged-elderly individuals with sun-damaged skin, particularly on the head and neck. Lentigo malignas resemble age spots and often go untreated, leading to a poor prognosis.	(Aquilato et al., 2014)
Acral lentiginous melanoma (ALM)	ALM is uncommon, as it accounts for 5% of all melanomas diagnosed. This subtype is largely observed in Asians and Africans and appears on the palms of hands, soles of feet and underneath fingernails and toenails.	(Aquilato et al., 2014)

(*Continued*)

TABLE 9.1 (Continued)

Subtypes	Descriptions	References
Amelanotic melanoma (AM)	AMs pose a challenge in terms of diagnosis, as some lesions are the same color as the skin but in other cases, the lesions appear to be pink or red. Thus, early detection is possible to improve the prognosis of melanoma.	(Aquilato et al., 2014; McClain et al., 2012)
Desmoplastic melanomas (DM)	DMs are a rare subtype characterized by spindle melanocytes and an abundance of collagen. This subtype differs from other subtypes as the lesions are atypical, colorless and aggressive, with a high local recurrence rate. Desmoplastic melanomas occur in elderly patients, especially males, and occurs on sun damaged skin, particularly on the head and neck.	(Shi et al., 2019)

9.3.4 Pathogenesis of Melanoma

Human cutaneous malignant melanoma arises from melanocytes in the stratum basale layer of the skin. The mortality rate for melanoma is significantly higher than in non-melanoma skin cancers (such as BCC and SCC). Patients with primary melanomas fare better than patients with metastatic melanoma. The main risk factor for melanoma is UV radiation, as it has severe mutagenic effects such as the formation of cyclobutene pyrimidine dimers (CPD) that alter genes that regulate normal physiological functions (growth, cell survival and metabolism) (Pfeifer and Besaratinia, 2012). Human cutaneous malignant melanoma is a growing concern, as UV radiation potentiates high mutation rates. These exceed mutation rates displayed by other solid cancers; however, primary cutaneous melanomas have shown to be treated effectively through surgery followed by radiation therapy (Moreno Nogueira, Valero Arbizu, and Pérez Temprano, 2013). Metastatic melanomas are challenging, as low survival rates have been observed after surgery and radiation therapy; thus, molecular markers need to be explored.

A proto-oncogene, BRAF, is a well-known target for targeted therapeutics, as BRAF mutations have been observed in 40–60% of all melanomas. BRAF forms part of the mitogen activated protein kinase (MAPK) pathway that controls cell growth and proliferation. The activation of this pathway occurs in response to growth signals (epidermal growth factor, G-proteins and vascular endothelial growth factor) resulting in the activation of BRAF and the activation of downstream kinases by BRAF that subsequently activate the cell cycle (Hannan et al., 2017). However, the mutated form of BRAF (oncogene) undergoes activation in the absence of growth signals resulting in uncontrolled proliferation. The missense mutation $BRAF^{V600E}$ is evident in 50% of metastatic melanoma cases, making BRAF an essential target for anti-cancer drug development (Ascierto et al., 2012).

Another key pathway in the pathogenesis of melanoma is the PI3K/Akt/mTOR pathway that regulates cell growth, survival and metabolism. This pathway is an emerging target for the treatment of melanoma because hyperactivation arises when the PIK3CA gene that encodes PI3K is mutated. The mutated form of PI3K undergoes activation in the absence of growth signals and activates downstream kinases such as protein kinase B (Akt) that enable the progression of the cell cycle and inhibit apoptosis (Davies, 2012). The hyperactivation of this pathway can occur when the tumor suppressor gene, phosphatase tensin homologue (PTEN), has been inactivated by loss of function mutations (deletions, frame-shift mutations, etc.). Phosphatase tensin homologue is a negative regulator of the PI3K/Akt/mTOR pathway and sustains homeostasis through apoptosis in damaged cells. Furthermore, the hyperactivation of the PI3K/Akt/mTOR pathway results in the loss of the apoptosis, inducing potential of chemotherapy and radiation therapy (Liu and Sheikh, 2014; Stahl et al., 2003). The deregulation of the MAPK and the PI3K/Akt/mTOR pathways are key elicitors of the transition of primary melanomas to metastatic melanomas. However, the transmembrane glycoprotein, epithelial cadherin (E-cadherin), implicates the pathogenesis of melanoma (Liu and Sheikh, 2014).

Melanocytes are connected to keratinocytes under normal physiological conditions through E-cadherin; the link between melanocytes and keratinocytes necessitates the normal functioning of melanocytes. An epithelial to mesenchymal transition enables the expression of neural cadherin (N-cadherin) instead of E-cadherin. Neural cadherin enables melanomas to migrate from the epidermis into the dermis enabling metastasis. Therefore, N-cadherin leads to the hyperactivation of the PI3K/Akt/mTOR pathway (Liu and Sheikh, 2014; Mrozik et al., 2018).

9.3.5 Conventional Treatment Approved by the Food and Drug Administration for Melanoma

Melanoma is effectively treated through surgery and the use of chemotherapy and radiation therapy as an adjuvant for the earlier stages of melanoma (stage I-IIIB). The later stages require the use of chemotherapy, targeted therapy and immunotherapy. Table 9.2 outlines the conventional treatment options approved by the U.S. Food and Drug Administration (FDA) for the treatment of melanoma.

TABLE 9.2
Conventional Treatments Approved by the FDA for the Treatment of Melanoma

Names	Common Names	Types of Therapy	Mechanisms of Action	Stages	References
Binimetinib	Mektovi	Targeted	An inhibitor of MEK[1] activated by $BRAF^{V600E}$[2] or $BRAF^{V600K}$[3] mutations	Stage III-IV, metastatic melanoma	(Koelblinger, Dornbierer, and Dummer, 2017)
Cobimetinib	Cotellic	Targeted	An inhibitor of MEK activated by $BRAF^{V600E}$ or $BRAF^{V600K}$ mutations	Stage III-IV, metastatic melanoma	(Signorelli and Shah Gandhi, 2017)
Dabrafenib mesylate	Tafinlar	Targeted	An inhibitor of mutated BRAF kinase, $BRAF^{V600E}$ or $BRAF^{V600K}$	Stage III-IV, metastatic melanoma	(Gibney and Zager, 2013)
Dacarbazine	DTIC-Dome	Chemotherapy	Alkylating agent that binds to DNA preventing cell division	Stage III-IV, metastatic melanoma	(Teimouri, Nikfar, and Abdollahi, 2013)
Encorafenib	Braftovi	Targeted	An inhibitor of mutated BRAF kinase, $BRAF^{V600E}$ or $BRAF^{V600K}$ mutations	Stage III-IV, metastatic melanoma	(Koelblinger, Thuerigen, and Dummer, 2018)
Interleukin-2	Aldesleukin, Proleukin	Immunotherapy	Increases the production of LAKs[4] that target cancer cells	Stage III-IV, metastatic melanoma	(Choudhry et al., 2018; West et al., 2011)
Interferon-α	Intron-A and Roferon-A	Immunotherapy	Upregulates the Major histocompatibility complexes found on cancer cells, thus rendering targeting by T cells more effective	Stage II-IV, metastatic melanoma	(Hwu, Lizée, and Liu, 2015)
Ipilimumab	Yervoy	Immunotherapy	A recombinant human monoclonal antibody that binds to CTLA-4[5], which blocks the interaction of CTLA-4 with its ligands, CD80[6] and CD86[7]; the blocking of this interaction enables the activation of the anti-tumor T-cell response	Stage III-IV, metastatic melanoma	(Festino et al., 2018)

(Continued)

TABLE 9.2 (Continued)

Names	Common Names	Types of Therapy	Mechanisms of Action	Stages	References
Nivolumab	Opdivo	Immunotherapy	An immune checkpoint inhibitor that blocks the interaction of programmed death protein 1 with its ligands PD-L1[8] and PD-L2[9]; the blockade of this interaction enhances anti-tumor immune responses and reduces tumor growth	Stage III-IV, metastatic melanoma	(Guo, Zhang, and Chen, 2017)
Peginterferon Alfa-2b	PegIntron	Target/ immunotherapy	Inhibits angiogenesis and increases the infiltration of $CD4^+$ cells into melanoma tumors	Stage II-IV, melanoma whereby surgery and lymph node removal occurred.	(Di Trolio et al., 2012)
Pembrolizumab	Keytruda	Immunotherapy	A targeted antibody that binds to programmed cell death protein 1 enabling activation of the immune response	Stage III-IV, metastatic melanoma	(Sahni, Valecha, and Sahni, 2018)
Talimogene laherparepvec	Imlygic	Oncolytic immunotherapy	Talimogene[10] multiplies in cancer cells causing lysis and stimulates the immune system through GM-CSF[11]	Stage III-IV, metastatic melanoma	(Andtbacka et al., 2015)
Trametinib	Mekinist	Targeted	An inhibitor of MEK activated by the BRAF mutation, $BRAF^{V600K}$ or $BRAF^{V600E}$	Stage III-IV, metastatic melanoma	(Chopra and Nathan, 2015)
Vemurafenib	Zelboraf	Targeted	An inhibitor of mutated BRAF kinase, $BRAF^{V600E}$	Stage III-IV, metastatic melanoma	(Chapman et al., 2011)

Notes: **1** Kinase activated in the MAPK pathway; **2** Single nucleotide substitution mutation resulting in the substitution of glutamic acid for valine; **3** Single nucleotide substitution mutation resulting in the substitution of lysine for valine; **4** Lymphokine activated killer cells (white blood cell stimulated to kill tumor cells); **5** Cytotoxic T-lymphocyte-associated protein 4; **6** Cluster of differentiation 80; **7** Cluster of differentiation 86; **8** Programmed death ligand 1; **9** Programmed death ligand 2; **10** Herpes simplex virus 1 whereby virulent genes have been replaced with GM-CSF; **11** Cytokine, granulocyte macrophage colony stimulating factor

9.3.6 The Significance of Inhibiting the PI3K/Akt/mTOR Pathway in Human Cutaneous Melanoma

The PI3K/Akt/mTOR pathway is upregulated in melanoma due to the loss of PTEN, a tumor suppressor gene that negatively regulates the PI3K pathway and has been identified in 30–40% of melanoma cell lines (Aguissa-Touré and Li, 2012). Phosphatidylinositol-3-kinase consists of a regulatory subunit (p85) and a catalytic subunit (p110) that is activated by epidermal growth factors that bind to growth factor receptors or by G-proteins that bind to G-protein–coupled receptors. The growth factors enable the phosphorylation of tyrosine kinases such as adenosine triphosphate (ATP) on the receptor subsequently activating PI3K, Akt and other downstream kinases (Jazirehi, Wenn, and Damavand, 2012). The loss of PTEN leads to the continuous activation of the cell cycle in the absence of apoptosis (Wu, Goel, and Haluska, 2003).

The MAPK pathway is closely linked to the PI3K pathway, and several inhibitors have been developed to target these pathways, e.g. BRAF inhibitors (dabrafenib and vemurafenib) and PI3K inhibitors (GSK2636771 and Burpalisib) (Aquilato et al., 2014; Eroglu and Ribas, 2016; Yang et al., 2019). The inhibitors largely target one tyrosine kinase, e.g. BRAF, PI3K or RAS; thus, the targeting of specific kinases leads to acquired resistance, resulting in the progression of cancer. Resistance to these inhibitors occurs due to the activation of parallel signaling pathways suggesting that the simultaneous inhibition of multiple signaling pathways minimizes the development of resistance (Inamdar, Madhunapantula, and Robertson, 2010). The indirect anti-cancer effects, i.e. the alteration of the tumor microenvironment by PI3K inhibitors, strengthens the anti-cancer response through the inhibition of the PI3K/Akt/mTOR pathway, in combination with other targeted therapies.

The tumor microenvironment is made up of cancer-associated fibroblasts, tumor associated macrophages, adipocytes, endothelial cells, immune cells and the extracellular matrix (Franco et al., 2020). PI3K inhibitors are largely used to target cancer cells specifically those that trigger apoptosis; however, the mechanism is largely affected by the development of resistance. Okkenhaug, Graupera, and Vanhaesebroeck (2016) suggested that isoform-selective PI3K inhibitors administered for a short period of time, i.e. three days, could enhance the functioning of blood vessels that supply nutrients and oxygen to the tumors. Improving the function of blood vessels through the downregulation of vascular endothelial growth factor (VEGF) enables effective drug delivery (chemotherapeutics, immunotherapy and targeted therapy) (Okkenhaug, Graupera, and Vanhaesebroeck, 2016).

The inhibitors of PI3K can be categorized as pan-PI3K inhibitors, isoform selective inhibitors and dual PI3K/mTOR inhibitors. The PI3K enzyme consists of four isoforms—namely PI3Kα, PI3Kβ, PI3Kδ and PI3Kγ—while pan-PI3K inhibitors inhibit all four isoforms and isoform selective inhibitors only inhibit one isoform, i.e. inhibition of PI3Kα (Vanhaesebroeck, Whitehead, and Piñeiro, 2016). The PI3Kβ isoform is associated with cancers that arise due to PTEN loss; therefore, this isoform is targeted in melanoma. Furthermore, melanomas are resistant to immune signaling; thus, immunotherapy has been suggested as a treatment option that subsequently sensitizes the melanomas to PI3Kβ inhibitors (Okkenhaug, Graupera, and Vanhaesebroeck, 2016). Table 9.3 highlights PI3K inhibitors undergoing clinical trials for use in the treatment of human cutaneous malignant melanoma.

Tumor-associated macrophages (TAMs) increase vasculature through enhanced production of vascular endothelial growth factor (VEGF) that promotes neo-angiogenesis in the tumor microenvironment. Tumor-associated macrophages secrete cytokines such as transforming growth factor beta (TGFβ) and interleukin-10 (IL-10) that dampen the anti-tumor effect of immune cells by enabling progression and survival of tumors. The dual isoform selective inhibitor of PI3Kδ/γ,

TABLE 9.3
PI3K Inhibitors in Clinical Trials for Melanoma

Drugs	Types of Inhibitors	Study Design/Mechanisms of Action	Phases of Clinical Trial and Identification	References
Copanlisib	Pan-class I PI3K[1] inhibitor	Melanoma and recurrent melanoma. PIK3CA[2] mutation and PTEN[3] loss	II NCT02465060	(Krause, Hassenrück, and Hallek, 2018; National Cancer Institute (NCI), 2020)
GSK2636771+ Pembrolizumab	PI3Kβ[4] inhibitor	Melanoma and other malignant neoplasms of the skin and metastatic melanoma	I/II NCT03131908	(Tawbi et al., 2018; M.D. Anderson Cancer Center, 2020)

Notes: **1** Phosphotidylinositol-3-kinase; **2** Phosphatidylinositol-4, 5-Bisphosphate 3-kinase catalytic subunit alpha; **3** Phosphatase tensin homologue; **4** isoform of PI3K related to skin cancer

RP6530, switches the TAMs to a M1 phenotype that favors anti-tumor responses such as a reduction in tumor vasculature and a decrease in the size of the tumor. The RP6530 inhibitor is particularly used in Hodgkin's lymphoma (Anfray et al., 2020).

9.4 JUSTIFICATION OF THE USE OF MEDICINAL PLANTS FOR THE TREATMENT OF CANCER

The treatment of cancer largely relies on the use of synthetic drugs that are administered in chemotherapy and targeted therapies. These synthetic drugs elicit adverse effects that may threaten the overall survival of patients. The use of phytochemicals derived from medicinal plants has been justified through pre-clinical studies that validate the efficacy of these phytochemicals for the treatment of cancer (Choudhari et al., 2020). The FDA has approved certain phytochemicals for the treatment of cancer, and 49% of the small molecule drugs approved between 1940 and 2014 for the treatment of cancer were natural products or derived from natural products (Newman and Cragg, 2016). Table 9.4 outlines the phytochemicals derived from medicinal plants approved by the FDA for the treatment of cancer. The chemical structures of phytochemicals derived from medicinal plants approved by the FDA for the treatment of cancer are displayed in Figure 9.1.

TABLE 9.4
Phytochemicals Derived from Medicinal Plants Approved by the FDA for the Treatment of Cancer

Drugs	Sources	Mechanisms of Action	References
Taxanes			
Paclitaxel **Cabazitaxel** **Docetaxel**	*Taxus brevifolia* Nutt.	Cell division is prevented as the depolymerization process in the microtubule is disrupted	(Abal, Andreu, and Barasoain, 2003)
Vinca alkaloids			
Vincristine **Vinblastine** **Vindesine** **Vinorelbine** **Vinflunine**	*Catharanthus roseus* (L.) G.Don	Binding of β-tubulin inhibiting polymerization into microtubules, resulting in cell death due to cell cycle arrest	(Moudi et al., 2013)
Camptothecins			
Irinotecan **Topotecan**	*Camptotheca acuminate* Decne.	Binds DNA[1] topoisomerase (type I), preventing cleavage and re-ligation of DNA, leading to double strand breaks in DNA, resulting in cell death	(Legarza and Yang, 2006)
Podophyllotoxins			
Etoposide **Teniposide**	*Podophyllum peltatum* L. and *Podophyllum emodi* Wall. ex Hook.f. & Thomson	Binds the tubulin that stabilizes microtubules that form the mitotic spindles during cell division, resulting in cell cycle arrest	(Ardalani, Avan, and Ghayour-Mobarhan, 2017)
Other classes			
Homoharringtone **Omacetaxine mepesuccinate**	*Cephalotaxus harringtonia* var fastigiate (Carriere) Rehder.	The interaction with the ribosomal A site is prevented; thus, the initial elongation step of protein synthesis is prevented, which affects proteins involved with cell division and growth, e.g. c-Myc[2], Cyclin D1[3], etc.	(Choudhari et al., 2020)

Notes: **1** Deoxyribonucleic acid; **2** cellular myelocytomatosis oncogene (human transcription factors); **3** regulator of cell cycle progression

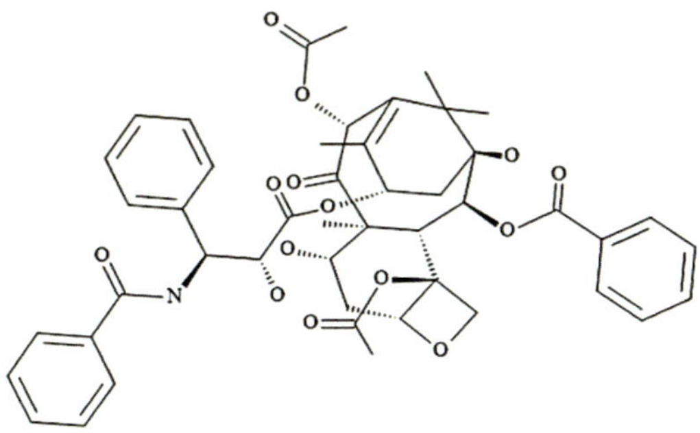

Paclitaxel

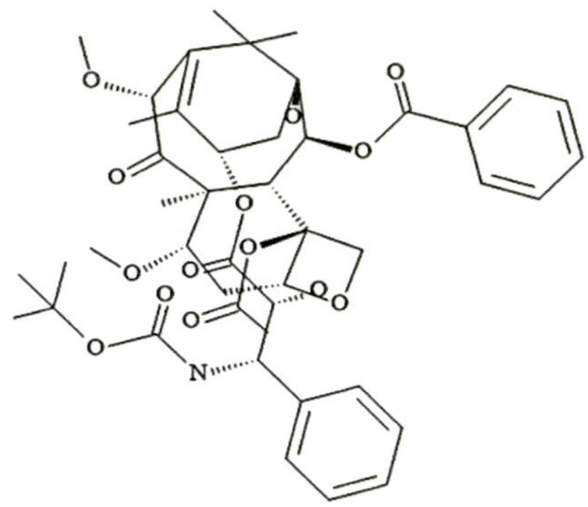

Cabazitaxel

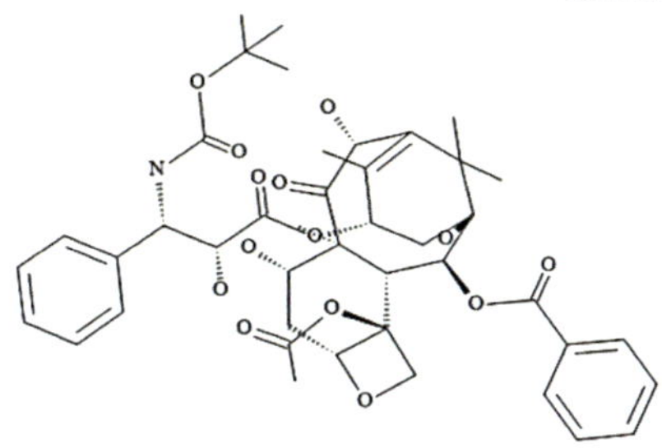

Docetaxel

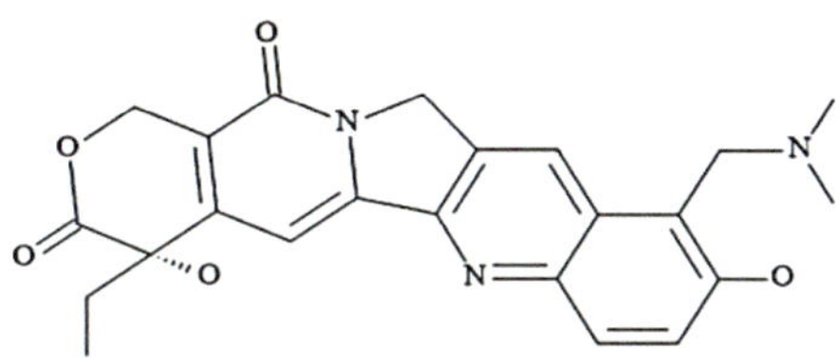

Topotecan

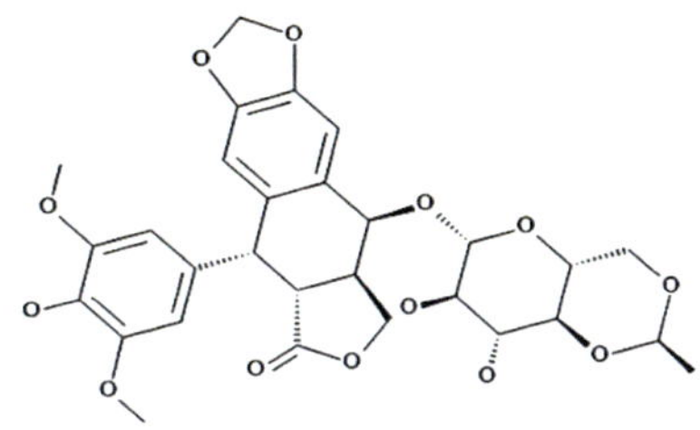

Etoposide

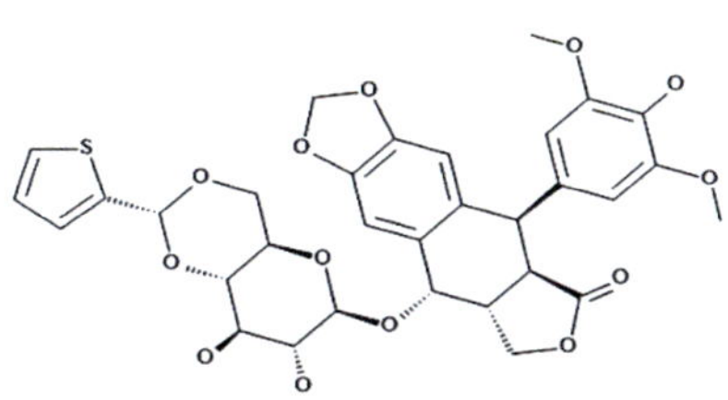

Teniposide

FIGURE 9.1 Chemical structures of phytochemicals isolated from medicinal plants approved by the FDA for the treatment of cancer (Drawn using BIOVIA Draw).

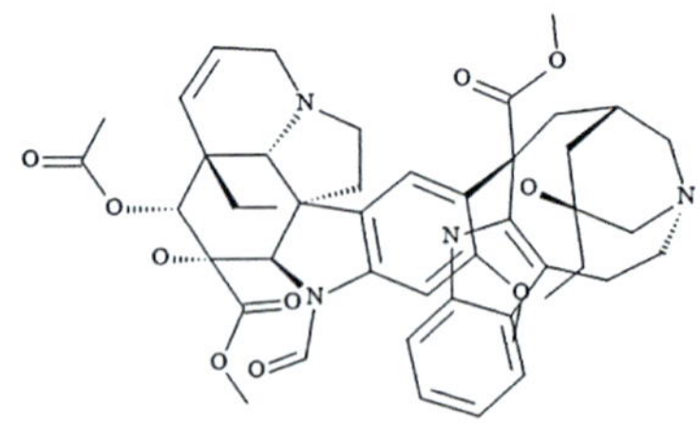

Vincristine

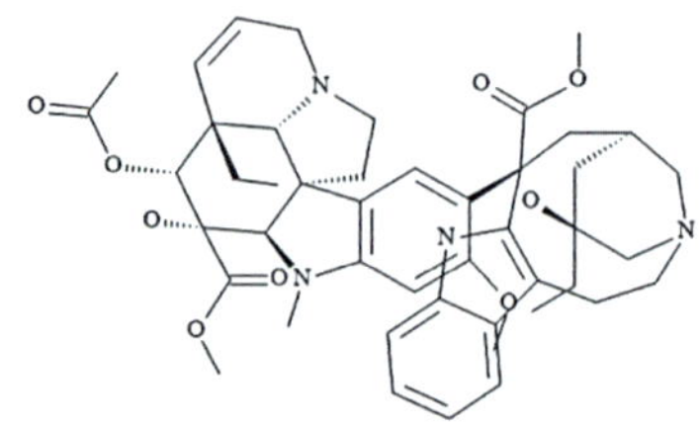

Vinblastine

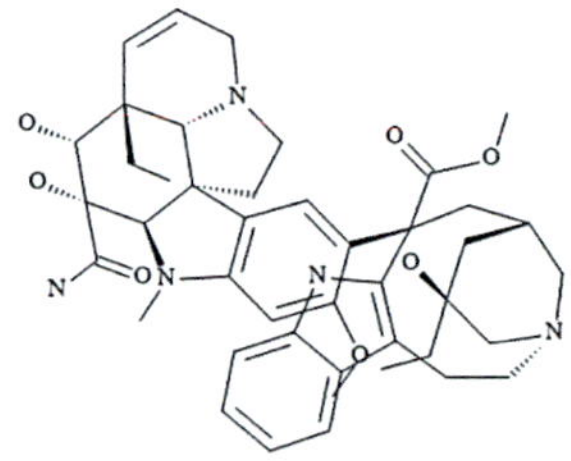

Vindesine

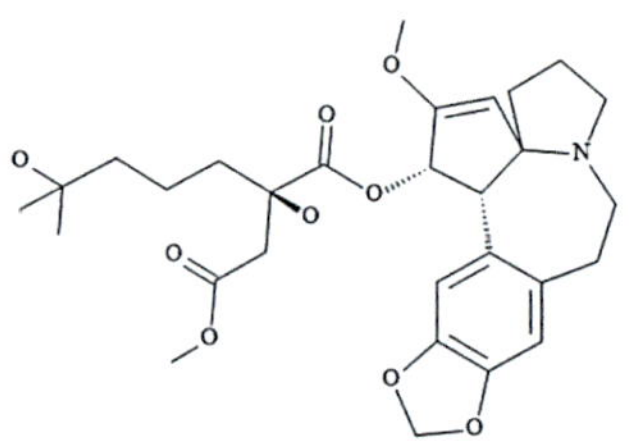

Homoharringtonine

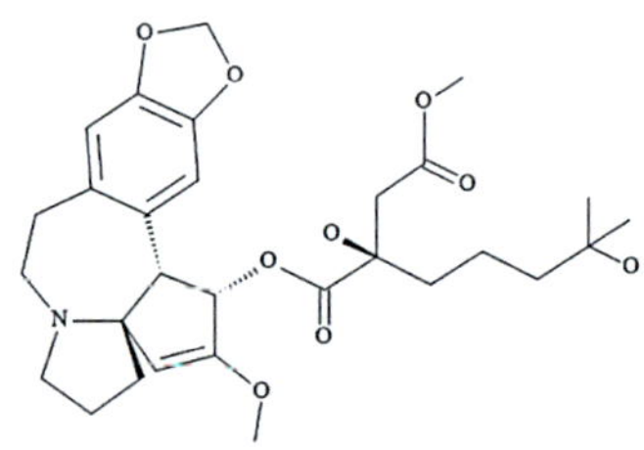

Omacetaxine mepesuccinate

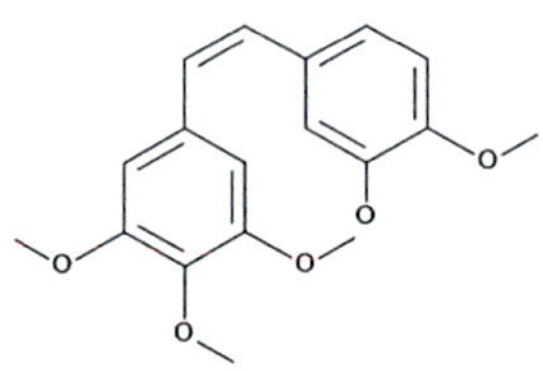

Combrestatin A4

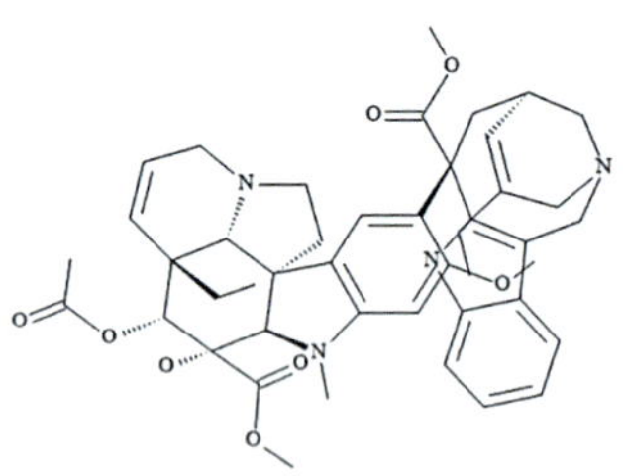

Vinorelbine

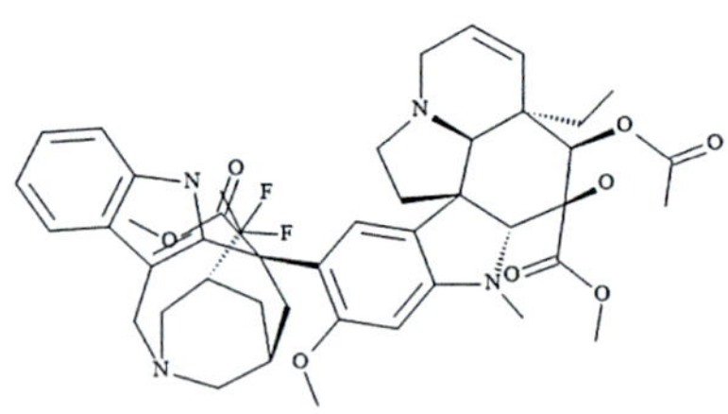

Vinflunine

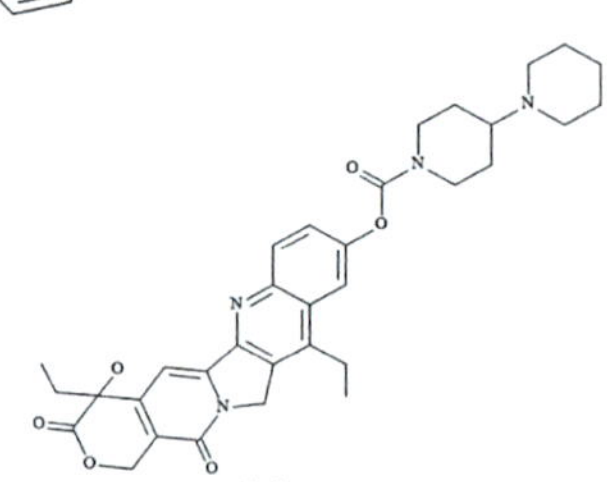

Irinotecan

FIGURE 9.1 (Continued)

9.5 PHYTOCHEMICALS DERIVED FROM MEDICINAL PLANTS AS INHIBITORS OF THE PI3K PATHWAY AND MAPK PATHWAY IN HUMAN CUTANEOUS MALIGNANT MELANOMA

The PI3K/Akt/mTOR pathway is key in the treatment of human cutaneous malignant melanoma, particularly in the hyperactivation of the PI3K/Akt/mTOR pathway due to PTEN loss (Chamcheu et al., 2019; Papa and Pandolfi, 2019). Several synthetic PI3K inhibitors have been approved by the FDA; however, these inhibitors cannot be used for long periods of time due to the adverse effects such as diarrhea, colitis, nausea, vomiting, hyperglycemia, fatigue, rash, anorexia and abnormal hepatic function (Buchanan, Lee, and Shepherd, 2019; Sanchez et al., 2019; Yang et al., 2019). Pan-PI3K inhibitors and dual PI3K/mTOR inhibitors have shown efficacy in clinical trials, but the adverse effects often lead to the discontinued use of these inhibitors (Rodon and Tabernero, 2017). In light of this, isoform-specific inhibitors such as p110β were favored for the inhibition of the PI3K pathway in melanoma; however, these inhibitors acquire resistance due to the activation of closely related signaling pathways, i.e. rat sarcoma (RAS)/rapidly accelerated fibrosarcoma (Raf)/ mitogen activated protein kinase (MAPK)/extracellular signal regulated kinase (ERK) pathway (Yang et al., 2019).

Thus, as mentioned previously, the targeting of the tumor microenvironment enhances the efficacy of PI3K inhibitors, i.e. the inhibition of VEGF, the downregulation of IL-10 and TGFβ activity. PI3K inhibitors can be combined with BRAF or MEK inhibitors to lower resistance and the occurrence of adverse effects mentioned previously. Thus, phytochemicals that can inhibit PI3K, modulate targets present in the tumor microenvironment (i.e. cancer associated fibroblasts, tumor associated macrophages, adipocytes, endothelial cells, immune cells and the extracellular matrix) and inhibit parallel signaling pathways such as the MAPK pathway which may yield better outcomes in clinical trials and decrease the likelihood of resistance, as well as adverse effects. In light of this, several phytochemicals isolated from medicinal plants or dietary sources that inhibit the PI3K/Akt/mTOR pathway and the MAPK pathway were explored.

Euplotin C (EC), isolated from *Euplotes crassus* Dujardin, induced apoptosis in three malignant melanoma cell lines (A375, MeWo and 501 Mel) with 50% minimum inhibitory concentrations (IC_{50}) of 3.53 ± 0.19, 2.68 ± 0.29 and 3.56 ± 0.38 μM, respectively. Western blotting analysis revealed that EC inhibited protein kinase B (p-Akt) by 68% and BRAF by 27% (Carpi et al., 2018). Bornyl cis-4-hydroxycinnamate (BCH), isolated from *Piper betle* L., reduced the viability of two malignant melanoma cell lines (A375 and A2058) at a concentration of 12 μM. Western blotting analysis revealed that the levels of p-PI3K and p-Akt were significantly reduced at 6 μM. Furthermore, BCH reduced the expression of extracellular signal regulated kinase (ERK), c-Jun N-terminal kinases (JNK) and p38 MAPK (MAPK pathway related proteins) (Yang et al., 2018). Isoangustone A (IA), isolated from *Glycyrrhiza glabra* L., inhibited the growth of the malignant melanoma cell line (SK-MEL28) by 67%. Through an *in vitro* PI3K assay and Western blotting, PI3K was significantly inhibited in comparison to the positive control, LY294002. Furthermore, IA inhibited p-Akt and molecular targets in the MAPK pathway (MKK4 and MKK7) (Song et al., 2013).

Magnolol, isolated from *Magnolia officinalis* Rehder & E.H. Wilson, induced cell death in NRAS-mutant WM1366 and BRAF mutant WM164 melanoma cells in a dose- and time-dependent manner. Western blotting analysis revealed significant downregulation of p-Akt, p-mTOR and p-ERK (Emran et al., 2019). Methylwogonin, isolated from *Scutellaria baicalensis* Georgi, significantly reduced the viability of human melanoma (A375) cells treated with methylwogonin with an IC_{50} of 72.9 μM (24 hours treatment) and 54.2 μM (48 hours treatment), respectively. Western blotting analysis revealed decreased expression of p-PI3K, p-Akt and p-mTOR (Chen et al., 2019). Plantamajoside, isolated from *Plantago asiatica* L., reduced the viability of the malignant melanoma cell line (A2058) in a dose- and time-dependent manner. Western blotting analysis revealed that Plantamajoside significantly inhibited the expression of p-Akt (Wang et al., 2020). Lycorine, isolated from *Lycoris radiate* (L'HÉR.) Herb. at 50 μM, significantly decreased the viability of A375 cells. Western blotting analysis revealed that lycorine significantly inhibited p-Akt and p-mTOR, suggesting that the anti-proliferative activity was due to the inhibition of the PI3K/Akt/mTOR pathway (Jiang and Liu, 2018).

Fisetin, a phytochemical present in strawberries, onions, grapes and persimmons, inhibited the proliferation of A375 cells at concentrations ranging from 20–80 μM. The IC_{50} values were 38.1 μM (24 h) and 20.3 μM (48 h), respectively. Furthermore, fisetin at concentrations of 40 μM and 80 μM were used to treat melanoma reconstructs. After 16 days, the fisetin-treated melanoma constructs had undetectable lesions compared to the control, where several tumor nests were evident. The inhibition of tumor growth in both the melanoma monolayer (A375) and 3-D melanoma constructs was through the downregulation of mTOR, Akt and PI3K. These were determined through immunohistochemical analysis of fisetin-treated melanoma constructs, Western blotting using whole-cell lysates from fisetin (20–40 μM) treated A375 and 451 Lu melanoma cells (Syed et al., 2014). Fisetin at concentrations ranging from 5–20 μM, significantly decreased the phosphorylation of MEK 1/2 and ERK 1/2 in A375 and RPMI-7951 cells when exposed for a 24-hour period (Pal et al., 2014).

Resveratrol, a phytochemical present in peanuts, pistachios, grapes, red wine, etc., inhibited the viability of A375 cells after a 24-hour period in a dose-dependent manner. Western blotting analysis revealed that resveratrol decreased the phosphorylation of p-PI3K, p-Akt and p-mTOR, suggesting that the induction of apoptosis in A375 cells was through the inhibition of the PI3K/Akt/mTOR pathway (Gong and Xia, 2020). Furthermore, resveratrol inhibited the proliferation of human malignant melanoma cells (HT-144 cells) in a dose-dependent manner. Western blotting analysis revealed that resveratrol significantly reduced the phosphorylation of MEK 1/2 and ERK 1/2 (Lei et al., 2017).

Curcumin, isolated from *Curcuma longa* L., inhibited the proliferation of human malignant melanoma cells (A375 and C8161) in a dose- and time-dependent manner. The whole-cell lysates of human cutaneous malignant melanoma cell lines (A375 and C8161) treated with curcumin were analyzed through Western blotting. This analysis revealed that p-Akt and p-mTOR were downregulated, suggesting that the induction of autophagy was through the inhibition of the PI3K/Akt/mTOR pathway (Zhao et al., 2016). The inhibition of JNK and ERK 1/2 was evaluated through Western blotting using whole-cell lysates from A375 cells treated with natural borneol and curcumin. The phosphorylation of JNK and ERK 1/2 was significantly reduced by the combination of natural borneol and curcumin (Chen et al., 2014).

The phytochemicals isolated from various medicinal plants and dietary sources (depicted in Figure 9.2) inhibited both the PI3K/Akt/mTOR and the MAPK pathways. The dual inhibition of these signaling pathways were advantageous, as monotherapies are often ineffective due to the development of resistance. Combination therapies that target more than one signaling pathway are often preferred, as resistance is less likely to occur and adverse effects are minimized (McCubrey et al., 2012).

9.6 CONCLUSION

The predominant hyperactivation of the PI3K/Akt/mTOR and MAPK pathways in human malignant melanoma are targeted with various synthetic inhibitors that acquire resistance and lead to adverse effects. Phytochemicals from medicinal plants or dietary sources display anti-proliferative activity in cancer; thus, they have been exploited in pre-clinical studies for the inhibition of the PI3K/Akt/mTOR and MAPK pathways. Some phytochemicals are in clinical trials for the inhibition of these signaling pathways; however, they have not yet been approved by the FDA. Phytochemicals could be used to simultaneously target tumor promoting factors in the tumor microenvironment, thereby enhancing the efficacy of synthetic inhibitors used to downregulate hyperactivated signaling pathways. The phytochemicals could be used to slightly decrease the production of VEGF to

FIGURE 9.2 Medicinal plants and dietary sources from which phytochemicals that inhibited the PI3K/Akt/mTOR and MAPK pathway in human cutaneous malignant melanoma. A, *Piper betle*; B, *Glycyrrhiza glabra*; C, *Magnolia officinalis*; D, *Scutellaria baicalensis*; E, *Plantago asiatica*; F, *Lycoris radiate*; G, *Curcuma longa* (Σ64, 2006; Cutler, 2009; Eugster, 2014; Han, 2013; Kenraiz, 2016; Stang, 2006; Shizhao, 2007).

normalize the vasculature leading to the tumor, thereby enhancing the delivery of FDA-approved targeted therapies or phytochemicals to the tumor.

REFERENCES

Abal, M, JM Andreu, and I Barasoain. 2003. "Taxanes: Microtubule and centrosome targets, and cell cycle dependent mechanisms of action." *Current Cancer Drug Targets* 3, 3: 193–203. doi: 10.2174/1568009033481967.

Aguissa-Touré, Almass-Houd, and Gang Li. 2012. "Genetic alterations of PTEN in human melanoma." *Cellular and Molecular Life Sciences* 69, 9: 1475–1491. doi: 10.1007/s00018-011-0878-0.

Andtbacka, RH, Howard L Kaufman, Frances Collichio, Thomas Amatruda, Neil Senzer, Jason Chesney, Keith A Delman, Lynn E Spitler, Igor Puzanov, and Sanjiv S Agarwala. 2015. "Talimogene laherparepvec improves durable response rate in patients with advanced melanoma." *Journal of Clinical Oncology* 33, 25: 2780–2788. doi: 10.1200/JCO.2014.58.3377.

Anfray, Clément, Aldo Ummarino, Fernando Torres Andón, and Paola Allavena. 2020. "Current strategies to target tumor-associated-macrophages to improve anti-tumor immune responses." *Cells* 9, 1: 46. doi: 10.3390/cells9010046.

Aquilato, Andrea, Victoria Lopez, Barbara Doonan, Tze-Chen Hsieh, John Thomas Pinto, Erxi Wu, and Joseph M Wu. 2014. "BRAF mutation in melanoma and dietary polyphenols as adjunctive treatment strategy." In *Polyphenols in Human Health and Disease*, 1353–1365. Elsevier. doi: 10.1016/B978-0-12-398456-2.00102-X.

Ardalani, Hamidreza, Amir Avan, and Majid Ghayour-Mobarhan. 2017. "Podophyllotoxin: A novel potential natural anticancer agent." *Avicenna Journal of Phytomedicine* 7, 4: 285. PMID: 28884079.

Ascierto, Paolo A, John M Kirkwood, Jean-Jacques Grob, Ester Simeone, Antonio M Grimaldi, Michele Maio, Giuseppe Palmieri, Alessandro Testori, Francesco M Marincola, and Nicola Mozzillo. 2012. "The role of BRAF V600 mutation in melanoma." *Journal of Translational Medicine* 10, 1: 1–9. doi: 10.1186/1479-5876-10-85.

Bray, Freddie, Jacques Ferlay, Isabelle Soerjomataram, Rebecca L Siegel, Lindsey A Torre, and Ahmedin Jemal. 2018. "Global cancer statistics 2018: GLOBOCAN estimates of incidence and mortality worldwide for 36 cancers in 185 countries." *CA: A Cancer Journal for Clinicians* 68, 6: 394–424. doi: 10.3322/caac.21492.

Buchanan, Christina M, Kate L Lee, and Peter R Shepherd. 2019. "For better or worse: The potential for dose limiting the on-target toxicity of PI 3-kinase inhibitors." *Biomolecules* 9, 9: 402. doi: 10.3390/biom9090402.

Burd, EM. 2003. "Human papillomavirus and cervical cancer." *Clinical Microbiology Reviews* 16, 1: 1–17. doi: 10.1128/cmr.16.1.1-17.2003.

Cancer Research UK. 2020a. "Stages and types." www.cancerresearchuk.org/about-cancer/melanoma/stages-types (Accessed 18 October 2020).

Cancer Research UK. 2020b. "TNM staging." www.cancerresearchuk.org/about-cancer/melanoma/stages-types/tnm-staging (Accessed 18 October 2020).

Carpi, Sara, Beatrice Polini, Giulio Poli, Gabriela Alcantara Barata, Stefano Fogli, Antonella Romanini, Tiziano Tuccinardi, Graziano Guella, Francesco Paolo Frontini, and Paola Nieri. 2018. "Anticancer activity of Euplotin C, isolated from the marine ciliate *Euplotes crassus*, against human melanoma cells." *Marine Drugs* 16, 5: 166. doi: 10.3390/md16050166.

Chamcheu, Jean Christopher, Tithi Roy, Mohammad Burhan Uddin, Sergette Banang-Mbeumi, Roxane-Cherille N Chamcheu, Anthony L Walker, Yong-Yu Liu, and Shile Huang. 2019. "Role and therapeutic targeting of the PI3K/Akt/mTOR signaling pathway in skin cancer: A review of current status and future trends on natural and synthetic agents therapy." *Cells* 8, 8: 803. doi: 10.3390/cells8080803.

Chapman, Paul B, Axel Hauschild, Caroline Robert, John B Haanen, Paolo Ascierto, James Larkin, Reinhard Dummer, Claus Garbe, Alessandro Testori, and Michele Maio. 2011. "Improved survival with vemurafenib in melanoma with BRAF V600E mutation." *New England Journal of Medicine* 364, 26: 2507–2516. doi: 10.1056/NEJMoa1103782.

Chen, Jianping, Lin Li, Jianyu Su, Bing Li, Tianfeng Chen, and Yum-Shing Wong. 2014. "Synergistic apoptosis-inducing effects on A375 human melanoma cells of natural borneol and curcumin." *PLoS One* 9, 6: e101277. doi: 10.1371/journal.pone.0101277.

Chen, Jiaorong, Chunmei Huang, Fangfang Liu, Zihui Xu, Li Li, Zheng Huang, and Hongfeng Zhang. 2019. "Methylwogonin exerts anticancer effects in A375 human malignant melanoma cells through apoptosis induction, DNA damage, cell invasion inhibition and downregulation of the mTOR/PI3K/Akt signalling pathway." *Archives of Medical Science: AMS* 15, 4: 1056. doi: 10.5114/aoms.2018.73711.

Chopra, Neha, and Paul D Nathan. 2015. "Trametinib in metastatic melanoma." *Expert Review of Anticancer Therapy* 15, 7: 749–760. doi: 10.1586/14737140.2015.1060127.

Choudhari, Amit S, Pallavi C Mandave, Manasi Deshpande, Prabhakar Ranjekar, and Om Prakash. 2020. "Phytochemicals in cancer treatment: From preclinical studies to clinical practice." *Frontiers in Pharmacology* 10: 1614. doi: 10.3389/fphar.2019.01614.

Choudhry, Hani, Nawal Helmi, Wesam H Abdulaal, Mustafa Zeyadi, Mazin A Zamzami, Wei Wu, Maged Mostafa Mahmoud, Mohiuddin Khan Warsi, Mahmood Rasool, and Mohammad S Jamal. 2018. "Prospects of IL-2 in cancer immunotherapy." *BioMed Research International*. doi: 10.1155/2018/9056173.

Colloby, Peter S, Kevin P West, and Alan Fletcher. 1991. "Observer variation in the measurement of Breslow depth and Clark's level in thin cutaneous malignant melanoma." *The Journal of Pathology* 163, 3: 245–250. doi: 10.1002/path.1711630310.

Cutler, Wendy. 2009. "*Magnolia officinalis* biloba." Obtained from Wikimedia commons under the Creative commons Attribution 2.0 Generic license. https://commons.wikimedia.org/wiki/File:Magnolia_officinalis_biloba.jpg (Accessed 18 October 2020).

Davies, Michael A. 2012. "The role of the PI3K-AKT pathway in melanoma." *The Cancer Journal* 18, 2: 142–147. doi: 10.1097/PPO.0b013e31824d448c.

Davis, Lauren E, Sara C Shalin, and Alan J Tackett. 2019. "Current state of melanoma diagnosis and treatment." *Cancer Biology & Therapy* 20, 11: 1366–1379. doi: 10.1080/15384047.2019.1640032.

Di Trolio, Rossella, Ester Simeone, Giuseppe Di Lorenzo, Antonio Maria Grimaldi, Anna Romano, Fabrizio Ayala, Corrado Caraco, Nicola Mozzillo, and Paolo A Ascierto. 2012. "Update on PEG-interferon α-2b as adjuvant therapy in melanoma." *Anticancer Research* 32, 9: 3901–3909. PMID: 22993335.

Dong, Jianli, Robert G Phelps, Rui Qiao, Shen Yao, Outhiriaradjou Benard, Zeev Ronai, and Stuart A Aaronson. 2003. "BRAF oncogenic mutations correlate with progression rather than initiation of human melanoma." *Cancer Research* 63, 14: 3883–3885. PMID: 12873977.

Emran, Abdullah Al, Brinda Reddy Chinna Chowdary, Farzana Ahmed, Heinz Hammerlindl, Antje Huefner, Nikolas K Haass, Wolfgang Schuehly, and Helmut Schaider. 2019. "Magnolol induces cell death through PI3K/Akt-mediated epigenetic modifications boosting treatment of BRAF-and NRAS-mutant melanoma." *Cancer Medicine* 8, 3: 1186–1196. doi: 10.1002/cam4.1978.

Eroglu, Zeynep, and Antoni Ribas. 2016. "Combination therapy with BRAF and MEK inhibitors for melanoma: Latest evidence and place in therapy." *Therapeutic Advances in Medical Oncology* 8, 1: 48–56. doi: 10.1177/1758834015616934.

Eugster, Simon A. 2014. "*Curcuma longa* roots." Obtained from Wikimedia commons under the creative commons Attribution-Share Alike 3.0 Unported license. https://commons.wikimedia.org/wiki/File:Curcuma_longa_roots.jpg (Accessed 18 October 2020).

Festino, Lucia, Vito Vanella, Martina Strudel, and Paolo A Ascierto. 2018. "Molecular mechanisms underlying the action of ipilimumab against metastatic melanoma." In *Immunology*, 85–96. Elsevier. doi: 10.1016/B978-0-12-809819-6.00007-1.

Franco, Pablo Igor Ribeiro, Arthur Perillo Rodrigues, Liliana Borges de Menezes, and Marina Pacheco Miguel. 2020. "Tumor microenvironment components: Allies of cancer progression." *Pathology-Research and Practice* 216, 1: 152729. doi: 10.1016/j.prp.2019.152729.

Gibney, Geoffrey T, and Jonathan S Zager. 2013. "Clinical development of dabrafenib in BRAF mutant melanoma and other malignancies." *Expert Opinion on Drug Metabolism & Toxicology* 9, 7: 893–899. doi: 10.1517/17425255.2013.794220.

Gong, Changhua, and Honglei Xia. 2020. "Resveratrol suppresses melanoma growth by promoting autophagy through inhibiting the PI3K/AKT/mTOR signaling pathway." *Experimental and Therapeutic Medicine* 19, 3: 1878–1886. doi: 10.3892/etm.2019.8359.

Greene, Victoria R, Marcella M Johnson, Elizabeth A Grimm, and Julie A Ellerhorst. 2009. "Frequencies of NRAS and BRAF mutations increase from the radial to the vertical growth phase in cutaneous melanoma." *Journal of Investigative Dermatology* 129, 6: 1483–1488. doi: 10.1038/jid.2008.374.

Guo, Liting, Haijun Zhang, and Baoan Chen. 2017. "Nivolumab as programmed death-1 (PD-1) inhibitor for targeted immunotherapy in tumor." *Journal of Cancer* 8, 3: 410. doi: 10.7150/jca.17144.

Han, Pharaoh. 2013. "*Glycyrrhiza glabra* inflorescence." Obtained from Wikimedia under the creative commons Attribution-Share Alike 3.0 Unported license. https://commons.wikimedia.org/wiki/File:Glycyrrhiza_glabra_inflorescence.jpg (Accessed 18 October 2020).

Hannan, Enda J, Donal P O'Leary, Stephen P MacNally, Elaine W Kay, Michael A Farrell, Patrick G Morris, Colm P Power, and Arnold DK Hill. 2017. "The significance of $BRAF^{V600E}$ mutation status discordance between primary cutaneous melanoma and brain metastases: The implications for BRAF inhibitor therapy." *Medicine* 96, 48. doi: 10.1097/MD.0000000000008404.

Hwu, Patrick, Gregory Lizée, and Yong-Jun Liu. 2015. "51 — cancer and the cellular immune response." In *The Molecular Basis of Cancer (Fourth Edition)*, edited by John Mendelsohn, Joe W Gray, Peter M Howley, Mark A Israel and Craig B Thompson, 695–708.e2. Philadelphia: W.B. Saunders.

Inamdar, Gajanan S, SubbaRao V Madhunapantula, and Gavin P Robertson. 2010. "Targeting the MAPK pathway in melanoma: Why some approaches succeed and other fail." *Biochemical Pharmacology* 80, 5: 624–637. doi: 10.1016/j.bcp.2010.04.029.

Jazirehi, Ali R, Peter B Wenn, and Mohsen Damavand. 2012. "Therapeutic implications of targeting the PI3Kinase/AKT/mTOR signaling module in melanoma therapy." *American Journal of Cancer Research* 2, 2: 178. PMID: 22485197.

Jiang, Qun-Qun, and Wei-Bing Liu. 2018. "Lycorine inhibits melanoma A375 cell growth and metastasis through the inactivation of the PI3K/AKT signaling pathway." *Médecine/Sciences* 34: 33–38. doi: 10.1051/medsci/201834f106.

Kenraiz. 2016. "*Scutellaria baicalensis*." Obtained from Wikimedia under the Creative commons Attribution-Share Alike 4.0 International license. https://commons.wikimedia.org/wiki/File:Scutellaria_baicalensis_kz2.jpg (Accessed 18 October 2020).

King, Roy, Katherine N Weilbaecher, Gaël McGill, Edward Cooley, Martin Mihm, and David E Fisher. 1999. "Microphthalmia transcription factor: A sensitive and specific melanocyte marker for melanoma diagnosis." *The American Journal of Pathology* 155, 3: 731–738. doi: 10.1016/S0002-9440(10)65172-3.

Koelblinger, Peter, Joelle Dornbierer, and Reinhard Dummer. 2017. "A review of binimetinib for the treatment of mutant cutaneous melanoma." *Future Oncology* 13, 20: 1755–1766. doi: 10.2217/fon-2017-0170.

Koelblinger, Peter, Olaf Thuerigen, and Reinhard Dummer. 2018. "Development of encorafenib for BRAF-mutated advanced melanoma." *Current Opinion in Oncology* 30, 2: 125. doi: 10.1097/CCO.0000000000000426.

Krause, Günter, Floyd Hassenrück, and Michael Hallek. 2018. "Copanlisib for treatment of B-cell malignancies: The development of a PI3K inhibitor with considerable differences to idelalisib." *Drug Design, Development and Therapy* 12: 2577. doi: 10.2147/DDDT.S142406.

Lawton, S. 2019. "Skin 1: The structure and functions of the skin." *Nursing Times* 115: 30–33.

Legarza, Kathleen, and Li-Xi Yang. 2006. "New molecular mechanisms of action of camptothecin-type drugs." *Anticancer Research* 26, 5A: 3301–3305. PMID: 17094444.

Lei, Ming-jun, Yao Dong, Chun-xia Sun, and Xiang-hong Zhang. 2017. "Resveratrol inhibits proliferation, promotes differentiation and melanogenesis in HT-144 melanoma cells through inhibition of MEK/ERK kinase pathway." *Microbial Pathogenesis* 111: 410–413. doi: 10.1016/j.micpath.2017.09.029.

Liu, Yuxin, and M Saeed Sheikh. 2014. "Melanoma: Molecular pathogenesis and therapeutic management." *Molecular and Cellular Pharmacology* 6, 3: 228. PMID: 25745537.

Mar, VJ, and HP Soyer. 2018. "Artificial intelligence for melanoma diagnosis: How can we deliver on the promise?" *Annals of Oncology* 29, 8: 1625–1628. doi: 10.1093/annonc/mdy191.

Matthews, NH, WQ Li, AA Qureshi, MA Weinstock, and E Cho. 2017. "Epidemiology of melanoma. Cutaneous melanoma: Etiology and therapy [Internet]." *Brisbane: Codon Publications*: 10.15586.

McClain, Susannah E, Kira B Mayo, Amber L Shada, Mark E Smolkin, James W Patterson, and Craig L Slingluff Jr. 2012. "Amelanotic melanomas presenting as red skin lesions: A diagnostic challenge with potentially lethal consequences." *International Journal of Dermatology* 51, 4: 420–426. doi: 10.1111/j.1365-4632.2011.05066.x.

McCubrey, James A, Linda S Steelman, William H Chappell, Stephen L Abrams, Richard A Franklin, Giuseppe Montalto, Melchiorre Cervello, Massimo Libra, Saverio Candido, and Grazia Malaponte. 2012. "Ras/Raf/MEK/ERK and PI3K/PTEN/Akt/mTOR cascade inhibitors: How mutations can result in therapy resistance and how to overcome resistance." *Oncotarget* 3, 10: 1068. doi: 10.18632/oncotarget.659.

M.D. Anderson Cancer Center. 2020. "Study of the selective PI3K-beta inhibitor GSK2636771 in combination with pembrolizumab in patients with metastatic melanoma and PTEN loss." https://clinicaltrials.gov/ct2/show/NCT03131908?term=NCT03131908&draw=2&rank=1.

Ming, Mei, and Yu-Ying He. 2009. "PTEN: New insights into its regulation and function in skin cancer." *Journal of Investigative Dermatology* 129, 9: 2109–2112. doi: 10.1038/jid.2009.79.

Moreno Nogueira, JA, M Valero Arbizu, and R Pérez Temprano. 2013. "Adjuvant treatment of melanoma." *International Scholarly Research Notices*. doi: 10.1007/s40257-019-00456-4.

Moudi, Maryam, Rusea Go, Christina Yong Seok Yien, and Mohd Nazre. 2013. "Vinca alkaloids." *International Journal of Preventive Medicine* 4, 11: 1231. PMID: 24404355.

Mrozik, Krzysztof Marek, Orest William Blaschuk, Chee Man Cheong, Andrew Christopher William Zannettino, and Kate Vandyke. 2018. "N-cadherin in cancer metastasis, its emerging role in haematological malignancies and potential as a therapeutic target in cancer." *BMC Cancer* 18, 1: 1–16. doi: 10.1186/s12885-018-4845-0.

Nachbar, Franz, Wilhelm Stolz, Tanja Merkle, Armand B Cognetta, Thomas Vogt, Michael Landthaler, Peter Bilek, Otto Braun-Falco, and Gerd Plewig. 1994. "The ABCD rule of dermatoscopy: High prospective value in the diagnosis of doubtful melanocytic skin lesions." *Journal of the American Academy of Dermatology* 30, 4: 551–559. doi: 10.1016/s0190-9622(94)70061-3.

National Cancer Institute (NCI). 2020. "Targeted therapy directed by genetic testing in treating patients with advanced refractory solid tumors, lymphomas, or multiple myeloma (The MATCH screening trial)." https://clinicaltrials.gov/ct2/show/NCT02465060?term=NCT02465060&draw=2&rank=1

Newman, David J, and Gordon M Cragg. 2016. "Natural products as sources of new drugs from 1981 to 2014." *Journal of Natural Products* 79, 3: 629–661. doi: 10.1021/acs.jnatprod.5b01055.

Okkenhaug, Klaus, Mariona Graupera, and Bart Vanhaesebroeck. 2016. "Targeting PI3K in cancer: Impact on tumor cells, their protective stroma, angiogenesis, and immunotherapy." *Cancer Discovery* 6, 10: 1090–1105. doi: 10.1158/2159-8290.CD-16-0716.

Pal, Asmita, and Rita Kundu. 2020. "Human papillomavirus E6 and E7: The cervical cancer hallmarks and targets for therapy." *Frontiers in Microbiology* 10: 3116. doi: 10.3389/fmicb.2019.03116.

Pal, Harish Chandra, Samriti Sharma, Leah Ray Strickland, Santosh K Katiyar, Mary E Ballestas, Mohammad Athar, Craig A Elmets, and Farrukh Afaq. 2014. "Fisetin inhibits human melanoma cell invasion through promotion of mesenchymal to epithelial transition and by targeting MAPK and NFκB signaling pathways." *PloS One* 9, 1: e86338. doi: 10.1371/journal.pone.0086338.

Papa, Antonella, and Pier Paolo Pandolfi. 2019. "The PTEN—PI3K axis in cancer." *Biomolecules* 9, 4: 153. doi: 10.3390/biom9040153.

Pfeifer, Gerd P, and Ahmad Besaratinia. 2012. "UV wavelength-dependent DNA damage and human non-melanoma and melanoma skin cancer." *Photochemical & Photobiological Sciences* 11, 1: 90–97. doi: 10.1039/c1pp05144j.

Rodon, Jordi, and Josep Tabernero. 2017. "Improving the armamentarium of PI3K inhibitors with isoform-selective agents: A new light in the darkness." *Cancer Discovery* 7, 7: 666–669. doi: 10.1158/2159-8290.CD-17-0500.

Sahni, Sakshi, Gautam Valecha, and Ankit Sahni. 2018. "Role of anti-PD-1 antibodies in advanced melanoma: The era of immunotherapy." *Cureus* 10, 12. doi: 10.7759/cureus.3700.

Sanchez, Vanessa Edna, Cydney Nichols, Hye Na Kim, Eun Ji Gang, and Yong-Mi Kim. 2019. "Targeting PI3K signaling in acute lymphoblastic leukemia." *International Journal of Molecular Sciences* 20, 2: 412. doi: 10.3390/ijms20020412.

Scheffner, Martin, Bruce A Werness, Jon M Huibregtse, Arnold J Levine, and Peter M Howley. 1990. "The E6 oncoprotein encoded by human papillomavirus types 16 and 18 promotes the degradation of p53." *Cell* 63, 6: 1129–1136. doi: 10.1016/0092-8674(90)90409-8.

Shi, Ping, Zhe Xu, Xin Lei, Feiluore Yibulayin, and Alimujiang Wushou. 2019. "Desmoplastic melanoma: Demographic and clinicopathological features and disease-specific prognostic factors." *Oncology Letters* 17, 6: 5619–5627. doi: 10.3892/ol.2019.10259.

Shizhao. 2007. "*Plantago asiatica*." Obtained from Wikimedia commons under the Creative Commons Attribution 3.0 Unported license. https://commons.wikimedia.org/wiki/File:Plantago_asiatica.jpg (Accessed 18 October 2020).

Signorelli, Jessie, and Arpita Shah Gandhi. 2017. "Cobimetinib: A novel MEK inhibitor for metastatic melanoma." *Annals of Pharmacotherapy* 51, 2: 146–153. doi: 10.1177/1060028016672037.

Soliman, Maher. 2018. "Cancer causing chemicals." In *Cancer Causing Substances*. Finland: IntechOpen.

Song, Nu Ry, Eunjung Lee, Sanguine Byun, Jong-Eun Kim, Madhusoodanan Mottamal, Jung Han Yoon Park, Soon Sung Lim, Ann M Bode, Hyong Joo Lee, and Ki Won Lee. 2013. "Isoangustone A, a novel licorice compound, inhibits cell proliferation by targeting PI3K, MKK4, and MKK7 in human melanoma." *Cancer Prevention Research* 6, 12: 1293–1303. doi: 10.1158/1940-6207.CAPR-13-0134.

Stahl, Jill M, Mitchell Cheung, Arati Sharma, Nishit R Trivedi, Sumathi Shanmugam, and Gavin P Robertson. 2003. "Loss of PTEN promotes tumor development in malignant melanoma." *Cancer Research* 63, 11: 2881–2890. PMID: 12782594.

Stang, David J. 2006. "*Piper betle*." Obtained from Wikimedia under the Creative commons Attribution-Share Alike 4.0 International license. https://commons.wikimedia.org/wiki/File:Piper_betle_3zz.jpg (Accessed 18 October 2020).

Syed, Deeba N, Jean-Christopher Chamcheu, Mohammad Imran Khan, Mario Sechi, Rahul K Lall, Vaqar M Adhami, and Hasan Mukhtar. 2014. "Fisetin inhibits human melanoma cell growth through direct binding to p70S6K and mTOR: Findings from 3-D melanoma skin equivalents and computational modeling." *Biochemical Pharmacology* 89, 3: 349–360. doi: 10.1016/j.bcp.2014.03.007.

Tawbi, Hussein Abdul-Hassan, Weiyi Peng, Denai Milton, Rodabe Navroze Amaria, Isabella Claudia Glitza, Wen-Jen Hwu, Sapna Pradyuman Patel, Michael KK Wong, Scott Eric Woodman, and Cassian Yee. 2018. "Phase I/II study of the PI3Kβ inhibitor GSK2636771 in combination with pembrolizumab (P) in patients (pts) with PD-1 refractory metastatic melanoma (MM) and PTEN loss." *Journal of Clinical Oncology* 36, 15_Suppl.: TPS9596. doi: 10.1200/JCO.2018.36.15_suppl.TPS9596

Teimouri, Fatemeh, Shekoufeh Nikfar, and Mohammad Abdollahi. 2013. "Efficacy and side effects of dacarbazine in comparison with temozolomide in the treatment of malignant melanoma: A meta-analysis consisting of 1314 patients." *Melanoma Research* 23, 5: 381–389. doi: 10.1097/CMR.0b013e3283649a97.

Vanhaesebroeck, Bart, Maria A Whitehead, and Roberto Piñeiro. 2016. "Molecules in medicine mini-review: Isoforms of PI3K in biology and disease." *Journal of Molecular Medicine* 94, 1: 5–11. doi: 10.1007/s00109-015-1352-5.

Wang, Yan, Mingzhu Liu, Shenglan Chen, and Qin Wu. 2020. "Plantamajoside represses the growth and metastasis of malignant melanoma." *Experimental and Therapeutic Medicine* 19, 3: 2296–2302. doi: 10.3892/etm.2020.8442.

Watson, Meg, Dawn M Holman, and Maryellen Maguire-Eisen. 2016. "Ultraviolet radiation exposure and its impact on skin cancer risk." *Seminars in Oncology Nursing*. doi: 10.1016/j.soncn.2016.05.005.

West, EJ, KJ Scott, VA Jennings, and AA Melcher. 2011. "Immune activation by combination human lymphokine-activated killer and dendritic cell therapy." *British Journal of Cancer* 105, 6: 787–795. doi: 10.1038/bjc.2011.290.

WHO. 2017a. "Ultraviolet (UV) index." www.who.int/news-room/q-a-detail/ultraviolet-(uv)-index (Accessed 18 October 2020).

WHO. 2017b. "Ultraviolet (UV) radiation and skin cancer." www.who.int/news-room/q-a-detail/ultraviolet-(uv)-radiation-and-skin-cancer (Accessed 18 October 2020).

Wu, Heng, Vikas Goel, and Frank G Haluska. 2003. "PTEN signaling pathways in melanoma." *Oncogene* 22, 20: 3113–3122. doi: 10.1038/sj.onc.1206451.

Yang, Jing, Ji Nie, Xuelei Ma, Yuquan Wei, Yong Peng, and Xiawei Wei. 2019. "Targeting PI3K in cancer: Mechanisms and advances in clinical trials." *Molecular Cancer* 18, 1: 1–28. doi: 10.1186/s12943-019-0954-x.

Yang, Tzu-Yen, Mei-Li Wu, Chi-I Chang, Chih-I Liu, Te-Chih Cheng, and Yu-Jen Wu. 2018. "Bornyl cis-4-hydroxycinnamate suppresses cell metastasis of melanoma through FAK/PI3K/Akt/mTOR and MAPK signaling pathways and inhibition of the epithelial-to-mesenchymal transition." *International Journal of Molecular Sciences* 19, 8: 2152. doi: 10.3390/ijms19082152.

Zhao, Guangming, Xiaodong Han, Siwen Zheng, Zhen Li, Yang Sha, Jing Ni, Zhe Sun, Song Qiao, and Zhiqi Song. 2016. "Curcumin induces autophagy, inhibits proliferation and invasion by downregulating AKT/mTOR signaling pathway in human melanoma cells." *Oncology Reports* 35, 2: 1065–1074. doi: 10.3892/or.2015.4413.

Σ64.2006. "*Lycoris radiate*." Obtained from Wikimedia under the Creative commons Attribution 4.0 International license. https://commons.wikimedia.org/wiki/File:Lycoris_radiate_01.jpg (Accessed 18 October 2020).

10 Anti-Proliferative Properties of Various South African *Buddleja* Species

Talita Turvey and Namrita Lall

CONTENTS

DOI: 10.1201/9781003108375-10

10.1 INTRODUCTION

Medicinal plants are extensively explored in traditional medicine and drug discovery; approximately one-third of the top-selling drugs that are currently on the market across all fields have been derived from plants. The broad diversity and wide range of known pharmacological activity observed in various plant species make them an invaluable natural resource (Strohl, 2000). Examples of phyto-derived drugs prevalent in cancer treatment include vinblastine, vincristine and vindesine (vinca alkaloids); etoposide and teniposide (epipodophyllotoxins); paclitaxel and docetaxel (taxanes); as well as camptothecin and irinotecan (camptothecin derivatives) (Desai et al., 2008). Cancer is a complex disease and its pathogenesis varies greatly from one patient to the next. There are over 200 types of cancers; due to the vast differences found in histopathological, genetic and clinical outcomes in different patients and different types of cancer, the search for effective treatment is immensely needed (Tesfaye et al., 2020).

There is a need for less toxic drugs to treat cancer since most cancer treatment results in adverse side effects that could include cardiac toxicity, renal toxicity, myelotoxicity, pulmonary toxicity, cutaneous toxicity and alopecia or immunosuppression (Desai et al., 2008; Yeh et al., 2019). Thus, the discovery of phyto-derived drugs that have anti-proliferative potential is vital for cancer treatment. Naturally occurring organic compound classes that have been reported to have anti-proliferative activity include alkaloids, phenylpropanoids and terpenoids (Kintzios, 2006). This chapter (contents illustrated below) focuses on the medicinal properties of *Buddleja* species and the anti-proliferative potential that some species might have based on the traditional usage across the world. The biological activity of the isolated compounds from plants have been reported. The use of various *Buddleja* species has been observed in different traditional medicine practices all over the world. Several species of this genus have been reported to have bioactive properties such as anti-inflammatory, antifungal and wound healing when used in traditional medicine (Asakawa, 1999; Khan et al., 2019).

The anti-inflammatory properties of *Buddleja* species and its isolated compounds have been confirmed in *in vitro* studies. Chronic inflammation has been associated with an increased risk of developing several types of cancers which were demonstrated by both epidemiological and experimental data (Munn, 2017). It was also reported that anti-inflammatory therapies are effective in

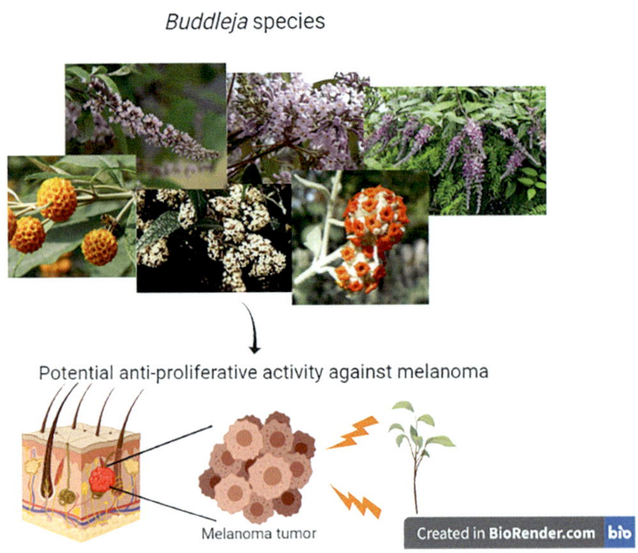

Graphical abstract representing the content of the chapter using Biorender (Efloraofindia, 2007; Joffe, 2003; Loader, 2015; Penganum, 2016; Salix, 2015; Shebs, 2005).

cancer treatment and prevention (Reuter et al., 2010). Furthermore, it has been reported that species from this genus have anti-microbial, anti-cataratic, anti-pyretic, anti-hepatotoxic, neuroprotective, molluscicidal and hypoglycemic activity (Khan et al., 2019).

Over the years, several chemical investigations have reported various compounds isolated from the species of this genus (Martínez-Pérez et al., 2020; Nguyen, 2020; Rocha-Guzmán et al., 2018). These include flavonoids, terpenoids and other compounds derived from the shikimate pathway (i.e. phenylethanoid glycosides such as verbascoside), along with monoterpene iridoids, which have been predicted to be present in the *Buddleja* genus (Houghton et al., 2003). This chapter explores the *Buddleja* genus as a result of several studies which have indicated the anti-mutagenic activity, anti-inflammatory properties and isolated compounds present in most species of this genus, more especially the flavonoids and terpenoids that have been reported to exhibit anti-cancer effects (Strohl, 2000; Houghton et al., 2003; Batra and Sharma, 2013). Particular focus on the anti-proliferative potential effects of the South African *Buddleja* species is discussed in this chapter, aw well as where further investigations should be considered.

10.2 THE *BUDDLEJA* GENUS

The *Buddleja* genus includes numerous plants that have a beautiful and fragrant inflorescence, comprised mostly of trees or shrubs and are known as the butterfly bush (Figure 10.1). Both evergreen and deciduous species occur in this genus and mostly occur in Asia, North and South America, and Africa (Pendota et al., 2014).

FIGURE 10.1 *Buddleja* species. A, *Buddleja auriculata*; B, *B. albiflora*; C, *B. asiatica*; D, *B. curviflora*; E, *B. cordata*; F, *B. dysophylla*; G, *B. globosa*; H, *B. incana*; I, *B. lindleyana*; J, *B. marrubifolia*; K, *B. madagascariensis*; L, *B. officinalis* (Efloraofindia, 2007; Elango, 2015; Flora, 2014; JMK, 2012; Joffe, 2003; Johansson, 2013; Kenraiz, 2017; Loader, 2015; Penganum, 2016; Porse, 2007; Salix, 2015; Shebs, 2005).

10.2.1 Phylogeny of the *Buddleja* Genus

The taxonomic history and classification of the *Buddleja* genus is complicated and has evolved considerably over the years. It was first grouped under Scrophulariaceae or Loganiaceae (Bentham, 1857; Bentham and Hooker, 1876). Later, it was separated into its own family known as Buddlejaceae (Takhtajan, 1980; Cronquist and Takhtadzhian, 1981; Oxelman et al., 2004). The *Buddleja* genus was finally grouped in the Scrophulariaceae family due to molecular data from plastid genomes as well as similarities observed in embryology, palynology and phytochemistry (Oxelman et al., 1999). Buddlejeae is classified as one of the eight tribes found in this family (Kornhall and Bremer, 2004; Oxelman et al., 2005).

The Buddlejeae tribe consists of 108 species (Tank et al., 2006); all the species in this tribe are woody and grow in arid areas, and are known as dicotyledonous flowering plants and not recognized as angiosperms. They consist of either trees, shrubs or subshrubs, and have a broad distribution spanning from Asia, South and North America, and Africa. There are seven species of *Buddleja* occurring in South Africa: *B. auriculata*, *B. dysophylla*, *B. glomerata*, *B. loricata*, *B. pulchella*, *B. saligna* and *B. salviifolia*, of which only *B. glomerata* is indigenous to South Africa (Raimondo et al., 2009).

10.2.2 South African *Buddleja* Species

10.2.2.1 *B. auriculata* Benth.

This species is a shrub or a small tree with dark green, glossy foliage with silver on the bottom of the leaves. It has spike inflorescence consisting of small, tubular cream; orange or pink flowers that have a sweet scent; and flowers from July to September. During this time, it bears fruit, which are minuscule, brown capsules that are split at the tip. The habitat ranges between mountain slopes, rocky ravines and forest margins in the Eastern Cape, KwaZulu-Natal, Limpopo, Mpumalanga and some regions of Zimbabwe (Joffe, 2001). The flowers of this plant attract many insects, such as butterflies, which can lead to the attraction of insectivorous birds (Joffe, 2001; Raimondo et al., 2009). This species is not considered a medicinal plant and is not used in traditional medicine. Therefore, no compounds have been isolated.

10.2.2.2 *B. glomerata* H.L.Wendl.

This species is characterized by the foul smell that the flowers possess, which is similar to the smell of a cockroach and causes one to sneeze; thus, it is often referred to as a sneezebush or as the Karoo sagewood. It has attractive foliage with silver, scalloped leaves and yellow flowers that bloom in spring and summer. The leaves are silvery-green when they are young and transition to blue-green when the leaves mature. These leaves have a rough texture, appear wrinkled and have sparsely distributed hairs. The fruit looks like small, hairy capsules. This plant is fast-growing and grows up to 4 meters in height and 3 meters in width to form a small tree or upright shrub. The distribution of this plant ranges from Karoo, Uniondale, Victoria West, Beaufort West, Graaff-Reinet, Pearson, Somerset East, Middelburg and Cradock; its natural growth habitat is among rocks and along dry hillsides (Joffe, 2003). There are no studies that report on the traditional usage or medicinal potential of this species thus far.

10.2.2.3 *B. loricata* Leeuwenb.

This species is commonly known as Mountain sagewood. It grows up to 4 meters high and is either a multi-stemmed, densely bushy shrub or a small tree. The leaves are small, rounded, dark green above and lighter in color below. It has cream-white or yellow flowers with an orange center and a sweet scent which occasionally attracts many insects, and flowers between October and December. The fruit are small capsules that are covered in fine, cream-gray hairs that form dense wooly heads. It is distributed throughout Lesotho, Eastern Cape, KwaZulu-Natal and the Free State, in habitats that range from rocky areas to moist, sheltered habitats on slopes, mountains and along stream banks situated at high elevations often above 1800 meters. *B. loricata* Leeuwenb is mostly used for horticultural purposes, though not often in South Africa (Pooley, 2003). The compounds that have been isolated from this species include 4-oxoisophorone, α-farnesene and linalool (which offer sedative, anti-depressant, anxiolytic, immune potentiating effects, analgesic and anticonvulsant effects) (Chen et al., 2012; McPartland and Russo, 2001; Russo and Marcu, 2017).

10.2.2.4 *B. pulchella* N.E.Br.

This species mostly forms a shrub that can grow up to 7 meters in height with leaves that are green on top, with small cream to light orange flowers that have orange centers. This plant is found in forests, mostly in the open parts or on the margins and it has been found in the Cape, Natal and east tropical Africa, as well as Swaziland (Roux, 2003). There is limited data on this species, specifically on compounds isolated from it.

10.2.2.5 *B. saligna* Willd.

This species is known as the false olive (English), umBatacwepe (Siswati), unGqepe (Zulu) and Lelothwane (South Sotho). It is an evergreen tree, which can range in size from small to medium, reaching heights up to 10 meters. It has a short trunk, with a domed-shaped, grayish-green foliage. The leaves are hairless, narrow and petiolate, with alternating colors, a dark green surface and a gray to white underside. The leaves are similar in appearance to that of olive tree leaves, although *B. saligna* leaves are more textured, and it bears inedible fruit, known as wild olives (Reid et al., 2006). The young stems are four-sided and have white velvet hairs that cover it, while older stems and trunks can range from gray or cream to dark brown. The plant flowers during spring and summer from August to January, with small, honey-scented, white flowers emerging from the ends of the branches. The miniature seeds found on this plant are present in the dried-out flowers, wrapped in a hairy capsule from October to March (Hutchings, 1996; Van Wyk and Gericke, 2000; Reid et al., 2006). It is distributed throughout South Africa and Zimbabwe. It can be found in dry and rocky hillsides, wooden valleys, forest and woodlands, along streams, in the coastal regions and the bushveld (Van Wyk and Gericke, 2000).

It is used for several ailments in traditional medicine. Globally, the fruit is used to treat high blood pressure. In South Africa, the bark and leaves are used as decoctions for the treatment of colds, coughs, colic, sore eyes, urinary problems and as an emetic (Hutchings, 1996; Reid et al., 2006). It has been reported to have anti-plasmodial, anti-mycobacterial, anti-bacterial, anti-mutagenic and antioxidant properties (Bamuamba et al., 2008; Chukwujekwu et al., 2014). Five compounds have been isolated from the stems and leaves of *B. saligna*, including α-spinasterol, betulonic acid, betulone, oleanolic acid and ursolic acid. However, no chemical compounds have been isolated from the flowers, and there have been no reports on the biological activity of the flowers (Bamuamba et al., 2008, Chukwujekwu et al., 2016).

10.2.2.6 *B. salviifolia* (L). Lam.

This species is a semi-evergreen bushy shrub that can grow up to 4 meters high when in optimal conditions, the leaves are dark green on top and white underneath with wrinkled surfaces. This plant has a spike inflorescence with small flowers that have a sweet scent and range from clear white to purple, flowering in spring, from August to October, and attracts bees, butterflies and birds. It is distributed widely in South Africa from the Western and Eastern Cape, Free State, Lesotho, KwaZulu-Natal, Northern Province and Mpumalanga, as well as Mozambique and Zimbabwe, occurring naturally in forest margins, rocky hillsides and along streams on escarpments. This plant is medicinal and used traditionally in South Africa; a decoction made from the roots is used as a remedy for coughs and to relieve colic, and an infusion made from the leaves is applied to the eyes as a lotion (Joffe, 1993; Van Wyk and Van Wyk, 1997; Venter and Venter, 2002).

10.3 MEDICINAL IMPORTANCE OF THE GENUS *BUDDLEJA*

10.3.1 Traditional Medicine Usage

Species in this genus have been used as a remedy against a variety of ailments in traditional medicine practices across the world for many years. The major ethnopharmacological uses of *Buddleja* species include wound healing and related skin conditions, treatment of liver diseases, sedative effects, diuretic activity, bronchial complaints, antirheumatic, analgesic activities and other medicinal uses (Table 10.1) (Khan et al., 2019).

TABLE 10.1
***Buddleja* Species in Traditional Medicine Practices across the World**

Traditional Medicines	Regions	Species	Preparations	Treatments	References
Traditional Chinese medicine	China	*B. officinalis* (Mi Meng Hua)	Unknown	Hypertension, diabetes, nephritis and cancer	(Smith and Stuart, 1911; Crevost and Petelot, 1929; Nikonov et al., 1961)
		B. officinalis	The flowers are soaked in wine and honey for three days	Painful eye conditions, cataracts and enhanced poor vision	
		B. officinalis	Mixed with the roots of various species of *Gentiana*	Skin diseases, ulcers and abscesses	
		B. curviflora Hook. Et Arn.	The flowers and leaves	Eye cataracts and additionally to treat chronic malaria-related symptoms	(Smith and Stuart, 1911)
		B. albiflora Hemsley and *B. davidii* Franchet	The leaves and roots	Rheumatism, arthralgia, hemorrhaging, traumatic injury and to treat skin ailments (i.e. eczema)	(Ying and Wan, 2012)
		B. davidii	The leafy stems are powdered and a hot water extract is prepared	Treat skin ulcers and to relieve the lesions caused by leprosy	(Houghton, 1984)
		B. asiatica	Various parts are used	Diarrhea and articular rheumatism	(Khan, 2013)
		B. albiflora	The leaves and flowers are simmered	Used as a cough remedy	(Cheo, 1947; Bhakuni et al., 1969)
		B. paniculata Wallich	An alcoholic extract was prepared	Central nervous system	
Mapuche medicine	Chile	*B. globose*	Leaves used as a decoction	Wash wounds	(Houghton, 1984)
		B. globose	Powdered leaves	Heal ulcers and wounds	
		B. globose	Infusion of the leaves	Administered internally to treat hemorrhoids and hepatitis; it is also used to treat callus ulcers and warts externally	
		B. incana Ruiz et Pav.	Decoction prepared from the leaves and flowers	It is used to treat callus ulcers and warts on the skin	(Hartwell, 1970)

Traditional Medicines	Regions	Species	Preparations	Treatments	References
African traditional medicine	South Africa	*B. saligna*	Decoction of the leaves Scrapings from the root	Cough and cold remedy Used as an emetic	(Watt and Breyer-Brandwijk, 1962)
		B. salviifolia	Various plant parts are prepared as a decoction	Used to treat coughs and colic	
		B. salviifolia	An infusion is prepared from various plant parts	Used as an eye lotion	
Mexican traditional medicine	Mexico	*B. americana* L.	An extract is prepared from the roots or bark	Used as a cure for a 'collapsed womb,' for increased urine production, and to detoxify the body	(de Barradas, 1957)
		B. americana	An extract prepared from leaves	Used to aid in good digestion	
		B. americana	A poultice of the leaves, roots or the bark	Used to relieve the pain in joints, aid in the treatment of tumors, cuts, ulcers and burns	
		B. americana	An extract of the plant	Administered orally to treat liver disease (i.e. cirrhosis) and to improve biliary function	(Díaz, 1976)
		B. marrubiifolia Benth.	An infusion of the leaves	Used to soak in as a treatment for rheumatic conditions and as a diuretic	(Martinez, 1959)
Philippine traditional medicine	Philippines	*B. asiatica* Lour.	Unknown	Used to treat skin diseases and aid in weight loss	(Houghton, 1984)
		B. asiatica	Roots and leaves	Used to treat tumor-like growths on the head	
		B. asiatica	Concentrated infusion of the roots	Used to treat malaria	
Bolivian traditional medicine	Bolivia	*B. incana*	The leaves	Used as an antirheumatic plaster	(Houghton, 1984)
		B. incana	A decoction of the bark	Taken orally for the treatment of genital infections	
		B. diffusa Ruiz et Pav.	A decoction of the bark	Taken orally for the treatment of genital infections	
		B. globose	The infusion of leaves	Used to treat hemorrhoids, hepatitis, cataracts and as a urinogenital antiseptic	

10.3.2 Chemical Composition of *Buddleja* Species

Interesting aspects of the phytochemical and ethnopharmacological composition of this genus prompted a further investigation on the phytochemistry of the *Buddleja* genus (Houghton, 1984). The flowers, leaves and the roots of various species of this genus are used globally in traditional medicine to treat a variety of ailments. A review written by Houghton et al. (2003), identified various compounds that have been isolated from the *Buddleja* genus (Houghton et al., 2003). This includes terpenes, phenylpropanoid, flavonoids, saponins and lignans (Table 10.2).

TABLE 10.2
Isolated Compounds from *Buddleja* Genus

Compounds	Species Compounds Are Isolated From	Functional Groups	Biological Activities	References
Aucubin	*B. globosa*	Terpene	Anti-inflammatory, anti-microbial, anti-analgesic and anti-tumor activity; decreased expression of various proteins in vivo (MMP-3[1], MMP-9, MMP-13, iNOS[2], COX-2[3])	(Wang et al., 2015; Xue et al., 2015).
Buddlejoside A	*B. crispa*	Terpene	-	(Ahmad et al., 2008)
Buddlejoside A1-A4	*B. japonica*	Terpene	-	(Ahmad et al., 2008)
Buddlejoside A5	*B. japonica*	Terpene	Antiangiogenic activity exhibited in two *in vivo* models (zebrafish embryos and chick embryo chorioallantoic membrane)	(Iannuzzi et al., 2019) (Miyase et al., 1991)
Buddlejoside A6–16	*B. japonica*	Terpene	-	(Miyase et al., 1991)
Buddlejoside B	*B. crispa*	Terpene	-	(Ahmad et al., 2008)
Buddlejoside C	*B. crispa*	Terpene	-	(Ahmad et al., 2008)
Methyl catalpol	*B. asiatica*	Terpene	Potential neuroprotective effects and anti-inflammation	(Zhang et al., 2013) (Khan, 2013)
Benzoyl catalpol	*B. davidii*	Terpene	Potential neuroprotective effects	(Zhang et al., 2013) (Houghton and Hikino, 1989)
ρ-methoxycinnamoyl catalpol	*B. davidii*	Terpene	-	(Yamamoto et al., 1993)
Dimethoxycinnamoyl catalpol	*B. davidii*	Terpene	-	(Yamamoto et al., 1993)
Buddlin	*B. asiatica*	Terpene	-	(Chen et al., 2005)
Buddledin A	*B. davidii*	Terpene	Antifungal activity, anti-inflammatory effects with inhibition exhibited against COX (IC_{50}[4] of 13.7μM) and 5-LOX[5] (IC_{50} of 50.4μM)	(Houghton et al., 2003) (Yoshida et al., 1978a)
Buddledin B-C	*B. davidii*	Terpene	Anti-inflammatory activity (inhibits COX and 5-LOX)	(Houghton et al., 2003) (Yoshida et al., 1978a)
Buddledin D-E	*B. davidii*	Terpene	-	(Houghton et al., 2003) (Yoshida et al., 1978b)
Zerumbone	*B. globosa*	Terpene	Anti-viral, anti-inflammatory, anti-oxidant, immunomodulatory and antiproliferative activity (cervical, pancreatic, lung, liver, breast, and leukemia cell lines)	(Abdul et al., 2008; Kalantari et al., 2017) (Liao et al., 1999)
Dihydroxy buddledin A	*B. globosa*	Terpene	-	(Liao et al., 1999)

Compounds	Species Compounds Are Isolated From	Functional Groups	Biological Activities	References
Buddledone A-B	*B. globosa*	Terpene	-	(Liao et al., 1999)
Buddlejone II	*B. crispa*	Terpene	-	(Ahmad et al., 2004)
Buddlindeterpene A-B	*B. lindleyana*	Terpene	-	(Lu et al., 2004)
Buddlejone	*B. albiflora*	Terpene	-	(Houghton et al., 1996)
Deoxy buddlejone	*B. globosa*	Terpene	-	(Mensah et al., 2000)
Maytenone	*B. globosa*	Terpene	Antimicrobial activity	(Mensah et al., 2000)
Crocetin-gentiobiose ester	*B. officinalis*	Terpene	Antiproliferative and antioxidant activity	(Singla and Giliyaru, 2011)
Buddlindeterpene C	*B. lindleyana*	Terpene	-	(Lu et al., 2004)
Saikosaponin A	*B. japonica*	Terpene	Anti-inflammatory, antitumor, antiviral, immunoregulation and neuroregulation	(Yuan et al., 2017)
Buddlejasaponin i -iii	*B. japonica*	Terpene	-	(Yamamoto et al., 1991)
Buddlejasaponin iv	*B. japonica*	Terpene	Anti-inflammatory and antinociceptive effects	(Won et al., 2006)
Mimengoside A-G	*B. officinalis*	Terpene	-	(Guo et al., 2004)
β-Amyrin	*B. globosa*	Terpene	Anti-inflammatory, hepatoprotective, analgesic	(Nogueira et al., 2019)
Lupeol	*B. globosa*	Terpene	Antiproliferative activity	(Saleem, 2009)
α-spinasterol	*B. saligna*	Terpene	Induces apoptosis in breast (MCF-7, MDA-MB-231) and ovarian cancer (SKOV-3) cells through an increased expression of pro-apoptotic proteins, Bax[6] and p53[7]; it was also reported to arrest the cell cycle (G_0-G_1) on both breast and ovarian cancer	(Mishra et al., 2007; Chukwujekwu et al., 2016; Socała and Wlaź, 2016; Sedky et al., 2018)
Betulonic acid	*B. saligna*	Terpene	Showed high cytotoxicity against prostate cancer cell lines	(Saxena et al., 2006; Chukwujekwu et al., 2016; Zhang et al., 2016)
Betulone	*B. saligna*	Terpene	Derivatives of betulone, antiproliferative effects against stomach, breast, and prostate cancer	(Hata et al., 2002; Liu et al., 2012).
Oleanolic acid	*B. saligna*	Terpene	Induces apoptosis in melanoma cells through upregulating the expression of pro-apoptotic proteins (p53, Bax and cytochrome c) and down-regulating the anti-apoptotic protein Bcl-2[8] in a dose-dependent manner	(Ghosh et al., 2014)
Ursolic acid	*B. saligna*	Terpene	Antiproliferative activity in human melanoma cells and induced apoptosis; downregulates angiogenesis-promoting proteins such as VEGF[9]	(Oprean et al., 2018)

(*Continued*)

TABLE 10.2 (Continued)

Compounds	Species Compounds Are Isolated From	Functional Groups	Biological Activities	References
α-Farnesene	*B. loricata*	Terpene	-	(Chen et al., 2012)
Linalool	*B. loricate*	Terpene	Sedative, antidepressant, anxiolytic, immune potentiating effects, analgesic and anticonvulsant properties	(McPartland and Russo, 2001; Russo and Marcu, 2017)
Verbascoside A	*B. japonica*	Phenylpropanoid	Antioxidant and antimicrobial effects	(Fazly Bazzaz et al., 2018)
Jionoside D	*B. davidii*	Phenylpropanoid	Antioxidant activity	(Chae et al., 2004)
2-Acetylmartynoside	*B. davidii* *B. albiflora*	Phenylpropanoid	Antioxidant activity	(Chae et al., 2007; Tao et al., 2009)
Martynoside	*B. davidii*	Phenylpropanoid	Anti-bacterial	(Pendota et al., 2014)
Isomartynoside	*B. davidii* *B. albiflora*	Phenylpropanoid	Lowers blood pressure	(Kang et al., 2003; Tao et al., 2009)
Ferulic acid methyl ester	*B. globosa*	Phenylpropanoid	Anti-inflammatory activity	(Lopez et al., 1979)
Coniferaldehyde	*B. davidii*	Phenylpropanoid	Anti-inflammatory activity by inducing heme oxygenase-1 (HO-1)	(Houghton, 1984; Nam et al., 2013)
Calceolarioside A	*B. davidii*	Phenylpropanoid	Immunosuppressant	(Chen et al., 2017)
Campneoside II	*B. davidii*	Phenylpropanoid	Antioxidant	(Gondal, 2020)
Forsythoside B	*B. davidii*	Phenylpropanoid	Inhibits inflammatory response, antioxidant, inhibits TNF-α[10], IL-6[11], IκB[12], and modulate NF-κB[13]	(Jiang et al., 2010; Jiang et al., 2012)
Acteoside	*B. salviifolia*	Phenylpropanoid	Anti-inflammatory agent, neuroprotective and antibacterial activity	(Qiao et al., 2019)
Leucosceptoside A and B	*B. davidii*	Flavonoid	-	(Yamamoto et al., 1991)
Phenylethyl glycoside	*B. officinalis*	Flavonoid	-	(Liao et al., 1999)
Luteolin	*B. officinalis* *B. albiflora*	Flavonoid	Anti-proliferative activity	(Lin et al., 2008; Tao et al., 2009)
Apigenin	*B. officinalis*	Flavonoid	Anti-inflammatory, anti-proliferative, and anti-viral activity	(Dormán et al., 2016; Salehi et al., 2019)
Apigenin-7-O-glucoside	*B. globose* *B. albiflora*	Flavonoid	Antifungal and anti-proliferative activity	(Tao et al., 2009; Smiljkovic et al., 2017)
Quercetin	*B. davidii* *B. albiflora* *B. salviifolia*	Flavonoid	Anti-inflammatory, antibacterial, antiviral, immune-modulatory activity	(Tao et al., 2009; Pendota et al., 2014; Kim et al., 2017)
6- Hydroxyluteolin	*B. globosa*	Flavonoid	Antioxidant activity	(Nkuété et al., 2015)
Echinacoside	*B. officinalis*	Flavonoid	Anti-inflammatory and neuroprotective activity	(Liu et al., 2018)
Asiatiside A-D	*B. asiatica*	Saponin	-	(Khan, 2013)
Buddlenol A-F	*B. davidii*	Lignan	-	(Houghton, 1984)
Neolignan 1–3	*B. davidii*	Neolignan	-	(Yamamoto et al., 1993)

Compounds	Species Compounds Are Isolated From	Functional Groups	Biological Activities	References
Methyl β-orcinolcarboxylate	*B. cordata*	Carboxyl group	-	(Acevedo et al., 2000)
β-orcinolcarboxylate	*B. cordata*	Carboxyl group	-	(Acevedo et al., 2000)
4-Oxoisophorone	*B. loricate*	Ketone	-	(Chen et al., 2012)

Notes: **1** Metalloproteinase (MMP); **2** Inducible nitric oxide synthase (iNOS); **3** Cyclooxygenase (COX); **4** 50% Inhibitory concentration (IC50); **5** Lipoxygenase (LOX); **6** Bcl-2-associated X protein (Bax); **7** Tumor protein p53 (p53); **8** B-cell lymphoma-2 (Bcl-2); **9** Vascular endothelial growth factor (VEGF); **10** Tumor necrosis factor (TNF-α); **11** Interleukin-6 (IL-6); **12** Inhibitor of nuclear factor kappa beta (IκB); **13** Nuclear factor kappa beta (NF- κB).

10.4 BIOLOGICAL ACTIVITY OF ISOLATED COMPOUNDS

10.4.1 Terpenes

Terpenes are part of the biggest class of secondary metabolites and mainly consist of five isoprene units assembled, which form simple hydrocarbons. Terpenoids are a modified class of terpenes that have different functional groups and oxidized methyl groups that are either moved or removed at different positions. Terpenoids can be divided into monoterpenes, sesquiterpenes, diterpenes, sesquiterpenes and triterpenes, dependent on the carbon units (Perveen and Al-Taweel, 2018). Terpenes can be primary and secondary metabolites that are produced by plants; some terpenes have anti-inflammatory, antiseptic, antibacterial and antiviral activity (Burt, 2004).

10.4.1.1 Monoterpenes

The major monoterpenes present in some of these species are iridoids (containing an ester or glycosidal link at C-6) such as aucubin and derivatives thereof, including conjugates with shikimate-derived moieties, catalpol and analogs (methoxy, benzoyl and ρ- methoxycinnamoyl) reported to be specifically present in *B. globosa* (Duff et al., 1965; Houghton and Hikino, 1989; Miyase et al., 1991). Other derivatives of catalpol have been reported to be present in *B. japonica* and congeners of ajugol and neolignans in the roots of *B. davidii* (Miyase et al., 1991; Yamamoto et al., 1993).

Although no extensive study has yet been carried out to determine which iridoids are present in various *Buddleja* species, it has been established that aucubin is present in several species, specifically in the aerial parts (Trim and Hill, 1952; Houghton et al., 1993). Of note are the prominent odors present in the flowers of most *Buddleja* species, be it pleasant or foul, may suggest the presence of volatile non-glycosylated monoterpenes in the aerial parts. However, no studies have been done to determine the chemical composition and volatiles present in the flowers (Houghton et al., 2003).

10.4.1.2 Sesquiterpenes

The presence of sesquiterpenes has been reported to be in the roots and bark of some *Buddleja* species. A series of caryophyllane compounds named buddledins A-E have been identified from the roots of *B. davidii*, while buddledins A and B have been reported to be present in *B. globosa* and *B. madagascarensis* (Yoshida et al., 1978a; Yoshida et al., 1978b; Mensah et al., 2000). Dehydrobuddledin A, zerumbone and two novel compounds, buddledones A and B, have been isolated from the roots of *B. globosa* (Liao et al., 1999). Two other compounds that have sesquiterpene skeletons were isolated from *B. cordata* H.B.K. and *B. sessiliflora* Kunth (de Vivar et al., 1995).

Buddledins A-E has been shown to have piscicidal effects, and buddledin A and buddledin B have been shown to have antifungal and antibacterial effects (Yoshida et al., 1978a; Yoshida et al.,

1978b; Mensah et al., 2000). Further investigation of the use of *Buddleja* species in traditional medicine for conditions that are related to inflammation revealed that buddledins A–C showed inhibition against cyclooxygenase (COX) and 5-lipoxygenase (5-LOX) under *in vitro* conditions. Buddledin A was identified as the most active against these two enzymes with an inhibitory concentration at 50% of the enzyme activity (IC_{50}) of 13.7 μM against COX and 50.4 μM against 5-LOX (Liao et al., 1999). The activity of buddledin D and E have not yet been tested.

10.4.1.3 Diterpenes

Buddlejone, which is an abietane diterpene, has been isolated from the root bark of *B. albiflora* Hemsl. After the first report of this compound present in the *Buddleja* genus, it was also isolated from the roots of *B. globosa* and in the bark of *B. yunanensis* (Houghton et al., 1996; Liao et al., 1999; Mensah et al., 2000). Maytenone was isolated from the roots and bark of *B. globosa* (Mensah et al., 2000). There is a considerable amount of non-cyclic crocetin-gentiobiose ester present in the flowers of *B. officinalis* Maxim., which is used as a yellow colorant in foods and is predicted to be responsible for the intense yellow color of the inner surface of the corollar tube in the flowers of various *Buddleja* species (Liao et al., 1999; Aoki, 2001; Houghton et al., 2003). Buddlejone, deoxybuddlejone and maytenone exhibited antifungal activity, but it was considerably less than that of sesquiterpenes (Mensah et al., 2000).

10.4.1.4 Triterpenoids

The leaves of various *Buddleja* species have been used for some time for cleansing and washing of wounds and as a method to poison fish. Due to these applications, it has been predicted that saponins could be present in these plants (Houghton, 1984). Four glycosides of saikosaponin A were isolated from the aerial parts of *B. japonica* Hemsl.; these four compounds were named buddlejasaponins I–IV. It has been reported that buddlejasaponin I are present in the leaves of *B. madagascariensis* Lam. (Yamamoto et al., 1991; Emam et al., 1997). Two other compounds, namely mimengoside A and mimengoside B, have been reported to be present in the flowers of *B. officinalis* Maxim., which is known as Mi Meng Hua in traditional Chinese medicine (TCM) to assist with various eye problems or disorders (Ding et al., 1992).

10.4.2 Phenylpropanoids

Phenylpropanoids are secondary metabolites that are commonly present in vegetables, fruits, cereal grains, beverages, herbs and spices. These metabolites are known to have various bioactive properties, including anti-oxidant, anti-inflammatory, anti-diabetic, and anti-cancer as well as being renoprotective, cardioprotective, hepatoprotective and neuroprotective (Neelam, 2020). These compounds are a diverse group of phenylalanine-derived metabolites (C_6–C_3) and play important roles in the interaction of plants with other living organisms (Soledade et al., 2010). The isolation of various phenylpropanoid glycosides from the ethanol extract of *B. davidii* includes calceolarioside A, isomartynoside, martynoside, 2-acetylmartynoside, jionoside D, plantainoside C (antimicro), campneoside II, angoroside C, forsythoside B and poliumoside (Ahmad and Sticher, 1988; Yamamoto et al., 1993; Liao et al., 1999).

10.4.3 Flavonoid

Flavonoids are polyphenol compounds that play an important role in defending plant cells against microorganisms, insects and ultraviolet (UV) radiation. These compounds can inhibit several pathways during the progression of cancer, such as invasion, metastasis and angiogenesis, by regulating the cell cycle, inducing apoptosis, inhibiting kinases and reducing transcription factors (Lin et al., 2008). These compounds have anti-oxidative, anti-inflammatory, anti-mutagenic and anti-carcinogenic activity, and could potentially inhibit several enzymes such as xanthine oxidase (XO), COX, lipoxygenase (LOX) and phosphoinositide 3-kinase (PI3K) (Panche et al., 2016).

10.4.4 Saponin

Saponins are comprised of a large family of structurally related compounds that contain a steroid or triterpenoid aglycone (sapogenin) that is linked to one or more oligosaccharide moieties. They are known for their foaming properties and hemolytic activity, and are also responsible for the bitter taste and astringency that can be present in plant material that contains a high concentration of saponins. Most of these compounds are insoluble and poorly absorbed (Mohan et al., 2016). Saponins act by permeabilizing plasma membranes and have been reported to interfere with cellular function and enzyme activity, and induce apoptosis. Some saponins, however, do not act by permeabilization of the membrane because it does not depend on amphipathicity (Mugford and Osbourn, 2012). Saponins isolated from *Buddleja* species are closely related to saikosaponins from *Bupleurum* species, which have exhibited anti-inflammatory activity (Houghton et al., 2003). However, the saponins isolated from *Buddleja* species have not been tested for anti-inflammatory properties *in vitro*. Furthermore, buddlejasaponin 1 showed antiprotozoal and moluscicidal activity (Emam et al., 1997).

10.4.5 Lignan

Lignans are 1,4-diarylbutan compounds that are derived from the shikimic acid biosynthetic pathway, and these compounds are known to exist as minor constituents in many plants (Imai et al., 2006). They form the building blocks for the formation of lignin in plant cell walls. Lignans derived from plants are metabolized to enterodiol and enterolactone, which are named enterolignans or mammalian lignans (Landete, 2012). These diphenolic compounds act as antioxidants and free radical scavengers, making them beneficial in preventing the development of cancer. Lignan has been shown to prevent colon cancer cell growth by inducing the production of various enzymes (Higuchi, 2014).

10.4.6 Iridoids

Iridoids are a type of monoterpenoid which are found in a variety of plants and some animals. They occur in plants typically as glycosides, most often bound to glucose. Limited research has been conducted on the biological activity of iridoids isolated specifically from the *Buddleja* species. However, aucubin and catalpol derived from other plant species have been investigated to some extent. The traditional usage of some *Buddleja* species to treat liver diseases resulted in a study to determine what the effects would be on liver cells; however, it was reported that aucubin had no protective effects on liver cells, whereas catalpol showed some hepatoprotective activity (Houghton and Hikino, 1989). Another study reported that the effects of aucubin *in vivo* showed conflicting results, indicating that this compound acted as hepatoprotective in mice; this could indicate the potential biotransformation of aucubin to an active compound due to enzyme systems in the body (Chang, 1998). Catalposide has been indicated to have inhibitory effects against nitric oxide (NO) synthesis in macrophages; however, there are no further reports of iridoids found in *Buddleja* species that have similar bioactivity (Oh et al., 2002).

10.5 POTENTIAL MEDICINAL VALUE OF *BUDDLEJA* SPECIES

10.5.1 General Biological Activity

10.5.1.1 Wound Healing

Many *Buddleja* species are used to aid in wound healing where the aerial parts, i.e. leaves, fruit and stem, are applied to the affected area in traditional medicine. Wound formation, regardless of the cause, results in damage to tissue and disrupts the surrounding tissue. When damage occurs to tissue in the body, a physiological response is initiated which leads to bleeding, vessel contraction and coagulation, as well as activation of an acute inflammatory and complement response. This can include regeneration, migration and proliferation of connective tissue and parenchyma cells, as well as the synthesis of extracellular matrix proteins and the remodeling of new connective tissue, parenchyma cells and collagen disposition (Rivera and Spencer, 2007).

The mechanism involved in wound healing includes inflammatory mediators and growth factors, cell-cell and cell-extracellular matrix interactions that control cell proliferation, migration and differentiation, events involved in epithelialization and angiogenesis, wound contraction and remodeling. These processes are initiated from the time of injury and proceed throughout the repair process (Velnar et al., 2009). There are various anti-inflammatory compounds present in the *Buddleja* genus; these include the iridoid, aucubin and flavonoids present in the leaves of *Buddleja* species (Liao et al., 1999).

Antioxidant effects can be attributed to the phenolics present in these species (Mensah et al., 2001). The flowers of *B. officinalis* are used in TCM and are known to have anti-inflammatory activity due to the high amount of crocetin monogentiobiose ester that has reported COX inhibition, thereby reducing the production of prostaglandins (which play an important role in pain and inflammation), and as a result could also aid in wound healing (Liao et al., 1999).

10.5.1.2 Antioxidant Properties

Antioxidants that scavenge active oxygen molecules (free radicals) play an important role in sustaining human health, because the antioxidants decrease the amount of oxygen stress that can be caused due to free radicals (i.e. air pollutants, ozone, exposure to X-rays, cigarette smoke, etc.). It has been identified that activated oxygen species are one of the major contributing factors of aging, diabetes, hardening of arteries and cancer. Approximately 90% of age-related diseases have been linked to some extent to activated oxygen (Ito and Hirose, 1989; Tarpey and White, 1994). Hexane, dichloromethane and methanol extracts of *B. globosa* display antioxidant activity (i.e., 1,1-diphenyl-2-picrylhydrazyl, DPPH, superoxide anion, lipid peroxidation and xanthine oxidase inhibition) (Backhouse et al., 2008). The phenylethanoid glycosides, which are a major class of compounds present in the flowers of *B. officinalis*, have been reported to have antioxidant activity utilizing a total oxidant scavenging capacity (TOSC) assay against peroxynitrite (Tai et al., 2009).

10.5.1.3 Topical Applications of Leaves and Flowers

The use of many Asiatic *Buddleja* species includes the treatment of skin conditions and as eye lotions. This could be related to the anti-inflammatory properties of the flavonoids that are present in this genus, which include the isolated compounds from *B. officinalis*, apigenin and echinacoside (Dormán et al., 2016; Liu et al., 2018). The saponins isolated from *B. asiatica* include asiatiside A–D, which assist in clearing the skin due to their ability to form a soap-like foam in an aqueous solution (Arslan, 2014). In the United States, the *Buddleja* species are applied topically to clean wounds and stimulate healing due to their antiseptic activity. It has been reported that aucubin is present in several drugs used to treat skin conditions and has antibiotic activity, which aids in clearing the skin of impurities (Rombouts and Links, 1956).

The leaves are applied to the skin topically as a poultice to treat rheumatic pains because they comprise various compounds that have analgesic and anti-inflammatory compounds present in the leaves that are being absorbed through the skin. However, no compounds isolated from *B. marrubiifolia* have exhibited any anti-inflammatory activity (Martínez-Pérez et al., 2020). Some compounds isolated from *B. globosa* exhibited analgesic and anti-inflammatory activity; this includes zerumbone, lupeol and aucubin that can inhibit matrix metalloproteinase (MMP-3, MMP-9 and MMP-13), inducible nitric oxide synthase (iNOS) and COX-2. These compounds are all related to anti-inflammation, which aids in the relief of various skin conditions. Dysregulated MMP activity leads to various pathological conditions, including arthritis, inflammation and cancer (Löffek et al., 2011).

COX-2 is an enzyme that synthesizes pro-inflammatory mediators, namely prostaglandins (Rawat et al., 2019), while iNOS is an important enzyme in the macrophage inflammatory response, is the source of NO and is induced in the response topro-inflammatory stimuli (McNeill et al., 2015). The anti-inflammatory-related compounds isolated from *B. officinalis* include apigenin and echinacoside. Buddledin A–C, quercetin and forsythoside B isolated from *B. davidii* express anti-inflammatory activity by inhibiting tumor necrosis factor-alpha (TNF-α), interleukin-6 (IL-6) and modulate nuclear factor kappa beta (NF-κB). Quercetin was also isolated from *B. albiflora* (Tao et al., 2009; Kim and Park, 2018).

TNF-α is an inflammatory cytokine that is produced by macrophages or monocytes during acute inflammation, is responsible for a cascade of events that leads to apoptosis or necrosis, and plays a role in the resistance to infection and cancer (Idriss and Naismith, 2000). IL-6 is an interleukin that acts as a pro-inflammatory cytokine and an anti-inflammatory myokine, acting as pro-inflammatory when trans-signaling occurs, while NF-κB represents a family of inducible transcription factors that regulate a large array of genes involved in different processes of inflammatory and immune responses; these responses could affect the development of cancer (Scheller et al., 2011; Liu et al., 2017).

10.5.2 Anti-Proliferative Activity

Various Buddleja species could potentially have anti-proliferative activity, based on their use in traditional medicine and the activity reported from the isolated compounds. The two species used to treat tumors in traditional medicine are *B. americana* and *B. asiatica*; however, the isolated compounds from these species have not been reported to have any anti-proliferative activity except for methyl catalpol isolated from *B. asiatica*, which exhibited anti-inflammatory activity (Khan, 2013; Zhang et al., 2013). Thus, based on the findings of the biological activity reported on the isolated compounds, as well as the traditional usage of these plants that have been summarized, the plants that would potentially have the best anti-proliferative activity of all the *Buddleja* species includin *B. globosa*, *B. davidii*, *B. salviifolia*, *B. japonica*, *B. officinalis*, *B. saligna* and *B. albiflora*.

10.5.2.1 *Buddleja albiflora*

The isolated compounds from this plant include flavonoids (apigenin-7-O-glucoside, quercetin and luteolin) and phenylpropanoids (2-acetylmartynoside and isomartynoside). The presence of inflammation within the body allows the development of cancer and aids in the progression of all stages of tumorigenesis. If the immune response is not restored to homeostasis, it could lead to chronic inflammation, immunosuppression and excessive proliferation. Apigenin-7-*O*-glucoside is a stable natural flavonoid that has anti-inflammatory activity and exhibits anti-proliferative activity against the colon cancer cell line HCT116 with an IC_{50} of 15 μM (Smiljkovic et al., 2017). Apoptosis in these cells was induced by the pro-apoptotic proteins Bcl-2-association X protein (Bax) and tumor protein p53 (Smiljkovic et al., 2017). Quercetin has anti-inflammatory activity, due to its ability to inhibit the COX-2 protein, and has immune-modulatory activity (Pendota et al., 2014).

Luteolin has been identified to have anti-proliferative activity due to interaction with several pathways involved in the progression of cancer, which includes inhibition of carcinogen activation, inhibition of cell proliferation, induction of apoptosis, the regulation of the cell cycle and the inhibition of angiogenesis and metastasis (Lin et al., 2008). Isomartynoside and 2-acetylmartynoside are other phenylpropanoids that have also been isolated from this plant; these two compounds both exhibit biological activity that indicates that these two compounds possess anti-proliferative activity (Wise, 2015). Isomartynoside has been reported to lower blood pressure, and 2-acetylmartynoside exhibits antioxidant activity. High blood pressure could increase the risk of developing some cancers (i.e., colorectal, lung, bladder and kidney), while an increase in ROS acts as a carcinogen which increases the risk of developing cancer, thus indicating the anti-proliferative activity of these two compounds (Wise, 2015).

10.5.2.2 *Buddleja davidii*

Several compounds are isolated from this species that display various biological activity, most of which only include inflammation, caused by the inhibition of COX and 5-lipoxygenase (5-LOX) induced by the terpenes, buddledin A–C. This relates to cancer because as mentioned before, inflammation contributes to the development of cancer (Yoshida et al., 1978a). The enzyme 5-LOX is vital in the biosynthesis of a variety of bioregulatory compounds such as hydroxyeicosatetraenoic acids (HETEs), leukotrienes, lipoxins and hepoxylines. It has been reported that 5-LOX products play important roles in a variety of disorders such as asthma, inflammation and tumor angiogenesis. Therefore, this is an important potential target for the treatment of bronchial asthma, cancer, autoimmune disease and inflammation, while the inhibition of COX can aid in decreasing inflammation within the body. Various phenylpropanoids have been isolated from this species, including jioniside D and campneoside II, which have antioxidant activity. Coniferaldehyde exhibits anti-inflammatory activity by inducing heme oxygenase-1 (HO-1), which has antioxidant, anti-inflammatory, anti-apoptotic, anti-proliferative and immune-modulating activity (Ahmad et al., 2006; Araujo et al., 2012; Nam et al., 2013). However, calceolarioside-A is known as an immunosuppressant, which could increase the risk of developing malignancy (Chen et al., 2017). Forsythoside B and quercetin have anti-inflammatory activity (Pendota et al., 2014; Kim et al., 2017).

10.5.2.3 *Buddleja globosa*

The flavonoid isolated from this species, apigenin-7-O-glucoside, exhibits anti-proliferative activity against colon cancer cell lines and can cause apoptosis by inducing pro-apoptotic proteins such as Bax and p53 (Smiljkovic et al., 2017). The terpenes lupeol and zerumbone have both been reported to have anti-proliferative activity (Abdul et al., 2008; Saleem, 2009). Zerumbone exhibited good anti-proliferative activity on laryngeal carcinoma cells (Hep-2) with an IC_{50} value of 15 μM and arrests the cell cycle of these cells *in vitro* (Jegannathan et al., 2016). The terpene aucubin exhibits anti-analgesic, anti-inflammation and anti-tumor activity by decreasing the expression of various proteins such as MMP-3, MMP-9 and MMP-13, as well as COX-2 and iNOS, which are all associated with inflammation. Finally, ferulic acid methyl ester and β-amyrin also both have anti-inflammatory activity (Nogueira et al., 2019).

10.5.2.4 *Buddleja japonica*

It was reported that two isolated terpenes exhibited anti-tumor activity. Buddlejoside A5, exhibited antiangiogenic activity in two *in vivo* assays, applied to zebrafish embryos and chick embryo chorioallantoic membrane (Iannuzzi et al., 2019). Saikosaponin A was reported to have antitumor and anti-inflammatory effects, as well as immunoregulatory activity (Yuan et al., 2017).

10.5.2.5 *Buddleja officinalis*

Three isolated compounds from this species have been reported to exhibit anti-proliferative activity, which includes apigenin, luteolin and crocetin-gentiobiose ester. Luteolin is also present in this species; the biological activity was mentioned in Section 10.5.2.1. The activity of apigenin includes the inhibition of dimethyl benzanthracene–induced skin tumors and UV light–induced cancers. It was reported that apigenin suppresses TNF-α and NF-κB, and when applied topically in a skin carcinogenesis model, it resulted in a higher frequency of tumor-free skin models (Shukla and Gupta, 2010). Finally, crocetin-gentiobiose ester has been reported to inhibit COX, thereby reducing the production of prostaglandins (Singla and Giliyaru, 2011).

10.5.2.6 *Buddleja saligna*

All the isolated compounds from this plant have some form of anti-proliferative activity; α-spinasterol induces apoptosis in breast cancer cell lines (MCF-7, MDA-MB-231) and ovarian cancer cells (SKOV-3) through an increased expression of pro-apoptotic proteins, Bax and p53.

It was reported to arrest the cell cycle (G0–G1) in both breast and ovarian cancer (Mishra et al., 2007; Chukwujekwu et al., 2016). Betulonic acid showed high cytotoxicity against prostate cancer cell lines after 24 hours; DU-145 (IC_{50}: 5.9 μg/mL), PC-3 (IC_{50}: 10.4 mg/mL) and LNCaP (IC_{50}: 38.0 μg/mL). Derivatives of betulone, such as betulone that has a carbonyl group at the C-3 position, exhibited anti-proliferative effects against the following cell lines; stomach (MGC-803, IC_{50}: 52.2 μg/mL), breast (Bcap-37, IC_{50}: 54.2 μg/mL and MCF-7, IC_{50}: 64.7 μg/mL), prostate cancer (PC3, IC_{50}: 52.3 μg/mL) (Hata et al., 2002; Saxena et al., 2006; Liu et al., 2012; Chukwujekwu et al., 2016; Zhang et al., 2016).

Oleanolic acid induces apoptosis in melanoma (A375) cells through upregulating the expression of pro-apoptotic proteins (p53, Bax and cytochrome c) and downregulating the anti-apoptotic protein Bcl-2 in a dose-dependent manner. It exhibited an IC_{50} of 40.7 μg/mL against A375 cells (Ghosh et al., 2014). Ursolic acid exhibited anti-proliferative activity in human melanoma cells (SK-MEL-2, IC_{50}: 58.4 μg/mL) and induced apoptosis in A2058 melanoma cells. It has been reported that ursolic acid downregulates angiogenesis-promoting proteins such as vascular endothelial growth factor (VEGF). Anti-proliferative effects were also observed in A375 cells (IC_{50}: 68.2 μg/mL) (Liu, 1995; Kashyap et al., 2016; Oprean et al., 2016; Caunii et al., 2017; Oprean et al., 2018).

10.6 DISCUSSION

Various *Buddleja* species have been used in traditional medicine for hundreds of years in different parts of the world. The prevalent use of these species led to further investigation of the chemical composition of many of the species, which explained the vast medicinal properties these plants have exhibited in traditional medicine practices across the world. This genus displayed anti-oxidant, anti-inflammatory activity, wound healing and the treatment of skin ailments or conditions. The isolated compounds of this genus exhibited significant and promising activity for further investigation of these species and the different isolated compounds of this genus. After compiling the biological activity of the isolated compounds from each species of this genus, it was concluded that *B. saligna*, *B. albiflora*, *B. davidii*, *B. officinalis* and *B. japonica* displayed significant anti-proliferative activity. However, due to the use of *B. americana* and *B. asiatica* in traditional medicine against tumor growths and the lack of research on the isolated compounds from these species, further investigation is justified. There is a lack of research on the chemical composition of the South African species except for *B. saligna* and *B. salviifolia*. Other species such as *B. auriculata*, *B. dysophylla*, *B. glomerata*, *B. loricata* and *B. pulchella* should be investigated further to determine whether these species could have any medicinal potential.

REFERENCES

Abdul, A. B., Abdelwahab, S. I., Al-Zubairi, A. S., Elhassan, M. M. & Murali, S. M. 2008. "Anticancer and antimicrobial activities of zerumbone from the rhizomes of *Zingiber zerumbut*." *International Journal of Pharmacology* 4: 301–304.

Acevedo, L., Martínez, E., Castañeda, P., Franzblau, S., Timmermann, B. N., Linares, E., Bye, R. & Mata, R. 2000. "New phenylethanoids from *Buddleja cordata* subsp. *cordata*." *Planta Medica* 66: 257–261. doi: 10.1055/s-2000-8570.

Ahmad, I., Afza, N., Anis, I., Malik, A. & Fatima, I. 2004. "Azhar-ul-Haq, RB Tareen." *Heterocycles* 63: 1875. doi: 10.1002/cbdv.200790080.

Ahmad, I., Chen, S., Peng, Y., Chen, S. & Xu, L. 2008. "Lipoxygenase inhibiting and antioxidant iridoids from *Buddleja crispa*." *Journal of Enzyme Inhibition and Medicinal Chemistry* 23: 140–143. doi: 10.1080/14756360701342532.

Ahmad, I., Malik, A., Afza, N., Fatima, I., Tareen, R., Nawaz, S. & Choudhary, M. 2006. "Iridoid glucoside and aryl ester from *Buddleja crispa*." *Polish Journal of Chemistry* 80: 1483.

Ahmad, M. & Sticher, O. 1988. "Isolation of acacetin-7-O-rutinoside and martynoside from *Buddleja davidii*." *Journal of the Chemical Society of Pakistan* 10: 117–123.

Aoki, H., Kuze, N., Ichi, T. & Koda, T. 2001. "Analytical method for *Buddleja* colorants in foods." *Journal of the Food Hygienic Society of Japan* 42, 2: 84–90. doi: 10.3358/shokueishi.42.84.

Araujo, J. A., Zhang, M. & Yin, F. 2012. "Heme oxygenase-1, oxidation, inflammation, and atherosclerosis." *Frontiers in Pharmacology* 3: 119. doi: 10.3389/fphar.2012.00119.

Arslan, I. 2014. "Saponins produced by gypsophila species enhance the toxicity of type I ribosome-inactivating proteins." *Studies in Natural Products Chemistry* 43: 375–380.

Asakawa, Y. 1999. "Phytochemicals in human health protection, nutrition, and defense." *Phytochemicals in human health protection, nutrition, and plant defense.* Romeo JT, Kluwer Academic/Plenum Publishers, New York: 319–342.

Backhouse, N., Rosales, L., Apablaza, C., Goty, L., Erazo, S., Negrete, R., Theodoluz, C., Rodriguez, J. & Delporte, C. 2008. "Analgesic, anti-inflammatory and antioxidant Bioactive constituents form *Buddleja* species: A review of the properties of *Buddleja globosa*, Buddlejaceae." *Journal of Ethnopharmacology* 116: 263–269.

Bamuamba, K., Gammon, D. W., Meyers, P., Dijoux-Franca, M. G. & Scott, G. 2008. "Anti-mycobacterial activity of five plant species used as traditional medicines in the Western Cape Province (South Africa)." *Journal of Ethnopharmacology* 117: 385–390. doi: 10.1016/j.jep.2008.02.007

Batra, P. & Sharma, A. K. 2013. "Anti-cancer potential of flavonoids: Recent trends and future perspectives." *Biotechnology* 3: 439–459.

Bentham, G. 1857. "Synopsis of the genus *Clitoria*." *Botanical Journal of the Linnean Society* 2: 33–44.

Bentham, G. & Hooker, J. D. 1876. *Genera plantarum.* Reeve, London. Volume 1: 57–102.

Bhakuni, D. S., Dhar, M., Dhar, M., Dhawan, B. & Mehrotra, B. 1969. "Screening of Indian plants for biological activity: Part II." *Indian Journal of Experimental Biology* 7: 398–402.

Burt, S. 2004. "Essential oils: Their antibacterial properties and potential applications in foods-a review." *International Journal of Food Microbiology* 94: 223–253. doi: 10.1016/j.ijfoodmicro.2004.03.022.

Caunii, A., Oprean, C., Cristea, M., Ivan, A., Danciu, C., Tatu, C., Paunescu, V., Marti, D., Tzanakakis, G. & Spandidos, D. A. 2017. "Effects of ursolic and oleanolic on SK-MEL-2 melanoma cells: *In vitro* and *in vivo* assays." *International Journal of Oncology* 51: 1651–1660. doi: 10.3892/ijo.2017.4160.

Chae, S. W., Kang, K. A., Kim, J. S., Kim, H. K., Lee, E. J., Hyun, J. W. & Kang, S. S. 2007. "Antioxidant activities of acetylmartynosides from *Clerodendron trichotomum*." *Journal of Applied Biological Chemistry* 50: 270–274.

Chae, S. W., Kim, J. S., Kang, K. A., Bu, H. D., Lee, Y., Hyun, J. W. & Kang, S. S. 2004. "Antioxidant activity of jionoside D from *Clerodendron trichotomum*." *Biological Pharmacological Bulletin* 27: 1504–1508. doi: 10.1248/bpb.27.1504.

Chang, I. M. 1998. "Liver-protective activities of aucubin derived from traditional oriental medicine." *Research Communications in Molecular Pathology and Pharmacology* 102, 2: 189–204. doi: 102(2):189–204.

Chen, G., Gong, W. C., Ge, J., Dunn, B. L. & Sun, W. B. 2012. "Floral scents of typical *Buddleja* species with different pollination syndromes." *Biochemical Systematics and Ecology* 44: 173–178. doi: 10.1016/j.bse.2012.05.010.

Chen, H., Xu, C., Liu, D. Q., An, S. Q. & Tan, R. X. 2005. "Buddlin, a new compound from *Buddleja asiatica*." *Fitoterapia* 76: 588–589. doi: 10.1016/j.fitote.2005.04.012.

Chen, Y., Xue, G., Liu, F. & Gongc, X. 2017. "Immunosuppressive effect of extracts from leaves of *Fraxinus Mandshurica* Rupr." *Bioengineered* 8: 212–216. doi: 10.1080/21655979.2016.1221023.

Cheo, T. 1947. "Medicinal plants of Omei-Shan." *Botanical Bulletin-Academia Sinica Taipei* 1: 298–308.

Chukwujekwu, J. C., Amoo, S. O., De Kock, C. A., Smith, P. J. & Van Staden, J. 2014. "Antiplasmodial, acetylcholinesterase and alpha-glucosidase inhibitory and cytotoxicity properties of *Buddleja saligna*." *South African Journal of Botany* 94: 6–8. doi: 10.1016/j.sajb.2014.05.003.

Chukwujekwu, J. C., Rengasamy, K. R., De Kock, C. A., Smith, P. J., Slavětínská, L. P. & Van Staden, J. 2016. "Alpha-glucosidase inhibitory and antiplasmodial properties of terpenoids from the leaves of *Buddleja saligna* Willd." *Journal of Enzyme Inhibition and Medicinal Chemistry* 31: 63–66. doi: 10.3109/14756366.2014.1003927.

Crevost, C. & Petelot, A. 1929. "Catalogue de produits de L'Indo-Chine (plantas medicinales)." *Bulletin Economic de L'Indo-Chine* 32: 325.

Cronquist, A. & TakhtadzhiaN, A. L. 1981. *An integrated system of classification of flowering plants.* Columbia University Press, New York.

de Barradas, J. P. 1957. *Plantas Magicas Americanas.* Consejo Superior de Investigaciones Cientgificas Instituto, Bernardino de Sahagun, Madrid, Spain. 153 p.

Desai, A. G., Qazi, G. N., Ganju, R. K., El-Tamer, M., Singh, J., Saxena, A. K., Bedi, Y. S., Taneja, S. C. & Bhat, H. K. 2008. "Medicinal plants and cancer chemoprevention." *Current Drug Metabolism* 9: 581–591. doi: 10.2174/138920008785821657.

De Vivar, A. R., Nieto, D. A., Gaviño, R. & Pérez, C. a.-L. 1995. "Isocapnell-9-en-8-one and 6α-hydroxyisocapnell-9-en-8-one, sesquiterpenes from *Buddleia* species." *Plantas magicas americanas Phytochemistry* 40: 167–170. doi: 10.1016/0031-9422(95)00261-5.

Díaz, J. L. 1976. Índice y sinonimia de las plantas medicinales de México (No. C/581.63472 I5/1).

Ding, N., Yahara, S. & Nohara, T. 1992. "Structure of mimengosides A and B, new triterpenoid glycosides from Buddlejae flos produced in China." *Chemical and Pharmaceutical Bulletin* 40: 780–782. doi: 10.1248/cpb.40.780.

Dormán, G., Flachner, B., Hajdú, I. & András, C. D. 2016. "Target identification and polypharmacology of nutraceuticals." *Nutraceuticals*: 263–286. doi: 10.1016/B978-0-12-821038-3.00023-9.

Duff, R., Bacon, J., Mundie, C., Farmer, V., Russell, J. & Forrester, A. 1965. "Catalpol and methylcatalpol: Naturally occurring glycosides in Plantago and Buddleia species." *Biochemical Journal* 96: 1–5. doi: 10.1042/bj0960001.

Efloraofindia. 2007. "*Buddleja asiatica*." Obtained from efloraofindia under the Creative Commons license. https://sites.google.com/site/efloraofindia/species/m—z/s/scrophulariaceae/buddleja/buddleja-asiatica (Accessed 19 September 2020).

Elango, Y. 2015. "*Buddleja madagascariensis*." Obtained from Wikimedia under the Creative Commons Attribution-Share Alike 4.0 International license. https://commons.wikimedia.org/wiki/File:Buddleja_madagascariensis-2-yercaud-salem-India.JPG (Accessed 5 November 2020).

Emam, A., Diaz-Lanza, A., Matellano-Fernandez, L., Faure, R. & Moussa, A. 1997. "Biological activities of buddlejasaponin isolated from *Buddleja madagascariensis* and *Scrophularia scorodonia*." *Pharmazie* 52: 76–77.

Fazly Bazzaz, B. S., Khameneh, B., Zahedian Ostad, M. R. & Hosseinzadeh, H. 2018. "*In vitro* evaluation of antibacterial activity of verbascoside, lemon verbena extract and caffeine in combination with gentamicin against drug-resistant *Staphylococcus aureus* and *Escherichia coli* clinical isolates." *Avicenna Journal of Phytomedicine* 8: 246–253.

Flora, A. 2014. "*Buddleia curviflora*. "www.asianflora.com/Buddleiaceae/Buddleia-curviflora-1.jpg (Accessed 22 September 2020).

Ghosh, S., Bishayee, K. & Khuda-Bukhsh, A. R. 2014. "Oleanolic acid isolated from ethanolic extract of *Phytolacca decandra* induces apoptosis in A375 skin melanoma cells: Drug-DNA interaction and signaling cascade." *Journal of Integrative Medicine* 12: 102–114. doi: 10.1016/S2095-4964(14)60015-7.

Gondal, H. Y., Zamir, R., Nisar, M. & Choudhary, M. I. 2020. "*Verbascum sinaiticum*: A rich source of antioxidant phenylethanoid glycosides." *The Natural Products Journal* 10: 158–162. doi: 10.2174/2210315509666190129160405.

Guo, H., Koike, K., Li, W., Satou, T., Guo, D. & Nikaido, T. 2004. "Saponins from the flower buds of *Buddleja officinalis*." *Journal of Natural Products* 67: 10–13. doi: 10.1021/np0300131.

Hartwell, J. L. 1970. "Plants used against cancer. A survey." *Lloydia* 33: 97–194.

Hata, K., Hori, K. & Takahashi, S. 2002. "Differentiation-and apoptosis-inducing activities by pentacyclic triterpenes on a mouse melanoma cell line." *Journal of Natural Products* 65: 645–648. doi: 10.1021/np0104673.

Higuchi, M. 2014. "Antioxidant properties of wheat bran against oxidative stress." *Wheat and Rice in Disease Prevention and Health*: 181–199. doi: 10.1016/B978-0-12-401716-0.00015-5.

Houghton, P. J. 1984. "Ethnopharmacology of some *Buddleja* species." *Journal of Ethnopharmacology* 11: 293–308. doi: 10.1016/0378-8741(84)90075-8.

Houghton, P. J., Aliancic, I. & Stefanovic, M. 1993. "Aucubin, an iridoid glycoside from *Buddleja americana* roots." *Journal of the Serbian Chemical Society* 58: 43–46.

Houghton, P. J. & Hikino, H. 1989. "Anti-hepatotoxic activity of extracts and constituents of *Buddleja* species." *Planta Medica* 55: 123–126. doi: 10.1055/s-2006-961903.

Houghton, P. J., Mensah, A. Y., Iessa, N. & Hong, L. Y. 2003. "Terpenoids in *Buddleja*: Relevance to chemosystematics, chemical ecology and biological activity." *Phytochemistry* 64: 385–393. doi: 10.1016/S0031-9422(03)00264-4.

Houghton, P. J., Woldemariam, T. Z., Candau, M., Barnardo, A., Khen-Alafun, O. & Shangxiao, L. 1996. "Buddlejone, a diterpene from *Buddleja albiflora*." *Phytochemistry* 42: 485–488. doi: 10.1016/0031-9422(96)00001-5.

Hutchings, A. 1996. "Zulu medicinal plants: An inventory." University of Natal Press. doi: 10.1021/np970084y.

Iannuzzi, A. M., Muñoz Camero, C., Beladjila, K. A., De Leo, M., D'Angelo, V., Germanò, M. P., Amira, S., Bader, A., De Tommasi, N. & Braca, A. 2019. "Antiangiogenic activity of iridoids from Lamiaceae and Plantaginaceae species." *67th International Congress and Annual Meeting of the Society for Medicinal Plant and Natural Product Research (GA) in cooperation with the French Society of Pharmacognosy AFERP* 85, 18: 1544. doi: 10.1055/s-0039-3400077.

Idriss, H. T. & Naismith, J. H. 2000. "TNF-α and the TNF receptor superfamily: Structure-function relationship (s)." *Microscopy Research and Technique*. 50: 184–195. doi: 10.1002/1097-0029(20000801)50:3<184::AID-JEMT2>3.0.CO;2-H.

Imai, T., Nomura, M. & Fukushima, K. 2006. "Evidence for involvement of the phenylpropanoid pathway in the biosynthesis of the norlignan agatharesinol." *Journal of Plant Physiology* 163, 5: 483–487. doi: 10.1016/j.jplph.2005.08.009.

Ito, N. & Hirose, M. 1989. "Antioxidants-carcinogenic and chemopreventive properties." *Advances in Cancer Research* 53: 247–302. doi: 10.1016/S0065-230X(08)60283-3.

Jegannathan, S. D., Arul, S. & Dayalan, H. 2016. "Zerumbone, a sesquiterpene, controls proliferation and induces cell cycle arrest in human laryngeal carcinoma cell line Hep-2." *Nutrition and Cancer* 68: 865–872. doi: 10.1080/01635581.2016.1159701.

Jiang, W. L., Fu, F. H., Xu, B. M., Tian, J. W., Zhu, H. B. & Jian, H. 2010. "Cardioprotection with forsythoside B in rat myocardial ischemia-reperfusion injury: Relation to inflammation response." *Phytomedicine* 17: 635–639. doi: 10.1016/j.phymed.2009.10.017.

Jiang, W. L., Yong, X., Zhang, S. P., Zhu, H. B. & Jian, H. 2012. "Forsythoside B protects against experimental sepsis by modulating inflammatory factors." *Phytotherapy Research* 26: 981–987. doi: 10.1002/ptr.3668.

JMK. 2012. "*Buddleja dysophylla*, bloeipluim." Obtained from Wikimedia under the Creative Commons Attribution-Share Alike 3.0 Unported license. https://commons.wikimedia.org/wiki/File:Buddleja_dysophylla,_bloeipluim,_Louwsburg.jpg (Accessed 19 November 2020).

Joffe, P. 1993. *The gardener's guide to South African plants*. Delos, Cape Town.

Joffe, P. 2001. *Creative gardening with indigenous plants*. Briza, Pretoria.

Joffe, P. 2003. *Easy guide to indigenous shrubs*. Briza Publications, Pretoria.

Johansson, C. T. 2013. "*Buddleja incana*." Obtained from Wikimedia under the Creative Commons Attribution-Share Alike 3.0 Unported license. https://commons.wikimedia.org/wiki/File:Buddleja_incana-IMG_9172.JPG (Accessed 5 November 2020).

Kalantari, K., Moniri, M., Boroumand Moghaddam, A., Abdul Rahim, R., Bin Ariff, A., Izadiyan, Z. & Mohamad, R. 2017. "A review of the biomedical applications of zerumbone and the techniques for its extraction from ginger rhizomes." *Molecules* 22, 10: 1645. doi: 10.3390/molecules22101645.

Kang, D. G., Lee, Y. S., Kim, H. J., Lee, Y. M. & Lee, H. S. 2003. "Angiotensin converting enzyme inhibitory phenylpropanoid glycosides from *Clerodendron trichotomum*." *Journal of Ethnopharmacoly* 89: 151–154. doi: 10.1016/S0378-8741(03)00274-5.

Kashyap, D., Tuli, H. S. & Sharma, A. K. 2016. "Ursolic acid (UA): A metabolite with promising therapeutic potential." *Life Sciences* 146: 201–213. doi: 10.1016/j.lfs.2016.01.017.

Kenraiz, K. Z. 2017. "*Buddleja cordata* in Hackfalls Arboretum, Hawkes's Bay, NZ." Obtained from Wikimedia under the Creative Commons Attribution-Share Alike 4.0 International license. https://commons.wikimedia.org/wiki/File:Buddleja_cordata_kz05.jpg (Accessed 30 November 2020).

Khan, F. A. 2013. "Phytochemical studies and biological activities of the constituents of *Buddleja Asiatica* Lour." Doctoral dissertation, Gomal University.

Khan, S., Ullah, H. & Zhang, L. 2019. "Bioactive constituents form *Buddleja* species." *Pakistan Journal of Pharmaceutical Sciences*. 32, 2. doi: 10.1016/j.ejmech.2020.112563.

Kim, G.-J., Jo, H.-J., Chung, K.-H., Lee, K.-J. & An, J. H. 2017. "Oleanolic acid induces p53 dependent apoptosis via the ERK/JNK/AKT pathway in cancer cell lines." *Europe PMC*. doi: 10.20944/preprints201705.0218.v1.

Kim, J. K. & Park, S. U. 2018. "Quercetin and its role in biological functions: An updated review." *Experimental and Clinical Sciences Journal* 17: 856–863. doi: 10.17179/excli2018-1538.

Kintzios, S. E. 2006. "Terrestrial plant-derived anticancer agents and plant species used in anticancer research." *Critical Reviews in Plant Sciences* 25: 79–113. doi: 10.1080/07352680500348824.

Kornhall, P. & Bremer, B. 2004. "New circumscription of the tribe Limoselleae (Scrophulariaceae) that includes the taxa of the tribe Manuleeae." *Botanical Journal of the Linnean Society* 146: 453–467. doi: 10.1111/j.1095-8339.2004.00341.x.

Landete, J. M. 2012. "Plant and mammalian lignans: A review of source, intake, metabolism, intestinal bacteria and health." *Food Research International* 46, 1: 410–424. doi: 10.1016/j.foodres.2011.12.023.

Liao, Y.-H., Houghton, P. J. & Hoult, J. 1999. "Novel and known constituents from *Buddleja* species and their activity against leukocyte eicosanoid generation." *Journal of Natural Products* 62: 1241–1245. doi: 10.1016/j.foodres.2011.12.023.

Lin, Y., Shi, R., Wang, X. & Shen, H. M. 2008. "Luteolin, a flavonoid with potential for cancer prevention and therapy." *Current Cancer Drug Targets* 8: 634–646. doi: 10.2174/156800908786241050.

Liu, J. 1995. "Pharmacology of oleanolic acid and ursolic acid." *Journal of Ethnopharmacology* 49: 57–68. doi: 10.1016/0378-8741(95)90032-2.

Liu, J., Yang, L., Dong, Y., Zhang, B. & Ma, X. 2018. "Echinacoside, an inestimable natural product in treatment of neurological and other disorders." *Molecules* 23: 1213. doi: 10.3390/molecules23051213.

Liu, M., Yang, S., Jin, L., Hu, D., Wu, Z. & Yang, S. 2012. "Chemical constituents of the ethyl acetate extract of *Belamcanda chinensis* (L.) DC roots and their antitumor activities." *Molecules* 17: 6156–6169. doi: 10.3390/molecules17056156.

Liu, T., Zhang, L., Joo, D. & Sun, S.-C. 2017. "NF-κB signaling in inflammation." *Signal Transduction and Targeted Therapy*. 2, pp. 1–9.

Loader, R. 2015. "*Buddleja albiflora* 16th July 2015." Obtained from The Buddleja Collection. www.buddlejacollection.com/plants/albiflora/ (Accessed 19 September 2020).

Löffek, S., Schilling, O. & Franzke, C.-W. 2011. "Biological role of matrix metalloproteinases: A critical balance." *European Respiratory Journal* 38: 191–208. doi: 1183/09031936.00146510.

Lopez, J., Sierra, J., Vegazo, M. & Cortes, M. 1979. "Chemical constituents of *Buddleja globosa* Lam." *Fitoterapia* 50, 5: 195–198.

Lu, J., Tu, G., Zhao, Y., Lv, Y., Liu, L. Y. & Wu, Y. 2004. "Structural determination of novel terpenes from *Buddleia lindleyana*." *Magnetic Resonance in Chemistry* 42: 893–897. doi: 10.1002/mrc.1408.

Martinez, M. M. 1959. *Las Plantas Medicinales de Mexico*. Ediciones Botas, Mexico City. Edition 4: 284.

Martínez-Pérez, E. F., Burgueño-Tapia, E., Roa-Flores, S., Bendaña-Piñeiro, A. E., Sánchez-Arreola, E., Bach, H. & Hernández, L. R. 2020. "A new diterpene and bioactivities of labdanes isolated from Buddleja marrubiifolia." *Natural Product Research*: 1–8. doi: 10.1080/14786419.2020.1762184.

McNeill, E., Crabtree, M. J., Sahgal, N., Patel, J., Chuaiphichai, S., Iqbal, A. J., Hale, A. B., Greaves, D. R. & Channon, K. M. 2015. "Regulation of iNOS function and cellular redox state by macrophage Gch1 reveals specific requirements for tetrahydrobiopterin in NRF2 activation." *Free Radical Biology and Medicine* 79: 206–216. doi: 10.1016/j.freeradbiomed.2014.10.575.

McPartland, J. M. & Russo, E. B. 2001. "Cannabis and cannabis extracts: Greater than the sum of their parts." *Journal of Cannabis Therapeutics* 1: 103–132. doi: 10.1300/J175v01n03_08.

Mensah, A. Y., Houghton, P. J., Bloomfield, S., Vlietinck, A. & Vanden Berghe, D. 2000. "Known and novel terpenes from *Buddleja globosa* displaying selective antifungal activity against dermatophytes." *Journal of Natural Products* 63: 1210–1213. doi: 10.1021/np0001023.

Mensah, A. Y., Sampson, J., Houghton, P., Hylands, P., Westbrook, J., Dunn, M., Hughes, M. & Cherry, G. 2001. "Effects of *Buddleja globosa* leaf and its constituents relevant to wound healing." *Journal of Ethnopharmacology* 77: 219–226. doi: 10.1021/np0001023.

Mishra, S. K., Sangwan, N. S. & Sangwan, R. S. 2007. "Plant review *Andrographis paniculata* (Kalmegh): A review." *Pharmacognosy Reviews* 1: 283–298.

Miyase, T., Akahori, C., Kohsaka, H. & Ueno, A. 1991. "Acylated iridoid glycosides from *Buddleja japonica* HEMSL." *Chemical and Pharmaceutical Bulletin* 39: 2944–2951. doi: 10.1248/cpb.39.2944.

Mohan, V. R., Tresina, P. S. & Daffodil, E. D. 2016. "Antinutritional factors in legume seeds: Characteristics and determination." *Encyclopedia of Food and Health*: 211–220.

Mugford, S. T. & Osbourn, A. 2012. "Saponin synthesis and function." *Isoprenoid Synthesis in Plants and Microorganisms* 22: 405–424.

Munn, L. L. 2017. "Cancer and inflammation." *Wiley Interdisciplinary Reviews: Systems Biology and Medicine* 9, 2: 1370. doi: 10.1002/wsbm.1370.

Nam, J. W., Kim, S. Y., Yoon, T., Lee, Y. J., Kil, Y. S., Lee, Y. S. & Seo, E. K. 2013. "Heat shock factor 1 inducers from the bark of *Eucommia ulmoides* as cytoprotective agents." *Chemistry & Biodiversity* 10: 1322–1327. doi: 10.1002/cbdv.201200401.

Neelam, Khatkar, A. & Sharma, K. K. 2020. "Phenylpropanoids and its derivatives: Biological activities and its role in food, pharmaceutical and cosmetic industries." *Critical Reviews in Food Science and Nutrition* 60, 16: 2655–2675. doi: 10.1080/10408398.2019.1653822.

Nguyen, A. T. & Kim, K. Y. 2020. "Inhibition of proinflammatory cytokines in *Cutibacterium acnes*-induced inflammation in HaCaT cells by using *Buddleja davidii* aqueous extract." *International Journal of Inflammation*. doi: 10.1155/2020/8063289.

Nikonov, G., Lou, T., Ch'u, C., Ma, L., Tung, L., Ming, C., Huo, T. & Lo, Y. 1961. "Materials in the study of drugs from Chinese folk medicine used in the treatment of hypertension, nephritis, diabetes and cancer." *Aptechnoe Delo*. 10: 71–83.

Nkuété, A. H. L., Migliolo, L., Wabo, H. K., Tane, P. & Franco, O. L. 2015. "Evaluation of multiple functions of polygonum genus compounds." *European Journal of Medicinal Plants* 6: 1–16. doi: 10.9734/EJMP/2015/15134.

Nogueira, A. O., Oliveira, Y. I. S., Adjafre, B. L., De Moraes, M. E. A. & Aragão, G. F. 2019. "Pharmacological effects of the isomeric mixture of alpha and beta amyrin from *Protium heptaphyllum*: A literature review." *Fundamental & Clinical Pharmacology* 33: 4–12. doi: 10.1111/fcp.12402.

Oh, H., Pae, H. O., Oh, G. S., Lee, S. Y., Chai, K. Y., Song, C. E., Kwon, T. O., Chung, H.-T. & Lee, H. S. 2002. "Inhibition of inducible nitric oxide synthesis by catalposide from *Catalpa ovata*." *Planta Medica* 68: 685–689. doi: 10.1055/s-2002-33810.

Oprean, C., Ivan, A., Bojin, F., Cristea, M., Soica, C., Drăghia, L., Caunii, A., Paunescu, V. & Tatu, C. 2018. "Selective *in vitro* anti-melanoma activity of ursolic and oleanolic acids." *Toxicology Mechanisms and Methods* 28: 148–156. doi: 10.1080/15376516.2017.1373881.

Oprean, C., Mioc, M., Csányi, E., Ambrus, R., Bojin, F., Tatu, C., Cristea, M., Ivan, A., Danciu, C. & Dehelean, C. 2016. "Improvement of ursolic and oleanolic acids' antitumor activity by complexation with hydrophilic cyclodextrins." *Biomedicine & Pharmacotherapy* 83: 1095–1104. doi: 10.1016/j.biopha.2016.08.030.

Oxelman, B., Backlund, M. & Bremer, B. 1999. "Relationships of the Buddlejaceae s. 1. Investigated using parsimony jackknife and branch support analysis of chloroplast ndhF and rbcL sequence data." *Systematic Botany* 24, 2: 164–182. doi: 10.2307/2419547.

Oxelman, B., Kornhall, P. & Norman, E. 2004. "Buddlejaceae." *Flowering plants*. Dicotyledons: 39–44. Springer, Berlin.

Oxelman, B., Kornhall, P., Olmstead, R. G. & Bremer, B. 2005. "Further disintegration of Scrophulariaceae." *Taxon* 54: 411–425. doi: 10.2307/25065369.

Panche, A., Diwan, A. & Chandra, S. 2016. "Flavonoids: An overview." *Journal of Nutritional Science* 5, 47: 1–15. doi: 10.1017/jns.2016.41.

Pendota, S., Ndhlala, A., Aremu, A., Aderogba, M. & Van Staden, J. 2014. "Anti-inflammatory, antioxidant and *in silico* studies of *Buddleja salviifolia* (L). Lam leaf constituents." *South African Journal of Botany* 93: 79–85. doi: 10.1016/j.sajb.2014.03.012.

Penganum. 2016. "*Buddleja lindleyana*." Obtained from Wikimedia under the license. https://commons.wikimedia.org/wiki/File:Buddleja_lindleyana_(20306938039).jpg (Accessed 5 November 2020).

Perveen, S. & Al-Taweel, A. 2018. "Introductory chapter: Terpenes and terpenoids." *Terpenes and Terpenoids* 1–12. doi: 10.5772/interchopen.79683.

Pooley, E. 2003. *Mountain flowers: A field guide to the flora of the Drakensberg and Lesotho*. The Flora Publications Trust, Durban.

Porse, S. 2007. "*Buddleja globosa*: Flower-heads." Obtained from Wikimedia under the Creative Commons Attribution-Share Alike 3.0 Unported license. https://commons.wikimedia.org/wiki/File:Buddleja-globosa-flowers.JPG (Accessed 5 November 2020).

Qiao, Z., Tang, J., Wu, W., Tang, J. & Liu, M. 2019. "Acteoside inhibits inflammatory response via JAK/STAT signaling pathway in osteoarthritic rats." *BMC Complementary and Alternative Medicine* 19: 1–8.

Raimondo, D., Staden, L. V., Foden, W., Victor, J., Helme, N., Turner, R., Kamundi, D. & Manyama, P. 2009. *Red list of South African plants 2009*. Gauteng: South African National Biodiversity Institute.

Rawat, C., Kukal, S., Dahiya, U. R. & Kukreti, R. 2019. "Cyclooxygenase-2 (COX-2) inhibitors: Future therapeutic strategies for epilepsy management." *Journal of Neuroinflammation* 16: 1–15.

Reid, K., Maes, J., Maes, A., Van Staden, J., De Kimpe, N., Mulholland, D. & Verschaeve, L. 2006. "Evaluation of the mutagenic and antimutagenic effects of South African plants." *Journal of Ethnopharmacology* 106: 44–50. doi: 10.1016/j.jep.2005.11.030.

Reuter, S., Gupta, S. C., Chaturvedi, M. M. & Aggarwal, B. B. 2010. "Oxidative stress, inflammation, and cancer: How are they linked?" *Free Radical Biology and Medicine* 49: 1603–1616. doi: 10.1016/j.freeradbiomed.2010.09.006.

Rivera, A. E. & Spencer, J. M. 2007. "Clinical aspects of full-thickness wound healing." *Clinics in Dermatology* 25: 39–48. doi: 10.1016/j.clindermatol.2006.10.001.

Rocha-Guzmán, N. E., Simental-Mendía, L. E., Barragán-Zúñiga, L. J., Ramírez-España, J. C., Gallegos-Infante, J. A., Lujan-Mendoza, C. I. & Gamboa-Gómez, C. I. 2018. "Effect of *Buddleja scordioides* K. leaves infusion on lipid peroxidation in mice with ultraviolet light-induced oxidative stress." *Medicinal Chemistry Research* 27, 10: 2379–2385.

Rombouts, J. & Links, J. 1956. "The chemical nature of the antibacterial substance present in *Aucuba japonica* thunbg." *Experientia* 12: 78–80.

Roux, J. 2003. "Flora of South Africa." Compton Herbarium: South African National Biodiversity Institute: Cape Town, South Africa. https://plants.jstor.org/collection/FLOSA?si=401&scope=plants&limit=50.

Russo, E. B. & Marcu, J. 2017. "Cannabis pharmacology: The usual suspects and a few promising leads." *Advances in Pharmacology*. Elsevier. 80: 67–134. Academic Press.

Saleem, M. 2009. "Lupeol, a novel anti-inflammatory and anti-cancer dietary triterpene." *Cancer Letters* 285: 109–115. doi: 10.1016/j.canlet.2009.04.033.

Salehi, B., Venditti, A., Sharifi-Rad, M., Kręgiel, D., Sharifi-Rad, J., Durazzo, A., Lucarini, M., Santini, A., Souto, E. B., Novellino, E., Antolak, H., Azzini, E., Setzer, W. N. & Martins, N. 2019. "The therapeutic potential of apigenin." *International Journal of Molecular Sciences* 20: 1305. doi: 10.3390/ijms20061305.

Salix. 2015. "*Buddleja officinalis*." Obtained from Wikimedia under the Creative Commons Attribution-Share Alike 4.0 International license. https://commons.wikimedia.org/wiki/File:Buddleja_officinalis_in_Jardin_des_Plantes_02.JPG (Accessed 5 November 2020).

Saxena, B. B., Zhu, L., Hao, M., Kisilis, E., Katdare, M., Oktem, O., Bomshteyn, A. & Rathnam, P. 2006. "Boc-lysinated-betulonic acid: A potent, anti-prostate cancer agent." *Bioorganic & Medicinal Chemistry* 14: 6349–6358. doi: 10.1016/j.bmc.2006.05.048.

Scheller, J., Chalaris, A., Schmidt-Arras, D. & Rose-John, S. 2011. "The pro-and anti-inflammatory properties of the cytokine interleukin-6." *Biochimica et Biophysica Acta (BBA)-Molecular Cell Research* 1813: 878–888. doi: 10.1016/j.bbamcr.2011.01.034.

Sedky, N. K., El Gammal, Z. H., Wahba, A. E., Mosad, E., Waly, Z. Y., El-Fallal, A. A., Arafa, R. K. & El-Badri, N. 2018. "The molecular basis of cytotoxicity of α-spinasterol from Ganoderma resinaceum: Induction of apoptosis and overexpression of p53 in breast and ovarian cancer cell lines." *Journal of Cellular Biochemistry* 119: 3892–3902. doi: 10.1002/jcb.26515.

Shebs, S. 2005. "*Buddleja marrubifolia* flowers." Obtained from Wikimedia under the Creative Commons Attribution-Share Alike 3.0 Unported license. https://commons.wikimedia.org/wiki/File:Buddleja_marrubifolia_flowers_3.jpg (Accessed 5 November 2020).

Shukla, S. & Gupta, S. 2010. "Apigenin: A promising molecule for cancer prevention." *Pharmaceutical Research* 27: 962–978.

Singla, R. K. & Giliyaru, V. 2011. "Crocin: An overview." *Indo Global Journal of Pharmaceutical Sciences* 1: 281–286.

Smiljkovic, M., Stanisavljevic, D., Stojkovic, D., Petrovic, I., Marjanovic Vicentic, J., Popovic, J., Golic Grdadolnik, S., Markovic, D., Sankovic-Babice, S., Glamoclija, J., Stevanovic, M. & Sokovic, M. 2017. "Apigenin-7-O-glucoside versus apigenin: Insight into the modes of anticandidal and cytotoxic actions." *Experimental and Clinical Sciences Journal* 16: 795–807. doi: 10.17179/excli2017-300.

Smith, F. P. & Stuart, G. A. 1911. "Chinese materia medica: Vegetable kingdom." American Presbyterian Mission Press. www.worldcat.org/title/chinese-materia-medica-vegetable-kingdom/oclc/14787160.

Socała, K. & Wlaź, P. 2016. "Evaluation of the antidepressant-and anxiolytic-like activity of α-spinasterol, a plant derivative with TRPV1 antagonistic effects, in mice." *Behavioural Brain Research* 303: 19–25. doi: 10.1016/j.bbr.2016.01.048.

Soledade, M., Pedras, C. & Zheng, Q. 2010. "The chemistry of *Arabidopsis thaliana*." In *Comprehensive Natural Products II: Chemistry and Biology*. Elsevier, Netherlands. 1297–1312.

Strohl, W. R. 2000. "The role of natural products in a modern drug discovery program." *Drug Discovery Today* 5: 39–41. doi: 10.1016/s1359-6446(99)01443-9.

Tai, B. H., Jung, B. Y., Cuong, N. M., Linh, P. T., Tung, N. H., Nhiem, N. X., Huong, T. T., Anh, N. T., Kim, J. A. & Kim, S. K. 2009. "Total peroxynitrite scavenging capacity of phenylethanoid and flavonoid glycosides from the flowers of *Buddleja officinalis*." *Biological and Pharmaceutical Bulletin* 32: 1952–1956. doi: 10.1248/bpb.32.1952.

Takhtajan, A. L. 1980. "Outline of the classification of flowering plants (Magnoliophyta)." *The Botanical Review* 46: 225–359.

Tank, D. C., Beardsley, P. M., Kelchner, S. A. & Olmstead, R. G. 2006. "Review of the systematics of *Scrophulariaceae* sp and their current disposition." *Australian Systematic Botany* 19: 289–307. doi: 10.1071/SB05009.

Tao, L., Huang, J., Zhao, Y. & Li, C. 2009. "Chemical constituents in *Buddleja albiflora*." *China Journal of Chinese Materia Medica* 34: 3043–3046.

Tarpey, M. M. & White, R. 1994. "Extensive nitration of protein tyrosines in human atherosclerosis detected by immunohistochemistry." *Biological Chemical Hoppe—Seyler* 375: 81–88. doi: 10.1515/bchm3.1994.375.2.81.

Tesfaye, S., Belete, A., Engidawork, E., Gedif, T. & Asres, K. 2020. "Ethnobotanical study of medicinal plants used by traditional healers to treat cancer-like symptoms in eleven districts, Ethiopia." *Evidence-Based Complementary and Alternative Medicine*. doi: 10.1155/2020/7683450.

Trim, A. & Hill, R. 1952. "The preparation and properties of aucubin, asperuloside and some related glycosides." *Biochemical Journal* 50: 310–319.

Van Wyk, B.-E. & Gericke, N. 2000. *People's plants: A guide to useful plants of Southern Africa*. Briza Publications, Pretoria.

Van Wyk, B.-E. & Van Wyk, P. 1997. *Trees of Southern Africa*. Africa Stuik Publication, Cape Town.

Velnar, T., Bailey, T. & Smrkolj, V. 2009. "The wound healing process: An overview of the cellular and molecular mechanisms." *Journal of International Medical Research* 37: 1528–1542.

Venter, F. & Venter, J. A. 2002. *Making the most of indigenous trees*. Briza Publications, Pretoria.

Wang, S.-N., Xie, G.-P., Qin, C.-H., Chen, Y.-R., Zhang, K.-R., Li, X., Wu, Q., Dong, W.-Q., Yang, J. & Yu, B. 2015. "Aucubin prevents interleukin-1 beta induced inflammation and cartilage matrix degradation via inhibition of NF-κB signaling pathway in rat articular chondrocytes." *International Immunopharmacology* 24: 408–415. doi: 10.1016/j.intimp.2014.12.029.

Watt, J. M. & Breyer-Brandwijk, M. G. 1962. *Medicinal and poisonous plants of Southern and Eastern Africa*. E. & S. Livingstone Ltd., Edinburgh. Edition 2: 1457.

Wise, J. 2015. "High blood pressure is linked to increased risk of diabetes." *British Medical Journal* 351: h5167. doi: 10.1136/bmj.h5167.

Won, J.-H., Im, H.-T., Kim, Y.-H., Yun, K.-J., Park, H.-J., Choi, J.-W. & Lee, K.-T. 2006. "Anti-inflammatory effect of buddlejasaponin IV through the inhibition of iNOS and COX-2 expression in RAW 264.7 macrophages via the NF-kappaB inactivation." *British Journal of Pharmacology* 148: 216–225. doi: 10.1038/sj.bjp.0706718.

Xue, B., Ma, B., Zhang, Q., Li, X., Zhu, J., Liu, M., Wu, X., Wang, C. & Wu, Z. 2015. "Pharmacokinetics and tissue distribution of Aucubin, Ajugol and Catalpol in rats using a validated simultaneous LC—ESI-MS/MS assay." *Journal of Chromatography B* 1002: 245–253. doi: 10.1016/j.jchromb.2015.08.026.

Yamamoto, A., Miyase, T., Ueno, A. & Maeda, T. 1991. "Buddlejasaponins I-IV, four new oleanane-triterpene saponins from the aerial parts of *Buddleja japonica* Hemsl." *Chemical and Pharmaceutical Bulletin* 39: 2764–2766.

Yamamoto, A., Nitta, S., Miyase, T., Ueno, A. & Li-Jun, W. 1993. "Phenylethanoid and lignan iridoid complex glycosides from roots of *Buddleja davidii*." *Phytochemistry* 32: 421–425.

Yeh, E. T., Ewer, M. S., Moslehi, J., Dlugosz-Danecka, M., Banchs, J., Chang, H. M. & Minotti, G. 2019. "Mechanisms and clinical course of cardiovascular toxicity of cancer treatment I. Oncology." *Seminars in Oncology* 46, 6 (December): 397–402. doi: 10.1053/j.seminoncol.2019.10.006.

Ying, C. & Wan, D. 2012. "Quantitative determination of total and individual flavonoids in stems and leaves of *Buddleja davidii* and *Buddleja albiflora*." *Pharmacognosy Magazine* 8, 32: 273–279. doi: 10.4103/0973-1296.103651.

Yoshida, T., Nobuhara, J., Fujii, N. & Okuda, T. 1978a. "Studies on the constituents of Buddleja species. II. Buddledin C, D and E, new sesquiterpenes from *Buddleja davidii* Franch." *Chemical and Pharmaceutical Bulletin* 26: 2543–2549.

Yoshida, T., Nobuhara, J., Uchida, M. & Okuda, T. 1978b. "Studies on the constituents of Buddleja species. I. Structures of buddledin A and B, two new toxic sesquiterpenes from *Buddleja davidii* Franch." *Chemical and Pharmaceutical Bulletin* 26: 2535–2542.

Yuan, B., Yang, R., Ma, Y., Zhou, S., Zhang, X. & Liu, Y. 2017. "A systematic review of the active saikosaponins and extracts isolated from Radix Bupleuri and their applications." *Pharmaceutical Biology* 55: 620–635.

Zhang, J., Chou, G., Liu, Z. & Liu, M. 2016. "Employing rubusoside to improve the solubility and permeability of antitumor compound betulonic acid." *Nanomedicine* 11: 2829–2844.

Zhang, X., Jin, C., Li, Y., Guan, S., Han, F. & Zhang, S. 2013. "Catalpol improves cholinergic function and reduces inflammatory cytokines in the senescent mice induced by D-galactose." *Food and Chemical Toxicology* 58: 50–55.

11 *Scutellaria* Species and Cancer Research
The Way Forward

Samantha H. Sherman, Lani Irvin, Prahlad Parajuli and Nirmal Joshee

CONTENTS

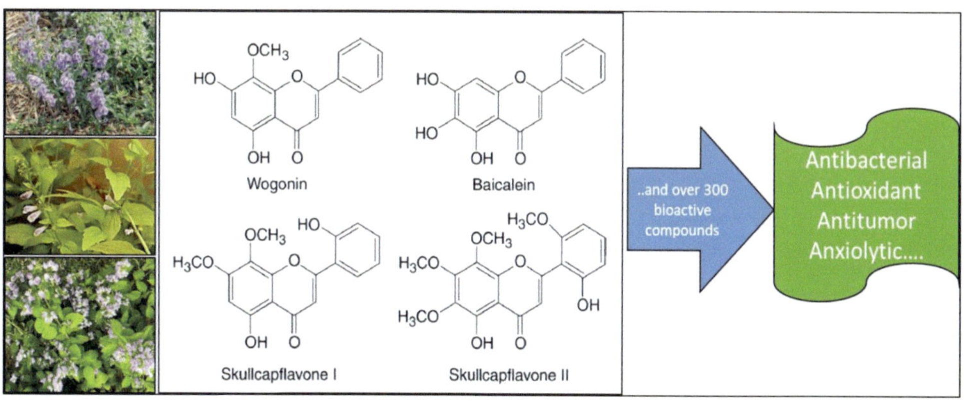

Graphical abstract representing the content of the chapter (Joshee et al., 2002).

11.1 INTRODUCTION

Cancer severely affects the human population globally, resulting in fatalities and enormous financial and psychological burden. Worldwide, cancer ranks as the second leading cause of death

DOI: 10.1201/9781003108375-11

(Nagai and Kim, 2017). A recent survey by the National Cancer Institute, USA, projected 1.8 million new cancer cases to be diagnosed worldwide in 2020; the incidence of lung, colon, breast and prostate cancers was predicted to be 22, 9, 7, and 5 respectively. The global cost of cancer care in 2020 was expected to exceed $150 billion (Sitarek et al., 2020). Although new treatment modalities—especially the advent of novel immunotherapeutic approaches—have considerably improved the clinical outcome in several hematologic cancers and some solid tumors, most tumors do not respond to these treatments or the disease recur after initial response. There is a continuous need for the development of more effective therapeutic approaches for this deadly disease (Ashraf, 2020; Shen et al., 2021).

Documentation of traditional medicine incorporating plants has triggered several in-depth investigations into alternative medications by researchers across the globe. It is estimated that bioactive secondary metabolites regulating signaling and metabolic pathways have been researched from up to 70,000 higher plant species (Vanisree et al., 2004). Plants possess phytochemicals with unlimited potential as agrochemicals, flavorings, fragrances, food additives and pharmaceuticals (Karuppusamy, 2009; Lahlou, 2013). Discovery of the anti-cancer alkaloids vinblastine and vincristine, and the isolation of cytotoxic podophyllotoxins, provided initial clues on the application of plant-based therapeutic compounds in cancer (Cragg and Newman, 2005). It was reported that 118 of the top 150 prescription drugs in the United States were traced to have their origin in a natural source, and approximately 60% of drugs used for cancer treatment have been based on compounds isolated from natural products (Chen et al., 2016; Gordaliza, 2007). Among these plants, derived compounds—such as *Taxus* diterpenes, *Camptotheca* alkaloids and *Podophyllum* lignans—are significant contributors. For example, paclitaxel derived from yew trees (*Taxus brevifolia* Nutt.) is used to treat ovarian, breast, lung, cervical and pancreatic cancers. In 2008 alone, 16 new plant-derived compounds were tested reaching phases I–III in clinical trials (Saklani and Kutty, 2008). Looking at the future therapeutic potential of *Scutellaria* compounds, *in vitro* plant tissue culture systems have been established to produce bioactive molecules (Grzegorcyzyk-Karolak et al., 2015).

Cancer starts with an accumulation of multiple mutations in oncogenes, tumor suppressor genes and apoptosis/cell death regulating genes, which provide the neoplastic cell with a capacity of uninhibited continuous proliferation (Hanahan and Weinberg, 2011). As the tumor progresses, the cancer cells acquire continuous growth signaling capability via overexpression or mutation of various growth factor receptors. This is supplemented by the induction of new blood vessels (angiogenesis) for nutrient supply, as well as for recruitment of pro-tumor inflammatory immune cells. There is a complex interaction among tumor cells and stromal components, mostly fibroblasts and immune cells, that drives a malignant phenotype (Hanahan and Weinberg, 2011).

Therapeutic interventions have targeted the molecules involved in tumor cell intrinsic proliferative pathways, as well the mechanisms driving the interactions between cancer and stromal cells (Patel et al., 2013). This chapter discusses phytochemicals derived from traditional medicinal plants *Scutellaria* spp., which may have the potential to interfere with the pathogenic factors of cancer development, progression and metastasis.

11.2 *SCUTELLARIA* SPECIES WITH POTENTIAL ANTI-CANCER COMPOUNDS

Scutellaria spp. is a member of the Lamiaceae family with approximately 400 species growing around the world. It has been used in the traditional medicinal system by the ancient Chinese, as well by the Native Americans (Shang et al., 2010). Some of the prominent *Scutellaria* species with well-documented anti-cancer activity are displayed in Table 11.1 and are discussed in what follows.

11.2.1 *Scutellaria baicalensis* Georgii

Scutellaria baicalensis (Chinese skullcap) is indigenous to East Asia, Korea, Japan, Russia and several European countries (Bochořáková et al., 2003; Bai et al., 2013) (Figure 11.1). *S. baicalensis*

TABLE 11.1
Status of Cancer Research in Clinical Trials Using *Scutellaria* Species

Study Titles	Conditions	Drugs Used	Statuses	Outcomes/Statuses	Locations	References
"Herbal Therapy in Treating Women with Metastatic Breast Cancer"	Breast cancer	Herba *Scutellaria barbata*	Completed	-	A. University of California, Comprehensive Cancer Center, San Francisco, USA B. Cancer Research Network, Incorporated, Plantation, Florida, USA	(Rugo, 2006)
"A Phase I/II Clinical Trial Assessing Safety and Efficacy of BZL101 for Metastatic Breast Cancer"	Metastatic Breast Cancer	BZL101 (Bezielle)	Completed with results	Primary: Toxicity Classified by the NCI Common Terminology Criteria Version 3 (Phase 2) [Time Frame: 2 Months] Response Evaluation Criteria in Solid Tumors (Phase 2) [Time Frame: 2 months]	-	(Hochester, 2006; Perez et al., 2010)
"Phase I Trial and Antitumor Effects of BZL101 for Patients with Advanced Breast Cancer"	Metastatic breast cancer	BZL101	Completed has results	Primary: Toxicity based upon Adverse Events Response Evaluation Criteria in Solid Tumors (Phase 1)	University of California, San Francisco, Carol Franc Buck Breast Cancer Center, USA	(Rugo et al., 2007)
"Phase I/II, Multi-Center, Open-Label, Dose-Escalation, Safety and Efficacy Study of PHY906 Plus Capecitabine in Patients with Advanced Pancreatic Carcinoma"	Pancreatic cancer	Capecitabine (Xeloda) PHY906 Other Name: traditional Chinese medicine: *Scutellaria baicalensis*, *Glycyrrhiza uralensis*, *Ziziphus jujuba*, and *Paeonia lactiflora*	Completed has results	Secondary: Median Overall Survival [Time Frame: Up to 100 weeks]	Yale Comprehensive Cancer Center, Yale University School of Medicine New Haven, Connecticut, USA	(Hochester, 2006)
"Baicalin from the Extract of *Scutellaria baicalensis* Affects the Innate Immunity and Apoptosis in Leukocytes in Children with Acute Lymphocytic Leukemia"	Lymphocytic leukemia	Baicalin	-	-	A. Ludwik Hiszfeld Institute of Immunology and Experimental Therapy (IIET), Polish Academy of Sciences, Wroclaw, Poland B. Department of Bone Marrow Transplantation, Oncology and Hematology, Medical University, Wroclaw, Poland C. Herbapol SA, WZZ, Wroclaw, Poland	(Orzechowska et al., 2014)

FIGURE 11.1 *Scutellaria* germplasm at Fort Valley State University. A, *S. montana*; B, *S. scandens*; C, *S. suffrutescens*; D, *S. lateriflora*; E, *S. ocmulgee*; F, *S. baicalensis*; G, *S. albida*; H, *S. elliptica*.

roots are used as herbal raw material in Korean, Japanese and Chinese pharmacopoeia (Shang et al., 2010). Dried roots of this plant, containing mainly flavonoids, are used to make Huang-Qin (golden herb) in traditional Chinese medicine and are used for the treatment of several health conditions, including diarrhea, tumors and inflammation (Han et al., 2007). Lately, the therapeutic role of *S. baicalensis* has undergone extensive evaluation for its anti-cancer activities (Scheck et al., 2006; Chiu et al., 2011; Park et al., 2011; Kim et al., 2013; Cheng et al., 2018).

Studies have been carried out to examine the anti-cancer activity of *S. baicalensis* via possible inhibition of prostaglandin E_2 (PGE_2) produced from arachidonic acid through cyclooxygenase-2 (COX-2) pathway. A dose-dependent growth inhibition assay using various cancer cell lines including MCF-7 (breast cancer), squamous cell carcinoma (SCC-25, KB), colon cancer (KM-12, HCT-15), hepatocellular carcinoma (HepG2) and prostate cancer (LNCaP, PC-3) was observed, with more prominent growth inhibition in prostate and breast cancer along with a negative effect on COX-2 activity (Ye et al., 2002). Another study performed on head and neck squamous cell carcinoma (HNSCC) reported significant *in vitro* and *in vivo* anti-cancer activity of *S. baicalensis* plant extract, along with a selective (celecoxib) and non-selective (indomethacin) inhibitor of COX-2. On the other hand, the isolated flavonoid baicalein from the plant has been reported to not show any tumor growth inhibition, suggesting a synergistic effect of different components present in the whole plant extract (Zhang et al., 2003).

Anti-cancer activity of *S. baicalensis* root extract has been studied on cell lines grown from the primary and recurrent glioblastoma multiforme (GBM), either alone or in combination with a chemotherapeutic agent 1,3-bis(2-chloroethyl)-1-nitrosourea (BCNU) (Scheck et al., 2006). Similarly, *S. baicalensis* extract and its major constituents (wogonin, baicalein, neobaicalein, skullcap flavone) efficiently inhibited the growth of prostate cancer cell lines LNCaP and PC-3, partly by inhibiting the androgen receptor (Bonham et al., 2005).

11.2.2 *Scutellaria barbata* D.Don

Scutellaria barbata is native to China and is known as Ban-Zhi-Lian in traditional Chinese medicine (TCM), where it is used for its diuretic, anti-cancer and anti-inflammatory activities (Yu et al., 2004). According to the Taiwan National Health Insurance Research Database, *S. barbata* is one of the most commonly used TCM-based herbs for the treatment of breast cancer (Yeh et al., 2014). The prominent phytochemicals identified in *S. barbata* include various flavonoids, alkaloids, essential oils, terpenoids and polysaccharides. Of these, flavonoids rank as the most important component of *S. barbata*, as these compounds contain antioxidant, anti-cancer and anti-inflammatory functions

(Gao et al., 2019). Studies suggest that flavonoids such as scutellarin and carthamidin serve as the main anti-cancer effector component of *S. barbata* aqueous extract, BZL101 (FDA IDN# 59521), which showed promising inhibitory activity against metastatic breast cancer in a phase I/II clinical trials (Rugo et al., 2007; Fong et al., 2008; Chen et al., 2012). The overall BZL101 efficacy is determined by the combined activity of its components acting synergistically toward selective cytotoxicity against cancer cells (Chen et al., 2012).

Similarly, *S. barbata* plant extract showed inhibitory effect against various cancers, including lung, breast, digestive system and liver cancers. However, the exact mechanism of its action remains unclear. A study evaluating the tumor inhibitory effect of ethanolic extracts of *S. barbata* on lung cancer cell line A549 reported induction of apoptosis and cytotoxicity following treatment with the extract. A cDNA microarray analysis showed a five-fold change in expression levels of 16 genes involved in DNA damage, cell cycle control, nucleic acid binding and protein phosphorylation that directly correlated with the cytolytic activity of *S. barbata*–derived compounds. An interesting finding was a 102-fold downregulation of CD209 (DC-SIGN), a molecule related to inhibition of dendritic cell (DC) function, indicating potential anti-cancer immune modulatory activity of *S. barbata*. Further functional studies in this area would elucidate the essential role of these genes in *S. barbata*–associated cytotoxic activity (Hossain and Wall, 2019; Yin et al., 2004).

S. barbata has a long history of use in Chinese medicine in the treatment of various liver diseases. A study evaluated anti-cancer activity of isolated active components of *S. barbata* on Hep3B hepatoma cell line in comparison to normal human liver cells WRL-68. The results showed that, among various constituents of *S. barbata*, pheophorbide induced sub-G1 cell cycle arrest and DNA fragmentation while suppressing anti-apoptotic protein Bcl-2, leading to specific inhibition of Hep3B cell proliferation (Chan et al., 2006). Another study reported isolation and identification of compounds named scutebarbatines from the *S. barbata* plant extract. These compounds showed significant cytotoxic activity against the cancer cell lines human nasopharyngeal carcinoma (HONE-1), oral epidermoid carcinoma (KB) and colorectal carcinoma (HT29) (Dai et al., 2006a). Similarly, a major alkaloid, scutebarbatine-A (SBT-A), isolated from *S. barbata*, has been shown to induce apoptosis in a non–small cell lung cancer (NSCLC) cell line known as A549. SBT-A upregulated the expression of caspase 3, 9 and cytochrome c, while down-regulating Bcl-2, as well as significantly inhibiting the growth of A549 tumor, transplanted in nude mice, via mitochondria-mediated apoptosis (Yang et al., 2014).

11.2.3 *Scutellaria lateriflora* L.

Scutellaria lateriflora, or American skullcap, is a perennial plant native to North America and has been used in treating kidney problems, digestive tract conditions, menstrual ailments, epilepsy, anxiety and insomnia (Figure 11.1). This plant contains a large number of flavonoids like wogonin, baicalin, and leteriflorin (Nishikawa et al., 1999; Gafner et al., 2000). The whole extract, as well as the phytochemicals present in *S. lateriflora*, showed strong anti-cancer activity against cancer cell lines, including U87-MG and human malignant glioma (U251), malignant breast cancer (MDA-MB-231) and human prostate cancer (PC3) (Parajuli et al., 2009).

Baicalin, a flavonoid isolated from the *S. lateriflora*, has been shown to suppress migration and invasion of an aggressive breast cancer cell line, MDA-MB-231, without any effect on cancer cell viability. Baicalin inhibits expression of markers for epithelial to mesenchymal transition (EMT), a phenomenon associated with metastatic events. In addition, *in vivo* study in xenograft mouse model illustrated that baicalin inhibits breast cancer metastasis to the lungs and liver (Zhou et al., 2017).

Another compound, scutellarein, demonstrated significant suppression of human fibrosarcoma cells (HT1080) by inducing apoptosis, while in *in vivo* experiments using Balb/c mice exhibited a decrease in tumor size and volume. Furthermore, scutellarein showed anti-metastatic activity as depicted by reduced migration and invasion of HT1080 cells, besides suppression of matrix metalloproteinase (MMP)-2, -9 and -14 (Shi et al., 2015).

11.2.4 *Scutellaria lindbergii* Rech.f.

Scutellaria lindbergii is a lesser studied skullcap that is reported to grow in Iran and surrounding areas. Methanolic extracts and fractions were tested on human Caucasian gastric adrenocarcinoma (AGS), human cervix epithelioid carcinoma (HeLa), human Caucasian breast adenocarcinoma (MCF-7) and rat adrenal phaeochromocytoma (PC12) cancer cell lines for cytotoxic properties. As detected by MTT assay, *S. lindbergii* extract inhibited the growth of malignant cells in a dose-dependent manner. Flow cytometric analysis of treated cells showed a sub-G1 peak indicating apoptotic cell death in tumor cells, while non-malignant cells were unaffected. Thus, *S. lindbergii*–derived active compounds may add to the list of potential chemotherapeutic agents in cancer treatment (Tayarani-Najaran et al., 2010, 2011, 2012a).

11.2.5 *Scutellaria litwinowii* Bornm. & Sint. ex Bornm.

Scutellaria litwinowii has been a part of traditional Iranian and Central Asian medicine (TICAM). Research on the cytotoxic activity of this plant on various cancer cell lines, including AGS, PC12, MCF-7 and HeLa exhibited a dose-dependent decrease in viability of the cell lines upon treatment with the root extract of *S. litwinowii* (Tayarani-Najaran et al., 2011). Two compounds, neobaicalein and wogonin, were isolated from the root extract of *S. litwinowii*, which showed dose-dependent cytotoxicity in HeLa cells, while normal lymphocytes were not killed (Tayarani-Najarani et al., 2012a). A study using *S. litwinowii* extracts on two myeloid cell lines, HL60 and K562, induced apoptosis in both types of cells, suggesting its potential use in various hematologic malignancy (Tayarani-Najaran et al., 2012b).

11.2.6 *Scutellaria ocmulgee* Small

Scutellaria ocmulgee is a native plant of North America, restricted to the states of Georgia and South Carolina (Figure 11.1). These plants are extremely rare and are placed under threatened status in Georgia. Following extensive studies on the anti-cancer activities of the leaf, stem and root extracts of 13 *Scutellaria* species, the leaf extracts of *S. ocmulgee* (SocL) demonstrated the highest anti-proliferative and apoptosis induction with no toxicity against non-cancerous cells derived from the same tissues in various breast and prostate cancers, which were tested on glioma cell lines (Parajuli et al., 2009). Subsequent studies showed that SocL extract inhibited glioma growth *in vitro* as well as *in vivo*, in rat models, via inhibition of AKT/GSK-3b/NF-κB signaling pathways (Parajuli et al., 2011).

Moreover, SocL extract inhibited induction/expansion of regulatory T cells in the tumor microenvironment via inhibition of transforming growth factor beta (TGF-β) secretion and signaling pathways (Dandawate et al., 2012). These results showed that the phytochemicals in *Scutellaria* could modulate tumor cell intrinsic signaling pathways while interfering with the pro-tumor interactions between the tumor and immune cells. Recent studies have demonstrated dose-dependent inhibition of proliferation and metastatic capacity of NSCLC cells and patient-derived brain-metastatic lung cancer cells *in vitro*, as well in zebrafish embryo xenograft models. The anti-tumor and anti-metastatic activity of SocL extract was associated with an inhibition of MMPs and activation of p38 MAPK signaling (Wright and Shahin et al., 2021).

Scutellaria flavonoids have been reported to have differential effects in the central nervous system, depending on the cell type and disease indicated, in a study that was reviewed by Patel et al. (2013). *Scutellaria* flavonoids inhibited the production of reactive oxygen species (ROS) in injured microglia and protect from neurotoxicity. On the other hand, the flavonoids have previously shown to enhance ROS production in malignant gliomas, leading to endoplasmic reticulum stress-related protein activation and caspase-mediated apoptosis. However, ROS production in normal astrocytes remains unaffected by flavonoids (Patel et al., 2013). Overall, it seems that using the active components of *Scutellaria*—in concert with modern medicines—may increase the efficacy of treatment for malignant gliomas.

11.2.7 *Scutellaria pinnatifida* A. Hamilt.

Scutellaria pinnatifida is native to Iran and it has been widely used in TCM (Mozaffarian, 1996). Few reports exist on the identification of phytochemicals of *S. pinnatifida* and its function. A recent report describes the isolation of phytochemicals from the *S. pinnatifida* including one phenylpropanoid, 1-*O*-feruloyl-β-D-glucose; two known flavonoids, luteolin and apigenin; and three known phenylethanoid glycosides: verbascoside, syringalide, and oleic acid. These compounds depicted free radical scavenging activity, especially with the methanol extract, while extract prepared in n-hexane showed dose-dependent insecticidal activity (Delazar et al., 2017). Like other members of the Lamiaceae family, *S. pinnatifida* possesses important flavonoids with cytotoxic activity, including neobaicalein (skullcapflavone II) and wogonin. Based on this knowledge, researchers later evaluated the cytotoxic activity of *S. pinnatifida* in two cancer cell lines human erythroleukemic cell line (K562) and HL-60, that revealed a significant decrease in the cell viability of malignant cell lines, with the dichloromethane plant extract showing higher cytotoxic activity compared to the n-hexane extract (Boozari et al., 2015).

11.3 *SCUTELLARIA* PHYTOCHEMICALS WITH ANTI-CANCER ACTIVITY

Flavonoids are one of the most important bioactive components in the root of *S. baicalensis*. Significant correlation has been reported between the dietary intake of flavonoids and a reduced risk of cancer, inflammation and heart disease (Li-Weber, 2009). Over 50 flavonoids have been isolated with baicalein as one of four major ones due to its wide spectrum of biological/physiological roles related to anti-cancer, anti-inflammation, anti-cardiovascular diseases, anti-bacterial or viral infection, and anti-neurodegenerative diseases (Kim et al., 2013). Studies have demonstrated that baicalein suppresses cancer cell proliferation and induces apoptosis and cell cycle arrest in human breast, prostate, hepatocellular, myeloma and T24 bladder cancer cells (Li et al., 2013). Baicalein inhibited migration and invasion of human breast cancer cells, hepatoma cells and skin cancer cells (Chiu et al., 2011; Wu et al., 2011).

Terpenoids are another class of phytochemicals with potential use in the prevention and therapy of cancer (Huang et al., 2012). Along with flavonoids, terpenoids constitute a major component of *Scutellaria* phytochemicals, and their cancer-specific cytotoxic activity has been demonstrated using various cancer cell lines and non-malignant cell controls (reviewed in Gao et al., 2019). Terpenes with anti-cancer activity have been reported from *Scutellaria havanensis* and *S. orientalis* subsp. Bornmuelleri (Table 11.2) (Delange et al., 2013; Gharari et al., 2021).

11.3.1 Anti-Tumor Molecular Mechanisms of *Scutellaria* Compounds

The anti-cancer properties of *Scutellaria* have mostly been attributed to its flavonoid and terpenoid content. While its antioxidant activity can be associated with potential prevention of cancer, the therapeutic benefits could most be attributed to its modulation of intracellular signaling pathways. *Scutellaria* flavonoids have been shown to have cytotoxic activity in cancer cells while sparing the normal epithelial, lymphoid/myeloid cells and astrocytes. This tumor selectivity is achieved by targeting signaling pathways that are specifically activated or over-activated in cancer cells. For example, the PI3K-AKT-GSK3β pathway is constitutively active in several tumor types. Research has shown that phytochemicals in SocL extract could directly bind to and inhibit the kinase activity of AKT, thereby modulating the downstream of GSK-3β activity in gliomas (Parajuli et al., 2011).

TABLE 11.2

Phytochemical Constituents of *Scutellaria* Species Tested for Anti-Cancer Activity on Cancer and Non-Cancer Cell Lines

No.	*Scutellaria* Species	Phytochemical Constituents	Formulations	Types of Cancer/Cell Lines Treated	References
1	*S. baicalensis*	Baicalin, baicalein, skullcap flavone, saikosaponins, scutellarin, scutellarien, esculin, scoparone, apigenincrysin, api-glcA, melatonin, serotonin, wogonoside, wogonin	Huang quin, Xiao-chai-hu-tang, and Sho-saiko-to/ TJ-9, PC-SPES	Prostate cancer PC-3[1], DU145[2], breast, gastric, liver, colon, pancreatic, human hepatocellular cancer, head and neck squamous carcinoma, human T- lymphoid leukemia, glioblastoma multiforme, 00wa[3], HL60[4], and LNCaP[5]	(Maher 2001; Zhang et al., 2003; Sonoda et al., 2004; Bonham et al., 2005; Horvath et al., 2005; Li and Khan 2006; Makino et al., 2006; Scheck et al., 2006; Ye et al., 2007; Cole et al., 2008)
2	*S. alpina*	Non clerodane diterpenoids: Scutalpin O. Scutalpin N, scutecyprol A, scutorientalin	Stem and leaf extract	-	(Malakov and Papanov, 1998)
3	*S. lateriflora*	Baicalin, baicalein, wogonin, melatonin, serotonin, viscidulin III-2'-O-glc, Chyrin-6-C-ara-glc, *trans*-Verbascoside, viscidulin, *trans*-Martynoside, Oroxylin A-7-O-glc, wogonoside, chitin, oroxylin A, scutellarin	Stem and leaf extract	-	(Nishikawa et al., 1999; Cole et al., 2008; Zhang et al., 2009)
4	*S. galecriculata*	Phenylethanoid glycosides: 2-(4-hydroxyphenyl)-ethyl-(6–*O*-caffeoyl)-β-D-glucopyanoside, calceolarioside B, osmanthuside E, martynoside	Stem and leaf extract	-	(Ersoz et al., 2002)
5	*S. barbata*	Scutebarbatine A, Scutebarbatine B, scutellarin, barbatins A-C, naringin, scutellarein, luteolin, apigenin, wogonon, pheophorbide a, 6–*O*-nictinoylscutebarbatine	Ban-Zhi-Lian	Ovarian cancer (A2780), human placental microsomes, myometrial and leiomyomal cells, HONE-1[6] nasopharangeal, KB[7] oral epidermoid carcinoma, and HT29[8] colorectal carcinoma, hepatoma H22 cells, HepG2[9] (HB-8065), Hep3B[10] (HB-8064), Calu-3 xenograft tumors, human hepatocarcinoma MHCC97H, C57BL/6 colorectal cancer, HT-29, human lung carcinoma A549, Human NSCLC[11], HI975[12], RAW264.7[13], Lewis lung carcinoma	(Powell et al., 2003; Lee et al., 2004; Chan et al., 2006; Dai et al., 2006a, 2006b, 2009; Dai et al., 2011, 2013; Wei et al., 2012; Yang et al., 2014; Yang et al., 2015; Kan et al., 2017; Liu et al., 2018)
6	*S. ocmulgee*	Apigenin, luteolin, scutapins, diterpenoids, neo-clerodanes, chrysin, iridoids, isoscutellarein, baicalin, baicalein, wogonin	Stem and leaf extract	F98[14] glioma in F344 rats	(Parajuli et al., 2009, 2011; Dandawate et al., 2012; Vaidya et al., 2013)

7	*S. ramosissima*	Scutellarin, chrysin, apigenin, apigenin-7-O-glucoside, cymaroside, pinocembrine	Stem, leaf, and root extract	HeLa[15], HepG-2, MCF-7[16]	(Mamadalieva et al., 2011)
8	*S. immaculata*	Scutellarin, chrysin, apigenin, apigenin-7-O-glucoside, cymaroside, pinocembrine	Stem, leaf, and root extract	HeLa, HepG-2, MCF-7	(Mamadalieva et al., 2011)
9	*S. litwinowii*	Wogonin, skullcapflavoneII, 6-hydroxyfavone, neobaicalein	Root extract	HeLa, PC12[17]	(Tayarani-Najaran et al., 2011; Afsharzadeh et al., 2012; Tayarani-Najaran et al., 2012b)
10	*S. incana*	Apigenin, Oroxylin, Chrysin Baicalein, Baicalin, Wogonoside, Wogonin, Vitexin, Isovitexin, Acetoside, Isoacetoside, Norwogonin, Scutellarien, Scutellarin	Stem, leaf, and root extract	-	(Islam et al., 2013)
11	*S. pinnatifida*	Neobaicalein (skullcapflavoneII), wogonin	Root extract	HL-60 and K562[18]	(Boozari et al., 2015)
12	*S. altissima*	Baicalin, wogonoside, luteolin, luteolin-7-gluconoside, verbascoside	Stem and leaf extract	-	(Grzegorcyzyk-Karolak et al., 2015)
13	*S. discolor*	Chrysin	Plant extract	HeLa	(Laishram et al., 2015)
14	*S. platystegia*	2(4- hydroxy phenyl) ethyl-O-β-D-glucopyranoside, apigenin 7-O-glucoside, verbascoside, martynoside	Stem and leaf extract	-	(Mandavi mousavi et al., 2015)
15	*S. colbrookiana*	Baicalein	Root extract	DLA[19] and EAC[20]	(Salini et al., 2013)
16	*S. violacea*	Baicalein	Root extract	DLA and EAC	(Salini et al., 2013)
17	*S. luteo-coerulea*	-	Stem and leaf extract	Human leukemia cancer cells (K562 and HL-60)	(Motaez et al., 2015)
18	*S. lindbergii*	-	Root extract	AGS[21], HeLa, MCF-7, PC12	(Tayarani-Najaran et al., 2010)
19	*S. orientalis*	-	Root extract	Metastatic breast cancer	(Ozmen et al., 2010)
20	*S. rivularis*	-	Plant extract	HL-60	(Sonoda et al., 2004)

Notes: **1** Prostate cancer; **2** Human prostate cancer cell line; **3** GBM tumor cell line, **4** Human caucasian promyelocytic leukemia cell line; **5** Androgen-sensitive human prostate adenocarcinoma cells; **6** Epithelial tumor cells; **7** Oral epidermal carcinoma; **8** Human colorectal adenocarcinoma; **9** Human hepatocellular carcinoma; **10** Hepatitis B containing hepatocellular carcinoma cell line; **11** Human non–small cell lung cancer; **12** Lung adenocarcinoma; **13** Mouse monocyte macrophage; **14** Rat gliomal cells; **15** Cervical cancer; **16** Human breast cancer; **17** Prostate cancer; **18** Mylogeneous leukemia; **19** Dalton's lymphoma Ascites; **20** Ehrlich ascites carcinoma; **21** Adenocarcinoma

11.3.2 *Scutellaria*-Based Formulations for Cancer Patients

Sho-saiko-to is a mixture of seven herbs, including *S barbata*, that has been traditionally used in Japan and China for the treatment of chronic hepatitis and liver fibrosis (Shimizu, 2000). Sho-saiko-to has been used for the treatment of liver cancer. A clinical study has been conducted to study the efficacy of Sho-saiko-to following removal of liver cancer by embolization in patients with surgically unresectable liver cancer; however, the result has not been published (DeMatteo, 2003).

BZL101/Bezielle is an aqueous extract from the herb *S. barbata*, which has undergone Phase 1B clinical studies for patients with metastatic breast cancer and its safety was determined at a dose of 40 g/day (Perez et al., 2010). Subsequent Phase I/II clinical studies showed encouraging results (1/7 stable disease, 3/7 progressive disease and 3/7 unknown), with nausea and diarrhea being major adverse events (Perez et al., 2010).

In 2017, the FDA issued a warning about potential hepatotoxicity and hypersensitivity pneumonitis conditions linked to Limbrel, a commercial supplement for osteoarthritis containing baicalin and catechin, which are isolated from *S. baicalensis* and *Acacia catechu* Lf., respectively (FDA, 2017). Although not related to cancer treatment, this incidence is a learning experience regarding necessary caution to adopt while designing therapeutics containing a combination of phytochemicals isolated from various sources. Adequate toxicity and efficacy studies need to be performed before bringing plant-based therapeutics to the patients. Historically and worldwide, *Scutellaria* has retained recognition as a powerful species of plant through its innate medicinal qualities. Further investigations of its so-far-untapped potential will enable current and future science to positively impact global health (Chen et al., 2016).

11.4 FUTURE PROSPECTS AND CONCLUSIONS

In conclusion, *Scutellaria* extracts and isolated phytochemicals have been shown to inhibit the growth of various types of cancer *in vitro* and in some *in vivo* studies, via induction of intracellular and intercellular anti-tumor mechanisms. Although the therapeutic efficacy of *Scutellaria* as a monotherapy against cancer in animal and human studies has been modest, it may have more potent effect if used in combination with standard radiation, chemotherapy or immunotherapy regimen. More *in vivo* studies in animal models and human clinical trials are needed to test the usefulness of *Scutellaria* extracts and phytochemicals in such combination therapies against cancer and other disorders. Future studies on the *in vitro* production of desirable active compounds and *in vivo* studies on the utilization of nanotechnology to improve their bioavailability will be of great interest.

REFERENCES

Afsharzadeh, M., Z. Tayarani-Najaran, A. Zare, and S. H. Mousavi. 2012. "Protective effect of *Scutellaria litwinowii* extract on serum/glucose-deprived cultured PC12 cells and determining the role of reactive oxygen species." *Journal of Toxicology* 413279. doi: 10.1155/2012/413279

Ashraf, M. A. 2020. "Phytochemicals as potential anti-cancer drugs: Time to ponder nature's bounty." *BioMed Research International* 1–7. doi: 10.1155%2F2020%2F8602879

Bai, C., M. Wen, L. L. Zhang, and G. Li. 2013. "Genetic diversity and sampling strategy of *Scutellaria baicalensis* germplasm resources based on ISSR." *Genetic Resources Crop Evolution* 60, 5: 1673–1685. doi: 10.1007/s10722-012-9949-9

Bochořáková, H., H. Paulová, J. Slanina, P. Musil, and E. Taborska. 2003. "Main flavonoids in the root of *Scutellaria baicalensis* cultivated in Europe and their comparative antiradical properties." *Phytotherapy Research* 17, 6: 640–644. doi: 10.1002/ptr.1216

Bonham, M., J. Posakony, I. Coleman, B. Montgomery, J. Simon, and P. S. Nelson. 2005. "Characterization of chemical constituents in *Scutellaria baicalensis* with antiandrogenic and growth-inhibitory activities toward prostate carcinoma." *Clinical Cancer Research* 11: 3905–3914. doi: 10.1158/1078-0432.CCR-04-1974

Boozari, M., A. Mohammadi, J. Asili, S. A. Emami, and Z. Tayarani-Najaran. 2015. "Growth inhibition and apoptosis induction by *Scutellaria pinnatifida* A. Ham. on HL-60 and K562 leukemic cell lines." *Environmental Toxicology Pharmacology* 39, 1: 307–312. doi: 10.1016/j.etap.2014.12.002

Chan, J. Y., P. M. Tang, P. Hon, S. W. Au, S. K. Tsui, M. M. Waye, S. Kong, T. C. Mak, and K. Fung. 2006. "Pheophorbide a, a major antitumor component purified from *Scutellaria barbata*, induces apoptosis in human hepatocellular carcinoma cells." *Planta Medica* 72, 1: 28–33. doi: 10.1055/s-2005-873149

Chen, S., H. Yu, H. Luo, Q. Wu, C. F. Li, and A. Steinmetz. 2016. "Conservation and sustainable use of medicinal plants: Problems, progress, and prospects." *Chinese Medicine* 11: 37. doi: 10.1186/s13020-016-0108-7

Chen, V., R. E. Staub, S. Baggett, R. Chimmani, M. Tagliafferi, I. Cohen, and E. Shtivelman. 2012. "Identification and analysis of the active phytochemicals from the anti-cancer botanical extract Bezielle." *PloS One* 7, 1: e30107. doi: 10.1371/journal.pone.0030107

Cheng, C.-S., J. Chen, H.-Y. Tan, N. Wang, Z. Chen, and Y. Feng. 2018. "*Scutellaria baicalensis* and cancer treatment: Recent progress and perspectives in biomedical and clinical studies." *The American Journal of Chinese Medicine* 46, 1: 25–54. doi: 10.1142/s0192415x18500027

Chiu, Y. W., T. H. Lin, W. S. Huang, C. Y. Teng, Y. S. Liou, W. H. Kuo, W. L. Lin, H. I. Huang, J. N. Tung, C. Y. Huang, J. Y. Liu, W. H. Wang, J. M. Hwang, and H. C. Kuo. 2011. "Baicalein inhibits the migration and invasive properties of human hepatoma cells." *Toxicology and Applied Pharmacology* 255: 316–326. doi: 10.1016/j.taap.2011.07.008

Cole, I. B., J. Cao, A. R. Alan, P. K. Saxena, and S. J. Murch. 2008. "Comparisons of *Scutellaria baicalensis, Scutellaria lateriflora*, and *Scutellaria racemosa*: Genome size, antioxidant potential and phytochemistry." *Planta Medica* 74: 474–481. doi: 10.1055/s-2008-1034358

Cragg, G. M., and D. J. Newman. 2005. "Plants as a source of anti-cancer agents." *Journal of Ethnopharmacology* 100, 1–2: 72–79. doi: 10.1016/j.jep.2005.05.011.

Dai, S. J., M. Chen, K. Liu, Y. T. Jiang, and L. Shen. 2006a. "Four new neo-clerodane diterpenoid alkaloids from *Scutellaria barbata* with cytotoxic activities." *Chemical and Pharmaceutical Bulletin* 54, 6: 869–872. doi: 10.1248/cpb.54.869

Dai, S. J., W. B. Peng, L. Shen, D. W. Zhang and Y. Ren. 2009. "Two new neo-clerodane diterpenoid alkaloids from *Scutellaria barbata* with cytotoxic activities." *Journal of Asian Natural Products Research* 11, 5: 451–456. doi: 10.1080/10286020902845652

Dai, S. J., J. Y. Tao, K. Liu, Y. T. Jiang, and L. Shen. 2006b. "Neo-Clerodanediterpenoids from *Scutellaria barbata* with cytotoxic activities." *Phytochemistry* 67: 1326. doi: 10.1080/14786419.2018.1514399

Dai, Z. J., J. Gao, Z. F. Li, Z. Z. Ji, H. F. Kang, H. T. Kang, H. T. Guan, Y. Diao, B. F. Wang, and X. J. Wang. 2011. "*In vitro* and *in vivo* antitumor activity of *Scutellaria barbata* extract on murine liver cancer." *Molecules* 16: 4389–4400. doi: 10.3390%2Fmolecules16064389

Dai, Z. J., B. F. Wang, W. F. Lu, Z. D. Wang, X. B. Ma, W. L. Min, H. F. Kang, X. J. Wang, and W. Y. Wu. 2013. "Total flavonoids of *Scutellaria barbata* inhibit invasion of hepatocarcinoma via MMP/TIMP *in vitro*." *Molecules* 18: 934–950. doi: 10.3390/molecules18010934

Dandawate, S., L. Williams, N. Joshee, A. M. Rimando, S. Mittal, A. Thakur, L. G. Lum, and P. Parajuli. 2012. "*Scutellaria* extract and wogonin inhibit tumor mediated induction of T_{reg} cells via inhibition of TGF-β1 activity." *Cancer Immunology Immunotherapy* 61, 5: 701–711. doi: 10.1007%2Fs00262-011-1130-3

Delange, D. M., C. L. M. Rico, V. G. Canavaciolo, E. A. R. Leyes, and R. S. Pérez. 2013. "Volatile constituents from leaves of endemic *Scutellaria havanensis* Jacq. in Cuba." *Journal of Essential Oil Bearing Plants* 16, 3: 368–371, doi: 10.1080/0972060X.2013.794045

Delazar, A., H. Nazemiyeh, F. H. Afshar, N. Barghi, S, Esnaashari, and P. Asgharian. 2017. "Chemical compositions and biological activities of *Scutellaria pinnatifida* A. Hamilt aerial parts." *Research in Pharmaceutical Science* 12, 3: 187. doi: 10.4103%2F1735-5362.207199

DeMatteo, R. 2003. "Sho-saiko-to following removal of liver cancer by embolization in treating patients with liver cancer that cannot be surgically removed." https://clinicaltrials.gov/ct2/show/NCT00040898 (Accessed: 3 May 2021).

Ersoz, T., D. Tasdemir, and I. Calis. 2002. "Phenylethanoid glycosides from *Scutellaria galericulata*." *Turkish Journal of Chemistry* 26: 465–471.

FDA. 2017. "FDA alerts consumers about potentially life-threatening health problems linked to Limbrel." www.fda.gov/food/alerts-advisories-safety-information/fda-alerts-consumers-about-potentially-life-threatening-health-problems-linked-limbrel (Accessed: 3 May 2021).

Fong, S., M. Shoemaker, J. Cadaoas, A. Lo, W. Liao, M Taglieferri, I. Cohen, and E. Shtivelmen. 2008. "Molecular mechanisms underlying selective cytotoxic activity of BZL101, an extract of *Scutellaria barbata* towards breast cancer cells." *Cancer Biology and Therapy* 7, 4: 577–586. doi: 10.4161/cbt.7.4.5535

Gafner, S., L. L. Batcha, C. Bergeron, and D. J. Carrier. 2000. "Comparison of different extracts of *Scutellaria lateriflora* by HPLC." *International Congress and 48th Annual Meeting of the Society of Medicinal Plant Research*, Switzerland.

Gao, J., W. Yin, and O. Corcoran. 2019. "From *Scutellaria barbata* to BZL101 in cancer patients: Phytochemistry, pharmacology, and clinical evidence." *Natural Product Communication* 14, 10: 1934578X19880645.

Gharari, Z., M, Aghajanzadeh., and A, Sharafi. 2021. "*Scutellaria orientalis* subsp. Bornmuelleri: Phytochemical composition and biological activities." *Natural Product Research*: 1–6. doi: 10.1080/14786419.2021.1875463

Gordaliza, M. 2007. "Natural products as leads to anti-cancer drugs." *Clinical and Translational Oncology* 9, 12: 767–776. doi: 10.1007/s12094-007-0138-9

Grzegorcyzyk-Karolak, I., L. Kuzma, and H. Wyokinska. 2015. "Study on the chemical composition and antioxidant activity of extracts from shoot culture and regenerated plants of *S. altissima*." *Acta Physiologiae Plant* 37: 1736. doi: 10.1007/s11738-014-1736-0

Han, J., Ye, M., Xu, M., Sun, J., Wang, B., and Guo, D. 2007. "Characterization of flavonoids in the traditional Chinese herbal medicine-Huang Qin by liquid chromatography coupled with electrospray ionization mass spectrometery." *Journal of Chromatography B: Analytical Technologies in the Biomedical and Life Sciences* 848, 2: 355–362. doi: 10.1016/j.jchromb.2006.10.061

Hanahan, D., and R. A. Weinberg. 2011. "Hallmarks of cancer: The next generation." *Cell* 144: 646. doi: 10.1016/j.cell.2011.02.013

Hochester, H. 2006. "Phase I/II, Multi-center, open-label, dose-escalation, safety and efficacy study of PHY906 plus Capecitabine in patients with advanced pancreatic carcinoma" https://clinicaltrials.gov/ct2/show/NCT00411762 (Accessed: 3 May 2021).

Horvath, C. R., P. A. Martos, and P. K. Saxena. 2005. "Identification and quantification of eight flavones in root and shoot tissues of the medicinal plant Huang-qin (*Scutellaria baicalensis* Georgi) using high performance liquid chromatography with diode array and mass spectrometric detection." *Journal of Chromatography A* 1062: 199–207. doi: 10.1016/j.chroma.2004.11.030

Hossain, M. K., and K. A. Wall. 2019. "Use of dendritic cell receptors as targets for enhancing anti-cancer immune responses." *Cancers (Basel)* 11, 3: 418. doi: 10.3390/cancers11030418

Huang, M., J.-J. Lu., M.-Q. Huang., J.-L. Bao., X.-P. Chen, and Y.-T. Wang. 2012. "Terpenoids: Natural products for cancer therapy." *Expert Opinion on Investigational Drugs* 21, 12: 1801–1818. doi: 10.1517/13543784.2012.727395

Islam, M. N., F. Downey, and C. K. Y. Ng. 2013. "Comprehensive profiling of flavonoids in *Scutellaria incana* L. using LC-Q-TOF-MS." *Acta Chromatographica* 25, 3: 1–24. doi: 10.1556/AChrom.25.2013.3.11

Joshee, N, T. S. Patrick, Rao, S. Mentreddy, and A. K. Yadav. 2002. "Skullcap: Potential medicinal crop." In: J. Janick and A. Whipkey (eds.). *Perspectives on New Crops and New Uses*. American Society for Horticultural Science Press, Alexandria, VA: 581–587.

Kan, X., W. Zhang, R. You, Y. Niu, J. Guo, and J. Xue. 2017. "*Scutellaria barbata* D, Don inhibits the tumor growth through down regulating of T_{regg} cells and manipulating Th1/Th17 immune response hepatoma H22 bearing mice." *BMC Complementary and Alternative Medicine* 17: 41. doi: 10.1186%2Fs12906-016-1551-9

Karuppusamy, S. 2009. "A review on trends in production of secondary metabolites from higher plants by *in vitro* tissue, organ and cell cultures." *Journal of Medicinal Plant Research* 3, 13: 1222–1239. doi: 10.5897/JMPR.9000026

Kim, D. H., M. A. Hossain, Y. J. Kang, J. Y. Jang, Y. J. Lee, E. Im, J. H. Yoon, H. S. Kim, H. Y. Chung, and N. D. Kim. 2013. "Baicalein, an active component of *Scutellaria baicalensis* Georgi, induces apoptosis in human colon cancer cells and prevents AOM/DSS-induced colon cancer in mice." *International Journal of Oncology* 43: 1652–1658. doi: 10.3892/ijo.2013.2086

Lahlou, M. 2013. "The success of natural products in drug discovery." *Journal of Pharmacy Pharmacology* 4: 17–31. doi: 10.4236/pp.2013.43A003

Laishram, S., D. S. Moirangthem, J. C. Borah, B. C. Pal, P. Suman, S. K. Gupta, M. C. Kalita, and K. N. Talukdar. 2015. "Chrysin rich *Scutellaria discolor* colebr. induces cervical cancer cell death via induction of cell cycle arrest and caspase-dependent apoptosis." *Life Science* 14: 105–113. doi: 10.1016/j.lfs.2015.10.035

Lee, T. K., D. H. Kim, J. Y. Han, and C. H. Kim. 2004. "Inhibitory effects of *Scutellaria barbata* D. Don. and *Euonymus alatus* Sieb. On aromatase activity of human leiomyomal cells." *Immunopharmacology and Immunotoxicology* 23, 3: 315–327. doi: 10.1081/iph-200026840

Li, H. L., S. Zang, Y. Wang, R. R. Liang, J. Li, P. An, Z. M. Wang, J. Yang, and Z. F. Li. 2013. "Baicalein induces apoptosis via a mitochondrial-dependent caspase activation pathway in T24 bladder cancer cells." *Molecular Medicine Reports* 7: 266–270.

Li, J., and I. Khan. 2006. "The advanced bioactivity studies of *Scutellaria baicalensis* Georgi and its phenolic compounds." *Proc. IVth on Quality and Safety in Botanicals* (Eds. I. A. Khan et al.,) *Acta Horticulture* ISHS. 157–169. doi: 10.17660/ActaHortic.2006.720.15

Liu, J., M. Jiang, Z. Li, X. Li, Y. Hao, X. Su, J. Zhu, C. Zheng, W. Xiao, and Y. Wang. 2018. "A novel systems pharmacology method to investigate molecular mechanisms of *Scutellaria barbata* D. Don for non-small cell lung cancer." *Frontiers in Pharmacology* 9: 1473. doi: 10.3389/fphar.2018.01473

Li-Weber, M. 2009. "New therapeutic aspects of flavones: The anti-cancer properties of *Scutellaria* and its main active constituents Wogonin, Baicalein and Baicalin." *Cancer Treatment Reviews* 35: 57–68. doi: 10.1016/j.ctrv.2008.09.005

Maher, T. J. 2001. "Complementary prostate cancer treatments continuing education test questions Natural Healing Track, New Hope Institute of Retailing." *Massachusetts College of Pharmacy and Health Sciences*: 1–15.

Makino, T., R. Tsubouchi, K. Murakami, M. Haneda, and M. Yoshino. 2006. "Generation of reactive oxygen species and induction of apoptosis of HL60 cells by ingredients of traditional herbal medicine, sho-saiko-to." *Basic and Clinical Pharmacology & Toxicology* 98: 401–405. doi: 10.1111/j.1742-7843.2006.pto_328.x

Malakov, P. Y., and G. Y. Papanov. 1998. "Neo-clerodane diterpenoids from *Scutellaria alpina*." *Phytochemisty* 49, 8: 2449–2452. doi: 10.1016/S0031-9422(98)00358-6

Mamadalieva, N. Z., F. Herrmann, M. Z. Elreadi, A. Tahrani, R. Hamound, D. R. Egamberdieva, S. S. Azimova, and M. Wink. 2011. "Flavonoids on *Scutellaria immaculata ramisissima* (Lamiaceae) and their biological activity." *Journal of Pharmacy and Pharmacology* 63, 10: 1346–1357. doi: 10.1111/j.2042-7158.2011.01336.x

Mandavi mousavi, S. N., A. Delazar, H. Nazemiyeh, and L. Khodaie. 2015. "Biological activity and phytochemical study of *Scutellaria platystegia*." *Iranian Journal of Pharmaceutical Research* 14, 1: 215–223.

Motaez, M., S. A. Emami, and Z. Tayarani-Najaran. 2015. "Growth inhibition and apoptosis induction of *Scutellaria luteo-coeulea Borm* and Sint. on leukemia cell lines K562 and HL60." *Environmental Toxicology and Pharmacology* 5, 6: 553–559. doi: 10.1016/j.etap.2014.12.002

Mozaffarian, V. A. 1996. "A dictionary of plant names." *Farhang Moa'ser*, Tehran: 497–499.

Nagai, H., and Y. H. Kim. 2017. "Cancer prevention from the perspective of global cancer burden patterns." *Journal of Thoracic Disease* 9, 3: 448–451. doi: 10.21037%2Fjtd.2017.02.75

Nishikawa, K., H. Furukawa, T. Fujioka, K. Mihashi, K. Shimomura, and K. Ishimaru. 1999. "Phenolics in tissue cultures of *Scutellaria*." *Recent Research Developments in Phytochemistry* 53, 4: 209–213.

Orzechowska, R., R. Chaber, A. Wisneiewska, E. Pajtaasz-Piasecka, B. Jatczak, I. Siemieniec, B. Gulanowski, A. Chybicka, and Z. Blach-Olszewska. "Baicalin from the extract of *Scutellaria baicalensis* affects the innate immunity and apoptosis in leukocytes in children with acute lymphocytic leukemia." *International Journal of Immunopharmacology* 23, 2: 558–567. doi: 10.1016/j.intimp.2014.10.005

Ozmen, A., S. Madlener, S. Bauer, S. Krasteva, C. Vonach, B. Giessrigl, M. Gridling, K. Viola, N. Stark, P. Saiko, B. Michel, M. Fritzer-Szkeres, T. Szekeres, T. Askin-Celik, L. Krenn, and G. Krupitza. 2010. "*In vitro* anti-leukemic activity of the ethnopharmacutical plant *Scutellaria orientalis* ssp. carica endemic to western Turkey." *Phytomedine*. 17, 1: 55–62. doi: 10.1016/j.phymed.2009.06.001

Parajuli, P., N. Joshee, A. M. Rimando, S. Mittal, and A. K. Yadav. 2009. "*In vitro* antitumor mechanisms of various *Scutellaria* extracts and constituent flavonoids." *Planta Medica* 75, 1: 41–48. doi: 10.1055/s-0028-1088364

Parajuli, P., N. Joshee, S. R. Chinni, A. M. Rimando, S. Mittal, S. Sethi, and A. K. Yadav. 2011. "Delayed growth of glioma by *Scutellaria* flavonoids involves inhibition of Akt, GSK-3 and NF-κB signaling." *Journal of Neuro-Oncology* 101, 1: 15–24. doi: 10.1007/s11060-010-0221-x

Park, K. I., H. S. Park, S. R. Kang, A. Magappan, D. H. Lee, J. A. Kim, D. Y. Han, and G. S. Kim. 2011. "Korean *Scutellaria baicalensis* water extract inhibits cell cycle G1/S transition by suppressing cyclin D1 expression and matrix-metalloproteinase-2 activity in human lung cancer cells." *Journal of Ethnopharmacology* 133, 2: 634–641. doi: 10.1016/j.jep.2010.10.057

Patel, P. S., N. Joshee, A. M. Rimando, and P. Parajuli. 2013. "Anti-cancer scopes and associated mechanisms of *Scutellaria* extract and flavonoid wogonin." *Current Cancer Therapy Reviews* 9, 1: 34–42. doi: 10.2174/1573394711309010004

Perez, B. Arun, D. Tripathy, M. A. Tagliaferri, H. S. Shaw, G. G. Kimmick, I. Cohen, E. Shtivelman, K. A. Caygill, D. Grady, M. Schactman, and C. L Shapiro. 2010. "A phase 1B dose escalation trial of *Scutellaria barbata* (BZL101) for patients with metastatic breast cancer." *Breast Cancer Research and Treatment* 120, 1: 111–118. doi: 10.1007/s10549-009-0678-5

Powell, C. B., P. Fung, J. Jackson, J. Dall'Era, D. Lewkowicz, I. Cohen, and K. Smith-McCune. 2003. "Aqueous extract of herbal *Scutellaria barbatae*, a Chinese herb used for ovarian cancer induces apoptosis of ovarian cancer cell lines." *Gynecologic Oncology* 91, 2: 332–340. doi: 10.1016/j.ygyno.2003.07.004

Rugo, H. 2006. "Herbal therapy in treating women with metastatic breast cancer." https://clinicaltrials.gov/ct2/show/NCT00028977 (Accessed: 3 May 2021).

Rugo, H., E. Shtivelman, A. Perez, C. Vogel, S. Franco, E. T. Chiu, M. Melisko, M. Tagliaferri, I. Cohen, M. Shoemaker, Z. Tran, and D. Tripathy. 2007. "Phase I trial and antitumor effects of BZL101 for patients with advanced breast cancer." *Breast Cancer Research and Treatment* 105, 1: 17–28.

Saklani, A., and S. K. Kutty. 2008. "Plant-derived compounds in clinical trials." *Drug Discovery Today* 13, 3–4: 161–171. doi: 10.1016/j.drudis.2007.10.010

Salini, S., T. Chubicka, N. Sasidharan, and T. D. Babu. 2013. "Cytotoxic and antioxidant properties of selected *Scutellaria* species from the Western Ghats Peninsular India." *Pharmacutical Biology* 51, 2: 152–159. doi: 10.3109/13880209.2012.715170

Scheck, A. C., K. Perry, N. C. Hank, and W. D. Clark 2006. "Anti-cancer activity of extracts derived from the mature roots of *Scutellaria baicalensis* on human malignant brain tumor cells." *BMC Complementary and Alternative Medicine* 6, 1: 27. doi: 10.1186%2F1472-6882-6-27

Shang, X., X. He, X. He, M. Li, R. Zhang, P. Fan, Q. Zhang, and Z. Jia. 2010. "The genus *Scutellaria* an ethnopharmacological and phytochemical review." *Journal of Ethnopharmacology* 128, 2: 279–313. doi: 10.1016/j.jep.2010.01.006

Shen, J., P. Li, S. Liu, L. Qing, L. Yue, and Y. Sun. 2021. "Traditional uses, ten years research progress on phytochemistry and pharmacology, and clinical studies of the genus *Scutellaria*." *Journal of Ethnopharmacology* 265: 1–23. doi: 10.1016/j.jep.2020.113198

Shi, X., G. Chen, X. Liu, Y. Qui, S. Yang, Y. Zhang, X. Fang, C. Zhang, and X. Liu. 2015. "Scutellarein inhibits cancer cell metastasis *in vitro* and attenuates the development of fibrosarcoma *in vivo*." *International Journal of Molecular Medicine* 35, 1: 31–38. doi: 10.3892/ijmm.2014.1997

Shimizu, I. 2000. "Sho-saiko-to: Japanese herbal medicine for protection against hepatic fibrosis and carcinoma." *Journal of Gastroenterology and Hepatology* 15, Suppl._1: 84–90. doi: 10.1046/j.1440-1746.2000.02138.x

Sitarek, P., A. Merecz-Sadowska, T. Sliwi´nski, R. Zajdel, and T. Kowalczyk. 2020. "An *in vitro* evaluation of the molecular mechanisms of action of medical plants from the Lamiaceae family as effective sources of active compounds against human cancer cell lines." *Cancers* 12: 2957. doi: 10.3390/cancers12102957

Sonoda, M., T. Nishiyama, Y. Matsukawa, and M. Moriyasu. 2004. "Cytotoxic activities of flavonoids from two *Scutellaria* plants in Chinese medicine." *Journal of Ethnopharmacology*. 91: 65–68. doi: 10.1016/j.jep.2003.11.014

Tayarani-Najarani, Z., J. Asili, H. Parsaee, S. H. Mousavi, N. V. Mashhadian, A. Merzaee, and S. A. Emami. 2012a. "Wogonin and neobaicalein from *Scutellaria litwinowii* roots are apoptotic for HeLa cells." *Revista Brasileira de Farmacognosia* 22, 2: 268–276. doi: 10.1590/S0102-695X2011005000161

Tayarani-Najaran, Z., S. A. Emami, J. Asili, M. Merzaei, and S. H. Mousavi. 2011. "Analyzing cytotoxic and apoptogenic properties of *Scutellaria litwinowii* root extract on cancer cell lines." *Evidence Based Complementary Alternative Medicine*. doi: 10.1093/ecam/nep214

Tayarani-Najaran, Z., S. H. Mousavi, J. Asilbi, and S. A. Emami. 2010. "Growth inhibitory effect of *Scutellaria lindbergii* in human cancer cell lines." *Food and Chemical Toxicology* 48, 2: 599–604. doi: 10.1016/j.fct.2009.11.038

Tayarani-Najaran, Z., S. H. Mousavi, N. V. Mashhadian, S. A. Emami, and H. Parsaee. 2012b. "*Scutellaria litwinowii* induces apoptosis through both extrinsic and intrinsic apoptotic pathways in human promyelocytic leukemia cells." *Nutrition and Cancer* 64, 1: 80–88. doi: 10.1080/01635581.2012.630162

Vaidya, B. N., T. A. Brearley, and N. Joshee. 2013. "Antioxidant capacity of fresh and dry leaf extracts of sixteen *Scutellaria* species." *Journal of Medicinally Active Plants* 2, 3: 42–49. doi: 10.7275/R5J9649K

Vanisree, M., C. Y. Lee, S. F. Lo, S. M. Nalawade, C. Y. Lin, and H. S. Tsay. 2004. "Studies on the production of some important secondary metabolites from medicinal plants by plant tissue cultures." *Botanical Bulleton Academia Sinica Tapei* 45, 1: 1–22.

Wei, L., J. Lin, W. Xu, Q. Cai, A. Shen, Z. Hong, and J. Peng. 2012. "*Scutellaria barbata* D. Don inhibits tumor angiogenesis via suppression of hedgehog pathway in a mouse model of colorectal cancer." *International Journal of Molecular Science* 13: 9419–9430. doi: 10.3390/ijms13089419

Wright, R. E., L. Shahin., V. Gogineni, Z. HussAin, A. Naeem[1], S. SadasIvan, I. Sinha, M. Neely, S. K. Michellhaugh, S. Mittal, N. Joshee, and P. Parajuli. 2021. "*Scutellaria* extract inhibits proliferation and migration of brain-metastatic lung cancer cells via regulation of multiple signaling pathways." *Journal of Medicinally Active Plants* 10, 1: 32–41. doi: 10.7275/ck6p-ej40

Wu, B., J. Li, D. Huang, W. Wang, Y. Chen, Y. Liao, X. Tang, H. Xie, and F. Tang. 2011. "Baicalein mediates inhibition of migration and invasiveness of skin carcinoma through Ezrin in A431 cells." *BMC Cancer* 11: 527. doi: 10.1186/1471-2407-11-527

Yang, J., G. Yang, G. Hou, Q. Liu, W. Hu, and P. He. 2015. "*Scutellaria barbata* D. Don polysaccharides inhibit the growth of Calu-3 xenograft tumors via suppression of HER2 pathway angiogenesis." *Oncology Letters* 9: 2721–2725. doi: 10.3892/ol.2015.3127

Yang, X. K., M. Y. Xu, G. S. Xu, Y. L. Zhang, and Z. X. Xu. 2014. "*In vitro* and *in vivo* antitumor activity of scutebarbatine A on human lung carcinoma A549 cell lines." *Molecules* 19, 7: 8740–8751. doi: 10.3390%2Fmolecules19078740

Ye, F., H. Volshonok, J. Wu, and D. Y. Zhang. 2007. "Molecular mechanism of anti-prostate cancer activity of *Scutellaria baicalensis* extract." *Nutrition and Cancer* 57, 1: 100–110. doi: 10.1080/01635580701268352

Ye, F., L. Xui, J. Yi, W. Zhang, and D. Y. Zhang. 2002. "Anti-cancer activity of *Scutellaria baicalensis* and its potential mechanism." *Journal of Alternative and Complementary Medicine* 8, 5: 567–572. doi: 10.1089/107555302320825075

Yeh, Y. C., H. Y. Chen, S. H. Yang, Y. H. Lin, J. H. Chui, Y. H. Lin, and J. L. Chen. 2014. "*Hedyotis diffusa* combined with *Scutellaria barbata* are the core treatment of Chinese herbal medicine used for breast cancer patients: A population-based study." *Evidence Based Complementary and Alternative Medicine* 2014: 1–9. doi: 10.1155/2014/202378

Yin, X., J. Zhou, C. Jie, D. Xing, and Y. Zhang. 2004. "Anti-cancer activity and mechanism of *Scutellaria barbata* extract on human lung cancer cell line A549." *Life Sciences* 75, 18: 2233–2244. doi: 10.1016/j.lfs.2004.05.015

Yu, J., J. Lei, H. Yu, X. Cai, and G. Zou. 2004. "Chemical composition and antimicrobial activity of the essential oil of *Scutellaria barbata*." *Phytochemistry* 65, 7: 881–884. doi: 10.1016/j.phytochem.2004.02.005

Zhang, D. Y., J. Wu, F. Ye, L. Zue, S. Jiang, J. Yi, W. Zhang, H. Wei, M. Sung, W. Wang, and X. Li. 2003. "Inhibition of cancer cell proliferation and prostaglandin E_2 synthesis by *Scutellaria baicalensis*." *Cancer Research* 63, 14: 4037–4043.

Zhang, Z., X. Lian, S. Li, and J. L. Stringer. 2009. "Characterization of chemical ingredients and anticonvulsant of American skullcap (*Scutellaria lateriflora*)." *Phytomedicine* 16: 485–493. doi: 10.1016/j.phymed.2008.07.011

Zhou, T., A. Zhang, G. Kuang, X. Gong, R. Jiang, D. Lin, J. Li, H. Li, X. Zhang, J. Wan, and H. Li. 2017. "Baicalin inhibits the metastasis of highly aggressive breast cancer cells by reversing epithelial-to-mesenchymal transition by targeting β-catenin signaling." *Oncology Reports* 38, 6: 3599–3607. doi: 10.3892/or.2017.6011

12 Lung Matters

Review on Lung Cancer and Therapeutic Possibilities of Medicinal Plants

Tenille Esmear and Namrita Lall

CONTENTS

12.1 INTRODUCTION INTO LUNG CANCER

Lung cancer is the leading cause of cancer-related death in the world, accounting for 188,679 and 607,465 reported deaths in 2020 in men and females, respectively (Ferlay et al., 2020). The global geographic distribution of lung cancer shows marked regional variation. In men, the highest incidence rates are observed in Central and Eastern Europe (53.5 per 100,000) and Eastern Asia (50.4 per 100,000). Considerably lower incidence rates are observed in Central and Western Africa (2.0 and 1.7 per 100,000, respectively). In women, the incidence rates are generally lower, with a different geographic pattern. This can mainly be reflected by historically different tobacco smoking

DOI: 10.1201/9781003108375-12

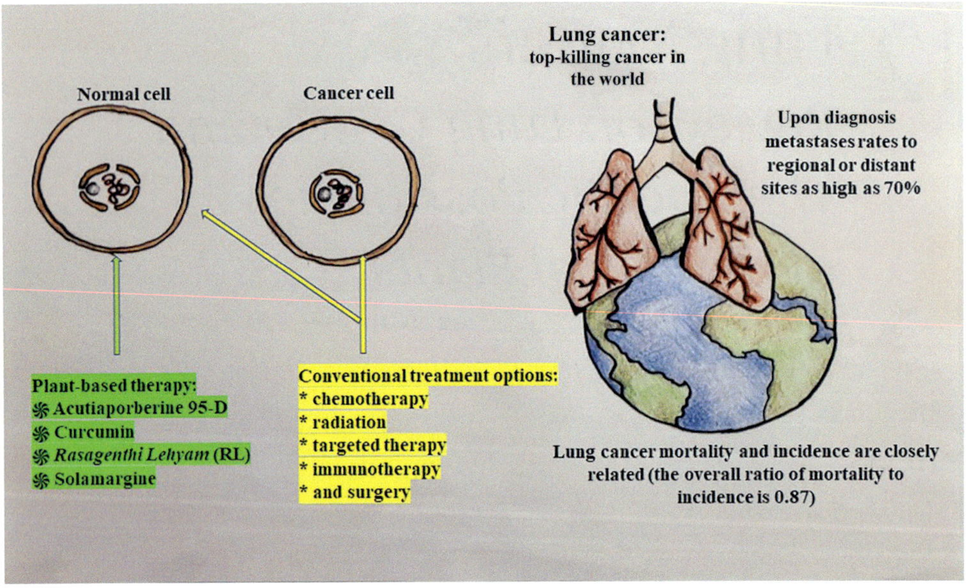

Graphical abstract representing the content of the chapter.

habits and exposure (Mao et al., 2016). The highest estimated rates are in North America (33.8 per 100,000) and Northern Europe (23.7 per 100,000), with a relatively high rate in East Asia (19.2 per 100,000) and the lowest rates again in Western and Central Africa (1.1 and 0.8 per 100,000, respectively). Lung cancer mortality and incidence are closely related (the overall ratio of mortality to incidence is 0.87), so the relative lack of variability in survival in different world regions and the geographic patterns in mortality closely follow those in incidence (Mao et al., 2016).

12.2 ANATOMY OF THE LUNGS

The lungs are the major organs of the respiratory system, with this pulmonary circulation aiding in gaseous exchange. Each of the lungs is paired and separated into lobes, with the left lung consisting of two lobes and the right lung consisting of three lobes. Blood circulation to the lungs plays a vital role, as blood is required to transport oxygen from the lungs to other tissues throughout the body. Deoxygenated blood from the body is delivered to the lungs via the pulmonary artery to the capillaries that form respiratory membranes with the alveoli, where oxygen is replenished. The pulmonary veins then return newly oxygenated blood to the heart for further transport throughout the body. Parasympathetic and sympathetic nervous systems throughout the lungs coordinate both bronchodilation and bronchoconstriction of the airways. The pleural membrane that is composed of visceral and parietal pleural layers encloses the lungs. These two layers have a space between them known as the pleural cavity. The pleural membrane has mesothelial cells which create pleural fluid, serving as a lubricant to reduce friction during breathing and as an adhesive to adhere the lungs to the thoracic wall, thereby facilitating the movement of the lungs during ventilation (Baile, 1996).

12.2.1 Causes of Lung Cancer

The causes of lung cancer are also called to the surface in this question. Epidemiologic studies have demonstrated that cigarette smoking is the major risk factor in the development of lung cancer and is dose-dependent. Approximately 80–90% of lung cancer cases are attributed to smoking, with the remaining 10–20% attributed to occupational exposure to agents such as uranium, ionizing radiation, asbestos, silica, arsenic, beryllium, chloromethyl, nickel chromates, indoor emissions from burning fuels, and polycyclic aromatic hydrocarbons (Bello et al., 2011).

Immunocompromised people such as those with the human immunodeficiency virus (HIV) or acquired immunodeficiency syndrome (AIDS) and other related diseases are more susceptible to the development of lung cancer. Although cancers of the lung are often difficult to diagnose early and treat successfully, the cause of most of these cancers is well known. There remains a significant exception in U.S. women, where more non-smoking women (never smoked and former smokers) die from lung cancer each year than women (non-smokers and smokers alike) who die from breast cancer (Bello et al., 2011).

12.2.2 Signs and Symptoms of Lung Cancer

Lung cancer can present with symptoms or may be found incidentally on chest imaging. Signs and symptoms often result from the location of the primary local tumor or due to compression of the adjacent thoracic structures, metastases in other vital organs, or due to paraneoplastic phenomena whereby symptoms mediated by humoral factors, such as hormones or cytokines, are secreted by tumor cells or by an immune response against the tumor.

The most common symptoms are as follows.

- Worsening cough.
- Shortness of breath.
- Dyspnea (difficult or labored breathing).

Other presenting symptoms include chest pain, hoarseness, malaise, anorexia, weight loss and hemoptysis (coughing up blood) (PDQ Adult Treatment Editorial Board, 2002).

12.2.3 Types of Lung Cancer

Lung cancer is classified into two types: small cell lung cancers (SCLC) and non–small cell lung cancers (NSCLC). They are uniformly aggressive tumors, with rates of regional or distant metastases upon diagnosis as high as 70%. Their classification is based upon the microscopic appearance of the tumor cells. It is important to make a distinction between these two types, as they grow, spread and are treated in different ways (Schiller, 2001) (Figure 12.1).

12.2.3.1 Small Cell Lung Cancer (SCLC)

SCLC is defined as a malignant epithelial tumor consisting of small cells with scant cytoplasm, ill-defined cell borders, finely granular nuclear chromatin and absent or inconspicuous nucleoli (Meerbeeck et al., 2011). Occurrence is less common than NSCLC and accounts for about 10–15% of lung cancers. However, this type of lung cancer is the most aggressive and rapidly growing of all types (Goldstraw et al., 2011). SCLC is strongly related to cigarette smoking (Meerbeeck et al., 2011). SCLCs metastasize rapidly to many sites within the body and are most often discovered after they have spread extensively (Kalemkerian et al., 2018).

Approximately 30% of patients with SCLC will have tumors confined to the hemithorax of origin, the mediastinum or the supraclavicular lymph nodes at the time of diagnosis, and are designated as having a limited-stage disease (LD). Patients with tumors that have spread beyond the supraclavicular areas are said to have an extensive-stage disease (ED) (PDQ Adult Treatment Editorial Board, 2002). SCLC is more responsive to chemotherapy and radiation therapy than other cell types of lung cancer. A cure, however, is difficult to achieve because SCLC has a greater tendency to have spread widely by the time of diagnosis (PDQ Adult Treatment Editorial Board, 2002).

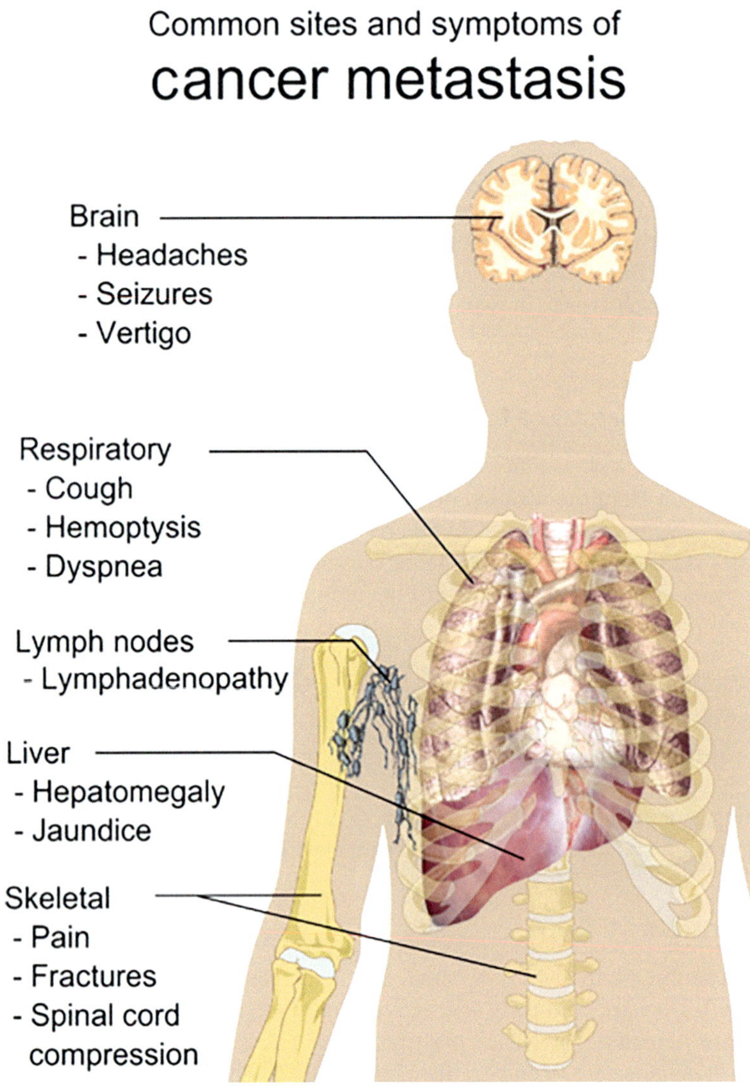

FIGURE 12.1 Depiction of Lung Cancer (Haggstrom, 2008).

12.2.3.2 Non–Small Cell Lung Cancers (NSCLC)

This is the most common lung cancer, accounting for about 85% of all cases (Baldwin and Popat, 2008). NSCLC is divided into three main types, which are determined by the type of cells found in the tumor. They are defined as follows.

12.2.3.2.1 Adenocarcinomas

- The most common type of NSCLC, accounting for up to 40% of lung cancer cases.
- Though found in smokers, this type is also seen particularly in non-smoking women who develop lung cancer.
- Occurrence is increasing in frequency and is more common in women.
- They arise mostly in the outer, or peripheral, areas of the lungs and tend to spread to the lymph nodes and beyond. They may develop at multiple sites in the lungs and spread along the pre-existing alveolar walls.

- They may present like pneumonia on a chest X-ray.
- Generally, the prognosis for this type of lung cancer is better than for the other types. (Wood et al., 2018)

12.2.3.2.2 Squamous Cell Carcinomas

- They were formerly more common than adenocarcinomas. Today, they account for about 25–30% of all lung cancer cases.
- They most frequently arise in the central chest area in the bronchi. This type of lung cancer stays most often within the lung, spreads to lymph nodes and grows quite large, forming a cavity. (Wood et al., 2018)

12.2.3.2.3 Large Cell Carcinomas (Also Known as Undifferentiated Carcinomas)

- These are the least common type of NSCLC, accounting for 10–15% of all lung cancers.
- This type of cancer has a high tendency to metastasize to the lymph nodes and distant sites in the body. (Wood et al., 2018)

12.2.3.3 Bronchial Carcinoids

Bronchial carcinoids account for up to 5% of lung cancers. These tumors are generally small (≤3 cm in diameter) when diagnosed and occur most commonly in persons under age 40. Unrelated to cigarette smoking, carcinoid tumors can metastasize, and a small proportion of these tumors secrete hormone-like substances. Carcinoids generally grow and spread more slowly than bronchogenic cancers, and many are detected early enough to be surgically removed. Cancers of supporting lung tissue such as smooth muscle, blood vessels or cells involved in the immune response are rare in the lungs (Wood et al., 2018).

12.2.3.4 Metastatic Cancers

Other primary tumors in the body are often found in the lungs. Tumors from anywhere in the body may spread to the lungs either through the bloodstream, through the lymphatic system or directly from nearby organs. Metastatic tumors are most often multiple, scattered throughout the lungs and concentrated in the outer areas rather than in the central areas of the organ (Wood et al., 2018).

12.3 OVERVIEW OF SYSTEMIC LUNG CANCER TREATMENTS

The efficiency of current treatment modalities for lung cancer depends strongly on the time of diagnosis, with better chances of survival if a tumor has been detected at an early stage (Risch and Plass, 2008). Treatment regimens are extensive and aggressive, since most lung cancers are detected after the tumor has grown considerably or metastasized elsewhere in the body (Kalemkerian et al., 2018; Baldwin and Popat, 2008).

12.3.1 Chemotherapy

Chemotherapy is administered in cycles. Often only one drug is used, but occasionally more than one drug is used (made with platinum), and these regimens are called platinum-doublet chemotherapy (Alder, 2017).

12.3.1.1 Chemotherapy Mechanisms

Table 12.1 provides the range of chemotherapy agents which are currently being used in the treatments of NSCLC and SCLC, and states the mechanisms of action of these drugs. Most often, a combination of drugs is used to target multiple mechanisms, since lung cancer is particularly resistant to treatment.

TABLE 12.1
Current Chemotherapeutic Agents Used in the Treatment of Lung Cancer (Alder, 2017; Kalemkerian et al., 2018)

Chemical Names	Mechanisms of Action	Lung Cancer Types
Carboplatin (platinum-based)	Inhibit DNA synthesis	NSCLC[1], SCLC[2]
Cisplatin (platinum-based)	Interfere with DNA replication	NSCLC, SCLC
Docetaxel	Inhibition of microtubule depolymerization Decrease of the effects of bcl-2[3] and bcl-xL[4] gene expression	NSCLC, SCLC
Etoposide: Etoposide phosphate	Causes DNA strands to break	NSCLC, SCLC
Gemcitabine hydrochloride	Inhibits processes for DNA synthesis	NSCLC, SCLC
Paclitaxel	Inhibits microtubule formation	NSCLC, SCLC
Paclitaxel, albumin-bound	Inhibits microtubule formation, albumin enhances the delivery of the drug	NSCLC, SCLC
Pemetrexed	Inhibits enzymes involved in folate metabolism and purine and pyrimidine synthesis	NSCLC
Vinblastine sulfate	Binds to microtubules proteins in the mitotic spindle, thereby preventing cell division during metaphase	NSCLC
Vinorelbine tartrate	Binds to tubulin and prevents the formation of the mitotic spindle, resulting in the arrest of tumor cell growth in metaphase	NSCLC, SCLC

Notes: **1** Non–small cell lung cancer; **2** small cell lung cancer; **3** B-cell Lymphoma 2; **4** B-cell lymphoma-extra-large

12.3.1.2 Chemotherapy Side Effects

Since chemotherapy does not specifically target cancer cells, many healthy cells are killed in the process of treatment, particularly fast-growing cells such as the hair follicles, gut, mouth and blood. Side effects include low red and white blood cell counts, loss in appetite, nausea, vomiting, diarrhea, hair loss and mouth sores, among many others (Pearce et al., 2017).

12.3.2 Targeted Therapy

Targeted therapy is a more desirable therapeutic approach, as these drugs target specific molecules found on cancer cells, inflicting less damage to normal healthy cells than chemotherapy agents (Alder, 2017).

12.3.2.1 Targeted Therapy Mechanisms

12.3.2.1.1 Anaplastic Lymphoma Kinase (ALK) Rearrangement

ALK is a surface receptor that triggers growth signals in lung cancer cells. When this receptor is overactive, it causes new cancer cells to proliferate quickly (Alder, 2017).

12.3.2.1.2 Proto-Oncogene Tyrosine-Protein Kinase (ROS1) Rearrangement

ROS1 is a surface receptor that triggers growth signals in lung cancer cells. When this receptor is overactive, it causes new cancer cells to proliferate quickly (Alder, 2017).

12.3.2.1.3 Serine/Threonine-Protein Kinase B-Raf (BRAF V600E) Mutation

BRAF is a kinase inside of cells that results in phosphate transfer. This stimulates the next growth signal in the pathway. When BRAF is overactive, it causes new cancer cells to proliferate quickly (Alder, 2017).

12.3.2.1.4 Vascular Endothelial Growth Factor (VEGF) Pathway

VEGF is not a mutation, but is made by cancer cells. It travels from cancer cells to endothelial cells, where it triggers growth signals to form new blood vessels (Alder, 2017).

12.3.2.1.5 Other Targets

These treatments have been approved for the treatment of lung cancer, though have not been extensively researched. They include ado-trastuzumab, which targets human epidermal growth factor receptor 2 (HER 2); and Cabozantinib and vandetanib, which both target rearranged transfection proto-oncogene (RET). Crizotinib targets anaplastic lymphoma kinase (ALK), while tyrosine kinase (ROSl) targets tyrosine-protein kinase (MET) (Alder, 2017). Table 12.2 summarizes the various targeted therapies which are being used, the receptors they exert their effect on, and the type of lung cancer these treatments are used for.

12.3.2.2 Targeted Therapy Side Effects

Though these therapies are more specific to cancer cells, many side effects remain, some severe. Side effects include diarrhea, skin rash, gut bleeding and severe eye, skin, lung, kidney and liver problems (Alder, 2017; Dholaria et al., 2016).

12.3.3 Immunotherapy

Immunotherapy is intended to increase the activity of the immune system, enabling the body to locate and destroy cancer cells. This is only used as second-line therapy.

12.3.3.1 Programmed Death-Ligand (PDL)-1

PDL-1 is found on normal cells and attaches to PD-1 found on T cells, thus turning off the immune response which occurs naturally when a response is no longer needed. However, cancer cells can also express PDL-1, thereby turning off the immune response and preventing the immune system from attacking the tumor. PDL-1 inhibitors attach to PD-1 on T cells, and PD-1 inhibitors attach to PD-1 on cancer cells. Both these inhibitors stop cancer cells with PDL-1 from attaching to T cells, thus enabling T cells to attack the cancer cells. Examples of PDL-1 inhibitors that are approved for use in

TABLE 12.2

Current Targeting Agents Used in the Treatment of Non–Small Cell Lung Cancer (NSCLC Lung Cancer) (Alder, 2017; Kalemkerian et al., 2018)

Chemical Names	Mechanisms of Action	Types of Lung Cancers
Ado-trastuzumab	HER2[1]	NSCLC[2]
Afatinib	Epidermal Growth Factor Receptor (EGFR) and HER2	NSCLC
Alectinib	ALK[3]	NSCLC
Bevacizumab	VEGF[4]	NSCLC
Brigatinib	ALK	NSCLC
Ceritinib	ALK	NSCLC
Cetuximab	EGFR	NSCLC
Crizotinib	ALK, ROS1[5] and MET[6]	NSCLC
Dabrafenib	BRAF V600E	NSCLC
Erlotinib hydrochloride	EGFR	NSCLC
Gefitinib	EGFR	NSCLC
Osimertinib	EGFR	NSCLC
Ramucirumab	VEGF	NSCLC
Trametinib	Mitogen-activated Protein Kinase (MEK1) and MEK2 (for BRAF mutation)	NSCLC
Vandetanib	RET [7]	NSCLC
Vemurafenib	BRAF V600E	NSCLC

Notes: **1** Human epidermal growth factor; **2** non–small cell lung cancer; **3** anaplastic lymphoma kinase; **4** receptor tyrosine kinase; **5** proto-oncogene; **6** braf proto-oncogene, serine/threonine kinase; **7** proto-oncogene tyrosine-protein kinase

NSCLC include Atezolizumab, Durvalumab, Nivolumab and Pembrolizumab (Alder, 2017). Studies have indicated that SCLC has a lower rate of PDL-1 expression than NSCLC (Sloan et al., 2016).

12.3.3.2 Immunotherapy Side Effects

Side effects seen are often less than experienced in chemotherapy. Common side effects include fatigue, constipation, nausea, lack of appetite and muscle or bone pain. Occasionally, an organ can become inflamed, leading to severe side effects involving the lung, gut, liver, kidney, skin or hormones (Dholaria et al., 2016).

12.3.4 Surgery

Despite complete surgical removal, a large proportion of patients with NSCLC have a recurrence of cancer, because most of the time, the cancer is advanced when they are diagnosed. Surgery is not widely used in SCLC. Because SCLC spreads widely and rapidly through the body, removing it all by surgery is usually impossible (Kalemkerian et al., 2018). Though there are many treatments available, many factors must be taken into consideration before a treatment option/s can be implemented.

12.4 CONVENTIONAL DRUG SETBACKS

Since chemotherapy does not specifically target cancer cells, many healthy cells are killed in the process of treatment, particularly fast-growing cells such as the hair follicles, gut, mouth and blood. The most common side effects include low red and white blood cell counts, loss in appetite, nausea, vomiting, diarrhea, hair loss and mouth sores, among many others (Alder, 2017). It is clear that many options are available to patients; however, their success is limited and comes at a price of often severe side effects. The liver is the primary site of metabolism for many of these drugs. All of the mentioned therapies can cause damage to the liver known as hepatotoxicity. The clinical presentations of hepatotoxicity can vary from asymptomatic to an increase of liver chemistries and evident cholestatic hepatitis. This damage can progress to fibrosis and cirrhosis, malignant transformation, veno-occlusive disease/sinusoidal obstruction and even fulminant hepatic failure (Grigorian and Brien, 2014).

Lung cancer patients are still plagued by rapid disease progression and recurrence, despite advances in molecular mechanisms and drug development. Overall prognosis and survival of these patients have seen no improvement with the introduction of new therapies. Another obstacle is the fact that lung cancer is particularly prone to develop resistance to chemotherapeutic agents. Identifying the molecular determinants of sensitivity and resistance to chemotherapy is expected to improve the therapeutic efficacy (Kummalue, 2005). The addition of plants as an adjuvant to the standard chemotherapeutic drug may improve the sensitivity and efficacy of the treatment.

12.5 THE COST OF CANCER

The World Health Organization (WHO) defines national cancer control programs as public health programs designed to reduce cancer incidence and mortality and improve quality of life of cancer patients, through the systematic and equitable implementation of evidence-based strategies for prevention, early detection, diagnosis, treatment, and palliation, making the best use of available resources (Song et al., 2008, p. 61).

Together with a functional cancer registry, more accurate measurement of the effect of malignant disease on population health and of the effectiveness of any preventive or curative interventions should now be possible. South Africa's establishment of a Ministerial Advisory Committee on the Prevention and Control of Cancer in 2012 is another positive step. However, funding for the national cancer control program was still lacking (Stefan et al., 2013). Due to the highly invasive, aggressive and resistant lung cancer, patients often must endure several drugs at the same time. However, a metanalysis regarding the addition of a third drug to the patients' regimen showed that this did not have affect with regards to the patients' survival rate (Delbaldo et al., 2004). Here, adding a plant-based adjuvant could improve the dismal outcomes for lung cancer patients. Figure 12.2 illustrates

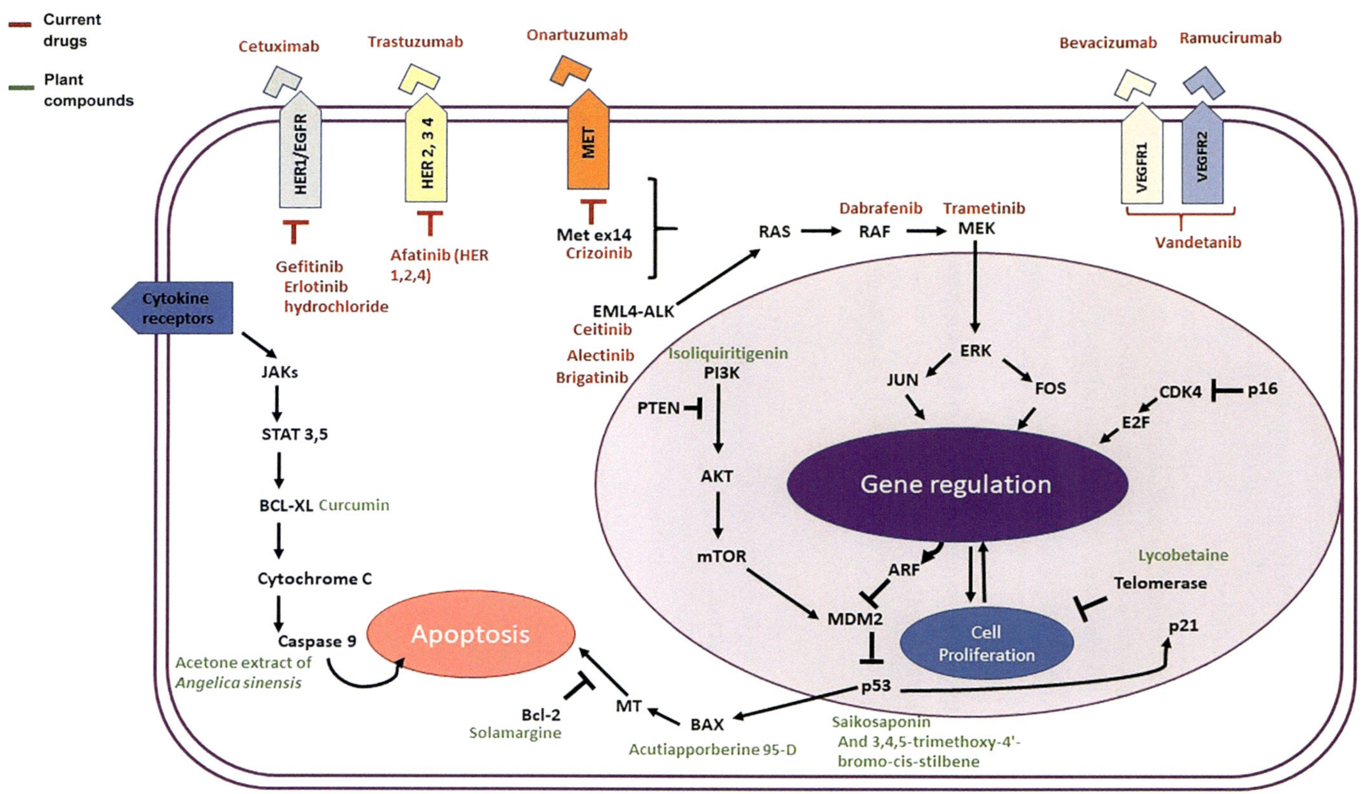

FIGURE 12.2 Illustration of cellular oncogenic pathways for immunotherapy and targeted therapy, and where the addition of an adjuvant plant-based compounds can benefit treatment outcomes. Adapted from Sundar et al. (2017).

where the newer therapies, immunotherapy and targeted therapy, can work with the addition of a plant-based compound.

12.6 NATURAL ALTERNATIVES FOR THE TREATMENT OF LUNG CANCER

Natural product molecules embody a promising approach for lung cancer therapy with fewer side effects, along with offering effective treatment outcomes. Various natural products have proven to be useful, and sensitizing is observed when used synergistically together with conventional therapies. Several natural product molecules can prevent the side effects of chemotherapy, as seen in Table 12.3. Of utmost importance, natural products can improve the quality of life and prolong the survival time of lung cancer patients. To date, anti-lung cancer herbs have included medicines with effective treatment effects (Figure 12.3) (Zhong et al., 2017). To verify the potential inhibitory effectiveness of the inhibitor, it is classically based on its half maximal inhibitory concentration (IC_{50}).

TABLE 12.3
Plants and Their Identified Compounds, along with Their Mechanisms of Action on Lung Cancer Cell Lines (Where IC Values Were Determined, They Were Expressed in μmol/mL)

Plants and Active Compounds	Cell Lines	Mechanisms/Results	References
Rasagenthi Lehyam (RL), a mixture of 48 plant parts and natural compounds from the chloroform fraction	A549[1] (IC_{50} 30μg/ml), H-460[2] (IC_{50} 40μg/ml), BEAS-2B[3] (no effect) cell lines	Upregulated p53[17] and Bax[18] Induced caspase-3 activation, down-regulated Bcl-2[19] and induced G2/M arrest	(Ranga et al., 2005)
Solanum incanum L. (gray bitter-apple) Solamargine	A549 (IC_{50} 2.9μM) H441[4], H520[5], H661[6] and H69[7] cells (IC_{50}'s of 3, 6.7, 7.2 and 5.8 μM, respectively)	Induced apoptosis of these cells determined by PS (phosphatidylserine, tumor necrosis factor receptor), release of cytochrome C, downregulation of Bcl-2 and Bcl-xL[20], increase of caspase-3 activity	(Li-Feng Liua et al., 2004)
Angelica sinensis (Oliv.) Diels Acetone extract of root	A549, HT29[8] and J5DBTRG-05MG[9] (IC_{50} between 35–50 μg/ml after 24hrs)	Activation of caspase 9 and 3 mediated via suppressing of Bcl-2 and cdk 4[21] expressions	(Cheng et al., 2004)
Bupleurum falcatum L. Saikosaponin D	A549 (IC_{50} 20 μM at 72 h)	Increased the expression of p53 and p21[22] proteins and induced of Fas/APO1[23]	(Hsu et al., 2004).
Curcuma longa L. Curcumin	A549 (IC_{50} 50 M) HT1299[10] (IC_{50} 40 M)	Decreased expression of p53, bcl-2, bcl-XL, inducing apoptosis after 24-hour exposure	(Pillai et al., 2004)
Lithospermum radix LR. β-Hydroxyisovalerylshikonin	DMS114[11] (IC_{50} 0.7 μM)	Decreased and suppressed the expression of tumor necrosis factor receptor-associated protein 1 (TRAP1)	(Masuda et al., 2004)
Thalictrum acutifolium (Hand. Mazz) Boivin (Root extract) Acutiaporberine 95-D	PLA-801[12] (IC_{50} 0.07) HepG2[13] (IC_{50} > 0.14) MCF-7[14] (IC_{50} > 0.14) NIH3T3[15] (IC_{50} 0.12)	Inhibited expression of bcl-2 gene and activated expression of bax gene and p53	(Chen et al., 2002).
Scutellaria barbata Ethanol extract	A549 (IC_{50} 0.21 mg/ml)	Activation through caspase cascade. Increase caspase 3/7	(Yin et al., 2004).
Liliaceae plants Isoliquiritigenin	A549 (IC_{50} 20 μM)	Downregulation of the PI3K/AKT/mTOR[24] signaling pathway proteins, and the downstream proteins P70 and CyclinD1	(Tian et al., 2018).

Plants and Active Compounds	Cell Lines	Mechanisms/Results	References
Grapes 3,4,5-trimethoxy-4'-bromo-cis-stilbene	A549 (IC_{50} 0.03 μM)	Elevated the expression levels of the pro-apoptotic protein p53, the cyclin-dependent kinase inhibitor p21, and the release of cytochrome c in the cytosol	(Lee et al., 2004).
Lycoris radiata Lycobetaine	LXFL 529[16] (IC_{50} 1.2 mM)	Inhibited topoisomerase I and II	(Chen et al., 2016)
Hedera colchica (K.Koch) Hedera Colchicoside	A549 (IC_{50} 10 mM)	Inhibited DNA synthesis	(Ancuceanu, 2004)

Notes: **1** Human non–small cell lung cancer; **2** large cell lung cancer; **3** human normal bronchial epithelial cell line; **4** normal alveolar epithelial cells; **5** human lung squamous carcinoma; **6** human large cell carcinoma; **7** human small cell lung cancer; **8** colorectal cancer cells; **9** human glioblastoma; **10** human non–small lung cancer cells; **11** human lung carcinoma; **12** human non-small cell cancer; **13** human liver cells; **14** human breast cancer; **15** normal fibroblast cells; **16** human non-small cell lung cancer; **17** tumor suppressor protein; **18** pro-apoptotic protein; **19** B-cell lymphoma, proteins regulate programmed cell death; **20** an anti-apoptotic protein; **21** cyclin-dependent kinase 4; **22** a potent cyclin-dependent kinase inhibitor; **23** a transmembrane receptor belonging to the tumor necrosis factor receptor superfamily; **24** phosphoinositide 3-kinase (PI3K)/serine-threonine protein kinase AKT (also known as protein kinase B (PKB)/ mammalian target of rapamycin (mTOR) pathway

FIGURE 12.3 Worldwide medicinal plants with lung cancer anti-proliferative activity. A, *Bupleurum falcatum*; B, *Thalictrum acutifolium*; C, *Ornithogalum umbellatum*; D, *Lycoris radiata*; E, *Indigofera tinctoria*; F, *Nigella sativa* (DenesFeri, 2014; Lefnaer, 2013; Stang, 2005; Stüber, 2004; Wae35244, 2000; Yamada, 2007).

12.7 CONCLUSION

There is a wealth of reputable sources exploring the efficacy of natural products for the development of new and novel drugs. As cancers become increasingly resistant to the classic drugs by which they are treated, it becomes vital to discover new, viable, natural candidates which demonstrate fewer toxic effects toward normal cells, thus decreasing side effects and improving patient quality of life. Many of these novel natural compounds, or plant extracts added as an adjuvant, could complement the existing therapeutic pathways by targeting multiple pathways simultaneously. This will allow for a more aggressive treatment plan, though with fewer side effects, and also allow for personalized treatment plans.

REFERENCES

Alder, Richard. 2017. "Lung Cancer." *Magill's Medical Guide (Online Edition).* doi: 10.1097/00063198-2001 07000-00009.

Ancuceanu, Robert. 2004. "Pharmacologically active natural compounds of lung cancer." *Alternative Medicine Review* 9, 4: 402–419.

Baile, Elisabeth M. 1996. "The anatomy and physiology of the bronchial circulation." *Journal of Aerosol Medicine and Pulmonary Drug Delivery* 9, 1: 1-6. doi: 10.1089/jam.1996.9.1

Baldwin, David R., and Sanjay Popat. 2008. "(NCCN-Guidelines) Non-Small Cell Lung Cancer." *PET-CT in Radiotherapy Treatment Planning* 140–152. doi: 10.1016/B978-1-4160-3224-3.50011-6.

Bello, Braimoh, Olufolawajimi Fadahun, Danuta Kielkowski, and Gill Nelson. 2011. "Trends in Lung Cancer Mortality in South Africa: 1995–2006." *BMC Public Health* 11: 209. doi: 10.1186/1471-2458-11-209.

Chen, Gui-Lin, Yong-Qiang Tian, Jian-Lin Wu, Na Li, and Ming-Quan Guo. 2016. "Antiproliferative Activities of Amaryllidaceae Alkaloids from Lycoris Radiata Targeting DNA Topoisomerase I." *Nature Publishing Group* 6: 38284. doi: 10.1038/srep38284.

Chen, Qi, Wenlie Peng, and Anlong Xu. 2002. "Apoptosis of a Human Non-Small Cell Lung Cancer (NSCLC) Cell Line, PLA-801, Induced by Acutiaporberine, a Novel Bisalkaloid Derived from Thalictrum Acutifolium (Hand.-Mazz.) Boivin." *Biochemical Pharmacology* 63, 8: 1389–1396. doi: 10.1016/ S0006-2952(02)00871-7.

Cheng, Yeung-Leung, Wen-Liang Chang, Shih-Chun Lee, Yau-Gye Liu, Cheng-Jueng Chen, Shinn-Zong Lin, Nu-Man Tsai, Dah-Shyong Yu, Chung-Yang Yen, and Horng-Jyh Harn. 2004. "Acetone Extract of Angelica Sinensis Inhibits Proliferation of Human Cancer Cells via Inducing Cell Cycle Arrest and Apoptosis." *Life Sciences* 75, 13: 1579–1594. doi: 10.1016/j.lfs.2004.03.009.

Delbaldo, C., S. Michiels, N. Syz, J. Soria, T. Le Chevalier, and J. Pignon. 2004. "Benefits of Adding a Drug to a Single-Agent or a 2-Agent Chemotherapy Regimen in Advanced Non–Small-Cell Lung Cancer: A Meta-analysis." *The Journal of the American Medical Association* 292, 4: 470–484. doi: 10.1001/ jama.292.4.470.

DenesFeri. 2014. "*Ornithogalum umbellatum*." Obtained from Wikimedia under the Creative Commons Attribution-Share Alike 4.0 International license. https://commons.wikimedia.org/wiki/File: Ornithogalum_umbellatum_-_1006.jpg (Accessed 3 October 2021)

Dholaria, Bhagirathbhai, William Hammond, Amanda Shreders, and Yanyan Lou. 2016. "Emerging Therapeutic Agents for Lung Cancer." *Journal of Hematology and Oncology* 9, 1: 138. doi: 10.1186/ s13045-016-0365-z.

Ferlay, J, M Ervik, F Lam, M Colombet, L Mery, M Piñeros, et al. 2020. "Lung Source: Globocan 2020 Number of New Cases in 2020, Both Sexes, All Ages." https://gco.iarc.fr/today/data/factsheets/cancers/15-Lung-fact-sheet.pdf.

Goldstraw, Peter, David Ball, James R Jett, Thierry Le Chevalier, Eric Lim, Andrew G Nicholson, and Frances A Shepherd. 2011. "Non-Small-Cell Lung Cancer." *The Lancet* 378: 1727–1740. doi: 10.1016/S0140.

Grigorian, Alla, and Christopher B O'Brien. 2014. "Hepatotoxicity Secondary to Chemotherapy." *Journal of Clinical and Translational Hepatology* 2, 2: 95–102. doi: 10.14218/jcth.2014.00011.

Haggstrom, Mikael. 2008. "Symptoms of Cancer Metastasis." Obtained from Wikimedia under the Public domain license. https://commons.wikimedia.org/wiki/File:Symptoms_of_cancer_metastasis.png (Accessed 6 November 2021)

Hsu, Ya-Ling, Po-Lin Kuo, and Chun-Ching Lin. 2004. "The Proliferative Inhibition and Apoptotic Mechanism of Saikosaponin D in Human Non-Small Cell Lung Cancer A549 Cells." *Life Sciences* 75, 10: 1231–1242. doi: 10.1016/j.lfs.2004.03.008.

Kalemkerian, Gregory P., Billy W. Loo, Wallace Akerley, Albert Attia, Michael Bassetti, Yanis Boumber, Roy Decker, et al. 2018. "NCCN Guidelines Insights: Small Cell Lung Cancer, Version 2.2018." *Journal of the National Comprehensive Cancer Network* 16, 10: 1171–1182. doi: 10.6004/jnccn.2018.0079.

Kummalue, Tanawan. 2005. "Molecular Mechanism of Herbs in Human Lung Cancer Cells." *Journal of the Medical Association of Thailand* 88, 11: 1725–1734.

Lee, Eun-Jin, Hye-Young Min, Hyen Joo Park, Hwa-Jin Chung, Sanghee Kim, Yong Nam Han, and Sang Kook Lee. 2004. "G2/M Cell Cycle Arrest and Induction of Apoptosis by a Stilbenoid, 3,4,5-Trime thoxy-4′-Bromo-Cis-Stilbene, in Human Lung Cancer Cells." *Life Sciences* 75, 23: 2829–2839. doi: 10.1016/j.lfs.2004.07.002.

Lefnaer, Stefan. 2013. "*Bupleurum falcatum*." Obtained from Wikimedia under the Creative Commons Attribution-Share Alike 3.0 Unported license. https://commons.wikimedia.org/wiki/File:Bupleurum_falcatum_sl3.jpg (Accessed 3 October 2021)

Li-Feng Liua, Chia-Hua Liangb, Li-Yen Shiub, Wei-Ling Linb, Chih-Chao Linb, and Kou-Wha Kuob. 2004. "Action of Solamargine on Human Lung Cancer Cells-Enhancement of the Susceptibility of Cancer Cells to TNFs." *FEBS Letters*, no. 577: 67–74. doi: 10.1016/j.febslet.2004.09.064.

Mao, Yousheng, Ding Yang, Jie He, and Mark J Krasna. 2016. "Epidemiology of Lung Cancer." *Surgical Oncology Clinics of North America* 25: 439–445. doi: 10.1016/j.soc.2016.02.001.

Masuda, Yutaka, Genryu Shima, Toshihiro Aiuchi, Masayo Horie, Kouichi Hori, Shigeo Nakajo, Sachiko Kajimoto, Toshiko Shibayama-Imazu, and Kazuyasu Nakaya. 2004. "Involvement of Tumor Necrosis Factor Receptor-Associated Protein 1 (TRAP1) in Apoptosis Induced by-Hydroxyisovalerylshikonin." *Journal of Biological Chemistry* 279, 41: 42503–42515. doi: 10.1074/jbc.M404256200.

Meerbeeck, Jan P van, Dean A Fennell, and Dirk KM De Ruysscher. 2011. "Small-Cell Lung Cancer." *The Lancet* 378: 1741–1755. doi: 10.1016/S0140.

PDQ Adult Treatment Editorial Board. 2002. "Small Cell Lung Cancer Treatment (PDQ®): Patient Version. PDQ Cancer Information Summaries." *National Cancer Institution*. www.ncbi.nlm.nih.gov/pubmed/26389478.

Pearce, Alison, Marion Haas, Rosalie Viney, Sallie-Anne Pearson, Philip Haywood, Chris Brown, and Robyn Ward. 2017. "Incidence and Severity of Self-Reported Chemotherapy Side Effects in Routine Care: A Prospective Cohort Study." *PLoS One* 12, 10: e0184360. doi: 10.1371/journal.pone.0184360.

Pillai, G Radhakrishna, Anand S Srivastava, Tarek I Hassanein, Dharam P Chauhan, and Ewa Carrier. 2004. "Induction of Apoptosis in Human Lung Cancer Cells by Curcumin." *Cancer Letter* 208, 2: 163–170. doi: 10.1016/j.canlet.2004.01.008.

Ranga, Rama S, Srinivasan Sowmyalakshmi, Ravshan Burikhanov, Mohammed A Akbarsha, and Damodaran Chendil. 2005. "A Herbal Medicine for the Treatment of Lung Cancer." *Molecular and Cellular Biochemistry* 280, 1–2: 125–133. doi: 10.1007/s11010-005-8518-3

Risch, Angela, and Christoph Plass. 2008. "Lung Cancer Epigenetics and Genetics." *International Journal of Cancer* 123, 1: 1–7. doi: 10.1002/ijc.23605.

Schiller, Joan H. 2001. "Current Standards of Care in Small-Cell and Non-Small-Cell Lung Cancer." *Oncology* 61, 1: 3–13. doi: 10.1159/000055386.

Sloan, Memorial, Scott J Antonia, José A López-Martin, Johanna Bendell, Patrick A Ott, Matthew Taylor, Joseph Paul Eder, et al. 2016. "Articles Nivolumab Alone and Nivolumab plus Ipilimumab in Recurrent Small-Cell Lung Cancer (CheckMate 032): A Multicentre, Open-Label, Phase 1/2 Trial." *Lancet Oncology* 17, 17: 883–895. doi: 10.1016/S1470-2045(16)30098-5.

Song, Sung, Yoo, Won Woo Lee, Jin-Haeng Chung, So Yeon Park, Yu Kyeong Kim, and Sang Eun Kim. 2008. "Correlation between FDG Uptake and Glucose Transporter Type 1 Expression in Neuroendocrine Tumors of the Lung." *Lung Cancer* 61. doi: 10.1016/j.lungcan.2007.11.012.

Stang, David. 2005. "*Thalictrum acutifolium*." Obtained from Wikimedia under the Creative Commons Attribution-Share Alike 4.0 international license. https://commons.wikimedia.org/wiki/File:Thalictrum_acutifolium_0zz.jpg (Accessed 3 October 2021)

Stefan, Daniela Cristina, Ahmed M Elzawawy, Hussein M Khaled, Fabien Ntaganda, Anita Asiimwe, Beatrice Wiafe Addai, and Seth Wiafe. 2013. "Series Cancer Control in Africa 7 Developing Cancer Control Plans in Africa: Examples from Five Countries." *Lancet Oncology* 14. doi: 10.1016/S1470-2045(13)70100-1.

Stüber, Kurt. 2004. "*Indigofera tinctoria*." Obtained from Wikimedia under the Creative Commons Attribution-Share Alike 3.0 Unported license. https://commons.wikimedia.org/wiki/File:Indigofera_tinctoria1.jpg (Accessed 3 October 2021)

Sundar, Raghav, Maxime Chénard-Poirier, Dearbhaile Catherine Collins, and Timothy A. Yap. 2017. "Imprecision in the Era of Precision Medicine in Non-Small Cell Lung Cancer." *Frontiers in Medicine* 4, April. doi: 10.3389/fmed.2017.00039.

Tian Tao, Sun Jinpeng, Wang Jianxin, Liu Yanchun, Liu Haitao. 2018. "Isoliquiritigenin Inhibits Cell Proliferation and Migration through the PI3K/AKT Signaling Pathway in A549 Lung Cancer Cells." *Oncology Letters* 16: 6133–6139. doi: 10.3892/ol.2018.9344

Wae35244.2000. "*Lycoris radiata*." Obtained from Wikimedia under the Creative Commons Attribution-Share Alike 4.0 International license. https://commons.wikimedia.org/wiki/File:Lycoris_radiata_in_Feb_2000.jpg (Accessed 3 October 2021)

Wood, Douglas E, Vice Chair, Dara L Aisner, Wallace Akerley, Jessica Bauman, Joe Y Chang, Lucian R Chirieac, et al. 2018. "NCCN Guidelines Version 3.2018 Panel Members Non-Small Cell Lung Cancer." *Official Journal of the National Comprehensive Cancer Network* 16, 4: 412–441. doi: 10.6004/jnccn.2018.0020

Yamada, K. 2007. "*Nigella sativa*." Obtained from Wikimedia under the Creative Commons Attribution-Share Alike 2.0 Generic license. https://commons.wikimedia.org/wiki/File:Nigella_sativa.jpg (Accessed 3 October 2021)

Yin, Xiaolu, Jiangbing Zhou, Chunfa Jie, Dongming Xing, and Ying Zhang. 2004. "Anticancer Activity and Mechanism of *Scutellaria barbata* Extract on Human Lung Cancer Cell Line A549." *Life Sciences* 75, 18: 2233–2244. doi: 10.1016/j.lfs.2004.05.015.

Zhong, Wei-long, Yuan Qin, Shuang Chen and, and Tao Sun. 2017. "Antitumor Effect of Natural Product Molecules against Lung Cancer." In *A Global Scientific Vision*: 201–220. doi: 10.5772/67241

13 Phytonanotechnology
Phytonanoformulations and Their Applications in Cancer Therapy

Tafadzwa J. Chiome and Asha Srinivasan

CONTENTS

13.1 PLANT-DERIVED DRUG DISCOVERY

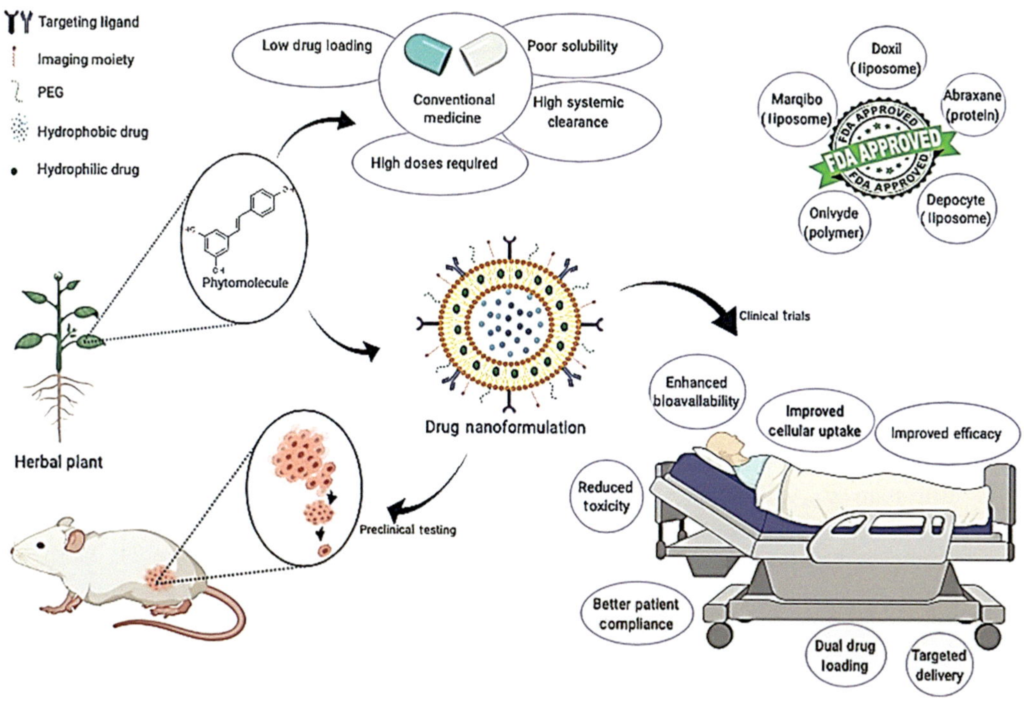

Graphical abstract representing the content of the chapter (created using Biorender).

DOI: 10.1201/9781003108375-13

Modern drug discovery is increasingly driven by scientific integration with plant-based medicines that have an ability to pull together seemingly unrelated branches of basic and applied sciences. In recent times, these combinations of basic and applied sciences have been capable of forming the elementary tools for innovative and efficacious solutions to multidisciplinary human health problems. Fossil records have revealed that human use of plants as traditional medicine dates back to the Middle Paleolithic age, approximately 60,000 years ago (Solecki, 1975).

Theophrastus (~300 BC) researched the medicinal qualities of herbs and noted their ability to change characteristics through cultivation. Similarly, Greeks, Romans and Chinese from 100–200 AD recorded most of the plant collections, their storage and use of medicinal herbs, and published more than 300 books on these subjects. The Indian Ayurveda, dates back to about 1000 BC, during which plants have formed the basis of sophisticated traditional medicine systems that have been in existence for thousands of years. Through all these records, it was noted that compounds extracted from medicinal plants are highly effective for the treatment of a variety of ailments (Li and Weng, 2017).

Some of the major examples are poppy (*Papaver somniferum* L.) and marijuana (*Cannabis sativa* L.), compounds in use for over 4,000 years. In 1806, Friedrich Serturner was able to isolate the alkaloid morphine from opium poppy seeds, which has led to the widespread investigations into the medicines derived from plants. Subsequently, in 1842 Pierre-Jean Robiquet was able to isolate codeine, and in 1848, the anti-spasmodic alkaloid papaverine was isolated from similar plant sources. These events led to a wide resurgence in plant-derived medicine research, and today, more than 95% of the plants are being used in traditional medicine and are collected from forest and other natural resources (Atanasov et al., 2015). There are countless natural active compounds that have been extracted and isolated from nature due to their unique chemical diversity. Some of the oldest plant-derived isolates that are of medicinal value include atropine, caffeine, digoxin and curare. Such compounds exhibit diverse biological activities and drug-like properties, thus becoming important resources for drug discovery and development of new lead compounds (Atanasov et al., 2015).

The last two decades have seen a dramatic rise in a wider selection of natural product development and drug discovery–related literature published, and shows different points of view in relation to use of natural compounds as drug leads and their potential (Table 13.1). The majority of the publications focused on plant-derived natural products and on compounds that are extracted from bacterial, fungal and marine hosts. These natural compounds are a good source of highly diverse structured elements that could be successfully marketed as new lead drugs in the forthcoming years (Guo, 2017).

There are several inherent advantages when using natural compounds in product development such as the following.

- A wide chemical diversity in natural compounds that influences the complexity of the structure and its potency.
- Capable of binding to biomolecules like proteins, giving them drug-like properties (Nisbet and Moore, 1997).
- The vast majority of natural products have not yet been explored to date, leading to the utilization of new drug discovery strategies of new bioactive compounds.
- Natural products are powerful biochemical tools that serve as pathfinders for molecular biology and chemistry, and in the investigation of cellular functions (Henkel et al., 1999).
- Research on natural products has led to the discovery of novel mechanisms of action.
- Natural products can guide the design of synthetic compounds and have led to a resurgence in plant-based drug discovery program tools and research (Breinbauer et al., 2002).

TABLE 13.1
Representation of Publications on Use of Natural Compounds in Modern Medicine as Drug Leads

Titles	Author(s), Year and Publications
"Natural Product Drug Discovery in the Next Millennium"	Cragg and Newman, 2001
"Natural Products in the Process of Finding New Drug Candidates"	Vuorela et al., 2004
"The Role of Natural Product Chemistry in Drug Discovery"	Butler, 2004
"Recent Natural Products Based Drug Development: A Pharmaceutical Industry Perspective"	Shu, 1998
"The Renaissance of Natural Products as Drug Candidates"	Paterson and Anderson, 2005
"Drug Discovery from Medicinal Plants"	Balunas and Kinghorn, 2005
"The Evolving Role of Natural Products in Drug Discovery"	Koehn and Carter, 2005
"Drug Discovery from Natural Products"	Gullo et al., 2006
"Drug Discovery from Natural Sources"	Chin et al., 2006
"Plant Natural Products: Back to the Future or into Extinction?"	McChesney et al., 2007
"Challenges and Opportunities in Drug Discovery from Plants"	Jachak and Saklani, 2007
"A Review of High Throughput Technology for the Screening of natural products"	Mishra et al., 2007
"New Aspects of Natural Products in Drug Discovery"	Lam, 2007
"The Value of Natural Products to Future Pharmaceutical Discovery"	Baker et al., 2007
"Molecular Understanding and Modern Application of Traditional Medicines: Triumphs and Trials"	Corson and Crews, 2007
"Natural Products in Drug Discovery"	Harvey, 2008
"Natural Products as a Robust Source of New Drugs and Drug Leads: Past Successes and Present-Day Issues"	Rishton, 2008
"Drug Discovery and Natural Products: End of an Era or an Endless Frontier?"	Li and Vederas, 2009
"Modern Natural Products Drug Discovery and Its Relevance to Biodiversity Conservation"	Kingston, 2011
"The Impact of the United Nations Convention on Biological Diversity on Natural Products Research"	Cragg et al., 2012
"The Pharmaceutical Industry and Natural Products: Historical Status and New Trends"	David et al., 2014
"The Re-Emergence of Natural Products for Drug Discovery in the Genomics Era"	Harvey et al., 2015

13.2 PLANT-DERIVED DRUG DISCOVERY PROCESSES

Drug discovery is a process by which potential and novel plant-based compounds are discovered after screening for therapeutic efficacy. The process through which a lead compound can be transformed into a commercial product can take up to 10–15 years for its development. Pharmaceutical companies have several approaches toward new drug discovery and development. However, two approaches are favored: phenotypic drug discovery and target-based drug discovery. In both these methods, the new drug discovery process involves rational drug design after initial target identification. In target-based drug discovery, the target is generally a gene or a protein that is linked with a disease. The success of new drug design is dependent on a target identification of a particular drug. The ability of a drug to bind to its target is assessed, with a thorough understanding of structural and thermodynamic basis on protein-ligand interactions, as it is this interaction that will initiate a cascade of biological events in response to the drug (Baker et al., 1995).

Once a target is identified, it is then validated for a response, which is critical in the disease process. The next step in the drug discovery process is identification, which involves discovery of the compound that will bind and interact with the target. A library of compounds is screened against the chosen validated target, and suitable hits are grown and developed into larger lead-like compounds. Several screening techniques such as high throughput screening, fragment-based screening and/or virtual screening can be used (Baker et al., 1995). Compounds in the hit series are optimized to improve potency and selectivity, which is achieved through functional group modifications. The identified lead compounds are modified to further improve the physicochemical and biological properties, making the compounds more effective and safer while transforming them into more viable drug candidates. The optimized compounds are then taken through the drug development process, starting with preclinical trials which lead to clinical testing (Baker et al., 1995).

13.3 MARKET VALUE FOR PLANT-BASED MEDICINES

The emergence in the use of natural products either as home remedies or as therapies for major ailments has augmented over the past 10–20 years. This sudden emergence has elevated the insatiable market for folkloric traditional medicine. Phyto-therapy is effective for some terminal diseases. According to the World Health Organization (WHO), it is estimated that the present demand for medicinal plants is about US $14 billion annually and by the year 2050 it would be about US $5 trillion (Nirmal et al., 2013). More than 80% of the population relies more on herbal drugs, and this trend is growing—not only in developing countries, but in developed countries, as well. It is predicted that this value will rise to US $59.4 million by 2025, while having 16.5% compound annual growth rate (CAGR) for the same period (Nirmal et al., 2013).

Follow-up studies have uncovered side effects related to the use of synthetic compounds, which have contributed to the wide acceptance of natural product as an alternative, thereby increasing its market value. In addition to their use in medicine, plant-based natural products are used as spices, essential oils, flavors, fragrance, dietary supplements, food and beverages, cosmetics and other personal care products. This has led to a huge rise in the market value of plant-based products. In addition, plant biotechnology is focused on drastically improving the yield and quality of these natural products for human purposes. Using the modern tools of genetic engineering, researchers are developing plant-based drugs or other active plant ingredients that are cost effective, easier to use and even more effective than their existing counterparts (Qiu et al., 2013).

13.4 PHYTONANOTECHNOLOGY IN CANCER

Plant-derived extracts are one of the richest sources of lead compounds in drug discovery. Plant extracts are screened and tested using *in vitro* targets, followed by bioassays and cytotoxicity analysis. Once a lead compound has been identified, it is isolated, purified and characterized. The compounds that show strong bioactivity are tested for *in vivo* toxicity, bioavailability and metabolite characterization in small animal models, and are taken further for drug development. Some of the earliest U.S. Food and Drug Administration (FDA)-approved plant-derived natural compounds used as therapy are the vinca alkaloids. These alkaloids were isolated from dry leaves of periwinkle plant (*Catharanthus roseus* [L.] G.Don), and two original alkaloids—vincristine (isolated in 1963) and vinblastine (isolated in 1965)—showed potential oral hypoglycemic agency (Marty et al., 1989). A detailed study on the two alkaloid extracts has indicated reduced white blood cell count in rats and extended the lives of mice with lymphocytic leukemia. As the original vinca-alkaloids had lower yields of the active compounds, semi synthetic vinca alkaloids with broad spectrum anti-cancer activity were approved for their anti-cancer activity in combination with other anti-cancer agents (Khayat and Antoine, 1997; Lilenbaum and Herbst, 2000). Plant-based medicine and related health products are being integrated into people's lives as a way to improve quality of life, improve immunity, prevent diseases and treat chronic illnesses. While developing modern cancer medicine, it is not only a matter of whether the drug can eliminate the cancer cells through apoptosis, but there is a need for advanced strategies which allow site-specific targeting, prevention of angiogenesis, immune response regulation or countering chronic inflammation associated with various types of cancers (Fridlender et al., 2015).

During the development of anti-cancer drugs, there has been a significant increase in the use of natural products, as they account for 60% of all anti-cancer drugs that have been developed (Seelinger et al., 2012). Such medicinal products originated from plants, marine organisms and microbes. However, over 3,000 species of plants alone are known to be used for the management of cancer (Song et al., 2014). Such plant-based medicines are developed from natural components or from their secondary derivatives, which further allow for the development of synthetic molecules inspired by nature's small molecules. The discovery of the key active ingredients from plant-based drugs is based on chemical, biochemical, pharmacological and clinical research data screening. Technological advancement has allowed for development of new synthesis strategies, which alter conventional methods used to deliver phyto-molecules, thereby creating a surge in the development of novel and more efficient herbal medicinal compounds. The majority of plant-based medicines are being screened for anti-cancer activity. In this context, several new primary or secondary plant-based derivatives of known anticancer agents are being subjected to clinical trials (Shukla and Mehta, 2015).

Some of the traditional examples of the translation from bench to bedside of a single type of natural product, particularly in pathological conditions such as cancer, include plant-derived compounds such as curcumin, genistein, reservetrol and silymarin. These compounds have entered clinical trials for various cancers such as colorectal cancer phase I, curcumin and genistein for pancreatic cancer phase II, resveratrol in phase I for colon cancer, silymarin in phase II for acute promyelocytic leukemia and liver cancer, and green tea for phase I studies in solid tumors (Carter et al., 2014; Dhillon, Aggarwal, Newman, et al., 2008; Ladas et al., 2010; Pisters et al., 2001; Sharma et al., 2004; Taylor et al., 2009).

Regardless of the immense knowledge known to date about medicinal plants, the FDA has approved little to no food or dietary supplements that can be used to treat or prevent cancer. This is due to the fact that even though a vast amount of products perform well in the phase II trials with promising results, a majority of them have not shown sufficient efficacy and safety, usually due to unfavorable pharmacokinetics (Pisters et al., 2001). Some of the bioactive compounds that are being used in medical research are shown in Table 13.2. The enormous diversity and chemical complexity in the structures of natural products, and their physicochemical and biological properties, are an important consideration to keep in mind when designing plant-based medicine (Weng et al., 2012).

13.5 HARNESSING NANOTECHNOLOGY IN PLANT-DERIVED MEDICINES

The therapeutic effectiveness of any plant-based medicine depends upon the ability of the dosage form to deliver a sufficient amount of the active compound to its site of action and elicit the expected pharmacological response. This attribute is commonly referred to as physiological bioavailability. Though plant-based bioactive compounds have shown impressive *in vitro* pharmacological activity, the *in vivo* effect has been less effective due to their poor solubility, poor systemic bioavailability, lipophilicity and inappropriate molecular size (Bilia et al., 2017). This has resulted in poor absorption and increased systemic clearance, which results in repeated administration of the drug, making the drug a poor candidate for therapeutic use. In addition to these features, the processes of absorption, distribution and metabolism lead to physicochemical modifications due to chemical interactions with the surrounding milieu. Thus, to successfully execute a therapeutic effect, a plant-derived bioactive compound must overcome the unfavorable physiological conditions *in vivo*. The challenge of delivering a bioactive drug molecule safely to a specific target site in a reproducible manner is of utmost importance in drug delivery systems (Laffleur and Keckeis, 2020).

TABLE 13.2
Phytochemicals Already Used in Cancer Research

Phytochemicals	Main Sources and Common Names	Cancer Types (*in vitro* and *in vivo*)	Clinical Trials	References
Ellagic acid	Pomegranate, grapes	Prostate, bladder, breast, colon	Prostate, follicular lymphoma	(Bell and Hawthorne, 2008) (Qiu et al., 2013) (Wang et al.,2012) (de Molina et al., 2015)
Epigallocatechin Gallate	Green tea	Prostate, breast, colon, pancreas, bone marrow, prostate	Postaste and breast	(Milligan et al., 2009) (Hwang et al., 2007) (Paschka et al., 1998) (Mittal et al., 2004) (Hoffmann et al., 2011) (Bettuzzi et al., 2006) (Ahn et al., 2003 Kumar et al., 2018)
Genistein	Soybean	Breast, cervical, colon	Breast, bladder	(Farina et al., 2006) (Shao et al., 1998) (Kim et al., 2009) (Nakamura et al., 2009)
Luteolin	Cabbages, celery, parsley	Prostate, breast, thyroid, cervical, ovary	-	(Yoo et al., 2009) (Lin et al., 2008) (Yin et al., 1999) (Park et al., 2013)
Silymarin	Thistle	Colon, lung, bladder skin, prostate	Upper gastrointestinal, leukemia	(Agarwal et al., 2006) (Zi et al., 1998) (Singh and Agarwal, 2002) (Deep et al., 2006)
Quercetin	Capers, apple	Pancreas, breast, colon, prostate, lung	Large bowel, ovary, pancreas, prostate, colorectal	(Vidya Priyadarsini et al., 2010) (Shan et al., 2009) (Ferry et al., 1996)
Resveratrol	Grape, berries	Breast, colorectal, hepatic melanoma, pancreas, prostate, skin, bladder, ovarian	Colorectal, colon, gastrointestinal tumors	(Alfaras et al., 2010) (Salado et al., 2011) (Wu et al., 2014) (Patel et al., 2010)
Sulforaphane	Brassica vegetables	Leukemia, breast, mammary, ovary, pancreas	Breast	(Gills et al., 2006) (Shen et al., 2007) (Singh et al., 2009) (Cornblatt et al., 2007)
Diallyl disulfide	Allyl vegetables	Neuroblastoma, prostate, thyroid	-	(Filomeni et al., 2003) (Yang et al., 2006) (Shin et al., 2010)
β-Sitosterol	Vegetal oils	Colon, breast, stomach, prostate, fibrosarcoma	-	(Baskar et al., 2010) (Zhao et al., 2009) (Moon et al., 2007) (Awad et al., 2007)
Curcumin	Curcuma longa l	Pancreas, prostate, cell carcinoma, leukemia, hepatoma,	Pancreas, colorectal, colon, pancreas, breast, head and neck	(Cheng et al., 2013) (Madden et al., 2009) (Bimonte et al., 2013) (Dhillon, Aggarwal and Newman, 2008)

Nanotechnology is on the threshold of providing unique delivery systems and equipping the materials with enhanced surface properties. These unique properties of nanomaterials are revolutionizing pharmaceutical industries. Several areas of medical research are already benefiting from the advantages of nanotechnology-based medicines. Transport of bioactive compounds to improve their bioavailability throughout the body, has greatly benefited by the advent of various nanocarriers. These nanocarriers release the bioactive in a controlled and sustained manner (Kariduraganavar et al., 2019).

The use of nanotechnology in the development of plant-based medicines has presented the pharmaceutical industries with a new platform for the development of efficient cancer drugs through repacking of conventional herbal molecules or plant-derived active components into various nano-sized structures. By encapsulating the active compound in a protective nanocarrier, potential physiochemical disruptions caused by enzymatic activity are reduced, thereby improving the bioavailability of the active compounds. In addition, nanoencapsulation of bioactive compounds helps reduce the frequency of dosing and provides physical protection during storage prior to their use for controlled cargo release. The popularity of nanomaterial-based delivery for anticancer agents is the result of the enhanced permeation and retention (EPR) effect, which is a passive tumor targeting phenomenon first described by Matsumura and Maeda (1986).

The rationale behind the EPR effect is that tumors possess leaky vascular architectures when compared with healthy blood vasculature. When nanoencapsulated active components are administered intravenously, their blood circulation time increases, leading it to accumulate at pathological sites, thereby enhancing the tumor concentration of anticancer therapeutics. However, recent studies have shown that, due to difficulties in reaching cancer cells deep inside malignant tissues, tumor targeting may not be achieved solely by the EPR effect, which highlights the need for synergistic passive and active targeting strategies. Therefore, several nanotechnology-based strategies have been sought to combine passive and active targeting through functionalization, conjugation and PEGylating of the nanomaterial surfaces for controlling the release of the drug payload in specific manner by engineering the nanomaterial to bind to the receptors expressed on the cancer cells (Singh et al., 2009).

Cancer therapy with the aid of active targeting has allowed for the conjugation of active compounds with the nanocarriers such that the designed molecules are capable of going forth and binding to the cancer target sites, which are overexpressed antigens or receptors. This targets specific features which make the drug safer when compared to other conventional chemotherapy treatments. The surface modification of the nanocarriers not only protects the neat drug from elimination, but will direct the drug to a specific tumor site where the drug will be released and induce a desired therapeutic effect with minimum systemic toxicity, with large drug concentrations successfully reaching the target sites (Patra et al., 2018).

13.6 IMPROVING BIOAVAILABILITY OF PLANT-BASED DRUG USING NANOTECHNOLOGY

With many plant-based medicines having poor bioavailability, the incorporation of nanotechnology in the development of herbal medicine has and is playing a pivotal role in countering the stability, solubility and delivery limitations involved with most conventional plant-based medicine. Use of phytochemicals in drug development is a complex process, as they have multiple precincts, which are a result of the molecular structure, physicochemical properties, hydrophobicity, size and complex formation in solvents via H-bond formation (Lipinski, 2000).

Poor bioavailability of bioactive compounds is especially noticeable when the route of delivery is oral. The gastrointestinal absorption forms a primary barrier between active component and systemic circulation. In addition, the enterocytes and hepatocytes lead to quick elimination of the phytomolecules. Due to these limitations in the oral route, most clinical trials adopt intravenous or intraperitoneal injections. A strategy to improve the bioavailability of the plant-based medicines is through the use of nanotechnology for the delivery of phytomolecules, which has made it possible to design drug systems capable of delivering amplified amounts of the therapeutic agent to the intended site while bypassing barriers such as the acid pH of the stomach and first pass metabolism, thereby increasing the residence time of the drug in the circulation (Kumari et al., 2019).

Herbal plants have become the perfect candidates for the development and reformulation of conventional drugs into nanoformulations, as they are highly feasible due to the following reasons.

- Most of the marketed formulations lack the capability of targeted delivery for chronic diseases like cancer.
- Some of the marketed formulations have side effects linked to their use.
- Lack of patient compliance due to large dosing requirements which are not directly proportional to the observed efficacy.

Nanosystems have shown more efficacy for the following reasons.

- High drug loading capacity, which allows for higher drug concentration accumulation in the circulatory system.
- Nanosizing the drug compound allows for a larger surface area, thereby increasing drug dissolution.
- An extended therapeutic effect is observed at the site of action.
- It has enhanced permeation and retention effect.
- Passive targeting to the site of action is observed without need of a ligand moiety.
- Reduces the dosage required for the formulation to give a therapeutic effect, thereby minimizing side effect.

There are several types of nanocarriers that are used to deliver bioactive compounds (Figure 13.1).

These nanocarriers in drug delivery systems aim to provide target-specific delivery, dosage miniaturization, better solubility and enhanced absorption, and to reduce or delay elimination of the drug. The use of nano-based delivery systems makes it possible to deliver phytomolecules to specific sites at predetermined rates, thereby resulting in site-specific action. Changing the structure of the compound at a molecular level and incorporating it into a nanocarrier system allows for a controlled drug release and distribution system. Taking into account timely and target-specific delivery, this helps to increase the efficacy of the compound while expanding its application range and thus ensuring efficient and optimal dosages. Currently, research is being done in the development of such carrier systems as they increase cellular uptake, drug solubility and stability, target-specific delivery and provide controlled release of the therapeutic payload (Yadav et al., 2011).

The use of plant extracts in conjunction with nanoformulations has shown to be advantageous, as nanoformulations can reduce the dose required while amplifying drug absorption. The activity of herbal medicines is highly dependent on the synergistic relationship between the constituents, which will determine the compound's overall function, as well as the therapeutic value (Ma et al., 2009). One of the major drawbacks of plant-based medicines is that most of them are insoluble in water, thereby resulting in poor absorption, translating to poor bioavailability. Such a property results in an increase in its systemic clearance, causing the drug to be administered at high dosages and making it a poor drug candidate due to its compromised therapeutic activity. The use of nanoformulations has shown to be effective in improving solubility, which results in higher bioavailability, enhanced pharmacological activity, better stability and reduced toxicity, and presents a controlled and sustained release, and drug protection from the host immune system, either through physical or chemical degradation. Some dosage forms used include polymeric nanoparticles, liposomes, nanoemulsions, proliposomes, liposomes and solid lipid nanoparticles. Thus, nanodrug delivery systems

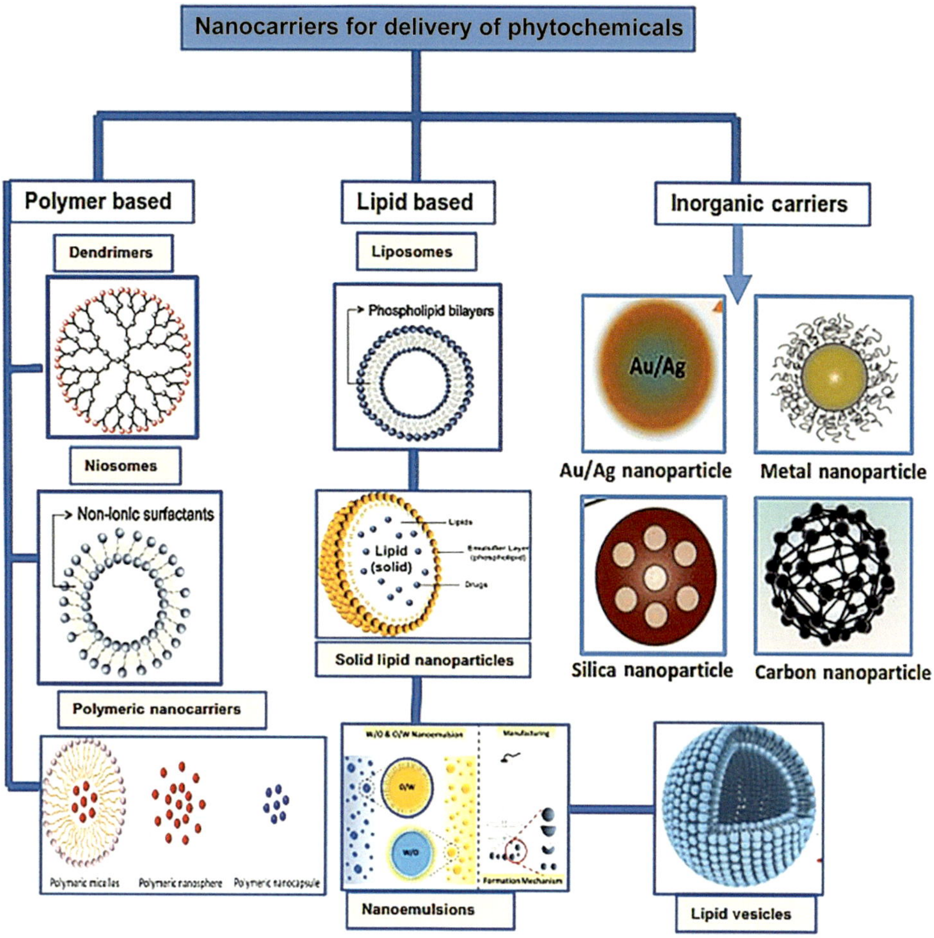

FIGURE 13.1 Types of nanocarriers used for the delivery of phytomolecules to improve drug efficacy, bioavailability and solubility.

are capable of enhancing the activity of the bioactive compound while resolving constraints associated with plant medicines (Tabassum and Pranav, 2018).

Currently, there are several nanocarriers that are used for encapsulating bioactive anticancer phytomolecules (Table 13.3). Among the several nanocarriers, research into polymer-based nanocarriers has been widely explored. Natural and selective synthetic polymers are preferred due to their biodegradability and biocompatibility nature. Polymers like poly(lactic-co-glycolic acid), poly vinyl alcohol and polyethylene glycol have been used in various studies for the delivery of bioactive molecule delivery to cancer cells, which resulted in improved drug uptake and efficiency (Pavlukhina et al., 2011).

Docetaxel is a well-known semi-synthetic poorly soluble taxane, which is being used in the treatment of breast, lung, oral and prostate cancers. The commercial form of Taxotere is formulated using high concentrations of Tween 80, which is linked to severe side effects like fluid retention, hypersensitivity reaction, mouth sores, hair loss, peripheral neuropathy and nausea (Immordino et al., 2003). Such a formulation resulted in lack of compatibility with poly-vinyl chloride administration sets. As a strategy to overcome multiple drug resistance during the treatment of breast cancer, docetaxel-loaded poly(ε-caprolactone)/Pluronic F68 nanoparticles was shown to be effective when used in docetaxel-resistant human breast cancer cell lines (Harush-Frenkel et al., 2008).

TABLE 13.3
List of Nanocarriers Used in Active Compounds and Their Advantages and Disadvantages

Types	Advantages	Disadvantages
Nanoparticles	Target delivery	Cytotoxicity
	Improved uptake by the cells because of their size	Nanoparticle accumulation
	Drug loading	Alveolar inflammation
	Better therapeutic efficacy	Broad size distribution
	Higher bioavailability	
	High surface to volume ratio	
Polymeric Nanoparticles	Administration through injection, oral and infusion	Difficult to scale up production
	Dual loading	Insufficient literature for toxicity assessment
	Improved stability	
	Better cellular uptake	
Liposomes	Allows for site targeting	Stability issues
	Efficient delivery	Short shelf life
	Reduces systemic toxicity	Elimination from the system by reticulo-endothelial system (RES)
	Better PK and PD control	Interaction with cells
Micelles	High drug loading capacity	Difficult to synthesize
	Highly stable	Stability issues
	Improved bioavailability	Slow release
	Enhanced circulatory time	
	Prevents drug degradation by immune system	
	Small diameter	
	Water soluble	
Gold nanoparticles	Less toxic	Toxicity
	Easy synthesis	Need for functionalization
	Size and morphology control	Biocompatibility
	High cellular uptake	
	Easy functionalization	
	High loading capacity	
Dendrimers	Can be tailor made	Complex to synthesize
	Enhanced permeability and retention	Limited drug loading capacity
	High drug entrapment	
	Low polydispersity index	
Carbon Nanotubes	Stable	Functionalization is necessary
	Possible to functionalize	Aggregate
	Large surface area	Lack of information on the health issues
	Easily taken up by cells	
	Hollow center for cargo loading	
Solid Lipid Nanoparticles	Cellular targeting is possible	Sudden gelation
	Controlled drug release	Burst release is usually observed
	Biocompatible	
	No use of organic solvents during synthesis	

This nanoformulation was prepared using the solvent displacement method to make porous spherical nanoparticles of 200 nm in diameter. The docetaxel-loaded nanoparticles exhibited a controlled release profile by which after 32 days, 67.91% of the drug had been released from the formulation. The developed nanoformulation had a higher cellular uptake efficiency in a breast cancer cell line (MCF-7), due to the nanoparticles' small size and with the aid of the clathrin endocytic machinery (Harush-Frenkel et al., 2008). The use of polycapri lactone/Pluronic F68 nanoparticles had a higher cytotoxicity as compared to that of PCL nanoparticles and commercial Taxotere, leading to the conclusion that such a nanoformulation could be used to overcome multiple drug resistance in human breast cancer cells (Mei et al., 2009).

Liposomes are nanocarriers that have spherical structures made up of concentric lipid bilayers separated by an aqueous medium and are used as nanocarriers. Liposomes are biocompatible, biodegradable and non-toxic, which has led to their use in the encapsulation of herbal biomolecules like curcumin and quercetin, thereby enhancing the therapeutic efficacy of the phytomolecules used. Their amphipathic nature makes it possible for encapsulation of both hydrophobic and hydrophilic bioactive compounds (Figure 13.2). The anticancer agent is encapsulated at the core of the liposome. Due to the nanosize and surface functionalization of the particle, the drug penetrates into the cancer cell, allowing the drug to be released, which induces a therapeutic effect. Surface functionalization of the liposome allows for targeted delivery to which most particles take advantage of the compromised structural integrity of the tissue to enter the tumor cells, while some require an endogenous or exogenous stimuli for targeting. The encapsulation of compounds in liposomal formulations help reduce peak valley fluctuations, provide a sustained release and offer specific targeting, while improving the solubility and bioavailability of the phyto-constituents (Singh et al., 2019).

Curcumin is one of the oldest substances extracted from multifunctional plants with numerous biological and therapeutic functions, such as wound healing, anti-inflammatory, anti-carcinogenic, antibacterial, antispasmodic, anticoagulant and anticancer activities. Studies have shown that curcumin strongly inhibits nuclear factor—Kappa B (NF-κB) activity and their related pathways, which plays a role in inflammation and cancer activation of cellular apoptosis (Olivera et al., 2012). Curcumin intake of up to 8 grams/day has not shown any toxic effect in humans, but poor bioavailability was observed. In order to overcome these limitations, a liposomal formulation of curcumin was developed with size ranging from 25–100 nm, which enhanced stability, bioavailability and site-specific targeting properties. Numerous studies have indicated that anticancer properties of curcumin alone and the liposomal curcumin are related to enzymes like cycloxygenase-2 (COX-2), AMP-activate protein kinase (AMPK), matrix metallopeptidases (MMPs) and nicotinamide adenine dinucleotide phosphate (NADPH); transcription factors such as NF-κB, AP-1, β-catenin and STAT-3; protein kinases; and certain growth factors (Figure 13.3) (Gera et al., 2017).

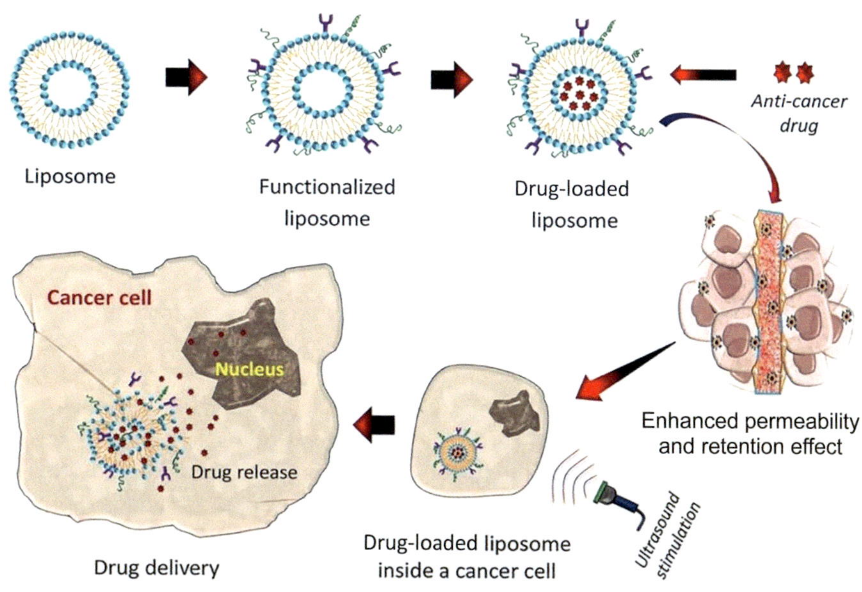

FIGURE 13.2 Schematic representation of drug delivery using liposomes (Hossen et al., 2019).

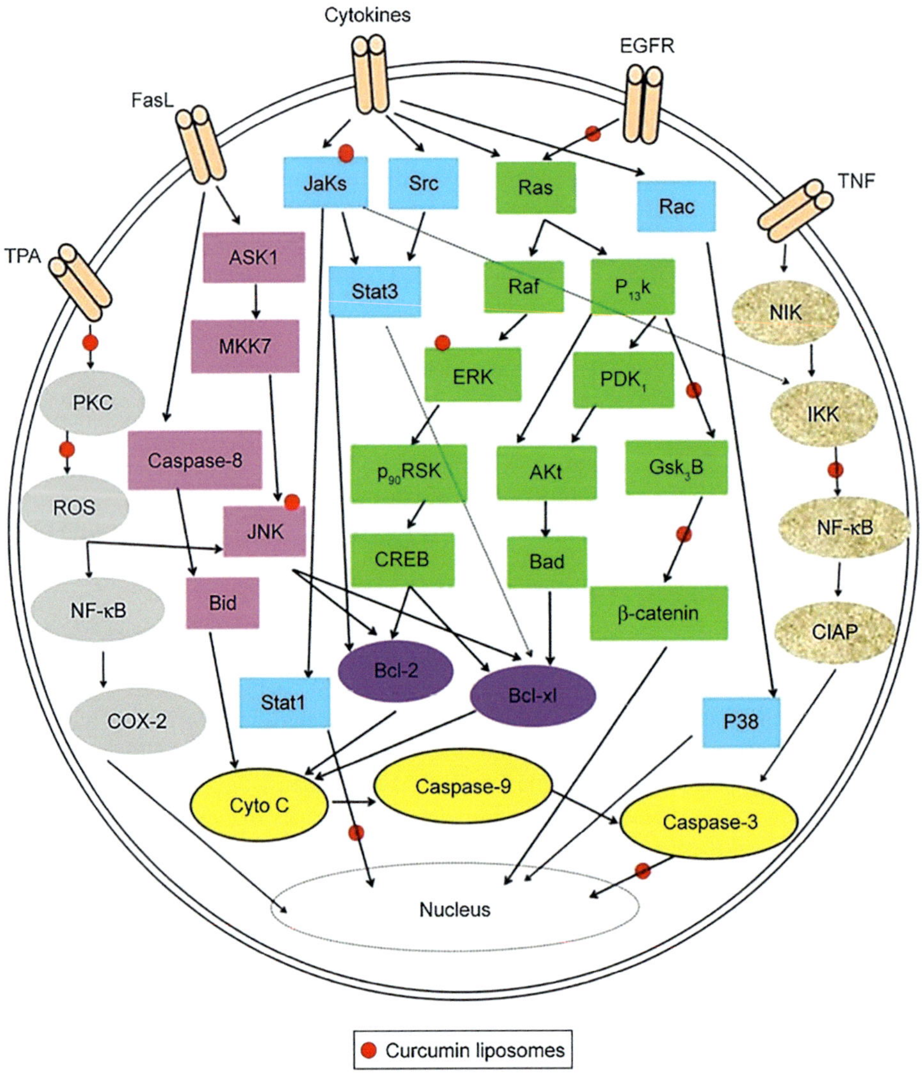

FIGURE 13.3 A schematic representation illustrating the multifunctional targets of liposomal curcumin in cancer by tumor promoter, extracellular growth factors, inflammatory cytokines and apoptotic markers (Feng et al., 2017).

Doxorubicin is a potent cytotoxic anticancer drug which is known to inhibit topoisomerase II and nucleic synthesis, leading to termination of cancer cell proliferation. Use of such a potent drug leads to the relaxation of supercoiled DNA structures during the transcription process whereby stabilization of DNA-topoisomerase II will prevent relaxation of the double helix and promotes termination. Regardless of the cytotoxicity nature of doxorubicin, its therapeutic use is limited due to adverse effects. It has been noted that it results in severe side effects like heart failure, dysrhythmia, leukocytopenia, nausea and hemorrhaging. Cardiotoxicity is the result of formation of cytotoxic free radicals in heart tissue. As a way to reduce the toxicity of doxorubicin, liposomes are being used to make doxorubicin nanoformulations. PEGylation of the liposomes resulted in a controlled release profile such that the circulation half-time lasted for up to four days, as compared to that of the neat drug, which is 20–25 hours (Boers-Sonderen et al., 2014).

Use of such novel formulations has shown clinical significance in decreasing cardiac events such as heart failure from taking place thereby leading to an increase in the survival rate of patients susceptible to cardiac risks. Other nanoformulations make use of micelles, which are spherical nanostructures of 10–100 nm formed through self-assembly of amphiphilic copolymers when in an aqueous environment. A semi-solid core is formed when the hydrophobic part is exposed to water, while a coronal layer is formed by the hydrophilic layer. Such a core-shell structure is important in drug delivery, as the hydrophobic core creates a reservoir for water insoluble drugs, while the outer shell acts as a protective layer to prevent rapid drug clearance (Torchilin, 2001).

13.7 COMBINATION BIOACTIVE MOLECULES IN NANOFORMULATIONS

Lack of durable response in monotherapies has led to development of combination therapies of two or more therapeutically active compounds to produce synergistic effect. Administering two or more drugs at a given time can be challenging, depending on mutually antagonistic effects of each drugs has on the other. However, as an alternative strategy, nanocarriers have facilitated the delivery of multiple drugs simultaneously. A combination of curcumin and resveratrol, polyphenolic compounds, is lipophilic in nature and has very low solubility in water. Several studies are in progress on these two phenolic compounds, as they are known for their many beneficial effects on human health (anticancer, anti-inflammatory, antimicrobial, antioxidant and chemo-preventive activity). It has already been reported that the use of these phytomolecules together is not only safe but is more efficient against breast cancer, with the only limitation being that they have poor solubility, which limits their application (Shindikar et al., 2016).

Rubusoside is a natural sweetener extracted from *Rubus suavissimus* S. Lee tea plant. Rubusoside is now increasingly being used in drug formulations due to its solubilizing properties, although very little is known about the specific mechanism behind its solubilizing nature (Zhang et al., 2017). Using coumarin-6 fluorescence markers, it was noted that the Rubusoside micelle formulation resulted in enhanced intracellular uptake of hydrophobic compounds. The nanoformulation showed greater cytotoxicity against MCF-7 cells when compared to the free compounds. The presence of rubusoside allowed for amplified cytotoxicity as compared to micelle formulations without the native compound. This study showed that it is possible to develop nanocarriers for the delivery of hydrophobic compounds while enhancing their efficacy in the process (Khoobchandani et al., 2013).

Due to the decrease in the diameter of particles from microscale to nanoscale, the particles start to exhibit unique physio-chemical and optical properties which are accompanied by a sharp increase in the drug loading capacity of the nanoparticles. Such properties have led to the use of materials like metal oxide and gold nanoparticles in cancer therapy, as they can be used in cancer diagnostics, as well as being functionalized for target-specific delivery use. As the world rallies for a greener environment, green synthesis of nanoparticles has made it possible to synthesize nanoparticles using natural or plant-based extracts, as they act as reducing agents whereby the bioactive materials found in the plant are coated onto the material (Khoobchandani et al., 2013).

Use of gold nanoparticles in cancer treatment has been forthcoming due to the particles having low cytotoxicity, great surface-to-volume ratio and are stable, which make them well suited for use in chemotherapy and immunotherapy. In order to improve the biocompatibility of material, researchers have turned to using proteins as drug carriers. Proteins are ideal natural molecules for the preparation of nanodrugs, since they are capable of interacting with both the solvents and the drug compounds. As carriers, proteins are both biocompatible and biodegradable such that they can be easily metabolized by the body without inducing any signs of cytotoxicity. They are known for their easy functionalization, which allows for attachment of drugs as well as ligands to facilitate targeted delivery (Kariduraganavar et al., 2019).

Abraxane is a commercially available nanoformulation of semi-synthetic paclitaxel whereby human serum albumin was used to synthesize 130 nm nanoparticles using nanoparticle albumin-bound (NAB) technology. This conjugation resulted in formation of a water-soluble compound which is currently being used worldwide for the treatment of metastatic breast cancer. The use of albumin as a carrier induces stability to the abraxane drug, but upon entry into the body, the compound rapidly dissolves to albumin-bound paclitaxel complex having a size similar to endogenous albumin. The transport of albumin-bound paclitaxel complexes across the endothelial barrier of tumor microvasculature is facilitated by GP60 receptor and caveolin-1 mediated transcytosis. Cellular uptake of this drug into cancer cells is mainly by the permeation effect during which the presence of albumin allows for facilitated uptake with the aid of glycoprotein (GP60). The presence of GP60 on the endothelial surface of tumor is responsible for the uptake of abraxane, which leads to the release of paclitaxel within the tumor mass. When the albumin-paclitaxel complex binds to the GP60 receptor molecule, it results in the clustering of GP60 and transcytosis of the drug complex, which is then released in the cancer cell Tiruppathi et al., 1997).

13.8 COMMERCIALLY AVAILABLE FDA-APPROVED NANODRUGS

Based on market research, the nanomedicine market is set to see a huge surge during the 2021–2027 period as compared to the period from 2012–2019. Regardless of the recession induced by the COVID-19 pandemic, the nanomedicine market is set to reach \$124.7 billion, with a compound annual growth rate (CAGR) of 22.6%, and the use of nanocrystals as carriers is predicted to set a market size of US \$83.1 billion (Research and Markets, 2020). Based on another market analysis for the 2017–2023 period, the CAGR is set to be 17.1% and determined based on applications and region of use. With an increase in significant investment in research and development from both the government and private sectors, cancer research has seen the most significant growth in nanomedicine, as it accounted for 35% of the total market share for the year 2016. The United States is set to see the largest growth, along with Europe, as it has recognized a profitable market for nanomedicine. Africa is also set to see an increase in the popularity and usage of nanodrugs over the coming years (Research and Markets, 2020).

The nanotechnology market has grown significantly over the past few years in the field of medicine, and has the potential to revolutionize drug delivery systems, medical diagnostics, gene therapies and other medical research areas. Impressive clinical results are being shown, as these nano-based delivery systems provide better diagnosis, reliable drug delivery and target specificity, and improve uptake by the tumor cells, leading to better disease management in cancer patients. Such nanotechnology platforms have led to development of drug delivery concepts and carrier properties which conventional drugs do not possess (Research and Markets, 2020). Recent decades have seen many nanoformulations entering the pharmaceutical market for the therapy of multiple diseases. Some of these FDA-approved formulations are listed in Table 13.4.

With still a lot of work being done in cancer research, co-delivery has shown to be a viable option when it comes to improving the efficiency of cancer treatment by using nanocarriers for the dual loading of both phytochemicals and other anti-tumor agents. Such a delivery has shown to be efficient at: (i) improving drug solubility; (ii) increasing bioavailability; (iii) minimizing chances of drug resistance development; (iv) delaying cellular adaptation; (v) inducing simultaneous therapeutic effect; (vi) delivery of optimum dosage; and (vii) reduced systemic cytotoxicity. Some studies have been done on several phytochemicals in conjunction with anti-cancer drugs used as first line treatment and some of the studies are listed in Table 13.5.

TABLE 13.4
FDA-Approved Nanodrugs and Their Active Compounds

Drug	Nanocarriers	Compounds	Applications	Year Approved
Ritalin	Nanocrystal		ADHD	1955
Abelcet	Liposome	Amphotericn B	Antifungal	1965
Adagen	Polymer	ADA	ADA deficiency	1990
Oncaspar	Polymer	Asparaginase	Acute lymphoblastic leukemia Chronic myelogenous leukemia	1994
Doxil	Liposome	Doxorubicin	Cancer	1995 (oral form) 2013 (injectable form)
Daunoxome	Liposome	Anthracycline	Cancer HIV associated Kaposi's Sarcoma Leukemia	1996
Ambisome	Liposome	Amphoeterin	Antifungal	1997
Taxol	Oil emulsion	Paclitaxel	Cancer	1998
Ontak	Polymer	Iterleukin 2	T-cell lymphoma	1999 (Discontinued 2014)
Depocyt	Liposome	Cytarabine	Cancer Lymphomatous meningitis	1999 2007
Visudyne	Liposome	Benzoporphyrin	Choroidal neovascularization	2001
Neulasta	Polymer	Filgrastim	Febrile neutropenia	2002
Pegasys	Polymer	Human alfa-2a interferon	Hepatisis C	2002
Somavert	Polymer	Human growth hormone	Acromegaly	2003
Macugen	Polymer	Pegatinib	Macular degeneration	2004
Tricor	Nanocrystal	Fenofibrate	Prevention of atherosclerosis	2004
Abraxane	Protein	Paclitaxel	Metastatic breast cancer Lung cancer Metastatic pancreatic adenocarcinoma	2005 2012 2013
Mirceta	Polymer	Epoetin	Anemia	2007
Cimzia	Polymer	IgG	Arthritis Crohn's disease	2008
Ferahene	Metal based	Ferumoxytol	Anemia	2009
Marqibo	Liposome	Vincristine	Cancer	2012
Onivyde	Liposome	Irinotecan	Metastatic pancreatic cancer	2015
Vyxeos	Liposome	Daunorubicin and cytarabine	Acute myeloid leukemia	2017
Rebinyn	Polymer	Factor IX	Hemophilia	2017

13.9 CONCLUSION

Natural products have been and will continue to be a main source of anticancer drug design and development. As discussed in this chapter, there are currently several phytochemical-based compounds undergoing clinical trials and drug development. The advent of nanotechnology has provided new strategies to overcome several limitations posed by plant-based active compounds. An effective drug discovery and development in phytochemicals is a long and tedious effort, which requires a multidisciplinary and collaborative approach for the discovery of lead active plant molecules. Nanotechnology has provided a unique method for drug delivery which scientists can manipulate in order to design and develop innovative nanocarrier molecules that effectively deliver the drug cargo to the site of action.

TABLE 13.5
Co-Administration of Phytochemical and Anti-Cancer Agents for Treatment of Various Cancers

Phytochemicals	Cancer Drugs	Type of Cancers	References
Genistein	Tamoxifen	Breast	(Ju et al., 2002)
	Gemcitabine	Pancreatic	(Banerjee et al., 2005)
	Interleukin-2	Osteosarcoma	(Karmanos, 2015)
		Kidney	(Zhang et al., 2010)
			(Kuzel, 2015)
Luteolin	Celecoxib	Breast	(Jeon and Suh, 2013)
Quercetin	Docetaxel	Prostate	(Wang et al., 2015)
	5-fluorouracil	Esophageal	(Chuang-Xin et al., 2012)
Resveratrol	Rapamycin	Breast	(Kim et al., 2014)
	Doxorubicin	Malignant glioma	(Lin et al., 2012)
	Temozolomide	Colon	(Chan et al., 2008)
	5-fluorouracil		(Ali and Braun, 2014)
	Mitomycin C		
Curcumin	Irinotecan	Colon	(Zhu et al., 2014)
	Dasatinib	Breast	(Nautiyal et al., 2011)
	Paclitaxel	Lung	(Aggarwal et al., 2005)
	Celecoxib		(Lev-Ari et al., 2005)
	Cyclodextrin		(Rocks et al., 2012)
Lycopene	Docetaxel	Prostate	(Tang et al., 2011)

Regardless of the advantages presented by the use of nano-based delivery systems, several challenges are encountered in the process of development of pharmaceuticals having maximum therapeutic effect while producing little to no side effects. The synthesis of nanoparticles is rather a complex process with economic product development procedures; however, other processes are expensive due to the several chemicals used during the initial setup. These chemicals are of concern toward the environment, as they may not be disposed of in the correct manner. On the other hand, the highly reactive nature of nanoparticles presents high risk in transportation and storage of such materials, as their stability is a concern.

REFERENCES

Agarwal, R., Agarwal, C., Ichikawa, H., Singh, R.P., and Aggarwal, B.B. 2006. "Anticancer potential of silymarin: From bench to bed side." *Anticancer Research* 26, 6: 4457–4498. PMID: 17201169

Aggarwal, B.B., Shishodia, S., Takada, Y. et al. 2005. "Curcumin suppresses the paclitaxel-induced nuclear factor-1κB pathway in breast cancer cells and inhibits lung metastasis of human breast cancer in nude mice." *Clinical Cancer Research* 11, 20: 7490–7498. doi: 10.1158/1078-0432.CCR-05-1192.

Ahn, W.S., Yoo, J., Huh, S.W., et al. 2003. "Protective effects of green tea extracts (polyphenon E and EGCG) on human cervical lesions." *European Journal of Cancer Prevention* 12, 5: 383–390. doi: 10.1097/00008469-200310000-00007.

Alfaras, I., Emìlia Juan, M., and Planas, J.M. 2010. "Trans-resveratrol reduces precancerous colonic lesions in dimethylhydrazine-treated rats." *Journal of Agricultural and Food Chemistry* 58, 13: 8104–8110. doi: 10.1021/jf100702x.

Ali, I., and Braun, D.P. 2014. "Resveratrol enhances mitomycin C mediated suppression of human colorectal cancer cell proliferation by up-regulation of p21WAF1/CIP1." *Anticancer Research* 34, 10: 5439–5446.

Atanasov, A.G., Waltenberger, B., Pferschy-Wenzig, E.M., Linder, T., Wawrosch, C., Uhrin, P., Temml, V., Wang, L., Schwaiger, S., Heiss, E.H., Rollinger, J.M., Schuster, D., Breuss, J.M., Bochkov, V.,

Mihovilovic, M.D., Kopp, B., Bauer, R., Dirsch, V.M., and Stuppner, H. 2015. "Discovery and resupply of pharmacologically active plant-derived natural products: A review." *Biotechnology Advances* 33, 8: 1582–1614. doi: 10.1016/j.biotechadv.2015.08.011.

Awad, B., Chinnam, M., Fink, C.S., and Bradford, P.G. 2007. "β-sitosterol activates Fas signaling in human breast cancer cells." *Phytomedicine: International Journal of Phytotherapy and Phytopharmacology* 14, 11: 747–754. doi: 10.1016/j.phymed.2007.01.003.

Baker, D.D., Chu, M., Oza, U., and Rajgarhia, V. 2007. "The value of natural products to future pharmaceutical discovery." *National Product Report* 24, 6: 1225–1244. doi: 10.1039/b602241n.

Baker, J.T., Borris, R.P., Carte´, B., Cordell, G.A., Soejarto, D.D., Cragg, G.M., Gupta, M.P., Iwu, M.M., Madulid, D.R., and Tyler, V.E. 1995. "Natural products drug discovery and development: New perspectives on international collaboration." *Journal of Natural Products* 58, 9: 1325–1357. doi: 10.1021/np50123a003.

Balunas, M.J., and Kinghorn, A.D. 2005. "Drug discovery from medicinal plants." *Life Sciences* 22, 5: 431–441. doi: 10.1016/j.lfs.2005.09.012.

Banerjee, S., Zhang, Y., Ali, S., et al. 2005. "Molecular evidence for increased antitumor activity of gemcitabine by genistein *in vitro* and *in vivo* using an orthotopic model of pancreatic cancer." *Cancer Research* 65, 19: 9064–9072. doi: 10.1158/0008-5472.CAN-05-1330.

Baskar, A.A., Ignacimuthu, S., Paulraj, G.M., and Al Numair, K.S. 2010. "Chemopreventive potential of beta-Sitosterol in experimental colon cancer model--an *in vitro* and *in vivo* study." *BMC Complementary and Alternative Medicine* 4; 10: 24. doi: 10.1186/1472-6882-10-24.

Bell, C., and Hawthorne, S. 2008. "Ellagic acid, pomegranate and prostate cancer: A mini review." *The Journal of Pharmacy and Pharmacology* 60, 2: 139–144. doi: 10.1211/jpp.60.2.0001.

Bettuzzi, S., Brausi, M., Rizzi, F., Castagnetti, G., Peracchia, G., and Corti, A. 2006. "Chemoprevention of human prostate cancer by oral administration of green tea catechins in volunteers with high-grade prostate intraepithelial neoplasia: A preliminary report from a one-year proof-of-principle study." *Cancer Research* 66, 2: 1234–1240. doi: 10.1158/0008-5472.CAN-05-1145.

Bilia, A., Piazzini, V., Guccione, C., Risaliti, L., Asprea, M., Capecchi, G., and Bergonzi, M. 2017. "Improving on nature: The role of nanomedicine in the development of clinical natural drugs." *Planta Medica* 83. doi: 10.1055/s-0043-102949.

Bimonte, S., Barbieri, A., Palma, G., LuciAno, A., Rea, D., and Arra, C. 2013. "Curcumin inhibits tumor growth and angiogenesis in an orthotopic mouse model of human pancreatic cancer." *Biomed Research International* 2013: 810423. doi: 10.1155/2013/810423.

Boers-Sonderen, M.J., van Herpen, C.M., van der Graaf, W.T., Desar, I.M., van der Logt, M.G., de Beer, Y.M., et al. 2014. "Correlation of toxicity and efficacy with pharmacokinetics (PK) of pegylated liposomal doxorubicin (PLD) (Caelyx®)." *Cancer Chemotherapy and Pharmacology* 74: 457–463. doi: 10.1007/s00280-014-2514-9.

Breinbauer, R., Vetter, I.R., and Waldmann, H. 2002. "Protein domains to drug candidates—Natural products as guiding principles in the design and synthesis of compound libraries." *Angewandte Chemie International Edition* 41: 2878–2890. doi: 10.1002/1521-3773(20020816)41:16<2878::AID-ANIE2878>3.0.CO;2-B.

Butler, M.S. 2004. "The role of natural product chemistry in drug discovery." *Journal of Natural Products* 67, 12: 2141–2153. doi: 10.1021/np040101y.

Carter, L.G., D'Orazio, J.A., and Pearson, K.J. 2014. "Resveratrol and cancer: Focus on in vivo evidence." *Endocrine Related Cancer* 21, 3: R209–R225. doi: 10.1530/ERC-13-1-71.

Chan, J.Y., Meng, S.P., Clement, M.-V., Pervaiz, S., and Shao, C.L. 2008. "Resveratrol displays converse dose-related effects on 5- fluorouracilevoked colon cancer cell apoptosis: The roles of caspase-6 and p53." *Cancer Biology and Therapy* 7, 8: 1305–1312. doi: 10.4161/cbt.7.8.6302.

Cheng, K.W., Wong, C.C., Mattheolabakis, G., Xie, G., Huang, L., and Rigas, B. 2013. "Curcumin enhances the lung cancer chemopreventive efficacy of phospho-sulindac by improving its pharmacokinetics." *International Journal of Oncology* 43, 3: 895–902. doi: 10.3892/ijo.2013.1995.

Chin, Y.W., Balunas, M.J., Chai, H.B., and Kinghorn, A.D. 2006. "Drug discovery from natural sources." *AAPS Journal* 14, 2: E239-E253. doi: 10.1007/BF02854894.

Chuang-Xin, L., Wen-Yu, W., Yao, C., Xiao-Yan, L., and Yun, Z. 2012. "Quercetin enhances the effects of 5-fluorouracil-mediated growth inhibition and apoptosis of esophageal cancer cells by inhibiting NF-1κB." *Oncology Letters* 4, 4: 775–778. doi: 10.3892/ol.2012.829.

Cornblatt, S., Ye, L., Dinkova-Kostova, A.T., et al. 2007. "Preclinical and clinical evaluation of sulforaphane for chemoprevention in the breast." *Carcinogenesis* 28, 7: 1485–1490. doi: 10.1093/carcin/bgm049.

Corson, T.W., and Crews, C.M. 2007. "Molecular understanding and modern application of traditional medicines: Triumphs and trials." *Cell* 7, 5: 769–774. doi: 10.1016/j.cell. 2007.08.021.

Cragg, G.M., Katz, F., Newman, D.J., and Rosenthal, J. 2012. "The impact of the United Nations Convention on Biological Diversity on natural products research." *Natural Products Report* 29, 12: 1407–1423. doi: 10.1039/c2np20091k.

Cragg, G.M., and Newman, D.J. 2001. "Natural product drug discovery in the next millennium." *Pharmaceutical Biology* 39, Suppl 1: 8–17. doi: 10.1076/phbi.39. s1.8.0009.

David, B., Wolfender, J., and Dias, D. A. 2014. "The pharmaceutical industry and natural products: historical status and new trends." *Phytochemistry reviews* 14, 2. doi: 10.1007/s11101-014-9367-z

Deep, G., Singh, R.P., Agarwal, C., Kroll, D.J., and Agarwal, R. 2006. "Silymarin and silibinin cause G1 and G2—M cell cycle arrest via distinct circuitries in human prostate cancer PC3 cells: A comparison of flavanone silibinin with flavanolignan mixture silymarin." *Oncogene* 25, 7: 1053–1069. doi: 10.1038/sj.onc.1209146.

de Molina, R., Vargas, T., Molina, S., et al. 2015. "The ellagic acid derivative 4,4′-Di-O-methylellagic acid efficiently inhibits colon cancer cell growth through a mechanism involving WNT16." *Journal of Pharmacology and Experimental Therapeutics* 353, 2: 433–444. doi: 10.1124/jpet.1114.221796.

Dhillon, N., Aggarwal, B.B., Newman, R.A., Wolff, R.A., Kunnumakkara, A.B., Abbruzzese, J.L., Ng, C.S., Badmaev, V., and Kurzrock, R. 2008. "Phase II trial of curcumin in patients with advanced pancreatic cancer." *Clinical Cancer Research* 15, 14: 4491–4499. doi: 10.1158/1078-0432.CCR-08-0024.

Farina, H.G., Pomies, M., Alonso, D.F., and Gomez, D.E. 2006. "Antitumor and antiangiogenic activity of soy isoflavone genistein in mouse models of melanoma and breast cancer." *Oncology Reports* 16, 4: 885–891.

Feng, T., et al. 2017. "Liposomal curcumin and its application in cancer." *International Journal of Nanomedicine* 12: 6027–6044. doi: 10.2147/IJN.S132434.

Ferry, D.R., Smith, A., Malkhandi, J., Fyfe, D.W., deTakats, P.G., Anderson, D., Baker, J., and Kerr, D.J. 1996. "Phase I clinical trial of the flavonoid quercetin: Pharmacokinetics and evidence for in vivo tyrosine kinase inhibition." *Clinical Cancer Research* 2, 4: 659–668.

Filomeni, G., Aquilano, K., Rotilio, G., and Ciriolo, M.R. 2003. "Reactive oxygen species-dependent c-Jun NH2-terminal kinase/c-Jun signaling cascade mediates neuroblastoma cell death induced by diallyl disulphide." *Cancer Research* 63, 18: 5940–5949.

Fridlender, M., Kapulnik, Y., and Koltai, H. 2015. "Plant derived substances with anti-cancer activity: From folklore to practice." *Frontiers in. Plant Science* 6: 799. doi: 10.3389/fpls.2015.00799.

Gera, M., Sharma, N., Ghosh, M., Luong Huynh, D., Lee, S. Jin, Min, T., Kwon, T., and Jeong, D. Kee. 2017. "Nanoformulations of curcumin: An emerging paradigm for improved remedial application." *Oncotarget* 8: 66680–66698. doi: 10.18632/oncotarget.19164.

Gills, J.J., Jeffery, E.H., Matusheski, N.V., Moon, R.C., Lantvit, D.D., and Pezzuto, J.M. 2006. "Sulforaphane prevents mouse skin tumorigenesis during the stage of promotion." *Cancer Letters* 236, 1: 72–79. doi: 10.1016/j.canlet.2005.05.007.

Gullo, V.P., McAlpine, J., Lam, K.S., Baker, D., and Petersen, F. 2006. "Drug discovery from natural products." *Journal of Industrial Microbiology and Biotechnology* 33, 7: 523–531. doi: 10.1007/s10295-006-0107-2.

Guo, Z. 2017. "The modification of natural products for medical use." *Acta Pharmaceutica Sinica B* 7, 2: 119–136. doi: 10.1016/j.apsb.2016.06.003.

Harush-Frenkel, O., Rozentur, E., Benita, S., and Altschuler, Y. 2008. "Surface charge of nanoparticles determines their endocytic and transcytotic pathway in polarized MDCK cells." *Biomacromolecules* 9: 435–443. doi: 10.1021/bm700535p.

Harvey, A.L. 2008. "Natural products in drug discovery." *Drug Discovery Today* 13, 19–20: 894–901. doi: 10.1016/j.drudis.2008.07.004.

Harvey, A.L., Edrada-Ebel, R., and Quinn, R. 2015. "The re-emergence of natural products for drug discovery in the genomics era." *National Reviews Drug Discovery* 14: 111–129. doi: 10.1038/nrd4510.

Henkel, T., Brunne, R.M., Müller, H., and Reichel, F. 1999. "Statistical investigation into the structural complementarity of natural products and synthetic compounds." *Angewandte Chemie International Edition* 38: 643–647. doi: 10.1002/(SICI)1521-3773(19990301)38:5<643: AID-Anie643>3.0.CO;2-G.

Hoffmann, J., Junker, H., Schmieder, A., et al. 2011. "EGCG downregulates IL-1RI expression and suppresses IL-1-induced tumorigenic factors in human pancreatic adenocarcinoma cells." *Biochemical Pharmacology* 82, 9: 1153–1162. doi: 10.1016/j.bcp.2011.07.063.

Hossen, S., Khalid Hossain, M., Basher, M.K., Mia, M.N.H., Rahman, M.T., and Jalal Uddin, M. 2019. "Smart nanocarrier-based drug delivery systems for cancer therapy and toxicity studies: A review." *Journal of Advanced Research* 15: 1–18. doi: 10.1016/j.jare.2018.06.005.

Hwang, J.T., Ha, J., Park, I.J., et al. 2007. "Apoptotic effect of EGCG in HT-29 colon cancer cells via AMPK signal pathway." *Cancer Letters* 247, 1–2: 115–121. doi: 10.1016/j.canlet.2006.03.030.

Immordino, M.L., Brusa, P., Arpicco, S., Stella, B., Dosio, F. and Cattel, L. 2003. "Preparation, characterization, cytotoxicity and pharmacokinetics of liposomes containing docetaxel." *Journal of Controlled. Release* 91: 417–429. doi: 10.1016/s0168-3659(03)00271-2.

Jachak, S., and Saklani, A. 2007. "Challenges and opportunities in drug discovery from plants." *Current Science* 92, 9: 1251–1257.

Jeon, Y.-W., and Suh, Y.J. 2013. "Synergistic apoptotic effect of celecoxib and luteolin on breast cancer cells." *Oncology Reports* 29, 2: 819–825. doi: 10.3892/or.2012.2158.

Ju, Y.H., Doerge, D.R., Allred, K.F., Allred, C.D., and Helferich, W.G. 2002. "Dietary genistein negates the inhibitory effect of tamoxifen on growth of estrogen-dependent human breast cancer (MCF-7) cells implanted in athymic mice." *Cancer Research* 62, 9: 2474–2477.

Kariduraganavar, M.Y., Heggannavar, G.B., Amado, S., and Mitchell, G.R. 2019. "Nanocarriers for drug delivery." *Micro and Nano Technologies*. Elsevier, Netherlands: 173–204. ISBN: 978-0-12-814033-8

Karmanos, B.A. 2015. "Gemcitabine hydrochloride and genistein in treating women with stage IV breast cancer." *Cancer Institute and National Cancer Institute (NCI)* NCT00244933.

Khayat, D., and Antoine, E. 1997. "Docetaxel in combination chemotherapy for metastatic breast cancer." *Seminars in Oncology* 24, 4 suppl 13: 19–26. PMID: 9335513

Khoobchandani, M., Zambre, A., Katti, K., Lin, C.H., and Katti, K.V. 2013. "Green nanotechnology from brassicaceae: development of broccoli phytochemicals–encapsulated gold nanoparticles and their applications in nanomedicine." *International Journal of Green Nanotechnology* 1: 1–15. doi: 10.1177/1943089213509474.

Kim, S.H., Kim, S.H., Lee, S.C., and Song, Y.S. 2009. "Involvement of both extrinsic and intrinsic apoptotic pathways in apoptosis induced by genistein in human cervical cancer cells." *Annals of the New York Academy of Sciences* 1171: 196–201. doi: 10.1111/j.1749-6632.

Kim, T.H., Shin, Y.J., Won, A.J., et al. 2014. "Resveratrol enhances chemosensitivity of doxorubicin in multidrug-resistant human breast cancer cells via increased cellular influx of doxorubicin." *Biochimica et Biophysica Acta (BBA)—General Subjects* 1840, 1: 615–625. doi: 10.1016/j.bbagen.2013.10.023.

Kingston, D.G. 2011. "Modern natural products drug discovery and its relevance to biodiversity conservation." *Journal of Natural Products* 25, 3: 496–511. doi: 10.1021/np100550t.

Koehn, F., and Carter, G. 2005. "The evolving role of natural products in drug discovery." *Natural Reviews Drug Discovery* 4: 206–220. doi: 10.1038/nrd1657.

Kumar, N.B., Dickinson, S.I., Schell, M.J., Manley, B.J., Poch, M.A., and Pow Sang, J. 2018. "Green tea extract for prevention of prostate cancer progression in patients on active surveillance." *Oncotarget* 9, 102: 37798–37806. doi: 10.18632/oncotarget.26519.

Kumari, P., Luqman, S., and Meena, A. 2019. "Application of the combinatorial approaches of medicinal and aromatic plants with nanotechnology and its impacts on healthcare." *Daru Journal of Pharmaceutical Sciences* 1: 475–489. doi: 10.1007/s40199-019-00271-6.

Kuzel, T. 2015. "Genistein and Interleukin-2 in treating patients with metastatic melanoma or kidney cancer." Northwestern University, National Cancer Institute (NCI). *Tech. Rep.* NCT00276835.

Ladas, E.J., Kroll, D.J., Oberlies, N.H., et al. 2010. "A randomized, controlled, double-blind, pilot study of milk thistle for the treatment of hepatotoxicity in childhood acute lymphoblastic leukaemia (ALL)." *Cancer* 116, 2: 506–513. doi: 10.1002/cncr.24723.

Laffleur, F., and Keckeis, V. 2020. "Advances in drug delivery systems: Work in progress still needed?" *International Journal of Pharmaceutics* 590: 119912. doi: 10.1016/j.ijpharm.2020.119912.

Lam, K.S. 2007. "New aspects of natural products in drug discovery." *Trends in Microbiology* 15, 6: 279–289. doi: 10.1016/j.tim.2007.04.001.

Lev-Ari, S., Strier, L., Kazanov, D., et al. 2005. "Celecoxib and curcumin synergistically inhibit the growth of colorectal cancer cells." *Clinical Cancer Research* 11, 18: 6738–6744. doi: 10.1158/1078-04322.CCR-05-0171.

Li, F.S., and Weng, J.K. 2017. "Demystifying traditional herbal medicine with modern approach." *Nature Plants* 3: 17109. doi: 10.1038/nplants.2017.109.

Li, J.W., and Vederas, J.C. 2009. "Drug discovery and natural products: End of an era or an endless frontier?" *Science* 10, 5937: 161–165. doi: 10.1126/science.1168243.

Lilenbaum, R.C., and Herbst, R.S. 2000. "Vinorelbine and gemcitabine combinations in advanced non-small-cell lung cancer." *Clinical Lung Cancer* 2, 2: 123–127. doi: 10.3816/clc. 2000. n.024.

Lin, C.J., Lee, C.C., Shih, Y.L., et al. 2012. "Resveratrol enhances the therapeutic effect of temozolomide against malignant glioma *in vitro* and *in vivo* by inhibiting autophagy." *Free Radical Biology and Medicine* 52, 2: 377–391. doi: 10.1016/j.freeradbiomed.2011.10.487.

Lin, Y., Shi, R., Wang, X., and Shen, H.M. 2008. "Luteolin, a flavonoid with potential for cancer prevention and therapy." *Current Cancer Drug Targets* 8, 7: 634–646. doi: 10.2174/156800908786241050.

Lipinski, A. 2000. "Drug-like properties and the causes of poor solubility and poor permeability." *Journal of Pharmacological and Toxicological Methods* 44, 1: 235–249. doi: 10.1016/s1056-8719(00)00107-6.

Ma, X.H., Zheng, C.J., Han, L.Y., Xie, B., Jia, J., Cao, Z.W., Li, Y.X., and Chen, Y.Z. 2009. "Synergistic therapeutic actions of herbal ingredients and their mechanisms from molecular interaction and network perspectives." *Drug Discovery Today* 14, 11–12: 579–588. doi: 10.1016/j.drudis.2009.03.012.

Madden, K., Flowers, L., Salani, R., et al. 2009. "Proteomics-based approach to elucidate the mechanism of antitumor effect of curcumin in cervical cancer." *Prostaglandins Leukotrienes and Essential Fatty Acids* 80, 1: 9–18. doi: 10.1016/j.plefa.2008.10.003.

Marty, M., Extra, J.M., Espie, M., Leandri, S., Besenval, M., and Krikorian, A. 1989. "Advances in vinca-alkaloids: Navelbine." *Nouvelle Revue Francaise d'Hematologie* 1, 2: 77–84.

Matsumura, Y., and Maeda, H. 1986. "A new concept for macromolecular therapeutics in cancer chemotherapy: Mechanism of tumoritropic accumulation of proteins and the antitumor agent smancs." *Cancer Research* 46: 6387–6392. PMID: 2946403

McChesney, J.D., Venkataraman, S.K., and Henri, J.T. 2007. "Plant natural products: Back to the future or into extinction?" *Phytochemistry* 68, 14: 2015–2022. doi: 10.1016/j.phytochem.2007.04.032.

Mei, L., Zhang, Y., Zheng, Y., Tian, G., Song, C., Yang, D., Chen, H., Sun, H., Tian, Y., Liu, K., Li, Z., and Huang, L. 2009. "A novel docetaxel-loaded poly (ε-caprolactone)/pluronic F68 nanoparticle overcoming multidrug resistance for breast cancer treatment." *Nanoscale Research Letters* 4, 12: 1530. doi: 10.1007/s11671-009-9431-6.

Milligan, S.A., Burke, P., Coleman, D.T., et al. 2009. "The green tea polyphenol EGCG potentiates the antiproliferative activity of c-Met and epidermal growth factor receptor inhibitors in non-small cell lung cancer cells." *Clinical Cancer Research* 15, 15: 4885–4894. doi: 10.1158/1078-0432.CCR-09-0109.

Mishra, K.P., Ganju, L., Sairam, M., Banerjee, P.K., and Sawhney, R.C. 2008. "A review of high throughput technology for the screening of natural products." *Biomedicine and Pharmacotherapy* 62, 2: 94–98. doi: 10.1016/j.biopha.2007.06.012.

Mittal, A., Pate, M.S., Wylie, R.C., Tollefsbol, T.O., and Katiyar, S.K. 2004. "EGCG down-regulates telomerase in human breast carcinoma MCF-7 cells, leading to suppression of cell viability and induction of apoptosis." *International Journal of Oncology* 24, 3: 703–710. PMID: 14767556

Moon, D.-O., Lee, K.-J., Choi, Y.H., and Kim, G.-Y. 2007. "β-Sitosterol-induced-apoptosis is mediated by the activation of ERK and the downregulation of Akt in MCA-102 murine fibrosarcoma cells." *International Immunopharmacology* 7, 8: 1044–1053. doi: 10.1016/j.intimp.2007.03.010.

Nakamura, Y., Yogosawa, S., Izutani, Y., Watanabe, H., Otsuji, E., and Sakai, T. 2009. "A combination of indol-3-carbinol and genistein synergistically induces apoptosis in human colon cancer HT-29 cells by inhibiting Akt phosphorylation and progression of autophagy." *Molecular Cancer* 8, 100: 1476–4598. doi: 10.1186/1476-4598-8-100.

Nautiyal, J., Banerjee, S., Kanwar, S.S. et al. 2011. "Curcumin enhances dasatinib-induced inhibition of growth and transformation of colon cancer cells." *International Journal of Cancer* 128, 4: 951–961. doi: 10.1002/ijc.25410.

Nirmal, S., Pal, S., Otimenyin, S., Aye, T., Mostafa, E., Kundu, S., Amirthalingam, R., and Mandal, S.C. 2013. "Contribution of herbal products in global market." *The Pharma Review*, 95–104.

Nisbet, L.J., and Moore, M. 1997. "Will natural products remain an important source of drug research for the future?" *Current Opinion in Biotechnology* 8, 6: 708–712. doi: 10.1016/s0958-1669(97)80124-3.

Olivera, A., et al. 2012. "Inhibition of the NF-κB signaling pathway by the curcumin analog, 3,5-Bis(2-pyridinylmethylidene)-4-piperidone (EF31): Anti-inflammatory and anti-cancer properties." *International Immunopharmacology* 12, 2: 368–377. doi: 10.1016/j.intimp.2011.12.009.

Park, S.H., Park, H.S., Lee, J.H., et al. 2013. "Induction of endoplasmic reticulum stress-mediated apoptosis and non-canonical autophagy by luteolin in NCI-H460 lung carcinoma cells." *Food and Chemical Toxicology* 56: 100–109. doi: 10.1016/j.fct.2013.02.022.

Paschka, G., Butler, R., and Young, C.Y.F. 1998. "Induction of apoptosis in prostate cancer cell lines by the green tea component, (-)-epigallocatechin-3-gallate." *Cancer Letters* 130, 1–2: 1–7. doi: 10.1016/s0304-3835(98)00084-6.

Patel, K.R., Brown, V.A., Jones, D.J.L., et al. 2010. "Clinical pharmacology of resveratrol and its metabolites in colorectal cancer patients." *Cancer Research* 70, 19: 7392–7399. doi: 10.1158/0008-5472.CAN-10-2027.

Paterson, I., and Anderson, E.A. 2005. "Chemistry: The renaissance of natural products as drug candidates." *Science* 310, 5747: 451–453. doi: 10.1126/science.1116364.

Patra, J.K., Das, G., Fraceto, L.F., et al. 2018. "Nano based drug delivery systems: Recent developments and future prospects." *Journal of Nanobiotechnology* 16, 1: 71. doi: 10.1186/s12951-018-0392-8.

Pavlukhina, S., and Sukhishvili, S. 2011. "Polymer assemblies for controlled delivery of bioactive molecules from surfaces." *Advanced Drug Delivery Reviews* 63, 9: 822–836. doi: 10.1016/j.addr.2011.03.017.

Pisters, K.M., Newman, R.A., Coldman, B., Shin, D.M., Khuri, F.R., Hong, W.K., Glisson, B.S., and Lee, J.S. 2001. "Phase I trial of oral green tea extract in adult patients with solid tumors." *Journal of Clinical Oncology* 15, 6: 1830–1838. doi: 10.1200/JCO.2001.19.6.1830.

Qiu, Z., Zhou, B., Jin, L., Yu, H., Liu, L., Liu, Y., Qin, C., Xie, S., and Zhu, F. 2013. "*In vitro* antioxidant and antiproliferative effects of ellagic acid and its colonic metabolite, urolithins, on human bladder cancer T24 cells." *Food and Chemical Toxicology* 59: 428–437. doi: 10.1016/j.fct.2013.06.025.

Research and Markets. 2020. "Global nanotechnology in drug delivery markets, 2020–2027 with profiles of 43 industry players." *PRNewswire*. 28 May. www.prnewswire.com/news-releases/global-nanotechnology-in-drug-delivery-markets-2020-2027-with-profiles-of-43-industry-players-301067245.html

Rishton, G.M. 2008. "Natural products as a robust source of new drugs and drug leads: Past successes and present-day issues." *The American Journal of Cardiology* 22, 10A: 43D–49D. doi: 10.1016/j.amjcard.2008.02.007.

Rocks, N., Bekaert, S., Coia, I., et al. 2012. "Curcumin-cyclodextrin complexes potentiate gemcitabine effects in an orthotopic mouse model of lung cancer." *British Journal of Cancer* 107, 7: 1083–1092. doi: 10.1038/bjc.2012.379.

Salado, C., Olaso, E., Gallot, N., et al. 2011. "Resveratrol prevents inflammation-dependent hepatic melanoma metastasis by inhibiting the secretion and effects of interleukin-18." *Journal of Translational Medicine* 9: 59. doi: 10.1186/1479-5876-9-59.

Seelinger, M., Popescu, R., Giessrigl, B., Jarukamjorn, K., Unger, C., Wallnöfer, B., Fritzer-Szekeres, M., Szekeres, T., Diaz, R., Jäger, W., Frisch, R., Kopp, B., and Krupitza, G. 2012. "Methanol extract of the ethnopharmaceutical remedy Smilax spinosa exhibits anti-neoplastic activity." *International Journal of Oncology* 41, 3: 1164–1172. doi: 103892//ijo.2012.1538.

Shan, B.E., Wang, M.X., and Li, R.Q. 2009. "Quercetin inhibit human SW480 colon cancer growth in association with inhibition of cyclin D1 and survivin expression through Wnt/β-catenin signalling pathway." *Cancer Investigation* 27, 6: 604–612. doi: 10.1080/07357900802337191.

Shao, Z.M., Wu, J., Shen, Z.Z., and Barsky, S.H. 1998. "Genistein exerts multiple suppressive effects on human breast carcinoma cells." *Cancer Research* 58, 21: 4851–4857.

Sharma, R.A., Euden, S.A., Platton, S.L., Cooke, D.N., Shafayat, A., Hewitt, H.R., Marczylo, T.H., Morgan, B., Hemingway, D., Plummer, S.M., Pirmohamed, M., Gescher, A.J., and Steward, W.P. 2004. "Phase I clinical trial of oral curcumin: Biomarkers of systemic activity and compliance." *Clinical Cancer Research* 15, 20: 6847–6854. doi: 10.1158/1078-0432.CCR-04-0744.

Shen, G., Tin, O.K., Hu, R., et al. 2007. "Chemoprevention of familial adenomatous polyposis by natural dietary compounds sulforaphane and dibenzoylmethane alone and in combination in Apc Min/+ mouse." *Cancer Research* 67, 20: 9937–9944. doi: 10.1158/0008-5472.CAN-07-1112.

Shin, H.A., Cha, Y.Y., Park, M.S., Kim, J.M., and Lim, Y.C. 2010. "Diallyl sulfide induces growth inhibition and apoptosis of anaplastic thyroid cancer cells by mitochondrial signaling pathway." *Oral Oncology* 46, 4: e15–e18. doi: 10.1016/j.oraloncology. 2009.10.012.

Shindikar, A., Singh, A., Nobre, M., and Kirolikar, S. 2016. "Curcumin and reserveratol as promising natural remedies with nanomedicine approach for the effective treatment of triple negative breast cancer." *Journal of Oncology*. doi: 10.1155/2016/9750785.

Shu, Y.Z. 1998. "Recent natural products based drug development: A pharmaceutical industry perspective." *Journal of Natural Products* 61, 8: 1053–1071. doi: 10.1021/np9800102.

Shukla, S., and Mehta, A. 2015. "Anticancer potential of medicinal plants and their phytochemicals: A review." *Brazilian Journal Botany* 38: 199–210. doi: 10.1007/s40415-015-0135-0.

Singh, M., Devi, S., Rana, V.S., Mishra, B.B., Kumar, J., and Ahluwalia, V. 2019. "Delivery of phytochemicals by liposome cargos: Recent progress, challenges and opportunities." *Journal of Microencapsulation* 36, 3: 215–235. doi: 10.1080/02652048.2019.1617361.

Singh, R., and Lillard, J.W. Jr. 2009. "Nanoparticle-based targeted drug delivery." *Experimental and Molecular Pathology* 86, 3: 215–223. doi: 10.1016/j.yexmp.2008.12.004.

Singh, R.P., and Agarwal, R. 2002. "Flavonoid antioxidant silymarin and skin cancer." *Antioxidants and Redox Signalling* 4, 4: 655–663. doi: 10.1089/15230860260220166.

Singh, S.V., Warin, R., Xiao, D., et al. 2009. "Sulforaphane inhibits prostate carcinogenesis and pulmonary metastasis in TRAMP mice in association with increased cytotoxicity of natural killer cells." *Cancer Research* 69, 5: 2117–2125. doi: 10.1158/0008-5472.CAN-08-3502.

Solecki, R. 1975. "Shanidar, IV, A Neanderthal flower burial in northern Iraq." *Science* 190: 880–881. doi: 10.1126/science.190.4217.880.

Song, Y., Sun, H., Zhang, A., Yan, G., Han, Y., and Wang, X. 2014. "Plant derived natural products as leads to anticancer drugs." *Journal of Medicinal Plants & Herbal Therapy Research* 2: 6–15.

Tabassum, K., and Pranav, G. 2018. "PhytoNanotechnology: Enhancing delivery of plant based anti-cancer drugs." *Frontiers in Pharmacology* 8: 1002. doi: 10.3389/fphar.2017.01002.

Tang, Y., Parmakhtiar, B., Simoneau, A.R., et al. 2011. "Lycopene enhances docetaxel's effect in castration-resistant prostate cancer associated with insulin-like growth factor I receptor levels." *Neoplasia* 13, 2: 108–119. doi: 10.1593/neo.101092.

Taylor, C.K., Levy, R.M., Elliott, J.C., and Burnett, B.P. 2009. "The effect of genistein aglycone on cancer and cancer risk: A review of in vitro, preclinical, and clinical studies." *Nutrition Reviews* 67: 398–415. doi: 10.1111/j.1753-4887.2009.00213. x.

Tiruppathi, C., Song, W., Bergenfeldt, M., Sass, P., and Malik, A.B. 1997. "Gp60 activation mediates albumin transcytosis in endothelial cells by tyrosine kinase-dependent pathway." *Journal of Biological Chemistry* 272: 25968–25975. doi: 10.1074/jbc.272.41.25968.

Torchilin, V.P. 2001. "Structure and design of polymeric surfactant-based drug delivery systems." *Journal of Controlled Release* 73: 137–172. doi: 10.1016/S0168-3659(01)00299-1.

Vidya Priyadarsini, R., Senthil Murugan, R., Maitreyi, S., Ramalingam, K., Karunagaran, D., and Nagini, S. 2010. "The flavonoid quercetin induces cell cycle arrest and mitochondria-mediated apoptosis in human cervical cancer (HeLa) cells through p53 induction and NF-κB inhibition." *European Journal of Pharmacology* 649, 1–3: 84–91. doi: 10.1016/j.ejpahr.2010.09.020.

Vuorela, P., Leinonen, M., Saikku, P., Tammela, P., Rauha, J.P., Wennberg, T., and Vuorela, H. 2004. "Natural products in the process of finding new drug candidates." *Current Medicinal Chemistry* 11: 1375. doi: 10.2174/0929867043365116.

Wang, N., Wang, Z.Y., Mo, S.L., et al. 2012. "Ellagic acid, a phenolic compound, exerts anti-angiogenesis effects via VEGFR-2 signalling pathway in breast cancer." *Breast Cancer Res Treat* 134, 3: 943–955. doi: 10.1007/s10549-012 1977-9.

Wang, P., Henning, S., Heber, D., and Vadgama, J. 2015. "Enhanced inhibition of PC-3 xenograft prostate tumor growth by combination of green tea and quercetin with docetaxel." *Cancer Research* 75, 15_ Suppl: 5345. doi: 10.1158/1538-7445.AM2015-5345.

Weng, J.K., Philippe, R.N., and Noel, J.P. 2012. "The rise of chemo-diversity in plants." *Science* 336, 6089: 1667–1670. doi: 10.1126/science.1217411.

Wu, M.L., Li, H., Yu, L.J., Chen, X.Y., Kong, Q.Y., Song, X., Shu, X.H., and Liu, J. 2014. "Short-term resveratrol exposure causes *in vitro* and *in vivo* growth inhibition and apoptosis of bladder cancer cells." *PLoS One* 25, 2: e89806. doi: 10.1371/journal.pone.0089806.

Yadav, D., Suri, S., Choudhary, A.A., Sikender, M., Hemant, Beg N.M., et al. 2011. "Novel approach: Herbal remedies and natural products in pharmaceutical science as nanodrug delivery systems." *International Journal Pharmacy and Technology* 3: 3092–3116.

Yang, J.S., Kok, L.F., Lin, Y.H., et al. 2006. "Diallyl disulfide inhibits WEHI-3 leukemia cells *in vivo*." *Anticancer Research* 26, 1: 219–225.

Yin, F., Giuliano, A.E., and Van Herle, A.J. 1999. "Growth inhibitory effects of flavonoids in human thyroid cancer cell lines." *Thyroid* 9, 4: 369–376. doi: 10.1089/thy.1999.9.369.

Yoo, D.R., Jang, Y.H., Jeon, Y.K., et al. 2009. "Proteomic identification of anti-cancer proteins in luteolin-treated human hepatoma Huh-7 cells." *Cancer Letters* 282, 1: 48–54. doi: 10.1016/j.canlet.2009.02.051.

Zhang, B., Shi, Z.L., Liu, B., Yan, X.B., Feng, J., and Tao, H.-M. 2010. "Enhanced anticancer effect of gemcitabine by genistein in osteosarcoma: The role of Akt and nuclear factor-IκB." *Anti-Cancer Drugs* 21, 3: 288–296. doi: 10.1097/CAD.0b013e328334da17.

Zhang, M., Dai, T., and Feng, N. 2017. "A novel solubility enhanced rubusocide based micelles for increased cancer therapy." *Nanoscale Research Letters* 12, 274: 1–10. doi: 10.1186/s11671-017-2054-4.

Zhao, Y., Chang, S.K.C., Qu, G., Li, T., and Cui, H. 2009. "β-Sitosterol inhibits cell growth and induces apoptosis in SGC-7901 human stomach cancer cells." *Journal of Agricultural and Food Chemistry* 57, 12: 5211–5218. doi: 10.1021/jf803878n.

Zhu, D.J., Chen, X.W., Wang, J., Ju, Y.L., Yang, M.Z.O., and Zhang, W.J. 2014. "Proteomic analysis identifies proteins associated with curcumin-enhancing efficacy of irinotecan-induced apoptosis of colorectal cancer LOVO cell." *International Journal of Clinical and Experimental Pathology* 7, 1: 1–15.

Zi, X., Feyes, D.K., and Agarwal, R. 1998. "Anticarcinogenic effect of a flavonoid antioxidant, silymarin, in human breast cancer cells MDA-MB 468: Induction of G1 arrest through an increase in Cipl/p21 concomitant with a decrease in kinase activity of cyclin-dependent kinases and associated cyclins." *Clinical Cancer Research* 4, 4: 1055–1064.

14 Plant-Based Adjunct Therapy for Tuberculosis

Lydia Gibango, Anna-Mari Reid, Jonathan L. Seaman and Namrita Lall

CONTENTS

14.1 INTRODUCTION

14.1.1 History of Tuberculosis

Tuberculosis (TB) is a disease known to humanity since ancient times, and it is believed that Clarissimus Galen, the Greek personal physician of the Roman Emperor Marcus Aurelius (174 AD), suspected its contagious nature. He suggested that the symptoms include sweating, fever, severe coughing and blood-stained sputum (Sakula, 1983; Daniel et al., 1994; Barberis et al., 2017). It has been hypothesized that the *Mycobacterium* genus originated more than 150 million years ago. Modern molecular genetics techniques allow for a more accurate estimation of the origin of mycobacteria. The low mutation rate of *Mycobacterium tuberculosis* facilitated this estimation and the conclusion was that an early progenitor of *M. tuberculosis* was present in East Africa as early as three million years ago. This suggests that early hominids at that time may have suffered from an infection, and it is likely that members of the *M. tuberculosis* complex—including the African variants *Mycobacterium africanum* and *Mycobacterium bovis*—had a common African ancestor about 15,000–35,000 years ago (Kapur et al., 1994; Brosch et al., 2002; Gutierrez et al., 2005). East Africa was believed to have been the ancestral home of both tubercle bacilli

DOI: 10.1201/9781003108375-14

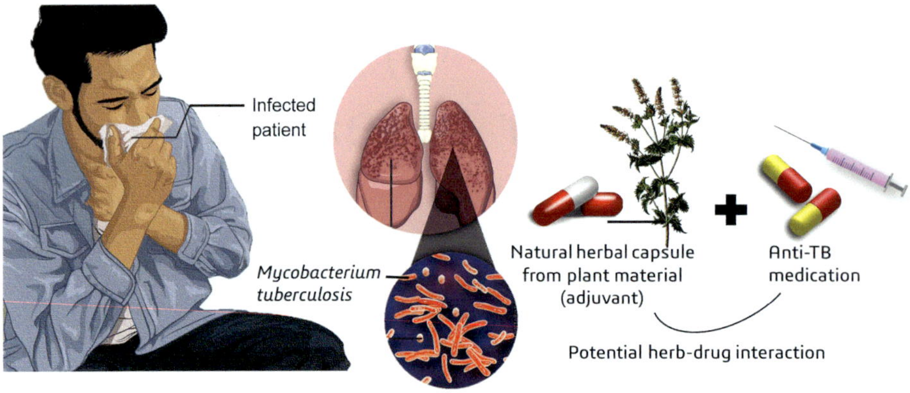

Graphical abstract representing the content of the chapter.

and its human hosts. Tuberculosis in Egypt was documented more than 5,000 years ago with the occurrence of peculiar skeletal abnormalities indicating Pott's disease found on Egyptian mummies (Cave and Demonstrator, 1939; Daniel, 2006).

Tuberculosis was well known to the Greeks and was called phthisis. Written records of tuberculosis became scarce during the Middle Ages, but that does not mean that the disease was not present. As the Renaissance emerged from Italy into Northern Europe, new knowledge of the disease emerged, as well. René Théophile Hyacinthe Laënnec is mostly remembered for his invention of the stethoscope and the elucidation of the pathogenesis of tuberculosis (Daniel, 2004, 2006). In 1819, Laënnec was convinced that the tubercle was the common factor in all forms of the disease, as he was surrounded by many patients who were infected with the disease, including his mother. The disease was christened "tuberculosis" in 1839 by Schönlein. By 1862, Pasteur's germ theory of infectious disease accelerated the search for the causative agent of the disease (Sakula, 1983). Demonstration of the infectious nature of tuberculosis occurred in 1865 by French military surgeon Jean-Antoine Villemin when he conducted animal experiments. He demonstrated that tuberculosis could be inoculated from humans or cattle to rabbits or guinea pigs, and that the sputum could be the source of infection of the rabbit with tuberculosis (Sakula, 1983; Daniel, 2006). The British government investigated these claims by setting up a commission, and both Sir John Burdon Sanderson and Sir John Simon confirmed Villemin's claims (Sakula, 1983).

A major event in the history of tuberculosis occurred on March 24, 1882, when Hermann Heinrich Robert Koch presented his presentation to the Berlin Physiological Society on the demonstrations of the tubercle bacillus and the identification of his famous postulates. These postulates, better known as Koch-Henle or Koch's postulates, is a standard for the demonstration of infectious etiology. In 1890, Koch presented to the Tenth International Medical Conference in Berlin, revealing that he had isolated a relatively harmless substance from the pathogenic tubercle bacilli, called tuberculin. He demonstrated that injections of it might be of use diagnostically by injecting himself with 0.25 cm^3 of the concentrated tuberculin and observed that he developed a rise in body temperature up to 39.6°C. Danish veterinarians soon followed suit and made use of tuberculin testing on cattle (Daniel, 2006).

More than half a century later, in 1948, a campaign to control tuberculosis with sponsorship by the United Nations International Children's Emergency Fund (UNICEF) and the Danish Red Cross was underway with Johannes Holm as its director. Initiation of the campaign was based on tuberculin testing followed by the bacilli Calmette-Guerin (BCG) vaccination. The program was the first disease control program 'undertaken' by an agency of the World Health Organization. The first oral mycobacterial drug, isoniazid, was released in 1952, followed by rifamycins in 1957. A new era of tuberculosis treatment had begun (Daniel, 2006).

14.2 TUBERCULOSIS AND ITS STAGES OF INFECTION

Tuberculosis is a very contagious and easily transmittable disease. The transmission cascade is initiated when an infected individual with active TB coughs, sneezes or engages in any other type of activity that would cause little aerosol droplets containing viable *M. tuberculosis* bacilli to be dispersed into the atmosphere. These droplets are then inhaled by a susceptible individual, who is then infected and may develop tuberculosis. The manifestations of tuberculosis may occur in various parts of the body (extrapulmonary tuberculosis), but pulmonary tuberculosis is the most common (Crevel et al., 2002).

The pathogenesis of tuberculosis is a complex interaction between the causative microorganism and the human host response to the invading threat (Schluger, 2005). Infected aerosol droplets are small in size (1–5 μm) and can bypass defense systems of the bronchi and penetrate the terminal alveoli, where engulfment of the bacilli by phagocytic immune cells occurs on the first contact with resident macrophages. Infection of non-phagocytic immune cells can occur, as well (Ahmad, 2011). Intracellular pathogens, including *M. tuberculosis*, initially reside in an endocytic phagosome after entry. Normal phagosomal processes involve a phagosome-lysosome fusion that creates a hostile environment including an acidic pH, toxic peptides, lysosomal enzymes and reactive oxygen intermediates. This would normally suffice in eliminating the threat within these newly formed phagolysosomes; however, *M. tuberculosis* has evolved ways to bypass this mechanism of defense by inhibiting the phagosome-lysosome fusion. This allows for the survival of the intracellular mycobacterial cells (Armstrong and Hart, 1975).

Infected macrophages in the lungs attract inactivated monocytes, neutrophils and lymphocytes, which are not effective in killing the surviving mycobacterial cells. Granulomas composed of macrophage-derived giant cells and lymphocytes then begin to form and become a means of containing the infection and preventing the spread of the mycobacterial cells (Fenton and Vermeulen, 1996). As the development of cellular immunity progresses, bacteria-laden macrophages are killed, which subsequently leads to the formation of the caseous center of the granuloma surrounded by lymphocytes, monocytes and fibroblasts. At this point, the *M. tuberculosis* bacilli, unable to replicate due to the toxic microenvironment created within the caseous tissue, will remain dormant (Smith, 2003).

The duration of the dormancy is dependent on the strength of the host's immune system and responses when the immune system is compromised in any way due to several factors such as aging, intake of immunosuppressant drugs or via infection of immune-weakening pathogens such as human immunodeficiency virus (HIV) (Daniel, 2006). Replication of the internalized *M. tuberculosis* cells results in bacteria-laden immune cells that may potentially cross the alveolar barrier and may disseminate systemically and manifest as various forms of extrapulmonary tuberculosis (Ahmad, 2011). Since dendritic cells are migratory cells, it is thought that they may play a role in dissemination (Smith, 2003).

14.3 TUBERCULOSIS TREATMENT

Prescribed tuberculosis treatment (Table 14.1) aims to cure, minimize disease transmission and prevent further complications, morbidity and mortality. For the prescribed medication to be effective, it should be able to reduce the bacterial load rapidly, prevent the selection of naturally resistant strains and be capable of sterilizing the lesion. Combinational therapy with several drugs is used to ensure all these requirements and aims are met (Sotgiu et al., 2016).

Patients infected and diagnosed with pulmonary tuberculosis for the first time have to be treated for six months. These patients are suspected to have drug-susceptible TB strains. For a two-month intensive phase, patients are administered a combined regimen that includes isoniazid, pyrazinamide, ethambutol and rifampicin. Thereafter, only isoniazid and rifampicin are prescribed for a four-month continuation phase. Effective regimens for patients with resistant strains or previously treated cases are tailored according to the phenotypic profile of the mycobacterial isolates (Sotgiu et al., 2016).

TABLE 14.1
Different Types of Antituberculosis Drug Treatments

Groups	Drugs	Requirements for Administration
First-line oral drugs	Isoniazid, Rifampicin, Pyrazinamide, Ethambutol, Rifapentine and Rifabutin	Patients are only prescribed this if they are new patients and generally considered to have susceptible TB.
Second-line oral drugs	Linezolid, Clofazimine, Amoxicillin/clavulanate, Imipenem/cilastatin, Meropenem, high-dose isoniazid, Thioacetazone, Clarithromycin, Bedaquiline and Delamanid	Prescribed to patients with MDR-TB[1] strains
Fluoroquinolones	Ofloxacin, Moxifloxacin, Levofloxacin and Gatifloxacin, Bedaquiline, and Linezolid	Prescribed to patients with MDR-TB strains
Injectables	Kanamycin, Streptomycin, Capreomycin and Amikacin	Prescribed to patients with drug-resistant TB strains but is slowly becoming less of a priority to be used in TB regimens but instead be replaced with oral drugs.

Note: [1]**MDR-TB**: Multidrug resistant tuberculosis

14.3.1 Biofilm

When *M. tuberculosis* infection develops within the host, chronic granulomatous inflammation is broadly characterized as the host's response (Canetti, 1968). Unregulated inflammation may result in failure to clear the infection and significantly impairs the host's immune responses, cellular and systemic metabolic homeostasis (Kiran et al., 2016). Macrophages are thought to play a role in TB granuloma formation as surviving intracellular mycobacterium bacilli establish initial *M. tuberculosis* infection within these macrophages (Armstrong and Hart, 1971). The infected macrophages and a failed host defense to eradicate or significantly slow the exponential growth of *M. tuberculosis* contribute to the recruitment of additional inflammatory cells that form infectious lesions. Following the release of intracellular bacilli from infected cells, extracellular organisms attach within a complex matrix of host and pathogen-derived macromolecules. The attachment and persistent growth of *M. tuberculosis* within these complex extracellular polymeric structures have been likened to biofilm formation (Orme, 2014; Kiran et al., 2016).

Bacteria have the ability to adhere to each other and grow on almost any surface (biotic and abiotic). They form complex communities of microbial cells that are enclosed in an extracellular matrix composed of polysaccharide material (Donlan, 2002). First reports of the concept of biofilms date back to 1978 (Costerton et al., 1978). Biofilms are of significance to human health as they can form in natural, industrial and medical environments. More specifically, in a medical setting, biofilms can form on medical devices such as catheters or implants, which may result in chronic infections that are harder to treat (López et al., 2010). Biofilm infections are associated with resistance and/or tolerance to antimicrobial compounds and persist despite prolonged host defenses (Hall-Stoodley and Stoodley, 2009).

The biofilm matrix dramatically reduces the rate of penetration of antibiotics into the biofilm, which simultaneously allows for the induction of expression of genes within the biofilm that could mediate resistance (Jefferson et al., 2005). Mycobacterial biofilms can be defined in the same manner as any other bacterial biofilm; however, some mycobacteria may develop biofilm structures on air-media interfaces *in vitro* (Ojha et al., 2008). Mycobacteria lack surface pili or fimbriae proteins, although other proteins have been described as potentially playing a role in the attachment to other cells (Menozzi et al., 1996; Esteban and García-Coca, 2017). Mycobacteria do not produce the usual exopolysaccharide constituents of the extracellular matrix, but still have the ability to adhere to various surfaces and develop fully mature biofilms (Zamora et al., 2007; Zambrano and Kolter, 2005; Esteban and García-Coca, 2017).

14.3.2 Adjuvant Therapy

The use of herbal medicine as primary treatment is feasible and may even be recommended if the risk-benefit ratio is favorable toward the alternative in comparison to conventional medicine. Herbal medicine may be the preferred choice in treating mild or self-limiting illnesses if it is known to be effective and safer, with fewer or minimal side effects than its conventional counterpart. For more severe illnesses, clinicians may be willing to suggest herbal medicine as a primary treatment only if conventional treatment was not working as anticipated or if the patient has succumbed to side effects that limited its use. For serious life-threatening illnesses, herbal medicine as primary treatment would not be the preferred option or would only be considered if there were no known effective conventional therapies or if sufficient compelling efficacy data were available for the proposed herbal treatment. It is also worth noting that herbal medicine is used complementary to conventional medicine to improve the quality of life of the patient (Girard and Vohra, 2011).

14.3.3 Adjuvants and Their Role in Vaccine Development

The only vaccine currently on the market for the treatment of tuberculosis is the live attenuated BCG vaccine derived from *Mycobacterium bovis* (Khademi et al., 2019). This vaccine has been in use for over 100 years up until today. The vaccine, however, does not have the ability to prevent active pulmonary tuberculosis, latent TB or adult infections (Kaufmann et al., 2010). The BCG vaccine has shown fluctuating efficacy levels in adults, in part said to be due to the different BCG strains used in vaccine design in different countries, as well as exposure to non-tuberculous mycobacteria that mask the efficacy (Davenne and McShane, 2016). Various vaccines are in the preclinical and clinical stages of development, and include genetically modified live attenuated vaccines, which are virally vectored and are considered as a subunit vaccine (Gengenbacher and Kaufmann, 2012; Kaufmann et al., 2010). Subunit vaccines are effectively co-administered with an adjuvant to strengthen this sought-after desired immune response. It is found that combinations of adjuvants with different modes of action are more effective in eliciting both a humoral and cellular response to subunit vaccine antigens by the infected individual (Garçon et al., 2007).

Several antigens that have been used in TB subunit vaccines include early secretory antigenic target-6 (ESAT6) and antigen 85B (Ag85B), which are conserved secreted proteins and have been shown to be immunogenic in animal models (Agger and Andersen, 2001; Stewart et al., 2019). Several hurdles have restricted the number of antigens used, such as the various different stages of infections that take place and the larger genome size of *Mycobacterium tuberculosis*. Other antigens that have been identified include sulfate adenylyltransferase subunit 2 (CysD), which is a non-secreted protein that forms part of the sulfur assimilation pathway and has been shown to be upregulated during latent infection and is conserved across several strains (Counoupas et al., 2016; Stewart et al., 2019).

During clinical trials and investigating various models, it was found that several adjuvants such as AS01, IC31 and CAF01 can increase a Th1-type cellular response in individuals infected with TB (Agger et al., 2006; Lindenstrom et al., 2009; van Dissel et al., 2011; Woodworth et al., 2017). According to Khademi et al. (2019), adjuvants are designed to stimulate the adaptive and innate immune system, to protect antigens from enzyme degradation *in vivo*, to improve antigen-presenting to the antigen-presenting cells (APCs) followed by processing by major histocompatibility complex class I and II (MHC-I and II) leading to induction of $CD4^+$ and $CD8^+$ T cells, and finally by controlling the particle size and controlling the induction of the cell-mediated responses (Th1 or Th2). Adjuvants are selected based on several criteria due to their different means of stimulating the immune system (Khademi et al., 2019).

14.3.4 Immunotherapeutic and Hepatoprotective Adjunct Therapies

Many immunotherapies have been proposed especially for cancer treatment. In this case, toll-like receptor (TLR) ligands are used as the main form of immunotherapy for cancer patients. Tumor-specific antigens were used to evoke the immune system to attack tumor cells in the body (Circelli et al., 2017). Subunit vaccines normally do not have the possibility of inducing an immune response. The addition of adjuvants has the ability to elicit this immune response when combined with the subunit vaccines. Allium salts are the only approved adjuvant used together with human vaccines (Circelli et al., 2017). The mechanism of action of allium salts is that they have the ability to induce a humoral response through the depot effect, by improving the antigen delivery to APCs, as well as the processing of the antigens and presentation and inducing cytokine production (Circelli et al., 2017). Eliciting a response from the immune system through hepatoprotective agents, normally evoking a response through cytochrome P450 (CYP P450) inhibition of certain CYP P450 enzymes, which prevent compounds from being metabolized in a certain manner. Another way through which an immune response can be elicited is through a compound's capability to either increase antioxidant or decrease anti-inflammatory responses (Knight, 2000).

14.3.5 Pattern Recognition Receptors (PRRs) as Adjuvants

14.3.5.1 NOD (Nucleotide-Binding Oligomerization Domain)–Like Receptors, RIG-I (Retinoic Acid-Inducible Gene-I)–Like Receptors and TLRs (Toll-Like Receptors)

TLRs have been identified in both humans (10 TLRs) and mice (12 TLRs). In humans, TLR2 is found on the cell membrane as a surface receptor and dimerizes with TRL1 or TLR6. The heterodimer formed between TLR1 and TLR2 is responsible for detecting triacylated lipopeptides from mycoplasma or gram-negative bacteria. The heterodimer formed between TLR2 and TLR6 could sense diacylated lipopeptides from mycoplasma or gram-positive bacteria (Circelli et al., 2017). TLR3 is a receptor found in the endosome and has the ability to recognize double-stranded RNA (dsRNA). TLR4 recognizes lipopolysaccharide (LPS), a component found in the outer membrane of gram-negative bacteria (Kawai and Akira, 2010). The only TLR4 agonist used as a cancer vaccine adjuvant that is approved is monophosphoryl lipid A (MPL), which has undergone many clinical trials to prove its safety and probability to elicit an immune response (Cluff, 2010). Both TLR7 and TLR8 are endosomal receptors that recognize single-stranded RNA (ssRNA); TLR7, Imiquimod, is the only approved ligand for use in the treatment of precancerous skin lesions (Vacchelli et al., 2012). TLR9 can identify unmethylated CpG dinucleotides of bacterial DNA origin (Kawai and Akira, 2010). TLRs that have had the ability to steer a Th1 elicited immune response include Poly:IC (TLR3), 3-O-desacyl-4'-monophosphoryl lipid A (MPLA: TLR4) or CpG oligonucleotides (TLR9). They have the ability to activate these specific TLRs and induce a downstream IL-12 response, which is important for the activation of a Th1 response (Stewart et al., 2019).

14.3.6 Antibacterial (Antibiotic) Adjunct Therapies

Due to the ongoing resistance against bacteria, antibacterial or antibiotic adjuvants have been proposed to prevent resistance from occurring. Several compounds are already in clinical development, such as β-lactamase inhibitors (González-Bello, 2017). These compounds are known as resistance breakers or antibiotic potentiators, and do not possess antibiotic activity of their own but can be co-administered with the main antibiotic of choice. Their mechanism of action includes blocking the main bacterial resistance mechanisms or enhancing the antimicrobial capabilities of the co-administered drug. There are four main mechanisms of how antibiotic resistance occurs (González-Bello, 2017), as follows.

- Inactivation of the antibiotic by enzymes.
- Activation of the efflux, causing drug excretion.

- Changes in the outer membrane, causing decreased uptake of the antibiotic.
- Any modifications of the drug target.

In response to this resistance mechanism, antibiotic adjuvants have been produced, namely β-lactamase inhibitors, efflux pump inhibitors and outer membrane permeabilizers (González-Bello, 2017).

β-lactamase inhibitors are mainly used to overcome the resistance to β-lactam antibiotics. As previously mentioned, these adjuvants protect against the inactivation of the antibiotic by enzymes. Bacterial inactivation involves β-lactamases that hydrolyze the β-lactam core of β-lactams through a process based on acylation-deacylation (González-Bello, 2017). In the 1980s, the first β-lactamase inhibitor was used in combination with penicillin. Diazabicyclooctanes have been under investigation since the mid-1990s as β-lactam mimics, and later were found to be efficient β-lactamase inhibitors (Bush, 2018). Boronic acids are used as transition state analogs. The development of these compounds was discovered when serine proteases were inhibited by boronic acids (Smoum et al., 2012).

14.3.7 Plant-Based Commercial Tuberculosis Adjuvant Therapy

Medicinal plants are nature's underlying gift used to treat numerous diseases. Herbal medicine has been used to increase the effectiveness of conventional medication or to minimize side effects (Girard and Vohra, 2011). A broad spectrum of *in vitro* and *in vivo* antimicrobial activity is considered when searching for viable plant-based adjuvant TB therapy. This includes phytochemicals that exhibit enhancement of combination therapy, the reversal of drug resistance or re-sensitizing activities, inhibition of biofilm formation or by induction and control of efflux pumps. These antimicrobial agents have two subgroups, each containing distinct adjuvant activities which are either pathogen-directed or host-directed (Thombre et al., 2019). With side effects such as induced toxicity from conventional medicine, it is beneficial for adjuvants to provide added hepatoprotective properties. There are various adjuvant natural herbal remedies for tuberculosis treatment (Figure 14.1). These evidence-based treatments are summarized in Table 14.2.

The development of plant-based adjuvant immunotherapy that may be used in conjunction with current TB treatment is ongoing in the respective research fields. Immunomodulatory approaches have received more attention over the years. T-cell responses during a TB infection may aid with the containment of *M. tuberculosis*; however, excessive responses may lead to inflammatory damage to the host (Cooper, 2009). In Ukraine, oral immunomodulators Dzherelo (Immunoxel), Svitanok, and Lizorm are used for the management of HIV infections and HIV and TB co-infection in patients (Nikolaeva et al., 2008).

Dzherelo is composed of various medicinal plants including purple coneflower, licorice, cottonweed, sage, thyme, juniper fruit, fennel, dandelion, St. John's wort, marigold, wormwood, elecampane, centuary, nettle, birch tree fungus, Siberian golden root, sweet-sedge, three-lobe beggartricks, aloe, common knotgrass, dog rose fruit, yarrow, greater plantain, snowball tree berries and dog rose fruit. The approval of the immunomodulatory supplement occurred in 1997 by the Ministry of Health of Ukraine (Nikolaeva et al., 2008). This herbal admixture, recently formulated into Honey Lozenges™, is used to aid immunocompromised HIV patients and co-administered with anti-TB drugs to TB patients (Khameneh et al., 2019). Previous investigations show an overall higher rate of *M. tuberculosis* clearance in sputum cultures with the co-administration Dzherelo and anti-TB drugs in comparison to the patients who received anti-TB drugs alone or with placebo. Improved liver function and decrease in body temperature was also noted (Kitenge et al., 2020). In TB/HIV co-infected patients, the addition to the herbal supplement to conventional treatment resulted in a decrease of the viral load and a burgeoning of total CD3+ lymphocytes and CD4 T cells (Nikolaeva et al., 2008). It has been suggested, however, that more studies with a larger sample size and more clinical trials would be beneficial to strengthen and establish the value of this adjunctive treatment for pulmonary TB (Kitenge et al., 2020).

FIGURE 14.1 Plants that have been used in plant-based tuberculosis adjuvants. A, Purple Coneflower (*Echinacea purpurea*); B, Licorice (*Helichrysum petiolare*); C, Fennel (*Foeniculum vulgare*); D, St. John's Wort (*Hypericum perforatum*); E, Wormwood (*Artemisia absinthium*); F, Elecampane (*Inula helenium*), G; Sweet sedge (*Acorus calamus*); H, Common knotgrass (*Polygonum aviculare*); I, Dog rose (*Rosa canina*); J, Chicory (*Cichorium intybus*); K, Coriander seeds (*Coriandrum sativum*); L, Wild strawberry (*Fragaria vesca*); M, Greater celandine (*Chelidonium majus*); N, French tamarisk (*Tamarix gallica*); O, Coffee senna (*Cassia occidentalis*); P, Arjun tree (*Terminalia arjuna*); Q, South African geranium (*Pelargonium sidoides*); R, Malabar nut (*Adathoda vasica*); S, Stonebreaker (*Phyllanthus niruri*); T, Indian mulberry (*Morinda citrifolia*) (Acharya, 2017; Aftabbanoori, 2014; AnRo0002, 2015a, 2015b; Blasutto, 2007; Challiyan, 2018; Cebeci, 2020; Dalgial, 2009; Gafford, 2004; Garg, 2008; Harper, 2008; Hillewaert, 2007; HQ, 2010; Leidus, 2013; Lily, 2014; Petruss, 2011; Stang, 2006; SterlingHerron, 2015; Vengolis, 2019; Ziarnek, 2016).

Svitanok is composed of the flowers of immortelle, barberry roots, chicory roots, coriander seeds, marigold, wormwood, and maize cores with stigmas. Svitanok is most often used for counteracting liver-damage effects of chemotherapy and hepatitis therapy. Lizorm is commonly used for alleviating symptoms of autoimmune disorders (Bodnar et al., 2002). Lizorm is composed of barberry roots, aronia berries, St. John's wort, centaury, immortelle, aronia berries, common knotgrass, nettle, wild strawberry leaves and greater celandine (Prihoda et al., 2007).

Several studies and clinical trials have been conducted to test the efficacy, safety and effects of Dzherelo, Svitanok, and Lizorm individually, in combination with each other and in combination with anti-tuberculosis drugs. In 2007, Prihoda et al. conducted open-label trials of anti-tuberculosis therapy in combination with the plant-derived oral immunomodulators. In this study, 14 Ukrainian patients—half of whom were co-infected with HIV and nine with MDR-TB, and two out of those nine individuals with XDR-TB—all received anti-tuberculosis therapy. In addition to this, the patients were administered with a daily dose of 30 drops of Dzherelo diluted in half a glass of water, and 30 drops of Lizorm and Svitanok administered in the same manner as Dzherelo. The

TABLE 14.2
Examples of Plant-Based Anti-Tuberculosis Adjuvants

Adjuvant Treatments	Key Herbs	Companies (Country of Origin)	References
Dzherelo (Immunoxel)	Purple coneflower, licorice, cottonweed, sage, thyme, juniper fruit, fennel, dandelion, St. John's wort, marigold, wormwood, elecampane, centuary, nettle, birch tree fungus, Siberian golden root, sweet-sedge, three-lobe beggartricks, aloe, common knotgrass, dog rose fruit, yarrow, greater plantain, snowball tree berries and dog rose fruit	Ekomed (Ukraine)	(Bodnar et al., 2002; Prihoda et al., 2007; Nikolaeva et al., 2008)
Svitanok	Immortelle, Barberry roots, chicory roots, coriander seeds, marigold, wormwood, and maize cores with stigmas	Ekomed (Ukraine)	(Prihoda et al., 2007)
Lizorm	Barberry roots, aronia berries, St. John's wort, centaury, immortelle, aronia berries, common knotgrass, nettle, wild strawberry leaves and greater celandine	Ekomed (Ukraine)	(Bodnar et al., 2002; Prihoda et al., 2007)
Liv52	*Capparis spinosa* L., *Solanum nigrum* L., *Achillea millefolium* L., *Tamarix gallica* L., *Cassia occidentalis* L., *Cichorium intybus* L. and *Terminalia arjuna* (Roxb.) Wight & Arn.	Himalaya Herbals (India)	(Dange, 2010)
Umckaloabo Spray	*Pelargonium sidoides* DC.	Medico Herbs (South Africa)	(Bladt and Wagner, 2007)
Lobodil	Banslochana *Adathoda vasica* L.	Salveo Life Sciences (India)	(Chaterjee, 2011)
Livful	*Picorhiza kurrora* Royle ex Benth *Phyllanthus niruri* L.	Salveo Life Sciences (India)	(Salveo Life Sciences, 2016)
Andaman Noni (Immune booster)	*Morinda citrifolia* L.	Salveo Life Sciences (India)	(Salveo Life Sciences, 2016)

results from this study demonstrated that when anti-tuberculosis therapy was combined with the plant-derived immunomodulatory drugs, the patients were cured after four months and showed no sign of *M. tuberculosis* from sputum culture when observed within a month of initiation of treatment (Prihoda et al., 2007).

Furthermore, in 2009, a study conducted by Zaitzeva et al., with 66 patients, all with active pulmonary TB, administered 50 drops of Dzherelo in a glass of water twice daily for three months in combination with anti-tuberculosis therapy drugs. Results included improvement in TB-associated symptoms such as night sweats, fatigue, nausea, dyspnea and general malaise within 2–3 weeks. Since anti-tuberculosis drugs can lead to hepatotoxicity, Dzherelo was found to be safe, improve liver function and even reverse liver damage caused by anti-tuberculosis drugs (Zaitzeva et al., 2009). An additional study by Nikolaeva et al. (2008) showed that Dzherelo is able to shorten the duration of anti-TB treatment and help achieve a better and higher response rate in TB infected patients, even those who harbor MDR and XDR forms of TB (Nikolaeva et al., 2008).

The frequent occurrence of liver toxicity during TB treatment is a concern. Withdrawal of treatment by patients and the development of multidrug resistant (MDR) TB is usually the result of hepatotoxicity (Ramappa and Aithal, 2013). The liver is responsible for regulating various physiological processes and a source of effective enzymes that transform foreign molecules to metabolize and detoxify several heterochthonous or autochthonous chemicals (Bedi et al., 2016). A natural plant-based anti-TB drug with hepatoprotective activity is Liv52, a polyherbal formulation that has been used for many years for various hepatic disorders. The main principle herbs in this preparation include *Capparis spinosa* L., *Solanum nigrum* L., *Achillea millefolium* L., *Tamarix gallica* L., *Cassia occidentalis* L., *Cichorium intybus* L. and *Terminalia arjuna* (Roxb. ex DC.) Wight & Arn. A meta-analysis conducted by Dange (2010) showed that the outcome of including Liv52 in the treatment of patients who suffered from various forms of TB, including pulmonary TB, extrapulmonary TB, miliary TB, tuberculous cervical lymphadenitis and meningitis resulted in significant improvements as a hepatoprotective drug. Furthermore, it aided in the weight gain of anorexic patients (Dange, 2010).

A more recent study conducted by Choijamts et al. (2018) tested the effects of Liv52 on 90 individuals in Mongolia. In the double-blind placebo-controlled two-arm study, the first arm was administered Liv52 as part of the treatment and placebo tablets in the second arm. Each individual from each arm was advised to take two tablets twice daily for a period of six months. Evaluation of the individuals took place after two-month intervals for the six months. Results correlated with previous studies in demonstrating the ability of Liv52 to alleviate symptoms of liver damage, increased weight gain, and hepatoprotective activity in individuals who had Liv52 as part of their treatment (Choijamts et al., 2018).

South African medicinal plants such as *Pelargonium sidoides* DC. and *Pelargonium reniforme* Curtis are traditionally used as infusions made from their tuberous, woody roots. *Pelargonium sidoides* DC. has a long history of use for medicinal purposes and for tuberculosis in particular. In 1897, an Englishman named Charles Henry Stevens who was suffering from tuberculosis was advised by his doctor to go to South Africa to cure the disease. In what is known as Lesotho today, he met Kagaitse, a Zulu man who was an expert on medicinal plants, and was given a boiled root preparation of the plant. After three months of treatment, Stevens was cured of tuberculosis (Bladt and Wagner, 2007). This plant, native to South Africa, was then taken to England by Stevens where today, the extract of the plant is sold all over Europe. In South Africa, it has long been used by the Zulu people to treat diarrhea, dysentery and gonorrhea (Drewes, 2012).

Along with the traditional usage, *P. sidoides* DC. is commercially available internationally in countries such as Mexico, Turkey, Brazil, Germany and South Africa. However, the antimycobacterial activity of *P. sidoides* DC. seems to have varying results against *M. tuberculosis*, depending on the strain of mycobacterium as well as the extract preparation and composition. The effectiveness of *P. sidoides* DC. in tuberculosis treatment may be attributed to its immunomodulatory activity (Kolodziej, 2011). The isolated compounds from the *P. sidoides* DC. root include umckalin (6,8-hydroxy-5,7-dimethoxycoumarin), gallic acid, catechin, tannins and certain fatty acids (White et al., 2008).

Salveo Life Sciences has a few products—Lobodil, Livful and Andaman Noni—that are notable for their use as potential adjuvants for tuberculosis treatment. Lobodil is mainly used for the management of bronchial asthma, Livful is used as a natural hepatoprotective agent and Andaman Noni is a strong antioxidant, immune booster and rejuvenator (Salveo Life Sciences, 2016). Lobodil consists of *Adhatoda vasica* (L.), which is considered beneficial for tuberculosis patients as it is used for soothing or eradicating incessant coughing (Sampath Kumar et al., 2010). With Livful being a hepatoprotective agent, it is expected that its main plant constituents contribute strong hepatoprotective properties to mitigate liver toxicity (Salveo Life Sciences, 2016).

14.4 PHARMACOKINETIC INTERACTIONS WITH ANTITUBERCULOSIS AGENTS

Drug–drug interactions occur when the effect of one drug is affected by the presence of a concomitant drug (Magro et al., 2012). These drug interactions may cause adverse drug reactions which, for the most part, are predictable and can be avoided. However, prevention is very complex in practice due to the high number of drugs that could potentially interact with one another (Létinier et al., 2019).

Prednisone is reported to reduce isoniazid concentrations in both rapid and slow acetylators (Sarma et al., 1980). Isoniazid is metabolized predominantly by acetylation. Genetic polymorphism determines whether patients are rapid or slow acetylators. Drug interactions may be more prominent in slow acetylators. Insulin enhances the absorption of isoniazid; however, the mechanism of that is unclear (Grange et al., 1994). Aluminum hydroxide lowers the bioavailability of isoniazid. When isoniazid is administered to patients who are stabilized on phenytoin or carbamazepine, plasma concentrations of the antiepileptic drugs increases and the potential of toxicity rises dramatically (Toman, 2004).

Rifampicin is a potent inducer of cytochrome isoenzymes CYP2C and CYP3A. Drugs that are metabolized by the same cytochromes or undergo the same route of metabolism are metabolized more rapidly, and their therapeutic effects are therefore futile. Instead, there is a higher risk of increased toxicity or therapeutic failure. Patients who are stabilized on the anticoagulant warfarin, while having rifampicin administered to them, may undergo over-anticoagulation if rifampicin is halted without adjusting warfarin dose (Grange et al., 1994). Drug–drug interactions occur between several antituberculosis drugs themselves, but some of their clinical significance is still unclear.

The use of herbal remedies is often used in combination with Western pharmaceuticals without considering the potential drug–herb interaction. The high polymorphic range of cytochrome P450 enzymes may pose as a problem and tailor the drug to meet an individual metabolic activity. The most desired interactions are those that result in increased efficacy and additional therapeutic benefit (Cordier and Steenkamp, 2011).

Whole plants or plant parts (i.e. leaves, flowers, fruits, roots, bark, stems, etc.) may provide medicinal, health-promoting properties. Herbalism is the use of alternative healthcare besides conventional pharmaceuticals. Herbal products are available in both commercial and crude preparations (Mehmood et al., 2018). Herbal products are herbal preparations produced by subjecting plant or herbal materials to extraction, fractionation, concentration, purification, or other physical and biological processes (Alissa, 2014).

Drug interactions can be defined as modifications caused as a result of interacting with exogenous chemicals (from other drugs, food or herbs) which have an effect on pharmacodynamics and/or pharmacokinetic profiles of the drug, which may inherently influence physiological responses in the body (Cordier and Steenkamp, 2011; Alissa, 2014).

14.4.1 Cytochromes and Drug Metabolism

Cytochrome P450 enzymes are heme-containing proteins found in humans and bacteria. Discovery and characterization of P450 in liver extracts occurred in the 1950s (Garfinkel, 1958; Klingenberg, 1958). The pigment displayed the ability to absorb maximally at 450 nm when reduced with carbon monoxide present. Hence, the term P450 from the indication 'pigment-450.' Expression of P450 enzymes occurs throughout nature. In mammals, about one-third of the P450 enzymes are expressed in the liver and play a role in the degradation of numerous xenobiotic compounds (Furge and Guengerich, 2006). Cytochrome P450 pathways are categorized according to similar gene sequences and assigned a family member (e.g. CYP1, CYP2) and a subfamily letter (e.g. CYP1A, CYP2C) and further differentiated by a number for the isoform or individual enzyme (e.g. CYP1A1, CYP2C9). Drugs that undergo common pathways have the potential for drug–drug interactions (McDonnell et al., 2013, Nelson, 2009).

Elimination of administered drugs can occur by one of three major routes. Polar drugs are often eliminated through urine via the kidneys, polar and nonpolar drugs can be eliminated through the bile and finally, drugs can be metabolized by several endogenous enzymes and then eliminated in urine or bile. Drug metabolism is the process of modification of xenobiotics to make them more readily available for excretion by several endogenous enzymes. Often, modifications are made to the parent drug molecule to make them more hydrophilic. The liver is one of the main organs that host/facilitate drug metabolism, followed by the intestine and kidneys, and several other organs to a lesser extent (Sane and Sinz, 2016).

Inhibition of cytochrome P450 enzymes results in a decreased rate of metabolism and excretion of drugs metabolized by these enzymes. This may concurrently result in the accumulation of the drug in the body, and toxicity may occur within 2–3 days, depending on the level of inhibition and the drug used. The clinical significance of enzyme inhibition interactions is dependent on the extent to which serum levels rise (Harris and Swart, 2005). Inhibitors inactivate specific CYP enzymes irreversibly. Metabolism will resume back to normality once the inhibitor has been cleared and new enzymes have been produced (Scripture et al., 2006).

Enzyme induction often affects phase I oxidation whereby the presence of the reduced form of nicotinamide adenine dinucleotide phosphate (NADPH) and heme-containing protein cytochrome P450 is required. The induction of enzymes increases the activity of microsomal enzymes (cytochrome P450 enzymes), therefore increasing the rate of metabolism and excretion. The outcome is that higher doses of the drug often need to be administered (Harris and Swart, 2005).

14.5 CONCLUSION

A wide variety of plants and herbs are used to treat tuberculosis or related symptoms, and have made a myriad of impacts on the maintenance of human health (Mangwani et al., 2019). Combining pathogen-directed and host-directed therapeutics offers a strategy to control the emergence of more tolerant and drug-resistant intracellular bacteria (Czyż et al., 2014). Herbal medicine is ubiquitous and its ingestion with conventional prescribed medication is becoming increasingly popular; however, the probability of herb-drug interactions increases, as well. Experimental data, case reports and case series are often limited, and therefore, more study of this topic is encouraged (Fugh-Berman and Ernst, 2001).

REFERENCES

Acharya, Sanjay. 2017. "Coriander Seeds." Obtained from Wikimedia under the Creative Commons Attribution-Share Alike 4.0 International license. https://commons.wikimedia.org/wiki/File:Coriander_Seeds.jpg (Accessed 1 July 2021)

Aftabbanoori. 2014. "Terminalia arjuna." Obtained from Wikimedia under the Creative Commons Attribution-Share Alike 4.0 International license. https://commons.wikimedia.org/wiki/File:Terminalia_arjuna_-_Bagh-e-Jinnah.jpg (Accessed 1 July 2021)

Agger, E. M., and P. Andersen. 2001. "Tuberculosis Subunit Vaccine Development: On the Role of Interferon-gamma." *Vaccine* 19, 17–19: 2298–2302. doi: 10.1016/s0264-410x(00)00519-3

Agger, E. M., I. Rosenkrands, A. W. Olsen, G. Hatch, A. Williams, C. Kritsch, K. Lingnau, A. von Gabain, C. S. Andersen, K. S. Korsholm, and P. Andersen. 2006. "Protective Immunity to Tuberculosis with Ag85B-ESAT-6 in a Synthetic Cationic Adjuvant System IC31." *Vaccine* 24, 26: 5452–5460. doi: 10.1016/j.vaccine.2006.03.072

Ahmad, Suhail. 2011. "Pathogenesis, Immunology, and Diagnosis of Latent Mycobacterium Tuberculosis Infection." *Clinical & Developmental Immunology*, December: 814943. doi: 10.1155/2011/814943.

Alissa, Eman M. 2014. "Medicinal Herbs and Therapeutic Drugs Interactions." *Therapeutic Drug Monitoring* 36, 4: 413–422. doi: 10.1097/FTD.0000000000000035.

AnRo0002. 2015a. "*Hypericum perforatum*." Obtained from Wikimedia Commons under the Creative Commons CC0 1.0 Universal Public Domain Dedication license. https://commons.wikimedia.org/wiki/File:20150605Hypericum_perforatum3.jpg (Accessed 30 June 2021).

AnRo0002. 2015b. "*Artemisia absinthium*." Obtained from Wikimedia Commons under the Creative Commons CC0 1.0 Universal Public Domain Dedication license. https://commons.wikimedia.org/wiki/File:20150821Artemisia_absinthium5.jpg (Accessed 30 June 2021).

Armstrong, J. A., and P. D'Arcy Hart. 1971. "Response of Cultured Macrophages to Mycobacterium Tuberculosis, with Observations on Fusion of Lysosomes with Phagosomes." *Journal of Experimental Medicine* 134, 3: 713–740. doi: 10.1084/jem.134.3.713.

Armstrong, J. A., and P. D'Arcy Hart. 1975. "Phagosome-lysosome Interactions in Cultured Macrophages Infected with Virulent Tubercle Bacilli. Reversal of the Usual Nonfusion Pattern and Observations on Bacterial Survival." *The Journal of Experimental Medicine*. 142, 1: 1–16. doi: 10.1084/jem.142.1.1.

Barberis, I., N. L. Bragazzi, L. Galluzzo, and M. Martini. 2017. "The History of Tuberculosis: From the First Historical Records to the Isolation of Koch's Bacillus." *Journal of Preventive Medicine and Hygiene* 58, 1: E9–12.

Bedi, Onkar, Krishna V. Reddy Bijjem, Puneet Kumar, Vinod Gauttam, M. Pharm, and Associate Professor. 2016. "Herbal Induced Hepatoprotection and Hepatotoxicity : A Critical Review." *Indian Journal of Physiology and Pharmacology* 60, 1: 6–21.

Bladt, Sabine, and Hildebert Wagner. 2007. "From the Zulu Medicine to the European Phytomedicine Umckaloabo®." *Phytomedicine* 14, Suppl. 1: 2–4. doi: 10.1016/j.phymed.2006.11.030.

Blasutto, Enrico. 2007. "*Chelidonium majus*." Obtained from Wikimedia under the Creative Commons Attribution-Share Alike 4.0 International license. https://commons.wikimedia.org/wiki/File:Chelidonium_majus_ENBLA03.jpg (Accessed 1 July 2021)

Bodnar, Petro, H. P. Mykhal'chyshyn, V. M. Reznichenko, Oleksandr Moshchych, and V. S. Pylypchuk. 2002. "Phytoconcentrates 'Dzherelo' and 'Lizorm' in Therapy of Autoimmune Thyroiditis." *Likars'ka Sprava/Ministerstvo Okhorony Zdorov'ia Ukraïny* 8: 127–129.

Brosch, R., S. V. Gordon, M. Marmiesse, P. Brodin, C. Buchrieser, K. Eiglmeier, T. Garnier, et al. 2002. "A New Evolutionary Scenario for the *Mycobacterium Tuberculosis* Complex." *Proceedings of the National Academy of Sciences of the United States of America* 99, 6: 3684–3689. doi: 10.1073/pnas.052548299.

Bush, K. 2018. "Past and Present Perspectives of β-lactamases." *Antimicrobial Agents and Chemotherapy* 62, 10: e01076–18. doi: 10.1128/AAC/01076-18.

Canetti, Georges. 1968. "Pathogenesis of Tuberculosis in Man." *Annals of the New York Academy of Sciences* 154, 1: 13–18. doi: 10.1111/j.1749-6632.1968.tb16691.x.

Cave, A. J. E., and Arnott Demonstrator. 1939. "The Evidence for the Incidence of Tuberculosis in Ancient Egypt." *British Journal of Tuberculosis* 33, 3: 142–152. doi: 10.1016/S0366-0850(39)80016-3.

Cebeci, Z. 2020. "*Cichorium intybus*." Obtained from Wikimedia Commons under the Creative Commons Share-Alike 4.0 license. https://commons.wikimedia.org/wiki/File:Cichorium_intybus_-_Chicory_02.jpg (Accessed 30 June 2021).

Challiyan. 2018. "*Phyllanthus niruri*." Obtained from Wikimedia under the Creative Commons Attribution-Share Alike 3.0 Unported license. https://commons.wikimedia.org/wiki/File:Phyllanthus_niruri_.jpg (Accessed 1 July 2021)

Chaterjee, Dr. Tapan Kumar. 2011. "A Novel Composition of Polyherbal TX Gel (Lobodil TX) for Respiratory Diseases and Disorders and Asthma and Process there off." New Delhi, October 7, 2011.

Choijamts, G., C. H. Batsyren, B. Orkhon, Otgontyr, D. Narantstsetseg, X. Oyunerdene, T. Zagirjav, Erdenesuvd, Enkhbayar, Pyrevbazar, Oyundelger, Tungalag, D. Erdenetsetseg, Enkhtsetseg, M. Bilegmaa, S. Sarantuya, and D. Palaniyamma. 2018. "Role of Liv.52 DS Tablets as a Hepatoprotective Agent in Tuberculosis Patients Receiving Antitubercular Drugs: A Double Blind Placebo Controlled Study." *Journal of Liver and Clinical Research* 5, 1: 1042.

Circelli, L., M. Tornesello, F. M. Buonaguro, and L. Buonaguro. 2017. "Use of Adjuvants for Immunotherapy." *Human Vaccines & Immunotherapeutics* 13, 8: 1774–1777. doi: 10.1080/21645515.2017.1321725.

Cluff, C. W. 2010. "Monophosphoryl Lipid A (MPL) as an Adjuvant for Anti-cancer Vaccines: Clinical Results." *Advances in Experimental Medicine and Biology* 667: 111–123. doi: 10.1007/978-1-4419-1603-7_10.

Cooper, A. M. 2009. "Cell-mediated Immune Responses in Tuberculosis." *Annual Review of Immunology* 27: 393–422. Doi: 10.1146/annurev.immunol.021908.132703.

Cordier, Werner, and Vanessa Steenkamp. 2011. "Drug Interactions in African Herbal Remedies." *Drug Metabolism and Drug Interactions* 26, 2: 53–63. doi: 10.1515/DMDI.2011.011.

Costerton, J. W., G. G. Geesey, and K. J. Cheng. 1978. "How Bacteria Stick." *Scientific American* 238, 1: 86–95. doi: 10.1038/scientificamerican0178-86

Counoupas, C., R. Pinto, G. Nagalingam, G. Hill-Cawthorne, C. G. Feng, W. J. Britton, and J. A. Triccas. 2016. "*Mycobacterium tuberculosis* Components Expressed During Chronic Infection of the Lung Contribute to Long-term Control of Pulmonary Tuberculosis in Mice." *NPJ Vaccines* 15, 1: 16012. doi: 10.1038/npjvaccines.2016.12

Crevel, Reinout van, Tom H. M. Ottenhoff, and Jos W. M. van der Meer. 2002. "Innate Immunity to *Mycobacterium tuberculosis*." *Clinical Microbiology Reviews* 15, 2: 294–309. doi: 10.1128/CMR.15.2.294.

Czyż, D. M., L. P. Potluri, N. Jain-Gupta, S. P. Riley, J. J. Martinez, T. L. Steck, S. Crosson, H. A. Shuman, and J. E. Gabay. 2014. "Host-directed Antimicrobial Drugs with Broad-spectrum Efficacy Against Intracellular Bacterial Pathogens." *mBio* 5, 4: e01534–14. Doi: 10.1128/mBio.01534-14

Dalgial. 2009. "*Polygonum aviculare*." Obtained from Wikimedia Commons under the Creative Commons Share-Alike 3.0 license. https://commons.wikimedia.org/wiki/File:Polygonum_aviculare_4.JPG (Accessed 30 June 2021).

Dange, S. V. 2010. "Liv.52 in the Prevention of Hepatotoxicity in Patients Receiving Antitubercular Drugs: A Meta-Analysis." *Indian Journal of Clinical Practice* 21, 2: 81–86.

Daniel, Thomas M. 2004. "René Théophile Hyacinthe Laënnec and the Founding of Pulmonary Medicine [Founders of Our Knowledge]." *The International Journal of Tuberculosis and Lung Disease* 8, 5: 517–518.

Daniel, Thomas M. 2006. "The History of Tuberculosis." *Respiratory Medicine* 100, 11: 1862–1870. Doi: 10.1016/J.RMED.2006.08.006

Daniel, Thomas M., Joseph H. Bates, and Katharine A. Downes. 1994. "History of Tuberculosis." *Tuberculosis: Pathogenesis, Protection, and Control*, 13–24. Doi: 10.1128/9781555818357.ch2.

Davenne, T., and H. McShane. 2016. "Why Don't We Have an Effective Tuberculosis Vaccine Yet?" *Expert Review of Vaccines* 15, 8: 1009–1013. Doi: 10.1586/14760584.2016.1170599/

Donlan, Rodney M. 2002. "Biofilms: Microbial Life on Surfaces." *Emerging Infectious Diseases* 8, 9: 881–890. Doi: 10.3201/eid0809.020063.

Drewes, Siegfried E. 2012. "Natural Products Research in South Africa : 1890–2010." *South African Journal of Science* 108: 1–8. Doi: 10.4102/sajs.v108i5/6.765.

Esteban, Jaime, and Marta García-Coca. 2017. "Mycobacterium Biofilms." *Frontiers in Microbiology* 8: 2651. Doi: 10.3389/fmicb.2017.02651.

Fenton, M. J., and M. W. Vermeulen. 1996. "Immunopathology of Tuberculosis: Roles of Macrophages and Monocytes." *Infection and Immunity* 64, 3: 683–690.

Fugh-Berman, A., and E. Ernst. 2001. "Herb-drug Interactions: Review and Assessment of Report Reliability." *British Journal of Clinical Pharmacology* 52, 5: 587–595. doi: 10.1046/j.0306-5251.2001.01469.x

Furge, Laura Lowe, and F. Peter Guengerich. 2006. "Cytochrome P450 Enzymes in Drug Metabolism and Chemical Toxicology: An Introduction." *Biochemistry and Molecular Biology Education* 34, 2: 66–74. doi: 10.1002/bmb.2006.49403402066.

Gafford, J. F. 2004. "*Acorus calamus*." Obtained from Wikimedia Commons under the Creative Commons Share-Alike 3.0 license. https://commons.wikimedia.org/wiki/File:Acorus_calamus1.jpg (Accessed 30 June 2021).

Garçon, N., P. Chomez, and M. Van Mechelen. 2007. "GlaxoSmithKline Adjuvant Systems in Vaccines: Concepts, Achievements and Perspectives." *Expert Review of Vaccines* 6, 5: 723–739. doi: 10.1586/14760584.6.5.723

Garfinkel, David. 1958. "Studies on Pig Liver Microsomes. I. Enzymic and Pigment Composition of Different Microsomal Fractions." *Archives of Biochemistry and Biophysics* 77, 2: 493–509. doi: 10.1016/0003-9861(58)90095-X.

Garg, J. M. 2008. "*Cassia occidentalis*." Obtained from Wikimedia under the Creative Commons Attribution-Share Alike 4.0 International, 3.0 Unported, 2.5 Generic, 2.0 Generic and 1.0 Generic license. https://commons.wikimedia.org/wiki/File:Cassia_occidentalis_in_Narshapur,_AP_W_IMG_1105.jpg (Accessed 1 July 2021)

Gengenbacher, M., and S. H. Kaufmann. 2012. "*Mycobacterium tuberculosis*: Success Through Dormancy." *FEMS Microbiology Reviews* 36, 3: 514–532. doi: 10.1111/j.1574-6976.2012.00331.x

Girard, Lauren, and Sunita Vohra. 2011. "Ethics of Using Herbal Medicine as Primary or Adjunct Treatment and Issues of Drug-Herb Interaction." In: *Herbal Medicine: Biomolecular and Clinical Aspects,* 2nd ed: 439–451. CRC Press. doi: 10.1201/b10787-22.

González-Bello C. 2017. "Antibiotic Adjuvants - A Strategy to Unlock Bacterial Resistance to Antibiotics." *Bioorganic & Medicinal Chemistry Letters* 27, 18: 4221–4228. doi: 10.1016/j.bmcl.2017.08.027

Grange, John M., Peter A. Winstanley, and Peter D. O. Davies. 1994. "Clinically Significant Drug Interactions with Antituberculosis Agents." *Drug Safety* 11, 4: 242–251. doi: 10.2165/00002018-199411040-00003.

Gutierrez, M. Cristina, Sylvain Brisse, Roland Brosch, Michel Fabre, Bahia Omaïs, Magali Marmiesse, Philip Supply, and Veronique Vincent. 2005. "Ancient Origin and Gene Mosaicism of the Progenitor of *Mycobacterium tuberculosis*." *PLoS Pathogens* 1, 1: e5. doi: 10.1371/journal.ppat.0010005.

Hall-Stoodley, Luanne, and Paul Stoodley. 2009. "Evolving Concepts in Biofilm Infections." *Cellular Microbiology* 11, 7: 1034–1043. doi: 10.1111/j.1462-5822.2009.01323.x.

Harper, D. 2008. "*Inula helenium*." Obtained from Wikimedia Commons under the Creative Commons Share-Alike 2.0 license. https://commons.wikimedia.org/wiki/File:Inula_helenium_-_Elecampane_-_geograph.org.uk_-_949099.jpg (Accessed 30 June 2021).

Harris, Vanessa, and Annoesjka Swart. 2005. "Drug Interactions With Tuberculosis Therapy" *Continuing Medical Education (CME)* 23, 2: 56–60.

Hillewaert Hans. 2007. "*Rosa canina*." Obtained from Wikimedia Commons under the Creative Commons Share-Alike 3.0 license. https://commons.wikimedia.org/wiki/File:Rosa_canina_.jpg (Accessed 30 June 2021).

HQ. 2010. "*Pelargonium sidoides*." Obtained from Wikimedia under the Creative Commons Attribution-Share Alike 2.0 Generic license. https://commons.wikimedia.org/wiki/File:%E5%A4%A9%E7%AB%BA%E8%91%B5%E5%B1%AC_Pelargonium_sidoides_-%E7%89%9B%E6%B4%A5%E5%A4%A7%E5%AD%B8%E6%A4%8D%E7%89%A9%E5%9C%92_Oxford_Botanic_Garden-_(9207603628).jpg (Accessed 1 July 2021)

Jefferson, K. K., D. A. Goldmann, and G. B. Pier. 2005. "Use of Confocal Microscopy To Analyze the Rate of Vancomycin Penetration through *Staphylococcus aureus* Biofilms." *Antimicrobial Agents and Chemotherapy* 49, 6: 2467–2473. doi: 10.1128/AAC.49.6.2467-2473.2005.

Kapur, V., T. S. Whittam, and J. M. Musser. 1994. "Is *Mycobacterium tuberculosis* 15,000 Years Old?" *Journal of Infectious Diseases* 170, 5: 1348–1349. doi: 10.1093/infdis/170.5.1348.

Kaufmann, S. H., G. Hussey, and P. H. Lambert. 2010. "New Vaccines for Tuberculosis." *Lancet* 375, 9731: 2110–2119. doi: 10.1016/S0140-6736(10)60393-5

Kawai, T., and S. Akira. 2010. "The Role of Pattern-recognition Receptors in Innate Immunity: Update on Toll-like Receptors." *Nature Immunology* 11, 5: 373–384. doi: 10.1038/ni.1863

Khademi, F., A. Yousefi, M. Derakhshan, A. Najafi, and M. Tafaghodi. 2019. "Enhancing Immunogenicity of Novel Multistage Subunit Vaccine of *Mycobacterium tuberculosis* Using PLGA:DDA Hybrid Nanoparticles and MPLA: Subcutaneous Administration." *Iranian Journal of Basic Medical Sciences* 22, 8: 893–900. doi: 10.22038/ijbms.2019.33962.8079

Khameneh, Bahman, Milad Iranshahy, Nasser Vahdati-Mashhadian, Amirhossein Sahebkar, and Bibi Sedigheh Fazly Bazzaz. 2019. "Non-Antibiotic Adjunctive Therapy: A Promising Approach to Fight Tuberculosis." *Pharmacological Research*. doi: 10.1016/j.phrs.2019.104289.

Kitenge, Marcel, Bessie Phiri, Sara M. Pheeha, Modupe Ogunrombi, and Peter Nyasulu. 2020. "Dzherelo (Immunoxel) as Adjunctive Therapy to Standard Antituberculosis Treatment in Patients with Pulmonary Tuberculosis: A Systematic Review and Meta-Analysis of Clinical Trials." *Preprint,* December. doi: 10.21203/rs.3.rs-121904/v1 (still under review).

Kiran, Dilara, Brendan K. Podell, Mark Chambers, and Randall J. Basaraba. 2016. "Host-Directed Therapy Targeting the *Mycobacterium tuberculosis* Granuloma: A Review." *Seminars in Immunopathology*. doi: 10.1007/s00281-015-0537-x.

Klingenberg, Martin. 1958. "Pigments of Rat Liver Microsomes." *Archives of Biochemistry and Biophysics* 75, 2: 376–386. doi: 10.1016/0003-9861(58)90436-3.

Knight, J. A. 2000. "Free Radicals, Antioxidants, and the Immune System." *Annals of Clinical & Laboratory Science* 30, 2: 145–158. PMID: 10807157.

Kolodziej, Herbert. 2011. "Antimicrobial, Antiviral and Immunomodulatory Activity Studies of Pelargonium Sidoides (EPs® 7630) in the Context of Health Promotion." *Pharmaceuticals* 4, 10: 1295–1314. doi: 10.3390/ph4101295.

Leidus, I. 2013. "*Fragaria vesca*." Obtained from Wikimedia Commons under the Creative Commons Share-Alike 3.0 license. https://commons.wikimedia.org/wiki/File:Fragaria_vesca_-_metsmaasikas.jpg (Accessed 30 June 2021).

Létinier, Louis, Sébastien Cossin, Yohann Mansiaux, Mickaël Arnaud, Francesco Salvo, Julien Bezin, Frantz Thiessard, and Antoine Pariente. 2019. "Risk of Drug-Drug Interactions in Out-Hospital Drug Dispensings in France: Results From the Drug Interaction Prevalence Study." *Frontiers in Pharmacology* 10, March: 265. doi: 10.3389/fphar.2019.00265.

Lily, Likhitha. 2014. "*Adathoda vasica*." Obtained from Wikimedia under the Creative Commons Attribution-Share Alike 3.0 Unported license. https://commons.wikimedia.org/wiki/File:Adathoda_vasica.JPG (Accessed 1 July 2021)

Lindenstrom, T., E. M. Agger, K.S. Korsholm, P. A. Darrah, C. Aagaard, R. A. Seder, I. Rosenkrands, and P. Andersen. 2009. "Tuberculosis Subunit Vaccination Provides Long-term Protective Immunity Characterized by Multifunctional CD4 Memory T Cells." *The Journal of Immunology* 182: 8047–8055. doi: 10.4049/jimmunol.0801592

López, Daniel, Hera Vlamakis, and Roberto Kolter. 2010. "Biofilms." *Cold Spring Harbor Perspectives in Biology* 2, 7: a000398. doi: 10.1101/cshperspect.a000398.

Mangwani, Neelam, Pawan Kumar Singh, and Vipin Kumar. 2020. "Medicinal Plants: Adjunct Treatment to Tuberculosis Chemotherapy to Prevent Hepatic Damage." *Journal of Ayurveda and Integrative Medicine*. doi: 10.1016/j.jaim.2019.02.004.

Magro, Lara, Ugo Moretti, and Roberto Leone. 2012. "Epidemiology and Characteristics of Adverse Drug Reactions Caused by Drug—Drug Interactions." *Expert Opinion on Drug Safety* 11, 1: 83–94. doi: 10.1517/14740338.2012.631910.

McDonnell, Anne M., and Cathyyen H. Dang. 2013. "Basic Review of the Cytochrome P450 System." *Journal of the Advanced Practitioner in Oncology*. doi: 10.6004/jadpro.2013.4.4.7.

Mehmood, Zafar, Mohammad Shavez Khan, Faizan Abul Qais, Samreen, and Iqbal Ahmad. 2018. "Herb and Modern Drug Interactions." *New Look to Phytomedicine*, 503–520. doi: 10.1016/b978-0-12-814619-4.00019-7.

Menozzi, F. D., J. H. Rouse, M. Alavi, M. Laude-Sharp, J. Muller, R. Bischoff, M. J. Brennan, and C. Locht. 1996. "Identification of a Heparin-Binding Hemagglutinin Present in Mycobacteria." *The Journal of Experimental Medicine* 184, 3: 993–1001.

Nelson, David R. 2009. "The Cytochrome P450 Homepage." *Human Genomics*. doi: 10.1186/1479-7364-4-1-59.

Nikolaeva, Lyudmila G., Tatyana V. Maystat, Volodymyr S. Pylypchuk, Yuri L. Volyanskii, Lilia A. Masyuk, and Galyna A. Kutsyna. 2008. "Effect of Oral Immunomodulator Dzherelo in TB/HIV Co-Infected Patients Receiving Anti-Tuberculosis Therapy under DOTS." *International Immunopharmacology* 8, 6: 845–851. doi: 10.1016/j.intimp.2008.01.029.

Ojha, Anil K., Anthony D. Baughn, Dhinakaran Sambandan, Tsungda Hsu, Xavier Trivelli, Yann Guerardel, Anuradha Alahari, et al. 2008. "Growth of *Mycobacterium tuberculosis* Biofilms Containing Free Mycolic Acids and Harbouring Drug-Tolerant Bacteria." *Molecular Microbiology* 69, 1: 164. doi: 10.1111/J.1365-2958.2008.06274.X.

Orme, Ian M. 2014. "A New Unifying Theory of the Pathogenesis of Tuberculosis." *Tuberculosis* 94, 1: 8–14. doi: 10.1016/j.tube.2013.07.004.

Petruss. 2011. "*Foeniculum vulgare* flowers." Obtained from Wikimedia Commons under the Creative Commons Share-Alike 3.0 license. https://commons.wikimedia.org/wiki/File:Foeniculum_vulgare_flowers.JPG (Accessed 30 June 2021).

Prihoda, N. D., O. V. Arjanova, L. V. Yurchenko, N. I. Sokolenko, L. a. Vihrova, V. S. Pylypchuk, and G. a. Kutsyna. 2007. "Open Label Trial of Adjuvant Immunotherapy with Dzherelo, Svitanok and Lizorm, in MDR-TB, XDR-TB and TB/HIV Co-Infected Patients Receiving Anti-Tuberculosis Therapy under DOT." *Journal of Medicinal Plants Research* 1, December: 117–122.

Ramappa, Vidyasagar, and Guruprasad P. Aithal. 2013. "Hepatotoxicity Related to Anti-Tuberculosis Drugs: Mechanisms and Management." *Journal of Clinical and Experimental Hepatology*. doi: 10.1016/j.jceh.2012.12.001.

Sakula, Alex. 1983. "Robert Koch: Centenary of the Discovery of the Tubercle Bacillus. 1882." *The Canadian Veterinary Journal* 24: 127–131.

Salveo Life Sciences. 2016. "LIVFUL® Syrup 200ML." https://salveolifesciences.com/product/livful-syrup-200ml/ (Accessed 21 April 2020).

Sampath Kumar, K. P., Debjit Bhowmik, Pankaj Tiwari, and Rakesh Kharel. 2010. "Indian Traditional Herbs Adhatoda Vasica and Its Medicinal Application." *Pharmaceutical Research* 2, 1: 240–245.

Sane, R., and M. Sinz. 2016. "Introduction of Drug Metabolism and Overview of Disease Effect on Drug Metabolism." *Drug Metabolism in Diseases*. Elsevier Inc. doi: 10.1016/B978-0-12-802949-7.00001-8.

Sarma, G. R., S. Kailasam, N. G. Nair, et al. 1980. "Effect of Prednisolone and Rifampin on Isoniazid Metabolism in Slow and Rapid Inactivators of Isoniazid." *Antimicrob Agents Chemother* 18, 5: 661–666. doi: 10.1128/aac.18.5.661

Schluger, Neil W. 2005. "The Pathogenesis of Tuberculosis." *American Journal of Respiratory Cell and Molecular Biology* 32, 4: 251–256. doi: 10.1165/rcmb.F293.

Scripture, C. D., W. D. Figg, and A. Sparreboom. 2006. "Peripheral Neuropathy Induced by Paclitaxel: Recent Insights and Future Perspectives." *Current Neuropharmacology* 4, 2: 165–172. doi: 10.2174/157015906776359568

Skeiky, Y. A., M. R. Alderson, P. J. Ovendale, J. A. Guderian, L. Brandt, D. C. Dillon, A. Campos-Neto, Y. Lobet, W. Dalemans, I. M. Orme, and S. G. Reed. 2004. "Differential Immune Responses and Protective Efficacy Induced by Components of a Tuberculosis Polyprotein Vaccine, Mtb72F, Delivered as Naked DNA or Recombinant Protein." *Journal of Immunology* 172, 12: 7618–7628. doi: 10.4049/jimmunol.172.12.7618

Smith, Issar. 2003. "*Mycobacterium tuberculosis* Pathogenesis and Molecular Determinants of Virulence." *Clinical Microbiology Reviews* 16, 3: 463–496. doi: 10.1128/CMR.16.3.463.

Smoum, Reem, Rubinstein, Abraham, Dembitsky, Valery, Srebnik, Morris. 2012. "Boron Containing Compounds as Protease Inhibitors." *Chemical Reviews* 112: 4156–4220. doi: 10.1021/cr608202m.

Sotgiu, Giovanni, Payam Nahid, Robert Loddenkemper, Ibrahim Abubakar, Marc Miravitlles, and Giovanni Battista Migliori. 2016. "The ERS-Endorsed Official ATS/CDC/IDSA Clinical Practice Guidelines on Treatment of Drug-Susceptible Tuberculosis." *European Respiratory Journal* 48, 4: 963–971. doi: 10.1183/13993003.01356-2016.

Stang, D. 2006. "*Helichrysum petiolare* Lime." Obtained from Wikimedia Commons under the Creative Commons Share-Alike 4.0 license. https://commons.wikimedia.org/wiki/File:Helichrysum_petiolare_Lime_3zz.jpg (Accessed 30 June 2021).

SterlingHerron. 2015. "Purple coneflower (*Echinacea purpurea*)." Obtained from Wikimedia Commons under the Creative Commons CC0 1.0 Universal Public Domain Dedication license. https://commons.wikimedia.org/wiki/File:Purple_coneflower_(Echinacea_purpurea)_01.jpg (Accessed 30 June 2021).

Stewart, E., J. A. Triccas, and N. Petrovsky. 2019. "Adjuvant Strategies for More Effective Tuberculosis Vaccine Immunity." *Microorganisms* 7, 8: 255. doi: 10.3390/microorganisms7080255

Thombre, R., Jangid, K., Shukla, R. and Dutta, N. K. 2019. "Editorial: Alternative Therapeutics Against Antimicrobial-Resistant Pathogens." *Frontiers in Microbiology* 10: 1–2. doi: 10.3389/fmicb.2019.02173

Toman, K. 2004. "Toman's Tuberculosis: Case Detection, Treatment, and Monitoring-questions and Answers." *Geneva* 2: 111.

Vacchelli, E., L. Galluzzi, A. Eggermont, W. H. Fridman, J. Galon, C. Sautès-Fridman, E. Tartour, L. Zitvogel, and G. Kroemer. 2012. "Trial Watch: FDA-approved Toll-like Receptor Agonists for Cancer Therapy." *Oncoimmunology* 1, 894–907. doi: 10.4161/onci.20931

van Dissel, J. T., D. Soonawala, S. A. Joosten, C. Prins, S. M. Arend, P. Bang, P. N. Tingskov, K. Lingnau, J. Nouta, S. T. Hoff, I. Rosenkrands, I. Kromann, T. H. Ottenhoff, T. M. Doherty, and P. Andersen. 2011. "Ag85B-ESAT-6 Adjuvanted with IC31® Promotes Strong and Long-lived *Mycobacterium tuberculosis* Specific T Cell Responses in Volunteers with Previous BCG Vaccination or Tuberculosis Infection." *Vaccine* 29, 11: 2100–2109. doi: 10.1016/j.vaccine.2010.12.135

Vengolis. 2019. "Morinda citrifolia." Obtained from Wikimedia under the Creative Commons Attribution-Share Alike 4.0 International license. https://commons.wikimedia.org/wiki/File:Morinda_citrifolia_0213.jpg (Accessed 1 July 2021)

White, A. G., M. T. Davies-Coleman, and B. S. Ripley. 2008. "Measuring and Optimising Umckalin Concentration in Wild-Harvested and Cultivated Pelargonium Sidoides (Geraniaceae)." *South African Journal of Botany* 74, 2: 260–267. doi: 10.1016/j.sajb.2007.11.015.

Woodworth, J. S., S. B. Cohen, A. O. Moguche, C. R. Plumlee, E. M. Agger, K. B. Urdahl, and P. Andersen. 2017. "Subunit Vaccine H56/CAF01 Induces a Population of Circulating CD4 T Cells that Traffic into the *Mycobacterium tuberculosis*-infected lung." *Mucosal Immunology* 10, 555–564. doi: 10.1038/mi.2016.70

Zambrano, María Mercedes, and Roberto Kolter. 2005. "Mycobacterial Biofilms: A Greasy Way to Hold It Together." *Cell* 123, 5: 762–764. doi: 10.1016/j.cell.2005.11.011.

Zamora, N., J. Esteban, T. J. Kinnari, A. Celdrán, J. J. Granizo, and C. Zafra. 2007. "*In-Vitro* Evaluation of the Adhesion to Polypropylene Sutures of Non-Pigmented, Rapidly Growing Mycobacteria." *Clinical Microbiology and Infection* 13, 9: 902–907. doi: 10.1111/j.1469-0691.2007.01769.x.

Zaitzeva, S. I., Matveeva, S. L., Gerasimova, T. G., Pashkov, Y. N., Butov, D. A., Pylypchuk, V. S., Frolov, V. M., and G. A. Kutsyna. 2009. "Treatment of Cavitary and Infiltrating Pulmonary Tuberculosis with and without the Immunomodulator Dzherelo." *Clinical Microbiology and Infection* 15: 1154–1162. doi: 10.1111/j.1469-0691.2009.02760.x

Ziarnek, K. 2016. "*Tamarix gallica*." Obtained from Wikimedia under the Creative Commons Attribution-Share Alike 4.0 International license. https://commons.wikimedia.org/wiki/File:Tamarix_gallica_kz2.jpg (Accessed 1 July 2021)

15 The Genus *Blumea*

Ethnomedicinal Uses, Phytochemistry and Pharmacology

Harish C. Upadhyay

CONTENTS

15.1 INTRODUCTION

Nature has blessed us with a versatile gift in terms of plants to cure a variety of diseases and debilities. Natural product–based remedies have been used since the beginning of human civilization in the form of folklore or traditional medicine (Cragg et al., 1997; Newman et al., 2000). Enhancement in experience-based knowledge led to orderly-regulated indigenous medicinal systems across the globe, viz. traditional Chinese medicine, Ayurveda, Siddha, Unani, Kampo, traditional Korean medicine and traditional African medicine (Yuan et al., 2016; Fitzgerald et al., 2020; Che et al., 2017).

Traditional Chinese medicine (TCM) is rooted in ancient Chinese philosophy, dates back over 3,000 years and has provided reliable, effective and affordable healthcare options for patients.

DOI: 10.1201/9781003108375-15

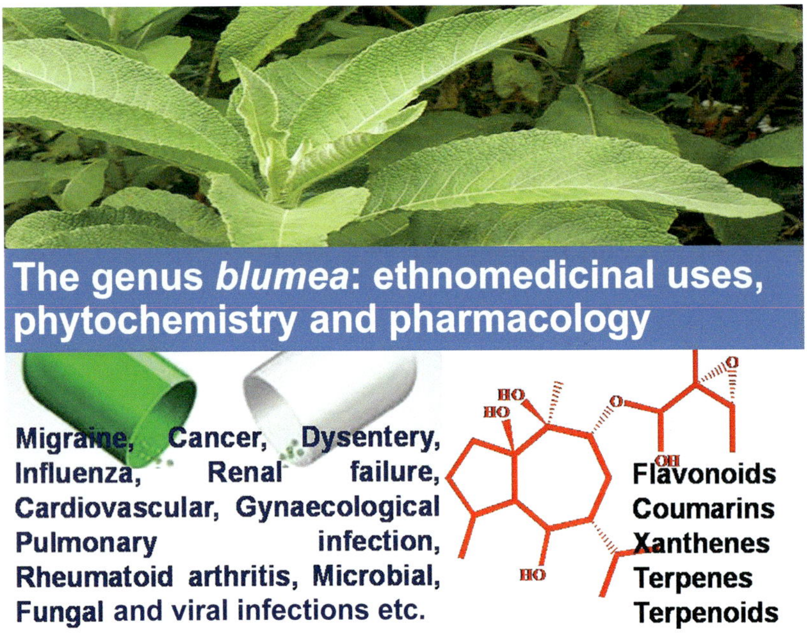

Graphical abstract representing the content of the chapter.

Similarly, Ayurveda, a prevention-oriented holistic science of natural healing developed by the great masters of India around 6000 BC, is reported to use more than 1,500 plant species for treatment (Che et al., 2017; Upadhyay, 2019; Yuan et al., 2016). These well-developed traditional medicines have been practiced either using a single plant, plant part or in the form of herbal formulations comprising more than one plant (Yuan et al., 2016; Pei, 2001; Bent, 2008).

Herbal treatments are the most popular form of traditional medicine and are considered as a holistic approach to prevent and treat diseases. According to the World Health Organization (WHO), in African countries, 70–80% of the population use some form of traditional herbal medicine for their primary healthcare because of their accessibility and affordability (Upadhyay, 2019; WHO, 2019). However, traditional herbal medicines are getting significant attention in the global health sector of developing as well as developed countries including China, India, Nigeria, and the United States (Tilburt & Kaptchuk, 2008).

Medicinal plants are rich in secondary metabolites produced as a defense mechanism to stress present in the environment. The identification of plant secondary metabolites as active principles of therapy became possible with the advancement in pharmaceutical science and technology, which ultimately led to the basis of 'modern drugs' at the beginning of the nineteenth century (Thomford et al., 2018).

Due to the existence of a diverse range of pharmacologically active multi-dimensional chemical entities, plants are still of continued interest to modern pharma scientists because compounds existing in plants have better compatibility with the human body, hence lesser side effects (Balunas & Kinghorn, 2005; Dias et al., 2012; Upadhyay et al., 2014). Indeed, of all the drugs licensed between 1981 and 2002, around 57–67% contain compounds originally identified from plants or their derivatives (Cragg et al., 2002). In the area of cancer, over the time frame from around the 1940s to 2010, of the 175 small molecules, 131 (74.8%) are non-synthetic, with 85 (48.6%) being either natural products or directly derived therefrom (Newman & Cragg, 2012). Therefore, besides direct use, the traditional folk medicine system is a significant and productive cradle of discovery and development of potent drug leads.

Asteraceae is one of the largest families of vascular plants represented by 30,000 species and over 1,100 genera across the world and 900 species under 167 genera in India (Baruah & Choudhury, 2012). Genus *Blumea* DC. is one of the largest genera of Asteraceae, comprising approximately 100 species distributed throughout the Old World tropics, with the highest diversity in tropical Asia and a few species in Australia and Africa (Pornpongrungrueng et al., 2007). Continental Southeast Asia—covering Myanmar, Thailand, Laos, Cambodia, Vietnam and peninsular Malaysia—has been reported to feature 27 species of *Blumea* (Pornpongrungrueng et al., 2016). The plants of genus *Blumea* are prized for their medicinal values and are being used as folklore and in traditional medicine to cure a variety of diseases and debilities. Among them, *B. balsamifera* (L.) DC, *B. riparia* DC, *B. lacera* DC, *B. membranacea* DC and *B. eriantha* DC are very common (Figure 15.1). One of the members of this genus having potent medicinal property, *B. balsamifera*, often grows in forests, river beds, valleys or grasslands, at 600–1,000 m elevation distributed in India, Pakistan, Burma, the Indo-China peninsula, Malaysia, Indonesia and the Philippines (Tolosa et al., 2020; Tan et al., 2020).

Blumea membranacea, *B. eriantha* and *B. lacera* are important medicinal plants of India, Indonesia, Malaysia, Myanmar, Nepal, Pakistan, Sri Lanka, Thailand and Vietnam (Mehta et al., 1986; Upadhyay et al., 2013). *Blumea eriantha* is an annual aromatic herb, which grows abundantly along roadsides and degraded forestlands (Upadhyay et al., 2012; Singh et al., 2012). Some species of this genus are components of patented Chinese herbal formulations to cure various types of skin diseases, menorrhagia, rheumatism, fever, urinary problem, bone fractures, wound treatment, malaria, microbial and fungal infections, cancer, etc. (Upadhyay et al., 2013).

The folklore principles have been verified by potent biological activities of plant extracts, which further led to the isolation and characterization of chemical constituents with their multifaceted activities (Chen et al., 2009; Pang et al., 2014a). The major bioactive phytomolecules isolated from *Blumea* consist of flavonoids, xanthenes, terpenes and terpenoids. Further, the essential oil of various *Blumea* species has been reported to possess promising bioactivities (Chen et al., 2009; Pang

FIGURE 15.1 Some medicinally important *Blumea* species. A, *Blumea lacera* L.; B, *B. membranacea* DC; C, *B. balasamifera* DC; D, *B. lanceolaria* Roxb.; E, *B. mollis* (D.Don) Merr.; F, *B. densiflora* DC (Kew, 2017; Mead, 2017; Valke, 2013, 2014; Vengolis, 2018; Vinayaraj, 2014).

et al., 2014a). Owing to the medicinal importance of the genus *Blumea*, this chapter presents a comprehensive review of ethnomedicinal uses, phytochemistry and pharmacology of various medicinally important *Blumea* species.

15.2 ETHNOMEDICINAL USES

Several species of this genus are used as folklore medicine in India, as well as in other countries. *B. balsamifera* is traditionally named 'Sambong' in many countries, including China and the Philippines, and has been reported as a medicinal plant of immense importance to cure a variety of diseases and illnesses. *Blumea balsamifera* has been mentioned in numerous publications to be used by females for childbirth recovery. The use of *B. balsamifera* helps in the treatment of weakness due to blood loss, tissue tearing and hormonal changes associated with labor (de Boer et al., 2011). The plant has been reported to be used to treat menstrual conditions such as amenorrhea and dysmenorrhea, pain, hemorrhage and infections; as an abortifacient, emmenagogue and lactagogue; and in baths mainly for postpartum women (de Boer & Cotingting, 2014).

A decoction of *B. wightiana* DC roots is taken twice a day for five days for urinary problems, while its leaf decoction is given for stomach aches (Verma & Chauhan, 2007). The leaf paste of *B. lanceolaria* (Roxb.) diluted in cold water is applied on the head and forehead to treat fever, while the leaf paste of *B. clarkei* Hook. F. is applied on the affected area to treat bone fracture (Rahman et al., 2007). In Shahjahanpur district of Uttar Pradesh (India), the plant *B. lacera* is locally named 'Kukrendu'; its leaf juice is given to relieve piles, and leaf paste is given in case of dog bites, while seed powder is given for leukorrhea (Sharma et al., 2010). The juice of *B. eriantha* is administered as a carminative. A warm infusion of *B. eriantha* leaves is given as a sudorific, while a cold infusion is considered diuretic and emmenagogue (Chen et al., 2009). The ethnomedicinal uses of common *Blumea* species are depicted in Table 15.1.

15.3 *BLUMEA* SPECIES IN PATENTED HERBAL FORMULATIONS

Many *Blumea* species have been patented as a component of herbal formulations to cure a variety of diseases in China. *Blumea balsamifera* (Sambong) is a component of various herbal formulations in TCM for treating liver and pancreatic cancer, cold dysentery, influenza, gynecological diseases and viral infections (Upadhyay et al., 2013; Li, 2004; Peng, 2016). It can be used as a component of herbal medicines for stomach and abdominal pain, alcoholic cardiomyopathy, migraine, cardiovascular diseases, gastrointestinal adverse reaction after chemotherapy and radiotherapy, bone tumors, knee osteoarthritis, oral cavity and throat diseases, sciatica, fractures and cerebrovascular disease (Chen, 2015a; Dou & Wei, 2014; Liu, 2014a, 2014b; Liu, 2015; Ouyang, 2015; Qian et al., 2014; Shi, 2010; Wang, 2015; Yin et al., 2015; Yu et al., 2014). *Blumea fistulosa* has been reported to be used in a soft capsule for relieving cough (Chen, 2015b).

Blumea laciniata is a component of TCM for the treatment of multiple neurofibromas, acute intestinal obstruction, acute renal failure, migraine, hyperosteogeny, knee osteoarthritis, rheumatoid arthritis, dermatitis and bone injury (Chen & Jiang, 2014; Ge & Wang, 2014; Gu et al., 2015; Liu & Sun, 2011; Meng & Zhang, 2014; Sun et al., 2014; Sun et al., 2015; Xu, 2014; Yu, 2014; Zhang, 2015; Zhou, 2015). *Blumea lacera* is a component of Chinese herbal formulation for the treatment of acne, viral infections, pulmonary infection, bed sores and oral cancer (Bian, 2015a, 2015b; Deng, 2015; Meng, 2015; Shi, 2015). *Blumea megacephala* is a component of external use in Chinese medicine for the treatment of eczema, female genital itching and psoriasis (Du, 2014; Lv, 2015; Shang, 2015). *Blumea oblongifolia* is a component of Chinese herbal preparation for the treatment of ulcerative colitis (Wang et al., 2014). *Blumea lanceolaria* is a component of herbal medicines for the treatment of cardiovascular disease, cough and mumps (Chen, 2014; Li, 2015; Wu, 2014). *Blumea riparia*, along with *B. lanceolaria*, is used in herbal medicine for treating chronic cholecystitis (Hao, 2014).

TABLE 15.1
Ethnomedicinal Uses of Some *Blumea* Species

Species	Ethnomedicinal Uses	References
B. aurita (L.f.) DC.	Leaf paste is used to stop bleeding from cuts.	(Vanila et al., 2008)
B. balsamifera	Powdered leaves and roots (oral and poultices) are used to stop bleeding and epistaxis; to cool down body temperature; for itching, bone and joint pain; and for refreshing juice.	(Panyaphu et al., 2011)
	Decoctions of the leaves and roots are used in herbal baths by mothers after childbirth. Leaf tea is taken for fevers and stomach pain, and leaf paste is applied to cuts and wounds.	(Ahmad & Ismail, 2003)
	Whole plant is used as a stomachic, expectorant, antispasmodic and sudorific, taken as a crude drug for beriberi, menorrhagia, leukorrhea, lumbago, rheumatism and a variety of ailments.	(Chen et al., 2009)
	Roots and leaves decoction for cough, urinary tract infection, stomach ulcer, fever and headache.	(Cordero et al., 2020)
	Hypertension, cough, cough with phlegm, urination difficulty, postpartum care and recovery, body ache, headache, fever, weakness and fatigue, gas pain and flatulence.	(Dapar et al., 2020)
	Burn/decoction/grinded poultices are used for Flank/axilla symptom, muscle pain, sprain/strain of joint.	(Kantasrila et al., 2020)
	Decoction of leaves for diarrhea, fever, heartburn, constipation.	(Sujarwo et al., 2015)
	Kidney problems, hypertension, cuts, infected wounds, diarrhea, rheumatism, spasms, colds and coughs, respiratory infection, stomach pain and improving the urinary tract.	(Alok et al., 2013; Boy et al., 2018)
	Entire plant is used as antiplasmodial. Decoction of the leaves is used for coughs, fever and influenza, sprains, pain, itching.	(Noor-Rain et al., 2007; Nguyen et al., 2020)
	The leaves as a stomachic, expectorant, antispasmodic and diaphoretic; in fever, lumbago, to increase appetite; in skin diseases, to treat wounds and liver cirrhosis and to treat gout.	(Nessa et al., 2010; Nguyen et al., 2004)
	Decoction is used in female genital disease/menstruation irregularity.	(Lee et al., 2019)
B. clarkei	Leaf paste is applied on the affected area to treat bone fracture.	(Rahman et al., 2007)
B. densiflora DC.	Juice of fresh leaves is used as insecticidal, mosquito repellant.	(Khare, 2007)
B. eriantha	The juice of the whole plant as a carminative. A warm infusion of leaves is given as a sudorific, while a cold infusion is used as a diuretic and emmenagogue.	(Benelli et al., 2017; Chen et al., 2009; Khare, 2007)
B. fastulosa (Roxb.) Kurz.	Diuretic.	(Khare, 2007)
B. lacera	Whole plant is used as antipyretic, bronchitis, blood diseases. Leaves as astringent, febrifuge, diuretic, deobstruent, anthelmintic (particularly in case of thread worm), astringent, stimulant, febrifuge, bleeding piles, anti-inflammatory repellent of flies and other insects. Roots are used as anticholerin.	(Khare, 2007; Mia et al., 2020; Akter et al., 2014)
	In Uttar Pradesh (India), its leaf juice is given to relieve piles and leaf paste is given for dog bites, while seed powder is given for leukorrhea.	(Sharma et al., 2010)
B. lanceolaria	A leaf paste diluted in cold water is applied on the head and forehead to treat fever.	(Rahman et al., 2007)
B. mollis (D.Don) Merr.	The boiled herb is used for treating diarrhea, inflammatory disease and fever. The leaf used for treatment of skin diseases.	(Guga- Bakshi et al., 1999; Brindha-Devi et al., 2011; Vanila et al., 2008)
B. riparia	Whole plant for puerperal metrorrhagia, puerperal edema, and barrenness.	(Chen et al., 2009)
B. wightiana	A decoction of roots is taken twice a day for five days for urinary problems, while its leaf decoction is given in stomachache.	(Verma & Chauhan, 2007)

Blumea species have received attention not only in China, but also from herbalists across the world as evident from some of the U.S. and Indian patented herbal formulations. In the Philippines, the tablet consisting of *B. balsamifera* is used as a clinically proven drug able to decrease the size or number of stones, passage of stones and disappearance or reduction of signs and symptoms, and hence is used for treating patients with kidney and urinary tract stones (Tolosa et al., 2020). A herbal medicinal formulation containing the whole plant of *B. lacera*, along with other plants and honey, has been patented and marketed as a body rejuvenating agent for physical and mental abilities against environmental toxins, as well as lifestyle disorders (Khan, 2014). *Blumea lacera* is used as a component of Indian herbal formulations for the treatment and prevention of neurological disorders (Yadav, 2015). *Blumea fistulosa* is a component of patented Indian herbal formulations for treating muscle cramps (Bibi, 2014).

15.4 PHYTOCHEMICAL AND PHARMACOLOGICAL INVESTIGATIONS OF ESSENTIAL OIL

Most of the *Blumea* species are pleasantly aromatic annual herbs, and hence, their aerial parts are rich in volatile constituents. The common traditional method for extraction of essential oil (EO) from aerial parts of various *Blumea* species is hydro-distillation using Clevenger apparatus. However, other methods like simultaneous distillation and extraction (SDE), CO_2 supercritical extraction and headspace solid-phase micro-extraction (HS-SPME) are used (Pang et al., 2014a; Wang et al., 2012).

The EO is reported to be analyzed by gas chromatography (GC) and gas chromatography-mass spectrometry (GC-MS), and compounds have been identified by mass spectral fragmentation patterns of peaks, matching with recorded mass spectral databases of compounds. Pang et al. (2014a) compiled the chemical constituents and biological activity of EO of *B. balsamifera* (Pang et al., 2014a). There are reports on EO of other species of *Blumea* viz. *B. eriantha*, *B. mollis*, *B. lacera*, *B. densiflora*, *B. martiniana*, *B. perrottetiana* and *B. megacephala* (Benelli et al., 2017; Owolabi et al., 2010; Satyal et al., 2015; Senthilkumar et al., 2008; Zhu et al., 2011; Zhu & Tian, 2011a, 2011b). The volatile chemical constituents of essential oil extracted from various *Blumea* species identified by GC-MS are viz. terpenes, terpenoids, fatty acids and alcohols, aldehydes, ethers, ketones, pyridines, and furans, as depicted in Table 15.2.

The major chemical components of EO—monoterpenes, sesquiterpenes and terpenoids like borneol, limonene, ocimene, allo-ocimene, myrcene, camphene, pinene, terpinene, terpineniol, sabinene, camphor, gurjunene, allo-aromadendrene, caryophyllene, caryophylleneoxide and murrolene—are common in all species examined (Table 15.2). Many of the phytochemical constituents reported that the EO of *Blumea* species has proven to have pharmacological effects. The EO of *B. mollis* showed significant anti-oxidant potential exhibiting the DPPH radical scavenging activity (50% inhibitory concentration (IC_{50}) 347.06 ± 0.001 µg/mL) as compared to methanol extract of micropropagated plant callus (IC_{50} 387.20 ± 0.001 µg/mL) (Tamilarasi & Thirugnanasampandan, 2014).

The EO of *B. balsamifera* exhibited moderate antitumor and anti-oxidant activities (Jiang et al., 2014). *B. lacera* EO was extracted by hydro-distillation and constituents were identified using GC-MS. One of the major components, (Z)-lachnophyllum ester, was isolated by preparative silica gel chromatography. The oil was found to be rich in (Z)-lachnophyllum ester (25.5%), (Z)-lachnophyllic acid (17.0%), Germacrene D (11.0%), (E)-β-farnesene (10.1%), Bicyclogermacrene (5.2%), (E)-caryophyllene (4.8%), (E)-nerolidol (4.2%), (E)-lachnophyllic acid (3.3%) and (E)-lachnophyllum ester (1.7%). Surprisingly, (Z)-lachnophyllic acid and (Z)-lachnophyllum ester were in much higher concentrations as compared with the (E) isomers. Both the EO and (Z)-Lachnophyllum ester showed promising activity against human mammary adenocarcinoma (MDA-MB-231 and MCF-7) and human urinary bladder carcinoma (5637) cell lines showing high cytotoxic effect toward MDAMB- 231 cells with IC_{50} value 7.2 ± 1.0 µg/mL (Satyal et al., 2015).

TABLE 15.2
Phyto-Constituents from Essential Oil of Various *Blumea* Species

Classifications	Sl. Nos.	Compound Names	Molecular Formulas	References
Monoterpenes	1	(+)-Limonene	$C_{10}H_{16}$	A, B, C, D, E, G
	2	(-)-Limonene		A
	3	Sabinene		A, B,C,D, E, F, G, H
	4	α-Thujene		A, B, C, E, G
	5	(E) Ocimene		A, C, D, E, G
	6	Allo-ocimine		A, C, F
	7	(Z)-β-Ocimene		A, E, G
	8	β-Myrcene		A, B, C, D, G
	9	Camphene		A, C, D, E, G
	10	α-Pinene		A, B, C,D, E, G, H
	11	β-Pinene		A, B, D, E, G, H
	12	α-Terpinene		B, C, H
	13	γ-Terpinene		C, E, F, G, H
	14	α-Phellandrene		C, E
	15	β-Phellandrene		E, G
	16	Terpinoline/iosterpine		C, E, H
	17	3-Carene		D, E, H
	18	Citronellol	$C_{10}H_{20}O$	E
	19	L-borneol	$C_{10}H_{18}O$	A, B, E, F, G
	20	Isoborneol		A, E, G
	21	Terpinen-4-ol		A, B, D, G, H
	22	Isopulegol		C
	23	α-Terpineol		A, D E, F, G, H
	24	γ-Terpineol		C, E
	25	Nerol		B, G
	26	Geraniol		D, E, G, H
	27	1,8-Cineole		A, D, H
	28	1,6 Dihydrocarveol		C
	29	Lavandulol		E
	30	Perillyl alcohol	$C_{10}H_{16}O$	A, D
	31	trans-Verbenol		B
	32	α-Thujone		D
	33	(Z)-Jasmone		D
	34	Camphor		A, D, E, G
	35	Cis-pinocamphone		C
	36	E-Pinocarveol		E
	37	Chrysanthenone	$C_{10}H_{14}O$	A
	38	Perilla aldehyde		A, G
	39	Myrtenal		A
	40	Cuminaldehyde	$C_{10}H_{12}O$	A,
	41	Linalool oxide	$C_{10}H_{18}O_2$	A, G

(Continued)

TABLE 15.2 (Continued)

Classifications	Sl. Nos.	Compound Names	Molecular Formulas	References
	42	Linalool	$C_{10}H_{18}O_2$	C, D, E, F, G, H
	43	Thymyl methyl ether	$C_{11}H_{16}O$	G, H
	44	Bornyl acetate	$C_{12}H_{20}O_2$	A, E,
	45	Thymohydroquinone dimethyl ether	$C_{12}H_{18}O_2$	A
Sesquiterpenes	46	α-Gurjunene	$C_{15}H_{24}$	A, B
	47	Alloaromadendrene		A, C, D, E, F, H
	48	(+)-Aromadendrene		A
	49	Aromadendrene		A
	50	Longifolene		A
	51	α-Caryophyllene		A, B, F
	52	β-Caryophyllene		A,C,D, E, F, H
	53	Guaia-3,9- diene		A
	54	γ-Cadinene		A, H
	55	δ-Cadinene		A,C,D, H
	56	β-Selinene		A
	57	(+)-β-Selinene(β-Selinene)		A
	58	β-Gurjunene		A, E
	59	(+)-γ-Gurjunene		A
	60	Thujopsene-13		A
	61	Viridiflorene		C
	62	β-Elemene		A,D
	63	γ -Elemene		A,D
	64	(−)-γ-Cadinene		A
	65	(−)-δ-Cadinene		A
	66	Cadina-1,4-diene		E, H
	67	(E)-β-Farnesene		D, H
	68	Copaene		C,D. F, H
	69	α-Cubebene		C
	70	α-Humulene		D, E, F, H
	71	Germacrene A		D
	72	Germacrene B		D
	73	Germacrene D		D, E, F
	74	Bisabolene		E
	75	Bicyclogermacrene		D
	76	α-trans-Bergamotene		D
	77	γ-Muurolene		A, B, C, H
	78	Cuparene	$C_{15}H_{22}$	C
	79	Aromadendrene, dehydro		A
	80	10-Epi-γ-Eudesmol	$C_{15}H_{26}O$	A, E
	81	7-Epi-γ-Eudesmol		B, E
	82	Globulol		A
	83	(−)-Guaiol		A
	84	Ledol		A, E
	85	E-Nerolidol		D, E,
	86	Elemol		A
	87	α-Eudesmol		A, E
	88	β-Eudesmol		A, D
	89	γ-Eudesmol		A
	90	Carotol		A

Classifications	Sl. Nos.	Compound Names	Molecular Formulas	References
	91	Cubenol		A
	92	Zingiberenol, Guaiol		E
	93	α-Muurolol		B, H
	94	α-cadinol		D
	95	δ—cadinol		H
	96	τ-cadinol		H
	97	Caryophyllene oxide	$C_{15}H_{24}O$	A,B,C,D, E, H
	98	Aromadendrene oxide		A
	99	Germacrene-D-4-ol		D
	100	α-Santalol		B
Diterpenes	101	16-Kaurene	$C_{20}H_{32}$	A
	102	Phytol	$C_{20}H_{40}O$	A, D
	103	1-Ang-4,7-dihydroxyeudesmane	$C_{20}H_{34}O_4$	A
Fatty acids and esters	104	Capric acid	$C_{10}H_{20}O_2$	A
	105	Xanthoxylin	$C_{10}H_{12}O_4$	A
	106	Eugenol	$C_{10}H_{12}O_2$	A, D, E
	107	E-Lachnophyllic acid		D
	108	Z-Lachnophyllic acid	$C_{10}H_{10}O_2$	D
	109	Trans-2-undecenoic acid	$C_{11}H_{20}O_2$	A
	110	Methyl eugenol	$C_{11}H_{14}O_2$	E
	111	(E)-Lachnophyllum ester	$C_{11}H_{12}O_2$	D
	112	Z-Lachnophyllum ester		D
	113	Bornyl acetate	$C_{12}H_{20}O_2$	G
	114	Dimethoxydurene	$C_{12}H_{18}O_2$	A
	115	Dodecyl acetate	$C_{14}H_{28}O_2$	B
			$C_{14}H_{24}O_2$	
	116	Neryl isobutanoate		D
	117	Palmitic acid	$C_{16}H_{32}O_2$	A
	118	n-Hexadecanoic acid		E
	119	9-Hexadecenoic acid	$C_{16}H_{30}O_2$	A
	120	(11Z)-11-hexadecenoic acid		A
Alcohol	121	n-Hexanol	$C_6H_{14}O$	B
	122	3-Hexen-1-ol	$C_6H_{12}O$	G
	123	3-Octanol	$C_8H_{18}O$	A
	124	1-Octen-3-ol	$C_8H_{16}O$	A, B, E, G
	125	2-phenylethanol	$C_8H_{10}O$	B, E, G
	126	Perilla alcohol	$C_{10}H_{16}O$	G
	127	Estragole	$C_{10}H_{12}O$	C
	128	10-Dodecyn-1-ol	$C_{12}H_{22}O$	D
Aldehydes and ketones	129	Heptanal	$C_7H_{14}O$	E
	130	3-Octanone	$C_8H_{16}O$	A
	131	Nonanal	$C_9H_{18}O$	H
	132	Isophorone	$C_9H_{14}O$	B
	133	Decanal	$C_{10}H_{20}O$	H
	134	(4Z)-Decenal	$C_{10}H_{18}O$	B
	135	Carvotanacetone	$C_{10}H_{16}O$	B
	136	Myrtenal	$C_{10}H_{14}O$	B
	137	3-Propyl benzaldehyde	$C_{10}H_{12}O$	A
			$C_{10}H_{12}O$	
	138	4-Isopropylbenzaldehyde		A

(Continued)

TABLE 15.2 (Continued)

Classifications	Sl. Nos.	Compound Names	Molecular Formulas	References
	139	2-Hydroxy-4,6-dimethoxyacetophenone	$C_{10}H_{12}O_4$	A
Miscellaneous compounds	140	3-Fluoro-5-amino-pyridine	$C_5H_5FN_2$	A
	141	α-Butenoic acid, 3-methoxy-4-nitro	$C_5H_7NO_5$	A
	142	O-acetyl-1-serine	$C_5H_9NO_4$	A
	143	α-Butenoic acid, 3-methoxy-4-nitro	$C_5H_7NO_5$	A
	144	p-Vinylguaiacol	$C_9H_{10}O_2$	D
	145	Furan,4,5-diethy-2,3-dihydro-2,3-dimethyl	$C_{10}H_{18}O$	A
	146	1,3,4,5,6,7-hexahydro-2,5,5-trimetehyl-2H-2,4A-et hanonaphthalene	$C_{10}H_{14}$	A
	147	P-cymene (4-Isopropyltoluene)		A, D, E, G, H
	148	o-cymene (2-Isopropyltoluene)		E, H
	149	Carvacrol	$C_{10}H_{14}O$	B, G
	150	2,5-Dimethoxy-*p*-cymene	$C_{12}H_{18}O_2$	H

Notes: A, (Pang et al., 2014a); B, (Benelli et al., 2017); C, (Senthilkumar et al., 2008); D, (Satyal et al., 2015); E, (Zhu & Tian, 2011a); F, (Zhu & Tian, 2011b); G, (Zhu et al., 2011); H, (Owolabi et al., 2010)

Essential oil of *B. balsamifera* and *B. mollis* have been reported to show significant antimicrobial activity, which can be further justified by the presence of antimicrobial phytoconstituents present in EO viz. β-pinene, camphor, bornyl acetate, borneol, linalool, D-limonene, terpinen-4-ol and α-terpinene (de Boer et al., 2011; Senthilkumar et al., 2009). Sakee et al. (2011) reported the EO of *B. balsamifera* to show potent antibacterial and antifungal activity with a minimal inhibitory concentration (MIC) of 150 μg/mL against *Bacillus cereus*, and 1.2 mg/mL against *Staphylococcus aureus* and *Candida albicans*. Notice that bioactivity was two-fold better than hexane extract (Sakee et al., 2011). *Blumea balsamifera* EO showed potent antibacterial activity against *Haemophilus parasuis*, a potentially pathogenic Gram-negative organism that colonizes the upper respiratory tract of pigs, causing Glasser's disease. The *in vitro* MIC and Minimum Bactericidal Concentration (MBC) of *B. balsamifera* EO were found to be 0.625 and 1.25 μg/mL, respectively (He et al., 2020).

In another study, the volatile oil of *B. balsamifera* showed potent antibacterial activity on plant pathogens *Rhizoctonia solani*, *R. solani* AG1-IA and *Sclerotinia sclerotiorum*, with MIC value of 0.195 μL/mL for all the strains (Jiang et al., 2014). *Blumea megacephala* EO was obtained through steam distillation and analyzed using gas chromatography with flame ionization detection (GC-FID) and GC-MS. A total of 65 compounds were identified with a dominance of borneol (13.6 %), β-caryophyllene (9.56 %), Germacrene D (9.09 %), Sabinene (6.37 %) and α-humulene (4.78%). The EO showed promising antimicrobial activity against Gram-positive *Bacillus subtilis*, *S. aureus* (MIC of 62.5 and 125 μg/mL, respectively) and Gram-negative *Pseudomonas aeruginosa*, *Escherichia coli* (MIC of 125 μg/mL for both) bacterial as well as yeast *Hansenula anomala* (MIC of 125 μg/mL) strains (Zhu et al., 2011). The antimicrobial activities may be justified due to the presence of bioactive components of EO. (Z)-Lachnophyllum ester, a major component of *B. lacera* EO, showed good activity against *B. cereus* (MIC of 312.5 μg/mL), *S. aureus* (MIC of 78 μg/mL), *E. coli* (MIC of 312.5 μg/mL), *C. albicans* (MIC of 78 μg/mL) and *Aspergillus niger* (MIC of 39 μg/mL) (Satyal et al., 2015).

The external application of *B. balsamifera* EO on the intact and damaged skin of rats improved wound contraction and closure without any acute toxicity or allergic reaction. Histopathology studies showed that application of EO increased wound-healing in rats by reducing the number of inflammatory cells, significantly increasing the hydroxyproline content, improving the formation of collagen and reducing the frequency of fibroblasts (Wang et al., 2013; Pang et al., 2014b). In another study, *B. balsamifera* EO was found highly effective for the acceleration of healing of burn injuries in deep second-degree burn models in rats (Fan et al., 2015). These results justify the traditional use of *B. balsamifera*, which is used to heal wounds and reduce itching (Wang, et al., 2013; Pang et al., 2014b).

The EO obtained from *B. membranacea* and *B. fistulosa* (Roxb.) Kurz. has been reported to possess central nervous system (CNS) depressant activity (Mehta et al., 1986; Khare, 2007). The *B. membranacea* EO produced a marked and long-lasting fall in blood pressure in anesthetized dogs by exerting a direct depressant action on frog heart and spasmolytic effect on rabbit ileum. The *B. membranacea* EO adversely affected rotarod performance and potentiated the pentobarbitone-induced hypnosis in rats (Mehta et al., 1986).

The essential oils of *B. balsamifera*, *B. eriantha*, *B. mollis*, *B. densiflora* and *B. martiniana* have been reported to possess potent larvicidal activity against various mosquito species. LC_{50} is the concentration of the chemical in the air or water that will kill 50% of the test animals with a single exposure. The *B. eriantha* EO showed high toxicity against larvae of six mosquito species: *Anopheles stephensi* (LC_{50} = 41.61 μg/mL), *Aedes aegypti* (LC_{50} = 44.82 μg/mL), *Culex quinquefasciatus* (LC_{50} = 48.92 μg/mL), *Anopheles subpictus* (LC_{50} = 51.21 μg/mL), *Aedes albopictus* (LC_{50} = 56.33 μg/mL) and *Culex tritaeniorhynchus* (LC_{50} = 61.33 μg/mL) (Benelli et al., 2017). The EO of *B. mollis* showed a significant toxic effect against early fourth-instar larvae of *C. quinquefasciatus* with an LC_{50} of 71.71 and an LC_{90} of 143.41 ppm (Senthilkumar et al., 2008). The *B. densiflora* EO has shown potent larvicidal activity against *A. anthropophagus* larvae with 100% mortality within six hours at 150 ppm, ten hours at 100 ppm, 30 hours at 50 ppm, and 30 hours with EO at a concentration of 25 ppm (Zhu & Tian, 2011a). Similarly, *B. martiniana* EO showed potent larvicidal activity against A. anthropophagus, the natural vector of malaria in a concentration-dependent manner (Zhu & Tian, 2011b). The larvicidal activity of EOs can be justified by the active components of the oil like linalool, Germacrene D, arvotanacetone, (4E,6Z)-allo-ocimene, allo-aromadendrene, borneol, β-caryophyllene, γ-terpinene, sabinene and β-bisabolene. Notice that (4E,6Z)-allo-ocimene and carvotanacetone achieved LC_{50} lower than 10 μg/mL on various tested mosquito species (Zhu & Tian, 2011b). The results favor the proposed *Blumea* EO as an alternative for developing eco-friendly mosquito control tools.

The EO from the aerial parts of *B. perrottetiana* was obtained by hydrodistillation. The GC-MS analysis of the EO revealed the presence of 2,5-dimethoxy-p-cymene (30.0%) and 1,8-cineole (11.0%), along with oxygenated monoterpenes as major constituents. The EO demonstrated notable insecticidal activity against the red flour beetle (*Tribolium castaneum*) in both dose- and exposure-dependent manner. EO at a dose concentration of 0.10 μL/cm2 increased the mortality rate to 73.3% after 12 hours and 93.3% after 24 hours of exposure, while a dose concentration of 0.4 μL/cm2 increased the mortality rate to 100% mortality after 12 hours (Owolabi et al., 2010). The significant insecticidal activity has been reported for EO of *B. balsamifera*, which showed fumigant toxicity against *Sitophilus zeamais*, *Aeromonas hydrophila*, *F. graminearum* and *Magnaporthe grisea* (Chu et al., 2013; Pang et al., 2014b). The results are consistent with traditional uses of the plant as an insecticide and anthelmintic, and suggest the possible use of these volatile oils to be a useful alternative to synthetic insecticides.

15.5 PHYTOCHEMICAL AND PHARMACOLOGICAL INVESTIGATIONS OF EXTRACTS

Various ethnomedicinal uses of *Blumea* species have been justified by phytochemical investigations. Plant materials have been extracted and fractionated with various suitable solvents and investigated for the presence of a major group of phytochemicals. The application of extensive chromatographic separation techniques on extracts and fractions led to the isolation of pure phytomolecules, which have been characterized by spectroscopic data analysis of compounds. Furthermore, the extracts, fractions and isolated compounds have been evaluated for their pharmacological effects on various debilities and disease-causing agents.

15.5.1 Preliminary Phytochemical Investigation of Extracts

The preliminary phytochemical screening consists of a set of tests following standard procedures to detect the major class of chemical constituents present in medicinal plants (Upadhyay et al., 2011). The preliminary phytochemical screening helps to find out the presence of important pharmacologically active principles which may be isolated further. There have been several reports on preliminary phytochemical screening of extracts of *Blumea* species which revealed the presence of carbohydrates, proteins, glycosides, phenols, tannins, flavonoids, triterpenoids, saponins, gums and mucilages (Ahmed et al., 2014; Rahman et al., 2013; Sreelekha et al., 2017; Khair et al., 2015; Khandekar et al., 2013). The alkaloids were found present either in trace amounts or absent in most of the species. During phytochemical screening studies, alkaloids were found to be absent in *B. balsamifera* water, as well as organic extracts though were present in *B. lacera* and *B. mollis* (Kumari et al., 2016; Rahman et al., 2013; Sreelekha et al., 2017).

In many studies, the total flavonoid contents of the different extracts of *Blumea* species was determined, which revealed the presence of high phenolic/flavonoid content, thereby justifying the anti-oxidant and superoxide radical scavenging activity (Fazilatun et al., 2004; Nessa et al., 2010; Gore & Desai, 2014). Under the efforts of primary phytochemical screening and quality control of herbal materials, there are reports on fluorescence profiling of the plant powder under visible, short and long ultraviolet (UV) light to signify characteristics of the plant material (Sreelekha et al., 2017). In a recent approach of primary phytochemical screening, there is a report on Fourier-transform infrared spectroscopy (FTIR) analysis of *B. lacera* plant powder or crude extracts, which may be the most powerful tool for identifying types of chemical bonds and functional groups (Khandekar et al., 2013).

15.5.2 Pharmacological Effects of Extracts

15.5.2.1 Antimicrobial Effects

In a study by Sakee et al., (2011), the dried leaves of *B. balsamifera* were extracted with hexane, dichloromethane and methanol. The extracts were evaluated for antibacterial and antifungal activities using the disc diffusion assay and agar microdilution method. The hexane extract showed a MIC of 9.6 mg/mL against Gram-positive *S. aureus* ATCC 25923, and a MIC of 4.8 mg/mL against Gram-negative *Enterobacter cloacae* DMST 17206. The dichloromethane extract was found inactive against *S. aureus*, but showed a MIC of 9.6 mg/mL against *E. cloacae*. No significant fungicidal activity was observed (Sakee et al., 2011).

Crude methanol extract of *B. lacera* was evaluated for its antibacterial activity against Gram-positive *B. cereus*, *Bacillus megaterium*, *B. subtilis*, *S. aureus* and Gram-negative *E. coli*, *P. aeruginosa*, *Salmonella typhi* and *Shigella dysenteriae* by disc diffusion method. The zones of inhibition were found within the range of 9.0–14.0 mm. The highest zone of inhibition (14.0 mm) was observed against *S. dysenteriae*, and the plant extract showed mild antibacterial activity against *S. typhi*, *B. cereus* and *S. aureus*. The extracts were further evaluated for their antifungal activities against *A. niger*, *Blastomyces dermatitidis* and *Cryptococcus neoformans*. The zones of inhibition were found within the range of 11.0–18.17 mm with the highest zone (18.17 mm) observed for *B. dermatitidis* (Khair et al., 2015).

In another study by Singh et al. (2010), the aqueous and methanolic extracts of *B. lacera* were screened for antimicrobial activity using the disc diffusion assay. The aqueous extract was found to possess potent antimicrobial activity against Gram-positive *B. subtilis*, *S. aureus* and Gram-negative *Serratia marcescens* bacterial strains. Both aqueous and methanolic extracts showed promising antifungal activity against *Candida albicans* in terms of zone of inhibition of 9–38 mm in agar well diffusion technique using serial dilutions (Singh et al., 2010). Further, in many studies, the methanol extract of *B. lacera* was reported to possess promising antibacterial activity against both Gram-positive and Gram-negative strains by disc diffusion assay (Mahida & Mohan, 2006; Khandekar et al., 2013; Jahan et al., 2014).

15.5.2.2 Antiplasmodial Effects

The antiplasmodial potential of *B. balsamifera* root and stem extracts were evaluated *in vitro* against *P. falciparum* chloroquine-sensitive D10 strain using the Plasmodium lactate dehydrogenase (pLDH) assay in which they showed an IC_{50} of 26.25 and 7.75 µg/mL, respectively. The extracts were confirmed to be non-toxic toward Madin Darby bovine kidney (MDBK) cells, which revealed that the antiplasmodial activity was not due to the toxic nature of extracts (Noor-Rain et al., 2007). In another study by Upadhyay et al. (2013), the aerial parts and roots of *B. membranacea* and *B. eriantha* were extracted with ethanol and were successively partitioned with n-hexane, ethyl acetate and n-butanol. These were later evaluated for their *in vitro* antiplasmodial activity against *P. falciparum* NF-54 and *in vitro* cytotoxicity against non-cancerous Vero cell line. The n-hexane, ethyl acetate fractions of aerial parts and n-butanol fraction of roots of *B. membranacea* showed IC_{50} values of 17.4, 19.0 and 3.3 µg/mL, respectively. N-hexane, ethyl acetate fractions of aerial parts and ethyl acetate fraction of roots of *B. eriantha* showed IC_{50} values of 25.0, 26.5 and 15.6 µg/mL, respectively, against *P. falciparum* NF-54. All these fractions were non-toxic to Vero cells (IC_{50} > 200 µg/mL), thus possessing a high degree of selective antiplasmodial activity (Upadhyay et al., 2013). Crude methanol extract of *B. lacera* showed good antiplasmodial activity against *P. ovale* with zone inhibition of 11.00 mm, as compared with the standard ciprofloxacin (15.33 mm) by disc diffusion assay (Khair et al., 2015).

15.5.2.3 Anticancer Effects

Norikura et al. (2008) evaluated the anticancer potential of methanol and ethanol extracts of *B. balsamifera*, which revealed that the 50% methanolic extract of *B. balsamifera* induced growth inhibitory activity in the rat (McA-RH7777) and human hepatocellular carcinoma (HepG2) cells without cytotoxicity to normal rat hepatocytes. Further studies were carried out to understand the mechanism of action, which revealed that the extract induces cell cycle arrest at the G1 phase via a decrease in expression of cyclin-E and phosphorylation of retinoblastoma (Rb) protein in both dose- and time-dependent manners. The extract depletes the level of a proliferation-inducing related ligand (APRIL), which stimulates tumor cell growth and thus may be an important mechanism on hepatoma cancer cells (Norikura et al., 2008). The excipient of *B. balsamifera*, along with other Chinese herbs, has been reported to have an anti-tumor effect on the primary or metastatic hepatoma and pancreatic cancer, and registered as a patented formulation by Lee (2006) for the treatment of hepatoma and pancreatic cancer (Lee, 2006).

In another study by Mutmainah et al. (2014), it was found that hot water extract of *B. balsamifera*, in combination with extracts of *Curcuma domestica* L. and *Amomum compactum* S., was effective in protecting the gastric mucosa and decreasing the number of mast cells and eosinophils in aspirin-induced gastric ulcer model in rats in a dose- and time-dependent manners. The results were comparable with sucralfate, a cytoprotective agent indicated for peptic ulcers (Mutmainah et al., 2014).

The cytotoxic effect of some Bangladeshi medicinal plants, including *B. lacera*, was studied by Uddin et al. (2011), in which the methanol and water extracts were screened for cytotoxic activity against healthy mouse fibroblasts (NIH3T3) and three human cancer cell lines: gastric (AGS), colon (HT-29) and breast (MDA-MB-435S) using the 3-(4,5-dimethylthiazol-2-yl)-2,5-diphenyl tetrazolium bromide (MTT) assay. Potent cytotoxic activity was observed for both methanol (IC_{50} 0.01–0.08 mg/mL) and aqueous (IC_{50} 0.39–0.67 mg/mL) extracts of *B. lacera* against all tested cell lines (Uddin et al., 2011). The methanol extract of *B. eriantha* showed potent cytotoxic activity in HeLa (IC_{50} = 782.25 µg/mL) and B16-F10 (IC_{50} = 841.84 µg/mL) cancer cell lines. The extract showed a potent antimetastatic effect and cell death by apoptosis (Gore & Desai, 2014). The crude methanolic extract of *B. lanceolaria* leaf was found to have mild antitumor activity against murine ascites Dalton's lymphoma (DL) up to a dose of 200 mg/kg body weight of mice *in vivo* (Rosangkima & Prasad, 2004).

15.5.2.4 Antiviral Effects

The aqueous extract of *B. laciniata* was evaluated for antiviral activities against human herpes simplex virus type 1 (HSV-1) and human respiratory syncytial virus (RSV) using a cytopathic effect (CPE) reduction assay. The extract was inactive against HSV-1 but showed potent activity (IC_{50} = 15.6 μg/mL) against RSV with high selectivity index (SI = 15.5). The results suggested that besides polyphenolic compounds, other kinds of constituents in the herb might contribute to their anti-RSV activity (Li, 2004). The results scientifically justify the traditional use of *B. laciniata* in the treatment of viral infection in the respiratory tract (Li et al., 2004). The hot water extract of *B. lacera* showed significant antiviral activity against HSV-1 and HSV-2, with IC_{50} values 83.2 and 43.3 μg/mL, respectively, without any cytotoxic effect on basal cell carcinoma cell line (BCC-1/KMC). Additionally, the extract showed potent *in vitro* antileukemic activity against leukemia cell lines K562 (ATCC CCL 243), L1210 (ATCC CCL 219), P3HR1 (ATCC HTB 62) and Raji (ATCC CCL 86) (Chiang et al., 2004).

15.5.2.5 Anti-Oxidant and Anti-Inflammatory Effects

Various organic extracts of *B. balsamifera* were evaluated for their anti-oxidant potential in terms of xanthine oxidase (XO) inhibitory activities. The methanol extract showed highest XO inhibitory activity (IC_{50} 0.111 mg/mL), followed by chloroform (IC_{50} 0.138 mg/mL) and petroleum ether (IC_{50} 0.516 mg/mL) extracts. The superoxide radical scavenging potential followed the same order, which may be due to a higher concentration of flavonoids in methanol extract (Nessa et al., 2010). The anti-oxidant potential of various extracts of *B. eriantha* was evaluated by 2,2-diphenyl-1-picrylhydrazyl (DPPH) radical scavenging and ferric reducing anti-oxidant power (FRAP) assay, which revealed that the methanol extract had the maximum reducing capacity, followed by the ethanol extract and aqueous extracts. The DPPH free radical scavenging activity (RSA) was 97.90% for methanol, 97.84% for ethanol and 89.67 % for aqueous extracts (Gore & Desai, 2014).

In another study by Singh et al. (2014), alcoholic and aqueous extracts of *B. eriantha* root and stem produced significant anti-oxidant activity. An evaluation was conducted in terms of total reducing power and DPPH, super oxide, nitric oxide, and hydrogen peroxide free radical scavenging power. The oxidizing power of extracts may be attributed to flavonoids and phenolic constituents present in the extract (Singh et al., 2014). Ethanol extract of *B. lacera* showed significant DPPH radical scavenging effect spectrophotometrically with IC_{50} of 19.9 μg/mL (Rahman et al., 2013). In another study, the crude methanol extract of *B. lacera* significantly inhibited the DPPH free radicals at the concentration ranging from 1,000–5,000 μg/mL (Khandekar et al., 2013).

In a study by Basnet et al. (2015) the ethanol extract of aerial parts of *B. lacera* was given to male Wistar rats at doses of 100 and 200 mg/kg for seven days. Enterocolitis was induced in experimental rats by two subsequent doses of indomethacin (7.5 mg/kg) on the seventh and eighth day, and the protective effect of extracts against indomethacin-induced enterocolitis was investigated based on macroscopic scores of ileum strips, histopathological parameters such as serum lactate dehydrogenase, tissue myeloperoxidase, lipid peroxidation and total thiols. The animals pre-treated with extracts did not show significant changes in tissue damage as well as histopathological parameters; hence, good protective effects against the development of indomethacin-induced enterocolitis was observed and may be attributed to its anti-oxidant, anti-inflammatory, antimicrobial and membrane-stabilizing properties. The animals showed good tolerance without any toxicity and mortality with a single dose of 2 g/kg (Basnet et al., 2015).

Similar results were obtained in another study whereby an aqueous-alcoholic extract of *B. lacera* was found to accelerate the healing of acetic acid-induced ulcerative colitis in rats by regulating oxidative stress and colonic inflammation (Adhikari et al., 2020). Crude methanol extract of *B. lacera* showed good red blood cell (RBC) membrane stabilizing activity, which may be indicative of their ability to exert anti-inflammatory activity (Khair et al., 2015).

15.5.2.6 Antidiabetic, Anti-Obesity and Antidyslipidemic Effects

To evaluate hypoglycemic potential, the effect of methanol extract of *B. lacera* was tested *in vivo* on a glucose-loaded hypoglycemic rat model. Results indicated that the leaf extract of *B. lacera* caused a significant decrease in blood glucose level in a dose-dependent manner. The effect was observed in all doses, but maximum hypoglycemic activity (46.85%) was observed at 400 mg/kg, while the standard drug, Glibenclamide, produced 47.53% at a dose of 10 mg/kg (Hasan et al., 2015). The extract of *B. lacera* showed an inhibitory effect on α-amylase, an enzyme responsible for breaking down of α-1,4-glycosidic bonds in starch, suggesting possible anti-hyperglycemic and antidiabetic activity of this plant (Khair et al., 2015).

In a study by Kubota et al. (2009) the ethanol extract of *B. balsamifera* caused a significant decrease in the activity of glycerol-3-phosphate dehydrogenase (GPDH) by down-regulation of key adipogenic transcription factors, including peroxisome proliferator-activated receptor (PPAR) gamma, CCAAT element-binding protein (C/EBPs) and leptin in 3T3-L1 adipocytes. However, the extract caused an up-regulation of adiponectin at the protein level in 3T3-L1 adipocytes. The results suggest that *B. balsamifera* extract may block adipogenesis, at least in part, along with antidiabetic, anti-atherogenic and anti-inflammatory properties (Kubota et al., 2009).

In vivo anti-hyperlipidemic potential of ethanol and aqueous extracts of roots and stem of *B. eriantha* was evaluated in triton WR-1339 (iso-octyl polyoxyethylene phenol)-induced hyperlipidemia in albino rats by determination of serum cholesterol, triglycerides, very-low-density lipoprotein (VLDL) and low-density lipoprotein (LDL) levels. The alcoholic and aqueous extracts of *B. eriantha* root and stem at doses of 200 and 400 mg/kg body weight showed significant reduction in cholesterol, triglyceride, LDL and VLDL levels, and an increase in plasma high-density lipoprotein (HDL) levels measured at 6 hours and 24 hours post-administration. The results were comparable with Fenofibrate at 20 mg/kg and Simvastatin at 4 mg/kg, considered as standard (Singh et al., 2014).

15.5.2.7 Wound Healing Effects

In an investigation by Pang et al. (2017) the effect of total flavonoid content (TFC) of *B. balsamifera* on skin excisional wounds on the back of Sprague-Dawley rats was evaluated. The powdered plant material was extracted with ethanol. Dried ethanolic extract was dissolved in hot water and fractionated with petroleum ether; the evaporated water part was subjected to polyamide macroporous resin column chromatography, eluting with a MeOH/H_2O gradient to afford three fractions (F1–F3); of these, F2, eluted with 80% methanol, containing 81.1% flavonoid content, was further considered as TFC for evaluation of wound healing property. Three different doses: high (2.52 g/kg), medium (1.26 g/kg), and low (0.63 g/kg) of TFC was applied for ten consecutive days post-wound. Significant increases of granulation tissue, fibroblast and capillary vessel proliferation were observed at day seven in the high dose and positive control groups, compared with the standard Jing Wan Hong (JWH) ointment (Pang et al., 2017).

15.5.2.8 Anti-Urolithic Effects

B. balsamifera showed potent anti-urolithic activity in ethylene glycol and ammonium chloride-induced urolithiasis male hooded Wistar rats (*Rattus norvegicus*). The 100% and 50% concentrations of the decoction inhibits calcium oxalate stone formation in the kidneys, effectively decreasing the number of stones and oxalate content of the kidney homogenate (Agdamag et al., 2020). Similar results were obtained in another study, whereby the extract of *B. balsamifera* was observed to have decreased crystal size, shifted crystal phase from calcium oxalate monohydrate to calcium oxalate dihydrate and thus prevented the aggregation of calcium oxalate crystals (Montealegre & De-Leon, 2017).

15.5.2.9 Hepatoprotective Effects

The methanol extract of *B. mollis* showed significant hepatoprotective activity by decreasing the activity of serum glutamate oxalate transaminase (SGOT), serum glutamate pyruvate transaminase (SGPT), alkaline phosphatase (ALP) and bilirubin, while it significantly increased the levels of

Hepatic glycogen reduced glutathione (GSH), Glutathione-S-transferase (GST), glutathione peroxidase (GPX) and GSH-R in a dose-dependent manner (Brindha-Devi & Revathi, 2012).

15.5.2.10 Other Pharmacological Effects

Crude methanol extract of *B. lacera* showed a statistically significant reduction in the incidence and severity of diarrhea in castor oil–induced diarrheal mice. The plant extract at 100–200 mg/kg body weight significantly lowered several typical parameters of diarrhea (Khair et al., 2015). The *B. lacera* methanolic extract showed anxiolytic activity in mice, as evident by open field and whole cross models (Khair et al., 2015). The ethanol extract of *B. densifolia* showed significant analgesic, antipyretic, antidiarrheal and anxiolytic activities. The analgesic effect was accessed by Eddy's hot plate model, whereby the extract at a dose of 400 mg/kg body weight decreased latency time significantly. The antipyretic activity was accessed by Brewer's yeast-induced pyrexia in mice, whereby the extract at a dose of 500 mg/kg body weight reduced rectal body temperature. The extract at the dose of 500 mg/kg body weight reduced the number of feces in castor oil-induced diarrheal mice. Furthermore, the extract at a dose of 500 mg/kg body weight decreased the number of field cross, whole cross and swing, which was statistically significant for anxiolytic activity (Ahsan, 2016).

15.5.3 Isolation of Chemical Compounds and their Pharmacology

The genus *Blumea* is a rich source of a diverse range of secondary metabolites. Owing to the medicinal importance of various species of this genus, many publications are available on detailed chemical investigation leading to isolation, characterization and evaluation of pharmacological effects of pure secondary metabolites. The major non-volatile constituents isolated from various plant extracts include flavonoids, flavanone, chalcone, steroids, triterpenes, sesquiterpene lactones and steroid glycosides. Owing to extensive pharmacological effects, the plant *B. balsamifera* has emerged as a center of attraction of the genus *Blumea*—and the majority of the compounds have been reported from this plant. The compounds have been reported to be isolated by extraction of dried plant material, fractionation of extracts and application of various chromatographic techniques on fractions like column chromatography and preparative thin-layer chromatography (TLC). The structure elucidation of isolated pure compounds has been carried out by infra-red, mass spectroscopic and various nuclear magnetic spectroscopic techniques (1H-NMR, 13C-NMR, DEPT, COSY, HMBC, HSQC, and NOESY), and UV spectral analysis. Furthermore, many of the isolated compounds are reported to possess significant biological activities evaluated *in vitro* and *in vivo*.

15.5.3.1 Flavonoids and Phenolic Compounds

Flavonoids, a class of plant secondary metabolites having a polyphenolic structure, are the major constituent of *Blumea* species, as evident from various reports on primary phytochemical investigations. Most of the flavonoids are reported to be isolated from *B. balsamifera*, justifying the immense pharmacological use of this plant. Plant flavonoids and phenolics are associated with a broad spectrum of health-promoting effects. They are a necessary component of a variety of nutraceutical, pharmaceutical, medicinal formulations, which may be attributed to their antioxidative, anti-inflammatory, antimutagenic and anticarcinogenic properties, coupled with their capacity to modulate key cellular enzymatic functions (Banjarnahor & Artanti, 2015; Rice-Evans et al., 1997).

Many of the phenolics and flavonoids isolated from *Blumea* species are reported to possess potent pharmacological effects against common debilities and disease-causing organisms. The flavanoids (sub-categorized as flavones, flavanones, flavonols, flavanonols and biflavonoids) and other phenolic compounds viz. chalcones, xanthenes, coumarins, lignanes, chromones, etc., reported from various *Blumea* species are summarized in Table 15.3, and their structures have been provided in Figure 15.2.

TABLE 15.3
Flavonoids and Phenolic Compounds Isolated from Genus *Blumea*

Compound Nos.	Compound Names	Sources	References
	Flavonoids (Flavones)		
1	Luteolin	*B. balsamifera* *B. riparia*	(Chen et al., 2009; Pang et al., 2014a)
2	Luteolin-7-O-β-D-glucopyranoside	*B. riparia*	(Chen et al., 2009)
3	Luteolin-7-methyl-ether	*B. balsamifera*	(Pang et al., 2014a)
4	Diosmetin (Luteolin-4'-methyl ether)	*B. balsamifera*	(Pang et al., 2014a)
5	Chrysoeriol (Luteolin-3'-methyl ether)	*B. balsamifera*	(Pang et al., 2014a)
6	4',5-dihydroxy-7-methoxyflavone	*B. balsamifera*	(Pang et al., 2014a)
7	4',5-dihydroxy-3',6,7-trimethoxyflavone	*B. lacera*	(Chen et al., 2009)
8	Nepitrin	*B. riparia*	(Chen et al., 2009)
9	Apigenin	*B. riparia*	(Chen et al., 2009)
10	Apigenin 7-O-β-D-glucopyranoside	*B. riparia*	(Chen et al., 2009)
11	Tricin	*B. riparia*	(Chen et al., 2009)
12	Trici-7-O-β-D—glucopyranoside	*B. riparia*	(Chen et al., 2009)
	Flavonoids (Flavanones)		
13	Eriodictyol	*B. balsamifera*	(Pang et al., 2014a)
14	5,7,3',5'-tetrahydroxyflavanone	*B. balsamifera*	(Pang et al., 2014a)
15	3',4',5-trihydroxy-7-metoxyflavanone	*B. balsamifera*	(Ali et al., 2005; Pang et al., 2014a)
16	Blumeatin (5,3',5'-trihydroxymethoxydihydroflavone)	*B. balsamifera*	(Pang et al., 2014a)
	Flavonoids (Flavonols)		
17	Quercetin	*B. riperia*	(Pang et al., 2014a)
18	3,5,3',4'-tetrahydroxy-7-methoxyflavone	*B. balsamifera*	(Pang et al., 2014a)
19	3,5,3'-trihydroxy-7,4-dimethoxyflavone	*B. balsamifera*	(Pang et al., 2014a)
20	Rhamnetin (7-methoxyquercetin)	*B. balsamifera*	(Pang et al., 2014a)
21	Tamarixetin	*B. balsamifera*	(Pang et al., 2014a)
22	Ombuine	*B. balsamifera*	(Pang et al., 2014a)
23	3,5,7-trihydroxy-3'4'-dimethoxyflavone	*B. balsamifera*	(Pang et al., 2014a)
24	3,3',4',5-tetrahydroxy-7-methoxyflavone	*B. balsamifera*	(Pang et al., 2014a)
25	3,5-dihydroxy-3',4',7-trimethoxyflavone	*B. balsamifera*	(Pang et al., 2014a)
26	4',5-dihydroxy-3,3',7-trimethoxy flavanone	*B. balsamifera*	(Pang et al., 2014a)
27	5,7-dihydroxy-3,3',4',-trimethoxyflavone	*B. balsamifera*, *B. aromatica*	(Pang et al., 2014a; Shen et al., 2019
28	Ayanin (5,3'-dihydroxy-3,7,4'-trimethoxyflavone)	*B. balsamifera*	(Pang et al., 2014a)
29	4',5,7-trihydroxy-3,3'-dimethoxyflavone	*B. balsamifera*	(Pang et al., 2014a)
30	5-hydroxy-3,3',4',7-tetramethoxyflavone	*B. riparia*	(Chen et al., 2009)
31	Isoquercitrin (Quercetin-3-O-glucoside)	*B. balsamifera*	(Pang et al., 2014a)
32	Hyperoside (Quercetin-3-O-galactoside)	*B. balsamifera*	(Pang et al., 2014a)
33	2',5',6-trihydroxy-3,5,7-trimethoxyflavone	*B. malcolmii*	(Chen et al., 2009)
34	5',6-dihydroxy-2',3,5,7-trimethoxyflavone	*B. malcolmii*	(Chen et al., 2009)
35	6-hydroxy-2',3,5,5',7-pentamethoxyflavone	*B. malcolmii*	(Chen et al., 2009)
36	6-hydroxy-3,4',5,7-tetramethoxyflavone	*B. malcolmii*	(Chen et al., 2009)
37	3,4',5-trihydroxy-3',6, 7-trimethoxyflavone	*B. lacera*	(Chen et al., 2009)
38	3,3',5-trihydroxy-4',7-dimethoxyflavone	*B. riparia*	(Chen et al., 2009)
39	3,3',4',5-tetrahydroxy-7-methoxyflavone	*B. riparia*	(Chen et al., 2009)
40	5-hydroxy-3,4',6,7-tetramethoxyflavone	*B. malcolmii*	(Chen et al., 2009)
41	Chrysosplenol C	*B. balsamifera*	(Chen et al., 2009)

(Continued)

TABLE 15.3 (Continued)

Compound Nos.	Compound Names	Sources	References
42	3,3',4' -trihydroxy- 6,7, 8-trimethoxyflavone	*B. eriantha*	(Tambewagh et al., 2019)
43	3-hydroxy-6,7,8,3',4'-pentamethoxyflavone	*B. eriantha*	(Tambewagh et al., 2019)
	Flavonoids (Flavanonols)		
44	Dihydroquercetin-4'-methylether	*B. balsamifera*	(Pang et al., 2014a)
45	Dihydroquercetin-7,4'-dimethylether	*B. balsamifera*	(Pang et al., 2014a)
46	3,4'5-trihydroxy-3'7-dimethoxyflavanone	*B. balsamifera*	(Ali et al., 2005; Pang et al., 2014a)
47	3,3',4',5-tetrahydroxy-7-methoxyflavanone	*B. balsamifera*	Pang et al., 2014a)
48	3,3',5-trihydroxy-4',7-dimethoxyflavanone	*B. balsamifera*	(Pang et al., 2014a)
49	3,3',5,5',7-pentahydroxyflavanone	*B. balsamifera*	(Pang et al., 2014a)
50	3',4',5-trihydroxy-3,7-dimetoxyflavanone	*B. balsamifera*	(Pang et al., 2014a)
51	Catechin	*B. balsamifera*	(Pang et al., 2014a)
52	(2R,3R)-(+)-7-*O*-methyldihydroquercetin	*B. balsamifera*	(Pang et al., 2014a)
	Biflavonoids		
53	3-O-7''-biluteolin	*B. balsamifera*	(Ali et al., 2005)
	Phenolic compounds other than Flavonoids		
54	Davidigenin	*B. balsamifera*	(Pang et al., 2014a)
55	Davidioside	*B. balsamifera*	(Pang et al., 2014a)
56	5,7-dihydroxychromone	*B. balsamifera*	(Pang et al., 2014a)
57	Umberlliferone (7-hydroxycoumarin)	*B. balsamifera*	(Pang et al., 2014a)
58	Hydranngetin	*B. balsamifera*	(Pang et al., 2014a)
59	Protocatechuic acid	*B. balsamifera; B. riparia*	(Chen et al., 2009; Tan et al., 2013)
60	Protocatechuic acid methyl ester	*B. riparia*	(Chen et al., 2009)
61	Caffeic acid	*B. riparia*	(Chen et al., 2009)
62	Caffeic acid methyl ester	*B. riparia*	(Chen et al., 2009)
63	2-isoprenyl-5-isopropylphenol 4-O-β-d-xylopyranoside	*B. riparia*	(Chen et al., 2009)
64	1-(4-angeloyloxy-3-methoxyphenyl)-3-(angeloyloxy) prop-1-ene	*B. riparia*	(Chen et al., 2009)
65	Syringaresinol	*B. balsamifera*	(Pang et al., 2014a)
66	Blumeaxanthene	*B. ripera*	(Cao et al., 2007)
67	Blumeaxanthene-II	*B. ripera*	(Huang et al., 2010)
68	2,4-dicumylphenol	*B. balsamifera*	(Tan et al., 2013)
69	p-hydroxybenzoic acid	*B. balsamifera*	(Tan et al., 2013)
70a	4,5-di-O-caffeoylquinic acid	*B. aromatica*	(Shen et al., 2019)
70b	4,5-di-O-caffeoylquinic methyl ester	*B. aromatica*	(Shen et al., 2019)

Pang et al. (2014a) in a review, cited 35 flavonoids and some other phenolic compounds isolated and characterized from *B. balsamifera* (Pang et al., 2014a). With the advancement in phytochemical studies in recent years, some more flavonoids have been reported from *B. balsamifera*. Ali et al. (2005) reported a new biflavanoid as 3-O-7''-biluteolin (**53**), along with 3',4',5-trihydroxy-7-methoxyflavanone (**15**) and 3,4',5-trihydroxy-3',7-dimethoxyflavanone (**46**) from various extracts of *B. balsamifera* (Ali et al., 2005). Tan et al., (2013) reported isolation of phenolic compounds gentisic acid (**59**), 2,4-dicumylphenol (**68**) and p-hydroxybenzoic acid (**69**) (Tan et al., 2013). Many flavonoids reported from *Blumea* species have been proven to have superoxide radical scavenging and anti-oxidant properties (Banjarnahor & Artanti, 2015; Rice-Evans et al., 1997; Nessa et al., 2010).

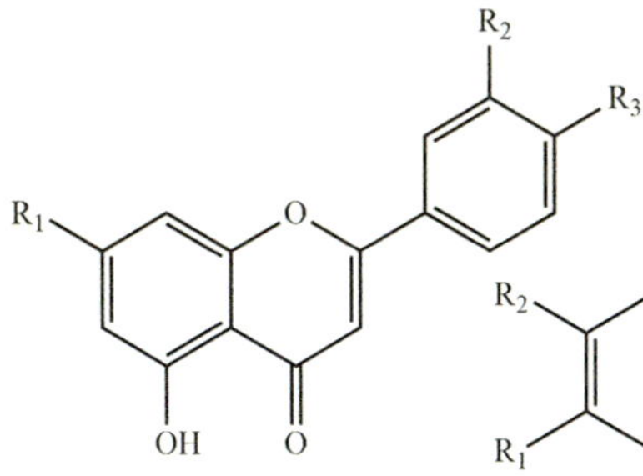

	R1	R2	R3
1	-OH	-OH	-OH
2	-OGlu	-OH	-OH
3	-OMe	-OH	-OH
4	-OH	-OH	-OMe
5	-OH	-OMe	-OH

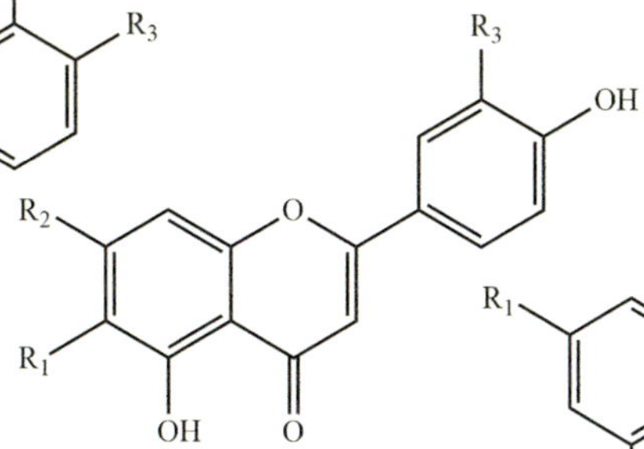

	R1	R2	R3
6	-H	-OMe	-H
7	-OMe	-OMe	-OMe
8	-OMe	-OGlu	-OH

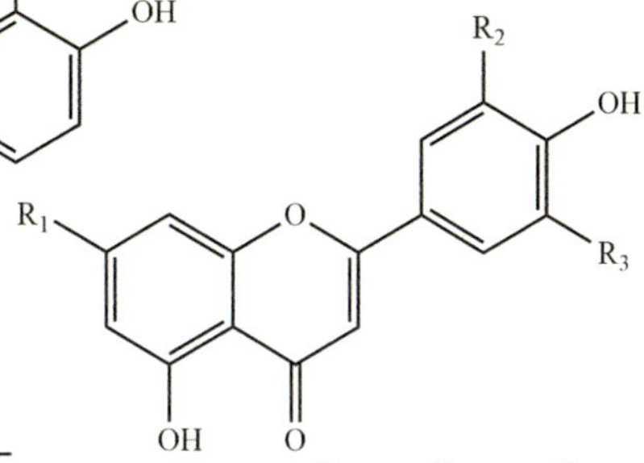

	R1	R2	R3
9	-OH	-H	-H
10	-OGlu	-H	-H
11	-OH	-OMe	-OMe
12	-OGlu	-OMe	-OMe

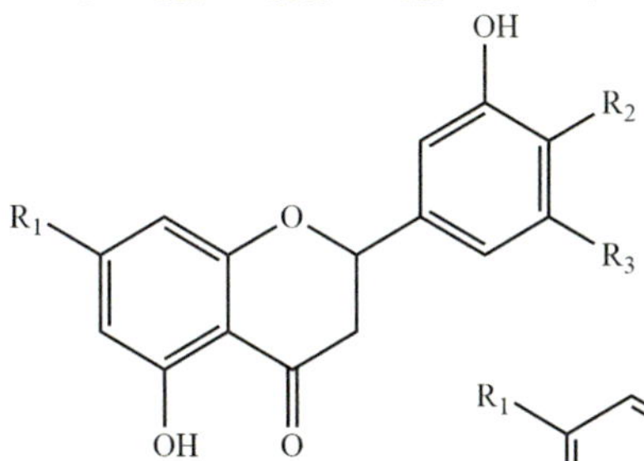

	R1	R2	R3
13	-OH	-OH	-H
14	-OH	-H	-OH
15	-OMe	-OH	-H
16	-OMe	-H	-OH

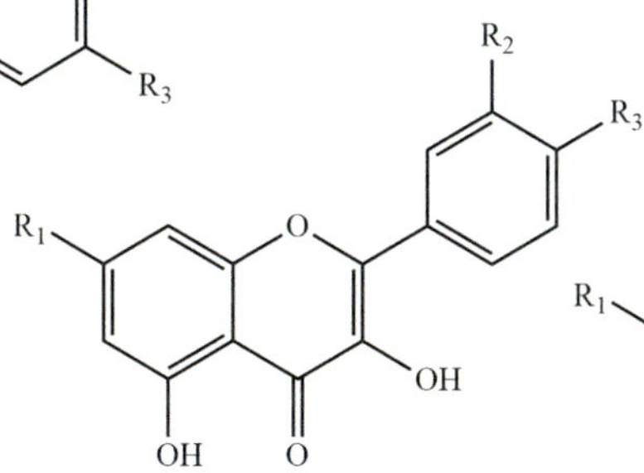

	R1	R2	R3
17	-OH	-OH	-OH
18	-OMe	-OH	-OH
19	-OMe	-OH	-OMe
20	-OMe	-OH	-OH
21	-OH	-OH	-OMe
22	-OMe	-OH	-OMe
23	-OH	-OMe	-OMe
24	-OMe	-OH	-OH
25	-OMe	-OMe	-OMe

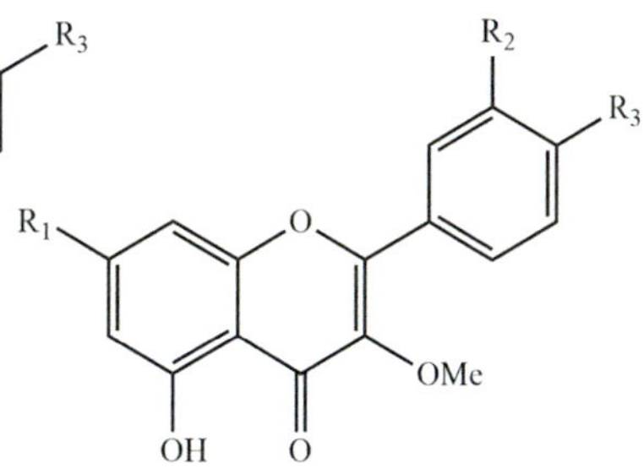

	R1	R2	R3
26	-OMe	-OMe	-OH
27	-OH	-OMe	-OMe
28	-OMe	-OH	-OMe
29	-OH	-OMe	-OH
30	-OMe	-OMe	-OMe

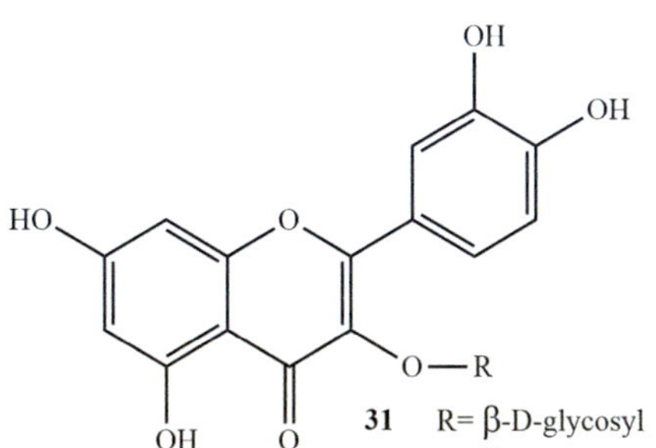

31 R= β-D-glycosyl
32 R= β-D-galactosyl

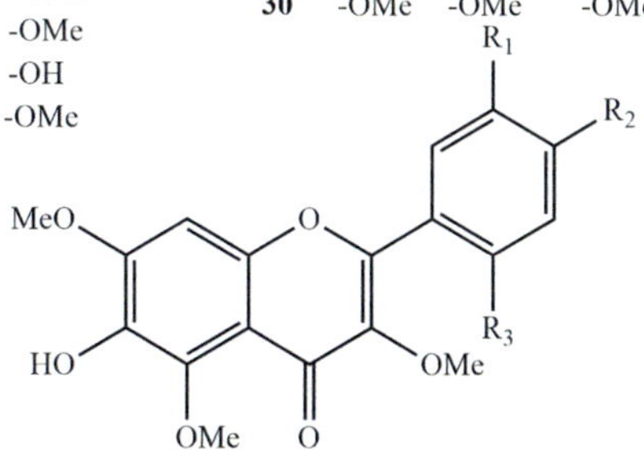

	R1	R2	R3
33	-OH	-H	-OH
34	-OH	-H	-OMe
35	-OMe	-H	-OH
36	-H	-OMe	-H

FIGURE 15.2 Molecular structure of flavonoids and phenolic compounds (**1–70**) from *Blumea* species.

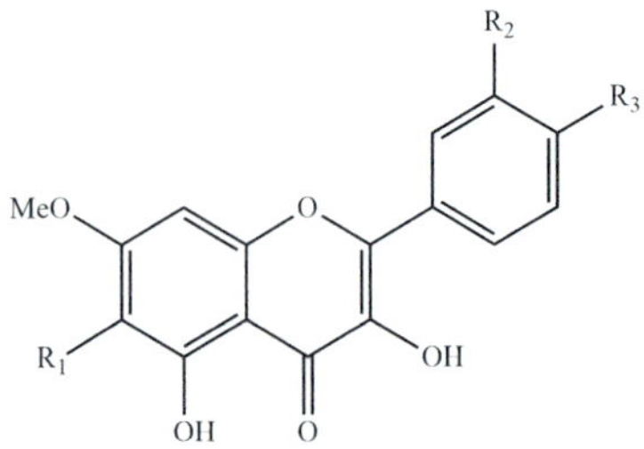

	R_1	R_2	R_3
37	-OMe	-OMe	-OH
38	-H	-OH	-OMe
39	-H	-OH	-OH

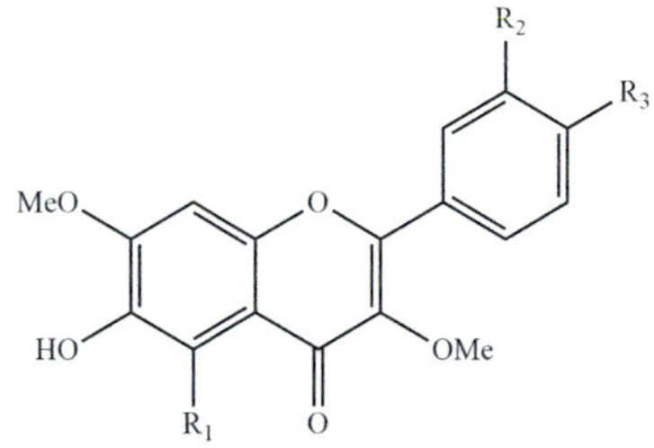

	R_1	R_2	R_3
40	-OMe	-H	-OMe
41	-OH	-OMe	-OH

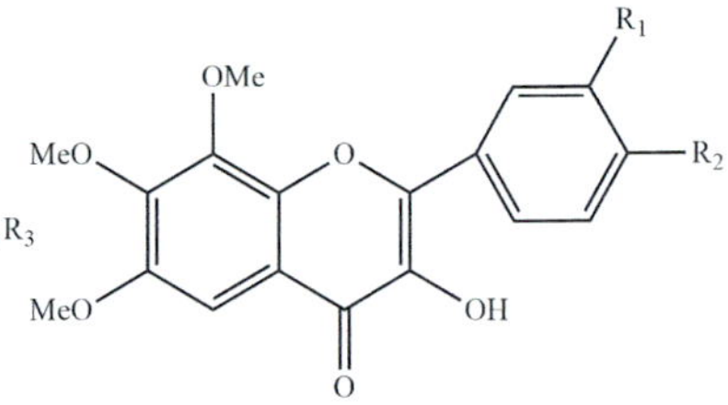

	R_1	R_2
42	-OH	-OH
43	-OMe	-OMe

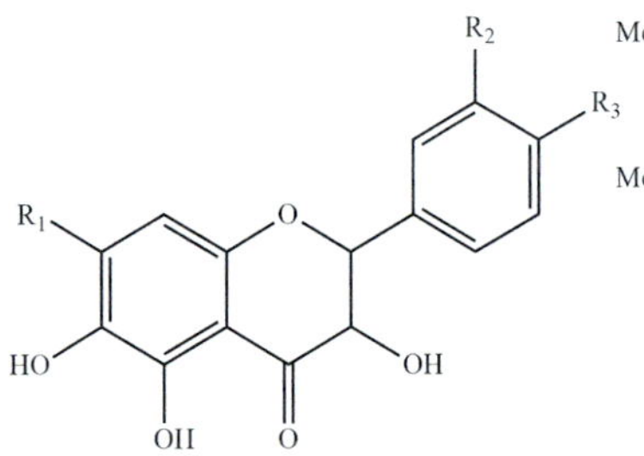

	R_1	R_2	R_3
44	-OH	-OH	-OMe
45	-OMe	-OH	-OMe
46	-OMe	-OMe	-OH
47	-OMe	-OH	-OH
48	-OMe	-OH	-OMe

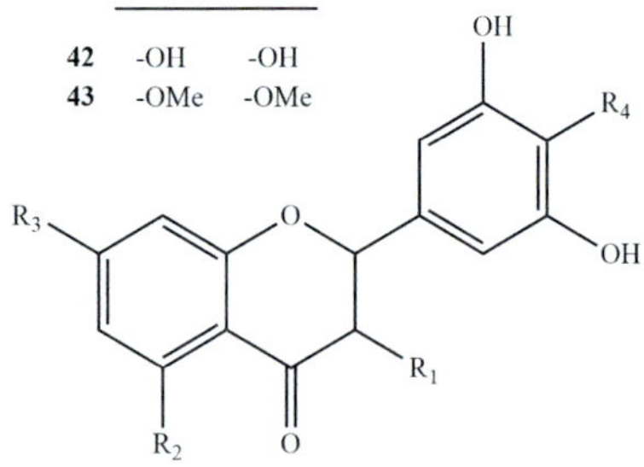

	R_1	R_2	R_3	R_4
49	-OH	-OH	-OH	-H
50	-OMe	-H	-OMe	-OH

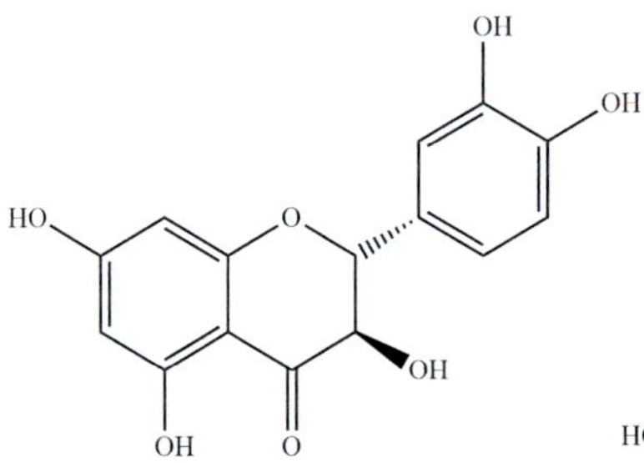

	R_1	R_2
51	-OH	-OH
52	-H	-OMe

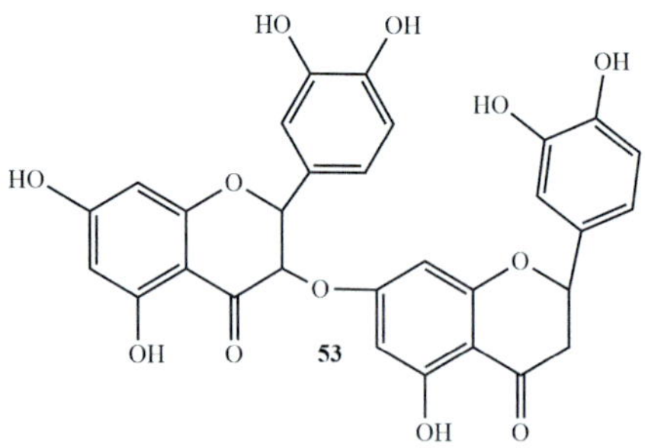

FIGURE 15.2A Continuation of Figure 15.2.

Nessa et al. (2004) evaluated the free radical scavenging activity of isolated flavonoids of *B. balsamifera* and reported the order of activity as Quercetin (**17**) > Luteolin (**1**) > 5,7,3',5'-tetrahydroxyflavanone (**14**) > Blumeatin (**16**) > Rhamnetin (**20**) > Tamarixetin > Luteolin-7-methyl ether (**3**) > Dihydroquercetin-4'-methyl ether (**44**) > Dihydroquercetin-4',7-dimethyl ether (**45**) (Nessa et al., 2004). Further in another study, the XO inhibitory activity of the

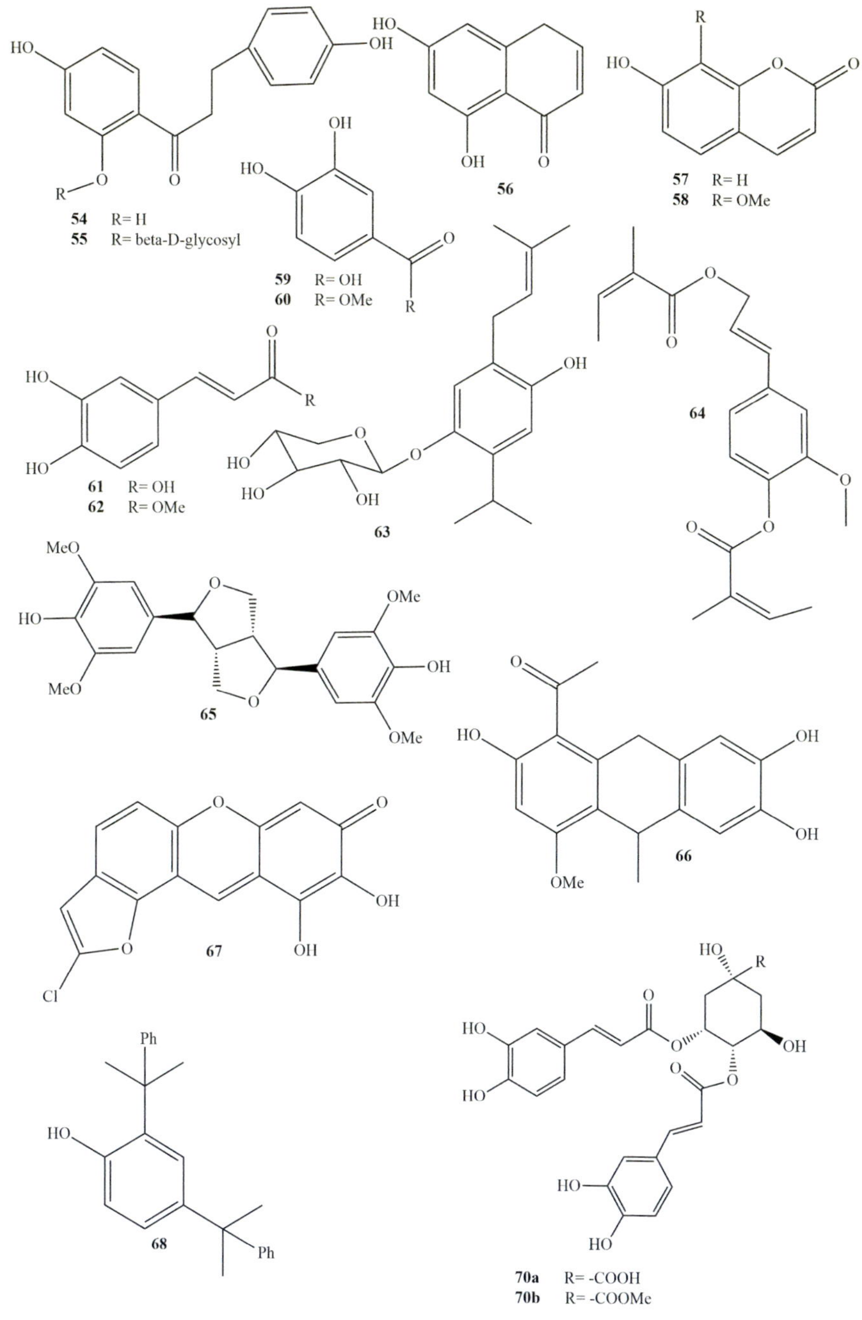

FIGURE 15.2B Continuation of Figure 15.2a.

isolated flavonoids was evaluated by Nessa et al. (2010) which reported XO inhibitory activity in the order of: Allopurinol (positive control) > Luteolin (**1**) > Quercetin (**17**) > Tamarixetin (**21**) > 5,7,3',5′-tetrahydroxyflavanone (**14**) > Rhamnetin (**20**) > Luteolin-7-methyl ether (**3**) > Blumeatin (**16**) > Dihydroquercetin-4′-methyl ether (**44**) > Dihydroquercetin-7,4′-dimethyl ether (**45**) > L-ascorbic acid (Nessa et al., 2010).

Nguyen and Nguyen (2012) reported the isolation of (2R,3S)-(-)-4′-O-methyldihydroquercetin, (2R,3R)-(+)-4′-O-methyldihydroquercetin, (2R,3R)-(+)-4′,7-di-O-methyldihydroquercetin, (2R,3R)-(+)-7-O-methyldihydroquercetin, 5,7,3′,5′-tetrahydroxy flavanone, Quercetin, Quercetin-3,7,3′-trimethyl ether and Quercetin-3,3′,4′-trimethyl ether from *B. balsamifera* extract. Further, they reported that the isolated flavonoids possess potent XO inhibitory activity, with IC_{50} values ranging from 0.23–1.91 mM, comparable to that of a positive control Allopurinol (IC_{50} 2.50 mM) (Nguyen and Nguyen, 2012).

The methanol extract of the aerial part of *B. eriantha* was subjected to repeated column chromatography and preparative TLC, which afforded 3, 3', 4' -trihydroxy- 6, 7, 8- trimethoxy flavone (**42**) and 3-hydroxy-6, 7, 8, 3', 4'-pentamethoxy flavone (**43**). The compounds **42** and **43** were evaluated for their antiproliferative activity on the human non–small cell lung cancer (NCI-H23) cell line. Compound **42** showed potent antiproliferative effect by a significant reduction in cell viability at concentrations 25,773 nM (89%) and 25,7731 nM (68%), respectively, comparable to standard drug Paclitaxel (56%) at 100 nM concentration after 24 hours of treatment. Compound **43** did not show any antiproliferative effect (Tambewagh et al., 2019). The chloroform extract of the aerial part of *B. riparia* afforded Blumeaxanthene (**66**) (Cao et al., 2007).

In further studies, a very rare halogenated xanthene was reported by Huang et al. (2010) from ethyl acetate fraction of *B. riperia* and was named Blumeaxanthene II (**67**). Compound **67** was found to be weakly cytotoxic against Bel-7404 liver cancer cells tested *in vitro* based on concentration (Huang et al., 2010). Saewan et al. (2011) isolated nine flavonoids from *B. balsamifera* and evaluated them for cytotoxicity against KB (oral cancer), MCF-7 (breast cancer) and NCI-H187 (lung cancer) cell lines, whereby most of the flavonoids were found to be active. Luteolin-7-methyl ether (**3**) showed strong cytotoxicity against human lung cancer (NCI-H187) cell lines with an IC_{50} value of 1.29 μg/mL and moderate toxicity against oral cavity cancer (KB) cell lines with an IC_{50} of 17.83 μg/mL. The isolated flavonoids—Quercetin (**17**), Rhamnetin (**20**), Tamarixetin (**21**), Dihydroquercetin-4'-methylether (**44**) and Dihydroquercetin-7,4'-dimethylether (**45**)—showed significantly higher anti-tyrosinase activity than Arbutin (positive control) on mushroom tyrosinase using levodopa (L-DOPA) as the substrate (Saewan et al., 2011).

As a common practice for quality control of herbal medicines, several high-performance liquid chromatography (HPLC) methods have been developed for the quantification of chemical constituents of *B. balsamifera*. The developed methods mainly include analysis of flavonoids and polyphenolic compounds of *B. balsamifera*. Reversed-phase HPLC method for the quantitative analysis of five flavonoids of *B. balsamifera*—namely 5,7,3',5'-tetrahydroxyflavanone (**14**), Blumeatin (**16**), Quercetin (**17**), Dihydroquercetin-4'-methyl ether (**44**) and Dihydroquercetin-7,4'-dimethyl ether (**45**), as well as methods to determine total phenolic contents—have been reported (Nessa et al., 2005; Pang et al., 2014a). Recently, in 2020, a new HPLC method has been reported for simultaneous quantification of 18 polyphenolic compounds from the leaves of *B. balsamifera*, including 17 flavonoids and one phenylethanone (Tan et al., 2020).

15.5.3.2 Terpenes, Terpenoids and other Non-Phenolic Compounds

Several volatile monoterpenes, sesquiterpenes and terpenoids have been reported from the essential oil of various *Blumea* species. The identified volatile chemical constituents and their pharmacological effects have been already discussed in 'Phytochemistry and Pharmacological Investigations of Essential Oils' Section 15.4 of this chapter. Herein, we discuss non-volatile sesquiterpenes, diterpenes, triterpenes, terpenoids and other non-phenolic compounds isolated from various *Blumea* species. The compounds of such category reported from various *Blumea* species are summarized in Table 15.4, and the structure of selected compounds are presented in Figure 15.3.

Some of the terpenoids isolated from *Blumea* (Figure 15.3) have been reported to possess anticancer properties. Ten new sesquiterpenoid esters, namely; Blumeaenes A–J (**83–82**), along with 13 flavonoids, were isolated from aerial parts of *B. balsamifera* and the structures were determined by analysis of 1D and 2D NMR spectroscopic data. All sesquiterpenoid esters were tested for their

TABLE 15.4
Terpenes, Terpenoids and Other Compounds Isolated from Genus *Blumea*

Compound Nos.	Compound Names	Sources	References
	Sesquiterpenes		
71	2-methylbut-2-enoic acid (3aR,4S,5R,7R,8aS)-decahydro-3a,4-dihydroxy-4-methyl-7-(1-methylethyl)-1-methylidene-8-oxoazulen-5-yl ester	*B. balsamifera*	(Chen et al., 2009)
72	2,3-dimethyloxiranecarboxylic acid (1R,3R,5E,10S)-10-hydroxy-6,10-dimethyl-3-(1-methylethyl)-4,9-dioxocyclodec-5-en-1-yl ester	*B. balsamifera*	(Chen et al., 2009)
73	Blumeaene A	*B. balsamifera*	(Chen et al., 2010; Pang et al., 2014a)
74	Blumeaene B	*B. balsamifera*	(Chen et al., 2010; Pang et al., 2014a)
75	Blumeaene C	*B. balsamifera*	(Chen et al., 2010; Pang et al., 2014a)
76	Blumeaene D	*B. balsamifera*	(Chen et al., 2010; Pang et al., 2014a)
77	Blumeaene E	*B. balsamifera*	(Chen et al., 2010; Pang et al., 2014a)
78	Blumeaene F	*B. balsamifera*	(Chen et al., 2010; Pang et al., 2014a)
79	Blumeaene G	*B. balsamifera*	(Chen et al., 2010; Pang et al., 2014a)
80	Blumeaene H	*B. balsamifera*	(Chen et al., 2010; Pang et al., 2014a)
81	Blumeaene I	*B. balsamifera*	(Chen et al., 2010; Pang et al., 2014a)
82	Blumeaene J	*B. balsamifera*	(Chen et al., 2010; Pang et al., 2014a)
83	Blumealactone A	*B. balsamifera*	(Pang et al., 2014a)
84	Blumealactone B	*B. balsamifera*	(Pang et al., 2014a)
85	Blumealactone C	*B. balsamifera, B. arfakiana*	(Pang et al., 2014a; Rücker et al., 1991)
86	Deacetylblumealactone C	*B. arfakiana*	(Rücker et al., 1991)
87	Deoxyblumealactone C	*B. arfakiana*	(Rücker et al., 1991)
88	7β,8β: 8α,12-diepoxyeremophil-11(13)-ene	*B. alata*	(Chen et al., 2009)
89	1-hydroxy-3-O-(2-methylbutanoyl)-2-oxoeudesma-3,11(13)-dien-12,8-olide	*B. densiflora*	(Chen et al., 2009)
90	1-hydroxy-3-(2-methylpropanoyl) eudesma-4(15),11(13)-dien-12,8-olide	*B. densiflora*	(Chen et al., 2009)
91	3-angeloyl-4-hydroxy-7β,11-epoxyeudesman-8-one	*B. alata*	(Chen et al., 2009)
92	4-acetyl-3-angeloyl-7β,11-epoxyeudesman-8-one	*B. alata*	(Chen et al., 2009)
93	3-(2,3-epoxy-2-methylbutanoyl)-4-hydroxy-7β,11-epoxyeudesman-8-one	*B. alata*	(Chen et al., 2009)
94	4-acetyl-3-(2,3-epoxy-2-methylbutanoyl)-7β,11-epoxyeudesman-8-one	*B. alata*	(Chen et al., 2009)
95	1-hydroxy-3-O-(2-methylbutanoyl)-2-oxoeudesma-3,11(13)-dien-12,8-olide	*B. densiflora*	(Chen et al., 2009)
96	1,3-dihydroxy-2-(2-methylbutanoyl) eudesma-4(15),11(13)-dien-12,8-olide	*B. densiflora*	(Chen et al., 2009)

(*Continued*)

TABLE 15.4 (Continued)

Compound Nos.	Compound Names	Sources	References
97	1,3-diacetyl-2-(2-methylbutanoyl) eudesma-4(15),11(13)-dien-12,8-olide	*B. densiflora*	(Chen et al., 2009)
98	1,2-dihydroxy-3-(2-methylbutanoyl) eudesma-4(15),11(13)-dien-12,8-olide	*B. densiflora*	(Chen et al., 2009)
99	1-acetyl-2-hydroxy-3-(2-methylbutanoyl) eudesma-4(15),11(13)-dien-12,8-olide	*B. densiflora*	(Chen et al., 2009)
100	4-acetoxy-3-(angeloyloxy)-11-hydroperoxyeudesm-6-en-8-one	*B. alata*	(Chen et al., 2009)
101	2-himachalene-7,12-diol	*B. gariepina*	(Chen et al., 2009)
102	Blumeaene E1	*B. balsamifera*	(Shirota et al., 2011)
103	Blumeaene E2	*B. balsamifera*	(Shirota et al., 2011)
104	Blumeaene K	*B. balsamifera*	(Shirota et al., 2011)
105	Blumeaene L	*B. balsamifera*	(Shirota et al., 2011)
106	Blumeaene M	*B. balsamifera*	(Shirota et al., 2011)
107	Samboginone	*B. balsamifera*	(Shirota et al., 2011)
108a	6*E*,10*E*,14*Z*-(3*S*)-17-hydroxygeranyllinalool-17-*O*-*β*-D-glucopyranosyl-(1→2)-[*α*-L-rhamnopyranosyl-(1→6)]-*β*-D-glucopyranoside	*B. lacera*	(Akter et al., 2016)
108b	6*E*,10*E*,14*Z*-(3*S*)-17-hydroxygeranyllinalool-3-O- *β*-D-glucopyranosyl-17-*O*-*β*-D-glucopyranosyl-(1→2)-[*α*-L-rhamnopyranosyl-(1→6)]-*β*-D-glucopyranoside	*B. lacera*	(Akter et al., 2016)
	Diterpenes		
109	Austroinulin	*B. glomerata*	(Chen et al., 2009)
110	Cryptomeridiol		(Pang et al., 2014a
111	trans-Phytol	*B. balsamifera*	(Tan and Tatsumura, 2015; Pang et al., 2014a)
112	6α-O-isovalerylnidorellol	*B. aromatica*	(Shen et al., 2019)
113	(12S)-blumdane	*B. aromatica*	(Shen et al., 2019)
114	(12R)-epiblumdane	*B. aromatica*	(Shen et al., 2019)
115	Sterebin A	*B. aromatica*	(Shen et al., 2019)
116	6α-O-(3-methyl-2-butenoyl)sterebin A	*B. aromatica*	(Shen et al., 2019)
117	6α-O-angeloylsterebin A	*B. aromatica*	(Shen et al., 2019)
118	6α-O-isovalerylsterebin A	*B. aromatica*	(Shen et al., 2019)
119	6α-hydroxynidorellol	*B. aromatica*	(Shen et al., 2019)
120	Acromatin A	*B. aromatica*	(He et al., 2019)
121	Acromatin B	*B. aromatica*	(He et al., 2019)
122	Acromatin C	*B. aromatica*	(He et al., 2019)
	Triterpenes		
123	Austroinulin	*B. balsamifera*	(Pang et al., 2014a)
124	3,13-Clerodadiene-6,15-diol	*B. balsamifera*	(Pang et al., 2014a)
125	β-Sitosterol/67	*B. balsamifera, B. aromatica*	(Chen et al., 2009; Shen et al., 2019)
126	Stigmasterol/68	*B. balsamifera*	(Chen et al., 2009)
127	19α-Hydroxyurs-12-ene-24,28-dioate-3-O-β-D-xylopyranoside/69	*B. lacera*	(Chen et al., 2009)
128	5*α*,8*α*-Epidioxyergosta-6,22-dien-3β-ol	*B. balsamifera*	(Pang et al., 2014a)
129	Daucosterol	*B. balsamifera*	(Pang et al., 2014a)
130	(25*R*)-3β-{*O*-β-D-glucopyranosyl-(1→4)-*O*- α-L-rhamnopyranosyl-(1→4)-[*O*- α -L-rhamnopyranosyl-(1→2)]-α-L-rhamnopyranosyl}-22α-N-spirosol-5-ene	*B. lacera*	(Akter et al., 2015)

Compound Nos.	Compound Names	Sources	References
	Other Non-Phenolic Compounds		
131	5',5''-Methylene-bis[5-(but-3-en-1-ynyl)-2,2'-bithiophene	*B. obliqua*	(Ahmad et al., 1998)
132	4-(5'-Methyl-[2,2'-bithiophen]-5-yl)but-3-ynyl 3-methylbutanoate	*B. obliqua*	(Ahmad et al., 1998)
133	2-Hydroxy-4-(5'-methyl-[2,2'-bithiophen]-5-yl) but-3-ynyl-3-methylbutanoate	*B. obliqua*	(Ahmad et al., 1998)
134	5-(But-3-en-1-ynyl)-5'-[(3-methylbutanoyloxy)-methyl]-2,2'-bithiophene	*B. obliqua*	(Ahmad et al., 1998)
135	Amplectol	*B. amplectens*	(Pathak et al., 1987)
136	Icthyothereol acetate	*B. balsamifera*	(Ragasa et al., 2005)
137	(Z)-2-(2,4-dihydroxy-2,6,6-trimethylcyclohexylidene) acetic acid	*B. aromatica*	(Shen et al., 2019)
138	Indole-3-carboxaldehyde	*B. aromatica*	(Shen et al., 2019)
139	Fumaric acid	*B. aromatica*	(Shen et al., 2019)
140	2-sec-butylmaleic acid	*B. aromatica*	(Shen et al., 2019)
141	2-sec-butylmaleic acid 4-methyl ether	*B. aromatica*	(Shen et al., 2019)

inhibitory activity against LPS-induced NO production in RAW264.7 macrophages. Blumeaene A (**83**), Blumeaene D (**86**) and Blumeaene E (**87**) showed a slight inhibitory effect on the production of NO, with IC_{50} values of 40.06 μg/mL, 46.35 μg/mL and 57.80 μg/mL, respectively while other sesquiterpenes were found inactive (IC_{50} > 100 μg/mL). Aminoguanidine (positive control) showed IC_{50} = 1.55 μg/mL under the same experimental conditions. All the sesquiterpenes showed no significant cytotoxicity toward normal RAW264.7 cells at their effective concentration for the inhibition of NO production accessed by MTT assay (Chen et al., 2010).

Fujimoto et al. (1988) isolated three sesquiterpene lactones, namely Blumealactone-A (**83**), Blumealactone-B (**84**) and Blumealactone-C (**86**), from *B. balsamifera* and evaluated for their antitumor activities. All three compounds exhibited growth inhibitory effects against Yoshida sarcoma cells at the concentration of 5–10 μg/mL (Fujimoto et al., 1988). Bioactivity-guided fractionation of the methanol extract of *B. lacera* led to isolation of steroidal glycoalkaloid, (25R)-3β-{O-β-D-glucopyranosyl-(1→4)-O- α-L-rhamnopyranosyl-(1→4)-[O- α -L-rhamnopyranosyl-(1→2)]-α-L-rhamnopyranosyl} -22α-N-spirosol-5-ene (**130**), which were evaluated *in vitro* for cytotoxicity potential against normal mouse fibroblast (NIH3T3) and monkey kidney (Vero) cells, as well as four human cancer cell lines. This included gastric adenocarcinoma (AGS), colon adenocarcinoma (HT-29) and breast ductal carcinoma cell lines (MCF-7 and MDA-MB-231) using the MTT assay. Compound **130** exhibited maximum cytotoxicity (IC_{50} 2.62 μM) and displayed the highest apoptotic potential on MCF-7 cells with significant cell cycle arrest (Akter et al., 2015).

Akter et al. (2016) isolated diterpenoid glycosides 6E,10E,14Z-(3S)-17-hydroxygeranyllinalool-17-O-β-D-glucopyranosyl-(1→2)-[α-L-rhamnopyranosyl-(1→6)]-β-D-glucopyranoside (**108a**) and 6E,10E,14Z-(3S)-17-hydroxygeranyllinalool-3-O- β-D-glucopyranosyl-17-O-β-D-glucopyranosyl-(1→2)-[α-L-rhamnopyranosyl-(1→6)]-β-D-glucopyranoside (**108b**), along with some known flavonoid glycosides from the methanol extract of *B. lacera* leaves. The compound **108a** showed potent cytotoxic activity with IC_{50} value 8.3 μM against MCF-7 breast cancer cells. In apoptosis and cell cycle analysis, **108a** revealed strong apoptotic activity against MCF-7 cells (45.5% AV+/PI–) after 24 hours, but showed no cell cycle arrest in of any of the cell cycle phases in MCF-7 cells (Akter et al., 2016).

Ragasa et al. (2005) isolated icthyothereol acetate (**136**), cryptomeridiol, lutein, and β-carotene from *B. balsamifera*. The isolated compounds icthyothereol acetate (**136**) and cryptomeridiol were

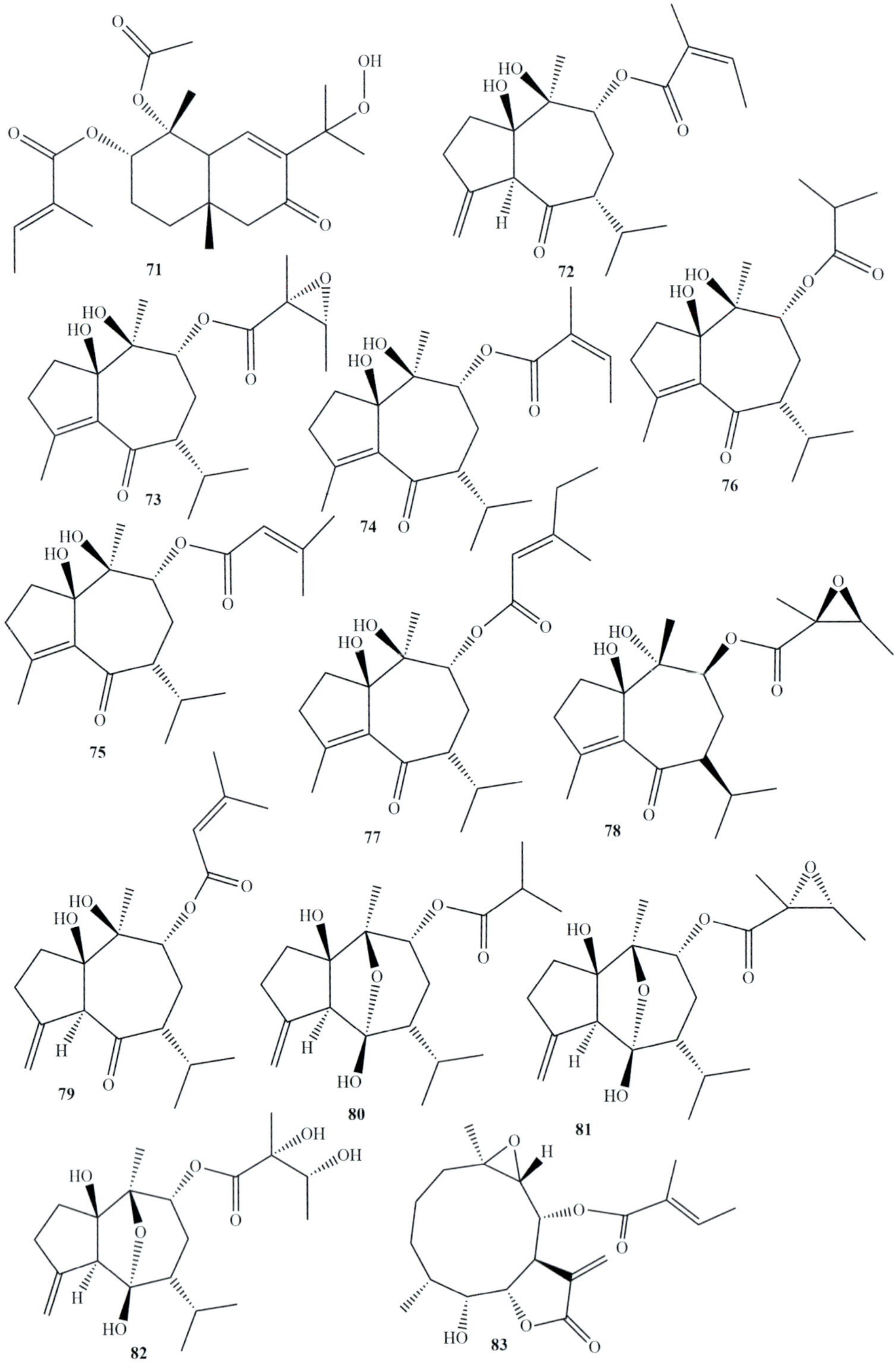

FIGURE 15.3 Molecular structure of terpenes, terpenoids and other non-phenolic compounds (**71–141**) from *Blumea* species.

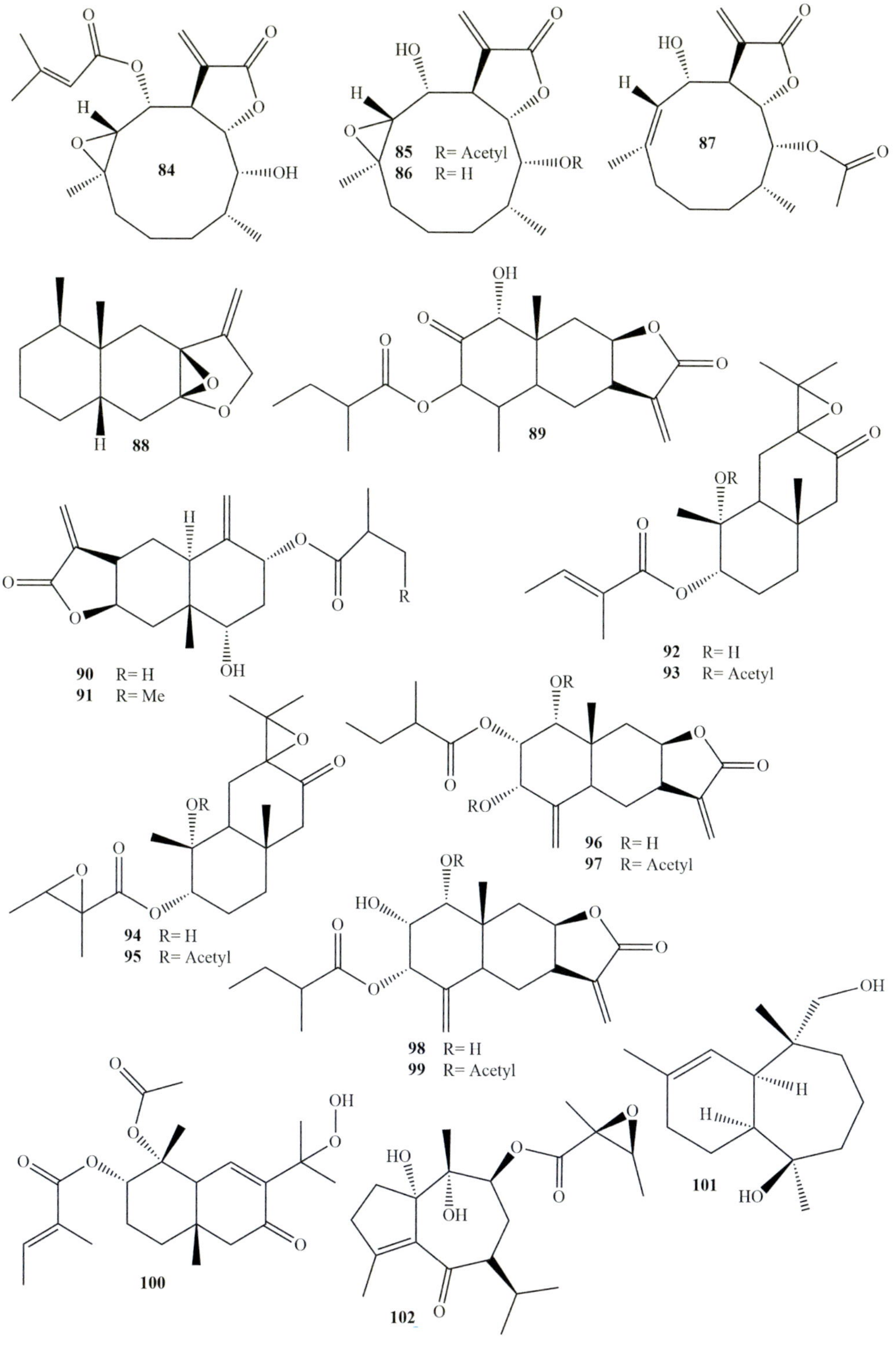

FIGURE 15.3A A and B Continuation of Figure 15.3.

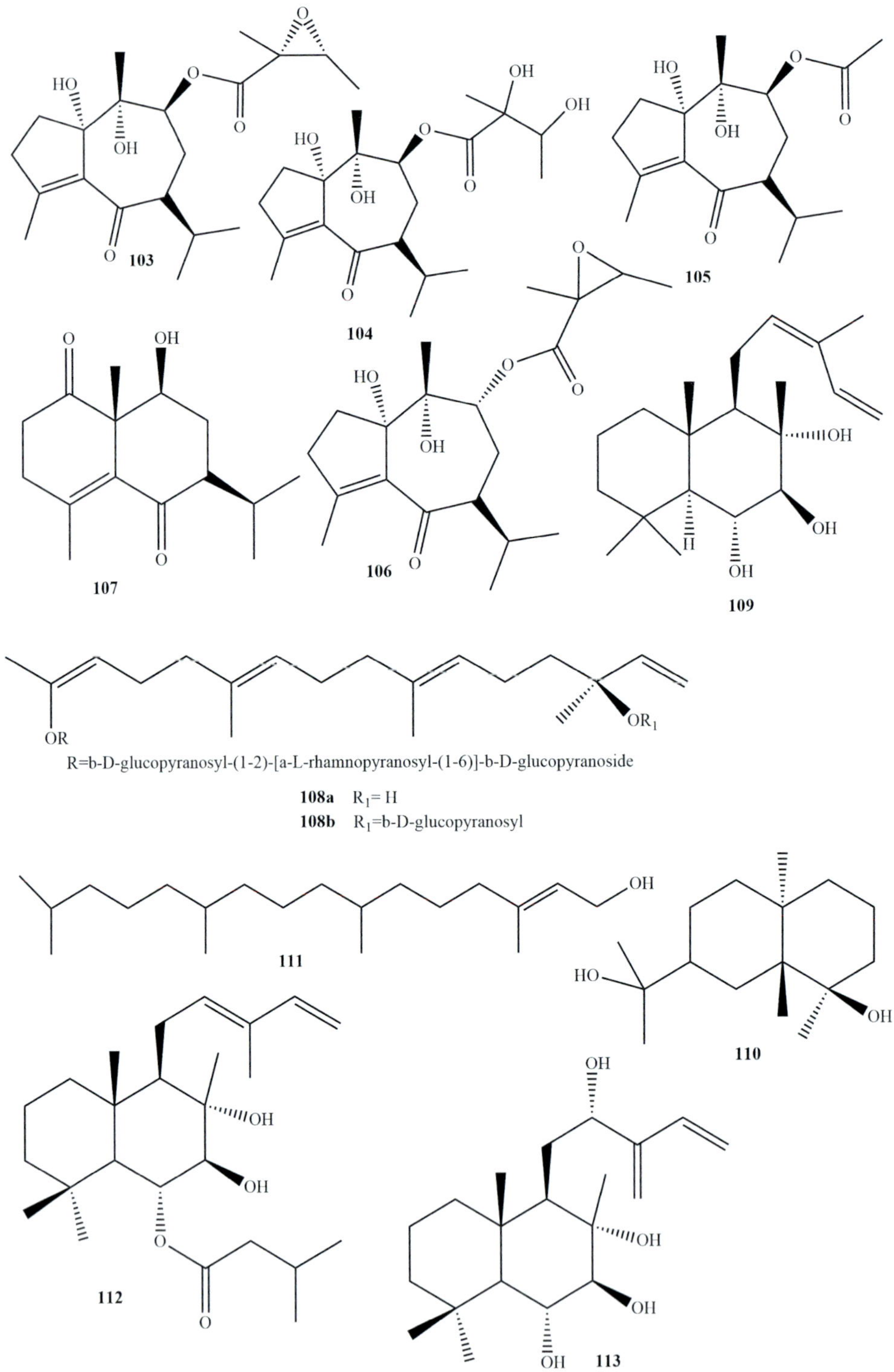

FIGURE 15.3A (Continued)

evaluated for their antimicrobial activity, in which compound **136** showed moderate activity while Cryptomeridiol showed low activity against the fungi *A. niger*, *T. mentagrophytes* and *C. albicans*. Both compounds did not show any antimicrobial activity against *P. aeruginosa*, *S. aureus*, *B. subtilis*, and *E. coli* (Ragasa et al., 2005).

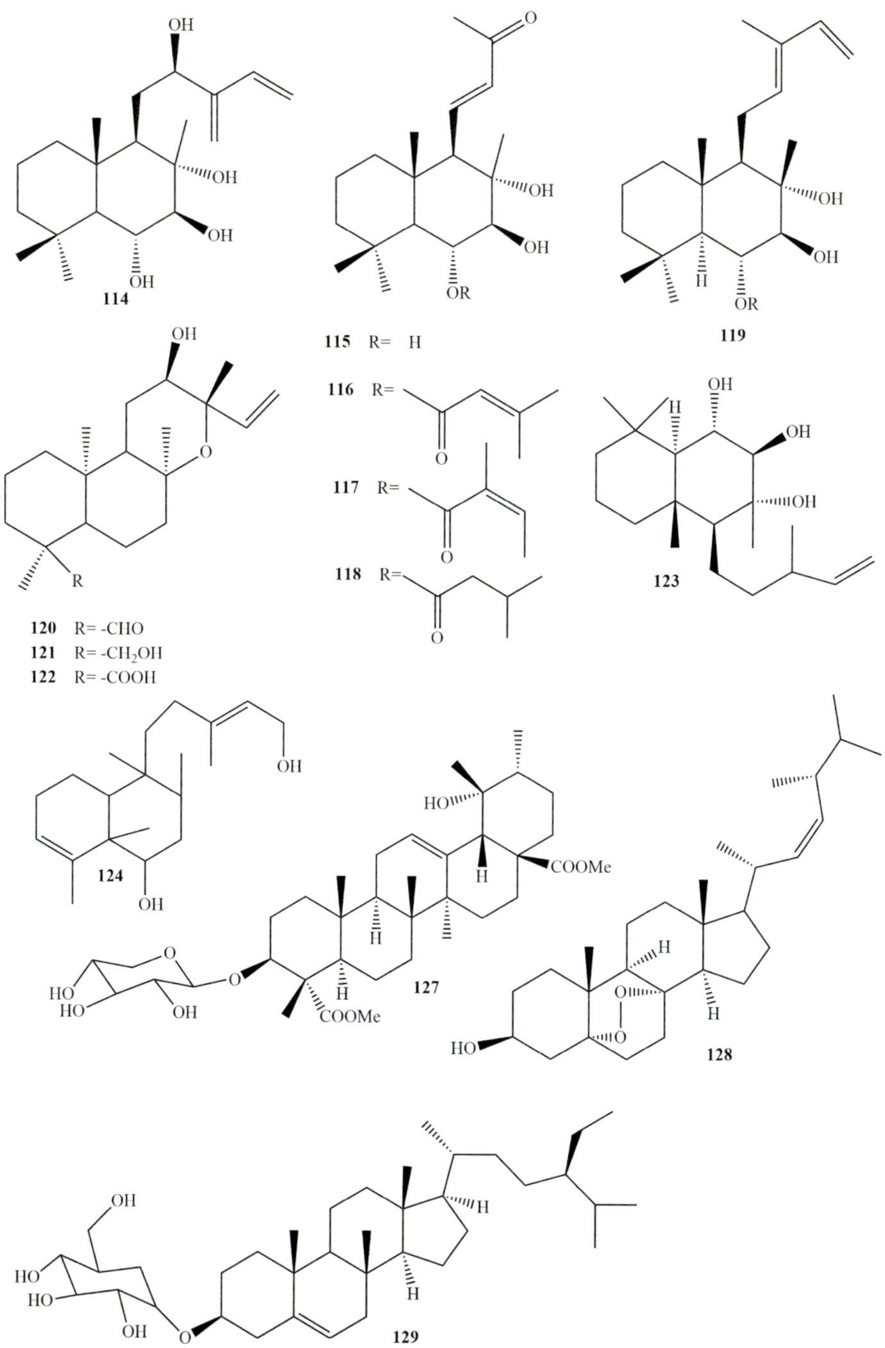

FIGURE 15.3B A and B Continuation of Figure 15.3a.

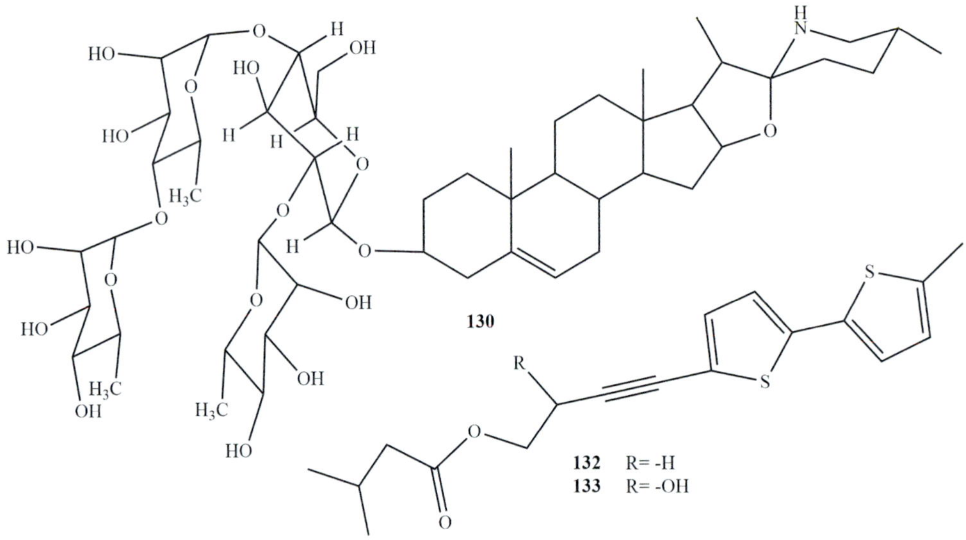

130

132 R= -H
133 R= -OH

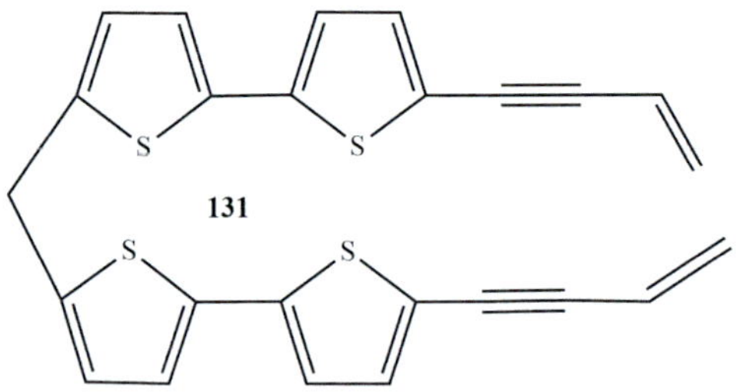

131

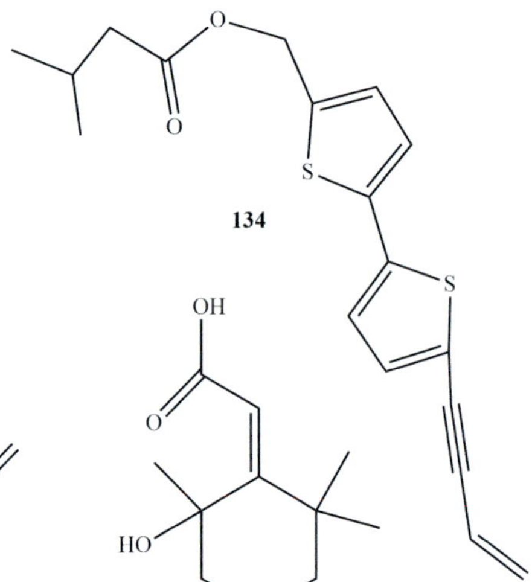

134

137

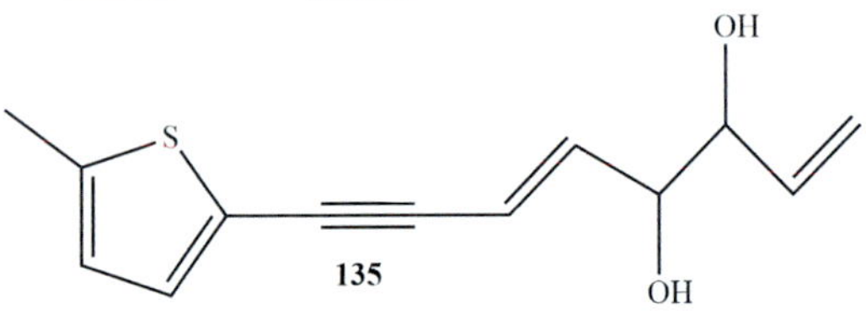

135

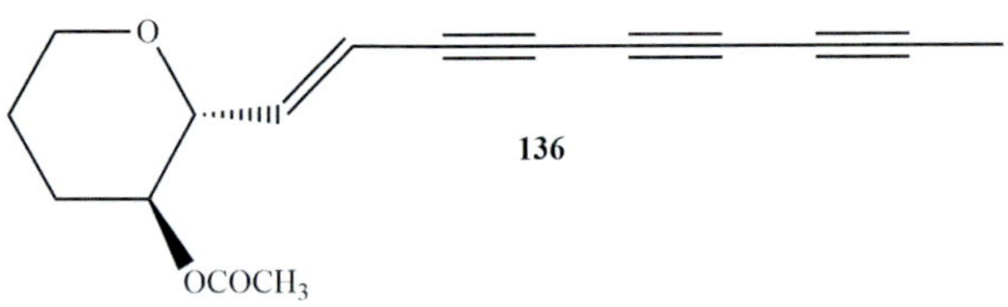

136

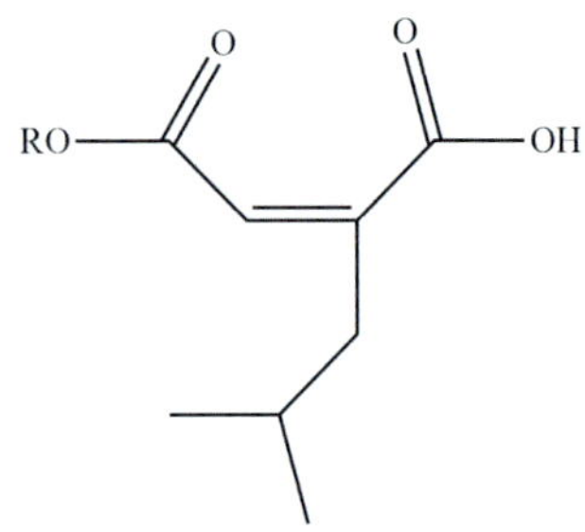

140 R= H
141 R= CH_3

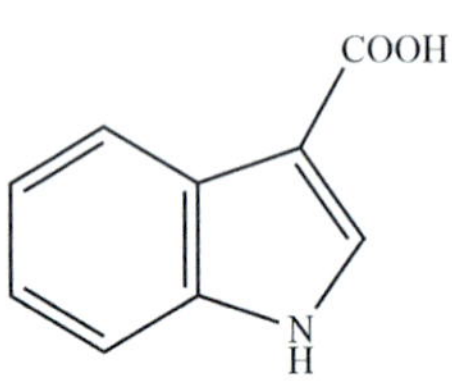

138

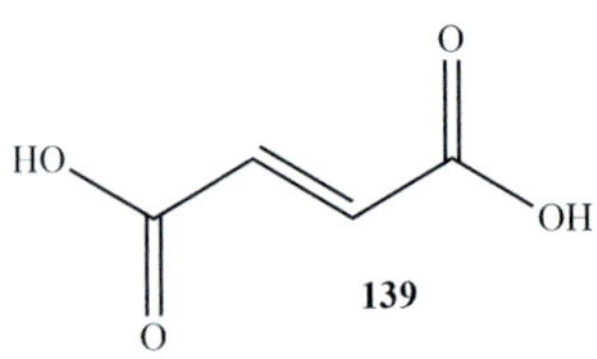

139

FIGURE 15.3B (Continued)

Shen et al. (2019) isolated 23 compounds from ethanol extract of *B. aromatica* including three labdane-type diterpenoids, 6α-O-isovalerylnidorellol (**112**), (12S)-blumdane (**113**) and (12R)-epiblumdane (**114**), and three bis-norditerpenoids, Sterebin A (**115**), 6α-O-(3-methyl-2-butenoyl)sterebin A (**116**), 6α-O-angeloylsterebin A (**117**) and 6α-O-isovalerylsterebin A (**118**). The isolated compounds were evaluated for their inhibitory effects on tumor necrosis factor-alpha (TNF-α) expression in LPS-stimulated RAW 264.7 murine macrophages. The results showed that compounds 4,5-di-O-caffeoylquinic acid, 5,7-dihydroxy-3,6,4'-trimethoxyflavone (**27a**), and (Z)-2-(2,4-dihydroxy-2,6, 6-trimethylcyclohexylidene) acetic acid (**137**) slightly suppressed the production of TNF-α with inhibition rates of 15.5 ± 0.7%, 24 ± 2.2%, and 27.5 ± 5.1%, respectively, at a concentration of 40 μM (Shen et al., 2019).

15.6 CONCLUSION

Genus *Blumea* is blessed with magical therapeutic potential, being used in folklore as well as in patented herbal formulations to treat various types of cancer, cold dysentery, influenza, cardiovascular diseases, gynecological diseases, pulmonary infection, rheumatoid arthritis, dermatitis and eczema, acute renal failure, viral infections, migraines and neurological disorders. The plants of this genus are a rich source of bioactive secondary metabolites of class flavonoids, coumarins, xanthenes, terpenes and terpenoids as a major component. The essential oil of these plants is rich in monoterpenes, sesquiterpenes and terpenoids as volatile components with proven pharmacological effects.

The potent mosquito larvicidal activity indicates that *Blumea*'s essential oil may be considered to develop into an alternative eco-friendly mosquito control tool. Many bioactive compounds have been isolated and characterized from *Blumea* species, justifying the ethnomedicinal use, as well as providing novel leads for drug development. A steroidal glycoalkaloid, (25*R*)-3β-{*O*-β-D-glucopyranosyl-(1→4)-*O*-α-L-rhamnopyranosyl-(1→4)-[*O*-α-L-rhamnopyranosyl-(1→2)]-α-L-rhamnopyranosyl}-22α-N-spirosol-5-ene (**130**), isolated from *B. balsamifera*, has been reported to possess potent *in vitro* cytotoxicity against breast cancer cell line MCF-7 (IC_{50} = 2.62 μM), displaying significant apoptotic potential and cell cycle arresting effect on MCF-7 cells. Further studies on this compound are suggested to develop a novel anticancer drug candidate of natural origin. Because of the enormous use of *Blumea* species in herbal formulations across the world, rigid steps for prevention of adulteration like detailed chemical investigation leading to the identification of marker compounds, DNA barcoding of various *Blumea* species, as well as the development of novel validated analytical procedures for quantification of bioactive constituents, are suggested.

REFERENCES

Adhikari, A., Basnet, S., Vijayakumar, S., *et al.*, 2020. "*Blumea lacera* DC, accelerates the healing of acetic acid induced ulcerative colitis in rats by regulating oxidative stress and colonic inflammation: *In-vivo* and *in silico* molecular docking experiments." *Advances in Traditional Medicine*. doi: 10.1007/s13596-020-00454-6.

Agdamag, A.S.C., Aggabao, L.H.C., Agudo, M.S.C., *et al.*, 2020. "Anti-urolithiatic activity of Sambong (*Blumea balsamifera*) extract in ethyleneglycol-induced urolithiatic Wistar Rats (*Rattus norvegicus*)." *Acta Medica Philippina* 54, 1: 31–35.

Ahmad, F.B. and Ismail, G., 2003. "Medicinal plants used by Kadazandusun communities around crocker range." *Asean Review of Biodiversity and Environmental Conservation* 1, 1–10.

Ahmad, V.U., Alam, N. and Qaisar, M., 1998."Acetylenic thiophenes from *Blumea obliqua*." *Phytochemistry* 49, 1: 259–261.

Ahmed, F., Rahman, A. and Mubassara, S., 2014. "Phytochemical composition, antioxidant activity and cytotoxicity of *Blumea lacera* Linn. from two different habitats." *Jahangirnagar University Journal of Biological Sciences* 3, 1: 37–45.

Ahsan, M.Q., 2016. "*In-vivo* investigation of analgesic, antipyretic anti-diarrheal and anxiolytic activity of *Blumea densiflora* DC." *European Journal of Pharmaceutical and Medical Research* 3, 11: 50–55.

Akter, R., Uddin, S., Grice, I. and Tiralongo, E., 2014. "Cytotoxic activity screening of Bangladeshi medicinal plant extracts." *Journal of Natural Medicines* 68, 1: 246–252.

Akter, R., Uddin, S., Tiralongo, J., *et al.*, 2015. "A new cytotoxic steroidal glycoalkaloid from the methanol extract of *Blumea lacera* leaves." *Journal of Pharmacy & Pharmaceutical Sciences* 18, 4: 616.

Akter, R., Uddin, S., Tiralongo, J., *et al.*, 2016. "A new cytotoxic diterpenoid glycoside from the leaves of *Blumea lacera* and its effects on apoptosis and cell cycle." *Natural Product Research* 30, 23: 2688–2693.

Ali, D., Wong, K. and Lim, P., 2005. "Flavonoids from *Blumea balsamifera*." *Fitoterapia* 76, 1: 128–130.

Alok, S., Jain, S., Verma, A., *et al.*, 2013. "Pathophysiology of kidney, gallbladder and urinary stones treatment with herbal and allopathic medicine: A review." *Asian Pacific Journal of Tropical Disease* 3, 6: 496–504.

Baruah, M.K. and Choudhury, D.M., 2012. "Ethno-medicinal uses of Asteraceae in Barak Valley, Assam." *International Journal of Plant Sciences* 7, 2: 220–223.

Balunas, M. and Kinghorn, A.D., 2005. "Drug discovery from medicinal plants." *Life Sciences* 78, 5: 431–441.

Banjarnahor, S.D. and Artanti, N., 2015. "Antioxidant properties of flavonoids." *Medical Journal of Indonesia* 23, 4: 239–244.

Basnet, S., Adhikari, A., Sachidananda, V., *et al.*, 2015. "Protective effect of *Blumea lacera* DC aerial parts in indomethacin-induced enterocolitis in rats." *Inflammopharmacology* 23, 6: 355–363.

Benelli, G., Govindarajan, M., Rajeswary, M., *et al.*, 2017. "Larvicidal activity of *Blumea eriantha* essential oil and its components against six mosquito species, including Zika virus vectors: The promising potential of (4E,6Z)-allo-ocimene, carvotanacetone and dodecyl acetate." *Parasitology Research* 116, 4: 1175–1188.

Bent, S., 2008. "Herbal medicine in the United States: Review of efficacy, safety, and regulation." *Journal of General Internal Medicine* 23, 6: 854–859.

Bian, Y., 2015a. "Anti-viral soft capsule and its preparation method and application." *Chinese Patent* 104435049, A 20150325.

Bian, Y., 2015b. "A soft capsule for resisting pulmonary infection." *Chinese Patent* 104435092, A 20150325.

Bibi, M., 2014. "A topical herbal composition for use in treating muscle cramps." *Indian Patent* 2012KO00554, A 20141212.

Boy, H., Rutilla, A., Santos, K., *et al.*, 2018. "Recommended medicinal plants as source of natural products: A review." *Digital Chinese Medicine* 1, 2: 131–142.

Brindha-Devi, G.B. and Amutha, K., 2011. "Anti-inflammatory and antipyretic activities of *Blumea mollis* (D. Don) Merr." *Research Journal of Pharmacognosy and Phytochemistry* 3, 14–16.

Brindha-Devi, G.B. and Revathi, K., 2012. "Evaluation of hepatoprotective activity of *Blumea mollis* D. Don Merr. on Paracetamol-induced hepatotoxicity in rats." *Research Journal of Pharmacology and Pharmacodynamics* 4, 4: 206–209.

Cao, J., Yao, Y., Chen, H., *et al.*, 2007. "A New xanthene from *Blumea riparia*." *Chinese Chemical Letters* 18, 3: 303–305.

Che, C.T., George, V., Ijinu, T.P., *et al.*, 2017. "Traditional medicine." In: *Pharmacognosy: Fundamentals, Applications and Strategies*, Eds. Badal McCreath, S. and Delgoda, R., Saint Louis: Elsevier Science, 15–30.

Chen, G., 2015a. "Chinese medicine composition for treatment of sciatica and its preparation method." *Chinese Patent* 104352616, A 20150218.

Chen, G., 2015b. "One kind of soft capsule for relieving cough and its preparation method and application. *Chinese Patent* 104510768, A 20150415.

Chen, H. and Jiang, X., 2014. "Traditional Chinese medicine for treatment of blood stasis migraine and its preparation method." *Chinese Patent* 103638367, A 20140319.

Chen, M., Jin, H., Zhang, W., *et al.*, 2009. "Chemical constituents of plants from the genus *Blumea*." *Chemistry & Biodiversity* 6, 6: 809–817.

Chen, M., Qin, J.J., Fu, J.J., *et al.*, 2010. "Blumeaenes A-J, sesquiterpenoid esters from *Blumea balsamifera* with NO inhibitory activity." *Planta Medica* 76, 9: 897–902.

Chen, Y., 2014. "Chinese medicine for treatment of cordiovascular disease." *Chinese Patent* 103690845, A 20140402.

Chiang, L.C., Cheng, H.Y., Chen, C.C. and Lin, C.C., 2004. "*In vitro* anti-leukemic and antiviral activities of traditionally used medicinal plants in Taiwan." *American Journal of Chinese Medicine* 32, 5: 695–704.

Chu, S.S., Du, S.S. and Liu, Z.L., 2013. "Fumigant compounds from the essential oil of Chinese *Blumea balsamifera* leaves against the maize weevil (*Sitophilus zeamais*)." *Journal of Chemistry*. doi: 10.1155/2013/289874.

Cordero, C.S., Ligsay, A. and Alejandro, G.J.D., 2020. "Ethnobotanical documentation of medicinal plants used by the Ati tribe in Malay, Aklan, Philippines." *Journal of Complementry Medicine Research* 11, 1: 170–198.

Cragg, G.M., Newman, D.J. and Snader, K.M., 1997. "Natural products in drug discovery and development." *Journal of Natural Products* 60, 52–60.

Cragg, G.M., Newman, D.J. and Snader, K.M., 2002. "Natural products as sources of new drugs over the period 1981–2002." *Journal of Natural Products* 66, 1022–1037.

Dapar, M.L.G., Alejandro, G.J.D., Meve, U., *et al.*, 2020. "Quantitative ethnopharmacological documentation and molecular confirmation of medicinal plants used by the Manobo tribe of Agusan del Sur." *Philippines Journal of Ethnobiology & Ethnomedicine.* doi: 10.1186/s13002-020-00363-7.

de Boer, H.J. and Cotingting, C., 2014. "Medicinal plants for women's healthcare in Southeast Asia: A meta-analysis of their traditional use, chemical constituents, and pharmacology." *Journal of Ethnopharmacology* 151, 2: 747–767.

de Boer, H.J., Lamxay, V. and Björk, L., 2011. "Steam sauna and mother roasting in Lao PDR: Practices and chemical constituents of essential oils of plant species used in postpartum recovery." *BMC Complementary and Alternative Medicine* 11, 128. doi: 10.1186/1472-6882-11-128.

Deng, R., 2015. "Medicine for treating bedsore with symptom of blood clotting toxin accumulation and its preparation method." *Chinese Patent* 104474452, A 20150401.

Dias, D.A., Urban, S. and Roessner, U., 2012. "A historical overview of natural products in drug discovery." *Metabolites* 2, 2: 303–336.

Dou, Q. and Wei, J., 2014. "Pharmaceutical composition for treating cardiovascular diseases, pharmaceutical preparations, application and manufacture method thereof." *Chinese Patent* 103768348, A 20140507.

Du, Y., 2014. "Chinese medicine for treating female genital itching." *Chinese Patent*, 103736068, A 20140423.

Fan, Z.W., Pang, Y.X., Wang, K., *et al.*, 2015. "*Blumea balsamifera* oil for the acceleration of healing of burn injuries." *Molecules* 20, 9: 17166–17179.

Fazilatun, N., Nornisah, M. and Zhari, I., 2004. "Superoxide radical scavenging properties of extracts and flavonoids isolated from the leaves of *Blumea balsamifera*." *Pharmaceutical Biology* 42, 6: 404–408.

Fitzgerald, M., Heinrich, M., Booker, A., 2020. "Medicinal plant analysis: A historical and regional discussion of emergent complex techniques." *Frontiers in Pharmacology* 10: 1480. doi: 10.3389/fphar.2019.01480.

Fujimoto, Y., Soemartono, A., Sumatra, M., 1988. "Sesquiterpenelactones from *Blumea balsamifera*." *Phytochemistry* 27, 4: 1109–1111.

Ge, W. and Wang, Q., 2014. "Chinese medicinal liquor for rapidly healing bone injury." *Chinese Patent* 103877453, A 20140625.

Gore, M. and Desai, N.S., 2014. "Characterization of phytochemicals and evaluation of anti-cancer potential of *Blumea eriantha* DC." *Physiology and Molecular Biology of Plants* 20, 4: 475–486.

Gu, W., Xue, W., Xue, *et al.*, 2015. "Chinese medicinal composition for treating rheumatoid arthritis and its preparation method." *Chinese Patent* 104288282, A 20150121.

Guga-Bakshi, D.N., Sensarma, P. and Pal, D.C., 1999. *A Lexicon of Medicinal Plants in India*, volume 1, Calcutta, India: Naya Prokash, 290.

Hao, H., 2014. "Chinese medicine for treatment of chronic cholecystitis." *Chinese Patent* 103610886, A 20140305.

Hasan, M.N., Rahman, M.H., Guo, R. and Hirashima, A., 2015. "Hypoglycemic activity of methanolic leaf extract of *Blumea lacera* in Swiss-albino mice." *Asian Pacific Journal of Tropical Diseases* 5, 3: 195–198.

He, C., Yang, P., Wang, L., *et al.*, 2020. "Antibacterial effect of *Blumea balsamifera* DC. essential oil against *Haemophilus parasuis*." *Archives of Microbiology* 202, 9: 2499–2508. doi: 10.1007/s00203-020-01946-4.

He, L., Zhang, Z., Yao, C., *et al.*, 2019. "Rapid screening of Forskolin-type diterpenoids of *Blumea aromatica* DC using ultra-high-performance liquid chromatography tandem quadrupole time-of-flight mass spectrometry based on the mass defect filtering approach." *Molecules* 24, 17: 3073. doi: 10.3390/molecules24173073.

Huang, L., Lei, T., Lin, C., *et al.*, 2010. "Blumeaxanthene II, a novel xanthene from *Blumea riparia* DC." *Fitoterapia* 81, 5: 389–392.

Jahan, K., Kundu, S.K. and Bake, M.A., 2014. "Evaluation of antimicrobial and cytotoxic activities of the methanolic and petroleum ether extract of *Blumea lacera* Burm.f in Bangladesh." *Journal of Pharmacognosy and Phytochemistry* 2, 6: 104–108.

Jiang, Z.L., Zhou, Y., Ge, W.C. and Yuan, K., 2014. "Phytochemical compositions of volatile oil from *Blumea balsamifera* and their biological activities." *Pharmacognosy Magagazine* 10, 39: 346–352.

Kantasrila, R., Pandith, H., Balslev, H., *et al.*, 2020. "Medicinal plants for treating musculoskeletal disorders among Karen in Thailand." *Plants (Basel)* 9, 7: e811. doi: 10.3390/plants9070811.

Kew, Royal Botanical Gardens, 2017. "*Blumea lanceolaria*." Obtained under the Creative Commons Attribution-Share Alike 3.0 license. www.plantsoftheworldonline.org/taxon/urn:lsid:ipni.org:names:185607-1 (Accessed 1 November 2021)

Khair, M.A., Ibrahim, Ahsan, Q., *et al.*, 2015. "Preliminary phytochemical screenings and pharmacological activities of *Blumea lacera* (Burn.f.) DC." *Dhaka University Journal of Pharmaceutical Sciences* 13, 1: 69–73.

Khan, M., 2014. "Body Revival herbal composition for cardiovascular use." *Indian Patent* 2011MU00540, A 20140321.

Khandekar, U., Tippat, S. and Ghongade, R., 2013. "Investigation on antioxidant, antimicrobial and phytochemical profile of *Blumea lacera* leaf." *International Journal of Biological & Pharmaceutical Research* 4, 11: 756–761.

Khare, C.P., 2007. *Indian Medicinal Plants, an Illustrated Dictionary. Blumea densiflora* DC. New York: Springer Publishing, 94–95. ISBN: 0387706372.

Kubota, H., Kojima-Yuasa, A., Morii, R., *et al.*, 2009. "Anti-obesity effect of *Blumea balsamifera* extract in 3T3-L1 preadipocytes and adipocytes." *American Journal of Chinese Medicine* 37, 5: 843–854.

Kumari, N., Shukla, K. and Sharma, H.P., 2016. "Phytochemical screening and biopesticidal efficacy of some plants of East Singhbhum, Jharkhand." *Indo American Journal of Pharmaceutical Research* 6, 10: 6716–6724.

Lee, C., Kim, S.Y., Eum, S., *et al.*, 2019. "Ethnobotanical study on medicinal plants used by local Van Kieu ethnic people of Bac Huong Hoa nature reserve, Vietnam." *Journal of Ethnopharmacology* 231, 283–294.

Lee, C.Y., 2006. "Medication combination for hepatoma and pancreatic cancer and preparation protocol." *U.S. Patent* US6998141 B2, A 20060214.

Li, J., 2004. "A pharmaceutical composition for treating liver cancer, pancreatic cancer and its preparation method." *Chinese Patent* 1491671 A 20040428.

Li, R., 2015. "Traditional Chinese medicine for treating mumps." *Chinese Patent* 104523895, A 20150422.

Li, Y., Ooi, L.S., Wang, H., *et al.*, 2004. "Antiviral activities of medicinal herbs traditionally used in southern mainland China." *Phytotherapy Research* 18, 9: 718–722.

Liu, A. and Sun, P., 2011. "Traditional Chinese medicine composition for treating multiple neurofibroma." *Chinese Patent* 102188654, A 20110921.

Liu, D., 2014a. "Chinese medicine for treatment of bone tumor." *Chinese Patent* 104055931, A 20140924.

Liu, D., 2014b. "Chinese medicinal fumigant for treating knee osteoarthritis and nursing method." *Chinese Patent* 103830325, A 20140604.

Liu, Z., 2015. "Chinese medicinal composition for treating fractures." *Chinese Patent* 104547794, A 20150429.

Lv, L., 2015. "A traditional Chinese medicine for treatment of psoriasis and its preparation method." Chinese *Patent* 104306831, A 20150128.

Mahida, Y. and Mohan, J.S.S., 2006. "Screening of Indian plant extracts for antibacterial activity." *Pharmaceutical Biology* 44, 8: 627–631.

Mead, D.E., 2017. "*Blumea balsamifera*." Obtained from Wikipedia under the Creative Commons CC0 1.0 Universal Public Domain Dedication license. https://en.m.wikipedia.org/wiki/File:Sambong_(Blumea_balsamifera).jpg (Accessed 1 November 2021)

Mehta, S.C., Vardhan, H. and Saxena, S.P., 1986. "Some pharmacological actions of the essential oil of *Blumea membranacea*." *Indian Journal of Physiology and Pharmacology* 30, 149–154.

Meng, H. and Zhang, J., 2014. "Chinese medicine powder for nursing infant diaper dermatitis." *Chinese Patent* 104173609, A 20141203.

Meng, L., 2015. "A pharmaceutical composition for treating oral cancer." *Chinese Patent* 104547956, A 20150429.

Mia, M.A., Khatun, M.M. and Sarwar, A.K.M.G., 2020. "Survey on Asteraceae weeds at Bangladesh Agricultural University campus and reviewed their ethno-botanical uses." *Journal of Bangladesh Agricultural University* 18, 1: 17–24.

Montealegre, C.M. and De-Leon, R.L., 2017. "Effect of *Blumea balsamifera* extract on the phase and morphology of calcium oxalate crystals." *Asian Journal of Urology* 4, 4: 201–207.

Mutmainah, R.S., Rahmawati, N. and Nugroho, A.E., 2014. "Gastroprotective effects of combination of hot water extracts of turmeric (*Curcuma domestica* L.), cardamom pods (*Ammomum compactum* S.) and Sembung leaf (*Blumea balsamifera* DC.) against aspirin-induced gastric ulcer model in rats." *Asian Pacific Journal of Tropical Biomedicine* 4, Suppl.1: S500–S504.

Nessa, F., Ismail, Z., Karupiah, S. and Mohamed, N., 2005. "RP-HPLC method for the quantitative analysis of naturally occurring flavonoids in leaves of *Blumea balsamifera* DC." *Journal of Chromatographic Science* 43, 8: 416–420.

Nessa, F., Ismail, Z. and Mohamed, N., 2010. "Xanthine oxidase inhibitory activities of extracts and flavonoids of the leaves of *Blumea balsamifera*." *Pharmaceutical Biology* 48, 12: 1405–1412.

Nessa, F., Ismail, Z., Mohamed, N. and Haris, M.R.H.M., 2004. "Free radical—scavenging activity of organic extracts and of pure flavonoids of *Blumea balsamifera* DC leaves." *Food Chemistry* 88, 243–252.

Newman, D.J. and Cragg, G.M., 2012. "Natural products as sources of new drugs over the 30 years from 1981 to 2010." *Journal of Natural Products* 75, 311–335.

Newman, D.J., Cragg, G.M., Snader, K.M., 2000. "The influence of natural products upon drug discovery." *Natural Product Reports* 17, 215–234.

Nguyen, M.T.T., Awale, S., Tezuka, Y., *et al.*, 2004. "Xanthine oxidase inhibitory activity of Vietnamese medicinal plants." *Biological and Pharmaceutical Bulletin* 27, 1414–1421.

Nguyen, M.T.T. and Nguyen, N.T., 2012. "Xanthine oxidase inhibitors from Vietnamese *Blumea balsamifera* L." *Phytotherapy Research* 26, 1178–1181.

Nguyen, X.M.A., Bun, S.S., Ollivier, E. and Dan, T.P.H., 2020. "Ethnobotanical study of medicinal plants used by K'Ho-Cil people for treatment of diarrhea in Lam Dong Province, Vietnam." *Journal of Herbal Medicine* 19, 100320. doi: 10.1016/j.hermed.2019.100320.

Noor-Rain, A., Khozirah, S. and Mohd-Ridzuan, M.A., *et al.*, 2007. "Antiplasmodial properties of some Malaysian medicinal plants." *Tropical Biomedicine* 24, 1: 29–35.

Norikura, T., Kojima-Yuasa, A., Shimizu, M., *et al.*, 2008. "Anticancer activities and mechanisms of *Blumea balsamifera* extract in hepatocellular carcinoma cells." *Amican Journal of Chinese Medicine* 36, 2: 411–424.

Ouyang, H., 2015. "Chinese medicine composition preparation for treating migraine and preparation method thereof." *Chinese Patent* 104857313, A 20150826.

Owolabi, M.S., Lajide, L., Villanueva, H.E. and Setzer, W.N., 2010. "Essential oil composition and insecticidal activity of *Blumea perrottetiana* growing in Southwestern Nigeria." *Natural Products Communications* 5, 7: 1135–1138.

Pang, Y., Wang, D. and Fan, Z., *et al.*, 2014a. "*Blumea balsamifera*-A phytochemical and pharmacological review." *Molecules* 9, 7: 9453–9477.

Pang, Y., Wang, D., Hu, X., *et al.*, 2014b. "Effect of volatile oil from *Blumea Balsamifera* (L.) DC. leaves on wound healing in mice." *Journal of Traditional Chinese Medicine* 34, 6: 716–724.

Pang, Y., Zhang, Y., Huang, L., *et al.*, 2017. "Effects and mechanisms of total flavonoids from *Blumea balsamifera* (L.) DC. on skin wound in rats." *International Journal of Molecular Sciences* 18, 12: 2766. doi: 10.3390/ijms18122766.

Panyaphu, K., On, T.V., Sirisa-ard, P., *et al.*, 2011. "Medicinal plants of the Mien (Yao) in Northern Thailand and their potential value in the primary healthcare of postpartum women." *Journal of Ethnopharmacology* 135, 2: 226–237.

Pathak, V.P., Jakupovic, J., Jain, S. and Bhakuni, D.S., 1987. "Amplectol: A novel acetylenic thiophene derivative from *Blumea amplectens* var. arenaria." *Planta Medica* 53, 1: 103–104.

Pei, S.J., 2001. "Ethnobotanical approaches of traditional medicine studies some experiences form Asia." *Pharmaceutical Biology* 39, 74–79.

Peng, G., 2016. "Traditional Chinese medicine composition for treating gynecological diseases, and its preparation method." *Chinese Patent* 105770050, A 20160720.

Pornpongrungrueng, P., Borchsenius, F., Englund, M., *et al.*, 2007. "Phylogenetic relationships in *Blumea* (Asteraceae: Inuleae) as evidenced by molecular and morphological data." *Plant Systematics and Evolution* 269, 223–243.

Pornpongrungrueng, P., Gustafsson, M.H.G., Borchsenius, F., *et al.*, 2016. "*Blumea* (Compositae: Inuleae) in continental Southeast Asia." *Kew Bulletin* 71, 1. doi: 10.1007/s12225-016-9612-2.

Qian, Y., Kang, J. and Wang, L., 2014. "A medical composition and a medical preparation for treating oral cavity and throat diseases and manufacture method thereof." *Chinese Patent* 104116734, A 20141029.

Ragasa, C. Y., Kristin, C., Co, A. L. and Rideout, J. A. 2005. "Antifungal metabolites from *Blumea balsamifera*." *Natural Product Research* 19, 3: 231–237. doi: 10.1080/14786410410001709773

Rahman, A., Bhuiyan, M.A. and Shahriar, M., 2013. "Preliminary *in vitro* biological, phytochemical screenings and *in vitro* antioxidant activities of *Blumea lacera* (Burm.f.) DC." *International Archive of Applied Sciences and Technology* 4, 2: 01–07.

Rahman, M.A., Uddine, S.B. and Wilcock, C.C., 2007. "Medicinal plants used by Chakma tribe in Hill tracts districts of Bangladesh." *Indian Journal of Traditional Knowledge* 6, 508–517.

Rice-Evans, C., Miller, N. and Paganga, G., 1997. "Antioxidant properties of phenolic compounds." *Trends in Plant Sciences* 2, 4: 152–159.

Rosangkima, G. and Prasad, S.B., 2004. "Antitumour activity of some plants from Meghalaya and Mizoram against murine ascites Dalton's lymphoma." *Indian Journal of Experimental Biology* 42, 10: 981–988.

Rücker, G., Paulini, G., Sakulas, H., Lawong, B. and Goeltenboth, F., 1991. "Medicinal plants from Papua New Guinea: Graniline-Esters and Germacranolides from *Blumea arfakiana*." *Planta Medica* 57, 3: 278–281.

Saewan, N., Koysomboon, S. and Chantraprommma, K., 2011. "Anti-tyrosinase and anti-cancer activities of flavonoids from *Blumea balsamifera* DC." *Journal of Medicinal Plants Research* 5, 1018–1025.

Sakee, U., Maneerat, S., Cushnie, T.P. and De-Eknamkul, W., 2011. "Antimicrobial activity of *Blumea balsamifera* (Lin.) DC. extracts and essential oil." *Natural Product Research* 25, 19: 1849–1856.

Satyal, P., Chhetri, B.K., Dosoky, N.S, *et al.*, 2015. "Chemical Composition of *Blumea lacera* essential oil from Nepal. Biological activities of the essential oil and (Z)-Lachnophyllum ester." *Natural Product Communications* 10, 10: 1749–1750.

Senthilkumar, A., Kannathasan, K. and Venkatesalu, V., 2008. "Chemical constituents and larvicidal property of the essential oil of *Blumea mollis* (D. Don) Merr. against *Culex quinquefasciatus*." *Parasitology Research* 103, 4: 959–962.

Senthilkumar, A., Kannathasan, K. and Venkatesalu, V., 2009. "Antibacterial activity of the leaf essential oil of *Blumea mollis* (D. Don) Merr." *World Journal of Microbiology and Biotechnology* 25, 1297–1300.

Shang, X., 2015. "External-use Chinese medicine for treatment of eczema and its preparation method." *Chinese Patent* 104906268, A 20150916.

Sharma, J., Painuli, R.M. and Gaur, R.D., 2010. "Plants used by rural communities of district Shahjahanpur, Uttar Pradesh." *Indian Journal of Traditional Knowledge* 9, 798–803.

Shen, C.C., Wei, W.C. and Lin, L.C., 2019. "Diterpenoids and Bisnorditerpenoids from *Blumea aromatica*." *Journal of Natural Products* 82, 11: 3181–3185.

Shi, H., 2010. "A Chinese medicinal soft capsule for the treatment of stomach and abdominal pain and its preparation method." *Chinese Patent* 101703573, A 20100512.

Shi, Q., 2015. "Topical ointment for treatment of acne and preparation method." *Chinese Patent* 104800429, A 20150729.

Shirota, O., Oribello, J.M., Sekita, S. and Satake, M., 2011. "Sesquiterpenes from *Blumea balsamifera*." *Journal of Natural Products* 74, 3: 470–476.

Singh, B., Singh, S., Singh, S., *et al.*, 2010. "Ethnobotanical Significance and Antimicrobial activity of *Blumea lacera* (Roxb.) DC." *International Journal of Pharmaceutical & Biological Archives* 1, 3: 314–316.

Singh, U.P., Parthasarathy, and Mohan, G., 2014. "*In-vitro* antioxidant and anti-hyperlipidemic activities of *Blumea eriantha* DC." *Pakistan Journal of Pharmaceutical Sciences* 27, 6: 1961–1965.

Singh, U.P., Singh, A.K. and Sarathy, R.P., 2012. "Effect of methanolic extracts of *Blumea eriantha* DC leaves on protein metabolism and marker enzymes in streptozotocin induced hyperglycemic animals." *International Journal of Pharmaceutical Sciences* 4, 1: 235–238.

Sreelekha, K.P., Ajeesh-Krishna, T.P., Adarsh-Krishna, T.P., *et al.*, 2017. "Pharmaco-chemical characterization of leaves of *Blumea mollis* (D. Don) merr. from Western Ghats of wayanad region of Kerala, India." *Journal of Pharmacognosy and Phytochemistry* 6, 4: 319–323.

Sujarwo, W., Keim, A.P., Savo, V., *et al.*, 2015. "Ethnobotanical study of Loloh: Traditional herbal drinks from Bali (Indonesia)." *Journal of Ethnopharmacology* 169, 34–48.

Sun, G., Zheng, X., Fan, X. and Yang, G., 2014. "Topical Chinese medicine for treatment of knee osteoarthritis." *Chinese Patent* 104116976, A 20141029.

Sun, Y., Yan, Q. and Wang, C., 2015. "Chinese medicine formulation for treatment of rheumatoid arthritis and preparation method thereof." *Chinese Patent* 104257908, A 20150107.

Tambewagh, U.U., Javir, G.R., Joshi, K.S. and Rojatkar, S.R., 2019. "Two new polyoxygenated flavonoids from *Blumea eriantha* DC, methanol extract and their anti-proliferative activity." *Natural Products Research*. doi: 10.1080/14786419.2019.1672063.

Tamilarasi, T. and Thirugnanasampandan, R., 2014. "Antioxidant activity evaluation of essential oil and RAPD analysis of in vitro regenerated *Blumea mollis* (D. Don) Merr." *Acta Physiologiae Plantarum* 36. doi: 10.1007/s11738-014-1535-7.

Tan, D., Yan, Q. and Kang, H., 2013. "Chemical constituents from *Blumea balsamifera*." *Chemistry of Natural Compounds* 48, 1072–1073.

Tan, D., Yang, Z., Zhang, Q., *et al.*, 2020. "Simultaneous quantitative determination of polyphenolic compounds in *Blumea balsamifera* (Ai-Na-Xiang, Sembung) by high-performance liquid chromatography with photodiode array detector." *International Journal of Analytical Chemistry* 9731327. doi: 10.1155/2020/9731327.

Tan, S.Y. and Tatsumura, Y., 2015. "Alexander Fleming (1881–1955): Discoverer of penicillin." *Singapore Medical Journal* 56, 7: 366–367.

Thomford, N.E., Senthebane, D.A., Rowe, A., *et al.*, 2018. "Natural products for drug discovery in the 21st century: Innovations for novel drug discovery." *International Journal of Molecular Sciences* 19, 6: 1578. doi: 10.3390/ijms19061578.

Tilburt, J.C. and Kaptchuk, T.J., 2008. "Herbal medicine research and global health: An ethical analysis." *Bulletin of the World Health Organization* 86, 8: 577–656.

Tolosa, E.N., Rodriguez, J.P. and Malamug, E.L.F., 2020. "A systematic review on the safety and efficacy of *Blumea balsamifera* (L.) DC (NIRPROMP Tablet) for the treatment of urinary tract stones." *Acta Medica Philippina* 54, 1: 53–61.

Uddin, S.J., Grice, I.D. and Tiralongo, E., 2011. "Cytotoxic effects of Bangladeshi medicinal plant extracts." *Evidence Based Complementary and Alternative Medicine* 578092. doi: 10.1093/ecam/nep111.

Upadhyay, H.C., 2019. "Medicinal chemistry of alternative therapeutics: Novelty and hopes with genus *Ammannia*." *Current Topics in Medicinal Chemistry* 19, 10: 784–794.

Upadhyay, H.C., Dwivedi, G.R., Roy, S., *et al*., 2014. "Novel phytol derivatives as drug resistance reversal agents." *CHEMMEDCHEM* 9, 1860–1868.

Upadhyay, H.C., Jaiswal, N., Tamrakar, A.K., *et al*., 2012. "Antihyperglycemic agents from *Ammannia multiflora*." *Natural Product Communications* 7, 7: 899–900.

Upadhyay, H.C., Saini, D.C. and Srivastava, S.K., 2011. "Phytochemical analysis of *Ammannia multiflora*." *Research Journal of Phytochemistry* 5, 170–176.

Upadhyay, H.C., Sisodia, B.S., Verma, R.K., *et al*., 2013. "Antiplasmodial potential of extracts from two species of genus *Blumea*." *Pharmaceutical Biology* 51, 10: 1326–1330.

Valke, D., 2013. "*Blumea membranacea* var. *jaquemontii*." Obtained from Wikimedia under the Creative Commons Attribution-Share Alike Generic license. https://commons.wikimedia.org/wiki/File:%C2%BF_Blumea_membranacea_var._jaquemontii_%3F_(11274557614).jpg (Accessed 1 November 2021)

Valke, D., 2014. "*Blumea mollis*." Obtained from Wikimedia under the Creative Commons Attribute-Share Alike 2.0 Generic license. https://commons.wikimedia.org/wiki/File:Blumea_%C2%BF_mollis_%3F_(11982741505).jpg (Accessed 1 November 2021)

Vanila, D., Ghanthikumar, S. and Manickam, V.S., 2008. "Ethnomedicinal uses of plants in the plains area of the Tirunelveli-District, Tamil Nadu, India." *Ethnobotanical Leaflets* 12, 1198–1205.

Vengolis, 2018. "*Blume* sp." Obtained from Wikimedia under the Creative Commons Attribution-Share Alike 4.0 International license. https://commons.wikimedia.org/wiki/File:Blumea_sp_20180827_103958.jpg (Accessed 1 November 2021)

Verma, S. and Chauhan, N.S., 2007. "Indigenous medicinal plants knowledge of Kunihar forest division, district Salon." *Indian Journal of Traditional Knowledge* 6, 494–497.

Vinayaraj, 2014. "*Blumea lacera*." Obtained from Wikimedia under the Creative Commons Attribution-Share Alike 4.0 International license. https://ayurwiki.org/Ayurwiki/File:Blumea_lacera_10.JPG (Accessed 1 November 2021)

Wang, D., Fu, W.J., Pang, Y.X., *et al*., 2013. "The Study of skin allergy and acute toxicity of *Blumea Balsamifera* oil." *Chinese Journal of Tropical Crops* 34, 2499–2502.

Wang, T., 2015. "Traditional Chinese medicine for treating alcoholic cardiomyopathy." *Chinese Patent* 105194623, A 20151230.

Wang, Y., Zhang, L. and Yang, M., 2014. "Chinese medicine for treating ulcerative colitis caused by retention of damp and heat in the interior." *Chinese Patent* 103800597, A 20140521.

Wang, Y.H., Tian, H.Y., He, S.J., *et al*., 2012. "Analysis of volatile components from leaf in *Blumea balsamifera* (L.) DC. with different extraction methods by gas chromatography—mass spectrometry." *Science and Technology of Food Industry* 33, 97–105.

WHO, 2019. "Traditional medicine." www.afro.who.int/health-topics/traditional-medicine (Accessed 14 August 2020).

Wu, Y., 2014. "Traditional Chinese medicine composition for treating cough due to lung-heat." *Chinese Patent* 104083646, A 20141008.

Xu, H., 2014. "Traditional Chinese medicine preparation for treatment of hyperosteogeny." *Chinese Patent*, 10405595,1 A 20140924.

Yadav, N., 2015. "Herbal compositions comprising *Blumea lacera*, *Celosia cristata* and, *Aristolochia indica* for treatment or prevention of neurological disorders." *Indian Patent* 2011KO00459, A 20150116.

Yin, A., Liu, D., Shao, W. and Zhang, W., 2015. "Chinese medicinal preparation for treating cerebrovascular disease." *Chinese Patent* 104587025, A 20150506.

Yu, L., Han, S., Han, S., 2014. "Chinese medicinal composition for treating gastrointestinal adverse reaction after chemotherapy and radiotherapy." *Chinese Patent* 103816393, A 20140528.

Yu, Q., 2014. "Traditional Chinese medicine composition for treating migraine." *Chinese Patent* 103610961, A 20140305.

Yuan, H., Ma, Q., Ye, L. and Piao, G., 2016. "The traditional medicine and modern medicine from natural products." *Molecules* 21, 5: 559. doi: 10.3390/molecules21050559.

Zhang, X., 2015. "A Chinese medicine formulation for treatment of acute intestinal obstruction." *Chinese Patent* 104825811, A 20150812.

Zhou, Y., 2015. "Application of traditional Chinese medicine composition in preparing medicament for the treatment of acute renal failure." *Chinese Patent* 104887910, A 20150909.

Zhu, L. and Tian, Y.J., 2011a. "Chemical composition and larvicidal activity of *Blumea densiflora* essential oils against *Anopheles anthropophagus*: A malarial vector mosquito." *Parasitology Research* 109, 5: 1417–1422.

Zhu, L. and Tian, Y.J., 2011b. "Chemical composition and larvicidal effects of essential oil of *Blumea martiniana* against Anopheles anthropophagus." *Asian Pacific Journal of Tropical Medicine* 4, 5: 371–374.

Zhu, L., Tian, Y.J., Yang, L. and Jiang, J.G., 2011. "Chemical composition and antimicrobial activities of essential oil of *Blumea megacephala*." *EXCLI Journal* 10, 62–68.

16 Natural Product Compounds from Plants in Neurodegenerative Diseases

Priya Darshani, Md TanjimAlam, Prem P. Tripathi and V.S. Pragadheesh

CONTENTS

16.1 DISEASE INTRODUCTION

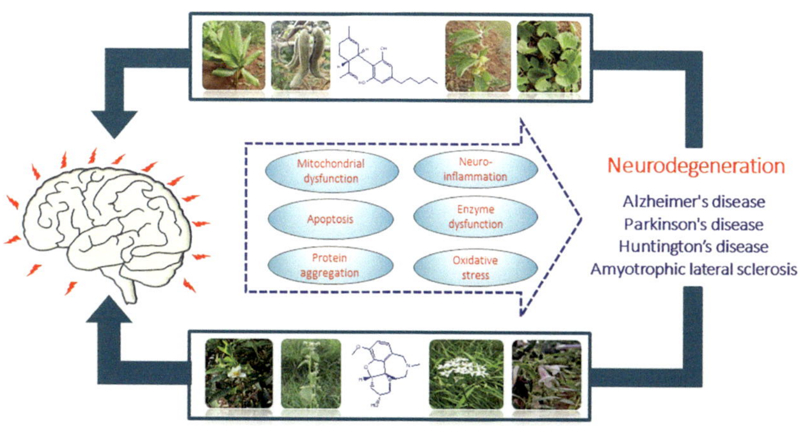

Graphical abstract representing the content of the chapter.

DOI: 10.1201/9781003108375-16

Neurodegenerative disorder or neurodegenerative disease is 'umbrella terminology' for the cluster of diseases that are characterized by progressive and irreversible damage to the central and peripheral nervous system. Neurodegeneration leads to a gradual loss of structure and function of the neuronal population of cells leading to impairments, thereby severely affecting the lives of patients (Wang et al., 2020). The most common factors which are intricately associated with neurodegeneration include aging, genetic mutations, epigenetics, environmental triggers, oxidative and endoplasmic reticulum (ER) stress, aberrant neuronal signaling, abnormal protein accumulation and programmed cell death (Pang et al., 2019; Taylor et al., 2002; Vila and Przedborski, 2003).

Epilepsy, Alzheimer's disease (AD), Parkinson's disease (PD), amyotrophic lateral sclerosis (ALS) and Huntington's disease (HD) are some of the most common neurodegenerative pathologies, which have garnered significant attention around the world due to their debilitative impact (Dugger and Dickson, 2017). Despite their increased prevalence, mainly among the elderly population, there has been diminutive progress in the therapeutic arena of research. The complex etiology of these neurological diseases has rendered them incurable. Therefore, researchers all around the world have gradually paced up their efforts to find the curative routes and effective treatment strategies for these diseases.

Chemical compounds of synthetic origin and their treatment effectiveness show major drawbacks in terms of their undesirable side effects in patients. For a long time, natural products and their derivative bioactive compounds have been used in the treatment of these diseases due to their attractive therapeutic potential. The medicinal use of herbal compounds was prevalent in ancient India and Chinese traditional medicine. Thus, natural products obtained from plant, animal or fungal sources have proven to be safe alternatives compared to synthetic pharmacological agents. The neuroprotective activity of natural products on neurodegenerative diseases has been delineated (Sharifi-Rad et al., 2020; Sairazi and Sirajudeen, 2020). A detailed analysis of the phytochemicals and their derivatives will aid the process of developing novel medicinal approaches. The cellular mechanisms that regulate neurodegenerative diseases and the interventional roles of various natural products in modulating the molecular machinery need to be comprehensively assayed for future translational research avenues.

16.2 THE FUNDAMENTALS OF THE DISEASE

Oxidative stress, irregular protein dynamics (misfolding, defective degradation, proteasomal dysfunction and aggregates) in nerve cells, aberrant synthesis of neurotransmitters, abnormal ubiquitination, mitochondrial dysfunction, disruption of cellular/axonal transport, neuroinflammatory processes and damage of the blood–brain barrier (BBB) are various biological mechanisms leading to the pathogenesis of neurodegenerative diseases (Rasool et al., 2014; Jellinger, 2010).

The basic underlying principle related to the onset and development of most neurological diseases is the progressive accumulation of protein aggregates, triggering inflammation and oxidative stress in the central nervous system (CNS). This can be attributed to extensive literature available regarding the roles of amyloid-β (Aβ) and tau proteins in AD, α-synuclein (AS) in PD, and TAR DNA-binding protein 43 (TDB-43) in ALS (Peng et al., 2020). The common cellular and subcellular features of these diseases underline the need to develop novel drugs and therapeutic avenues that will effectively target such diseases. The key features and causes of a few notable neurodegenerative disorders are summarized here.

16.2.1 Alzheimer's Disease

AD is a chronic neurodegenerative ailment affecting adults in their later stages of life due to memory and cognitive impairment, combined with gradual neuronal death (Hodson, 2018). AD pathogenesis is categorized by neuropathological hallmarks that include formation and deposition of the extracellular Aβ aggregates and formation of neurofibrillary tangles in the brain due to the accumulation of intracellular hyperphosphorylated tau proteins (Sairazi and Sirajudeen, 2020).

In some AD patients, deficiency of the neurotransmitter acetylcholine (ACh) can occur. The Aβ plaques and oligomers are potent synaptotoxins that block the functions of proteasomes and affect neuronal functions. AD therapy involves addressing the aggregation of Aβ through drugs that cause disaggregation of Aβ in neurons and by using strong antioxidant and anti-inflammatory drugs. However, no definitive drug exists which will prevent the progressive loss of neurons in AD patients (Uddin et al., 2020). Promising preclinical results have failed repeatedly to yield successful treatment results.

16.2.2 Parkinson's Disease

PD is characterized by the formation of abnormal proteinaceous bodies (α-synuclein rich inclusion) called Lewy bodies containing branching Lewy neurites in the cell body of affected neurons. It affects the motor neurons adversely. Degeneration of dopaminergic nigrostriatal neurons containing Lewy bodies is the key neuropathological adversity in PD, though similar damage may be associated with glutamatergic, cholinergic and GABA-ergic nerve cells (Sveinbjornsdottir, 2016).

Early symptoms include shaking, rigidity and difficulty in motor movements, with dementia being common in advanced stages. This disease is caused by a multifaceted interplay of genetic and environmental factors that affect the basic cellular processes (Kalia and Lang, 2015). PD is incurable; the available dopaminergic drugs act only upon the motor disturbance symptoms and do not alleviate the disease's onset and development. The revolutionary clinical treatment of PD with Levodopa (L-Dopa) has become relatively ineffective in the advanced stages of the disease (Chou et al., 2018).

16.2.3 Epilepsy

Epilepsy is characterized by the manifestation of epileptic seizures and linked social, psychological and cognitive complications. It involves the occurrence of two unprovoked seizures, a single unprovoked seizure with a high risk of relapse and consequent appearance of an epileptic syndrome caused by abnormal coordinated neuronal cell firing (Devinsky et al., 2018). Though the causes of epilepsy are relatively unknown, genetic mutations, brain strokes, traumatic injury, inflammatory and auto-immune diseases are linked with epileptic development in many patients (Vossler et al., 2018).

A growing line of evidence has shown the involvement of α-synuclein, Aβ and tau protein-mediated neurodegeneration to be the causative agents behind epilepsy. Upregulated α-synuclein levels, elevated expression of amyloid precursor protein (APP) and its processing enzymes (β-APP). Aβ1–40 and Aβ1–42 levels in epilepsy models have highlighted their importance in mediating neurodegeneration in epilepsy. Further studies have shown that an increase in tau hyperphosphorylation stimulates the activity of glycogen synthase kinase 3 beta (GSK3β). This further activates the phosphatidylinositol-3-kinase (PI3K/AKT) pathways, which stimulates epileptogenesis (Paudel et al., 2020). Anti-seizure drugs (ASDs) are the first-line treatment for epilepsy. Even though there are almost 30 U.S. Food and Drug Administration (FDA)-approved anti-epileptic drugs, they only provide transient symptomatic relief and do not halt the disease (Vossler et al., 2018).

16.2.4 Huntington's Disease

HD is a neurodegenerative genetic disorder caused by autosomal dominance mutation in either copy of the genes called huntingtin. It is a rare, inherited, autosomal-dominant neurodegenerative disease that progresses to death in 10–30 years after its onset. It involves a spectrum of motor, cognitive and psychiatric impairments (Rieux et al., 2020). The huntingtin gene provides genetic information to the huntingtin protein. It is associated with the accumulation of a mutant huntingtin protein (mHTT), which consists of an extended polyglutamine (poly Q) stretch. It provokes pathological changes in both the central and peripheral nervous systems. mHTT is formed due to the expansion of CAG (cytosine-adenine-guanine) triplet repeats in the gene coding for huntingtin protein. However, the exact pathophysiological mechanism remains unknown to date (Jimenez-Sanchez et al., 2016).

The symptoms include lack of coordination, unsteady gait, jerky body movements, inability to talk during the advanced stage of the disease and finally mental abilities declining into dementia. Being a complicated neurodegenerative inherited disorder, its treatment modalities are limited and have not offered much success (Dayalu and Albin, 2015). However, scientists have been in continuous search of novel drugs that can effectively target the disease at the cellular or genetic levels.

16.2.5 Amyotrophic Lateral Sclerosis

ALS is an irreversible and progressive neurodegenerative disease associated with selective motor neuron death in the spinal cord, brain stem and motor cortex. The progressive loss of motor neurons leads to muscle weakening, atrophy and dysfunction, resulting in final respiratory failure. ALS affects the nerve cells in the CNS and PNS (peripheral nervous system), causing loss of muscle control. Protein aggregates of TAR DNA-binding protein 43 (TDP-43) have been reported in patients with ALS (Van Es et al., 2017). Mutation of the superoxide dismutase-1 (SOD1) gene has been attributed to the development of this disease (Ciesler and Youssef, 2013).

The symptoms include stiff muscles, muscle twitching, weakness in the arms or legs, and difficulty in speaking, ultimately culminating into an inability to walk, talk, eat or breathe. Genetic and environmental factors might trigger this disease (Van Es et al., 2017; Hardiman et al., 2017). The causal mechanism involves damage of both the motor neurons (Hardiman et al., 2017). Treatment options for this fatal disease are limited; a medication called riluzole exists with limited efficacy (Ciesler and Youssef, 2013). Several neuroprotective factors and chemicals are being researched as novel drug candidates in ALS treatment, and advancement in the study of molecular mechanistic aspects and different genetic models of ALS will gradually open up doors for a successful therapeutic regime.

16.3 ASPECTS OF PHYTOMEDICINE FOR TREATMENT OF NEURODEGENERATIVE DISEASE

Several of today's drugs present in the market are either directly from natural sources or derivatives of the natural products, e.g. paclitaxel and etoposide as anticancer drugs, artemisinin and chloroquine as antimalarial drugs, reserpine and digoxin as cardiovascular drugs, ephedrine as antiasthmatic drugs and galantamine and physostigmine in treatment of neurodegenerative disorders like AD (Cragg and Newman, 2013). Patients still count on phytomedicines due to several issues such as dissatisfaction with the conventional treatment, desire to achieve healthcare solutions and a perception that phytomedicines are harmonious due to their beliefs (Akhondzadeh and Abbasi, 2006). Several age-old studies and documents indicated that herbal medicine plays a crucial and unique role in neuroprotection. The potential natural product compounds (NPCs) from plants in treating different neurodegenerative diseases are shown in Figure 16.1.

16.3.1 Alzheimer's Disease

It has been more than 100 years since the German neuropathologist Alois Alzheimer described the distinct clinical and pathological characteristics of AD. Since then, several diagnostic markers and hypotheses have been found, but it is still unclear whether these are causative or just the ultimate products of a complex cascade of events. These hypothesis and biomarkers have

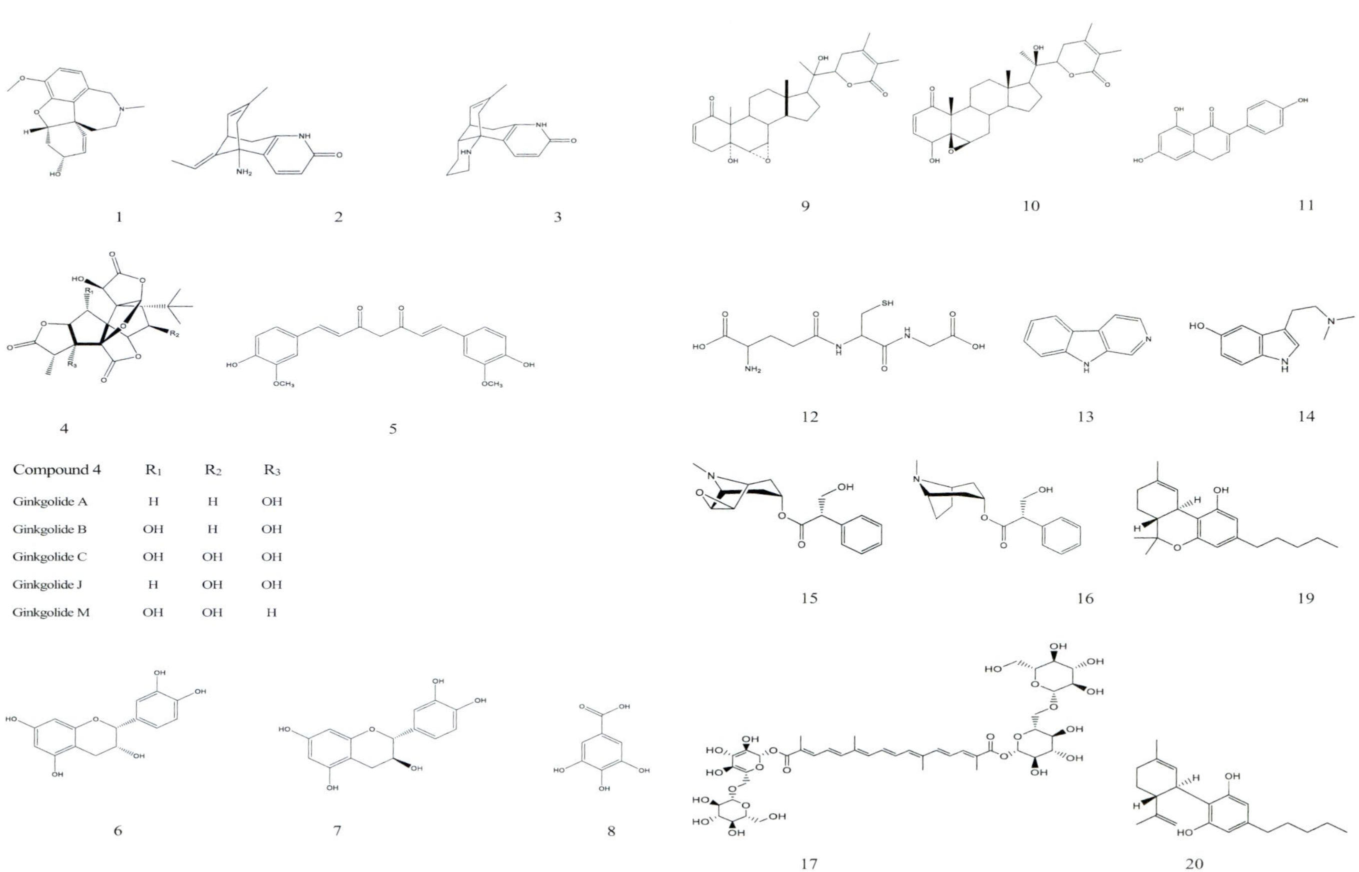

Compound 4	R1	R2	R3
Ginkgolide A	H	H	OH
Ginkgolide B	OH	H	OH
Ginkgolide C	OH	OH	OH
Ginkgolide J	H	OH	OH
Ginkgolide M	OH	OH	H

FIGURE 16.1 Potential natural product compounds from plants for the treatment of neurodegenerative diseases. 1, Galantamine; 2, Huperzine A; 3, Huperzine B; 4, Ginkgolides; 5, Curcumin; 6, (-)-Epicatechin; 7, (+)-Catechin; 8, Gallic acid; 9, Withanolide A; 10, Withanolide D; 11, Genistein; 12, Glutathione; 13, *β*-carboline; 14, Bufotenine; 15, Scopolamine; 16, Hyoscyamine; 17, Crocin; 18, Crocetin; 19, THC; 20, Cannabidiol; 21, Nabilone; 22, Resveratrol; 23, *β*-phenethylamine; 24, (*S*)-(+)-N_b-methyltryptophan methylester; 25, Vasicinone; 26, 5,7-dihydroxy-3-isoprenyl flavone; 27, 5-hydroxy-3-isoprenyl flavone; 28, Baicalein; 29, Stepholidine; 30, Harmine; 31, Harmaline; 32, Ginsenosides; 33, Embelin; 34, Madecassoside; 35, Diallyl trisulfide; 36, Epigallocatechin gallate; 37, Ampelopsin; 38, Morroniside; 39, β-Asarone; 40, Celastrol.

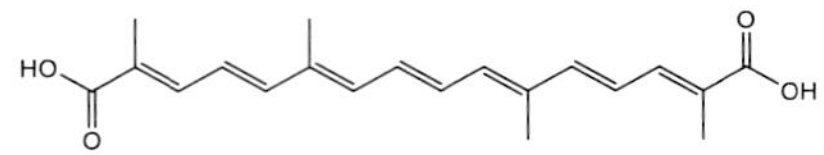

18

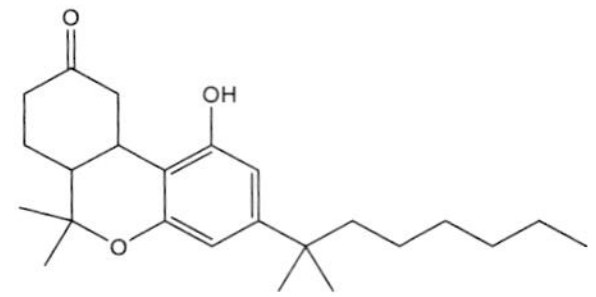

21

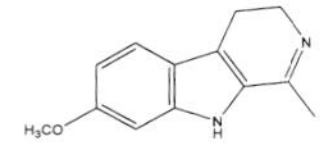

31

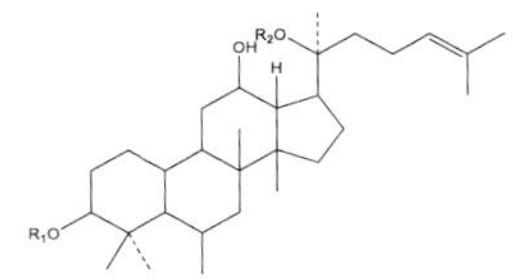

32

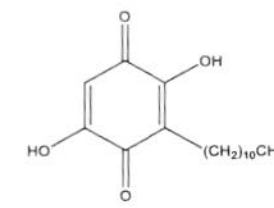

33

Compounds of 32	R_1	R_2
Ginsenoside Rb_1	glc(2-1)glc	glc(6-1)glc
Ginsenoside Rb_2	glc(2-1)glc	glc(6-1)ara(p)
Ginsenoside Rc	glc(2-1)glc	glc(6-1)ara(f)
Ginsenoside Rd	gla(2-1)glc	glc
Ginsenoside Rg_3	glc(2-1)glc	H

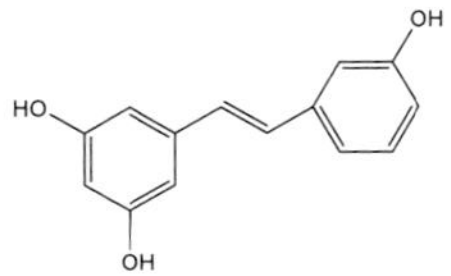

22

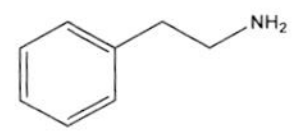

23

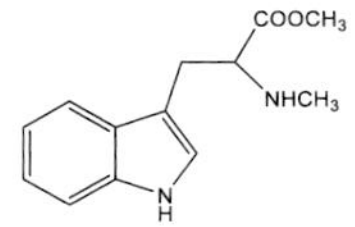

24

Note:

glc – glucopyranose; ara(p) – α-L-arabinose(pyranose); ara(f) – α-L-arabinose(furanose)

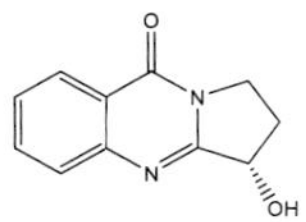

25

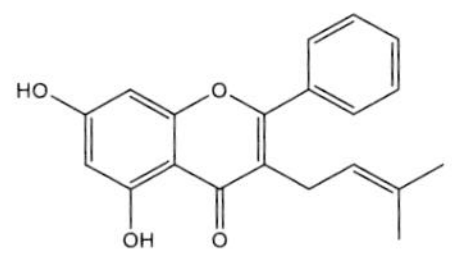

26

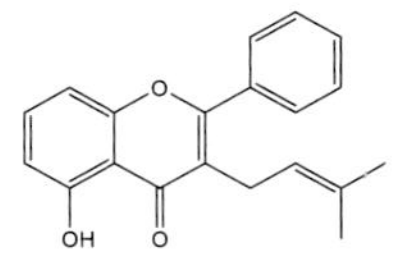

27

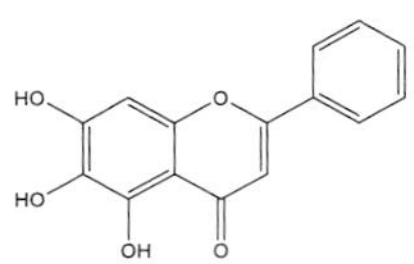

28

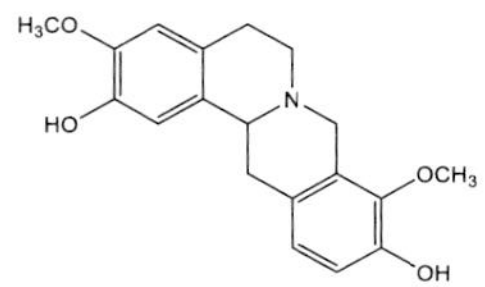

29

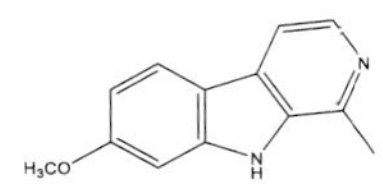

30

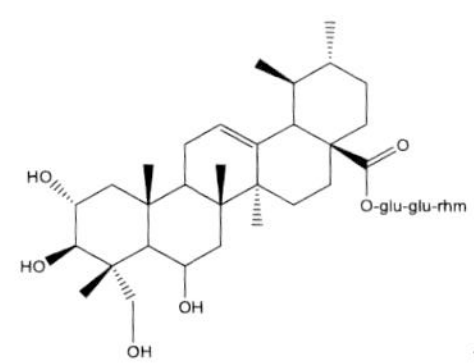

34

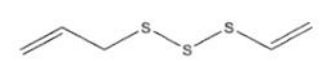

35

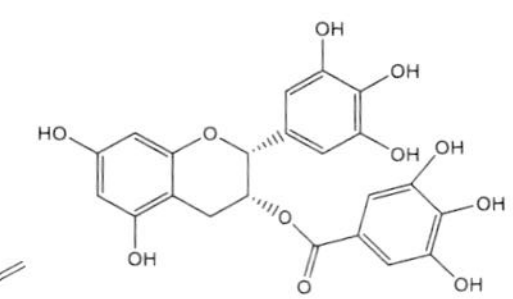

36

FIGURE 16.1 (Continued)

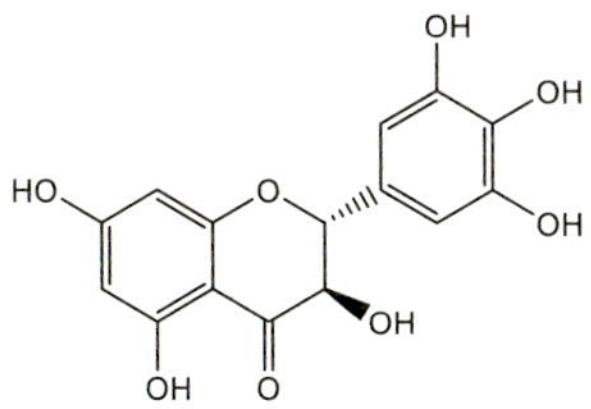

37

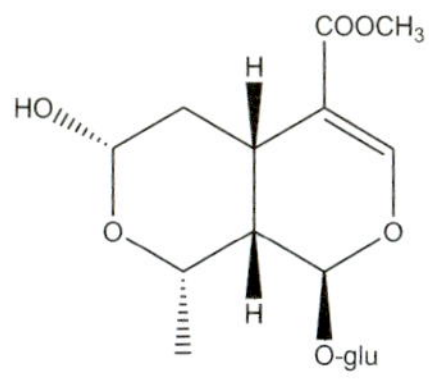

38

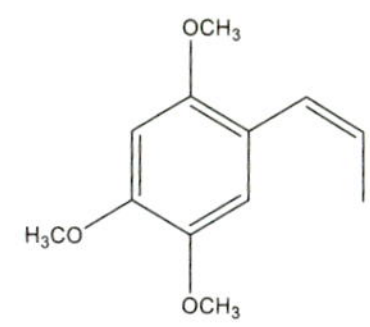

39

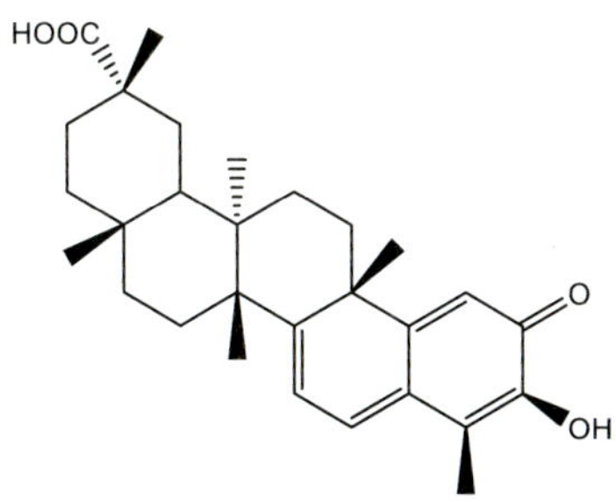

40

FIGURE 16.1 (Continued)

assisted in the development of different therapeutic approaches which can be generally categorized as follows.

- Standard strategies: Acetylcholinesterase inhibitors (AChEI), N-methyl-D-aspartate (NMDA) antagonists and antioxidants.
- Newer and non-standard strategies: anti-tau agents, anti-amyloid agents (β and γ-secretase inhibitor), apolipoprotein (APOE) targeting agents, α-secretase, anti-inflammatory drugs, hormonal replacement therapy, metal chelators immunotherapeutic agents, etc. (Pietrzik and Behl, 2005; Masters et al., 2015).

Several synthetic, semisynthetic and phytochemicals have shown significant effects on different therapeutic strategies. Among the strategies mentioned earlier, only anticholinesterase inhibitors like galantamine (phytocompound), rivastigmine (semisynthetic) and donepezil (synthetic) were approved by the FDA for the treatment of AD (Masters et al., 2015).

Various phytoconstituents have shown promising potential in the treatment of AD and are in different phases of clinical trials. Galantamine is an isoquinoline alkaloid found in the plants of the Amaryllidaceae family. The compound was first isolated in 1956 by a Bulgarian chemist and his team from the bulb of *Galanthus nivalis* L. (commonly known as snowdrop) (Chrusciel and Varagić, 1965). Currently, galantamine is being extracted from the bulb of *Galanthus woronowi* Losinsk and *Galanthus alpinus* Sosns., and also synthesized chemically (Loy and Schneider, 2006). Galantamine acts on the brain cholinergic system by a dual mechanism of action i.e., a reversible competitive inhibitor of acetylcholinesterase (AChE) enzyme and allosteric enhancement of nicotinic acetylcholine receptor (nAChR), which potentiate the presynaptic response to acetylcholine (Heinrich and Teoh, 2004; Albuquerque et al., 2001).

Apart from these modes of action, galantamine has been shown to affect mitochondrial dysfunction by modulating mitochondrial membrane potential (Alexi et al., 2000). Galantamine is metabolized (75%) by hepatic cytochrome P450 enzymes into inactive metabolites and has a half-life of 5–6 hours. CYP2D6 and CYP3A4 are the key isoenzymes that play a critical role in the metabolism of galantamine (Kumar and Kumar, 2009). According to a study conducted on patients with mild AD, when galantamine and placebo at a dose of 16–25 mg were given for six months, patients treated with galantamine had improved cognition and behavioral changes, as compared to those who received only the placebo (Tariot et al., 2000).

Huperzine A is a lycopodium alkaloid derived from the Chinese herb *Huperzia serrata* (Thnub.) Trevis. (Liu et al., 1986). *H. serrata*, belonging to the Lycopodiaceae family, has been used for centuries in the Chinese system of medicine for the treatment of fever, blood disorders, inflammation and schizophrenia. In 1986, a Chinese scientist discovered Huperzine A as a competitive potent reversible AChE inhibitor (Wang et al., 1986). According to a Cochrane systematic review published in 2008, which included six trials conducted on 454 patients of AD, the study showed that huperzine A seemed to have a beneficial effect on cognitive functions and clinical status (Li et al., 2008). A similar study published in 2013 involving 20 randomized clinical trials (RCTs) on 1,823 participants confirmed the beneficial effect of huperzine A on improving cognitive functions, during the clinical assessment in patients having AD (Yang et al., 2013). In China, huperzine A is clinically approved as a symptomatic agent for the treatment of AD, whereas in the United States, it is available as a dietary supplement for dementia and mental deficiency (Zangara, 2003).

Ginkgo biloba L. extract has been used to treat various maladies in different parts of the world for ages. The standardized extract termed as EGb 761 contains 6% terpene trilactone (3.1% ginkgolides and 2.9% bilobalide) and 24% flavonoids (flavonol-*O*-glycosides, a combination of the phenolic aglycones quercetin, kaempferol, or isorhamnetin, with glucose or rhamnose or both at different positions of the flavonol moiety) (Stromgaard and Nakanishi, 2004). A randomized, double-blind, placebo-controlled examination of *G. biloba* dried leaves extract (EGb 761) was conducted and evaluated for clinical efficacy in 216 patients with presenile and senile degenerative dementia for 24 weeks (oral dose of 240 mg). The study showed the significant efficacy of EGb 76, with $p < 0.005$ (Kanowski et al., 1997). In another study, 120 mg of *G. biloba* extract was given for six months and it was found that 27% of patients who received the extract showed moderate improvement in cognitive tests without any specific adverse effects (Le Bars et al., 1997).

Researchers have reported that *G. biloba* extract normalizes the ACh receptor in the brain, which increases cholinergic activity and provides improvement. EGb 761 restrains β-amyloid toxicity in brain cells, thereby playing a crucial part in the progression of AD (Thancharoen et al., 2019, Kanowski and Hoerr, 2002). Recent studies on *Rhodiola crenulata* Hook. f. & Thomson extracts (RCE) demonstrated its cognitive action in the treatment of AD by mediating protection against cholinergic system insufficiency; GSK3β activation and the neuroprotective role of RCE was highlighted in AAβ1–42 models of AD (Zhang et al., 2019).

Vinpocetine, a semi-synthetic chemical derived from the vincamine, present in the leaves of *Vinca minor* L., is used for the treatment of dementia and mental impairment. Vinpocetine is sold as a drug in Europe named Cavinton, whereas, in the United States, it is sold as a dietary supplement to treat memory loss. Vinpocetine has been reported to enhance neurotransmitter production in the brain and improve cerebral metabolism by enhancing the glucose and oxygen consumption in the brain and elevating cerebral cAMP and ATP levels. It has been shown to ameliorate microcirculation and cerebral blood flow by inhibiting platelet aggregation. Vinpocetine has been evaluated in several double-blind clinical trials that demonstrated its significant benefits in patients of AD (Szatmari and Whitehouse, 2003; Balestreri et al., 1987).

Curcumin, a diarylheptanoid-type compound from *Curcuma longa* L., crosses the BBB and reduces the cerebral inflammation of AD by inhibiting the interleukin-1β and oxidative stress (Figure 16.2) (Lim et al., 2001). Curcumin binds with the β-amyloid and inhibits its aggregation, as well as reduces the existing plaque. Neuroprotective effects of curcumin and aerobic yoga reduce the amyloid production and decrease neuroglial cell proliferation, this approach to treat AD is currently in phase II of clinical trials (Cummings et al., 2019).

FIGURE 16.2 Representative medicinal plants having therapeutic potential on various neurodegenerative diseases. a, *Centella asiatica* (L.) Urb.; b, *Sesamum indicum* L.; c, *Allium* species; d, *Withania somnifera* (L.) Dunal; e, *Curcuma longa* L.; f, *Sida cordifolia* L.; g, *Camellia sinensis* (L.) Kuntze.; h, *Mucuna pruriens* (L.) DC; i, *Bacopa monnieri* (L.) Wettst.; j, *Clitoria ternatea* L.

Anthocyanin-rich elderberry juice, which exhibits anti-inflammatory and antioxidant properties and improves mitochondrial function, stands as a therapeutic agent in phase II of clinical trials. Grape seed extract, which contains polyphenolic compounds and acts as an anti-oligomerization agent and prevents aggregation of amyloid and tau protein, is another therapeutic agent in the pipeline for the treatment of AD (Cummings et al., 2019). Korean red ginseng is used as an adjuvant treatment because of its neuroprotective and anti-inflammatory effects (Fei et al., 2020). Ginseng extracts play diversified roles in modulating the signaling pathways through their multidimensional effect on nuclear factor kappa beta (NF-κB), MAP kinase and LPS-induced activation of the STAT signaling pathway. The antioxidant activity of soy isoflavone glycitein, which is the most recognized biologically active compound in soybeans and soy products, may inhibit Aβ aggregate deposition, consequently affecting Aβ-related neurodegenerative disease (Fei et al., 2020).

Evidence from the thioflavin T assay and transgenic mouse models of AD showed the importance of plant secondary metabolites in binding the soluble Aβ in blood and Aβ peptides in the CNS, thereby reducing their aggregation and inhibiting AD development. Medicinal plants containing inhibitors of Aβ fibrillogenesis and synthesis of pharmacologically active compounds may aid in characterizing the mode of action and efficacy of each of these compounds (Stefanescu et al., 2020).

16.3.2 Parkinson's Disease

The main therapeutic approach for PD involves the administration of L-Dopa, the precursor of dopamine to elevate its level in the brain, inhibition of monoamine oxidase (MAO), and catechol-*O*-methyltransferase (COMT). Ayurveda, the traditional Indian system of medicine, describes the use of four plants—*Mucuna pruriens* (L.) DC, *Hyoscyamus reticulatus* L. seeds, *Withania somnifera* (L.) Dunal, and *Sida cordifolia* L. roots—as a concoction in cow's milk for the treatment of PD (Figure 16.2). The concoction has been reported to contain compounds like L-Dopa, neuroactive agents like hyoscyamine, somniferin and ephedrine. A prospective clinical study demonstrates the efficacy of Ayurveda treatment, in which 18 patients of PD received a mixture of warm cow's milk (200 mL) and powdered *M. pruriens* (4.5 g), *H. reticulatus* seeds (0.75 g), *W. somnifera* (14.5 g) and *S. cordifolia* roots (14.5 g). The symptoms like tremor, bradykinesia, stiffness and cramp-like pain were improved by 61.5%, 69.2%, 100% and 75%, respectively, in patients who followed the Ayurvedic treatment (Nagashayana et al., 2000).

W. somnifera extract has been reported to lessen neuronal injuries seen in Parkinson's by normalizing lipid peroxidation, modulating glutathione content and activities of several enzymes like glutathione-S-transferase (GST), glutathione reductase (GR), glutathione peroxidase (GP), superoxide dismutase (SOD) and catalase (Ahmad et al., 2005). *W. somnifera* root extract has been reported to increase dopamine and other catecholamines. The root extract of the plant was in the clinical trials in healthy volunteers (Ahmad et al., 2005).

C. longa has been extensively studied in neuroprotection in the 6-OHDA animal model of PD; curcuminoids inhibited α-synuclein oligomerization into higher molecular weight aggregates, increased dopamine content and moderate iron chelation in dopaminergic neurons (Pandey et al., 2008). In the MPTP-induced model of PD, curcumin has been found to reduce the loss of neurons in substantia nigra, inhibited pro-inflammatory cytokines and NF-kB, decreased the exhaustion of dopamine and reversed the glial fibrillary acidic protein. Curcumin inhibited the enzyme MAO, responsible for breaking down dopamine (Gandhi and Vaidya, 2017).

Root ethanolic extract of *Nardostachys jatamansi* (D.Don) DC. delayed the neuronal injury induced in the 6-OHDA rat model of PD. Animals treated with this extract significantly attenuated the neurotoxin-induced lipid peroxidation led to the loss of catecholamine and increased the glutathione (GSH) content (Ahmad et al., 2006). Phytoconstituents like crocin and crocetin of *Crocus sativus* L. executed neuroprotective activity through antioxidant, anti-inflammatory and immunoregulatory pathways. Cannabinoids from the *Cannabis sativa* L. such as delta-9-tetrahydrocannabinol (THC) and cannabidiol showed neuroprotection through their antioxidant activity. Cannabinoid derivatives, namely AM404 and WIN-55212–2, showed therapeutic activity by recovering affected MAO dopaminergic neurons and constrain excitotoxicity, glial activation and oxidative injury that prevented the deterioration of nigrostriatal neurons. Additionally, cannabinoid compounds such as CE-178253, oleoylethanolamide, nabilone and HU210 have shown their consistent effects against bradykinesia and levodopa-induced dyskinesia in PD (Carrera and Cacabelos, 2019; Pertwee, 2015; Concannon et al., 2015).

Harmine and harmaline are two bioactive compounds of *Banisteriopsis caapi* (Spruce ex Griseb.) Morton which exhibited dose-dependent MAO-A and MAO-B inhibitory activity. The extract of *B. caapi* also increased the dopamine release from rat striatal slices. The antioxidant, anti-inflammatory, free radical scavenging, iron-chelating and DA uptake inhibitory properties of *Camellia sinensis* (L.) Kuntze. showed its neuroprotective potentials in PD (Figure 16.2) (Sengupta et al., 2016).

Resveratrol, a polyphenol antioxidant found in cranberries, peanuts, grapes and red wine, has shown its beneficial role in treating PD. Resveratrol protects the SH-SY5Y cells by progressively inducing the collapse of α-synuclein. A recent *in vivo* study reported that resveratrol possesses the therapeutic capability for rotenone, 6-OHDA and MPTP-induced PD by preventing ER stress-mediated apoptosis, over-expressing antioxidant enzymes and suppressing α-synuclein expression, respectively. Nanoencapsulation of resveratrol increased its bioavailability, thus prolonging neuronal survival against oxidative stress. An emulsion of resveratrol and Vitamin E increased the availability of resveratrol in the brain and thus reduced the oxidative stress of PD, improved coordination movements and memory performance in the mouse model. Atremorine, a phytobioactive agent obtained from *Vicia faba* L., has demonstrated protection against MPTP-induced dopaminergic neurodegeneration, inhibition of MPTP-induced microglial activation and neurotoxicity in substantia nigra along with improvement of motor function in MPTP-induced mice. According to preclinical and clinical studies, atremorine is an effective nutraceutical and manifests anti-inflammatory, neuroprotective and antioxidant effects, presenting the ability to regulate neuroendocrine function in PD (Carrera and Cacabelos, 2019).

Korean red ginseng was shown to prevent dopaminergic neuron cell death by decreasing the expression of cyclin-dependent kinase 5 (Cdk5) in the substantia nigra and striatum. In the cellular model on rat PC12 cells, it was shown that Korean red ginseng treatment prevented 1-methyl-4-phenylpyridinium ion- (MPP+-)–induced cell death. Decreased cell viability and apoptosis were consequently reported. Another plant, *Uncaria rhynchophylla* (Miq.) Miq. ex Havil., reduced apoptosis and reactive oxygen species (ROS) generation, therefore exhibiting anti-Parkinson's activity (Sairazi and Sirajudeen, 2020). Flavonoids such as baicalein, hesperidin, and quercetin were reported to increase dopamine and serotonin levels in the striatum and downregulate oxidative stress and the astroglial response, thus highlighting their anti-neurodegenerative roles in PD mice models (Corona, 2018). Administration of monosialotetrahexosylganglioside (GM1) can alleviate the effects of various neurodegenerative processes. GM1-deficient animals showed Parkinson-like symptoms, suggesting the potential therapeutic role of GM1 in PD treatment (Wang et al., 2019).

16.3.3 Huntington's Disease

W. somnifera has been known to increase vitality and longevity in Ayurveda. Two-week treatment with root extract of *W. somnifera* improved the 3-nitropropionic acid (3-NP)–induced behavioral, biochemical and mitochondrial dysfunction in animal models of HD. Biochemical analysis showed that normalizing the alteration is due to the antioxidant activity of *W. somnifera* (Kumar and Kumar, 2009).

Curcuminoids of *C. longa* exhibited various direct mechanisms like scavenging activity of superoxides and hydroxyl radical, induction of antioxidant enzymes like SOD, catalase, GP and GST, and inhibition of cyclooxygenase (COX-2), which indicated its potential to protect the degradation of neurons. According to Choudhary et al. (2013), chronic administration of curcumin improved the motor deficit, improved memory and increased SDH activity in 3-NP–treated animal models. Ginsenosides from the *Panax ginseng* C.A.Mey. have been reported to exhibit neuroprotective effects against 3-NP-induced striatal neuronal damage. Scientists report that the neuroprotective effect of these ginsenosides could be because of their potential to inhibit glutamate-induced Ca^{2+} responses in spinal neuronal cultures (Huang et al., 2019).

Centella asiatica (L.) Urb. attenuated the 3-NP–induced depletion of GSH levels, total thiols and endogenous antioxidants in the striatum and other brain regions. It also displayed protection against 3-NP-induced mitochondrial dysfunctions, reduced activity of SDH, electron transport chain enzymes and decreased mitochondrial viability (Figure 16.2). Various flavonoids like naringin, hesperidin, kaempferol and epigallocatechin gallate (EGCG) have been reported to provide beneficial effects against 3-NP–induced neurotoxicity. Studies of these plants indicated the potential of phytoconstituents in the treatment of HD (Choudhary et al., 2013).

Embelin from *Embelia ribes* Burm.f. decreased the 3-NP induced malondialdehyde (MDA) and increased the antioxidant enzymes (GST, GSH and CAT); hence improving motor coordination and body balance. The efficacy of embelin could be attributed to its antioxidant potential and the potential of embelin to regulate Ca^{2+} influx linked with increased brain glutamate levels (Kundap et al., 2017). *Psoralea corylifolia* L. increased mitochondrial respiration and decreased cell apoptosis associated with the management of HD (Brahmachari, 2017). Various flavonoids, triterpenoids (celastrol), carotenoids (lycopene) and *Sesamum indicum* L. derivatives showed anti-HD effects in suitable mice model studies (Choudhary et al., 2013).

16.3.4 Amyotrophic Lateral Sclerosis

The etiology of ALS is yet not determined; however, inheritance, mitochondrial dysfunction, a mutation in superoxide dismutase 1 (SOD1), neuro-inflammation, TAR DNA binding protein (TDP-43), glutamate excitotoxicity, translocation in liposarcoma/fused in sarcoma (FUS/TLS) and misfolded protein aggregation are some pathological hallmarks found in the development of ALS. The only approved treatment for ALS is antiglutamatergic riluzole, which has shown the life span extension in patients of ALS in clinical trials (Rowland and Shneider, 2001). Research is ongoing to develop a valid drug candidate for the treatment of ALS.

Administration of withaferin A in mouse model overexpressing the TDP-43 protein has demonstrated improvement in disease pathology. Withaferin A was used as an antagonist of NF-κB, which is a crucial factor in ALS development. It has markedly reduced the inflammatory responses and improved motor responses in transgenic mice (Swarup et al., 2011). In another mouse model FALS, which arose from the mutation in SOD1, withaferin A–treated mice survived longer than the non-treated. In a similar study, where mice expressing G93A mutant form of SOD1 were treated with *W. somnifera* extract and exhibited improved motor performance, increased motor neurons in the lumbar spinal cord and increased longevity by reducing levels of misfolded SOD1, glial activation and prevented phosphorylation of NF-κB (Dutta et al., 2018).

KCHO-1, an herbal combination of 30% ethanol and extract of nine herbs (*Curcuma longa, Salvia miltiorrhiza* Bunge, *Gastrodia elata* Blume, *Chaenomeles sinensis* (Dum. Cours.) Koehne, *Polygala tenuifolia* Willd., *Paeonia japonica* (Makino) Miyabe & H. Takeda, *Glycyrrhiza uralensis* Fisch., *Atractylodes japonica* Koidz. ex Kitam and processed *Aconitum carmichaelii* Debeaux) was administered in the ALS mice model. The results demonstrated improved motor function and delayed onset of the disease. KCHO-1 reduced the oxidative stress in activated microglia and the spinal cord of hSOD1G93a Tg transgenic mice model (Kook et al., 2017). PYM50018 (Myogane), a saponin-type natural product–derived small molecule, has demonstrated neuroprotective activity in several clinical models and is currently in phase I of a clinical trial for the treatment of ALS (Dunkel et al., 2012).

Natural compounds targeting oxidative stress and other cellular aberrations are either effective in themselves or used together with other natural/chemical substances, such as novel therapeutic modalities for ALS and other neurodegenerative disease management. The safety and efficacy of nanocurcumin in ALS adults rendered positive results in a study conducted in Iran recently. Nanocurcumin, together with riluzole, significantly enhanced the survival rate in ALS (Momtaz et al., 2020). Melittin obtained from bee venom, suppressed iNOS (inducible nitric oxide synthase), COX-2 and expression of phosphorylated p38 MAPKin ALS mice models. These, in turn, have been shown to decrease motor neuron injury in ALS models (Jin et al., 2019).

Ginsenoside, a saponin, is the most potent phytochemical used for the treatment of ALS. It reduces iNOS and COX-2 expression levels in LPS-induced BV-2 cells, indicating its effective role in inhibiting the M1 polarization of microglia (Jin et al., 2019). Apart from these, phytoconstituents like madecassoside, epigallocatechin gallate, diallyl trisulfide, ampelopsin, morroniside, picroside-II and astragaloside IV have been reported by researchers for their beneficial activity against oxidative stress in different ALS animal models (Zhang et al., 2013).

16.4 POTENTIAL PLANTS AND NATURAL PRODUCTS WITH SPECIFIC ACTIVITY AGAINST NEURODEGENERATIVE DISEASES

Alkaloids isolated from different plants such as berberine, stylopine, epiberberine, pseudodehydrocorydaline, pseudocopsitine, pseudoberberine, 19,20-dihydrotabernamine and 19,20-dihydroervahanine A exhibited strong AChE inhibitory activity. *Bacopa monnieri* (L.) Wettst., *Convolvulus pluricaulis* Choisy, *Acorus calamus* L. and *Clitoria ternatea* L., which are well known for their memory-enhancing properties, have shown significant inhibition of AChE. Salvianolic acid A, a polyphenol from *S. miltiorrhiza* and rosmarinic acid from *Salvia* species, were shown to inhibit Aβ aggregation. Flavonoids such as myricetin, kaempferol, quercetin, morin, apigenin, kuwanon C, kuwanon A, morusin, and norartocarpetin have shown potent β-secretase inhibition. Diarylheptanoid, a derivative root extract of *Myrica cerifera* L., (-)-epicatechin-3-gallate from *C. sinensis,* myricetin and (+)-aR, 11S-myricanol have demonstrated tau protein inhibition activity (Malik, 2017).

Alkaloids possess vigorous acetylcholinesterase inhibitory activity, while polyphenolic also exhibits significant AChE inhibition for AD. Curcuminoids, ginkgolides, ginsenosides and withanolides have shown encouraging results against potential targets of neurodegeneration such as oxidative stress, Aβ aggregation, tau protein aggregation, α-synuclein, MAO and inflammatory mediators like TNF-α and IL-1β. Neuroprotective properties of flavonoids such as baicalein, kaempferol, quercetin and genistein are attributed to their antioxidant properties. The diverse therapeutic properties of curcumin such as anti-inflammatory, antioxidant, amyloid β-binding, tau inhibition, increased neurogenesis and enhanced synaptogenesis help in the management of AD, PD and other neurological disorders. Novel nanoformulations are being developed to overcome their bioavailability limitations (Salehi et al., 2020).

Polyphenolic compounds such as phenolic acids, flavonoids, stilbenes, alkaloids, catechins, carotenoids, and terpenes have important roles in treating neurodegenerative diseases. Glycan-based therapies have found their way in treating ALS and other neurodegenerative disorders. Lesser amounts of fucosylated glycans and higher amounts of sialylated glycans have been reported in the serum of ALS patients. Furthermore, cerebrospinal fluid IgG N-glycosylation has been indicated as a potential biomarker for ALS. Also, a carbohydrate sulfotransferase, GlcNAc6ST1, is found to be overexpressed and belongs to one of the topmost 40 ALS relevant genes in microglia (Wang et al., 2019).

Medicinal plants cited in the traditional system of medicine have beneficial effects on neurodegenerative diseases via several cellular and molecular mechanisms. Plants listed in Ayurveda and Chinese medicine systems are at the center of most of the research being carried out in neuroprotection. Plants and their phytochemicals offer numerous options for treating the pathological process involved in several neurodegenerative diseases such as AD, PD, ALS and HD (Table 16.1).

TABLE 16.1
Potential Plants and Their Phytochemicals for Neuroprotection

Plants	Compound Numbers	Compound Names	Diseases	Specific Activities and Targets Studied	References
Galanthus nivalis (bulbs)	1	Galantamine	AD	Acetylcholinesterase inhibitor Allosteric enhancement of nAChR Prevent mitochondrial dysfunction	(Albuquerque et al., 2001; Cabral-Costa and Kowaltowski, 2020)
Huperzia serrata	2	Huperzine A	AD	Acetylcholinesterase inhibitor	(Liu et al., 1986)
	3	Huperzine B	AD	Acetylcholinesterase inhibitor	(Liu et al., 1986)
Ginkgo biloba	4	Ginkgolides	AD, PD, HD and ALD	Increase cholinergic activity in the brain Reduce Aβ toxicity in brain cells Reduce oxidative stress Regulation of GST, GS, GT and GP Tau protein inhibition	(Stromgaard and Nakanishi, 2004; Thancharoen et al., 2019; Kanowski and Hoerr, 2002; Malik, 2017)
Curcuma longa	5	Curcumin	AD, PD, HD and ALD	Prevent Aβ aggregation Inhibit COX-2. Reduce oxidative stress Prevent α-synuclein aggregation Induce antioxidant activity Scavenge ROS	(Corona, 2018; Choudhary et al., 2013; Sengupta et al., 2016)
Grape seed extract	6 7 8	(-)-Epicatechin (+)-Catechin Gallic acid	AD	Prevent Aβ deposition Attenuate inflammation on the brain	(Ferruzzi et al., 2009; Wang et al., 2009)

(Continued)

TABLE 16.1 (Continued)

Plants	Compound Numbers	Compound Names	Diseases	Specific Activities and Targets Studied	References
Withania somnifera	9 10	Withanolide A Withanolide D	AD, PD, HD and ALD	Inhibits AChE and butyrylcholinesterase (BChE) Increase activity of GS, GP, GR and GST Improve mitochondrial dysfunction	(Ahmad et al., 2005; Jin et al., 2019; Swarup et al., 2011; Malik, 2017)
Mucuna pruriens	11 12 13 14	Genistein Glutathione β-carboline Bufotenine	PD	Inhibits dopa decarboxylase Free radical scavenger MAO inhibitor activity of DMT 5-HT_{1A} agonist reduces dyskinesia	(Kasture et al., 2013)
Hyoscyamus niger	15 16	scopolamine Hyoscyamine	PD	Anticholinergic activity Ca^{2+} channel blocker Decrease MPTP-induced DA depletion	(Sengupta et al., 2016)
Crocus sativus	17 18	crocin crocetin	PD	Inhibits of LPS-induced nitric oxide Reduce TNF-α and IL-1β Reduces LPS-elicited NF-κB activation	(Nam et al., 2010)
Cannabis sativa	19 20 21	THC Cannabidiol Nabilone	PD	Attenuate bradykinesia Reduce calcium influx and glial activation Anti-oxidant and anti-inflammation activity	(Carrera and Cacabelos, 2019; Pertwee, 2015)
Morus sp.	22	Resveratrol	PD	Upregulation of antioxidant enzymes Downregulating α-synuclein expression Reduce oxidative stress Improve coordination movement	(Carrera and Cacabelos, 2019; Sengupta et al., 2016)
Sida cordifolia	23 24 25 26 27	*β*-phenethylamine (S)-(+)-N_b-methyltryptophan methylester Vasicinone 5,7-dihydroxy-3-isoprenyl flavone 5-hydroxy-3-isoprenyl flavone	PD	Protects dopaminergic neurons against oxidative stress	(Sengupta et al., 2016; Ahmed et al., 2015)
Scutellaria baicalensis	28	Baicalein	PD	Decrease LPS-induced dopamine uptake Reduces production of TNF-α and NO Attenuates oxidative stress Inhibit Aβ aggregation	(Li et al., 2017)
Stephania intermedia	29	Stepholidine	PD	Partial D_1 receptor agonist D_2 receptor antagonist Lessened neuronal damage by an antioxidant mechanism	(Kechun et al., 2007)

Plants	Compound Numbers	Compound Names	Diseases	Specific Activities and Targets Studied	References
Banisteriopsis caapi	30 31	Harmine Harmaline	PD	Inhibit of MAO-A Increase the level of dopamine Antioxidant property	(Samoylenko et al., 2010)
Panax ginseng	32	Ginsenosides	AD, PD, HD	Promote neuronal growth Increase expression of growth-promoting kinases Antiapoptotic activity inhibitor of α-synuclein fibrillation Inhibits glutamate-induced changes in Ca^{2+} concentrations Reduce behavioral disorders	(Huang et al., 2019)
Embelia ribes	33	Embelin	HD	Increase the activity of enzymes GST, GSH and CAT Reduce malonaldehyde Modulate Ca^{2+} influx	(Kundap et al., 2017)
Centella asiatica	34	Madecassoside	HD and ALS	Decrease the level of malondialdehyde Increase the activities of SOD Protect motor neuron from degradation	(Choudhary et al., 2013; Kumar et al., 2019)
Allium sativum	35	Diallyl trisulfide	ALS	Antioxidant and anti-inflammatory Downregulate the expression of glial fibrillary acidic protein	(Zhang et al., 2013)
Camellia sinensis	36	Epigallocatechin gallate	HD and ALS	Upregulation of PI3K/Akt and GSK-3 pathways Downregulation of caspase-3, PARP and mitochondrial damage	(Kumar et al., 2019)
Ampelopsis grossedentata	37	Ampelopsin	ALS	Attenuate ROS production Induction of HO-1 protein expression Inhibit of caspase-3	(Zhang et al., 2013)
Cornus officinalis	38	Morroniside	ALS	Reduce the level of lactate dehydrogenase Stabilize the cell and mitochondrial membrane potential Inhibit the formation of ROS	(Zhang et al., 2013)
Acorus tatarinowii	39	β-Asarone	ALS	Inhibit NMDA or glutamate-induced excitotoxicity Decrease LDH leakage Reduce the apoptosis ratio and the intracellular calcium concentration	(Zhang et al., 2013)
Tripterygium wilfordii	40	Celastrol	ALS	Inhibits phosphorylation of the mitogen-activated protein kinase/ERK1/2 signaling pathway and NF-κB Block the microglial activation Block TNF-α and IL-1β	(Zhang et al., 2013)

16.5 CONCLUSION

Natural products hold tremendous potential to carve out their niche in the field of neurodegenerative disease therapy. Structurally and functionally diversified roles played by numerous phytochemicals and other derivatives will benefit in preventive and therapeutic use. Since they are derived from living organisms, natural products demonstrate naturally optimized biological properties for binding to specific biomolecules or targeting proteins within a cell. They exhibit diversified multi-target efficacy and physiologically broad safety profiles. Despite enormous advantageous roles played by natural products, the limitations in their applicability are attributed to low bioavailability, poor water solubility, physicochemical instability, fast metabolism and susceptibility.

The question regarding the degree to which such natural compounds traverse across the blood–brain barrier remains elusive. There have been several reports of them being metabolized and producing secondary metabolites with different degrees of bioavailability at the target site. The isolation and purification process of bioactive compounds from natural plant sources is also time consuming. In the absence of a requisite treatment for neurodegenerative diseases, it is indisputable that there is an urgent need to identify new targets and approaches for comprehensively assessing the role of natural products in basal cellular mechanisms causing neurodegenerative disorders.

Thorough *in vitro* and *in vivo* studies of these compounds will help in enhancing evaluation of their therapeutic potential and help in monitoring their bioavailability. Proper channelization of pharmacological studies will assist in designing effective multi-target drugs against neurodegenerative diseases. The use of high throughput screening and combinatorial chemistry of natural compounds can overcome these issues. Modern drug delivery approaches such as nanotechnology and nanoencapsulation have to be employed to avail the bioavailability of plant-derived compounds for neuroprotection.

Despite all said limitations, a major advantage of natural product–based therapy is its multidimensional roles across multiple neurodegenerative diseases. Hence there is an urgent need for extensive research of phytochemicals—and their mechanisms of action, safety and efficacy—to explore the real benefits underneath.

REFERENCES

Ahmad, Muzamil, Sofiyan Saleem, Abdullah Shafique Ahmad, Mubeen Ahmad Ansari, SeemaYousuf, Md Nasrul Hoda, and Fakhrul Islam. 2005. "Neuroprotective Effects of *Withania somnifera* on 6-Hydroxydopamine Induced Parkinsonism in Rats." *Human & Experimental Toxicology* 24, 3: 137–147. doi: 10.1191/0960327105ht509oa.

Ahmad, Muzamil, Seema Yousuf, M. Badruzzaman Khan, Md Nasrul Hoda, Abdullah Shafique Ahmad, Mubeen Ahmad Ansari, Tauheed Ishrat, Ashok Kumar Agrawal, and Fakhrul Islam. 2006. "Attenuation by *Nardostachys jatamansi* of 6-Hydroxydopamine-Induced Parkinsonism in Rats: Behavioral, Neurochemical, and Immunohistochemical Studies." *Pharmacology Biochemistry and Behavior* 83, 1: 150–160. doi: 10.1016/j.pbb.2006.01.005.

Ahmed, Galal, Vijayasankar Raman, and Ikhlas A. Khan. 2015. "*Sida cordifolia*, a Traditional Herb in Modern Perspective—A Review." *Current Traditional Medicine* 1, 1: 5–17. doi: 10.2174/22150838016 66141226215639.

Akhondzadeh, Shahin, and Seyed Hesameddin Abbasi. 2006. "Herbal Medicine in the Treatment of Alzheimer's Disease." *American Journal of Alzheimer's Disease & Other Dementias* 21, 2: 113–118. doi: 10.1177/153331750602100211.

Albuquerque, E.X., M.D. Santos, M. Alkondon, E.F.R. Pereira, and A. Maelicke. 2001. "Modulation of Nicotinic Receptor Activity in the Central Nervous System." *Alzheimer Disease & Associated Disorders* 15, Supplement 1: S19–S25. doi: 10.1097/00002093-200108001-00004.

Alexi, Tajrena, Cesario V. Borlongan, Richard L.M. Faull, Chris E. Williams, Ross G. Clark, Peter D. Gluckman, and Paul E. Hughes. 2000. "Neuroprotective Strategies for Basal Ganglia Degeneration: Parkinson's and Huntington's Diseases." *Progress in Neurobiology* 60, 5: 409–470. doi: 10.1016/s0301-0082(99)00032-5.

Balestreri, Roberto, Luigi Fontana, and Federico Astengo. 1987. "A Double-Blind Placebo Controlled Evaluation of the Safety and Efficacy of Vinpocetine in the Treatment of Patients with Chronic Vascular Senile Cerebral Dysfunction." *Journal of the American Geriatrics Society* 35, 5: 425–430. doi: 10.1111/j.1532-5415.1987.tb04664.x.

Brahmachari, Goutam. 2017. *Neuroprotective Natural Products: Clinical Aspects and Mode of Action*. 1st ed. Wiley-Vch. Verlag GmbH & Co. KGaA, Weinheim, Germany, 1–5.

Cabral-Costa, J.V., and A.J. Kowaltowski. 2020 "Neurological Disorders and Mitochondria." *Molecular Aspects of Medicine* 71: 100826. doi: 10.1016/j.mam.2019.10.003.

Carrera, Ivan, and Ramon Cacabelos. 2019. "Current Drugs and Potential Future Neuroprotective Compounds for Parkinson's Disease." *Current Neuropharmacology* 17, 3: 295–306. doi: 10.2174/1570159x1766618 1127125704.

Chou, L. Kelvin, Stacy Mark, Simuni Tanya, Miyasaki Janis, Oertel H. Wolfgang, Sethi Kapil, Fernandez H. Hubert, and Stocchi Fabrizio. 2018. "The Spectrum of 'Off' in Parkinson's Disease: What Have We Learned over 40 Years?" *Parkinsonism & Related Disorders* 51: 9–16. doi: 10.1016/j.parkreldis.2018.02.001

Choudhary, Sunayna, Kumar Puneet, and Malik Jai. 2013. "Plants and Phytochemicals for Huntington's Disease." *Pharmacognosy Reviews* 7, 14: 81–91. doi: 10.4103/0973-7847.120505.

Chrusciel, M., and V. Varagić. 1965. "The Effect of Galantamine on the Blood Pressure of the Rat." *British Journal of Pharmacology and Chemotherapy* 26, 2: 295–301. doi: 10.1111/j.1476-5381.1966.tb01908.x.

Ciesler, Jessica, and Sari Youssef. 2013. "Neurotrophic Peptides: Potential Drugs for Treatment of Amyotrophic Lateral Sclerosis and Alzheimer's Disease." *Open Journal of Neuroscience* 3, 2: 1–13.

Concannon, Ruth, David P. Finn, and Eilís Dowd. 2015. "Cannabinoids in Parkinson's Disease." In book *Cannabinoids in Neurologic and Mental Disease* 1: 35–59. doi: 10.1016/b978-0-12-417041-4.00003-5.

Corona, Juan Carlos. 2018. "Natural Compounds for the Management of Parkinson's Disease and Attention-Deficit/Hyperactivity Disorder." *BioMed Research International* 1–12. doi: 10.1155/2018/4067597.

Cragg, M. Gordon, and David J. Newman. 2013. "Natural Products: A Continuing Source of Novel Drug Leads." *Biochimica et Biophysica Acta (BBA)-General Subjects* 1830, 6: 3670–3695. doi: 10.1016/j.bbagen.2013.02.008.

Cummings, Jeffrey, Garam Lee, Aaron Ritter, Marwan Sabbagh, and Kate Zhong. 2019. "Alzheimer's Disease Drug Development Pipeline: 2019." *Alzheimer's & Dementia: Translational Research & Clinical Interventions* 5, 1: 272–293. doi: 10.1016/j.trci.2019.05.008.

Dayalu, Praveen, and Roger L. Albin. 2015. "Huntington Disease." *Neurologic Clinics* 33, 1: 101–114. doi: 10.1016/j.ncl.2014.09.003.

Devinsky, Orrin, Annamaria Vezzani, Terence J. O'Brien, Nathalie Jette, Ingrid E. Scheffer, Marco De Curtis, and Piero Perucca. 2018. "Epilepsy." *Nature Reviews Disease Primers* 4: 18024. doi: 10.1038/nrdp.2018.24.

Dugger, N. Brittany, and Dennis W. Dickson. 2017. "Pathology of Neurodegenerative Diseases." *Cold Spring Harbor Perspectives in Biology* 9, 7: 1–23. doi: 10.1101/cshperspect.a028035.

Dunkel, Petra, Christina Ll Chai, Beáta Sperlágh, Paul B. Huleatt, and Péter Mátyus. 2012. "Clinical Utility of Neuroprotective Agents in Neurodegenerative Diseases: Current Status of Drug Development for Alzheimer's, Parkinson's and Huntington's Diseases, and Amyotrophic Lateral Sclerosis." *Expert Opinion on Investigational Drugs* 21, 9: 1267–1308. doi: 10.1517/13543784.2012.703178.

Dutta, Kallol, Priyanka Patel, and Jean-Pierre Julien. 2018. "Protective Effects of *Withania somnifera* Extract in SOD1G93A Mouse Model of Amyotrophic Lateral Sclerosis." *Experimental Neurology* 309: 193–204. doi: 10.1016/j.expneurol.2018.08.008.

Fei, Fei, Ning Su, Xia Li, and Zhou Fei. 2020. "Neuroprotection Mediated by Natural Products and Their Chemical Derivatives." *Neural Regeneration Research* 15, 11: 2008–2015. doi: 10.4103/1673-5374.282240.

Ferruzzi, G. Mario, Jessica K. Lobo, Elsa M. Janle, Bruce Cooper, James E. Simon, Qing-Li Wu, Cara Welch, Lap Ho, Connie Weaver, and Giulio M. Pasinetti. 2009. "Bioavailability of Gallic Acid and Catechins from Grape Seed Polyphenol Extract is Improved by Repeated Dosing in Rats: Implications for Treatment in Alzheimer's Disease." *Journal of Alzheimer's Disease* 18, 1: 113–124. doi: 10.3233/jad-2009-1135.

Gandhi, Namyata Pathak, and Ashok D.B. Vaidya. 2017. "Management of Parkinson's Disease in Ayurveda: Medicinal Plants and Adjuvant Measures." *Journal of Ethnopharmacology* 197, Special issue on Ayurveda: 46–51. doi: 10.1016/j.jep.2016.08.020.

Hardiman, Orla, Ammar Al-Chalabi, Adriano Chio, Emma M. Corr, Giancarlo Logroscino, Wim Robberecht, Pamela J. Shaw, Zachary Simmons, and Leonard H. Van Den Berg. 2017. "Amyotrophic Lateral Sclerosis." *Nature Reviews Disease Primers* 3: 170711. doi: 10.1038/nrdp.2017.71.

Heinrich, Michael, and Hooi L. Teoh. 2004. "Galanthamine from Snowdrop—the Development of a Modern Drug against Alzheimer's Disease from Local Caucasian Knowledge." *Journal of Ethnopharmacology* 92, 2–3: 147–162. doi: 10.1016/j.jep.2004.02.012.

Hodson, Richard. 2018. "Alzheimer's Disease." *Nature* 559, 7715: S1. doi: 10.1038/d41586-018-05717-6.

Huang, Xing, Ning Li, Yiqiong Pu, Tong Zhang, and Bing Wang. 2019. "Neuroprotective Effects of Ginseng Phytochemicals: Recent Perspectives." *Molecules* 24, 16: 1–20. doi: 10.3390/molecules24162939.

Jellinger, A. Kurt. 2010. "Basic Mechanisms of Neurodegeneration: A Critical Update." *Journal of Cellular and Molecular Medicine* 14, 3: 457–487. doi: 10.1111/j.1582-4934.2010.01010.x.

Jimenez-Sanchez, Maria, Floriana Licitra, Benjamin R. Underwood, and David C. Rubinsztein. 2016. "Huntington's Disease: Mechanisms of Pathogenesis and Therapeutic Strategies." *Cold Spring Harbor Perspectives in Medicine* 7, 7: 1–22. doi: 10.1101/cshperspect.a024240.

Jin, Xin, Ming-Yan Liu, Dong-Fang Zhang, Xin Zhong, Ke Du, Ping Qian, Hua Gao, and Min-Jie Wei. 2019. "Natural Products as a Potential Modulator of Microglial Polarization in Neurodegenerative Diseases." *Pharmacological Research* 145: 104253. doi: 10.1016/j.phrs.2019.104253.

Kalia, V. Lorraine, and Anthony E. Lang. 2015. "Parkinson's Disease." *The Lancet* 386, 9996: 896–912. doi: 10.1016/s0140-6736(14)61393-3.

Kanowski, S., W.M. Herrmann, K. Stephan, W. Wierich, and R. Hörr. 1997. "Proof of Efficacy of the *Ginkgo biloba* Special Extract EGb 761 in Outpatients Suffering from Mild to Moderate Primary Degenerative Dementia of the Alzheimer Type or Multi-Infarct Dementia." *Phytomedicine* 4, 1: 3–13. doi: 10.1016/s0944-7113(97)80021-9.

Kanowski, S., and R. Hoerr. 2002. "*Ginkgo biloba* Extract EGb 761® in Dementia: Intent-to-Treat Analyses of a 24-Week, Multi-Center, Double-Blind, Placebo-Controlled, Randomized Trial." *Pharmacopsychiatry* 36, 6: 297–303. doi: 10.1055/s-2003-45117.

Kasture, Sanjay, Mahalaxmi Mohan, and Veena Kasture. 2013. "*Mucuna pruriens* Seeds in Treatment of Parkinson's Disease: Pharmacological Review." *Oriental Pharmacy and Experimental Medicine* 13, 3: 165–174. doi: 10.1007/s13596-013-0126-2.

Kechun, Yang, Guozhang Jin, and Jie Wu. 2007. "The Neuropharmacology of (-)-Stepholidine and Its Potential Applications." *Current Neuropharmacology* 5, 4: 289–294. doi: 10.2174/157015907782793649.

Kook, MyungGeun, Soon Won Choi, YoojinSeo, Dong Woung Kim, Bong Keun Song, Ilhong Son, Sungchul Kim, and Kyung-Sun Kang. 2017. "KCHO-1, a Novel Herbal Anti-Inflammatory Compound, Attenuates Oxidative Stress in an Animal Model of Amyotrophic Lateral Sclerosis." *Journal of Veterinary Science* 18, 4: 487–497. doi: 10.4142/jvs.2017.18.4.487.

Kumar, Puneet, and Anil Kumar. 2009. "Possible Neuroprotective Effect Of *Withania somnifera* Root Extract Against 3-Nitropropionic Acid-Induced Behavioral, Biochemical, and Mitochondrial Dysfunction in an Animal Model of Huntington's Disease." *Journal of Medicinal Food* 12, 3: 591–600. doi: 10.1089/jmf.2008.0028.

Kumar, Vijay, Preeti Gupta, and Md Imtaiyaz Hassan. 2019. "Mechanism and Implications of Traditional Chinese Medicine in Amyotrophic Lateral Sclerosis Therapy." *Journal of Proteins and Proteomics* 10, 2: 131–147. doi: 10.1007/s42485-019-00009-7.

Kundap, P. Uday, Saatheeyavaane Bhuvanendran, Yatinesh Kumari, Iekhsan Othman, and Mohd. Farooq Shaikh. 2017. "Plant Derived Phytocompound, Embelin in CNS Disorders: A Systematic Review." *Frontiers in Pharmacology* 8, 76: 1–13. doi: 10.3389/fphar.2017.00076.

Le Bars, Pierre L. 1997. "A Placebo-Controlled, Double-Blind, Randomized Trial of an Extract of *Ginkgo biloba* for Dementia." *JAMA: The Journal of the American Medical Association* 278, 16: 1327–1332. doi: 10.1001/jama.1997.03550160047037.

Li, Jun, Hong Mei Wu, Rongle L. Zhou, Guan Jian Liu, and Bi Rong Dong. 2008. "Huperzine A for Alzheimer's Disease." *Cochrane Database of Systematic Reviews* 2: 1–36. doi: 10.1002/14651858.CD005592.pub2.

Li, Yanwei, Jinying Zhao, and Christian Hölscher. 2017. "Therapeutic Potential of Baicalein in Alzheimer's Disease and Parkinson's Disease." *CNS Drugs* 31, 8: 639–652. doi: 10.1007/s40263-017-0451-y.

Lim, P. Giselle, Teresa Chu, Fusheng Yang, Walter Beech, Sally A. Frautschy, and Greg M. Cole. 2001. "The Curry Spice Curcumin Reduces Oxidative Damage and Amyloid Pathology in an Alzheimer Transgenic Mouse." *The Journal of Neuroscience* 21, 21: 8370–8377. doi: 10.1523/jneurosci.21-21-08370.2001.

Liu, Jia-Sen, Yuan-Long Zhu, Chao-Mei Yu, You-Zuo Zhou, Yan-Yi Han, Feng-Wu Wu, and Bao-Feng Qi. 1986. "The Structures of Huperzine A and B, Two New Alkaloids Exhibiting Marked Anticholinesterase Activity." *Canadian Journal of Chemistry* 64, 4: 837–839. doi: 10.1139/v86-137.

Loy, Clement, and Lon Schneider. 2006. "Galantamine for Alzheimer's Disease and Mild Cognitive Impairment." *Cochrane Database of Systematic Reviews* 1: 1–47. doi: 10.1002/14651858.cd001747.pub3.

Malik, Jai. 2017. "Phytomedicine A Possible Tool for Alzheimer's Disease Therapeutics." Essay in *Neuroprotective Effects of Phytochemicals in Neurological Disorders*, edited by Tahira Farooqui and Akhlaq A. Farooqui, 1st ed. Wiley-Blackwell, 167–198. doi: 10.1002/9781119155195.ch8

Masters, L. Colin, Randall Bateman, Kaj Blennow, Christopher C. Rowe, Reisa A. Sperling, and Jeffrey L. Cummings. 2015. "Alzheimer's Disease." *Nature Reviews Disease Primers* 1, 1: 1–18. doi: 10.1038/nrdp.2015.56.

Momtaz, Saeideh, Zahra Memariani, Fardous F. El-Senduny, Nima Sanadgol, Fereshteh Golab, Majid Katebi, Amir Hossein Abdolghaffari, Mohammad Hosein Farzaei, and Mohammad Abdollahi. 2020. "Targeting Ubiquitin-Proteasome Pathway by Natural Products: Novel Therapeutic Strategy for Treatment of Neurodegenerative Diseases." *Frontiers in Physiology* 11: 1–24. doi: 10.3389/fphys.2020.00361.

Nagashayana, N., P. Sankarankutty, M.R.V. Nampoothiri, P.K. Mohan, and K. Mohanakumar. 2000. "Association of l-DOPA with Recovery Following Ayurveda Medication in Parkinson's Disease." *Journal of the Neurological Sciences* 176, 2: 124–127. doi: 10.1016/s0022-510x(00)00329-4.

Nam, Kyong Nyon, Young-Min Park, Hoon-Ji Jung, Jung Yeon Lee, Byung Duk Min, Seong-Uk Park, Woo-Sang Jung, et al. 2010. "Anti-Inflammatory Effects of Crocin and Crocetin in Rat Brain Microglial Cells." *European Journal of Pharmacology* 648, 1–3: 110–116. doi: 10.1016/j.ejphar.2010.09.003.

Pandey, Neeraj, Jeffrey Strider, William C. Nolan, Sherry X. Yan, and James E. Galvin. 2008. "Curcumin Inhibits Aggregation of α-Synuclein." *Acta Neuropathologica* 115, 4: 479–489. doi: 10.1007/s00401-007-0332-4.

Pang, Shirley Yin-Yu, Philip Wing-LokHo, Hui-Fang Liu, Chi-Ting Leung, Lingfei Li, Eunice EunSeo Chang, David Boyer Ramsden, and Shu-Leong Ho. 2019. "The Interplay of Aging, Genetics and Environmental Factors in the Pathogenesis of Parkinson's Disease." *Translational Neurodegeneration* 8, 1: 1–11. doi: 10.1186/s40035-019-0165-9.

Paudel, Yam Nath, Efthalia Angelopoulou, Christina Piperi, Iekhsan Othman, and Mohd. Farooq Shaikh. 2020. "Revisiting the Impact of Neurodegenerative Proteins in Epilepsy: Focus on Alpha-Synuclein, Beta-Amyloid, and Tau." *Biology* 9, 6: 122. doi: 10.3390/biology9060122.

Peng, Chao, John Q. Trojanowski, and Virginia M.-Y. Lee. 2020. "Protein Transmission in Neurodegenerative Disease." *Nature Reviews Neurology* 16, 4: 199–212. doi: 10.1038/s41582-020-0333-7.

Pertwee, R.G. 2015. "Pharmacological Actions of Cannabinoids." *Handbook of Experimental Pharmacology Cannabinoids* 168: 1–51. doi: 10.1007/3-540-26573-2_1.

Pietrzik, Claus, and Christian Behl. 2005. "Concepts for the Treatment of Alzheimer's Disease: Molecular Mechanisms and Clinical Application." *International Journal of Experimental Pathology* 86, 3: 173–185. doi: 10.1111/j.0959-9673.2005.00435.x.

Rasool, Mahmood, Arif Malik, Muhammad Saeed Qureshi, Abdul Manan, Peter Natesan Pushparaj, Muhammad Asif, Mahmood Husain Qazi, et al. 2014. "Recent Updates in the Treatment of Neurodegenerative Disorders Using Natural Compounds." *Evidence-Based Complementary and Alternative Medicine*, 1–7. doi: 10.1155/2014/979730.

Rieux, Marie, Melanie Alpaugh, Giacomo Sciacca, Martine Saint-Pierre, Maria Masnata, Hélèna L. Denis, Sébastien A. Lévesque, et al. 2020. "Shedding a New Light on Huntington's Disease: How Blood Can Both Propagate and Ameliorate Disease Pathology." *Molecular Psychiatry* 26, 5441–5463. doi: 10.1038/s41380-020-0787-4.

Rowland, P. Lewis, and Neil A. Shneider. 2001. "Amyotrophic Lateral Sclerosis." *New England Journal of Medicine* 344, 22: 1688–1700. doi: 10.1056/nejm200105313442207.

Sairazi, Nur Shafika Mohd, and K.N.S. Sirajudeen. 2020. "Natural Products and Their Bioactive Compounds: Neuroprotective Potentials against Neurodegenerative Diseases." *Evidence-Based Complementary and Alternative Medicine*, 1–30. doi: 10.1155/2020/6565396.

Salehi, Bahare, Daniela Calina, Anca Oana Docea, Niranjan Koirala, Sushant Aryal, Domenico Lombardo, Luigi Pasqua, et al. 2020. "Curcumin's Nanomedicine Formulations for Therapeutic Application in Neurological Diseases." *Journal of Clinical Medicine* 9, 2: 1–35. doi: 10.3390/jcm9020430.

Samoylenko, Volodymyr, Md Mostafizur Rahman, Babu L. Tekwani, Lalit M. Tripathi, Yan-Hong Wang, Shabana I. Khan, Ikhlas A. Khan, Loren S. Miller, Vaishali C. Joshi, and Ilias Muhammad. 2010. "*Banisteriopsis caapi*, a Unique Combination of MAO Inhibitory and Antioxidative Constituents for the Activities Relevant to Neurodegenerative Disorders and Parkinson's Disease." *Journal of Ethnopharmacology* 127, 2: 357–367. doi: 10.1016/j.jep.2009.10.030.

Sengupta, T., J. Vinayagam, R. Singh, P. Jaisankar, and K.P. Mohanakumar. 2016 "Plant-Derived Natural Products for Parkinson's Disease Therapy." In book *Advances in Neurobiology* 12, *The Benefits of Natural Products for Neurodegenerative Diseases*. Springer, Switzerland, 415–496. doi: 10.1007/978-3-319-28383-8_23.

Sharifi-Rad, Mehdi, Chintha Lankatillake, Daniel A. Dias, Anca Oana Docea, Mohamad Fawzi Mahomoodally, Devina Lobine, Paul L. Chazot, et al. 2020. "Impact of Natural Compounds on Neurodegenerative Disorders: From Preclinical to Pharmacotherapeutics." *Journal of Clinical Medicine* 9, 4: 1–19. doi: 10.3390/jcm9041061.

Stefanescu, Raluca, Gabriela Dumitrița Stanciu, Andrei Luca, Luminita Paduraru and Bogdan-Ionel Tamba. 2020. "Secondary Metabolites from Plants Possessing Inhibitory Properties against Beta-Amyloid Aggregation as Revealed by Thioflavin-T Assay and Correlations with Investigations on Transgenic Mouse Models of Alzheimer's Disease." *Biomolecules* 10, 6: 1–21. doi: 10.3390/biom10060870.

Stromgaard, Kristian, and Koji Nakanishi. 2004. "Chemistry and Biology of Terpene Trilactones from *Ginkgo biloba*." *AngewandteChemie International Edition* 43, 13: 1640–1658. doi: 10.1002/anie.200300601.

Sveinbjornsdottir, Sigurlaug. 2016. "The Clinical Symptoms of Parkinson's Disease." *Journal of Neurochemistry* 139, Suppl. 1: 318–324. doi: 10.1111/jnc.13691.

Swarup, Vivek, Daniel Phaneuf, Nicolas Dupre, Susanne Petri, Michael Strong, Jasna Kriz, and Jean-Pierre Julien. 2011. "Deregulation of TDP-43 in Amyotrophic Lateral Sclerosis Triggers Nuclear Factor KB—Mediated Pathogenic Pathways." *The Journal of Experimental Medicine* 208, 12: 2429–2447. doi: 10.1084/jem.20111313.

Szatmari, Szabolcs, and Peter Whitehouse. 2003. "Vinpocetine for Cognitive Impairment and Dementia." *Cochrane Database of Systematic Reviews* 1: CD003119. doi: 10.1002/14651858.cd003119.

Tariot, P.N., P.R. Solomon, J.C. Morris, P. Kershaw, S. Lilienfeld, and C. Ding. 2000. "A 5-Month, Randomized, Placebo-Controlled Trial of Galantamine in AD." *Neurology* 54, 12: 2269–2276. doi: 10.1212/wnl.54.12.2269.

Taylor, J. Paul, John Hardy, and Kenneth H. Fischbeck. 2002. "Toxic Proteins in Neurodegenerative Disease." *Science* 296, 5575: 1991–1995. doi: 10.1126/science.1067122.

Thancharoen, Onnita, Chulaporn Limwattananon, Onanong Waleekhachonloet, Thananan Rattanachotphanit, Phumtham Limwattananon, and Panita Limpawattana. 2019. "*Ginkgo biloba* Extract (EGb761), Cholinesterase Inhibitors, and Memantine for the Treatment of Mild-to-Moderate Alzheimer's Disease: A Network Meta-Analysis." *Drugs & Aging* 36, 5: 435–452. doi: 10.1007/s40266-019-00648-x.

Uddin, Md Sahab, Md Tanvir Kabir, Devesh Tewari, Abdullah Al Mamun, Bijo Mathew, Lotfi Aleya, George E. Barreto, May N. Bin-Jumah, Mohamed M. Abdel-Daim, and Ghulam Md Ashraf. 2020. "Revisiting the Role of Brain and Peripheral Aβ in the Pathogenesis of Alzheimer's Disease." *Journal of the Neurological Sciences* 416. doi: 10.1016/j.jns.2020.116974.

Van Es, Michael A, Orla Hardiman, Adriano Chio, Ammar Al-Chalabi, R. Jeroen Pasterkamp, Jan H. Veldink, and Leonard H. Van Den Berg. 2017. "Amyotrophic Lateral Sclerosis." *The Lancet* 390, 10107: 2084–2098. doi: 10.1016/s0140-6736(17)31287-4.

Vila, M., and S. Przedborski. 2003. "Targeting Programmed Cell Death in Neurodegenerative Diseases." *Nature Reviews Neuroscience* 4, 5: 365–375. doi: 10.1038/nrn1100

Vossler, G. David, Mindl Weingarten, and Barry E. Gidal. 2018. "Summary of Antiepileptic Drugs Available in the United States of America." *Epilepsy Currents* 18, Suppl. 1: 1–26. doi: 10.5698/1535-7597.18.4s1.1.

Wang, Jie, Wei-Wei Hu, Zhi Jiang, and Mei-Jiang Feng. 2020. "Advances in Treatment of Neurodegenerative Diseases: Perspectives for Combination of Stem Cells with Neurotrophic Factors." *World Journal of Stem Cells* 12, 5: 323–338. doi: 10.4252/wjsc.v12.i5.323.

Wang, Wenyue, Sandeep Gopal, Roger Pocock, and Zhicheng Xiao. 2019. "Glycan Mimetics from Natural Products: New Therapeutic Opportunities for Neurodegenerative Disease." *Molecules* 24, 24: 1–23. doi: 10.3390/molecules24244604.

Wang, Y.E., D.X. Yue, and X.C. Tang. 1986. "Anti-cholinesterase Activity of Huperzine A." *Actapharmacologica Sinica* 7, 2: 110–113.

Wang, Yan-Jiang, Philip Thomas, Jin-HuaZhong, Fang-Fang Bi, ShanthaKosaraju, Anthony Pollard, Michael Fenech, and Xin-Fu Zhou. 2009. "Consumption of Grape Seed Extract Prevents Amyloid-β Deposition and Attenuates Inflammation in Brain of an Alzheimer's Disease Mouse." *Neurotoxicity Research* 15: 3–14. doi: 10.1007/s12640-009-9000-x.

Yang, Guoyan, Yuyi Wang, Jinzhou Tian, and Jian-Ping Liu. 2013. "Huperzine A for Alzheimer's Disease: A Systematic Review and Meta-Analysis of Randomized Clinical Trials." *PLOS One* 8, 9: 1–8. doi: 10.1371/journal.pone.0074916.

Zangara, Andrea. 2003. "The Psychopharmacology of Huperzine A: An Alkaloid with Cognitive Enhancing and Neuroprotective Properties of Interest in the Treatment of Alzheimer's Disease." *Pharmacology Biochemistry and Behavior* 75, 3: 675–686. doi: 10.1016/s0091-3057(03)00111-4.

Zhang, Xiaoxue, Xue Wang, Xinhua Hu, Xiaowen Chu, Xintong Li, and Fei Han. 2019. "Neuroprotective Effects of a *Rhodiola crenulata* Extract on Amyloid-β Peptides (Aβ1–42)—Induced Cognitive Deficits in Rat Models of Alzheimer's Disease." *Phytomedicine* 57: 331–338. doi: 10.1016/j.phymed.2018.12.042.

Zhang, Xue, Yan-Long Hong, De-Sheng Xu, Yi Feng, Li-Jie Zhao, Ke-FengRuan, and Xiu-Juan Yang. 2013. "A Review of Experimental Research on Herbal Compounds in Amyotrophic Lateral Sclerosis." *Phytotherapy Research* 28, 1: 9–21. doi: 10.1002/ptr.4960.

17 The Potential of Plants as Treatments for Venous Thromboembolism

A Review of Traditional Use, Biological Activity and Chemical Constituents

Lilitha L. Denga and Namrita Lall

CONTENTS

DOI: 10.1201/9781003108375-17

17.1 INTRODUCTION

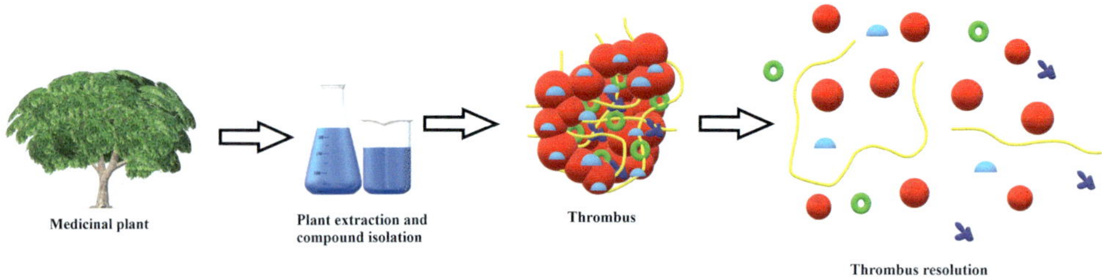

Graphical abstract that represents the content of the chapter.

Venous thromboembolism (VTE) is a collective term that consists of two cardiovascular diseases: deep vein thrombosis (DVT) and pulmonary embolism (PE). DVT is the development of a thrombus (blood clot) in the large veins of the leg. When a part of this thrombus breaks away and travels through the circulatory system, it is known as an embolus. The blood flow carries the embolus to the pulmonary circulation in the lungs, where it can block an artery. This results in PE, which occurs in up to one-third of DVT cases and is the main contributor to mortality (Mackman 2008). Cardiovascular diseases cause the most deaths globally, with more than 75% of these deaths occurring in low- and middle-income countries (World Health Organisation 2017).

This chapter, illustrated in graphic above, focuses on the role of coagulation and platelets in the progression of VTE, as well as the different risk factors and current treatments. Plants are used as medicine by a large percentage of the population in developing countries (Ekor 2014). Hence, this chapter reviews plants for their potential for VTE, highlighting the traditional use and their biological activity.

17.2 PATHOGENESIS OF VTE

Pathogenesis is the biological mechanism by which a disease develops. Understanding pathogenesis provides key factors and conditions that contribute to disease development and can be targeted when developing treatments. The *Global Burden of Diseases, Injuries, and Risk Factors Study* (GBD study), conducted in 2010, revealed that ischemic heart disease and stroke collectively caused 1 in 4 deaths globally (Horton 2012). This is relevant because, like VTE, the key pathologic mechanism underlying these cardiovascular diseases is thrombosis, indicating that thrombosis is a major contributor to the pathology of cardiovascular diseases (Raskob et al. 2014). Therefore, targeting factors that cause thrombosis is an effective way to treat VTE.

17.2.1 Blood Coagulation

Coagulation is a physiological mechanism that prevents excessive bleeding by converting blood from a liquid to a gel state. Coagulation is regulated in a physiological pathway known as the coagulation cascade (Schenone, Furie, and Furie 2004). VTE is a condition where coagulation is upregulated, resulting in thrombosis and subsequently PE (Koupenova et al. 2017).

17.2.2 Structure of a Venous Thrombus

A venous thrombus (Figure 17.1) is composed of erythrocytes and immune cells encased in an interlaced structure of platelets and fibrin. Fibrin is a protein which provides elasticity and a structural backbone to the thrombus (Kattula, Byrnes, and Wolberg 2017). Platelets are small disc-shaped anucleate blood cells that function as procoagulant surfaces, which recruit additional platelets and leukocytes, such as macrophages and neutrophils, to the growing thrombus (Monroe, Hoffman, and Roberts 2002).

17.2.3 The Coagulation Cascade

The coagulation cascade (Figure 17.2) is a series of reactions that regulate hemostasis. The plasma portion of blood carries a set of serine proteases that exist in an inactivated state. These serine proteases are known as coagulation factors or procoagulants (Schenone, Furie, and Furie 2004). Inactive coagulation factors are represented by Roman numerals and are appended with a lowercase "a" once activated. The activity of the coagulation factors can increase by as much as five orders of magnitude when activated (Smith, Travers, and Morrissey 2015). The coagulation cascade consists of two pathways: the intrinsic and extrinsic.

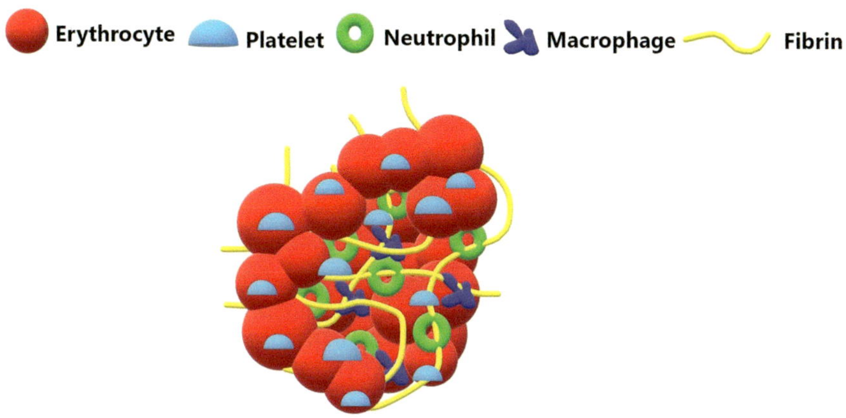

FIGURE 17.1 Structure of a venous thrombus (Paint 3D).

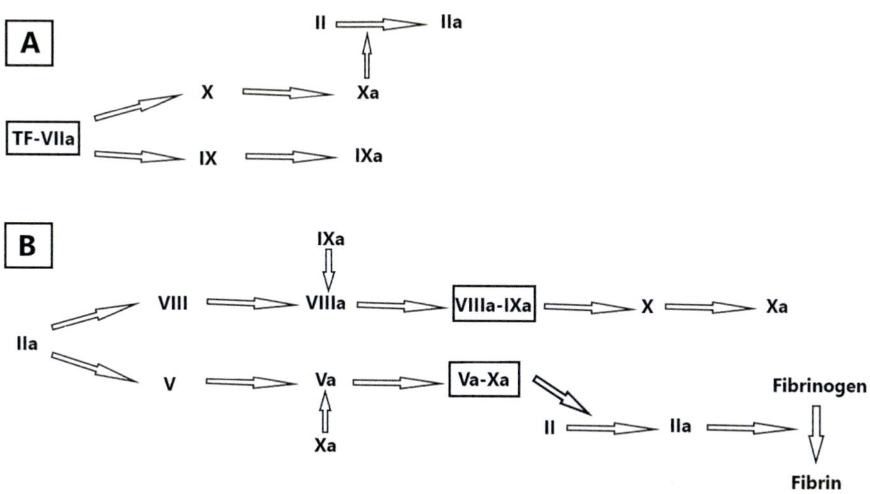

FIGURE 17.2 The coagulation cascade (Paint 3D). A, Initiation of the coagulation cascade. Tissue factor (TF) forms a complex with activated coagulation factor VII (VIIa). The complex (TF-VIIa) activates coagulation factors X (X) and IX (IX), resulting in activated coagulation factors X (Xa) and IX (IXa). Xa converts prothrombin (II) to thrombin (IIa); B, Propagation of the coagulation cascade. Thrombin activates coagulation factors VIII (VIII) and V (V), resulting in activated coagulation factors VIII (VIIIa) and V (Va). VIIIa forms the tensase complex (VIIIa-IXa) with IXa, resulting in additional Xa which forms the prothrombinase complex (Va-Xa) with Va. The prothrombinase complex converts II to IIa, which converts fibrinogen to fibrin.

17.2.3.1 The Extrinsic Pathway

The extrinsic pathway is the initiation step of coagulation, and it requires an integral membrane protein known as tissue factor (TF). TF is expressed on endothelial cells surrounding blood vessels, as well as on monocytes (Schecter et al. 2000). Expression of TF is normally induced when a blood vessel wall is damaged; however, active TF can be expressed by smooth muscle and adventitial cells in the absence of gross vessel wall injury (Manly, Boles, and Mackman 2011). Furthermore, under pathological conditions, expression of TF by endothelial cells and monocytes can be induced by inflammatory cytokines such as tumor necrosis factor (TNF-α) and interleukin-1 (IL-1) (Chu 2006).

The coagulation cascade (Figure 17.2) is initiated when TF forms a complex with the activated serine protease, coagulation factor VII (FVIIa). The tissue factor and activated coagulation factor VII (TF-FVIIa) complex is a potent activator of coagulation because it activates the downstream substrates, coagulation factors IX (FIX) and X (FX), which become activated coagulation factors IX (FIXa) and X (FXa), respectively (Smith, Travers, and Morrissey 2015). The small amount of FXa, then converts initial amounts of prothrombin (FII) to thrombin (FIIa). Thrombin is a serine protease that is responsible for the amplification of the coagulation cascade and conversion of fibrinogen to fibrin in the intrinsic pathway (Brass 2003).

17.2.3.2 The Intrinsic Pathway

The intrinsic pathway is the amplification and propagation step of the coagulation cascade. It creates numerous positive feedback loops, which accelerate the coagulation cascade. The initial creation of thrombin (Figure 17.2) generates additional thrombin by activating factors V (FV) and VIII (FVIII), which become activated factors V (FVa) and VIII (FVIIIa), respectively (Monroe, Hoffman, and Roberts 2002). FVIIIa then forms the tenase complex with FIXa.

The tenase complex proteolytically creates more FXa by activating additional FX (Autin et al. 2004). FXa then forms the prothrombinase complex with FVa. The prothrombinase complex rapidly converts prothrombin to thrombin, which further amplifies the cascade and converts soluble fibrinogen to insoluble fibrin (Smith, Travers, and Morrissey 2015).

17.2.3.3 The Role of Platelets in the Coagulation Cascade

In addition to cascade amplification and fibrinogen conversion, thrombin is a potent activator of platelets (Brass 2003). Studies have shown that the protein components of platelets provide virtually all the specificity required for procoagulant activity and subsequent thrombin generation (Monroe, Hoffman, and Roberts 2002). Thrombin activates platelets and localizes them near a site of exposed TF, resulting in the release of partially active FV, FVIII and FIX (Manly, Boles, and Mackman 2011). Thrombin then cleaves the partially activated coagulation factors to fully active forms, which rapidly bind to the activated platelet (Figure 17.3).

The initial FIXa formed by the TF-VIIa complex diffuses to the platelet surface. Additionally, FXIa on the platelet surface activates FIX and provides additional FIXa, which forms the tenase complex with FVIIIa (Monroe, Hoffman, and Roberts 2002). FX is then activated by the tenase complex and forms the prothrombinase complex with FVa on the surface of the platelet, resulting in a burst of thrombin (Smith, Travers, and Morrissey 2015). Fibrinogen binds to activated platelets via activated fibrinogen receptor and is converted to fibrin by thrombin (Dorsam and Kunapuli 2004).

17.2.4 Anticoagulant Mechanisms

Physiological anticoagulant mechanisms exist to regulate coagulation in the bloodstream. Coagulation is regulated by enzyme inhibition or modification of the activity of procoagulants (De Caterina et al. 2013). A deficiency in any of these anticlotting mechanisms can cause VTE. Understanding the natural anticoagulant mechanisms aids in the development of treatments for

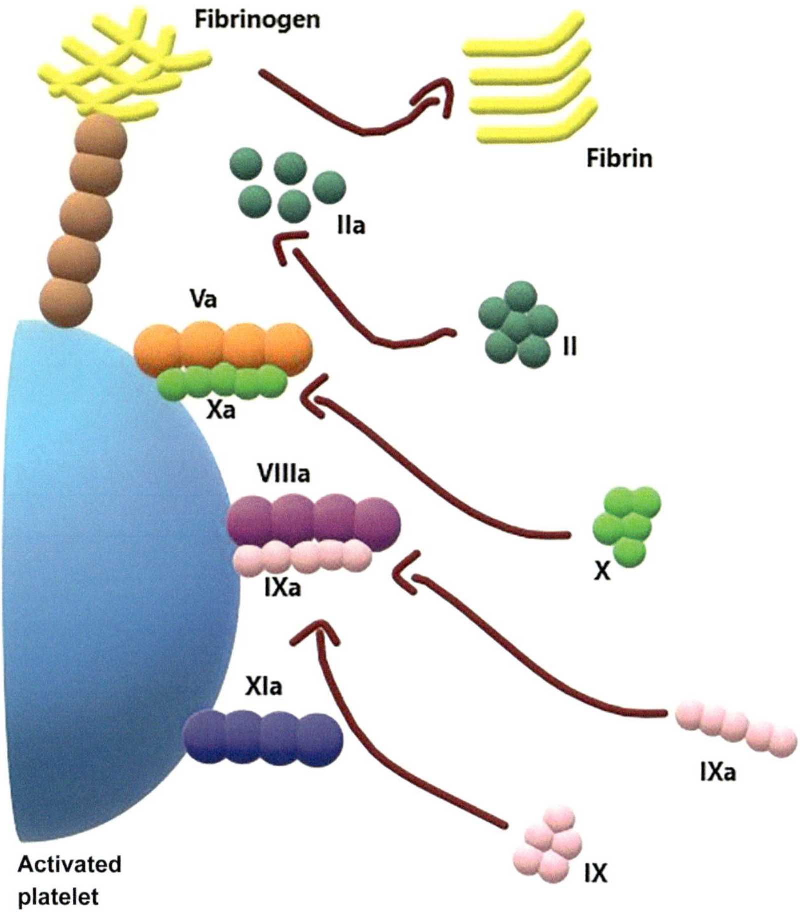

FIGURE 17.3 Activated platelet (Paint 3D). Activated platelets rapidly bind activated coagulation factors V (Va), VIII (VIIIa) and XI (XIa). Activated coagulation factor IX (IXa) diffuses to the platelet surface, and XIa on the platelet surface provides more IXa by activating coagulation factor IX (IX). IXa forms the tenase complex with VIIIa, which activates coagulation factor X(X). Activated coagulation factor X (Xa) forms prothrombinase complex with Va, The prothrombinase complex converts prothrombin (II) to thrombin (IIa), which converts fibrinogen to fibrin.

VTE. Anticoagulant mechanisms can be divided into two systems, the fibrinolytic system and the anticoagulant system.

17.2.4.1 Anticoagulant System

The anticoagulant system aims to reduce the formation of fibrin by reducing the production and effects of thrombin. The different proteins involved in the anticoagulant system are summarized in Table 17.1.

17.2.4.2 Fibrinolytic System

The aim of the fibrinolytic system is to resolve existing thrombi by cleaving fibrin to soluble degradation products. Plasmin is the enzyme that degrades fibrin, while plasminogen is the inactive precursor of plasmin (Anwar et al. 2011). Plasmin is created when plasminogen is released into the bloodstream and activated by tissue plasminogen activator (TPA) or urokinase plasminogen

TABLE 17.1
Proteins Involved in the Anticoagulant System and Mechanisms of Action

Anticoagulant Proteins	Mechanisms of Action	References
Antithrombin (AT)	Binds to thrombin and irreversibly inhibits action on fibrinogen, platelets, FV[1] and FVIII[2]. Directly inhibits FXa[3], FIXa[4] and TF-VIIa[5] complex	(Ezihe-ejiofor and Hutchinson 2013; Thomas 2001)
Tissue factor pathway inhibitor (TFPI)	Forms a complex with FXa, which subsequently forms a quaternary complex with TF-FVIIa complex leading to inhibition of FXa and FVIIa[6]	(Dahm, Sandset, and Rosendaal 2008)
Activated Protein C (APC)	Proteolytically degrades FVa[7] and FVIIIa[8] once activated by thrombin	(Maqbool et al. 2013)
Thrombomodulin (TM)	Forms a complex with thrombin that accelerates the activation of protein C	(Conway 2012)
Protein S	Functions as a cofactor for APC to enhance the degradation of FVa and FVIIIa, as well as direct inhibition of the prothrombinase complex	(Van Cott, Khor, and Zehnder 2016)
Protein Z–dependent protease inhibitor (ZPI)	Inhibits FXa in the presence of protein Z as a cofactor Inhibits FIXa and FXIa[9] in the absence of protein Z	(Ezihe-ejiofor and Hutchinson 2013)

Notes: **1** Coagulation factor V; **2** Coagulation factor VIII; **3** Activated coagulation factor X; **4** Activated coagulation factor IX; **5** Tissue factor and activated coagulation factor VII complex; **6** Activated coagulation factor VII; **7** Activated coagulation factor V; **8** Activated coagulation factor VIII; **9** Activated coagulation factor XI

activator (UPA). TPA binds to clot-bound plasminogen and converts it to plasmin. UPA binds to the UPA receptors located on cell surfaces and converts cell-bound plasminogen to plasmin (Ezihe-ejiofor and Hutchinson 2013).

17.2.5 Hypercoagulability

Hypercoagulability (thrombophilia) is a state of increased risk for thrombosis. A hypercoagulable state describes a pathologic state whereby coagulation is exaggerated in the bloodstream due to hyperactivity of pro-coagulant factors or a deficiency in anticoagulants (Schafer 2007). Hypercoagulability is caused by genetic factors which are inherited or it can be acquired through physiological responses to stress or trauma (Khan and Dickerman 2006; Thomas 2001). In many VTE cases, a hypercoagulable state develops due to an interaction between inherited and acquired factors.

17.2.5.1 Inherited Hypercoagulability

Inherited hypercoagulability occurs when an inherited DNA mutation results in the production of lower concentrations or mutations of anticoagulant proteins (Khan and Dickerman 2006). Table 17.2 summarizes the most common inherited hypercoagulability disorders that pose as risk factors for VTE.

17.2.5.2 Acquired Hypercoagulability

Acquired hypercoagulability (Table 17.3) refers to medical conditions or physiological responses that create an environment that favors thrombus formation (Thomas 2001). Medical conditions such as obesity, pregnancy and prolonged hospitalization after injury, as well as physiological responses such as inflammation and oxidative stress, increase the risk of VTE (Previtali et al. 2011; Katz and Beilin 2015; Mackman 2012; Chu 2006; Herkert et al. 2002).

TABLE 17.2
Inherited Hypercoagulability Disorders That Pose as Risk Factors for VTE

Disorders	Descriptions	Risks	References
AT[1] deficiency	Low concentration or activity of AT	Heterozygous individuals critically predisposed to DVT and tenfold increased risk of thrombosis	(Thomas 2001)
Protein C deficiency	Low concentration of Protein C	Homozygous individuals develop thrombotic disorders and heterozygous individuals are at increased risk for VTE[2]	(Maqbool et al. 2013)
Factor V Leiden	Mutation in FV[3]	Causes APC[4] resistance resulting in failure of APC-mediated degradation of FVa[5] and FVIIIa[6], found in 15–25% of patients with DVT[7]	(Van Cott, Khor, and Zehnder 2016)
Protein S deficiency	Low concentration of Protein S	Twofold to 11-fold increased risk for developing VTE in comparison with those without a deficiency	(Lipe and Ornstein 2011)
Prothrombin G20210A	Higher levels of prothrombin	Prothrombin levels of more than 110% increases risk of VTE	(Jadaon 2011)
AT deficiency	Low concentration or activity of AT	Heterozygous individuals critically predisposed to DVT and tenfold increased risk of thrombosis	(Thomas 2001)
Protein C deficiency	Low concentration of Protein C	Homozygous individuals develop thrombotic disorders and heterozygous individuals are at increased risk for VTE	(Maqbool et al. 2013)

Notes: **1** Antithrombin; **2** Venous thromboembolism; **3** Coagulation factor V; **4** Activated Protein C; **5** Activated coagulation factor V; **6** Activated coagulation factor FVIII; 7, Deep vein thrombosis

TABLE 17.3
Acquired Hypercoagulability Risk Factors

Risk Factors	Contribution to VTE	References
Inflammation	TNF-α[1] and IL-1[2] induce expression of TF[3]; TNF-α and IL-1 downregulate TM[4]	(Chu 2006) (Esmon 2003)
Reactive oxygen species (ROS)	ROS induce expression of TF	(Herkert et al. 2002)
	Inhibit TFPI[5], TM, ZPI[6] and APC[7] by oxidative modifications	(Gutmann et al. 2020)
Obesity	Lack of mobility reduces blood flow	(Previtali et al. 2011)
	Adipose tissue secretes TNF-α, and IL-1	(Blokhin and Lentz 2013)
Pregnancy	Increase of FVII[8], FVIII[9], FIX[10], FX[11] and FXI[12]	(Katz and Beilin 2015)
	Increased fibrinogen	(Prisco, Ciuti, and Falciani 2005)
Injury, surgery and hospitalization	Reduced blood flow results in accumulation of procoagulant factors	(Mackman 2012)
	Vessel trauma exposes TF	(Schenone, Furie, and Furie 2004)

Notes: **1** Tumor necrosis factor alpha; 2 Interleuken-1; **3** Tissue factor; **4** Thrombomodulin; **5** Tissue factor pathway inhibitor; **6** Protein Z–dependent protease inhibitor; **7** Activated Protein C; **8** Coagulation factor VII; **9** Coagulation factor FVIII; **10** Coagulation factor IX; 11 Coagulation factor X; **12** Coagulation factor XI

17.3 CURRENT TREATMENTS

Antithrombotic drugs (Table 17.4) display anticoagulant, antiplatelet or thrombolytic activity. Drugs containing anticoagulant activity prevent further development of blood clots and allow an individual's natural fibrinolytic system to lyse an existing clot (Bauer 2008). Other drugs that contain

TABLE 17.4
Antithrombotic Drugs

Types	Mechanisms of Action	Examples	Side Effects	References
Cyclooxygenase-1 (COX-1) Inhibitors[a]	Inhibit COX-1, which prevents production of thromboxane A_2 and reduces platelet aggregation	Aspirin	Gastrointestinal erosions	(Armstrong et al. 2011) (Süleyman, Demircan, and Karagöz 2007)
GPIIb/IIIa receptor antagonists[a]	Bind to receptor and prevent binding of fibrinogen to platelet	Abciximab Eptifibatide Tirofiban	Bleeding	(Peyvandi, Garagiola, and Baronciani 2011) (Stangl and Lewis 2010)
$P2Y_{12}$ receptor antagonists[a]	Prevent binding of ADP[1] to the receptor. Lowers ADP induced platelet aggregation. Reduces reactions of platelets to stimuli of thrombosis.	Clopidogrel Ticagrelor Prasugrel	Nausea Vomiting Diarrhea Severe skin rashes	(Damman et al. 2012) (Kam and Nethery 2003)
Phosphodiesterase inhibitors[a]	Bind to phosphodiesterase. Prevents hydrolysis of cAMP[2] and cGMP[3], resulting in elevated levels of cAMP, cGMP; increases platelet inhibition	Dipyridamole	Nausea Headaches	(Kim and Liao 2008) (Lee et al. 2017)
Thrombin inhibitors[b]	Bind to the active site of thrombin and prevent interaction with substrates	Epirudin Desirudin Bivalirudin Agratroban	Bleeding	(Lee and Ansell 2011) (Kong et al. 2014)
FXa[4] inhibitors[b]	Selectively bind to active site of FXa and prevent interaction with substrates	Rivaroxaban Apixaban Edoxaban	Headache Nausea Bleeding	(Eriksson, Quinlan, and Weitz 2009) (Abdalwahab, Mansour, and El-dib 2015)
Vitamin K antagonists[b]	Inhibit Vitamin K reductase, which prevents formation of thrombin, FVII[5], FIX[6] and FX[7]	Warfarin	Bleeding	(Bevans et al. 2013) (Pirmohamed 2006)
UFH[8] and LMWH[9] [b]	Bind to AT[10], which enhances the anticoagulant effects of AT on thrombin and FXa	N/A	Thrombocytopenia	(Hirsh et al. 2001) (Merli and Groce 2010)
Thrombolytics	Bind to plasminogen and convert it to plasmin	Anistreplase Streptokinase	Nausea Bleeding	(Anwar et al. 2011) (Ezihe-ejiofor and Hutchinson 2013)

Notes: **a** Antiplatelet; **b** Anticoagulant; **1** Adenosine diphosphate; **2** Cyclic adenosine monophosphate; **3** Cyclic guanosine monophosphate; **4** Activated coagulation factor X; **5** Coagulation factor VII; **6** Coagulation factor IX; **7** Coagulation factor X; **8** Unfractionated heparin; **9** Low molecular weight heparin; **10** Antithrombin

thrombolytic activity dissolve thrombi by activating plasminogen, while those with anti-platelet activity decrease platelet aggregation by interfering with platelet receptors and signaling molecules (Anwar et al. 2011; Hovens et al. 2006).

17.4 PLANTS FOR THE TREATMENT OF VTE

It is estimated that 80% of the world's population living in developing countries rely on herbal medicinal products as a primary source of healthcare (Ekor 2014). This is mainly due to the ease of accessibility and availability of medicinal plants as a health resource in these communities

(Mahomoodally 2013). Plants produce secondary compounds with medicinal properties; therefore, plants are being investigated to explore their potential for VTE.

17.4.1 Natural Antioxidants

Natural antioxidants and polyphenolic compounds are believed to inhibit platelet activity and thrombosis via free radical scavenging (Santhakumar 2015). Resveratrol is a polyphenolic phytoalexin that was first isolated from *Veratrum grandiflorum* (Maxim. ex Miq.) O.Loes. (white hellebore) roots in 1940 (Salehi et al. 2018). The best described medicinal property of resveratrol is its ability to act as a potent antioxidant (Brown et al. 2009). Resveratrol has been detected in more than 70 plant species; however, resveratrol is present at high concentrations in the seeds and skin of grapes, the fruits of *Vitis vinifera* L. (Mehta, Branford, and Rolfe 2016).

An *in vitro* study found that resveratrol inhibited platelet aggregation by scavenging free radicals produced after collagen-induced activation of platelets (Sobotková et al. 2009). Another study found that resveratrol downregulated the expression of TF in human umbilical vein endothelial cells and mononuclear cells by removal of highly reactive free radicals such as superoxide (Santo et al. 2003). In another study, the grape seed extract of *V. vinifera* displayed dual antiplatelet and anticoagulant activity (Bijak et al. 2019). Plants with high concentrations of polyphenolic compounds such as resveratrol and significant antioxidant activity should be further explored for their potential for VTE.

17.4.2 Natural COX-1 Inhibitors

Plants traditionally used for pain and inflammatory disorders often have the ability to inhibit cyclooxygenase enzyme (Figure 17.4). This is due to the role that cyclooxygenase plays in the production of pain and inflammation (Bushra and Aslam 2010).

Plant extracts with these properties can be used as an effective treatment for VTE as inflammation promotes thrombosis and COX-1 is directly involved in platelet aggregation (Armstrong et al. 2011; Esmon 2003). Table 17.5 summarizes some plants that are traditionally used for pain and inflammatory disorders, which displayed COX-1 inhibition in *in vitro* studies.

17.4.3 Natural Phosphodiesterase Inhibitors

There are plants that have displayed phosphodiesterase inhibitory activity in various studies. A study found that ethanolic extracts of *Hiptage benghalensis* (L.) Kurz (hiptage) and *Leea indica* (Burm.f.) Merr. (bandicoot berry), which are traditionally used for inflammation and body ache, inhibited phosphodiesterase enzyme at a 50% inhibitory concentration (IC_{50}) of 2.75 μg/mL and 2.62 μg/mL, respectively (Temkitthawon et al. 2008). The phytochemicals that were responsible for this activity were flavonoids (Temkitthawon et al. 2008). In another study, the ethanolic extracts of the South African plant *Sceletium tortuosum* (L.) N.E.Br. (kanna) reduced the activity of phosphodiesterase (PDE_3) by 88% at a concentration of 750 μg/mL (Harvey et al. 2011). The plant is traditionally used as a psychotropic in the form of a tincture (Terburg et al. 2013).

17.4.4 Natural GPIIb/IIIa and P2Y12 Receptor Antagonists

Millettia vine (*Spatholobus suberectus* Dunn) is a plant traditionally used in China for anemia, rheumatism and abnormality in menstruation (Zhang et al. 2016). A study revealed that the ethanolic extract of *S. suberectus* inhibited fibrinogen from binding to the GPIIb/IIIa receptor (Lee, Wu, and Fried 2011). Lobed actinostemma (*Actinostemma lobatum* Maxim.) is a plant traditionally used for cardiovascular diseases in Korea. A study found that the methanolic extract of *A. lobatum* inhibited the binding of fibrinogen to platelet surface GP IIb/IIIa in a dose dependent manner (Kim et al. 2008).

FIGURE 17.4 Medicinal plants with COX-1 inhibitory activity. A, *Merwilla plumbea* (Lindl.) Speta; B, *Harpagophytum procumbens* (Burch.) DC ex Meisn.; C, *Aloe ferox* Mill.; D, *Siphonochilus aethiopicus* (Schweinf.) B.L.Burtt.; E, *Ocotea bullata* (Burch.) Baill.; F, *Leonotis leonurus* (L.) R.Br.; G, *Eucomis autumnalis* (Mill.) Chitt. (Barrison 2012; BotBln 2010; Gregorydavid 2007; Pidoux 2005; Rulkens 2013, 2020; Shawka 2010).

TABLE 17.5
Plants Traditionally Used for Pain and Inflammatory Disorders with Reported COX-1 Activity

Plants	Traditional Uses	Reported COX-1 Activity	Phytochemicals Responsible for Activity	References
Merwilla plumbea (Lindl.) Speta (Blue hyacinth)	Sprains and fractures, menstrual pains, support woman when giving birth, boils, veld sores	DCM[1] extract showed 80% inhibition of COX-1[2] and hexane extract showed 71% inhibition of COX-1 at a concentration of 250 µg/mL	Saponins and Bufadienolides	(Street and Prinsloo 2013) (Sparg, Van Staden, and Jäger 2002)
Siphonochilus aethiopicus (Schweinf.) B.L.Burtt. (Wild ginger)	Rhizomes are chewed fresh for pain relief and dysmenorrhea	Ethanolic extract of leaves showed 80% inhibition of COX-1 at a concentration of 250 µg/mL	Sesquiterpenes	(van Wyk, van Oudsthoorn, and Gericke 1997) (Light et al. 2002) (Igoli and Obanu 2011)

Plants	Traditional Uses	Reported COX-1 Activity	Phytochemicals Responsible for Activity	References
Eucomis autumnalis (Mill.) Chitt. (Pineapple lily)	Leaves and bulbs administered for backache, post-operative recovery, fevers and fractures	Ethanolic extract of leaves, bulbs and roots displayed IC_{50}[3] of 15 µg/mL, 72 µg/mL and 27 µg/mL, respectively, against COX-1	Flavonoids	(Masondo, Finnie, and van Staden 2014) (Taylor and van Staden 2002)
Aloe ferox Mill. (Bitter aloe)	Bruises, burns, inflamed skin	Methanolic extract displayed 80% inhibition of COX-1	Cinnamic acid	(van Wyk, van Oudsthoorn, and Gericke 1997) (Lindsey et al. 2002) (Lawrence, Tripathi, and Jeyakumar 2009)
Ocotea bullata (Burch.) Baill. (Black Stinkwood)	Bark is taken as a snuff for headaches	Methanolic extracts of dried leaves showed 70% inhibition of COX-1 at a concentration of 100 µg/mL	Sesquiterpenes	(Zschocke et al. 2000)
Harpagophytum procumbens (Burch.) DC ex Meisn. (Devil's claw)	Analgesic and remedy for fever and allergies	Fraction decreased the activity of COX-1 by 37.2% at a concentration of 30 µg/mL	Harpagoside	(Mncwangi et al. 2012) (Anauate, Torres, and de Mello 2010)
Leonotis leonurus (L.) R.Br. (Lion's ear)	Inflammation, headaches, skin rash	Ethanolic extracts of fresh plant material totally inhibited COX-1	Diterpenoids, marrubiin	(Bienvenu et al. 2002) (Stafford, Jäger, and Van Staden 2005) (Scott, Springfield, and Coldrey 2004)

Notes: **1** Dichloromethane; **2** Cyclooxygenase-1; 3 50% inhibitory concentration

Lion's ear (*Leonotis leonurus* [L.] R. Br.) is a South African plant traditionally used to treat coughs, colds, influenza, bronchitis, inflammation and headaches (Bienvenu et al. 2002). A study revealed that the ethanolic extract of *L. leonurus* and one of the diterpenoid isolates known as marrubiin inhibited platelet aggregation by inhibiting the binding of fibrinogen to GPIIb/IIIa (Mnonopi et al. 2011).

Ginseng (*Panax ginseng* C.A.Mey.) is a medicinal plant traditionally used in China and Korea for fatigue and healing. The main active phytochemicals in *P. ginseng* have been reported to be triterpene saponins called ginsenosides (Kim 2018). A study on Korean red ginseng revealed that the extract significantly reduced Adenosine diphospahte (ADP)-induced platelet aggregation (Jin et al. 2007).

17.4.5 Natural FXa and Thrombin Inhibitors

The development of direct, potent and orally bioavailable FXa and thrombin inhibitors has become the focus of many medicinal chemists (Patel et al. 2016). However, due to polyfunctional characters of thrombin in the human body, directly inhibiting thrombin may result in many undesirable effects (Ibrahim et al. 2020). An active molecule of FXa can generate more than 1,000 thrombin molecules; therefore, inhibiting FXa will effectively block coagulation without interfering with other processes in the body (Toschi and Lettino 2011).

Several studies have revealed that plant extracts can inhibit thrombin. White mangrove (*Laguncularia racemose* [L.] C. F. Gaertn.) is a plant traditionally used for healing sores and fever

(Bandaranayake 1998). A study found that glycosylated flavonoids isolated from methanolic extracts of *L. racemosa* inhibited human thrombin activity (Rodrigues et al. 2015). Garden nasturtium (*Tropaeolum majus* L) is a plant that is traditionally used to treat cardiovascular and renal diseases (Gasparotto et al. 2011). In a study on medicinal plants from the Azores, the methylene chloride extract of garden nasturtium inhibited the activity of thrombin by 98% (Manuel et al. 2000). In another study, the methylene chloride extract of *Capsella bursa-pastoris* (L.) Medik (shepherd's purse), a plant traditionally used to improve blood flow, presented antithrombin activity by 87% (Goun et al. 2002).

Plant extracts and isolated compounds have displayed inhibitory activity of FXa in different studies. One study found that the flavonoids procyanidin B2, cyanidin, quercetin and silybin had inhibitory effects on FXa activity at an IC_{50} of 1.2, 3.0, 5.5 and 35 μM, respectively (Bijak, Ponczek, and Nowak 2014). In another study, the two flavonoids, persicarin and isorhamnetin, inhibited the activity of FXa in a dose-dependent manner without potentiation by antithrombin (Ku, Hoon, and Bae 2013). The anti-FXa activity of ethanolic extracts of some plants traditionally used to treat pain and blood flow were determined (Ibrahim et al. 2020). The IC_{50} values were 0.363 mg/mL, 0.729 mg/mL and 0.866 mg/mL for *Glycyrrhiza glabra* L. (Liquorice), *Trifolium alexandrinum* L. (Egyptian clover) and *Olea europaea* L. (Olive), respectively. An earlier study found that polysulfated flavonoids could directly inhibit FXa (Correia da Silva et al. 2011). These studies showed that plants with a high concentration of flavonoids can be considered as potential FXa or thrombin inhibitors.

17.4.6 Natural Vitamin K Antagonists

Coumarins are a family of compounds that bear the typical benzopyrone (benzene ring fused to α-pyrone ring) framework of coumarin (Venugopala, Rashmi, and Odhav 2013). Coumarins are present in many plants and are produced in response to environmental stress and pathogen attacks. Coumarin was first isolated from *Dipteryx odorata* (Aubl.) Willd. (Tonka bean) in 1820; the name coumarin is derived from 'Coumarou,' the vernacular name of the tonka bean (Jain and Joshi 2012).

Coumarin derivatives have attractive anticoagulant Vitamin K antagonist activity (Lei et al. 2015). Dicoumarol is a Vitamin K antagonist produced by fungal species such as *Penicillium nigricans* and *Penicillium jensi* on *Melilotus alba* Medik. and *Melilotus officinalis* (L.) Pall. (sweet clover) (Madari et al. 2003). Dicoumarol is the building block for modern-day Vitamin K antagonists, and its derivatives are hydroxylated in position 4 and have a substitution in position 3 (Lei et al. 2015). These are the minimum requirements for the anticoagulant activity of coumarins. Warfarin is a synthetic derivative of dicoumarol synthesized by the Michael addition of 4-hydroxycoumarin and benzolactone (Jain and Joshi 2012).

17.4.7 Natural Thrombolytic Plants

The thrombolytic potential of plant extracts has been tested in different studies. However, these studies monitored clot lysis after the addition of plant extract to coagulated blood in test tubes. Hence, the mechanism of clot lysis is not clear. Neem (*Azadirachta indica* A. Juss.) is a plant that is considered to be one of the most powerful blood purifiers and detoxifiers in the Ayurvedic system of medicine (Hussain, Islam, and Hossain 2020). A study found that at a concentration of 5 mg/mL, the n-hexane root extracts of *A. indica* was able to lyse blood clots by 21.26% (Hussain, Islam, and Hossain 2020). In another study, the whole plant methanolic extracts of *Averrhoa bilimbi* L. (bilimbi), traditionally used for hypertension, reduced blood clots by 23.93% (Ali et al. 2014).

17.5 CONCLUSION

Whole plant extracts and isolated compounds from plants have shown *in vitro* activity against the various factors that contribute to the development of VTE. With regards to thrombolytic plant extracts and compounds, more targeted *in vitro* studies need to be conducted. Furthermore, *in vivo* studies on plant extracts and isolated compounds need to be conducted to confirm the results obtained from *in vitro* investigations. It is evident that studies have focused on European and Asian plants species for thrombotic disorders. The potential of medicinal plants in other regions of the globe is underexplored. Therefore, future thrombosis research can focus on medicinal plants from these underexplored regions.

REFERENCES

Abdalwahab, M., Y. Mansour, and A. El-dib. 2015. "A Study on the Role of Rivaroxaban in Management of Venous Thromboembolism." *Egyptian Journal of Chest Diseases and Tuberculosis* 64, 4: 893–896. doi: 10.1016/j.ejcdt.2015.04.002.

Ali, R., M. Hossain, J.F. Runa, M. Hasanuzzaman, and M. Islam. 2014. "Evaluation of Thrombolytic Potential of Three Medicinal Plants Available in Bangladesh, as a Potent Source of Thrombolytic Compounds." *Avicenna Journal of Phytomedicine* 4, 6: 430–436. doi: 10.22038/ajp.2014.3226.

Anauate, M.C., L.M. Torres, and S.B.V. de Mello. 2010. "Effect of Isolated Fractions of *Harpagophytum Procumbens* D.C. (Devil's Claw) on COX-1, COX-2 Activity and Nitric Oxide Production on Whole-Blood Assay." *Phytotherapy Research* 24, 9: 1365–1369. doi: 10.1002/ptr.3124.

Anwar, M.S., I.N. Khan, S. Barua, A.T.M. Mostafa Kamal, S.M. Zahid Hosen, and M.H. Kawsar. 2011. "Assessment of Thrombolytic and Cytotoxic Activity of Herbal Preparations Originated from Botanical Source of Bangladesh." *Journal of Applied Pharmaceutical Science* 1, 7: 77–80.

Armstrong, P.C., N.S. Kirkby, Z.N. Zain, M. Emerson, J.A. Mitchell, and T.D. Warner. 2011. "Thrombosis Is Reduced by Inhibition of COX-1, but Unaffected by Inhibition of COX-2, in an Acute Model of Platelet Activation in the Mouse." *PLoS ONE* 6, 5: 5–10. doi: 10.1371/journal.pone.0020062.

Autin, L., W.H. Lee, M.A. Miteva, K. Mertens, K.P. Radtke, and B.O. Villoutreix. 2004. "Molecular Models of the Procoagulant Intrinsic Tenase Complex." *Blood* 104, 11: 1723. doi: 10.1182/blood.V104.11.1723.1723.

Bandaranayake, W.M. 1998. "Traditional and Medicinal Uses of Mangroves." *Mangroves and Salt Marshes* 2, 3: 133–148. doi: 10.1023/A:1009988607044.

Barrison, H. 2012. "Kirstenbosch National Botanical Gardens/Lion's Ear (*Leonotis Leonurus*)." Obtained from Wikimedia under the Creative Commons Share-Alike 2.0 license. https://commons.wikimedia.org/w/index.php?curid=21469335 (Accessed 5 June 2020).

Bauer, K.A. 2008. "New Anticoagulants." *Current Opinion in Hematology* 15, 5: 509–515. doi: 10.1093/ndt/gft293.

Bevans, C.G., C. Krettler, C. Reinhart, H. Tran, K. Koßmann, M. Watzka, and J. Oldenburg. 2013. "Determination of the Warfarin Inhibition Constant Ki for Vitamin K 2,3-Epoxide Reductase Complex Subunit-1 (VKORC1) Using an *in Vitro* DTT-Driven Assay." *Biochimica et Biophysica Acta* 1830: 4202–4210. doi: 10.1016/j.bbagen.2013.04.018.

Bienvenu, E., G.J. Amabeoku, P.K. Eagles, G. Scott, and E.P. Springfield. 2002. "Anticonvulsant Activity of Aqueous Extract of *Leonotis Leonurus*." *Phytomedicine* 9: 217–223. doi: 10.1078/0944-7113-00103.

Bijak, M., M.B. Ponczek, and P. Nowak. 2014. "Polyphenol Compounds Belonging to Flavonoids Inhibit Activity of Coagulation Factor X." *International Journal of Biological Macromolecules* 65, 2014: 129–135. doi: 10.1016/j.ijbiomac.2014.01.023.

Bijak, M., A. Sut, A. Kosiorek, J. Saluk-Bijak, and J. Golanski. 2019. "Dual Anticoagulant/Antiplatelet Activity of Polyphenolic Grape Seeds Extract." *Nutrients* 11, 1: 1–9. doi: 10.3390/nu11010093.

Blokhin, I.O., and S.R. Lentz. 2013. "Mechanisms of Thrombosis in Obesity." *Current Opinion in Hematology* 20, 5: 437–444. doi: 10.1016/j.arth.2019.10.026.

BotBln. 2010. "At Kirstenbosch National Botanical Gardens Species Merwilla Plumbea." Obtained from Wikimedia under the Creative Commons Share-Alike 3.0 license. https://commons.wikimedia.org/w/index.php?curid=12695830 (Accessed 5 June 2020).

Brass, L.F. 2003. "Thrombin and Platelet Activation." *Chest* 124, 3: 18S–25S. doi: 10.1378/chest.124.3_suppl.18S.

Brown, L., P.A. Kroon, D.K. Das, S. Das, A. Tosaki, V. Chan, M.V. Singer, and P. Feick. 2009. "Alcoholic Beverages Industries." *Alcoholism: Clinical and Experimental Research* 33, 9: 1513–1523. doi: 10.1111/j.1530-0277.2009.00989.x.

Bushra, R., and N. Aslam. 2010. "An Overview of Clinical Pharmacology of Ibuprofen." *Oman Medical Journal* 25, 3: 155–161. doi: 10.5001/omj.2010.49.

Chu, A.J. 2006. "Tissue Factor Upregulation Drives a Thrombosis-Inflammation Circuit in Relation to Cardiovascular Complications." *Cell Biochemistry and Function* 24: 173–192. doi: 10.1002/cbf.1200.

Conway, E.M. 2012. "Thrombomodulin and Its Role in Inflammation." *Seminars in Immunopathology* 34: 107–125. doi: 10.1007/s00281-011-0282-8.

Correia da Silva, M., E. Sousa, B. Duarte, F. Marques, F. Carvalho, L.M. Cunha-Ribeiro, and M.M.M. Pinto. 2011. "Flavonoids with an Oligopolysulfated Moiety : A New Class of Anticoagulant Agents." *Journal of Medical Chemistry* 54: 95–106. doi: 10.1021/jm1013117.

Dahm, A.E.A., P.M. Sandset, and F.R. Rosendaal. 2008. "The Association between Protein S Levels and Anticoagulant Activity of Tissue Factor Pathway Inhibitor Type 1." *Thrombosis and Haemostasis* 6, 2: 393–395. doi: 10.1111/j.1538-7836.2008.02859.x.

Damman, P., P. Woudstra, W.J. Kuijt, R.J. De Winter, and S.K. James. 2012. "P2Y12 Platelet Inhibition in Clinical Practice." *Journal of Thrombosis and Thrombolysis* 33, 2: 143–153. doi: 10.1007/s11239-011-0667-5.

De Caterina, R., S. Husted, L. Wallentin, F. Andreotti, H. Arnesen, F. Bachmann, C. Baigent, et al. 2013. "General Mechanisms of Coagulation and Targets of Anticoagulants (Section I)." *Thrombosis and Haemostasis* 109, 4: 569–579. doi: 10.1160/TH12-10-0772.

Dorsam, R.T, and S.P. Kunapuli. 2004. "Central Role of the P2Y12 Receptor in Platelet Activation." *Journal of Clinical Investigation* 113, 3: 340–345. doi: 10.1172%2FJCI20986.

Ekor, M. 2014. "The Growing Use of Herbal Medicines: Issues Relating to Adverse Reactions and Challenges in Monitoring Safety." *Frontiers in Pharmacology* 4: 1–10. doi: 10.3389/fphar.2013.00177.

Eriksson, B.I., D.J. Quinlan, and J.I. Weitz. 2009. "Comparative Pharmacodynamics and Pharmacokinetics of Oral Direct Thrombin and Factor Xa Inhibitors in Development." *Clinical Pharmacokinetics* 48, 1: 1–22. doi: 10.2165/0003088-200948010-00001.

Esmon, C.T. 2003. "Inflammation and Thrombosis." *Journal of Thrombosis and Haemostasis* 1: 1343–1348. doi: 10.1046/j.1538-7836.2003.00261.x.

Ezihe-ejiofor, J.A., and N. Hutchinson. 2013. "Anticlotting Mechanisms 1 : Physiology and Pathology." *Continuing Education in Anaesthesia, Critical Care & Pain* 13, 3: 87–92. doi: 10.1093/bjaceaccp/mks061.

Gasparotto, A., F.M. Gasparotto, E.L.B. Lourenço, S. Crestani, M.E.A. Stefanello, M.J. Salvador, J.E. Da Silva-Santos, M.C.A. Marques, and C.A.L. Kassuya. 2011. "Antihypertensive Effects of Isoquercitrin and Extracts from *Tropaeolum majus* L.: Evidence for the Inhibition of Angiotensin Converting Enzyme." *Journal of Ethnopharmacology* 134, 2011: 363–372. doi: 10.1016/j.jep.2010.12.026.

Goun, E.A, V.M Petrichenko, S.U Solodniko, T.V Suhinina, A. Kline, G. Cunningham, C. Nguyen, and H. Miles. 2002. "Anticancer and Antithrombin Activity of Russian Plants." *Journal of Ethnopharmacology* 81: 337–342. doi: 10.1016/S0378-8741(02)00116-2.

Gregorydavid. 2007. "*Aloe ferox* between Cofimvaba and Ngcobo." Obtained from Wikimedia under the Creative Commons Share-Alike 2.5 license. https://commons.wikimedia.org/w/index.php?title=File:Aloe_Ferox_between_Cofimvaba_and_Ngcobo.jpg&oldid=464576744 (Accessed 5 June 2020).

Gutmann, C., R. Siow, A.M. Gwozdz, P. Saha, and A. Smith. 2020. "Reactive Oxygen Species in Venous Thrombosis." *International Journal of Molecular Sciences* 21, 6: 1–28. doi: 10.3390/ijms21061918.

Harvey, A.L., L.C. Young, A.M. Viljoen, and N.P. Gericke. 2011. "Pharmacological Actions of the South African Medicinal and Functional Food Plant *Sceletium tortuosum* and Its Principal Alkaloids." *Journal of Ethnopharmacology* 137, 3: 1124–1129. doi: 10.1016/j.jep.2011.07.035.

Herkert, O., I. Diebold, R.P. Brandes, J. Hess, R. Busse, and A. Görlach. 2002. "NADPH Oxidase Mediates Tissue Factor-Dependent Surface Procoagulant Activity by Thrombin in Human Vascular Smooth Muscle Cells." *Circulation* 105, 17: 2030–2036. doi: 10.1161/01.CIR.0000014611.28864.1E.

Hirsh, J., S.S. Anand, J.L. Halperin, and V. Fuster. 2001. "Mechanism of Action and Pharmacology of Unfractionated Heparin." *Arteriosclerosis, Thrombosis, and Vascular Biology* 21: 1094–1096. doi: 10.1161/hq0701.093686.

Horton, R. 2012. "GBD 2010: Understanding Disease, Injury, and Risk." *The Lancet* 380: 2053–2054. doi: 10.1016/S0140-6736(12)62133-3.

Hovens, M.M.C., J.D. Snoep, J.T. Tamsma, and M.V. Huisman. 2006. "Aspirin in the Prevention and Treatment of Venous Thromboemblism." *Journal of Thrombosis and Haemostasis* 4: 1470–1475. doi: 10.1111/j.1538-7836.2006.01928.x.

Hussain, F., A. Islam, and M.S. Hossain. 2020. "In Vitro Thrombolytic Potential of Root Extracts of Four Medicinal Plants Available in Bangladesh." *Ancient Science of LIfe* 33: 162–164. doi: 10.4103/0257-7941.144620.

Ibrahim, R.S., R. Mahrous, H.M. Fathy, A.A. Omar, and R.M. Abu EL-Khair. 2020. "Anticoagulant Activity Screening of an In-House Database of Natural Compounds for Discovering Novel Selective Factor Xa Inhibitors; a Combined *in Silico* and *in Vitro* Approach." *Medicinal Chemistry Research* 29: 707–726. doi: 10.1007/s00044-020-02516-5.

Igoli, N.P, and Z. Obanu. 2011. "The Volatile Components of Wild Ginger (*Siphonochilus aethiopicus* (Schweinf) B.l Burtt)." *African Journal of Food Science* 5, 9: 541–549. doi: 10.5897/AJFS.9000298.

Jadaon, M.M. 2011. "Epidemiology of Prothrombin G20210A Mutation in the Mediterranean Region." *Meditteranenan Jourrnal of Hematology and Infectious Diseases* 3, 1: 1–7. doi: 10.4084/mjhid.2011.054.

Jain, P.K., and H. Joshi. 2012. "Coumarin: Chemical and Pharmacological Profile." *Journal of Applied Pharmaceutical Science* 2, 6: 236–240. doi: 10.7324/JAPS.2012.2643.

Jin, Y., J. Yu, J. Lee, S. You, J. Chung, J. Noh, J. Im, et al. 2007. "Antithrombotic and Antiplatelet Activities of Korean Red Ginseng Extract." *Basic and Clinical Pharmacology and Toxicology* 100, 3: 170–175. doi: 10.1111/j.1742-7843.2006.00033.x.

Kam, P.C.A., and C.M. Nethery. 2003. "The Thienopyridine Derivatives (Platelet Adenosine Diphosphate Receptor Antagonists), Pharmacology and Clinical Developments." *Anaesthesia* 58: 28–35. doi: 10.1046/j.1365-2044.2003.02960.x.

Kattula, S., J.R. Byrnes, and A.S. Wolberg. 2017. "Fibrinogen and Fibrin in Hemostasis and Thrombosis." *Arteriosclerosis, Thrombosis, and Vascular Biology* 37, 3: 13–21. doi: 10.1161/ATVBAHA.117.308564.

Katz, D., and Y. Beilin. 2015. "Disorders of Coagulation in Pregnancy." *British Journal of Anaesthesia* 115: 75–88. doi: 10.1093/bja/aev374.

Khan, S., and J.D. Dickerman. 2006. "Hereditary Thrombophilia." *Thrombosis Journal* 4, 15: 1–17. doi: 10.1186/1477-9560-4-15.

Kim, H., and J.K. Liao. 2008. "Translational Therapeutics of Dipyridamole." *Arteriosclerosis, Thrombosis, and Vascular Biology* 28, 3: 39–42. doi: 10.1161/ATVBAHA.107.160226.

Kim, J. 2018. "Pharmacological and Medical Applications of *Panax ginseng* and Ginsenosides: A Review for Use in Cardiovascular Diseases." *Journal of Ginseng Research* 42: 264–269. doi: 10.1016/j.jgr.2017.10.004.

Kim, K., H. Lee, J. Lee, Y. Jang, D. Kim, B. Shim, K. Cho, S. Ko, K. Ahn, and S. Kim. 2008. "Blockade of Glycoprotein IIb/IIIa Mediates the Antithrombotic Activity of Butanol Fraction of *Actinostemma lobatum* Maxim." *Journal of Ethnopharmacology* 116: 431–438. doi: 10.1016/j.jep.2007.12.012.

Kong, Y., H. Chen, Y. Wang, L. Meng, and J. Wei. 2014. "Direct Thrombin Inhibitors: Patents 2002–2012 (Review)." *Molecular Medicine Reports* 9, 5: 1506–1514. doi: 10.3892/mmr.2014.2025.

Koupenova, M., B.E. Kehrel, H.A. Corkrey, and J.E. Freedman. 2017. "Thrombosis and Platelets: An Update." *European Heart Journal* 38: 785–791. doi: 10.1093/eurheartj/ehw550.

Ku, S., T. Hoon, and J. Bae. 2013. "Anticoagulant Activities of Persicarin and Isorhamnetin." *Vascular Pharmacology* 58, 4: 272–279. doi: 10.1016/j.vph.2013.01.005.

Lawrence, R., P. Tripathi, and E. Jeyakumar. 2009. "Isolation, Purification and Evaluation of Antibacterial Agents from *Aloe vera*." *Brazilian Journal of Microbiology* 40: 906–915. doi: 10.1590/s1517-838220090004000023.

Lee, C.J., and Jack E. Ansell. 2011. "Direct Thrombin Inhibitors." *British Journal of Clinical Pharmacology* 72, 4: 581–592. doi: 10.1111/j.1365-2125.2011.03916.x.

Lee, M., Y. Wu, and S.K. Fried. 2011. "Adipose Tissue Remodeling." *Current Opinion in Clinical Nutrition and Metabolic Care* 13, 4: 371–376. doi: 10.1097/mco.0b013e32833aabef.

Lee, S., W. Huang, N. Peng, and C. Hu. 2017. "Dipyridamole-Induced Adverse Effects in Myocardial Perfusion Scans: Dynamic Evaluation." *IJC Heart and Vasculature* 14: 14–19. doi: 10.1016/j.ijcha.2016.11.002.

Lei, L., Y. Xue, Z. Liu, S. Peng, Y. He, Y. Zhang, and R. Fang. 2015. "Coumarin Derivatives from *Ainsliaea fragrans* and Their Anticoagulant Activity." *Scienctific Reports* 5: 1–9. doi: 10.1021/jm1013117.

Light, M.E., L.J. Mcgaw, T. Rabe, S.G. Sparg, M.B. Taylor, D.G. Erasmus, A.K. Jäger, and J. Van Staden. 2002. "Investigation of the Biological Activities of *Siphonochilus aethiopicus* and the Effect of Seasonal Senescence." *South African Journal of Botany* 68: 55–61. doi: 10.1016/S0254-6299(16)30455-0.

Lindsey, K.L., A.K. Jäger, A.M. Viljoen, and B. Van Wyk. 2002. "Cyclooxygenase Inhibitory Activity of Aloe Species." *South African Journal of Botany* 68: 47–50. doi: 10.1016/S0254-6299(16)30453-7.

Lipe, B., and D.L. Ornstein. 2011. "Deficiencies of Natural Anticoagulants, Protein C, Protein S, and Antithrombin." *Circulation* 124: 365–368. doi: 10.1161/CIRCULATIONAHA.111.044412.

Mackman, N. 2008. "Triggers, Targets and Treatments for Thrombosis." *Nature* 451, 7181: 914–918. doi: 10.1038/nature06797.

Mackman, N. 2012. "New Insights into the Mechanisms of Venous Thrombosis Find the Latest Version." *The Journal of Clinical Investigation* 122, 7: 2331–2336. doi: 10.1172%2FJCI60229.

Madari, H., D. Panda, L. Wilson, and R.S. Jacobs. 2003. "Dicoumarol : A Unique Microtubule Stabilizing Natural Product That Is Synergistic with Taxol." *Cancer Research* 63: 1214–1220.

Mahomoodally, M.F. 2013. "Traditional Medicines in Africa : An Appraisal of Ten Potent African Medicinal Plants." *Evidence-Based Complementary and Alternative Medicine*, 1–10. doi: 10.1155/2013/617459.

Manly, D.A., J. Boles, and N. Mackman. 2011. "Role of Tissue Factor in Venous Thrombosis." *Annual Review of Physiology* 73: 515–525. doi: 10.1146/annurev-physiol-042210-121137.

Manuel, J., R. De Medeiros, M. Macedo, J.P Contancia, C. Nguyen, G. Cunningham, and D.H. Miles. 2000. "Antithrombin Activity of Medicinal Plants of the Azores." *Journal of Ethnopharmacology* 72: 157–165. doi: 10.1016/s0378-8741(00)00226-9.

Maqbool, S., V. Rastogi, A. Seth, S. Singh, V. Kumar, and A. Mustaqueem. 2013. "Protein-C Deficiency Presenting as Pulmonary Embolism and Myocardial Infarction in the Same Patient." *Thrombosis Journal* 11, 19: 1–4. doi: 10.1186%2F1477-9560-11-19.

Masondo, N.A., J.F. Finnie, and J. van Staden. 2014. "Pharmacological Potential and Conservation Prospect of the Genus Eucomis (Hyacinthaceae) Endemic to Southern Africa." *Journal of Ethnopharmacology* 151: 44–53. doi: 10.1016/j.jep.2013.11.002.

Mehta, M., O.A. Branford, and K.J. Rolfe. 2016. "The Evidence for Natural Therapeutics as Potential Anti-Scarring Agents in Burn-Related Scarring." *Burns & Trauma* 4, 15: 1–12. doi: 10.1186/s41038-016-0040-1.

Merli, G.J., and J.B. Groce. 2010. "Pharmacological and Clinical Differences Between Low-Molecular-Weight Heparins: Implications for Prescribing Practice and Therapeutic Interchange." *Pharmacy and Therapeutics* 35, 2: 95–105.

Mncwangi, N., W. Chen, I. Vermaak, A.M. Viljoen, and N. Gericke. 2012. "Devil's Claw—A Review of the Ethnobotany, Phytochemistry and Biological Activity of *Harpagophytum Procumbens*." *Journal of Ethnopharmacology* 143: 755–771. doi: 10.1016/j.jep.2012.08.013.

Mnonopi, N., R.A. Levendal, M.T. Davies-Coleman, and C.L. Frost. 2011. "The Cardioprotective Effects of Marrubiin, a Diterpenoid Found in *Leonotis leonurus* Extracts." *Journal of Ethnopharmacology* 138: 67–75. doi: 10.1016/j.jep.2011.08.041.

Monroe, D.M., M. Hoffman, and H.R. Roberts. 2002. "Platelets and Thrombin Generation." *Arteriosclerosis, Thrombosis, and Vascular Biology* 22: 1381–1389. doi: 10.1161/01.ATV.0000031340.68494.34.

Patel, N.R., D.V. Patel, P.R. Murumkar, and M.R. Yadav. 2016. "European Journal of Medicinal Chemistry Contemporary Developments in the Discovery of Selective Factor Xa Inhibitors : A Review." *European Journal of Medicinal Chemistry* 121: 671–698. doi: 10.1016/j.ejmech.2016.05.039.

Peyvandi, F., I. Garagiola, and L. Baronciani. 2011. "Role of von Willebrand Factor in the Haemostasis." *Blood Transfusion* 9, 2: 3–8. doi: 10.2450/2011.002s.

Pidoux, H. 2005. "Harpagophytum." Obtained from Wikimedia under the Public domain license. https://commons.wikimedia.org/w/index.php?curid=3267827 (Accessed 5 June 2020).

Pirmohamed, M. 2006. "Warfarin: Almost 60 Years Old and Still Causing Problems" *British Journal of Clinical Pharmacology* 62, 5: 509–511. doi: 10.1111/j.1365-2125.2006.02806.x.

Previtali, E., P. Bucciarelli, S.M. Passamonti, and I. Martinelli. 2011. "Risk Factors for Venous and Arterial Thrombosis." *Blood Transfusion* 9, 2: 120–138. doi: 10.2450/2010.0066-10.

Prisco, D., G. Ciuti, and M. Falciani. 2005. "Hemostatic Changes in Normal Pregnancy." *Hematology Meeting Reports* 1, 10: 1–5. doi: 10.1016/j.thromres.2004.08.004.

Raskob, G.E., P. Angchaisuksiri, A.N. Blanco, H. Buller, A. Gallus, B.J. Hunt, E.M. Hylek, et al. 2014. "Thrombosis: A Major Contributor to Global Disease Burden." *Arteriosclerosis, Thrombosis, and Vascular Biology* 34, 11: 2363–2371. doi: 10.1111/jth.12698.

Rodrigues, C.F.B., H.H. Gaeta, M.N. Belchor, M.J.P. Ferreira, M.V.T. Pinho, D.D. Toyama, and M.H. Toyama. 2015. "Evaluation of Potential Thrombin Inhibitors from the White Mangrove (*Laguncularia racemosa* (L.) C.F. Gaertn.)." *Marine Drugs* 13: 4505–4519. doi: 10.3390/md13074505.

Rulkens, T. 2013. "Eucomis autumnalis in Its Natural Habitat, Close to Namaacha in Mozambique." Obtained from wikimedia under the Creative Commons Share-Alike 2.0. license. https://commons.wikimedia.org/w/index.php?curid=27832158 (Accessed 5 June 2020).

Rulkens, T. 2020. "In the Wild in Metuge District in Northern Mozambique, about 50 m.a.s. in Heavy Clay Soil. Early Rainy Season." Obtained from Wikimedia under the Creative Commons Share-Alike 2.0 license. https://commons.wikimedia.org/w/index.php?curid=42212756 (Accessed 5 June 2020).

Salehi, B., A.P. Mishra, M. Nigam, B. Sener, M. Kilic, M. Sharifi-Rad, P.V.T. Fokou, N. Martins, and J. Sharifi-Rad. 2018. "Resveratrol: A Double-Edged Sword in Health Benefits." *Biomedicines* 6, 91: 1–20. doi: 10.3390/biomedicines6030091.

Santhakumar, A.B. 2015. "Hematology & Blood Disorders." *Journal of Blood Disorders & Transfusion* 6, 5: 38–39. doi: 10.4172/2155-9864.C1.010%0A3rd.

Santo, A.D., A. Mezzetti, E. Napoleone, R.D. Tommaso, M.B. Donati, G. De Gaetano, and R. Lorenzet. 2003. "Resveratrol and Quercetin Down-Regulate Tissue Factor Expression by Human Stimulated Vascular Cells." *Journal of Thrombosis and Haemostasis* 1: 1089–1095. doi: 10.1046/j.1538-7836.2003.00217.x.

Schafer, Andrew I. 2007. "Hypercoagulable State." In *Cardiovascular Medicine*, 2423–2424. London: Springer.

Schecter, A.D., B. Spirn, M. Rossikhina, P.L.A. Giesen, V. Bogdanov, J.T. Fallon, E.A. Fisher, L.M. Schnapp, Y. Nemerson, and M.B. Taubman. 2000. "Release of Active Tissue Factor by Human Arterial Smooth Muscle Cells." *Circulation Research* 87: 126–132. doi: 10.1161/01.RES.87.2.126.

Schenone, M., B.C. Furie, and B. Furie. 2004. "The Blood Coagulation Cascade." *Current Opinion in Hematology* 11: 272–277. doi: 10.1097/01.moh.0000130308.37353.d4.

Scott, G., E.P. Springfield, and N. Coldrey. 2004. "A Pharmacognostical Study of 26 South African Plant Species Used as Traditional Medicines." *Pharmaceutical Biology* 42, 3: 186–213. doi: 10.1080/13880200490514032.

Shawka, A. 2010. "The Clean, Straight Trunk of a Young Stinkwood." Obtained from Wikimedia under the Public domain license. https://commons.wikimedia.org/w/index.php?curid=12339506 (Accessed 5 June 2020).

Smith, S.A., R.J. Travers, and J.H. Morrissey. 2015. "Initiation of Clotting Cascade." *Critical Reviews in Biochemistry and Molecular Biology* 50, 4: 326–336. doi: 10.3109/10409238.2015.1050550.

Sobotková, A., L. Mášová-Chrastinová, J. Suttnar, J. Štikarová, P Májek, Z. Reicheltová, R. Kotlín, J.W. Weisel, M. Malý, and E.D. Jan. 2009. "Antioxidants Change Platelet Responses to Various Stimulating Events." *Free Radical Biology and Medicine* 47, 12: 1707–1714. doi: 10.1016%2Fj.freeradbiomed.2009.09.015.

Sparg, S.G., J. Van Staden, and A.K. Jäger. 2002. "Pharmacological and Phytochemical Screening of Two Hyacinthaceae Species: *Scilla natalensis* and *Ledebouria ovatifolia*." *Journal of Ethnopharmacology* 80: 95–101. doi: 10.1016/S0378-8741(02)00007-7.

Stafford, G.I., A.K. Jäger, and J. Van Staden. 2005. "Effect of Storage on the Chemical Composition and Biological Activity of Several Popular South African Medicinal Plants." *Journal of Ethnopharmacology* 97: 107–115. doi: 10.1016/j.jep.2004.10.021.

Stangl, P.A., and S. Lewis. 2010. "Review of Currently Available GP IIb/IIIa Inhibitors and Their Role in Peripheral Vascular Interventions." *Seminars in Interventional Radiology* 27, 4: 412–421. doi: 10.1055%2Fs-0030–1267856.

Street, R.A., and G. Prinsloo. 2013. "Commercially Important Medicinal Plants of South Africa: A Review." *Journal of Chemistry*, 1–16. doi: 10.1155/2013/205048.

Süleyman, H., B. Demircan, and Y. Karagöz. 2007. "Anti-Inflammatory and Side Effects of Cyclooxygenase Inhibitors." *Pharmacological Reports* 59: 247–258.

Taylor, J.L.S., and J. van Staden. 2002. "COX-1 and COX-2 Inhibitory Activity in Extracts Prepared from Eucomis Species, with Further Reference to Extracts from *E. autumnalis autumnalis*." *South African Journal of Botany* 68: 80–85. doi: 10.1016/S0254-6299(16)30460-4.

Temkitthawon, P., J. Viyoch, N. Limpeanchob, and W. Pongamornkul. 2008. "Screening for Phosphodiesterase Inhibitory Activity of Thai Medicinal Plants." *Journal of Ethnopharmacology* 119: 214–217. doi: 10.1016/j.jep.2008.05.034.

Terburg, D., S. Syal, L.A. Rosenberger, S. Heany, N. Phillips, N. Gericke, D.J. Stein, and J. Van Honk. 2013. "Acute Effects of *Sceletium tortuosum* (Zembrin), a Dual 5-HT Reuptake and PDE4 Inhibitor, in the Human Amygdala and Its Connection to the Hypothalamus." *Neuropsychopharmacology* 38: 2708–2716. doi: 10.1038/npp.2013.183.

Thomas, R.H. 2001. "Hypercoagulability Syndromes." *Archives of Internal Medicie* 161: 2433–2439. doi: 10.1001/archinte.161.20.2433.

Toschi, V., and M. Lettino. 2011. "Inhibitors of Propagation of Coagulation : Factors V and X." *British Journal of Clinical Pharmacology* 72, 4: 563–580. doi: 10.1111/j.1365-2125.2011.04001.x.

Van Cott, E.M., B. Khor, and J.L. Zehnder. 2016. "Factor V Leiden." *American Journal of Hematology* 91, 1: 46–49. doi: 10.1002/ajh.24222.

Van Wyk, B.E., B. van Oudsthoorn, and N Gericke. 1997. *Meidicinal Plants of South Africa*. Pretoria: Briza Publications.

Venugopala, K.N., V. Rashmi, and B. Odhav. 2013. "Review on Natural Coumarin Lead Compounds for Their Pharmacological Activity." *BioMed Research International*, 1–14. doi: 10.1155/2013/963248.

World Health Organisation. 2017. "Cardiovascular Diseases (CVDs)." www.who.int/news-room/fact-sheets/detail/cardiovascular-diseases-(cvds) (Accessed 5 June 2020).

Zhang, R., C. Liu, X. Liu, and Y. Guo. 2016. "Protective Effect of *Spatholobus suberectus* on Brain Tissues in Cerebral Ischemia." *American Journal of Translational Research* 8, 9: 3963–3969.

Zschocke, S., S.E. Drewes, K. Paulus, R. Bauer, and J. Van Staden. 2000. "Analytical and Pharmacological Investigation of *Ocotea bullata* (Black Stinkwood) Bark and Leaves." *Journal of Ethnopharmacology* 71, 1–2: 219–230. doi: 10.1016/S0378-8741(00)00159-8.

18 A Pharmacological Appraisal of Antimalarial Plant Species

Mahwahwatse J. Bapela, Precious B. Ramontja and Mcebisi J. Mabuza

CONTENTS

18.1 INTRODUCTION

Malaria remains the leading protozoan infection that is killing millions of people in low-income tropical and subtropical countries (Ashley et al., 2018; Sachs and Malaney, 2002). Its recurrent resistance has devastating global health implications in the endemic countries. According to the World Health Organization (WHO), malaria is transmitted in 108 countries, causing an estimated 216 million clinical cases and 655,000 deaths annually (WHO, 2019). The severity of mortality and transmissions vary in malaria-endemic areas. The African region alone accounts for most cases (93%), followed by Southeast Asia (3.4%) and the Eastern Mediterranean (2.1%) (WHO, 2019). Historical fruitful developments in the fight for malaria eradication have been well documented, with a notable decline in death rates. The estimated number of global malaria deaths was 839,000 in 2010 and has decreased to 405,000 in 2018 (Dhiman, 2019; WHO, 2019). However, the malaria-related child mortality rate is increasing and was estimated to be 67% in 2018, with 272,000 deaths (WHO, 2019). The battle to control and eradicate malaria infection has been hampered by many factors, including plasmodial resistance to the conventional antimalarial drugs, the emergence of vector resistance and the lack of an effective vaccine offering a high level of protection (Menard and Dondorp, 2017; Parija and Praharaj, 2011).

Antimalarial drugs have always been part of the indispensable tools used at all phases of malaria elimination along the path toward eradication. Chemotherapy and chemoprevention of malaria are dependent on a few and less chemically diverse therapeutic agents, which present some limitations. The variable efficacy of the drugs between strains or species, their toxicity and the required treatment regimens are undermining the control approach to the disease (Witkowski et al., 2010). The main challenge in malaria chemotherapy is to discover safe and selective therapeutic agents with a potency that will not be compromised by plasmodial resistance and with more convenient dosing regimens (White et al., 2014).

DOI: 10.1201/9781003108375-18

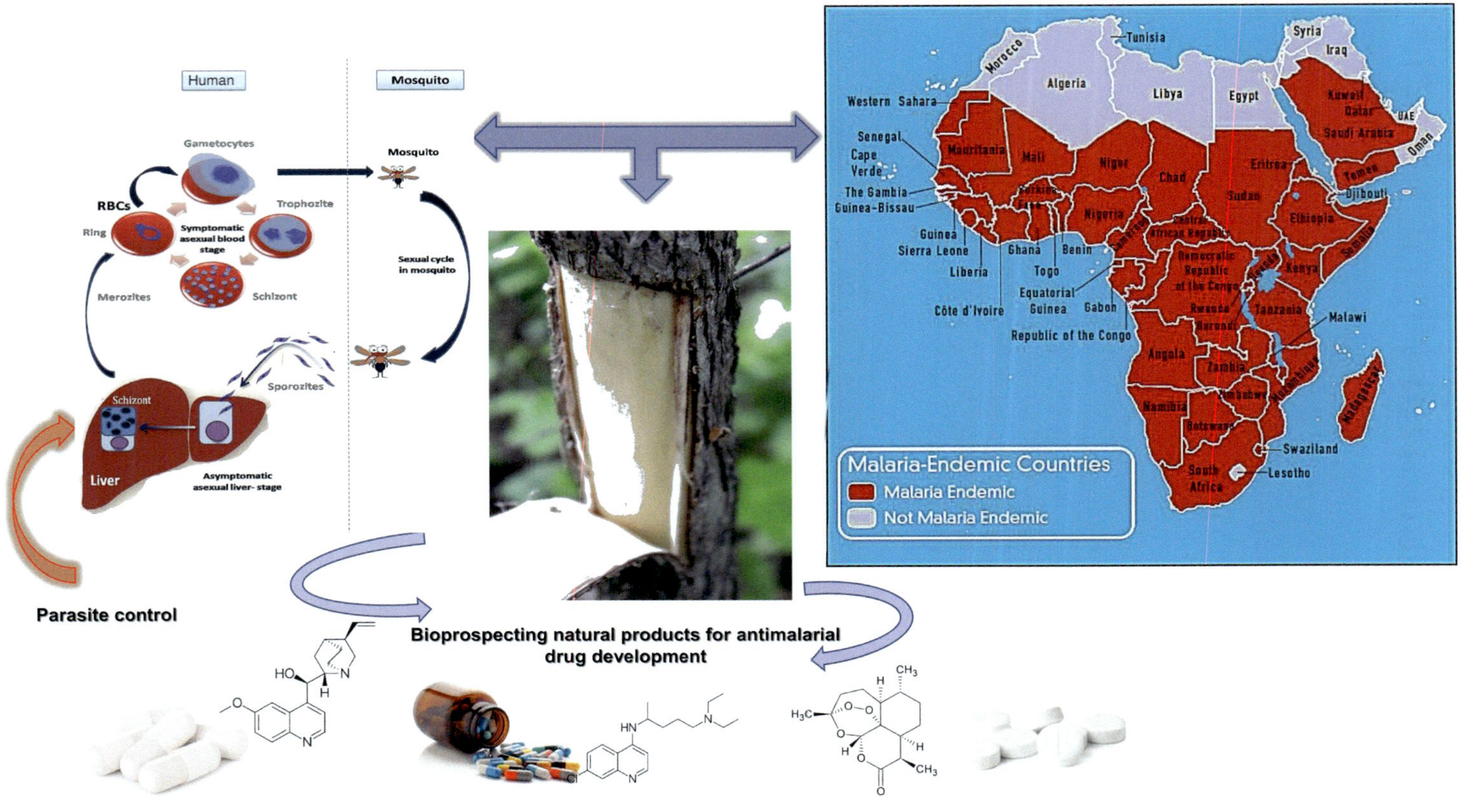

Graphical abstract representing the content of the chapter (Davis, 2011; Rees, 2021; White et al., 2014; Williams et al., 2007).

Plants have been the basis of the current antimalarial drugs and continue to be the promising source of novel molecular templates for malaria therapy. Screening plants based on ethnopharmacological data seems to increase the likelihood of finding novel compounds due to their long history of safe use, and it may therefore increase the prospects of finding new antiplasmodial plant leads with better pharmacokinetic profiles (Ntie-Kang et al., 2014; Zimmermann et al., 2012). This is also based on the immense anecdotal evidence of their efficacy by the indigenous people.

18.2 THE LIFE CYCLE OF PLASMODIUM PARASITES

Malaria is a vector-borne disease caused by parasitic protozoans of the genus *Plasmodium* and transmitted to human hosts by infected female *Anopheles* mosquitoes. Only five plasmodial protists—*P. falciparum*, *P. vivax*, *P. ovale*, *P. malariae* and *P. knowlesi*—are infectious to humans (Rondón et al., 2019). These species differ in geographical distribution, microscopic morphology, clinical presentation and susceptibility to antimalarial drugs. Of all these infectious human parasites, *P. falciparum*—which predominates in Africa—is the most virulent and causes the vast majority of deaths (Diakité et al., 2019). Infection by *P. falciparum* usually results in an uncomplicated disease; however, in some cases, the disease becomes severe and may lead to death. *Plasmodium vivax* is the most geographically widespread human malaria parasite and is common in tropical areas outside Africa. Although *P. vivax* malaria is considered to be benign, with a very low case-to-fatality ratio, it may effect morbidity, severe disease and death as a result of splenomegaly (Dayananda et al., 2018).

Plasmodium ovale is a protozoal parasite that causes tertian malaria in humans and comprises a relatively limited distribution range, which is restricted to West Africa, the Philippines and eastern Indonesia (Sitali et al., 2019; Zhou et al., 2019). The parasitic *P. malariae* species is characterized by low global prevalence and milder clinical manifestations, and is widespread throughout sub-Saharan Africa, Southeast Asia, Indonesia, the Western Pacific and in areas of the Amazon Basin of South America (Rutledge et al., 2017). The fifth species, *P. knowlesi*, is a primate malaria parasite that is endemic in some Southeast Asian areas and can cause severe malaria in humans (Amir et al., 2018). These five *Plasmodium* species are transmitted by *Anopheles* mosquitoes with diverse breeding and feeding habits, resulting in different disease spectra in various population groups (Berenger and Parola, 2017). Humans often harbor multiple *Plasmodium* species, but the clinical implications of mixed infections remain unclear.

The *Plasmodium* parasite exhibits a complex, invasive multistage life cycle occurring within the vector mosquitoes and the vertebrate hosts. Infection of the human host with a malaria parasite begins when an infected *Anopheles* mosquito inoculates microscopic motile sporozoites during a blood meal (Yap et al., 2019). Injected sporozoites are carried by the circulatory system to the liver, where they invade hepatocytes and then undergo exoerythrocytic schizogony (Figure 18.1). The asymptomatic liver stage of infection yields merozoites that are liberated into the bloodstream (Jaskiewicz et al., 2019). Hypnozoites are parasites of *P. vivax* and *P. ovale* that go through a latent period instead of immediate asexual multiplication. On the contrary, *P malariae* does not form hypnozoites, but it can persist for decades as an asymptomatic blood-stage infection. Unlike *P. vivax*, which binds to the Duffy antigen to invade reticulocytes, *P. falciparum* uses multiple pathways to infect erythrocytes, further complicating the effort to develop anti-invasion vaccines (Mikolajczak et al., 2015).

Merozoites invade erythrocytes, enlarge and develop into trophozoites. The early trophozoite stage is referred to as the ring form, whereas the late form is known as the schizont. Merozoites bud from schizonts and initiate another round of the blood-stage replicative cycle. The blood stages of infection include asexual forms of the parasite that undergo repeated cycles of mitotic multiplication. They also include male and female sexual forms, called micro- and macro-gametocytes, respectively, which await ingestion by mosquitoes before developing further (Marin-Mogollon et al., 2018; Yeoh et al., 2017).

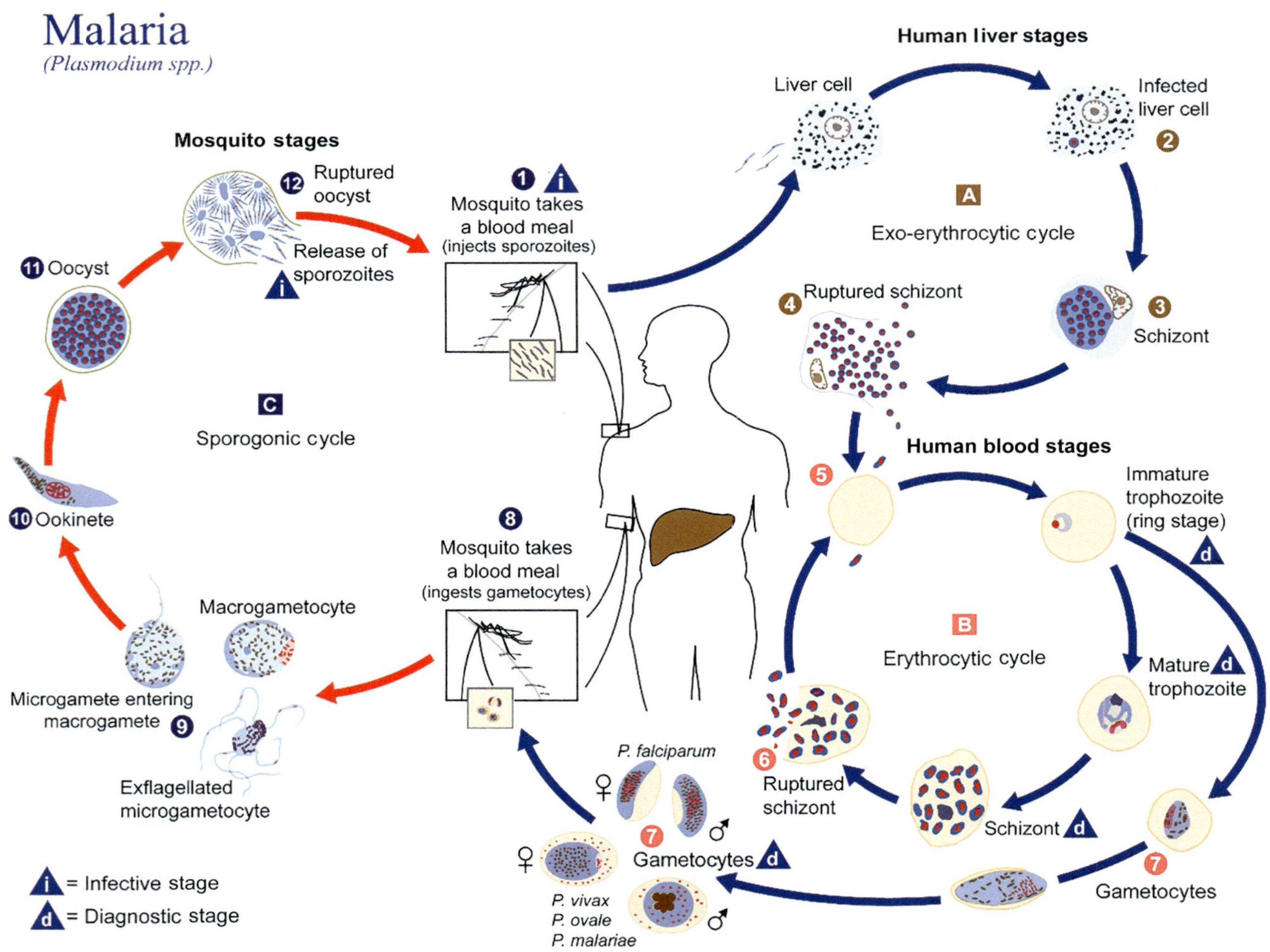

FIGURE 18.1 The life cycle of *Plasmodium falciparum* (da Silva and Moser, 2002).

The asexual stage is responsible for the pathogenesis of the disease. Intermittent fever results from the ingestion of host cytoplasm, proteolysis of hemoglobin, lysis of infected erythrocytes and a suite of host reactions. *Plasmodium falciparum* often exhibits continuous fever, higher levels of parasitemia and sequestration of infected erythrocytes in tissues, which explain its increased virulence, compared to the other *Plasmodium* species (Abdi et al., 2016; Bachmann et al., 2019). Sexual stage parasites are non-pathogenic but transmissible to the *Anopheles* vector. Ingestion of the gametocytes by the mosquito vector induces gametogenesis in the mid-gut of the mosquito. Microgametes fertilize macrogametes and form zygotes, which develop into ookinetes that penetrate the intestinal walls of the mid-gut and develop into oocysts. Asexual replication occurs within the oocyst and produces sporozoites. The sporozoites migrate to invade the salivary glands of the mosquitoes, thereby completing the life cycle (Yeoh et al., 2017).

18.2.1 Potential Targets for Chemotherapeutic Interventions

New targets for malaria chemotherapy keep changing to surpass the continuous parasite resistance. The parasites have proven to be resistant against the sulfa compounds and 4-aminoquinolines (Baldwin et al., 2007; White, 1999). Currently, artemisinin-based combination therapies (ACTs) are considered effective for uncomplicated falciparum malaria (Lunev et al., 2016; Mavoko et al., 2017). Some potential chemotherapy targets include the inhibition of unique processes found in either the host or parasite. Examples of such processes include phospholipid metabolism, folate metabolism and shikimate pathway (Kumar et al., 2018; Tse et al., 2019). The shikimate pathway does not exist in mammals and therefore possesses multiple targets for the growth of new antiparasitic agents, which can interfere with biosynthesis of amino acids by malaria parasites (Kumar et al., 2018; Mcconkey, 1999; Rosenthal, 1998).

The primary source of energy for *P. falciparum* is the continuous availability of glucose, making the glycolytic pathway an excellent target (Kumar et al., 2018; Van Niekerk et al., 2016). This parasite cannot salvage pyrimidines, and the inhibition of dihydroorotate and dehydrogenase will result in the lack of crucial metabolites necessary for DNA and RNA biosynthesis (Kumar et al., 2018; Tse et al., 2019). With *P. falciparum*, the proteins are exported quicker compared to other malaria-carrying parasites (Kumar et al., 2018; Lindner et al., 2013). Consequently, the apicoplast can also be the target, since it has the ability to traffic about 500 proteins (Tse et al., 2019; Wilson et al., 1996).

Each of the developmental stages represents a potential target at which the life cycle can be interrupted with vaccines, antivectors and chemotherapeutic agents. Proteases are potential malaria chemotherapy targets (Alam, 2014; Rosenthal, 1998). They are necessary for the rapturing of erythrocytes, as well as hemoglobin degradation (Yadav et al., 2019). Malaria protease appears to play an active part in breaching merozoites to erythrocyte cytoskeletons (Blackman, 2008; Rosenthal and Taylor, 2003).

Nevertheless, the complex life cycles of the five *Plasmodium* species infecting humans exhibit diverse challenges. All these stages are characterized by unique shapes, distinct structures and many protein complements. *Plasmodium* parasites have approximately 16 chromosomes, yet their specialized proteins help the parasites to invade, multiply and grow within multiple cell types of their hosts (Patgiri et al., 2019). Their surface proteins and metabolic pathways keep changing during these different stages, enabling the parasites to evade the immune clearance while also creating problems for the development of drugs and vaccines (Belachew, 2018).

Plasmodium vivax, *P. ovale* and *P. malariae* can persist for years without causing symptoms and at low levels that cannot be detected by standard diagnostic methods. They can, however, still propagate the transmission of malaria. Apart from its persistence, *P. vivax* is less responsive to control interventions than *P. falciparum* infections because of its wider ecological range and the ability to produce infectious gametocytes soon after it emerges from the liver (Ome-Kaius et al., 2019). In contrast to the other *Plasmodium* species, the appearance of gametocytes in peripheral circulation is delayed in *P. falciparum*, making recrudescence common in clinically cured individuals (Kagaya et al., 2019). Most of these malaria parasites attach and enter host cells via several pathways, and their species-specific characteristics present varied complications to chemotherapy.

18.3 CONVENTIONAL DRUGS FOR THE TREATMENT OF MALARIA

Antimalarial drugs are used to prevent the infection caused by *Plasmodium* species, treat the disease, eliminate latent parasites and prevent the transmission of malaria. The major purpose of antimalarial treatment in severe malaria is to prevent death, whereas in uncomplicated malaria it is to cure and to avoid the development of severe disease (Conroy et al., 2019). While insecticides are aptly credited for much of the success in reducing the burden of malaria, treatment and prevention with drugs have always been the essential components of all successful malaria elimination programs. Similarly, while many diseases have been eradicated primarily through the use of vaccines, it is very unlikely that malaria could be eradicated with a vaccine alone without the integration of antivector methods and drugs (Molyneux, 2020; The malERA Group, 2017).

Antimalarial drugs possess selective actions at different stages of the parasite life cycle. The ideal target profile for malaria eradication is a co-formulated drug combination that can be administered in a single dose and result in a radical cure of all life cycle stages of the five *Plasmodium* species infecting humans (Cohee and Laufer, 2018). However, the desirable drug might have limitations such as targeting only one or two parasite species, the priorities being *P. falciparum* and *P. vivax*. Given the differences in the life cycle stages of the five *Plasmodium* species, it is reasonable to assume that drugs which target *P. falciparum* will be generally effective against *P. malariae*, and those targeting *P. vivax* may be efficacious against *P. ovale* and *P. knowlesi* (The malERA Group, 2017).

A limited number of antimalarials are currently in clinical use, and they fall into seven different pharmacological classes. They include 4-aminoquinolines, aryl-amino alcohols, 8-aminoquinolines, antifolates, hydroxynaphthoquinones, antibiotics and artemisinins (Nqoro et al., 2017; Tse et al., 2019). Classification of these antimalarial drugs into groups is based on the differences in their pharmacophores, mechanisms of action and selective actions on the developmental stages of the parasites. 4-Aminoquinolines, chloroquine and its derivative amodiaquine are quinoline-based molecules containing a primary amine group in the 4-position (Figure 18.2) (Elderfield et al., 1946). They accumulate in vacuoles of susceptible parasites, thereby interfering with the parasite's detoxification pathway (Witkowski et al., 2010).

Aryl-amino alcohols include quinine, mefloquine, halofantrine and lumefantrine, and their anti-malarial activity seem to require the presence of an aromatic portion and an amino alcohol portion (Figure 18.2). Like the 4-aminoquinolines, aryl-amino alcohols are assumed to act primarily on the erythrocyte stage of the malaria parasite by inhibiting the formation of hemozoin (Anderson et al., 2006). 8-Aminoquinolines are derived quinoline molecules with an amine group at the 8-position of quinoline. Primaquine is the only 8-aminoquinoline used in malaria therapy, and studies have shown that it interferes with the parasite's DNA structure and disrupts its mitochondrial membranes (Miller et al., 2013). It is also the only available drug that can prevent transmission of mature gametocytes, although it can cause intravascular hemolysis in patients with glucose-6-phosphate dehydrogenase deficiency (Sinha et al., 2014). The precise mechanisms of action of the above-mentioned quinoline-based antimalarial drugs are not yet fully understood, and are still under investigation.

Antifolates include two classes of antimalarial drugs that interfere with folate metabolism, a pathway essential for DNA replication and protein synthesis of malaria parasites. They specifically target two enzymes involved in the biosynthesis of tetrahydrofolate, the dihydropteroate synthase (DHPS) and dihydrofolate reductase (DHFR) (Rudrapal, 2011). Sulphonamides, sulphadoxine (Figure 18.3), sulphones, dapsone and sulphalene act as competitive inhibitors against DHPS, thereby preventing the formation of dihydropteroate. The second class includes proguanil, chloroproguanil, cycloguanil and pyrimethamine, which inhibit DHFR, consequently preventing the reduction of dihydrofolate (DHF) to tetrahydrofolic acid (THF). The combination of both classes of

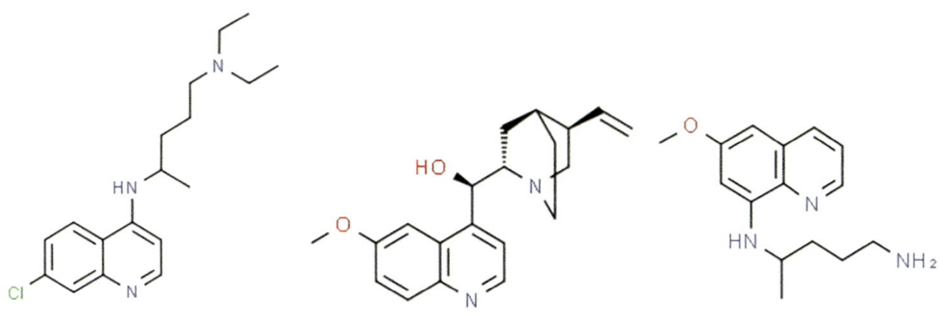

Chloroquine Quinine Primaquine

FIGURE 18.2 Chemical structures of a few quinolone based antimalarial drugs.

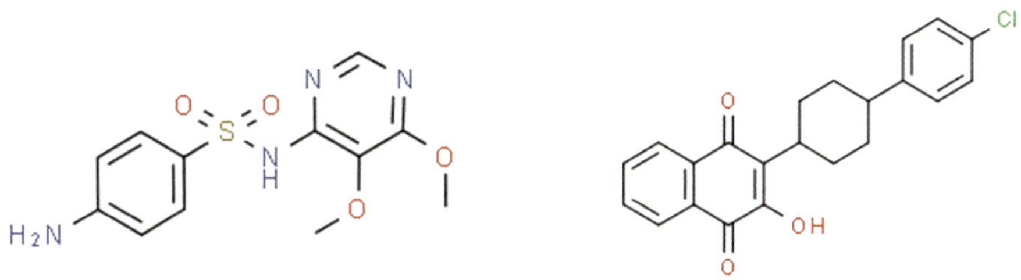

Sulphadoxine Atovaquone

FIGURE 18.3 Structures of sulphadoxine and atovaquone.

antifolates proved to have a synergistic effect against malaria, though clinical resistance developed quickly, compromising their use in many malaria-endemic areas (Heinberg and Kirkman, 2015). Atovaquone (Figure 18.3) is a hydroxynaphthoquinone that is used in combination with proguanil for prophylaxis and therapy of asymptomatic malaria. It is active against all *Plasmodium* species, inhibiting the exoerythrocytic stage in humans and the oocyst development in the mosquitoes. It acts by inhibiting the mitochondrial electron transport chain, thereby leading to a breakdown of the mitochondrial membrane potential (Barton et al., 2010). The prevalence of mutations of the cytochrome b gene has resulted in therapy failure of atovaquone (Siregar et al., 2015).

Certain antibiotics, such as doxycycline, clindamycin, tetracycline and chloramphenicol, play a major role in the treatment and prophylaxis of malaria. Antibiotics display an antimalarial effect by inhibiting mitochondrial protein synthesis and blocking the biosynthetic pathway of nucleic acids in *Plasmodium* species (Rudrapal, 2011). However, monotherapy of antibiotics for malaria treatment is ineffective, with fever and parasite clearance taking longer than with classical antimalarials. Artemisinin is the active constituent of *Artemisia annua* L., and is a sesquiterpene lactone consisting of three isoprene units bound to cyclic organic peroxide ester (Figure 18.4) (Maude et al., 2010). While the endoperoxide bond is now known to be necessary for antimalarial activity, the exact mechanism of action is still a topic of debate (O'Neill et al., 2010; Paddon and Keasling, 2014). The relatively poor pharmacokinetics of artemisinin had resulted in its semi-synthetic and synthetic structural variations of clinical use. They include dihydroartemisinin, artemether, arteether, artesunate and artelinic acid. Artemisinins display rapid parasite clearance, are active against all species of malaria that invade humans, are potent blood schizontocide and inhibit the early gametocytic stage of the parasite (Yahiya et al., 2019). The advantage of their short half-lives is that selection for

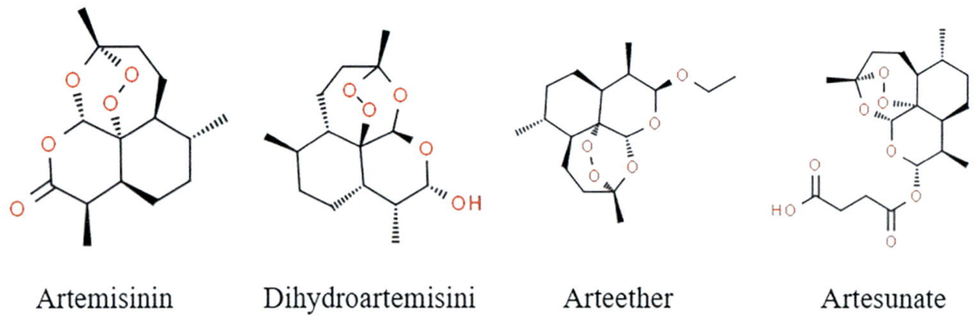

FIGURE 18.4 Artemisinin and its derivatives.

drug-resistant parasites is reduced; nevertheless, there is a higher associated risk of recrudescence when these drugs are administered as monotherapeutic regimens. Recrudescence is an indication of treatment failure, and has been thought to arise in response to the selection of drug-resistant parasites (Narula et al., 2019).

In an attempt to reduce the occurrence of recrudescence and therefore the development of resistance, WHO advocates the administration of ACTs, which provide symptomatic relief, prevent severe disease and reduce mortality rates (WHO, 2018). The ACTs combine a derivative of artemisinin with a longer-lasting partner drug that continues to reduce the parasite biomass after the short-lived artemisinin has dropped below therapeutic levels (Phyo and Von Seidlein, 2017).

Recommended ACTs currently under clinical use include artemether-lumefantrine (AL), artesunate-mefloquine (ASMQ), artesunate-amodiaquine (ASAQ), artesunate-sulfadoxine-pyrimethamine (AS-SP) and dihydroartemisinin-piperaquine (DHA-PQ) (Banek et al., 2014; Pousibet-Puerto et al., 2016). Reports of delayed responses to some ACTs, as well as supporting *in vitro* drug-susceptibility data, suggest the possibility of clinical artemisinin resistance along the Thai–Cambodian border (Dondorp et al., 2017). Significant failure rates of these combinations have also been observed in African countries where resistance to one drug has been previously encountered. Since artemisinins have very short half-lives, this loss of potency will render the partner drugs more vulnerable to resistance. These findings present a significant challenge to efforts aimed at preventing, controlling and eliminating malaria in recent years.

18.3.1 Drug Resistance by *Plasmodium* Parasites

Antimalarial drug resistance is the persistence or recurrence of *Plasmodium* parasites after appropriate drug treatment. Most of the chromosomal mutations that confer antimalarial drug resistance occur in nature, are spontaneous, were found to be independent of the kind of drug used and are not necessarily a result of treatment failure (Koenderink et al., 2010). These chromosomal mutations are selected by different concentrations of antimalarial drugs that are sufficient to reduce susceptible parasites but not adequate to inhibit the multiplication of mutants (Adderley et al., 2020). Presently, antimalarial drug resistance has been documented in *P. falciparum*, *P. vivax* and *P. malariae* strains. Resistance to antimalarials presents several survival advantages to the malaria parasite. They include recrudescence, the proliferation of resistant parasites and enhancement of the transmission of gametocytes harboring the resistant genes (Abdul-Ghani and Beier, 2014).

Recrudescent infections as a result of resistant strains of the parasite tend to have higher gametocyte carriage rates, with higher infectivity, than primary infections. Gametocytes are therefore an important mechanism for resistance and drugs that could block transmission by eliminating the gametocytes are much needed to counteract resistance (Richter et al., 2016). Antimalarial drug resistance is complicated by cross-resistance, which mostly occurs among groups of drugs that belong to a similar chemical family or which have the same mode of action. Furthermore, multiple drug resistance of *P. falciparum* has been observed when the parasite is resistant to more than two operating antimalarial compounds of different chemical classes and mechanisms of action (Sinha et al., 2014). The prevalence of resistance to the well-established antimalarial drugs has resulted in the expansion of antimalarial drug discovery efforts.

There is growing global collaboration between academic institutions, non-profit organizations and pharmaceutical companies in an attempt to develop new antimalarials, including compounds with novel properties (Shanks and Möhrle, 2017). Several new antimalarial agents are currently in advanced stages of clinical studies. They include mainly those compounds that were resurrected from previous antimalarial drug discovery programs or structural analogs of existing antimalarials. Some of these compounds are developed based on existing chemical scaffolds but with improved pharmacological properties. The advantage of this approach is that the safety and pharmacokinetic profiles of the modified compounds are well established; however, cross-resistance remains a threat. Additionally, while the global pipeline of new medicines for the control and elimination of malaria is broadening, there are still several gaps in the current portfolio (Anthony et al., 2012). Simpler single-dose regimens that can prevent transmission inhibit relapse and possess better safety profiles than current medicines are still needed.

18.4 ANTIMALARIAL AGENTS FROM ETHNOMEDICINAL PLANT SPECIES

Higher plants have for centuries been used to cater to humanity's healthcare needs (James et al., 2018). There exists a large cannon of human interaction with the botanicals with regards to health and wellness. Many different cultural groups around the globe are notable for their rich indigenous knowledge systems (IKSs) which form the backbone of their primary healthcare systems (Pan et al., 2018). Modern scientific research on medicinal plants owes its continued existence to many IKSs, wherefrom it draws inspiration for further development of chemotherapeutic agents. Some of the best prescribed drugs used to treat malaria were derived from plant species (Uzor, 2020). Screening medicinal plants based on ethnopharmacological data seems to increase the likelihood of finding novel compounds, due to their long history of safe use, and may therefore increase the prospects of finding new antiplasmodial plant leads with better pharmacokinetic profiles. This is also based on the immense anecdotal evidence of their efficacy by the indigenous people, particularly when the same herbal preparation is used by different ethnic groups with the same results (Cock et al., 2019; Pillay et al., 2008).

In recent years, many studies have reported on the antiplasmodial effectiveness of extracts and isolated compounds from plant species where their selection was based on an ethnopharmacological approach (Zimmermann et al., 2012; Ntie-Kang et. al., 2014). Most of these data together tend to justify the use of an ethnopharmacological approach when bioprospecting for antimalarial plant products. Quinine, which was the first antimalarial drug, was discovered early in 1820 from the stem bark of a *Cinchona* tree (Achan et al., 2011).

Quinine and its derivatives are still considered to be effective antimalarial drugs, even though in most cases, issues of parasite resistance have been reported to be a major problem. Another important antimalarial drug derived from a plant species is artemisinin (Haldar et al., 2018). Artemisinin and its derivatives are currently used as first-line drug regimens for complicated malaria cases (Oboh et al., 2018). Artemisinin was derived from *Artemisia annua* L., and has since contributed largely to the reduction of the malaria burden in many parts of the world (Tilley et al., 2016). Since then further studies and research in medicinal plants have been ongoing, to discover new antimalarial drugs with different chemical structures and mechanisms of action. Both crude plant extracts and isolated pure compounds have been studied and are continuously being investigated for the possible discovery of novel antimalarial drugs that will be effective.

18.4.1 Crude Plant Extracts with Antimalarial Activity

Ethnobotanical surveys have confirmed the presence of a plethora of medicinal plants that are traditionally used by indigenous people residing in malaria-endemic areas (Adebayo and Krettli, 2011; Cock et al., 2019; Dike et al., 2012; Odugbemi et al., 2007; Ojeh and Aworinde, 2016). The pharmacological significance of some of these plant species has not been validated for their antiplasmodial properties. Both polar and non-polar crude extracts from several indigenous medicinal plants have been investigated for their antimalarial activities, and some positive results have been reported. These ethnobotanical surveys and the subsequent *in vitro* and *in vivo* antiplasmodial studies conducted on the plants have opened up huge possibilities for further exploration of antimalarial plant species as sources for new drug discovery.

Plants have vast arrays of secondary metabolites with different structures, modes of action and pharmacological responses. Further phytochemical studies on these plants obtained from different IKS will ensure a constant supply of new effective drug interventions. Selected crude plant extracts that displayed significant antiplasmodial activity (IC_{50} < 1.5 µg/mL) will be discussed. Additionally, the selectivity index (SI), which determines the therapeutic potential of the crude plant extracts, will be highlighted.

Kenya is among the 15 most malaria-burdened countries in sub-Saharan Africa, with 70% of its population living in high-risk areas (Bashir et al., 2019). Malaria was the second leading cause of death in 2016, and this was continuously exacerbated by socioeconomic and environmental factors (Gopal et al., 2019). Kenya's tropical climate provides suitable conditions for the breeding of *Anopheles* vector and the proliferation of *Plasmodium* species. A large number of malaria cases in Kenya are caused by *P. falciparum* parasite, which is transmitted by *A. gambiae* (Niu et al., 2017). The shortage of proper health facilities and other socioeconomic factors further complicate and worsen the malaria burden in Kenya (Bashir et al., 2019). Kenyan people have a rich IKS on the different medicinal plant species that have been reported and vetted by indigenous people to have antiplasmodial activity. These plants have a long-standing history of being used in different parts of the country, and most have been validated for their ethnopharmacological relevance against malaria (Kimondo et al., 2015; Kiringe, 2006; Makule et al., 2014; Muregi et al., 2003; Muthaura et al., 2011, 2007a, 2007b; Nguta et al., 2010).

Boscia angustifolia A. Rich., *Schkuhria pinnata* (Lam.) Kuntze ex Thell., *Fuerstia africana* T.C.E. Fries, *Maytenus arbutifolia* (A. Rich.) Wilczek, *Warburgia ugandensis* Warburgia salutaris (G. Bertol.) Chiov. and *Vernonia lasiopus* O.Hoffm., are some of the common plant species that are traditionally used by Kenyan locals to remedy malaria and its symptoms (Table 18.1). The water extract of *B. angustifolia* stem bark showed high selectivity (SI = 4732) to *P. falciparum* (D6) strain and displayed an IC_{50} value of 1.43 µg/mL and cytotoxicity value of 304.92 µg/mL on Vero E6 cells. *Schkuhria pinnata* methanol (MeOH) extract demonstrated high selectivity (SI = 124) for *P. falciparum* (D6) strain and displayed an IC_{50} value of 1.30 µg/mL and a cytotoxicity value of 161.5 µg/mL on Vero E6 cells. The MeOH extract (whole plant) of *F. africana* showed a high selectivity (SI = 974) for the *P. falciparum* (D6) strain and displayed an IC_{50} value of 0.98 µg/mL and cytotoxicity of 124.2 µg/mL on Vero E6 cells (Muthaura et al., 2007a).

The leaf water extract of *Maytenus arbutifolia* (L.) displayed a significantly high parasite selectivity (SI = 3838) for *P. falciparum* (D6) strain. It showed an IC_{50} value of 0.95 µg/mL and a cytotoxicity value of 3645.7 µg/mL when tested on Vero E6 cells (Muthaura et al., 2007b). *Warburgia ugandensis* from the Canellaceae family was reported by Muthaura et al. (2011) to have significant antiplasmodial activity. Its MeOH stem bark extract displayed an IC_{50} value of 1.4 µg/mL on the KI strain of *P. falciparum*, with a cytotoxicity value of 0.34 µg/mL on L6 cells. *Warburgia ugandensis* showed a very low parasite selectivity (SI = 0.24), which does not fall within the acceptable range of potential drug hits (Muthaura et al., 2011). However, there are many different derivatization procedures, techniques and approaches that can be applied to chemically manipulate the plant extract to be clinically safe and more selective.

TABLE 18.1
The 50% Inhibitory Concentration (IC_{50}) Values (µg/mL) of Plant Extracts That Exhibited High Antiplasmodial Activity and Their Antiproliferative Activity

Species	Solvents	Antiplasmodial Activity (µG/ML)	Cytoxicity (µG/ML)	References
Alchornea cordifolia	EtOH[1]	0.08	6.2	(Adebayo and Krettli, 2011)
Azadirachta indica	EtOH	0.02	2.3	(Adebayo and Krettli, 2011)
Boscia angustifolia	H_2O[2]	1.43	304.92	(Muthaura et al., 2007a)
Combretum erythrophyllum	EtOH	0.87	925.72	(Van Dyk et al., 2009)
	DCM[3]	0.70	70.51	(Van Dyk et al., 2009)
Crinum bulbispermum	EtOAc[4]	0.02	9.60	(Van Dyk et al., 2009)
Drypetes gerrardii	DCM:MeOH[5] (1:1)	0.50	50.80	(Mokoka et al., 2013)
Fuerstia africana	MeOH[6]	0.98	1242	(Muthaura et al., 2007a)
Guiera senegalensis	H_2O	0.79	>25	(Adebayo and Krettli, 2011)
Maytenus arbutifolia	H_2O	0.95	3645.7	(Muthaura et al., 2007b)
Nauclea latifolia	H_2O	0.6	400	(Adebayo and Krettli, 2011)
Pavetta gardeniifolia	DCM	1.30	153.43	(Van Dyk et al., 2009)
Schkuhria pinnata	MeOH	1.30	161.5	(Muthaura et al., 2007a)
Tabernaemontana	DCM:50% MeOH (1:1)	0.331/0.834	4.68/38.2	(Bapela et al., 2014)
Vernonia lasiopus	$CHCl_3$[7]	1.2	ND [8]	(Muregi et al., 2003)
	EtOAc	1.0	ND	(Muregi et al., 2003)
Warburgia ugandensis	MeOH	1.4	0.34	(Muthaura et al., 2011)

Notes: **1** Ethanol; **2** Water; **3** Dichloromethane; **4** Ethyl acetate; **5** Dichloromethane: Methanol; **6** Methanol; **7** Chloroform; 8 Not detected

Vernonia lasiopus (Asteraceae) displayed high antiplasmodial activity when screened against the K39 strain of *P. falciparum*. The chloroform ($CHCl_3$) and ethyl acetate (EtOAc) leaf extracts displayed IC_{50} values of 1.2 and 1.0 µg/mL, respectively. The cytotoxicity values were not reported; however, the bioactivity indicated the potential candidates worthy of pursuit for further isolation of antiplasmodial phytoconstituents (Muregi et al., 2003; Sadiq et al., 2017). The leaf extract from *Acacia nilotica* (Fabaceae) displayed an IC_{50} value of 1.3 µg/mL on the 3D7 strain of *P. falciparum*. The cytotoxicity values could not be determined; however, their respective antiplasmodial activities are considered to be within the range of hits for drug discovery and further validate their traditional use against malaria (Muregi et al., 2003).

About 85% of the global malaria burden is carried by 19 African countries, with Nigeria contributing the highest (25%) percentage of cases. A total of about 84 million cases were recorded in 2018, and Nigeria recorded the highest increase in the number of cases from the previous year (Akpan et al., 2018). The most dominant anopheles vector in Nigeria is *A. gambiae*, which transmits the deadly *P. falciparum* parasite. Despite efforts to curb the disease through vector and parasite control measures, the burden is still alarming (Dhiman, 2019). Socioeconomic factors are contributing immensely to the high burden of the disease (Oladepo et al., 2019). In response to the high prevalence of malaria, especially in rural communities, many Nigerians have always relied on some indigenous plant species for the treatment of malaria and its symptoms. A large part of the population prefers the use of natural herbal remedies over synthetically produced pharmaceutical drugs, and this trend has made medicinal plants popular in many parts of the country (Judith et al., 2016). Traditional IKS has revealed antimalarial plants that have been used over the centuries by local people, and anecdotal evidence has been used to guide further ethnopharmacological studies.

Alchornea cordifolia (Schumach. & Thonn.) Muell. Arg., *Azadirachta indica* A. Juss., *Guiera senegalensis* J.F.Gmel. and *Nauclea latifolia* (Sm.) are antimalarial plant species that have demonstrated significant antiplasmodial activity against different *Plasmodium* parasites (Table 18.1). *Alchornea cordifolia* extract displayed a high selectivity to the FcM29 strain of *P. falciparum* (SI = 78) with an IC_{50} value of 0.08 μg/mL and a cytotoxicity value of 6.2 μg/mL on Hela cells. *Azadiracta indica* showed selectivity (SI = 115) for the W2 strain of *P. falciparum* with an IC_{50} of 0.02 μg/mL and cytotoxicity of 2.3 μg/mL on KB cells. *Guiera senegalensis* displayed selectivity (SI = 38) for the W2 strain of *P. falciparum* with a cytotoxicity value of 49.0 μg/mL on THP-1 cells. The water extract of *N. latifolia* showed selectivity (SI = 667) of the FcB1 strain of *P. falciparum* with an IC_{50} of 0.6 μg/mL and cytotoxicity of 400 μg/mL on human melanoma cells (> 1500 NBMH cells) (Adebayo and Krettli, 2011).

Ethiopia is among the few African countries where both *P. falciparum* and *P. vivax* are dominating (Deress and Girma, 2019; Lo et al., 2017). *Anopheles arabiensis* is the most prevalent vector, while *A. pharoensis*, *A. counstani* and *A. nili* are considered benign (Kenea et al., 2016). Many parts of Ethiopia vary in elevation and climate, causing seasonal and unstable malaria cases (Taffese et al., 2018; Yalew et al., 2017).

Ethiopia has done better than many sub-Saharan African countries in reducing the burden of malaria over the last decade. In 2018, the total number of cases dropped by almost 8.4 % since 2010 (Dabaro et al., 2020). Recently, the emergence of plasmodium resistance—particularly to the currently used malaria therapy—has exacerbated the economic as well as health impacts of malaria (Alemu et al., 2012; Taffese et al., 2018). About 80% of Ethiopians in rural societies depend solely on traditional medicinal remedies because of the cost and inaccessibility of modern health services and medications (Alebie et al., 2017). Traditional medicinal herbs used for malaria by Ethiopians include *Allium sativum* (Liliaceae), *Lepidium sativum* (Brassicaceae), *Carica papaya* (Caricaceae), *Croton macrostachyus* (Euphorbiaceae), *Vernonia amygdalina* (Asteraceae), *Artemisia afra* (Asteraceae) and *Moringa stenopetala* (Moringaceae). Crude plant extracts from these plant species are well documented for their antimalarial activities (Alebie et al., 2017).

In Mali, medicinal herbs are mostly taken concurrently for the treatment of malaria and liver diseases (Haidara et al., 2016). In the Malian local language, "Sumaya" is used to describe these two major health concerns (Haidara et al., 2016; Winch et al., 2008). *Cochlospermum tinctorium* A. Rich., *Entada Africana* Guill. & Perrott, *Argemone mexicana* L. and *Terminalia macroptera* Guill. & Perr., among other traditionally used medicinal plants, have great health benefits in fighting the spread of malaria in Mali (Diarra et al., 2015; Haidara et al., 2016) (Table 18.1). A study conducted in Burkina Faso reported on the antiplasmodial activity of the water extracts of *T. macroptera* with IC_{50} of 1 μg/mL for the W2 strain of *P. falciparum* (Sanon et al., 2003). Furthermore, the study Haidara et al. (2018) reported an IC_{50} of 1.2 μg/mL for the leaf extracts of *T. macroptera* against FcB1 *P. falciparum* strain (Haidara et al., 2018). The results show that *T. macroptera* is a potential herb with good antiplasmodial activity; however, the cytotoxicity of the assayed extracts was not reported. In Ivory Coast and Mali, *Nauclea latifolia* is commonly used as a therapy for malaria (Kokwaro, 2009). Its water extracts were shown to have antiplasmodial activity of IC_{50} = 0.6 g/mL against FcB1, with cytotoxicity of 400 μg/mL on human melanoma cells (Haidara et al., 2016).

The East Asia region is home to approximately one-quarter of the world's population, with 1.6 billion of its population at risk of malaria, a preventable and curable parasitic disease (WHO, 2019). Many East Asian countries rely on traditional medicine for the management and prevention of different types of diseases. The Japanese use a combination of modern and traditional systems (Kampo) (Witt et al., 2011). This integration helps conserve the historical traditional uses of Kampo that have triggered scientific research, as well as its relevance for the current healthcare system. Kampo herbal extracts have been used against malaria for many years (Teklemichael et al., 2020). Coptis chloride of the berberine alkaloids is a major active compound of coptis rhizome, and it has displayed good activity against *P. falciparum* 3D7 strain with IC_{50} of 1.1 µg/mL with cytotoxicity of 41.1 µM on atypical memory B (AMB) cells (Teklemichael et al., 2020).

India contributes a greater share of malaria cases in the East Asia region (Kumar et al., 2007). In 1958, the National Malaria Eradication Programme was implemented, resulting in a massive decline from one million to no death cases of malaria (Narain and Nath, 2018). The success story of dichlorodiphenyltrichloroethane (DDT) lost meaning when vectors and parasites developed resistance. Malaria re-emerged, with about 6.5 million cases that time (Sharma and Mehrotra, 1986). In 2015, the country accounted for 47% of global malaria deaths (WHO, 2017). India has plenty of medicinal plants used in traditional medicine for the treatment of diseases. *Curcuma longa* L., a popular spice in India and many other Asian countries, has shown significant antiplasmodial effects and has also been used in combination with artemisinin to fight malaria (Araújo and Leon, 2001; Nandakumar et al., 2006).

Thalictrum foliolosum DC is used by the Mompa tribes for fever, pain relief and stomach problems (Das et al., 2016). Many studies have reported on the antimalarial activity of *T. foliolosum* against Plasmodium species (Bhat and Surolia, 2001). Chloroform extracts of *T. foliolosum* showed good inhibition activity on malaria-sensitive (3D7) and -resistant (LS1) *P. falciparum* strains, with IC_{50} of 0.5 ± 0.0 and 1.1 ± 0.0 µg/mL, respectively. The study of Das et al. (2016) did not report on the cytotoxicity; however, leaves and roots ethanol extracts administration were tested on animals (2,000 mg/kg), and no harm was observed.

All these plants present a vast array of possibilities for discovering novel drug leads with effective antiplasmodial properties. Studies have shown that non-polar extracts display the highest antiplasmodial activity when compared to polar extracts. However, there are special cases when the polar extracts exhibit significantly high antiplasmodial activity. Herbal medicinal crude extract preparation involves the use of highly polar solvents like water. The presence of bioactive polar extracts strongly supports the traditional uses of these medicinal plants. A large proportion of these medicinal plants have been sourced out from communities with notable high malaria burdens. Ethnobotanical surveys have revealed the relation occurring between the presence of plants with antimalarial properties in different communities and the prevalence of the malaria burden in those communities. These observations become expedient when searching for plants with antimalarial properties; preference should be given to malaria-endemic communities where there is a high likelihood of finding indigenous knowledge on plants with antimalarial bioactivity.

18.4.2 Phytoconstituents with Antiplasmodial Activity

The potential of medicinal plant-derived antimalaria drugs has received great attention from researchers around the globe. *Plasmodium*'s ability to build up resistance over different treatment measures has fueled different drug discovery endeavors to study and discover more drugs that can be used to wedge war against malaria. The botanical history shared by some of the earliest antimalarial drugs has led many researchers to study the botanicals in depth for more solutions. Validation of antiplasmodial activity of different crude extracts is not sufficient, in some instances; the specific phytochemicals responsible for the antimalarial pharmacological responses need to be characterized.

Generally, if the observed bioactivity in a crude extract is attributed to a specific phytoconstituent, then there is a chance that the phytoconstituent has an even better antiplasmodial activity compared to the crude extract. In such a case, the pure compound will be more beneficial for drug discovery and further clinical studies. Two recently published review articles document a plethora of potential antiplasmodial phytoconstituents isolated from different plants with a history of traditional use (Bekono et al., 2020; Tajuddeen and Van Heerden, 2019). Only those which displayed significantly high antiplasmodial activity ($IC_{50} < 1.5$ µg/mL) will be discussed.

18.4.2.1 Terpenes

The sesquiterpenes mustakone, muskone and corymbolone constituents isolated from *Cyperus articulatus* L. Muskone (Table 18.2), displayed IC_{50} values of 0.14 and 0.25 μg/mL on *P. falciparum* NF54 and ENT30 strains, respectively. Their chemical structures have been elucidated; however, the cytotoxicity was not determined (Rukunga et al., 2008). Gedunin, a terpenoid isolated from *Khaya grandifoliola* (Welw) C.D.C., displayed significant antiplasmodial activity with an IC_{50} of 1.25 μg/mL when tested against *P. falciparum* (Amoa Onguéné et al., 2013). The molecule 13α-epi-dioxiabiet-8(14)-en-18-ol, a type of diterpene endoperoxide, was isolated from *Hyptis suaveolens* (L.) and displayed high antiplasmodial activity, with an IC_{50} of 0.1 μg/mL against the D10 strain of *P. falciparum* (Chukwujekwu et al., 2005).

A sesquiterpene lactone, tagitinin C was isolated from *Tithonia diversifolia* (Hemsl) A. Grey. and displayed significant bioactivity (IC_{50} = 0.33 μg/mL) against the FCA strain of *Plasmodium* (Goffin et al., 2003). Pristimerin, a quinonemethide triterpenoid isolated from *Maytenus senegalensis* (Lam.), showed high antiplasmodial activity, recording an IC_{50} of 0.5 μg/mL against *P. falciparum* Dd2 strain (Amoa Onguéné et al., 2013). Biological fractionation from *Kigelia africana* (Lam.) Benth. yielded 2β,3β,19α-trihydroxy-urs-12–20-en-28-oic acid (triterpenoid), which recorded an IC_{50} of 0.933 μg/mL when assayed on *P. falciparum* W2 strain (Zofou et al., 2012).

18.4.2.2 Alkaloids

Five alkaloidal phytoconstituents were isolated from *Albizia gummifera* (J.F. Gmel.) C.A. Sm. (Table 18.2). Budmunchiamine k1 demonstrated significant antiplasmodial activity of IC_{50} = 0.09 and 0.25 μg/mL against NF54 and ENT30 *P. falciparum* strains, respectively. 6-hydroxybudmunchiamine showed high antiplasmodial activity against the NF54 and ENT30 strains of *P. falciparum* (IC_{50} = 0.15 and 0.73 μg/mL, respectively). 5-Normethylbudmanchiamine k3 alkaloid was tested on *P. falciparum* NF54 and ENT30 strains and displayed a significantly high antiplasmodial activity of IC_{50} = 0.17 and 0.88 μg/mL, respectively. 6-Hydroxy-5-normethybudmunchiamine k4 (IC_{50} = 0.12 μg/mL on *P. falciparum* NF54) and 9-Normethylbudmunchiamine k5 (IC_{50} = 0.12 and 0.89 μg/mL on NF54 and ENT30 *P. falciparum*) displayed high antiplasmodial activity (Rukunga et al., 2007).

The phenylanthraquinone jazikniphone A was isolated from the acetone crude extract of *Bulbine frutescens* (L.) and recorded an IC_{50} value of 0.14 μg/mL against the k1 strain of *P. falciparum*. The cytotoxicity value was 16.3 μg/mL on L6 cells (Bringmann et al., 2008). The compound 5-hydroxynoracronycine was isolated from *Citropsis articulata* (Willd. ex Spreng.) Swingle & M.Kellerm. and displayed an IC_{50} value of 0.9 μg/mL on *P. falciparum*. The cytotoxicity value of 9.3 μg/mL was recorded when screening on Vero E6 cells (Lacroix et al., 2011).

Phytoconstituents that display significantly higher antiplasmodial activity generally occur in high concentrations in the plant metabolome structure and exhibit higher IC_{50} values compared to the parent crude extract. In some instances, isolated pure compounds display increased antiplasmodial activity compared to crude extract, and the trade-off is cytotoxicity (Gontijo et al., 2019). Most pure compounds, especially those isolated from non-polar extracts, display significantly high cytotoxicity levels, thus compromising the phytoconstituent's clinical safety and selectivity (Bapela et al., 2014).

The high cytotoxicity observed is—to some extent—related to the solvent used in extraction, as in many polyherbal formulations prepared by indigenous people; water is the main solvent, alcohol to a lesser extent. Some isolated compounds display reduced bioactivity compared to the crude extract, and this alludes to a synergistic relationship between more than one compound acting to produce the observed antiplasmodial activity in the crude extract (Chukwujekwu et al., 2016). Further studies are required in an attempt to determine the identity and ratio of the plant metabolites that are responsible for the synergistic interactions.

TABLE 18.2
Antiplasmodial Activity of Selected Phytoconstituents

Plant Species	Compounds	IC_{50}[1] of Different *P. falciparum* Strains (μG/ML)	References
Albizia gummifera	Budmunchiamine k1	0.09 (NF54[2]) and 0.25 (ENT30[3])	(Rukunga et al., 2007)
	6-Hydroxybudmunchiamine	0.15 (NF54) and 0.73 (ENT30)	(Rukunga et al., 2007)
	5-Normethylbudmanchiamine k3	0.17 (NF54) and 0.88 (ENT30)	(Rukunga et al., 2007)
	6-Hydroxy-5-normethybudmunchiamine k4	0.12 14 (NF54)	(Rukunga et al., 2007)
	9-Normethylbudmunchiamine k5	0.12 (NF54) and 0.89 (ENT30)	(Rukunga et al., 2007)
Bulbine frutescens	Jazikniphone A	0.14 (KI[4])	(Bringmann et al., 2008)
Citropsis articulate	5-Hydroxynoracronycine	0.9 (NF54)	(Lacroix et al., 2011)
Cyperus articulatus	Muskone	0.14 (NF54) and 0.25 (ENT30)	(Rukunga et al., 2008)
Hyptis suaveolens	13α-Epi-dioxiabiet-8(14)-en-18-ol	0.1 (D10[5])	(Chukwujekwu et al., 2005)
Cyperus articulatus	Corymbolone	1.07 (NF54)	(Rukunga et al., 2008)
Kigelia africana	2β,3β,19α-Trihydroxy-urs-12–20-en-28-oic acid	0.9 33 (W2[6])	(Zofou et al., 2012)
Maytenus senegalensis	Pristimerin	0.5 (Dd2[7])	(Amoa Onguéné et al., 2013)
Tithonia diversifolia	Tagitinin C	0.33 (FCA[8])	(Goffin et al., 2003)

Notes: **1** Half-maximal inhibitory concentration; **2** Chloroquine-sensitive strain; **3** Chloroquine-resistant strain; **4** Chloroquine-resistant strain; **5** Chloroquine-sensitive strain; **6** Chloroquine-resistant strain; **7** Chloroquine-resistant strains; **8** Chloroquine-sensitive strain

Plant constituents that dissolve in polar solvents have a high chance of not harming the body and have a higher possibility of not being cytotoxic. Unfortunately, most of these polar substances have lower bioactivity and their pharmacological significance is irrelevant. On the contrary, pure compounds from non-polar extracts display high antiplasmodial activity and reduced selectivity. Further post-isolation techniques to derivatize the compounds for reduced cytotoxicity could the preferred route to develop phytoconstituents with the desired drug-like properties.

18.5 CONCLUSION

The chapter highlights the therapeutic potential of ethnomedicinal plants that are traditionally used for the treatment of malaria (Figure 18.5). Validity assessments of the crude extracts and phytoconstituents derived from the reviewed antimalarial plant species substantiate their traditional use and ethnopharmacology as a reliable approach to bioprospecting for novel antiplasmodial chemotherapeutic agents. However, isolation of relatively few compounds and their subsequent screening represents the culmination of most ethnopharmacological investigations of medicinal plants. Further research is required for the biochemical exploration of the bioactive compounds in an attempt to develop novel drugs for the eradication of malaria.

FIGURE 18.5 Plants used for the treatment of malaria. a, *Cinchona officinalis*; b, *Kigelia africana* (Lam.) Benth.; c, *Gymnosporia senegalensis* (Lam.) Loes; d, *Bulbine frutescens* (L.) Willd.; e, *Cyperus articulatus* L.; f, *Khaya anthotheca* (Welw.) C.DC.; g, *Tithonia diversifolia* (Hemsl) A. Grey.; h, *Hyptis suaveolens* (L.) Poit.; i, *Artemisia annua* L.; j, *Albizia gummifera* (J.F. Gmel.) C.A. Sm. (Brisse, 2007; Donald, 2020; Ferreira, 2008; Gargg, 2020; JMK, 2015; Navez, 2006; Steakley, 2011, 2012; Vinayaraj, 2018; Zell, 2020).

REFERENCES

Abdi, A.I., Warimwe, G.M., Muthui, M.K., Kivisi, C.A., Kiragu, E.W., Fegan, G.W., Bull, P.C., 2016. "Global selection of *Plasmodium falciparum* virulence antigen expression by host antibodies." *Scientific Reports* 6: 1–9. doi: 10.1038/srep19882.

Abdul-Ghani, R., Beier, J.C., 2014. "Strategic use of antimalarial drugs that block falciparum malaria parasite transmission to mosquitoes to achieve local malaria elimination." *Parasitology Research* 113: 3535–3546. doi: 10.1007/s00436-014-4091-6.

Achan, J., Talisuna, A.O., Erhart, A., Yeka, A., Tibenderana, J.K., Baliraine, F.N., Rosenthal, P.J., D'Alessandro, U., 2011. "Quinine, an old anti-malarial drug in a modern world: Role in the treatment of malaria." *Malaria Journal* 10: 144. doi: 10.1186/1475-2875-10-144.

Adderley, J., John, von F.S., Jackson, S., Bird, M., Burns, A., Anar, B., Metcalf, T., Semblat, J., Billker, O., Wilson, D., Doerig, C., 2020. "Analysis of erythrocyte signalling pathways during *Plasmodium falciparum* infection identifies targets for host-directed antimalarial intervention." *Nature Communications* 11: 1–13. doi: 10.1038/s41467-020-17829-7.

Adebayo, J.O., Krettli, A.U., 2011. "Potential antimalarials from Nigerian plants: A review." *Journal of Ethnopharmacology* 133: 289–302. doi: 10.1016/j.jep.2010.11.024.

Akpan, G.E., Adepoju, K.A., Oladosu, O.R., Adelabu, S.A., 2018. "Dominant malaria vector species in Nigeria: Modelling potential distribution of Anopheles gambiae sensu lato and its siblings with MaxEnt." *PLOS ONE* 13: e0204233. doi: 10.1371/journal.pone.0204233.

Alam, A., 2014. "Serine proteases of malaria parasite *Plasmodium falciparum*: Potential as antimalarial drug targets." *Interdisciplinary Perspectives on Infectious Diseases* 2014: 453186. doi: 10.1155/2014/453186.

Alebie, G., Urga, B., Worku, A., 2017. "Systematic review on traditional medicinal plants used for the treatment of malaria in Ethiopia: Trends and perspectives." *Malaria Journal* 16: 307. doi: 10.1186/s12936-017-1953-2.

Alemu, A., Muluye, D., Mihret, M., Adugna, M., Gebeyaw, M., 2012. "Ten year trend analysis of malaria prevalence in Kola Diba, North Gondar, Northwest Ethiopia." *Parasites and Vectors* 5: 6. doi: 10.1186/1756-3305-5-173.

Amir, A., Cheong, F.W., de Silva, J.R., Liew, J.W.K., Lau, Y.L., 2018. "*Plasmodium knowles*i malaria: Current research perspectives." *Infection and Drug Resistance* 11: 1145–1155. doi: 10.2147/IDR.S148664.

Amoa Onguéné, P., Ntie-Kang, F., Lifongo, L.L., Ndom, J.C., Sippl, W., Mbaze, L.M.A., 2013. "The potential of anti-malarial compounds derived from African medicinal plants, part I: A pharmacological evaluation of alkaloids and terpenoids." *Malaria Journal* 12: 1–26. doi: 10.1186/1475-2875-12-449.

Anderson, M.O., Sherrill, J., Madrid, P.B., Liou, A.P., Weisman, J.L., DeRisi, J.L., Guy, R.K., 2006. "Parallel synthesis of 9-aminoacridines and their evaluation against chloroquine-resistant *Plasmodium falciparum*." *Bioorganic and Medicinal Chemistry* 14: 334–343. doi: 10.1016/j.bmc.2005.08.017.

Anthony, M.P., Burrows, J.N., Duparc, S., Jmoehrle, J., Wells, T.N.C., 2012. "The global pipeline of new medicines for the control and elimination of malaria." *Malaria Journal* 11: 1–25. doi: 10.1186/1475-2875-11-316.

Araújo, C.A.C., Leon, L.L., 2001. "Biological activities of *Curcuma longa* L." *Memorias do Instituto Oswaldo Cruz* 96: 723–728. doi: 10.1590/s0074-02762001000500026.

Ashley, E.A., Pyae Phyo, A., Woodrow, C.J., 2018. "Malaria." *The Lancet* 391: 1608–1621. doi: 10.1016/S0140-6736(18)30324-6.

Bachmann, A., Bruske, E., Krumkamp, R., Turner, L., Wichers, J.S., Petter, M., Held, J., Duffy, M.F., Sim, B.K.L., Hoffman, S.L., Kremsner, P.G., Lell, B., Lavstsen, T., Frank, M., Mordmüller, B., Tannich, E., 2019. "Controlled human malaria infection with *Plasmodium falciparum* demonstrates impact of naturally acquired immunity on virulence gene expression." *PLOS Pathogens* 15: e1007906. doi: 10.1371/journal.ppat.1007906.

Baldwin, S.A., Mcconkey, G.A., Cass, C.E., Young, J.D., 2007. "Nucleoside transport as a potential target for chemotherapy in malaria." *Bentham Science* 13: 569–580. doi: 10.2174/138161207780162845.

Banek, K., Lalani, M., Staedke, S.G., Chandramohan, D., 2014. "Adherence to artemisinin-based combination therapy for the treatment of malaria: A systematic review of the evidence." *Malaria Journal* 13: 1–13. doi: 10.2174/138161207780162845.

Bapela, M.J., Meyer, J.J.M., Kaiser, M., 2014. "*In vitro* antiplasmodial screening of ethnopharmacologically selected South African plant species used for the treatment of malaria." *Journal of Ethnopharmacology* 156: 370–373. doi: 10.1016/j.jep.2014.09.017.

Barton, V., Fisher, N., Biagini, G.A., Ward, S.A., O'Neill, P.M., 2010. "Inhibiting *Plasmodium cytochrome* bc1: A complex issue." *Current Opinion in Chemical Biology* 14: 440–446. doi: 10.1016/j.cbpa.2010.05.005.

Bashir, I.M., Nyakoe, N., Van Der Sande, M., 2019. "Targeting remaining pockets of malaria transmission in Kenya to hasten progress towards national elimination goals: An assessment of prevalence and risk factors in children from the Lake endemic region." *Malaria Journal* 18: 233. doi: 10.1186/s12936-019-2876-x.

Bekono, B.D., Ntie-Kang, F., Onguéné, P.A., Lifongo, L.L., Sippl, W., Fester, K., Owono, L.C.O., 2020. "The potential of anti-malarial compounds derived from African medicinal plants: A review of pharmacological evaluations from 2013 to 2019." *Malaria Journal* 19: 183. doi: 10.1186/s12936-020-03231-7.

Belachew, E.B., 2018. "Immune response and evasion mechanisms of *Plasmodium falciparum* parasites." *Journal of Immunology Research*, 1–6. doi: 10.1155/2018/6529681.

Berenger, J.-M., Parola, P., 2017. "Arthropod vectors of medical importance." In: *Infectious Diseases*. Elsevier, New York: 104–112. doi: 10.1155/2018/6529681.

Bhat, G.P., Surolia, N., 2001. "*In vitro* antimalarial activity of extracts of three plants used in the traditional medicine of India." *American Journal of Tropical Medicine and Hygiene* 65: 304–308. doi: 10.1155/2018/6529681.

Blackman, M.J., 2008. "Malarial proteases and host cell egress: An 'emerging' cascade." *Cellular Microbiology* 10: 1925–1934. doi: 10.1111/j.1462-5822.2008.01176.x.

Bringmann, G., Mutanyatta-Comar, J., Maksimenka, K., Wanjohi, J.M., Heydenreich, M., Brun, R., Müller, W.E.G., Peter, M.G., Midiwo, J.O., Yenesew, A., 2008. "The roots of *Bulbine frutescens*." *Chemistry-A European Journal* 14: 1420–1429. doi: 10.1002/chem.200701328.

Brisse, H., 2007. "*Albizia adiantifolia*." Obtained from Wikimedia commons under the creative commons Attribution-Share Alike 3.0 Unported license. https://commons.wikimedia.org/wiki/File:Albizia_adiantifolia_12102003_Afrique_du_sud.JPG (Accessed 30 June 2021).

Chukwujekwu, J.C., de Kock, C.A., Smith, P.J., Van Heerden, F.R., Van Staden, J., 2016. "Antiplasmodial activity of compounds isolated from *Erythrina caffra*." *South African Journal of Botany* 106: 101–103. doi: 10.1016/j.sajb.2016.05.019.

Chukwujekwu, J.C., Smith, P., Coombes, P.H., Mulholland, D.A., Van Staden, J., 2005. "Antiplasmodial diterpenoid from the leaves of *Hyptis suaveolens*." *Journal of Ethnopharmacology* 102: 295–297. doi: 10.1016/j.jep.2005.08.018.

Cock, I.E., Selesho, M.I., van Vuuren, S.F., 2019. "A review of the traditional use of southern African medicinal plants for the treatment of malaria." *Journal of Ethnopharmacology* 245: 1–23. doi: 10.1016/j.jep.2019.112176.

Cohee, L., Laufer, M., 2018. "Tackling malaria transmission in sub-Saharan Africa." *The Lancet Global Health* 6: e598–e599. doi: 10.1016/S2214-109X(18)30197-9.

Conroy, A.L., Datta, D., John, C.C., 2019. "What causes severe malaria and its complications in children? Lessons learned over the past." *BMC Medicine* 17: 52. doi: 10.1186/s12916-019-1291-z.

Dabaro, D., Birhanu, Z., Yewhalaw, D., 2020. "Analysis of trends of malaria from 2010 to 2017 in Boricha District, Southern Ethiopia." *Malaria Journal* 19: 1–10. doi: 10.1186/s12936-020-03169-w.

Das, N.G., Rabha, B., Talukdar, P.K., Goswami, D., Dhiman, S., 2016. "Preliminary in vitro antiplasmodial activity of *Aristolochia griffithii* and *Thalictrum foliolosum* DC extracts against malaria parasite *Plasmodium falciparum*." *BMC Research Notes* 9: 1–5. doi: 10.1186/s13104-016-1862-4.

da Silva, A., Moser, M., 2002. "Malaria." Obtained from the Centers for Disease Control and Prevention. https://phil.cdc.gov/details.aspx?pid=3405 (Accessed 18 June 2021).

Davis, L., 2011. "The best and worst drugs for women." Obtained from Prevention. www.prevention.com/life/a20463350/the-best-and-worst-drugs-for-women/ (Accessed 19 October 2021).

Dayananda, K.K., Achur, R.N., Gowda, D.C., 2018. "Epidemiology, drug resistance, and pathophysiology of *Plasmodium vivax* malaria." *Journal of Vector Borne Diseases* 55: 1–8. doi: 10.4103/0972-9062.234620.

Deress, T., Girma, M., 2019. "*Plasmodium falciparum* and *Plasmodium vivax* Prevalence in Ethiopia: A systematic review and meta-analysis." *Malaria Research and Treatment* 2019: 1–12. doi: 10.1155/2019/7065064.

Dhiman, S., 2019. "Correction to: Are malaria elimination efforts on right track? An analysis of gains achieved and challenges ahead." *Infectious Diseases of Poverty* 8: 1–19. doi: 10.1186/s40249-019-0527-7.

Diakité, S.A.S., Traoré, K., Sanogo, I., Clark, T.G., Campino, S., Sangaré, M., Dabitao, D., Dara, A., Konaté, D.S., Doucouré, F., Cissé, A., Keita, B., Doumbouya, M., Guindo, M.A., Toure, M.B., Sogoba, N., Doumbia, S., Awandare, G.A., Diakité, M., 2019. "A comprehensive analysis of drug resistance molecular markers and *Plasmodium falciparum* genetic diversity in two malaria endemic sites in Mali." *Malaria Journal* 18: 361. doi: 10.1186/s12936-019-2986-5.

Diarra, N., Klooster, C.V.T., Togola, A., Diallo, D., Willcox, M., Jong, J. De, 2015. "Ethnobotanical study of plants used against malaria in Sélingué subdistrict, Mali." *Journal of Ethnopharmacology* 166: 352–360. doi: 10.1016/j.jep.2015.02.054.

Dike, I.P., Obembe, O.O., Adebiyi, F.E., 2012. "Ethnobotanical survey for potential anti-malarial plants in south-western Nigeria." *Journal of Ethnopharmacology* 144: 618–626. doi: 10.1016/j.jep.2012.10.002.

Donald, M.R., 2020. "*Cyperus iria*." Obtained from Wikimedia commons under the creative commons Attribution-Share Alike 4.0 Unported license. https://commons.wikimedia.org/wiki/File:Cyperus_iria_P6140677.jpg (Accessed 7 August 2021).

Dondorp, A.M., Smithuis, F.M., Woodrow, C., Seidlein, L. von, 2017. "How to contain artemisinin- and multidrug-resistant falciparum malaria." *Trends in Parasitology* 33: 353–363. doi: 10.1016/j.pt.2017.01.004.

Elderfield, R.C., Gensler, W.J., Birstein, O., Kreysa, F.J., Maynard, J.T., Galbreath, J., 1946. "Synthesis of certain simple 4-aminoquinoline derivatives." *Journal of the American Chemical Society* 68: 1250–1251. doi: 10.1021/ja01211a032.

Ferreira J., 2008. "*Artemisia annua*." Obtained from Wikimedia under the Public domain license. https://commons.wikimedia.org/wiki/File:Artemisia_annua.jpg (Accessed 30 June 2021).

Gargg, J.M., 2020. "*Hyptis suaveolens*." Obtained from Wikimedia commons under the creative commons Attribution-Share Alike 3.0 Unported license. https://commons.wikimedia.org/wiki/File:Hyptis_suaveolens_(Vilayti_Tulsi)_in_Hyderabad,_AP_W_IMG_0114.jpg (Accessed 30 June 2021).

Goffin, E., da Cunha, A.P., Ziemons, E., Tits, M., Angenot, L., Frederich, M., 2003. "Quantification of tagitinin C in *Tithonia diversifolia* by reversed-phase high-performance liquid chromatography." *Phytochemical Analysis* 14: 378–380. doi: 10.1002/pca.732.

Gontijo, D.C., Brandão, G.C., Nascimento, M.F.A. do, Oliveira, A.B. de, 2019. "Antiplasmodial activity and cytotoxicity, isolation of active alkaloids, and dereplication of *Xylopia sericea* leaves ethanol extract by UPLC-DAD-ESI-MS/MS." *Journal of Pharmacy and Pharmacology* 71: 260–269. doi: 10.1111/jphp.13029.

Gopal, S., Ma, Y., Xin, C., Pitts, J., Were, L., 2019. "Characterizing the spatial determinants and prevention of malaria in Kenya." *International Journal of Environmental Research and Public Health* 16: 1–19. doi: 10.3390/ijerph16245078.

Haidara, M., Bourdy, G., De Tommasi, N., Braca, A., Traore, K., Giani, S., Sanogo, R., 2016. "Medicinal plants used in Mali for the treatment of malaria and liver diseases." *Natural Product Communications* 11: 339–352. doi: 10.1016/j.jep.2015.02.054.

Haidara, M., Haddad, M., Denou, A., Marti, G., Bourgeade-Delmas, S., Sanogo, R., Bourdy, G., Aubouy, A., 2018. "*In vivo* validation of anti-malarial activity of crude extracts of *Terminalia macroptera*, a Malian medicinal plant." *Malaria Journal* 17: 68. doi: 10.1186/s12936-018-2223-7.

Haldar, K., Bhattacharjee, S., Safeukui, I., 2018. "Drug resistance in plasmodium." *Nature Reviews Microbiology* 16: 156–170. doi: 10.1038/nrmicro.2017.161.

Heinberg, A., Kirkman, L., 2015. "The molecular basis of antifolate resistance in *Plasmodium falciparum*: Looking beyond point mutations." *Annals of the New York Academy of Sciences* 1342: 10–18. doi: 10.1111/nyas.12662.

James, P.B., Wardle, J., Steel, A., Adams, J., 2018. "Traditional, complementary and alternative medicine use in sub-Saharan Africa: A systematic review." *The BMJ Global Health* 3: 1–18. doi: 10.1136/bmjgh-2018-000895.

Jaskiewicz, E., Jodłowska, M., Kaczmarek, R., Zerka, A., 2019. "Erythrocyte glycophorins as receptors for *Plasmodium* merozoites." *Parasites and Vectors* 12: 317. doi: 10.1186/s13071-019-3575-8.

JMK., 2015. "*Khaya anthotheca*." Obtained from Wikimedia commons under the creative commons Attribution-Share Alike 3.0 Unported license. https://commons.wikimedia.org/wiki/File:Khaya_anthotheca,_habitus,_Manie_van_der_Schijff_BT,_a.jpg (Accessed 30 June 2021)

Judith, O., Ijeoma, O., Ekere, K., Isaac, A., Christianah, I., Olubunmi, O., Martins, E., 2016. "Popularity and customer preferences for herbal medicines in Nigeria: A questionnaire based survey." *Humanities and Social Sciences Letters* 4: 69–76. doi: 10.18488/journal.73/2016.4.3/73.3.69.76.

Kagaya, W., Gitaka, J., Chan, C.W., Kongere, J., Md Idris, Z., Deng, C., Kaneko, A., 2019. "Malaria resurgence after significant reduction by mass drug administration on Ngodhe Island, Kenya." *Scientific Reports* 9: 1–11. doi: 10.1038/s41598-019-55437-8.

Kenea, O., Balkew, M., Tekie, H., Gebre-Michael, T., Deressa, W., Loha, E., Lindtjørn, B., Overgaard, H.J., 2016. "Human-biting activities of Anopheles species in south-central Ethiopia." *Parasites and Vectors* 9: 1–12. doi: 10.1016/j.actatropica.2021.105837.

Kimondo, J., Miaron, J., Mutai, P., Njogu, P., 2015. "Ethnobotanical survey of food and medicinal plants of the Ilkisonko Maasai community in Kenya." *Journal of Ethnopharmacology* 175: 463–469. doi: 10.1016/j.jep.2015.10.013.

Kiringe, J.W., 2006. "A survey of traditional health remedies used by the Maasai of Southern Kaijiado District, Kenya." *Ethnobotany Research & Applications* 4: 61–73. doi: 10.17348/ERA.4.0.61-74.

Koenderink, J.B., Kavishe, R.A., Rijpma, S.R., Russel, F.G.M., 2010. "The ABCs of multidrug resistance in malaria." *Trends in Parasitology* 26: 440–446. doi: 10.1016/j.pt.2010.05.002.

Kokwaro, O.J., 2009. "Medicinal plants of East Africa." *Contemporary Sociology* 24: 359. doi: 10.2307/2076505.

Kumar, A., Valecha, N., Jain, T., Dash, A.P., 2007. "Burden of malaria in India: Retrospective and prospective view." *American Journal of Tropical Medicine and Hygiene* 77: 69–78. doi: 10.1186/1475-2875-8-281.

Kumar, S., Bhardwaj, T.R., Prasad, D.N., Singh, R.K., 2018. "Drug targets for resistant malaria: Historic to future perspectives." *Biomedicine and Pharmacotherapy* 104: 8–27. doi: 10.1016/j.biopha.2018.05.009.

Lacroix, D., Prado, S., Kamoga, D., Kasenene, J., Bodo, B., 2011. "Structure and *in vitro* antiparasitic activity of constituents of *Citropsis articulata* root bark." *Journal of Natural Products* 74: 2286–2289. doi: 10.1021/np2004825.

Lindner, J., Meissner, K.A., Schettert, I., Wrenger, C., 2013. "Trafficked proteins-druggable in *Plasmodium falciparum*?" *International Journal of Cell Biology* 2013, 13. doi: 10.1155/2013/435981.

Lo, E., Hemming-Schroeder, E., Yewhalaw, D., Nguyen, J., Kebede, E., Zemene, E., Getachew, S., Tushune, K., Zhong, D., Zhou, G., Petros, B., Yan, G., 2017. "Transmission dynamics of co-endemic *Plasmodium vivax* and *P. falciparum* in Ethiopia and prevalence of antimalarial resistant genotypes." *PLoS Neglected Tropical Diseases* 11: 1–25. doi: 10.1371/journal.pntd.0005806.

Lunev, S., Batista, F.A., Bosch, S.S., Wrenger, C., Groves, M.R., 2016. "Identification and validation of novel drug targets for the treatment of *Plasmodium falciparum* malaria: New insights." In: Current Topics in Malaria. InTech. doi: 10.5772/65659.

Makule, E., Heilmann, J., Kraus, B., 2014. "Ethnobiological survey of Maasai traditional medicinal plants." *Planta Medica* 80: P2B46. doi: 10.1016/j.jep.2015.10.013.

Marin-Mogollon, C., van de Vegte-Bolmer, M., van Gemert, G.J., van Pul, F.J.A., Ramesar, J., Othman, A.S., Kroeze, H., Miao, J., Cui, L., Williamson, K.C., Sauerwein, R.W., Janse, C.J., Khan, S.M., 2018. "The Plasmodium falciparum male gametocyte protein P230p, a paralog of P230, is vital for ookinete formation and mosquito transmission." *Scientific Reports* 8: 1–13. doi: 10.1038/s41598-018-33236-x.

Maude, R.J., Woodrow, C.J., White, L.J., 2010. "Artemisinin antimalarials: Preserving the 'magic bullet'." *Drug Development Research* 71: 12–19. doi: 10.1002/ddr.20344.

Mavoko, H.M., Nabasumba, C., da Luz, R.I., Tinto, H., D'Alessandro, U., Kambugu, A., Baraka, V., Rosanas-Urgell, A., Lutumba, P., Van geertruyden, J.P., 2017. "Efficacy and safety of re-treatment with the same artemisinin-based combination treatment (ACT) compared with an alternative ACT and quinine plus clindamycin after failure of first-line recommended ACT (QUINACT): A bicentre, open-label, phase 3, randomise." *The Lancet Global Health* 5: e60–e68. doi: 10.1016/S2214-109X(16)30236-4.

Mcconkey, G.A., 1999. "Targeting the shikimate pathway in the malaria parasite *Plasmodium falciparum*." *Antimicrobial Agents and Chemotherapy* 43: 175–177. doi: 10.1128/AAC.43.1.175.

Menard, D., Dondorp, A., 2017. "Antimalarial drug resistance: A threat to malaria elimination." *Cold Spring Harbor Perspectives in Medicine* 7: 1–24. doi: 10.1101/cshperspect.a025619.

Mikolajczak, S.A., Vaughan, A.M., Kangwanrangsan, N., Roobsoong, W., Fishbaugher, M., Yimamnuaychok, N., Rezakhani, N., Lakshmanan, V., Singh, N., Kaushansky, A., Camargo, N., Baldwin, M., Lindner, S.E., Adams, J.H., Sattabongkot, J., Kappe, S.H.I., 2015. "*Plasmodium vivax* liver stage development and hypnozoite persistence in human liver-chimeric mice." *Cell Host and Microbe* 17: 526–535. doi: 10.1016/j.chom.2015.02.011.

Miller, L.H., Ackerman, H.C., Su, X.Z., Wellems, T.E., 2013. "Malaria biology and disease pathogenesis: Insights for new treatments." *Nature Medicine* 19: 156–167. doi: 10.1038/nm.3073.

Mokoka, T.A., Xolani, P.K., Zimmermann, S., Hata, Y., Adams, M., Kaiser, M., Moodley, N., Maharaj, V., Koorbanally, N.A., Hamburger, M., Brun, R., Fouche, G., 2013. "Antiprotozoal screening of 60 South African plants, and the identification of the antitrypanosomal germacranolides schkuhrin I and II." *Planta Med* 79: 1380–1384. doi: 10.1055/s-0033-1350691.

Molyneux, D., 2020. "Malaria eradication." *The Lancet* 395: e71. doi: 10.1016/S0140-6736(20)30240-3.

Muregi, F.W., Chhabra, S.C., Njagi, E.N.M., Lang'at-Thoruwa, C.C., Njue, W.M., Orago, A.S.S., Omar, S.A., Ndiege, I.O., 2003. "*In vitro* antiplasmodial activity of some plants used in Kisii, Kenya against malaria and their chloroquine potentiation effects." *Journal of Ethnopharmacology* 84: 235–239. doi: 10.1016/s0378-8741(02)00327-6.

Muthaura, C.N., Keriko, J.M., Derese, S., Yenesew, A., Rukunga, G.M., 2011. "Investigation of some medicinal plants traditionally used for treatment of malaria in Kenya as potential sources of antimalarial drugs." *Experimental Parasitology* 127: 609–626. doi: 10.1016/j.exppara.2010.11.004.

Muthaura, C.N., Rukunga, G.M., Chhabra, S.C., Omar, S.A., Guantai, A.N., Gathirwa, J.W., Tolo, F.M., Mwitari, P.G., Keter, L.K., Kirira, P.G., Kimani, C.W., Mungai, G.M., Njagi, E.N.M., 2007a. "Antimalarial activity of some plants traditionally used in Meru district of Kenya." *Phytotherapy Research* 21: 860–867. doi: 10.1002/ptr.2170.

Muthaura, C.N., Rukunga, G.M., Chhabra, S.C., Omar, S.A., Guantai, A.N., Gathirwa, J.W., Tolo, F.M., Mwitari, P.G., Keter, L.K., Kirira, P.G., Kimani, C.W., Mungai, G.M., Njagi, E.N.M., 2007b. "Antimalarial activity of some plants traditionally used in treatment of malaria in Kwale district of Kenya." *Journal of Ethnopharmacology* 112: 545–551. doi: 10.1016/j.jep.2007.04.018.

Nandakumar, D.N., Nagaraj, V.A., Vathsala, P.G., Rangarajan, P., Padmanaban, G., 2006. "Curcumin-artemisinin combination therapy for malaria." *Antimicrobial Agents and Chemotherapy* 50: 1859–1860. doi: 10.1128/AAC.50.5.1859-1860.2006.

Narain, J.P., Nath, L.M., 2018. "Eliminating malaria in India by 2027: The countdown begins!" *Indian Journal of Medical Research* 148: 123–126. doi: 10.4103/ijmr.IJMR_1175_18.

Narula, A.K., Azad, C.S., Nainwal, L.M., 2019. "New dimensions in the field of antimalarial research against malaria resurgence." *European Journal of Medicinal Chemistry* 181: 111353. doi: 10.1016/j.ejmech.2019.05.043.

Navez, B., 2006. "*Tithonia diversifolia* (leaves and flowers)." Obtained from Wikimedia commons under the creative commons Attribution-Share Alike 3.0 Unported license https://commons.wikimedia.org/wiki/File:Tithonia_diversifolia.JPG (Accessed 30 June 2021).

Nguta, J.M., Mbaria, J.M., Gakuya, D.W., Gathumbi, P.K., Kiama, S.G., 2010. "Antimalarial herbal remedies of Msambweni, Kenya." *Journal of Ethnopharmacology* 128: 424–432. doi: 10.1016/j.jep.2010.01.033.

Niu, G., Zhang, G., Franca, C., Cui, Y., Munga, S., Afrane, Y., Li, J., 2017. "FBN30 in wild Anopheles gambiae functions as a pathogen recognition molecule against clinically circulating *Plasmodium falciparum* in malaria endemic areas in Kenya." *Scientific Reports* 7: 1–9. doi: 10.1038/s41598-017-09017-3.

Nqoro, X., Tobeka, N., Aderibigbe, B.A., 2017. "Quinoline-based hybrid compounds with antimalarial activity." *Molecules* 22: 1–22. doi: 10.3390/molecules22122268.

Ntie-Kang, F., Onguéné, P.A., Lifongo, L.L., Ndom, J.C., Sippl, W., Mbaze, L.M., 2014. "The potential of antimalarial compounds derived from African medicinal plants, part II: A pharmacological evaluation of non-alkaloids and non-terpenoids." *Malaria Journal* 13: 81. doi: 10.1186/1475-2875-13-81.

Oboh, M.A., Ndiaye, D., Antony, H.A., Badiane, A.S., Singh, U.S., Ali, N.A., Bharti, P.K., Das, A., 2018. "Status of artemisinin resistance in malaria parasite *Plasmodium falciparum* from molecular analyses of the Kelch13 gene in Southwestern Nigeria." *BioMed Research International* 2018: 1–5. doi: 10.1155/2018/2305062.

Odugbemi, T.O., Akinsulire, O.R., Aibinu, I.E., Fabeku, P.O., 2007. "Medicinal plants useful for malaria therapy in Okeigbo, Ondo State, Southwest Nigeria." *African Journal of Traditional, Complementary and Alternative Medicines* 4: 191–198. doi: 10.4314/ajtcam.v4i2.31207.

Ojeh, V.N., Aworinde, S., 2016. "Climate variation and challenges of human health in Nigeria: Malaria in perspective." In: *Climate Change Management*. Springer: 171–185. doi: 10.1016/j.jclepro.2017.12.144.

Oladepo, O., Oyeyemi, A.S., Titiloye, M.A., Adeyemi, A.O., Burnett, S.M., Apera, I., Oladunni, O., Alliu, M., 2019. "Malaria testing and treatment knowledge among selected rural patent and proprietary medicine vendors (PPMV) in Nigeria." *Malaria Journal* 18: 103. doi: 10.1186/s12936-019-2732-z.

Ome-Kaius, M., Kattenberg, J.H., Zaloumis, S., Siba, M., Kiniboro, B., Jally, S., Razook, Z., Mantila, D., Sui, D., Ginny, J., Rosanas-Urgell, A., Karl, S., Obadia, T., Barry, A., Rogerson, S.J., Laman, M., Tisch, D., Felger, I., Kazura, J.W., Mueller, I., Robinson, L.J., 2019. "Differential impact of malaria control interventions on *P. falciparum* and *P. vivax* infections in young Papua New Guinean children." *BMC Medicine* 12: 1–13. doi: 10.1186/s12916-019-1456-9.

O'Neill, P.M., Barton, V.E., Ward, S.A., 2010. "The molecular mechanism of action of artemisinin-The debate continues." *Molecules* 15: 1705–1721. doi: 10.3390/molecules15031705.

Paddon, C.J., Keasling, J.D., 2014. "Semi-synthetic artemisinin: A model for the use of synthetic biology in pharmaceutical development." *Nature Reviews Microbiology* 12: 355–367. doi: 10.1038/nrmicro3240.

Pan, W.H., Xu, X.Y., Shi, N., Tsang, S.W., Zhang, H.J., 2018. "Antimalarial activity of plant metabolites." *International Journal of Molecular Sciences* 19: 1–40. doi: 10.3390/ijms19051382.

Parija, S.C., Praharaj, I., 2011. "Drug resistance in malaria." *Indian Journal of Medical Microbiology* 29: 243–248. doi: 10.4103/0255-0857.83906.

Patgiri, S.J., Sarma, K., Sarmah, N., Bhattacharyya, N., Sarma, D.K., Nirmolia, T., Bhattacharyya, D.R., Mohapatra, P.K., Bansal, D., Bharti, P.K., Sehgal, R., Mahanta, J., Sultan, A.A., 2019. "Characterization of drug resistance and genetic diversity of *Plasmodium falciparum* parasites from Tripura, Northeast India." *Scientific Reports* 9: 1–10. doi: 10.1038/s41598-019-50152-w.

Phyo, A.P., Von Seidlein, L., 2017. "Challenges to replace ACT as first-line drug." *Malaria Journal* 16: 296. doi: 10.1186/s12936-017-1942-5.

Pillay, P., Maharaj, V.J.J., Smith, P.J.J., 2008. "Investigating South African plants as a source of new antimalarial drugs." *Journal of Ethnopharmacology* 119: 438–454. doi: 10.1016/j.jep.2008.07.003.

Pousibet-Puerto, J., Salas-Coronas, J., Sánchez-Crespo, A., Molina-Arrebola, M.A., Soriano-Pérez, M.J., Giménez-López, M.J., Vázquez-Villegas, J., Cabezas-Fernández, M.T., 2016. "Impact of using artemisinin-based combination therapy (ACT) in the treatment of uncomplicated malaria from *Plasmodium falciparum* in a non-endemic zone." *Malaria Journal* 15: 339. doi: 10.1186/s12936-016-1408-1.

Rees, P., 2021. "Vaccinations and malaria prophylaxis for travelers to Africa" Obtained from The Travel Doctor. www.traveldoctor.co.uk/africa.htm (Accessed 18 October 2021).

Richter, J., Franken, G., Holtfreter, M.C., Walter, S., Labisch, A., Mehlhorn, H., 2016. "Clinical implications of a gradual dormancy concept in malaria." *Parasitology Research* 115: 2139–2148. doi: 10.1007/s00436-016-5043-0.

Rondón, S., León, C., Link, A., González, C., 2019. "Prevalence of *Plasmodium parasites* in non-human primates and mosquitoes in areas with different degrees of fragmentation in Colombia." *Malaria Journal* 18: 276. doi: 10.1186/s12936-019-2910-z.

Rosenthal, P.J., 1998. "Proteases of malaria parasites: New targets for chemotherapy." *Emerging Infectious Diseases* 4: 49–57. doi: 10.3201/eid0401.980107.

Rosenthal, P.J., Taylor, W.R.J., 2003. "Antimalarial drug discovery: Old and new approaches." *Journal of Experimental Biology* 206: 3735–3744. doi: 10.1242/jeb.00589.

Rudrapal, M., 2011. "A brief review on malaria and current antimalarial drugs." *Current Pharma Research* 1: 286–292.

Rukunga, G.M., Muregi, F.W., Omar, S.A., Gathirwa, J.W., Muthaura, C.N., Peter, M.G., Heydenreich, M., Mungai, G.M., 2008. "Anti-plasmodial activity of the extracts and two sesquiterpenes from *Cyperus articulatus*." *Fitoterapia* 79: 188–190. doi: 10.1016/j.fitote.2007.11.010.

Rukunga, G.M., Muregi, F.W., Tolo, F.M., Omar, S.A., Mwitari, P., Muthaura, C.N., Omlin, F., Lwande, W., Hassanali, A., Githure, J., Iraqi, F.W., Mungai, G.M., Kraus, W., Kofi-Tsekpo, W.M., 2007. "The antiplasmodial activity of spermine alkaloids isolated from *Albizia gummifera*." *Fitoterapia* 78: 455–459. doi: 10.1016/j.fitote.2007.02.012.

Rutledge, G.G., Böhme, U., Sanders, M., Reid, A.J., Cotton, J.A., Maiga-Ascofare, O., Djimdé, A.A., Apinjoh, T.O., Amenga-Etego, L., Manske, M., Barnwell, J.W., Renaud, F., Ollomo, B., Prugnolle, F., Anstey, N.M., Auburn, S., Price, R.N., McCarthy, J.S., Kwiatkowski, D.P., Newbold, C.I., Berriman, M., Otto, T.D., 2017. "*Plasmodium malariae* and *P. ovale* genomes provide insights into malaria parasite evolution." *Nature* 542: 101–104. doi: 10.1038/nature21038.

Sachs, J., Malaney, P., 2002. "The economic and social burden of malaria." *Nature* 415: 680–685. doi: 10.1038/415680a.

Sadiq, M.B., Tharaphan, P., Chotivanich, K., Tarning, J., Anal, A.K., 2017. "*In vitro* antioxidant and antimalarial activities of leaves, pods and bark extracts of *Acacia nilotica* (L.) Del." *BMC Complementary and Alternative Medicine* 17: 372. doi: 10.1186/s12906-017-1878-x.

Sanon, S., Ollivier, E., Azas, N., Mahiou, V., Gasquet, M., Ouattara, C.T., Nebie, I., Traore, A.S., Esposito, F., Balansard, G., Timon-David, P., Fumoux, F., 2003. "Ethnobotanical survey and *in vitro* antiplasmodial activity of plants used in traditional medicine in Burkina Faso." *Journal of Ethnopharmacology*: 143–147. doi: 10.1016/S0378-8741(02)00381-1.

Shanks, G.D., Möhrle, J.J., 2017. "Treating malaria: New drugs for a new era." *The Lancet Infectious Diseases* 17: 1223–1224. doi: 10.1016/S1473-3099(17)30475-9.

Sharma, V.P., Mehrotra, K.N., 1986. "Malaria resurgence in India: A critical study." *Social Science and Medicine*. doi: 10.1016/0277-9536(86)90238-8.

Sinha, S., Medhi, B., Sehgal, R., 2014. "Challenges of drug-resistant malaria." *Parasite* 21: 61. doi: 10.1051/parasite/2014059.

Siregar, J.E., Kurisu, G., Kobayashi, T., Matsuzaki, M., Sakamoto, K., Mi-ichi, F., Watanabe, Y. ichi, Hirai, M., Matsuoka, H., Syafruddin, D., Marzuki, S., Kita, K., 2015. "Direct evidence for the atovaquone action on the *Plasmodium* cytochrome bc1 complex." *Parasitology International* 64: 295–300. doi: 10.1016/j.parint.2014.09.011.

Sitali, L., Miller, J.M., Mwenda, M.C., Bridges, D.J., Hawela, M.B., Hamainza, B., Chizema-Kawesha, E., Eisele, T.P., Chipeta, J., Lindtjørn, B., 2019. "Distribution of *Plasmodium* species and assessment of performance of diagnostic tools used during a malaria survey in Southern and Western Provinces of Zambia." *Malaria Journal* 18: 130. doi: 10.1186/s12936-019-2766-2.

Steakley, J., 2011. "*Bulbine frutescens*." Obtained from Wikimedia commons under the creative commons Attribution-Share Alike 3.0 Unported license. https://commons.wikimedia.org/wiki/File:Bulbine_frutescens_-_Missouri_Botanical_Garden.jpg (Accessed 30 June 2021)

Steakley, J., 2012. "*Cinchona officinalis*." Obtained from Wikimedia commons under the creative commons Attribution-Share Alike 3.0 Unported license https://commons.wikimedia.org/w/index.php?title=File:Cinchona_officinalis_(DITSL).JPG&oldid=416648658 (Accessed 2 July 2021)

Taffese, H.S., Hemming-Schroeder, E., Koepfli, C., Tesfaye, G., Lee, M.C., Kazura, J., Yan, G.Y., Zhou, G.F., 2018. "Malaria epidemiology and interventions in Ethiopia from 2001 to 2016." *Infectious Diseases of Poverty* 7: 1–9. doi: 10.1186/s40249-018-0487-3.

Tajuddeen, N., Van Heerden, F.R., 2019. "Antiplasmodial natural products: An update." *Malaria Journal* 18: 1–62. doi: 10.1186/s12936-019-3026-1.

Teklemichael, A.A., Mizukami, S., Toume, K., Mosaddeque, F., Kamel, M.G., Kaneko, O., Komatsu, K., Karbwang, J., Huy, N.T., Hirayama, K., 2020. "Anti-malarial activity of traditional Kampo medicine *Coptis rhizome* extract and its major active compounds." *Malaria Journal* 19: 1–10. doi: 10.1186/s12936-020-03273-x.

The malERA Group, 2017. "malERA: An updated research agenda for diagnostics, drugs, vaccines, and vector control in malaria elimination and eradication." *PLoS Medicine* 14: 1–35. doi: 10.1371/journal.pmed.1002455.

Tilley, L., Straimer, J., Gnädig, N.F., Ralph, S.A., Fidock, D.A., 2016. "Artemisinin action and resistance in *Plasmodium falciparum*." *Trends in Parasitology* 32: 682–696. doi: 10.1016/j.pt.2016.05.010.

Tse, E.G., Korsik, M., Todd, M.H., 2019. "The past, present and future of anti-malarial medicines." *Malaria Journal* 18: 1–21. doi: 10.1186/s12936-019-2724-z.

Uzor, P.F., 2020. "Alkaloids from plants with antimalarial activity: A review of recent studies." *Evidence-based Complementary and Alternative Medicine* 2020. doi: 10.1155/2020/8749083.

Van Dyk, S., Griffiths, S., Van Zyl, R.L., Malan, S.F., 2009. "The importance of including toxicity assays when screening plant extracts for antimalarial activity." *African Journal of Biotechnology* 8: 5584–5590. doi: 10.4314/ajb.v8i20.66014.

Van Niekerk, D.D., Penkler, G.P., du Toit, F., Snoep, J.L., 2016. "Targeting glycolysis in the malaria parasite *Plasmodium falciparum*." *FEBS Journal* 283: 634–646. doi: 10.1111/febs.13615.

Vinayaraj, V.R., 2018. "*Maytenus senegalensis-Gymnosporia senegalensis* from Masinagudi." Obtained from Wikimedia under the Creative Commons Attribution-Share Alike 4.0 international license. https://commons.wikimedia.org/wiki/File:Maytenus_senegalensis-Gymnosporia_senegalensis_from_Masinagudi_(4).jpg (Accessed 30 June 2021).

White, N., 1999. "Antimalarial drug resistance and combination chemotherapy." *Philosophical Transactions of the Royal Society B: Biological Sciences* 354: 739–749. doi: 10.1098/rstb.1999.0426.

White, N.J., Pukrittayakamee, S., Hien, T.T., Faiz, M.A., Mokuolu, O.A., Dondorp, A.M., 2014. "Malaria." *Lancet* 383: 723–735. doi: 10.1016/S0140-6736(13)60024-0.

WHO, 2017. "World malaria report: 2017." Geneva, WHO Press.

WHO, 2018. "Artemisinin resistance and artemisinin-based combination therapy efficacy." Geneva, WHO Press.

WHO, 2019. "World malaria report: 2019." Geneva, WHO Press.

Williams, V.L., Witkowski, E.T.F., Balkwill, K., 2007. "Relationship between bark thickness and diameter at breast height for six tree species used medicinally in South Africa." *South African Journal of Botany* 73: 449–465. doi: 10.1016/J.SAJB.2007.04.001

Wilson, R.J.M. Iain., Denny, P.W., Preiser, P.R., Rangachari, K., Roberts, K., Roy, A., Whyte, A., Strath, M., Moore, D.J., Moore, P.W., Williamson, D.H., 1996. "Complete gene map of the plastid-like DNA of the malaria parasite *Plasmodium falciparum*." *Journal of Molecular Biology* 261: 155–172. doi: 10.1006/jmbi.1996.0449.

Winch, P.J., Doumbia, S., Kanté, M., Malé, A.D., Swedberg, E., Gilroy, K.E., Ellis, A.A., Cissé, G., Sidibé, B., 2008. "Differential community response to introduction of zinc for childhood diarrhea and combination therapy for malaria in Southern Mali." *Journal of Nutrition* 138: 642–645. doi: 10.1093/jn/138.3.642.

Witkowski, B., Lelièvre, J., Barragán, M.J.L., Laurent, V., Su, X.Z., Berry, A., Benoit-Vical, F., 2010. "Increased tolerance to artemisinin in *Plasmodium falciparum* is mediated by a quiescence mechanism." *Antimicrobial Agents and Chemotherapy* 54: 1872–1877. doi: 10.1128/AAC.01636-09.

Witt, C.M., Watanabe, K., Matsuura, K., Gao, P., Hottenbacher, L., Tokunaga, H., Nishimura, K., Imazu, Y., Reissenweber, H., 2011. "Traditional Japanese Kampo medicine: Clinical research between modernity and traditional medicine-The state of research and methodological suggestions for the future." *Evidence-based Complementary and Alternative Medicine* 2011: 1–20. doi: 10.1093/ecam/neq067.

Yadav, D.K., Kumar, S., Teli, M.K., Yadav, R., Chaudhary, S., 2019. "Molecular targets for malarial chemotherapy: A review." *Current Topics in Medicinal Chemistry* 19: 861–873. doi: 10.2174/1568026619666190603080000.

Yahiya, S., Rueda-Zubiaurre, A., Delves, M.J., Fuchter, M.J., Baum, J., 2019. "The antimalarial screening landscape—looking beyond the asexual blood stage." *Current Opinion in Chemical Biology* 50: 1–9. doi: 10.1016/j.cbpa.2019.01.029.

Yalew, W.G., Pal, S., Bansil, P., Dabbs, R., Tetteh, K., Guinovart, C., Kalnoky, M., Serda, B.A., Tesfay, B.H., Beyene, B.B., Seneviratne, C., Littrell, M., Yokobe, L., Noland, G.S., Domingo, G.J., Getachew, A., Drakeley, C., Steketee, R.W., 2017. "Current and cumulative malaria infections in a setting embarking on elimination: Amhara, Ethiopia." *Malaria Journal* 16: 242. doi: 10.1186/s12936-017-1884-y.

Yap, X.Z., Lundie, R.J., Feng, G., Pooley, J., Beeson, J.G., O'Keeffe, M., 2019. "Different life cycle stages of *Plasmodium falciparum* induce contrasting responses in dendritic cells." *Frontiers in Immunology* 10: 32. doi: 10.3389/fimmu.2019.00032.

Yeoh, L.M., Goodman, C.D., Mollard, V., McFadden, G.I., Ralph, S.A., 2017. "Comparative transcriptomics of female and male gametocytes in *Plasmodium berghei* and the evolution of sex in alveolates." *BMC Genomics* 18: 734. doi: 10.1186/s12864-017-4100-0.

Zell, H. 2020. "*Kigelia africana*." Obtained from Wikimedia commons under the creative commons Attribution-Share Alike 3.0 Unported license. https://commons.wikimedia.org/wiki/File:Kigelia_africana_-_Jardim_Municipal_do_Funchal_01.jpg (Accessed 30 June 2021).

Zhou, R., Li, S., Zhao, Y., Yang, C., Liu, Y., Qian, D., Wang, H., Lu, D., Zhang, H., 2019. "Characterization of *Plasmodium ovale* spp. imported from Africa to Henan Province, China." *Scientific Reports* 9: 1–8. doi: 10.1038/s41598-019-38629-0.

Zimmermann, S., Thomi, S., Kaiser, M., Hamburger, M., Adams, M., 2012. "Screening and HPLC-based activity profiling for new antiprotozoal leads from European plants." *Scientia Pharmaceutica* 80: 205–213. doi: 10.3797/scipharm.1111-13.

Zofou, D., Tene, M., Tane, P., Titanji, V.P.K., 2012. "Antimalarial drug interactions of compounds isolated from *Kigelia africana* (Bignoniaceae) and their synergism with artemether, against the multidrug-resistant W2mef *Plasmodium falciparum* strain." *Parasitology Research* 110: 539–544. doi: 10.1007/s00436-011-2519-9.

19 The Potential of Medicinal Plants as Treatments for Infections Caused by *Aspergillus* spp.

Tefo K. Pule, Marco N. De Canha, Namrita Lall and Quenton Kritzinger

CONTENTS

19.1 INTRODUCTION

Fungi are one of the most ubiquitous groups of eukaryotic microorganisms on earth, and manifest typically as powdery molds in favorable environmental conditions. One attribute that contributes to their prevalence in various types of environments is their ability to produce spores (Shah and Panjabi, 2014). Several fungal species, such as *Fusarium* and *Aspergillus* spp., have been isolated from both indoor and outdoor environments (Latgé and Chamilos, 2020). *Aspergillus* spp. (Figure 19.1A) are saprophytic fungi that contribute to the recycling of carbon and nitrogen in nature, surviving and proliferating on organic debris. Of over 250 *Aspergillus* spp., *Aspergillus fumigatus* and a number of counterparts in this genus have an abundance of airborne conidia (Figure 19.1B) (Latgé and Chamilos, 2020). Each conidial head can produce a great number of conidia that sporulate efficiently.

No known specialized mechanisms are employed by the fungi to disseminate into the air; they rely simply on wind and other dispersion methods such as physical interactions with insects, animals and humans (Li, Lu, and Meng, 2019). Once released into the air, the buoyancy of the conidia keeps them airborne and prevalent in various environments where disturbances have occurred.

DOI: 10.1201/9781003108375-19

Graphical abstract representing the content of the chapter (Anthony, 2018).

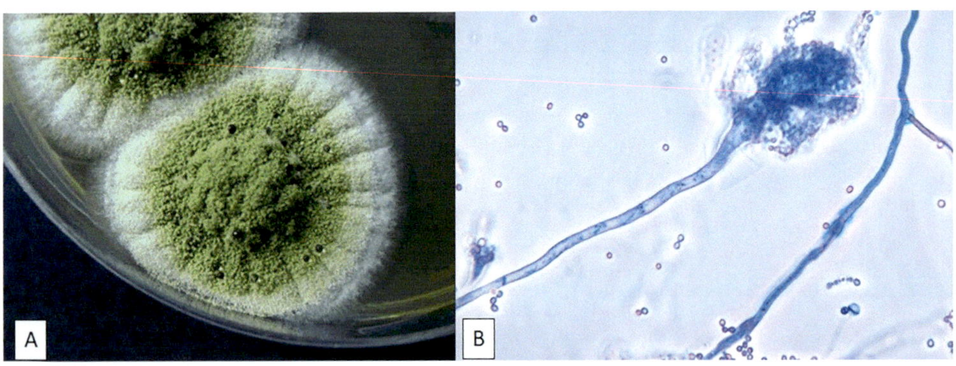

FIGURE 19.1 A, Colony morphology of *Aspergillus fumigatus*; B, light micrograph of the conidiophore of *A. fumigatus* (Centers for Disease Control and Prevention, 2014; Center for Disease Control Public Health Image Library, 1963).

Inhalation is the common route of entry for *Aspergillus* conidia to enter the human body; therefore, it is inevitable that conidia are inhaled via the human respiratory pathway, with a high probability of reaching the lung alveoli. The characteristic small size and shape of the conidia allow them to travel through the bronchus into the bronchioles until they reach their final destination, which is the lung alveoli (Li, Lu, and Meng, 2019).

Alternative routes of entry include skin wounds and ingestion of contaminated foods, and in some cases, the fungi become opportunistic and pathogenic by infecting individuals during surgery (Xavier et al., 2008). Although *Aspergillus* spp. are usually harmless to healthy individuals, there have been some reports which suggest that immunocompromised individuals have succumbed to *Aspergillus* infections (Li, Lu, and Meng, 2019; Latgé and Chamilos, 2020). This occurs when *Aspergillus* strains parasitize the human body. Depending on the route of entry of the fungus, symptoms can manifest as skin lesions, hypersensitive reactions, chronic pulmonary infections, and other acute life-threatening infections (Li, Lu, and Meng, 2019).

Aspergillus fumigatus is reported to be the main etiological agent of infections. However, in some instances, *A. flavus* has also been isolated from diseased individuals. The collection of respiratory diseases caused by various *Aspergillus* spp. is called aspergillosis (Xavier et al., 2008). Since the U.S. Centers for Disease Control and Prevention (CDC) classifies aspergillosis as a non-notifiable infection in the United States, it is challenging to report on the statistics and global impact of this fungal infection. There have, however, been reports that show 2.5% of adults who suffer from asthma also have aspergillosis, which is estimated to be around 4.8 million people who could potentially be suffering from the infection (Centers for Disease Control and Prevention, 2014).

While treatments for aspergillosis exist, they are not particularly desirable due to their systemic toxicity (Jenks and Hoenigl, 2018). This warrants the re-evaluation of currently available treatments and the investigation of less explored fields of study as sources of potential treatments. One field which has caught the attention of many scientists is the use of medicinal plants in finding and possibly pioneering new treatments. The ultimate goal would be the development of treatments that are more effective, less toxic to the human body and, hopefully, more promising than the commercially available treatments. This chapter aims to critically review the current treatments of *Aspergillus*-related infections and highlight the potential of medicinal plants as a treatment for respiratory infections in humans caused by *Aspergillus* spp. The chapter provides an overview of the various types of aspergillosis, the mode of action of the fungus in disease progression and the advantages and disadvantages of current treatments and medicinal plants used traditionally to treat fungal diseases, with emphasis on *Aspergillus* spp.

19.2 ASPERGILLOSIS

Aspergillosis refers to infections caused by fungi from the genus *Aspergillus*. *Aspergillus* infections are not only limited to humans, and can cause aspergillosis in birds as well (avian aspergillosis) (Suleiman et al., 2012). However, this chapter will focus only on aspergillosis affecting humans. The most common type of aspergillosis is known as pulmonary aspergillosis, which affects the respiratory pathway of the host. Other types include cutaneous aspergillosis, which affects the skin, and otomycosis, which affects the ears (Centers for Disease Control and Prevention, 2014). The three types of *Aspergillus* infections (pulmonary [with subtypes], cutaneous and otomycosis) are discussed in detail based on reports that have implicated *Aspergillus* spp. as the causative pathogens, confirmed by the isolation and identification of these fungi from afflicted individuals.

19.2.1 Pulmonary Aspergillosis

Pulmonary aspergillosis is an infection where *Aspergillus* spp. compromises the respiratory pathway of individuals with pre-existing complications, either lung infections or because of a feeble immune system (Lat and Thompson, 2011). Pulmonary aspergillosis can be categorized into distinct types, which include invasive pulmonary aspergillosis (IA or IPA), allergic bronchopulmonary aspergillosis (ABPA) and chronic pulmonary aspergillosis (CPA) (Dogra et al., 2016). The main etiological agent of this disease is widely reported to be *A. fumigatus*; however, in rare instances, *A. flavus* has been reported and isolated concurrently with *A. fumigatus*, which suggests that a mixed infection could be responsible for aspergillosis (Rudramurthy et al., 2019).

19.2.1.1 Invasive Pulmonary Aspergillosis (IA)

Invasive pulmonary aspergillosis is a result of hyphal penetration by *A. fumigatus* into lung tissues (Jenks and Hoenigl, 2018). The lungs are affected the most, followed by the paranasal sinuses and then the central nervous system (CNS) (Rudramurthy et al., 2019). Symptoms that follow manifest in the forms of chest pains, fever and hemoptysis (coughing up blood) (Jenks and Hoenigl, 2018).

19.2.1.2 Allergic Bronchopulmonary Aspergillosis (ABPA)

A hypersensitive reaction to *A. fumigatus* known as allergic bronchopulmonary aspergillosis (ABPA) (Agarwal and Road, 2012), is characterized by the immediate-type cutaneous hypersensitivity to *A. fumigatus* antigens. Susceptibility to ABPA is common among asthmatic patients and individuals with cystic fibrosis (Greenberger, 2002), but remains highly variable between patients. Variability of ABPA in patients is generally characterized by the part of the lung which is affected (Agarwal and Road, 2012).

19.2.1.3 Chronic Pulmonary Aspergillosis (CPA)

Chronic pulmonary aspergillosis is a lung disease within the scope of IA. It refers to long-term infection of aspergillosis associated with cavitary lung disease, chronic respiratory symptoms and *Aspergillus* serum antibody precipitation (Thompson and Patterson, 2008). Once the conidia successfully evade the immune system of immunocompromised individuals, they germinate and form hyphal networks in the lung's interior surface cavity, consequently damaging the surrounding parenchyma tissue. In some cases, the growth may progress into the formation of a fungal ball also known as an aspergilloma (Figure 19.2) (Maghrabi and Denning, 2017).

19.2.2 Cutaneous Aspergillosis

In addition to individuals who are immunocompromised, the risk of pulmonary aspergillosis is also strongly associated with debilitating respiratory illnesses. Cutaneous aspergillosis is no exception for individuals with impaired immune systems. The keratin and epidermal barriers of the skin play a pivotal role in preventing foreign particles or microorganisms, including fungi, from entering

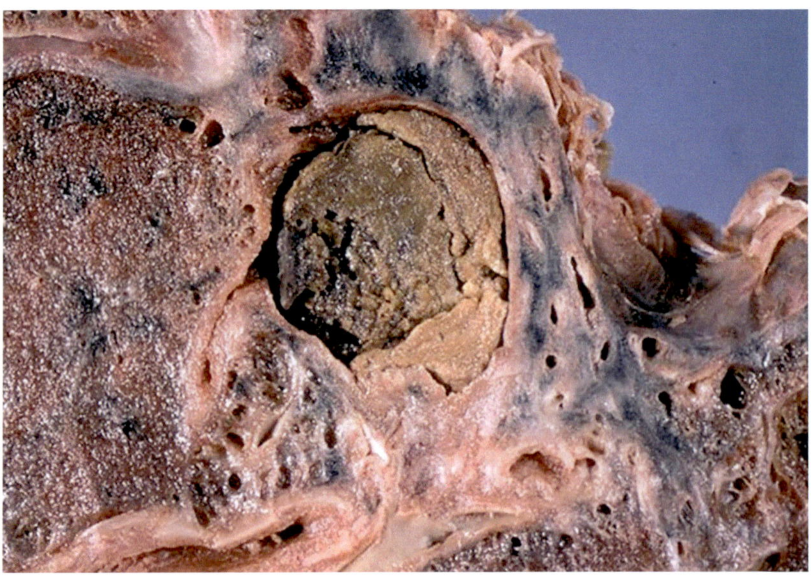

FIGURE 19.2 An aspergilloma present in part of the lung of a patient following a lobectomy (Rosen, 2011).

the human body (Avkan-Oğuz et al., 2020). When the skin is damaged by burns, cuts or when an individual suffers from skin conditions such as skin cancer, the skin barrier weakens. This results in a compromised skin barrier, and the chances of fungal invasion become more prevalent. Long periods of exposure to high conidial counts of *Aspergillus* with a damaged cutaneous barrier increase the likelihood of cutaneous aspergillosis manifestation (Liu, Yang, and Ma, 2017). Cutaneous aspergillosis is defined as primary or secondary, depending on the etiological mechanisms and the initial location of involvement. Primary cutaneous aspergillosis (PCA) occurs commonly at burns, catheter inlets and plaster regions that interfere with skin integrity. It can also appear in areas of maceration, repeated skin trauma to the same site and surgical wounds through direct inoculation (Tahir et al., 2011).

In contrast, secondary cutaneous aspergillosis (SCA) occurs by direct spread from organs that have been already infected with the fungi through cell damage or via hematogenous spread (Avkan-Oğuz et al., 2020). SCA predominantly affects immunocompromised individuals. It is usually characterized by the presence of lesions on the skin which may appear in various forms such as papules, nodules, plaques, or macules (Liu, Yang, and Ma, 2017). In a case study by Avkan-Oğuz et al (2020), it was reported that both *A. fumigatus* and *A. flavus* were isolated from a middle-aged and elderly male, while *A. flavus* was isolated from an elderly woman (Avkan-Oğuz et al., 2020). Such cases do not provide a clear insight into whether *A. fumigatus* independently causes the infection, but *A. flavus* appears to be a possible etiological agent in some instances.

19.2.3 Otomycosis

Otomycosis is defined as a fungal infection that usually affects the external ear canal (Ali et al., 2018). In rare cases, it may also involve the middle ear, provided that the eardrum is perforated (Righetti, Dall'Asta, and Hajslova, 2016). Primary otomycosis occurs in the external auditory canal in an immunocompetent individual, while secondary otomycosis is present in the middle ear alongside the history of existing otitis externa, trauma or postoperative ears. Patients can be immunocompromised or healthy (Prasad et al., 2014). Common causative agents of this disease are *A. fumigatus*, *A. flavus*, *A. niger* and *Candida* spp., with *A. niger* being the most prevalent (Prasad et al., 2014). Otomycosis can be characterized by inflammation, scaling, itching, pain, feeling of blocked ears, tinnitus, deafness and otoscopy revealing masses of hyphae, spores or curd-like gray/white discharge (Anwar and Gohar, 2014).

19.3 MODE OF ACTION OF INFECTION

Aspergillus spp. are primarily transmitted through inhalation of the conidia. They begin to infect epithelial cells in several ways, targeting damaged cells. For the fungus to establish an infection, it must evade recognition by the immune system, which is achieved through alterations of the β-1,3 glucan and chitin with different molecules (Hernández-Chávez et al., 2017). A weakened immune system generally fails to recognize the fungi. Fumagillin and gliotoxin are compounds produced by the fungi and are known to be involved in the pathogenicity of the fungi (Dagenais and Keller, 2009).

Aspergillus fumigatus conidia that survive mucociliary clearance attach to damaged respiratory epithelial cells and the basement membrane, using sialic acid residues (Latgé and Chamilos, 2020). The conidia are then engulfed into a phagosome, where the conidia can germinate. With the additional risk associated with immunocompromised individuals, chances of macrophage evasion are highly probable. When this occurs, *Aspergillus* spp. triggers an inflammatory response after germination in the phagosome. The dissemination of *Aspergillus* spp. involves hyphal growth and the invasion of host endothelial cells that line the blood cells near the lungs (Dagenais and Keller, 2009). Hyphae move through the cells into the interior of the blood vessels. Fragments of the fungi then adhere to the luminal side of other endothelial cells in the blood (Latgé and Chamilos, 2020).

The differential hyphal growth away and toward the lumen is accomplished by using various mechanisms that are currently not well understood.

19.4 CURRENT TREATMENTS AND SIDE EFFECTS

Current treatments available for the treatment of aspergillosis are mostly synthetic drugs, including triazoles, liposomal polyenes and echinocandins (Jenks and Hoenigl, 2018). These drugs have been reported to have antifungal activity and play significant roles in mitigating the effects of *Aspergillus*-related infections (Kousha, Tadi, and Soubani, 2011). With the exception of otomycosis, treatments for all the previously mentioned infections are similar. This includes the use of broad-spectrum antifungal drugs and surgical procedures in severe cases such as lung or skin surgery, depending on the type of aspergillosis (Alastruey-Izquierdo et al., 2018).

Amphotericin B (AmB) is a polyene drug that binds to ergosterol in fungal membranes. The binding action forms pores in the membrane, which eventually leads to death due to cell lysis (Jenks and Hoenigl, 2018). It has been a preferred antifungal agent for most patients, having a recommended dose of less than 2 mg/kg per day (Jenks and Hoenigl, 2018). Triazoles such as voriconazole, itraconazole and posaconazole have been introduced as part of aspergillosis treatments. These are heterocyclic compounds that bind to lanosterol-14-α-demethylase, which prevents the conversion of lanosterol to ergosterol. This in turn inhibits the synthesis of ergosterol (Sherif and Segal, 2010). The triazole compound, voriconazole, is more effective than AmB; however, this drug shows significant inter-individual variability (Alastruey-Izquierdo et al., 2018). In some cases, the voriconazole dosage may need to be increased over time. Finally, echinocandins are lipopeptides that act as fungistatic by inhibiting fungal wall synthesis. This occurs by echinocandins binding to the enzyme β-(1,3)-D-glucan synthases, thereby preventing hyphal tip growth (Maghrabi and Denning, 2017).

These drugs are administered orally in various doses. The dosage can have different effects on different individuals. In some instances, the previously mentioned compounds can be taken together as mixtures for better efficacy, especially when one type of treatment shows no sign of symptom amelioration (Alastruey-Izquierdo et al., 2018).

The treatment of otomycosis involves eliminating predisposing factors. This includes nail inspection that ensures that there will not be continuous infection by unintentional interaction of the hands. The thorough removal of the debris in the ear canal must also be performed. This ensures that there is no remaining debris that may re-infect the ear canal (Prasad et al., 2014). Syringing is generally practiced, alongside the application of fungicidal drops (containing clotrimazole) (Prasad et al., 2014). A more popular and traditional approach also involves the application of fungicidal creams containing ketoconazole or fluconazole (Prasad et al., 2014).

Surgery for the previously mentioned infections (especially pulmonary, cutaneous and otomycosis aspergillosis) is usually employed when long-term antifungal treatments are ineffective, intolerable or unattainable. It involves the removal of aspergillomas forming at an infected site, fungal debris that precipitates or removal of infected tissue. However, postoperative complications have been reported to become more frequent without any insight to elaborate on the possible reasons (Maghrabi and Denning, 2017).

19.4.1 Side Effects

Although AmB is a popular treatment against aspergillosis, Jenks and Hoenigl (2018) reported that there have been a number of adverse side effects. These include nephrotoxicity, electrolyte disturbances and hypersensitivity. These are side effects which need to be considered carefully before the use of AmB. Although triazoles have a better effect than AmB treatments, patients tend to show tolerance to triazoles, and this leads to an increase in dosage (Jenks and Hoenigl, 2018). This suggests that triazoles not only might become less effective after prolonged use, but the patients who are prescribed long-term use of these treatments might become more tolerant to the drug. Moreover, the extended use of triazoles increases renal toxicity and nephrotoxicity (Alastruey-Izquierdo et al., 2018). Echinocandins are merely fungistatic for *Aspergillus* by preventing hyphal tip growth, and are thus reserved for patients who have failed prior treatments or have a renal impairment that would lead to the exclusion of drugs causing renal and nephrotoxicity (Maghrabi and Denning, 2017).

The prolonged use of synthetic antifungal agents has been reported to be less effective for patients and/or cause increased toxicity to internal organs (Afolayan et al., 2002; Bobbarala et al., 2009; Njoki, Okoth, and Wachira, 2017). In some instances, the etiological fungi confer resistance to prescribed treatments. The prospect of investigating medicinal plants with antifungal activity against *Aspergillus* has thus garnered great interest (Njoki, Okoth, and Wachira, 2017).

19.5 NATURAL TREATMENTS

Medicinal plants have been the basis for medicinal treatment throughout much of human history. To this day, traditional medicine is widely practiced by people across the world for various health challenges (Dikhoba et al., 2019). They offer alternative treatments to conventional drug use. Fungal infections have widely been treated by medicinal plants traditionally—not only for their ease of access to the community, but also because they are less expensive (Dikhoba et al., 2019). For example, *Aloe ferox* is considered the most common *Aloe* species in South Africa, and has been well documented as an alternative medicine (Mahomoodally, 2013). It is commonly used as a laxative; however, a study has reported that *Aloe* species showed anti-inflammatory and antimicrobial properties (Mahomoodally, 2013). *Aloe* species have also been reported to be taken orally with a mixture of other medicinal plants (Mahomoodally, 2013). Another endemic South African fynbos species, *Aspalathus linearis* (Brum.f.), also known as rooibos, has been reported to be used beyond its value as tea such as a food additive and in the cosmetic industry (Mahomoodally, 2013).

The tested compounds from rooibos such as chrysoeriol have been reported to show a selective bronchodilator effect, showing potential for respiratory treatments (Khan and Gilani, 2006). In addition, several plant species such as *Artemisia afra* (W.), *Bulbine* spp., *Carpobrotus edulis* L., *Leonotis leonurus* L., *Lippia javanica* (Brum.f.) Spreng, *Melianthus comosus* Vahl, *Pentanisia prunelloides* (Eckl. & Zeyh.) Walp., *Rauvolfia caffra* (Sond), *Rothmannia capensis* (T.) and *Scilla natalensis* (Lindl.) are used to treat several respiratory ailments such as coughs and difficulty with breathing (Mabona and Van Vuuren, 2013). Furthermore, they are used to treat skin ailments including but not limited to skin boils, abscesses, ringworm and fungal acne (Mabona and Van Vuuren, 2013). These plants are reported to have antimicrobial active compounds including essential oils, phenolic compounds and proteins which have been used as antifungal agents (Mabona and Van Vuuren, 2013).

There are secondary metabolites in some plants which have been known to not only be beneficial to the human body but also have anti-*Aspergillus* activity (Njoki, Okoth, and Wachira, 2017). The antifungal activity may be attributed to secondary metabolites produced by the plant such as flavonoids (for example, ρ-coumaric acid), glycosides, saponins, tannins, alkaloids and aromatics, which are known to protect plants from fungi and bacteria (Njoki, Okoth, and Wachira, 2017). These phytochemicals may act individually or in synergy to improve human health (Mahomoodally, 2013). In some instances, the compounds can be further developed into biofungicides (Dikhoba et al., 2019). The studies in the following section report on plants possessing *in vitro* anti-*Aspergillus* activity; however, scientific evidence is yet to be found that reports insight on how to treat aspergillosis.

19.5.1 Medicinal Plants with the Potential to Treat Aspergillosis

Several studies have been carried out which aim to investigate the biological activities of plants against microorganisms. Table 19.1 provides a summary of several plant species reported to have antifungal activity against *Aspergillus* spp. The *in vitro* antifungal activity of *Melaleuca alterifolia* F. Muell. (tea tree) has been investigated (Homeyer et al., 2015), and the extract showed significant antifungal activity against *Aspergillus niger* and *A. fumigatus*, with as little as 4% v/v TTO and 2%

TABLE 19.1

Plant Extracts and Essential Oils from Medicinal Plants with Reported Anti-Aspergillus Activity (Sepahvand et al., 2016)

Scientific Names	Families	Traditional Uses	Plant Parts Used/ Extracts	Fungal Species/ Strain Tested Against	*In vitro* Antifungal Activity			References
					MIC[1] (mg/mL)	MFC[2] (mg/mL)	ZOI[3] (mm)	
Artemisia spp.	Asteraceae	Antimalaria, measles, common cold, cough, muscle weakness, roundworms	Leaves, root, stem/ ethanoic, methanolic, hexane extract	*A. flavus*, *A. niger*	6.25	12.5; 25; 50	6.25–25	(Jhansi Rani et al., 2015)
Thymus eriocalyx (Ronniger) Jalas	Laminaceae	Antiviral, anti-inflammatory	Essential oil extract	*A. flavus*, *A. niger*, *A. parasiticus*	0.5	N/A	22	(Silva et al., 2012)
Aloe vera (L.) Burm.f.	Liliaceae	Acne, promotes oral and digestive health	Leaves/acetone, aqueous, methanol extract	*A. flavus*, *A. niger*, *A. fumigatus*	1,000	N/A	15	(Saniasiaya et al., 2017)
Myrtus communis L.	Myrtaceae	Diarrhea, dysentery, cancer, sinuses, insomnia, diabetes, hypertension, pulmonary disorders	Leaves/acetone, ethyl acetate, essential oil extract	*A. fumigatus*, *A. flavus*	0.1	N/A	40; 20; 10	(Nebras, Al-Muttarri and Mohammed, 2016)
Rosa damascena Herrm.	Rosaceae	Abdominal, chest pains; menstrual bleeding, digestive problems; skin problems and headaches	Petals/essential oil, ethyl acetate	*A. fumigatus*, *A. niger*	437.5 ± 87.8	500 ± 20.41	11–17	(Shohayeb et al., 2014)
Cinnamomum zeylanicum Blume	Lauraceae	Parasiticide, aphrodisiac, haemostatic	Leaves/essential oil	*A. flavus*, *A. niger*, *A. fumigatus*	0.012	N/A	N/A	(Carmo et al., 2008)
Melaleuca alterifolia F.Muell.	Myrtaceae	Skin infections	Bark, leaves/tree tea oil	*A. fumigatus*, *A. niger*	20	N/A	N/A	(Homeyer et al., 2015)
Sarcococca saligna Müll.Arg	Buxacaea	Blood purifier, laxative, muscular analgesic	Leaves/ethanolic extract	*A. niger*, *A. treus*, *A. flavus*, *A. fumigatus*	4	N/A	N/A	(Moghaddam et al., 2010)

Notes: **1** Minimum inhibition concentration; **2** Minimum fungal concentration; **3** Zone of inhibition

v/v TTO against *A. niger* and *A. fumigatus*, respectively, showing a log decline in the relative count after eight hours. Chloroform extracts of seaweeds *Sargassum tenerrimum* and *Turbinaria ornate* have also been reported to show antifungal activity against *A. niger* (Kumar et al., 2015).

The chloroform extract of *S. tenerrimum* showed a significant minimum inhibitory concentration (MIC) of 50 mg/µL with the zone of inhibition (ZOI) of 19 mm, while that of *T. ornate* showed a MIC of 50 mg/µL with the ZOI of 11 mm. The controls fluconazole and amphotericin B showed ZOIs of 11 mm and 18.5 mm, respectively, at MIC concentrations of 0.0001 mg/µL and 0.00018 mg/µL (Kumar et al., 2015). Another plant species, *Sarcococca saligna* Müll. Arg, was investigated against clinical isolates of *A. niger, A. treus, A. flavus* and *A. fumigatus* (Moghaddam et al., 2010). Not only did the ethanolic extract of *S. saligna* show antifungal activity at concentrations of 4 mg/µL against the clinical isolates, but they also showed enhanced antifungal activity when combined with the antifungal agent, fluconazole, showing a larger ZOI as compared to without the extract or the antifungal agent alone (Moghaddam et al. 2010).

This may suggest that there may be synergistic effects between some plants exhibiting antifungal activity and commercial antifungal agents. *Artemisia* spp. (Figure 19.3A), *Solanum aculeastrum* (Figure 19.3B) and *Myrtus communis* (Figure 19.3C) are examples of medicinal plants reported to

FIGURE 19.3 Selected plant species with antifungal activity. A, *Artemisia afra*; B, *Solanum aculeastrum*; C, *Myrtus communis* and emerging species; D, *Emblica officinalis* (Franz, 2020; Kenraiz, 2020; Manuel Anastacio, 2006; Sengai Phoduvhan, 2012).

have anti-*Aspergillus* activity. Leaf extracts from *Artemisia* spp. have been reported to be effective against *Aspergillus* spp., with the ZOI ranging from 6.25–25 mm (Njoki, Okoth, and Wachira, 2017). *Solanum aculeastrum* also showed antifungal activity against *A. flavus* with the positive control, Apron Star (a class III blue active ingredient fungicide containing 20% thiamethoxam + 20% metalaxyl-M + 2% difenoconazole) having a larger ZOI (± 6 mm difference) (Njoki, Okoth, and Wachira, 2017).

A study by Mongalo et al. (2018) showed that *Warburgia salutaris*, *Bauhinia galpinii* and *Harpephyllum caffrum* have *in vitro* antifungal activity against not only *Aspergillus*, but *Fusarium* spp., as well. The organic extracts of these plants were reported to have better anti-proliferative effects than that of most plant species, which may indicate their prolonged use as traditional medicines in treating patients having illnesses associated with malignant tumors (Mongalo et al., 2018). The antifungal activities of these plants were ascribed to compounds from a class of phenols, tannins, terpenoids and flavonoids, which is consistent with the findings by Dikhoba et al. (2019). A study on the antifungal potential of *Emblica officinalis* (Figure 19.3D), *Terminalia bellirica* and *T. chebula* was performed by Gautam et al. (2012). It was observed that *T. chebula* had the most promising antifungal activity against *A. flavus*, *A. fumigatus*, *A. versicolor* and *A. terreus* (Gautam et al., 2012).

While these mentioned plants have been reported to have antifungal activity, current studies have not indicated their potential as a treatment for human aspergillosis. One study, however, has highlighted *Loxostylis alata* as a potential treatment against avian aspergillosis but with no further insight as to how the compounds can be used for further treatment (Suleiman et al., 2012). Literature that provides insight into the potential to treat pulmonary aspergillosis in humans is currently very limited (Suleiman et al., 2012).

19.6 CONCLUSION

Aspergillus infections remain an important cause of morbidity and mortality, especially in immunocompromised patients. The fact that *Aspergillus* spp. are also present in healthy individuals in high numbers warrants a better understanding of the infection. While *in vitro* antifungal tests showed promising results after a certain period, the lack of information on antifungal treatments *in vivo* with both immunodeficient and immunocompetent states necessitates several approaches. The studies may attempt to augment or restore the integrity of the immune system of individuals and ameliorate the symptoms of the infections. With reports showing several plants having antifungal activities *in vitro*, they can be used in conjunction with other plants to induce the better activity of the immune system, as they may have the potential to elicit immune-boosting effects. The immune system may then become strong enough to deal with the invasion of conidia that is especially hard to eliminate with the use of antifungal agents alone.

With several plants showing antifungal activity, it is shown the potential that some of the plants may have in pioneering better treatments against infections. Since trials have not yet been done that determine their efficacy on patients suffering from aspergillosis, further investigations—including the mode of action of how infection occurs—would need to be carried out. In addition, studies on the cytotoxicity of antifungals also need to be carried out to determine the selectivity and need to identify fungal targets. This is to avoid lung tissues as collateral damage. Moreover, multi-drug resistance has been reported from several studies, which is a medical problem faced throughout the world (Moghaddam et al., 2010). The introduction of medicinal plant-based treatments warrants more research on poorly investigated infections with little to no background on plant-based treatments.

REFERENCES

Afolayan, A. J., D. S. Grierson, L. Kambizi, I. Madamombe, and P. J. Masika. 2002. "In Vitro Antifungal Activity of Some South African Medicinal Plants." *South African Journal of Botany* 68, 1: 72–76. doi: 10.1016/s0254-6299(16)30458-6.

Agarwal, Ritesh, and Dundee Road. 2012. "Allergic Bronchopulmonary Aspergillosis: Lessons for the Busy Radiologist." *World Journal of Radiology* 3, 7: 178–181. doi: 10.1378/chest.08-2586.

Alastruey-Izquierdo, Ana, Jacques Cadranel, Holger Flick, Cendrine Godet, Christophe Hennequin, Martin Hoenigl, Chris Kosmidis, et al. 2018. "Treatment of Chronic Pulmonary Aspergillosis: Current Standards and Future Perspectives." *Respiration* 96, 2: 159–170. doi: 10.1159/000489474.

Ali, Khaled, Mahmood A. Hamed, Hameda Hassan, Amira Esmail, and Abeer Sheneef. 2018. "Identification of Fungal Pathogens in Otomycosis and Their Drug Sensitivity: Our Experience." *International Archives of Otorhinolaryngology* 22, 4: 400–403. doi: 10.1055/s-0038-1626702.

Anthony, R. C. 2018. "Severe Influenza Tied to Increased Invasive Pulmonary Aspergillosis Risk." Obtained from MIMS Respirology. https://specialty.mims.com/topic/severe-influenza-tied-to-increased-invasive-pulmonary-aspergillosis-risk- (Accessed 10 October 2021).

Anwar, Khurshid, and Muhammad Shahid Gohar. 2014. "Otomycosis; Clinical Features, Predisposing Factors and Treatment Implications." *Pakistan Journal of Medical Sciences* 30, 3: 2–5. doi: 10.12669/pjms.303.4106.

Avkan-Oğuz, Vildan, Muammer Çelik, Ismail S Satoglu, Mahmut Cem Ergon, and Ahmet emrah Açan. 2020. "Primary Cutaneous Aspergillosis in Immunocompetent Adults: Three Cases and a Review of the Literature." *Cureus* 12, 1: 1–8. doi: 10.7759/cureus.6600.

Bobbarala, Varaprasad, Prasanth Kumar Katikala, K. Chandrasekhar Naidu, and Somasekhar Penumajji. 2009. "Antifungal Activity of Selected Plant Extracts against Phytopathogenic Fungi *Aspergillus niger* F2723." *Indian Journal of Science and Technology* 2, 4: 87–90. doi: 10.17485/ijst/2009/v2i4/29438.

Carmo, Egberto Santos, Edeltrudes De Oliveira Lima, Evandro Leite De Souza, and Frederico Barbosa De Sousa. 2008. "Effect of *Cinnamomum zeylanicum* Blume Essential Oil on the Growth and Morphogenesis of Some Potentially Pathogenic *Aspergillus* Species." *Brazilian Journal of Microbiology* 39, 1: 91–97. doi: 10.1590/S1517-83822008000100021.

Center for Disease Control Public Health Image Library. 1963. Obtained from Wikimedia under the Creative Commons Attribution-Share Alike 2.0 licence https://commons.wikimedia.org/wiki/File:Aspergillus_fumigatus_01.jpg (Accessed 7 July 2020).

Centers for Disease Control and Prevention. 2014. Clinical Features of Aspergillosis – Aspergillosis. February 13. https://www.cdc.gov/fungal/diseases/aspergillosis/symptoms/clinical-features.html (Accessed 24 June 2020).

Dagenais, Taylor R. T., and Nancy P. Keller. 2009. "Pathogenesis of *Aspergillus fumigatus* in Invasive Aspergillosis." *Clinical Microbiology Reviews* 22, 3: 447–465. doi: 10.1128/CMR.00055-08.

Dikhoba, P. M., N. I. Mongalo, E. E. Elgorashi, and T. J. Makhafola. 2019. "Antifungal and Anti-Mycotoxigenic Activity of Selected South African Medicinal Plants Species." *Heliyon* 5, 10: e02668. doi: 10.1016/j.heliyon.2019.e02668.

Dogra, Vikas, Ankit Kumar Sinha, Rajat Saxena, and Deepak Talwar. 2016. "Aspergillus March: From ABPA to Aspergilloma to Subacute Invasive Aspergillosis." *Allergy, Asthma and Clinical Immunology* 12, 1: 1–5. doi: 10.1186/s13223-016-0170-9.

Franz, X. 2020. "*Artemisia afra*." Obtained from Wikimedia under the Creative Commons Attribution-Share Alike 4.0 license. https://commons.wikimedia.org/wiki/File:Artemisia_annua_kz05.jpg (Accessed 7 July 2020)

Gautam, A. K., S. Avasthi, A. Sharma, and R. Bhadauria. 2012. "Antifungal Potential of Triphala Churna Ingredients Against Aspergillus Species Associated with Them During Sstorage." *Pakistan Journal of Biological Sciences* 15, 5: 244–249. doi: 10.3923/pjbs.2012.244.249

Greenberger, Paul A. 2002. "Allergic Bronchopulmonary Aspergillosis." *Journal of Allergy and Clinical Immunology* 110, 5: 685–692. doi: 10.1067/mai.2002.130179.

Hernández-Chávez, Marco, Luis Pérez-García, Gustavo Niño-Vega, and Héctor Mora-Montes. 2017. "Fungal Strategies to Evade the Host Immune Recognition." *Journal of Fungi* 3, 4: 51. doi: 10.3390/jof3040051.

Homeyer, Diane C., Carlos J. Sanchez, Katrin Mende, Miriam L. Beckius, Clinton K. Murray, Joseph C. Wenke, and Kevin S. Akers. 2015. "In Vitro Activity of *Melaleuca alternifolia* (Tea Tree) Oil on Filamentous Fungi and Toxicity to Human Cells." *Medical Mycology* 53, 3: 285–294. doi: 10.1093/mmy/myu072.

Jenks, Jeffrey D., and Martin Hoenigl. 2018. "Treatment of Aspergillosis." *Journal of Fungi* 4, 3: 1–17. doi: 10.3390/jof4030098.

Jhansi Rani, S., P. Supraja, P. Sujitha, A. Kiranmayee, and R. Usha. 2015. "Evaluation of Antibacterial and Antifungal Activity of Evaluation of *Artemisia annua* during Pre and Post Flowering Stages." *International Journal of Current Research* 7, 10: 21581–21587.

Kenraiz. 2020. "Solanum aculeastrum." Obtained from Wikimedia under the Creative Commons Attribution-Share Alike 4.0 license. https://commons.wikimedia.org/wiki/File:Solanum_aculeastrum_1.jpg (Accessed 7 July 2020)

Khan, Arif Ullah, and Anwarul Hassan Gilani. 2006. "Selective Bronchodilatory Effect of Rooibos Tea (*Aspalathus linearis*) and Its Flavonoid, Chrysoeriol." *European Journal of Nutrition* 45, 8: 463–469. doi: 10.1007/s00394-006-0620-0.

Kousha, M., R. Tadi, and A. O. Soubani. 2011. "Pulmonary Aspergillosis: A Clinical Review." *European Respiratory Review* 20, 121: 156–174. doi: 10.1183/09059180.00001011.

Kumar, Isaiah Nirmal, Megha Barot, Rita Kumar, and Nirmal Kumar. 2015. "Phytochemical Analysis and Antifungal Activity of Selected Seaweeds from Okha Coast, Gujarat, India." *Journal of Coastal Life Medicine* 3, 7: 520–525. doi: 10.12980/jclm.3.201514j26.

Lat, Asma, and George R. Thompson. 2011. "Update on the Optimal Use of Voriconazole for Invasive Fungal Infections." *Infection and Drug Resistance* 4, 1: 43–53. doi: 10.2147/IDR.S12714.

Latgé, Jean Paul, and Georgios Chamilos. 2020. "*Aspergillus fumigatus* and Aspergillosis in 2019." *Clinical Microbiology Reviews* 33, 1: 310–350. doi: 10.1128/CMR.00140-18.

Li, Zhi, Gen Lu, and Guangxun Meng. 2019. "Pathogenic Fungal Infection in the Lung." *Frontiers in Immunology* 10, JUL: 1–20. doi: 10.3389/fimmu.2019.01524.

Liu, Xiaoyan, Jun Yang, and Weiyuan Ma. 2017. "Primary Cutaneous Aspergillosis Caused by *Aspergillus fumigatus* in an Immunocompetent Patient." *Medicine (United States)* 96, 48: 3–5. doi: 10.1097/MD.0000000000008916.

Mabona, U., and S. F. Van Vuuren. 2013. "Southern African Medicinal Plants Used to Treat Skin Diseases." *South African Journal of Botany* 87: 175–193. doi: 10.1016/j.sajb.2013.04.002.

Maghrabi, Firas, and David W. Denning. 2017. "The Management of Chronic Pulmonary Aspergillosis: The UK National Aspergillosis Centre Approach," *Current Management of Fungal Infections* 11: 242–251.

Mahomoodally, M. F. 2013. "Traditional Medicines in Africa: An Appraisal of Ten Potent African Medicinal Plants." *Traditional Medicines in Africa: An Appraisal of Ten Potent African Medicinal Plants* 2013: 14. doi: 10.1155/2013/617459.

Manuel Anastacio. 2006. "Myrtus communis." Obtained from Wikimedia under the Creative Commons Attribution-Share Alike 4.0 license. https://commons.wikimedia.org/wiki/File:Myrtus_communis.jpg (Accessed 7 July 2020)

Mongalo, N. I., P. M. Dikhoba, S. O. Soyingbe, and T. J. Makhafola. 2018. "Antifungal, Anti-Oxidant Activity and Cytotoxicity of South African Medicinal Plants against Mycotoxigenic Fungi." *Heliyon* 4, 11: e00973. doi: 10.1016/j.heliyon.2018.e00973.

Nebras, S., A. Al-Muttarri, and H. A. Mohammed. 2016. "Study Effects of Plant Extracts from *Conocarpus erectus* and *Mytrus communis* on the Growth of Some Fungi Isolated from Different Types of Insects." *International Journal of PharmTech* 9, 12: 599–606.

Njoki, Loise M., Sheila A. Okoth, and Peter M. Wachira. 2017. "Effects of Medicinal Plant Extracts and Photosensitization on Aflatoxin Producing *Aspergillus flavus* (Raper and Fennell)." *International Journal of Microbiology* 2017, 2001. doi: 10.1155/2017/5273893.

Prasad, Sampath Chandra, Subbannayya Kotigadde, Manisha Shekhar, Nikhil Dinaker Thada, Prashanth Prabhu, Tina D'Souza, and Kishore Chandra Prasad. 2014. "Primary Otomycosis in the Indian Subcontinent: Predisposing Factors, Microbiology, and Classification." *International Journal of Microbiology,* 5–8. doi: 10.1155/2014/636493.

Righetti, Laura, Chiara Dall'Asta, Jana Hajslova, and Joseph Rubert. 2016. "Metabolomics Approaches and Their Hidden Potential for Explaining the Mycotoxin Contamination Problem." In *Metabolomics Fundamentals and Applications*, i: 120–150. doi: 10.5772/65647.

Rosen, Y. 2011. "Aspergilloma in Old Tuberculous Cavity." Obtained from Wikimedia under the Creative Commons Attribution-Share Alike 2.0 license. https://commons.wikimedia.org/wiki/File:Aspergilloma_in_old_tuberculous_cavity_(6570247863).jpg (Accessed 7 July 2020).

Rudramurthy, Shivaprakash M., Raees A. Paul, Arunaloke Chakrabarti, Johan W. Mouton, and Jacques F. Meis. 2019. "Invasive Aspergillosis by *Aspergillus flavus*: Epidemiology, Diagnosis, Antifungal Resistance, and Management." *Journal of Fungi* 5, 3: 1–23. doi: 10.3390/jof5030055.

Saniasiaya, Jeyasakthy, Rosdan Salim, Irfan Mohamad, and Azian Harun. 2017. "Antifungal Effect of Malaysian *Aloe vera* Leaf Extract on Selected Fungal Species of Pathogenic Otomycosis Species in *in Vitro* Culture Medium." *Oman Medical Journal* 32, 1: 41–46. doi: 10.5001/omj.2017.08.

Sengai Phoduvhan. 2012. "Emblica officinalis." Obtained from Wikimedia under the Creative Commons Attribution-Share Alike 4.0 license. https://commons.wikimedia.org/wiki/File:Emblica_officinalis_fruits.JPG (Accessed 7 July 2020)

Sepahvand, Asghar, Zohre Eftekhari, Mahmoud Rafieian-Kopaei, and Setareh Soroush. 2016. "Phytotherapy in *Aspergillus*: An Overview of the Most Important Medicinal Plants Affecting *Aspergillus*." *International Journal of PharmTech Research* 9, 6: 274–281.

Shah, Ashok, and Chandramani Panjabi. 2014. "Allergic Aspergillosis of the Respiratory Tract," *European Respiratory Update* 1: 8–29. doi: 10.1183/09059180.00007413.

Sherif, R., and B. H. Segal. 2010. "Pulmonary Aspergillosis: Clinical Presentation, Diagnostic Tests, Management and Complications." *Current Opinion in Pulmonary Medicine* 16, 3: 242–250. doi: 10.1097/MCP.0b013e328337d6de

Shohayeb, Mohamed, El Sayed S. Abdel-Hameed, Salih A. Bazaid, and Ibrahim Maghrabi. 2014. "Antibacterial and Antifungal Activity of *Rosa damascena* MILL. Essential Oil, Different Extracts of Rose Petals." *Global Journal of Pharmacology* 8, 1: 1–7. doi: 10.5829/idosi.gjp.2014.8.1.81275.

Silva, Fernanda C. da, Sara M. Chalfoun, Virgínia M. de Siqueira, Deila M.dos S. Botelho, Nelson Lima, and Luís R. Batista. 2012. "Evaluation of Antifungal Activity of Essential Oils against Potentially Mycotoxigenic *Aspergillus flavus* and *Aspergillus parasiticus*." *Brazilian Journal of Pharmacognosy* 22, 5: 1002–1010. doi: 10.1590/S0102695X2012005000052.

Suleiman, Mohammed M., Neil Duncan, Jacobus N. Eloff, and Vinny Naidoo. 2012. "A Controlled Study to Determine the Efficacy of *Loxostylis alata* (Anacardiaceae) in the Treatment of *Aspergillus* in a Chicken (*Gallus domesticus*) Model in Comparison to Ketoconazole." *BMC Veterinary Research* 8: 210. doi: 10.1186/1746-6148-8-210.

Tahir, Chubado, Musa Garbati, Haruna A. Nggada, Edith H.Terna Yawe, and Auwal M. Abubakar. 2011. "Primary Cutaneous Aspergillosis in an Immunocompetent Patient." *Journal of Surgical Technique and Case Report* 3, 2: 94–96. doi: 10.4103/2006-8808.92802.

Thompson, George R. III, and Thomas F. Patterson. 2008. "Pulmonary Aspergillosis." *Seminars in Respiratory and Critical Care Medicine* 28, 2: 103–110. doi: 10.1055/s-2008-1063849.

Xavier, O. Melissa, Alessandro Comarú Pasqualotto, Maria Da Penha Uchoa Sales, Cecília Bittencourt Severo, José J. Peixoto Camargo, and Luiz Carlos Severo. 2008. "Invasive Pulmonary Aspergillosis Due to a Mixed Infection Caused by *Aspergillus flavus* and *Aspergillus fumigatus*." *Revista Iberoamericana de Micologia* 25, 3: 176–178. doi: 10.1016/S1130-1406(08)70041-X.

20 Naturally Occurring Alkaloids with Anti-HIV Activity

Surjeet Verma and Namrita Lall

CONTENTS

20.1 DISEASE INTRODUCTION

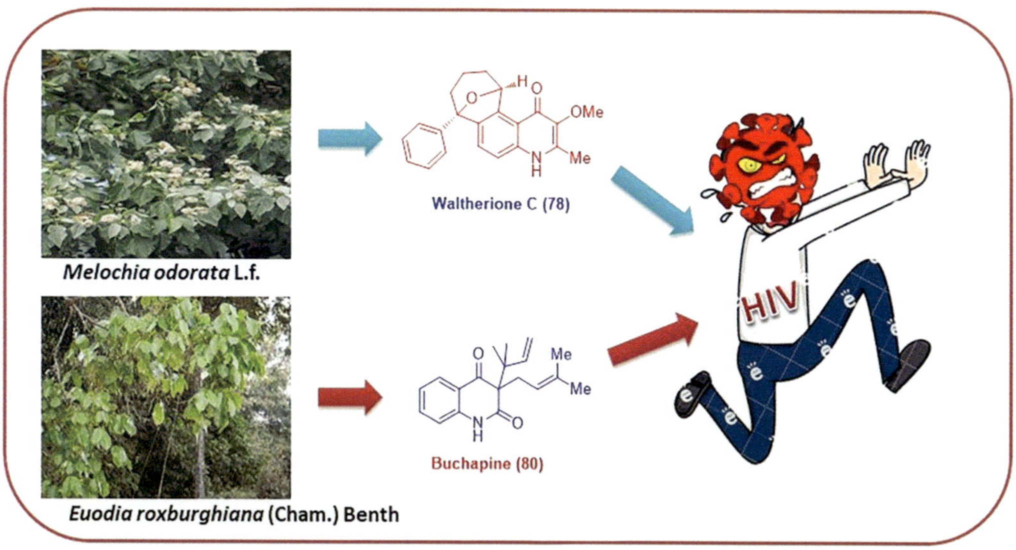

Graphical abstract representing the content of the chapter (Griffiths, 2016; Valke, 2012).

DOI: 10.1201/9781003108375-20

Human immunodeficiency virus (HIV) is the etiological agent of acquired immunodeficiency syndrome (AIDS) and continues to spread through the world population at an alarming rate. HIV continues to be a major global public health issue, having claimed almost 33 million lives so far. An estimated 38 million people were living with HIV at the end of 2019 (WHO, 2020). Few specific groups of the population (key population)—including homosexuals, intravenous drug users, prisoners and others living in closed settings, sex workers and their clients, transgender, adolescent girls and young women in Southern and Eastern Africa—are at high risk of infection. These groups and their sexual partners accounted for more than 60% of all new HIV infections globally among the age group 15–49 years (an estimated 62%) in 2019. According to the World Health Organization (WHO) estimation, more than two-thirds of all people living with HIV (25.7 million) live in the African region (Bedoya et al., 2008, WHO, 2020).

HIV spreads through parenteral, vertical or sexual transmission, and can be transmitted via the exchange of a variety of body fluids from infected people, such as blood, breast milk, semen and vaginal secretions. HIV can be transmitted from a mother to her child during pregnancy and delivery. Currently, more than 90% of all adolescent and adult HIV infections are a consequence of heterosexual intercourse. Three mechanisms of viral spread have been described. The first mechanism involves entry in which cell-free virions bind to host cells by receptor interactions. Second, an infected cell can directly infect another cell, without virion release; and third, dendritic cells (DCs) can capture viral particles by binding to C-type lectins, presenting the virus to surrounding lymphocytes in a cell environment known as a viral synapse. This last mechanism increases the efficiency of infection and prevents viral neutralization by antibodies or complements, and should therefore be considered as an important target to virucides (Bedoya et al., 2008, WHO, 2020).

20.2 THE FUNDAMENTALS OF HIV

HIV targets the immune system and weakens the patient's defense against many infections and some types of cancer. The immunity and immune function are typically measured using the amount of CD4 cells present. The CD4 cells—more commonly known as T cells, T lymphocytes or helper cells—are white blood cells that play an important role in the immune system. The CD4 cell count indicates the health of the immune system and the body's natural defense system against pathogens, infections and illnesses. The CD4 cell count of a person who does not have HIV ranges between 500 and 1,500. People living with HIV who have a CD4 count over 500 are considered usually in good health, while those who have a CD4 cell count below 200 are at high risk of developing serious illnesses. HIV treatment is especially important for people with low CD4 counts (Greta, 2017).

Immunodeficiency results in increased susceptibility to a wide range of infections, cancers and other diseases. The symptoms of HIV vary depending on the stage of infection. In the first few weeks after the initial infection, people may experience no symptoms or an influenza-like illness including fever, headache, rash or sore throat. However, they tend to be most infectious during this time. In severe cases, the people have symptoms including swollen lymph nodes, weight loss, fever, diarrhea and cough. In case of improper diagnosis and treatment, they could develop severe illnesses such as tuberculosis (TB), cryptococcal meningitis, severe bacterial infections and cancers such as lymphomas and Kaposi's sarcoma. However, the most advanced stage of HIV infection is AIDS, which can take many years to develop if not treated, depending on the individual. AIDS is defined by the development of certain cancers, infections or other severe long-term clinical manifestations (WHO, 2020).

20.3 PHYTOMEDICINAL ASPECTS FOR TREATMENT/PREVENTION

Due to various adverse effects, the emergence of drug resistance and the exorbitant prices of antiretroviral drugs and treatment, most of the afflicted population in developing countries are unable to use the currently approved anti-HIV drugs. Thus, the search for novel, effective, safe and affordable anti-HIV agents is an urgent need of the hour. Natural products are the important sources of new drugs and leads, and the best option in this situation. Many plant species and natural products with

Ancistrocladus heyneanus *Argemone Mexicana* *Tripterygium hypoglaucum* *Tripterygium wilfordii* *Schumanniophyton magnificum*

Flueggea virosa *Melochia odorata* *Daphniphyllum longeracemosa* *Euodia roxburghiana* *Leitneria floridana*

FIGURE 20.1 Selected medicinal plants containing anti-HIV alkaloids (BotBin, 2011; Daderot, 2011; Dupont, 2015; Griffiths, 2016; Plantdrew, 2017; Valke, 2012; Vinayaraj, 2012; Qwert1234, 2008).

anti-HIV activity have been isolated and identified. The most promising naturally occurring anti-HIV compounds are flavonoids, coumarins, terpenoids, alkaloids, polyphenols, polysaccharides or proteins (Asres et al., 2005).

In this chapter, only alkaloids and their anti-HIV potential will be discussed. Although the efficacy and side effects of these medicines are not well researched and scientifically validated, herbal medicines are often used in Africa as primary treatment for HIV/AIDS and HIV-related problems such as dermatological disorders, nausea, depression, insomnia and weakness. The ministries of health of several African nations are currently promoting traditional medicines for the treatment of HIV and opportunistic infections. The Ministry of Health in South Africa has recommended Garlic (*Allium sativum* L.), African Potato (*Hypoxis hemerocallidea* Fisch., C.A.Mey. & Avé-Lall.) and Sutherlandia (*Sutherlandia Frutescens* [L.] R.Br., subspecies Microphylla) for its use against HIV (Mills et al., 2005).

The 14 member states of the South African Development Community (SADC)—which includes Angola, Botswana, Democratic Republic of Congo, Lesotho, Malawi, Mauritius, Mozambique, Namibia, Seychelles, South Africa, Swaziland, Tanzania, Zambia and Zimbabwe—support their use (Mills et al., 2005). However, the concurrent use of antiretroviral drugs and some medicinal plants have been reported to have potentially harmful effects, such as, the harmful interactions between garlic, St. John's wort (*Hypericum perforatum* L.), *Sutherlandia frutescens* (L.) R.Br., and vitamins and cannabis (Peltzer et al., 2008). A few selected medicinal plants containing anti-HIV alkaloids are shown in Figure 20.1.

20.4 SCIENTIFIC RESULTS

Although not everyone is able to access HIV testing, treatment and care, by the end of 2019, an estimated 81% of people living with HIV knew their status. Currently, there is no cure for HIV infection. However, effective antiretroviral drugs (ARVs) can control the virus and help prevent onward transmission to other people. At the end of 2019, 67% (25.4 million) of the people living with HIV were receiving antiretroviral therapy (ART) and 59% had achieved suppression of the HIV with no risk of infecting others. ARVs act by binding/inhibiting the different viral target proteins (WHO, 2020). A few selected targets and respective ARVs approved by the U.S. Food and Drug Administration (FDA) are summarized in Table 20.1.

TABLE 20.1

FDA-Approved Anti-HIV Drugs (Jiang et al., 2011)

Target-Specific Drugs	Basic Chemical Moieties	Inhibitory Drugs	Brand Names	FDA Approval Dates
Reverse transcriptase inhibitors (RTIs): inhibit *reverse transcriptase*, an enzyme HIV needs to replicate (make copies of itself)	Nucleoside-based drugs	Abacavir (Abacavir sulfate)*	Ziagen	Dec. 17, 1998
		Emtricitabine (FTC)*	Emtriva	July 2, 2003
		Lamivudine (3TC)	Epivir	Nov. 17, 1995
		Tenofovir disoproxil Fumarate (Tenofovir DF or TDF)	Viread	Oct. 26, 2001
			Retrovir	March 19, 1987
		Aidovudine (Azidothymidine, AZT or ZDV)		
	Non-Nucleoside drugs	Doravirine (DOR)	Pifeltro	Aug. 30, 2018
		Efavirenz (EFV)	Sustiva	Sept. 17, 1998
		Etravirine (ETR)	Intelence	Jan. 18, 2008
		Nevirapine (NVP)	Viramune	Jun. 21, 1996
		Nevirapine (extended-release nevirapine)	Viramune XR (extended-release)	March 25, 2011 May 20, 2011
		Rilpivirine (Rilpivirine hydrochloride, RPV)	Edurant	
Protease Inhibitors (PIs): inhibit *protease*, an enzyme HIV needs to replicate.		Atazanavir (atazanavir sulfate, ATV)	Reyataz Prezista	June 20, 2003 June 23, 2006
		Darunavir (darunavir ethanolate, DRV)	Lexiva Norvir	Oct. 20, 2003 March 1, 1996
		Fosamprenavir (fosamprenavir calcium, FOS-APV, FPV)	Invirase Aptivus	Dec. 6, 1995 June 22, 2005
		Ritonavir (RTV)		
		Saquinavir (saquinavir mesylate, SQV)		
		Tipranavir (TPV)		
Fusion inhibitors: block HIV from entering the CD4 cells of the immune system.		Enfuvirtide (T-20)	Fuzeon	March 13, 2003
CCR5 antagonists: block CCR5 co-receptors on the surface of certain immune cells that HIV needs to enter the cells.		Maraviroc (MVC)	Selzentry	Aug. 6, 2007
Integrase inhibitors: block HIV integrase, an enzyme HIV needs to replicate.		Dolutegravir (dolutegravir sodium, DTG)	Tivicay	Aug. 13, 2013
		Raltegravir (raltegravir potassium, RAL)	Isentress Isentress HD	Oct. 12, 2007 May 26, 2017
Attachment inhibitors: bind to the gp120 protein on the outer surface of HIV, preventing HIV from entering CD4 cells.		Fostemsavir (fostemsavir tromethamine, FTR)	Rukobia	July 2, 2020
Post-attachment inhibitors: block CD4 receptors on the surface of certain immune cells that HIV needs to enter the cells.		Ibalizumab-uiyk (Hu5A8, IBA, Ibalizumab, TMB-355, TNX-355)	Trogarzo	March 6, 2018
Pharmacokinetic enhancers: used in HIV treatment to increase the effectiveness of an HIV medicine included in an HIV regimen.		Cobicistat (COBI, c)	Tybost	Sep. 24, 2014

Note: *These are the synonyms or short names of the drugs

Apart from the aforementioned drugs, FDA has approved several different combinations of these drugs, as well (https://aidsinfo.nih.gov/understanding-hiv-aids/fact-sheets/21/58/fda-approved-hiv-medicines). Despite the anti-HIV activity of several medicinal plants and natural compounds, all the approved drugs are human-made synthetic compounds. Since the natural compounds are synthesized by biological processes in the presence of biological macromolecules, these are safer, less toxic or have fewer side effects because the effect of the drug in the human body is mediated by specific interactions of the drug molecule with biological macromolecules like proteins or nucleic acids. Furthermore, natural compounds are more complex in structure than synthetic molecules and thus possess more novelty and uniqueness. This complexity and novelty allow for more selective binding to targets, such as proteins and other biological macromolecules, which leads to more effective drugs with fewer side effects. However, a nucleoside (precursors of nucleoside-based RTIs), adenosine has been isolated from the aqueous extract of the fruiting bodies of *Cordyceps militaris* (L.) Fr. Interestingly, adenosine displayed HIV-1 protease inhibiting activities even though nucleosid-based anti-HIV drugs are known to inhibit reverse transcriptases (Jiang et al., 2011).

20.5 POTENTIAL PLANTS AND NATURAL ALKALOIDS WITH ANTI-HIV ACTIVITY

20.5.1 Carbazole and Carboline Alkaloids

Two carbazole alkaloids, siamenol (**1**) and mahanimbilol (**2**), exhibited anti-HIV activity with EC_{50} (concentration that inhibited viral replication by 50%) values of 2.6 μM and 8.6 μM, respectively (Figure 20.2). These alkaloids were isolated from the organic extract of *Murraya siamensis* Craib (Meragelman et al., 2000). A *β*-carboline alkaloid, 1-methoxycanthinone (**3**), has been reported to

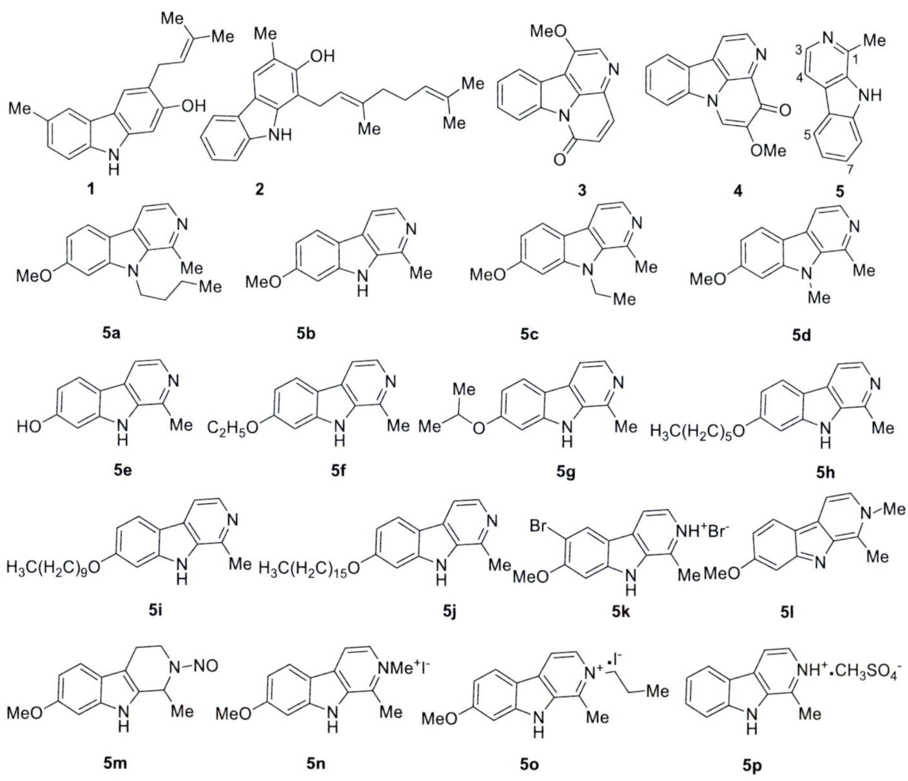

FIGURE 20.2 Anti-HIV carbazole and carboline alkaloids.

suppress the HIV-infected H9 cell growth significantly and had an impressive therapeutic index (TI = IC_{50} of toxicity study divided by EC_{50}) of more than 391. The corresponding IC_{50} (concentration that inhibited uninfected H9 lymphocyte cell growth by 50%) and EC_{50} values were more than 100 µg/mL and 0.256 µg/mL, respectively. This alkaloid was isolated from the MeOH extract of the whole plant material of *Leitneria floridana* Chapm (Xu et al., 2000). Drymaritin (**4**), an isomer of alkaloid **3**, exhibited anti-HIV effects in H9 lymphocytes with an EC_{50} value of 0.699 µg/mL and a TI of 20.6. Drymaritin was isolated from the methanolic extract of *Drymaria diandra* Bl. (Hsieh et al., 2004). Another β-carboline alkaloid, harman (**5**), has been reported for its inhibitory activity against HIV-1 replication in H9 lymphocytic cells with EC_{50} and TI values of 10.7 µM and 10.4 µM, respectively. This alkaloid was isolated from the 95% alcoholic extract of the *Symplocos setchuensis* Brand (Ishida et al., 2001).

In the same study, 28 synthetic analogs of alkaloid **5** were evaluated for their inhibitory activity against HIV-1 replication in H9 lymphocytic cells. Out of these 28 analogs, compound **5a** displayed the most potent activity with EC_{50} and TI values of 0.037 µM and 210, respectively. The EC_{50} value of **5a** was even lower than that of the positive control azidothymidine (AZT) used in that study. The EC_{50} and TI values for AZT were found to be 0.045 µM and 41667, respectively. A similar analog **5b**, with free NH group, displayed 14 times poorer activity (EC_{50} = 0.52 μM, TI: 50.8). This reduction in activity shows that the N-alkyl group plays a significant role in improving the activity. However, the analogs containing N-ethyl (**5c**) and N-methyl (**5d**) groups showed poor activity and TI values (EC_{50} = 0.35 and 1.1 µM, TI: 24.2 and 9.2 respectively) than **5a**, which contains N-butyl group. Similarly, the analog containing free OH group at 7-position (**5e**) displayed almost 22 times poorer activity (EC_{50} = 11.1 µM, TI: 9.7) than **5b**, which contains OMe group at 7-position. However, it is noted that an increase in the alkyl chain length results in a decrease within the activity. For example, analog **5f** (EC_{50} = 6.5 µM, TI: 4.0) and **5g-j** are inactive. The other analogs that showed better activity than harman (**5**) were **5k, 5l, 5m, 5n, 5o,** and **5p** (EC_{50} = 0.22 µM –9.6 µM, TI: 26.8–5.8) (Ishida et al., 2001).

20.5.2 Chromone and Flavonoid Alkaloids

Houghton et al. (1994) reported the anti-HIV activity of nine naturally occurring chromone alkaloids, isolated from the methanolic extract of root bark of *Schumanniophyton magnificum* (K.Schum.) Harms (Figure 20.3). Of those, schumannificine (**6**) displayed the best activity with EC_{50} value of 1.6 µM and selectivity index (SI) of 62.5 µM i.e. with the concentration of compound, which reduced the viability of uninfected cells by 50% (TC_{50} of 100 µM). However, the best EC_{50} value (0.4 µM) was displayed by anhydroschumannificine (**7**) but the SI value was found to be 10. Two other alkaloids, *N*-methylschumannificine (**8**) and schumanniophytine (**9**) showed considerable activity with EC_{50} values of 5 µM and 8 µM respectively and SI values of 20 and 12.5, respectively (Houghton et al., 1994).

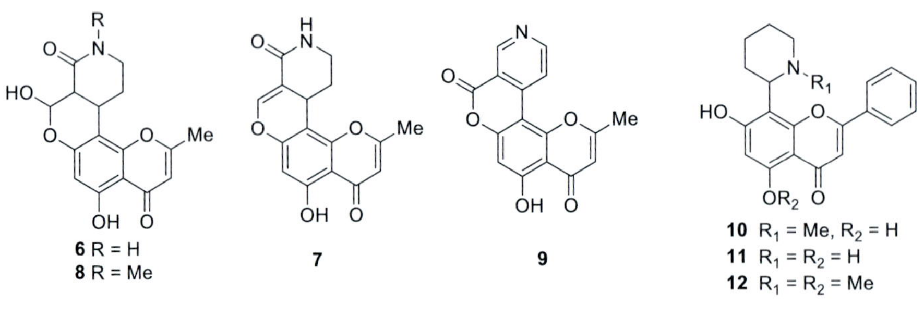

FIGURE 20.3 Anti-HIV chromone and flavonoid alkaloids.

A few *N*- and/or *O*-acyl derivatives of **6** were synthesized in this study and evaluated against HIV, but none of those showed promising activity compared to the parent compound. This shows that the NH and OH groups in **6** are essential for the activity and/or increasing lipophilicity to reduce the anti-HIV potential of **6** (Houghton et al., 1994). Three flavonoid alkaloids—*O*-demethylbuchenavianine (**10**), *N*, *O*-didemethylbuchenavianine (**11**) and buchenavianine (**12**)—were extracted and isolated with organic solvents from the leaves of *Buchenavia capitata* (Vahl) Eichler. Of these, alkaloids **10** exhibited prominent anti-HIV activity with EC_{50} value of 0.26 μM. However, the TI value was around 2.5 (IC_{50}, 0.66 μM). Alkaloid **11** showed poor activity with equal EC_{50} and IC_{50} values of 142 μM (TI = 1). Alkaloid **12** showed less than 50% inhibition at the highest test concentration (Beutler et al., 1992).

20.5.3 Guanidine Alkaloids

Seven polycyclic guanidine alkaloids, batzelladines L, M and N (**13–15**), ptilomycalin A (**16**), crambescidine 800 (**17**), batzelladine C (**18**), and dehydrobatzelladine C (**19**), isolated from the MeOH extract of Jamaican sponge *Monanchora unguifera* (De Laubenfels), displayed remarkable anti–HIV-1 activity in human peripheral blood mononuclear cells (PBMCs) (Figure 20.4).

The EC_{50} values for alkaloids **16** and **17** were found to be quite low (0.011 μM and 0.04 μM, respectively), while for other compounds (**13–15, 18, 19**), it was in the range of 1.6–7.7 μM. The SI

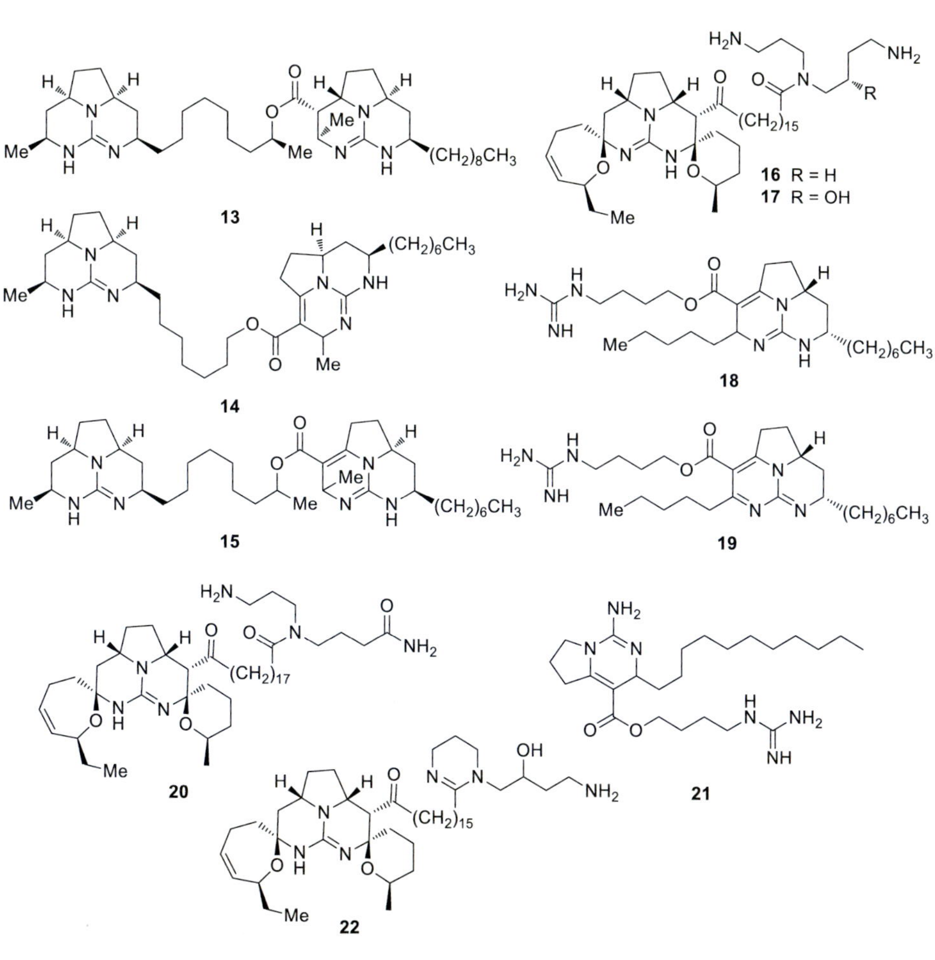

FIGURE 20.4 Anti-HIV guanidine alkaloids.

or TI values for these compounds were not evaluated in this study. However, alkaloids **16** and **17** exhibited cytotoxicity against Vero cells with TC_{50} values of more than 0.67 μM, while other compounds did not show any cytotoxicity against these cells. All these compounds showed prominent cytotoxicity (at 0.03–10 μg/mL concentrations) against an array of cancer cell lines (Hua et al., 2007).

Four other guanidine alkaloids, crambescidin 826 (**20**), dehydrocrambine A (**21**) and fromiamycalin (**22**) have shown anti–HIV-1 activity. Alkaloids **20** and **22** showed promising *in vitro* activity and inhibited HIV-1 envelope-mediated fusion with IC_{50} of 1–3 μM, while compound **21** showed a moderate inhibition, with an IC_{50} of 35 μM. These alkaloids were isolated from the CH_2Cl_2-MeOH (1:1) and MeOH extracts of the marine sponge *Monanchora* sp. (Chang et al., 2003).

20.5.4 Indole Alkaloids

Three unprecedented indole alkaloids, trigonoliimines A-C (**23–25**), with a unique polycyclic system, were isolated from the EtOH extract of leaves of *Trigonostemon lii* Y. T. Chang. Of these, alkaloid **23** showed anti–HIV-1 activity with EC_{50} and TI values of 0.95 μg/mL (2.67 μM) and 7.9, respectively (Figure 20.5). Alkaloid **25** was inactive while **24** was not tested. However, alkaloids **23** and **24** are isomers of each other and differ only in the position methoxy (OMe) group (Tan et al., 2010).

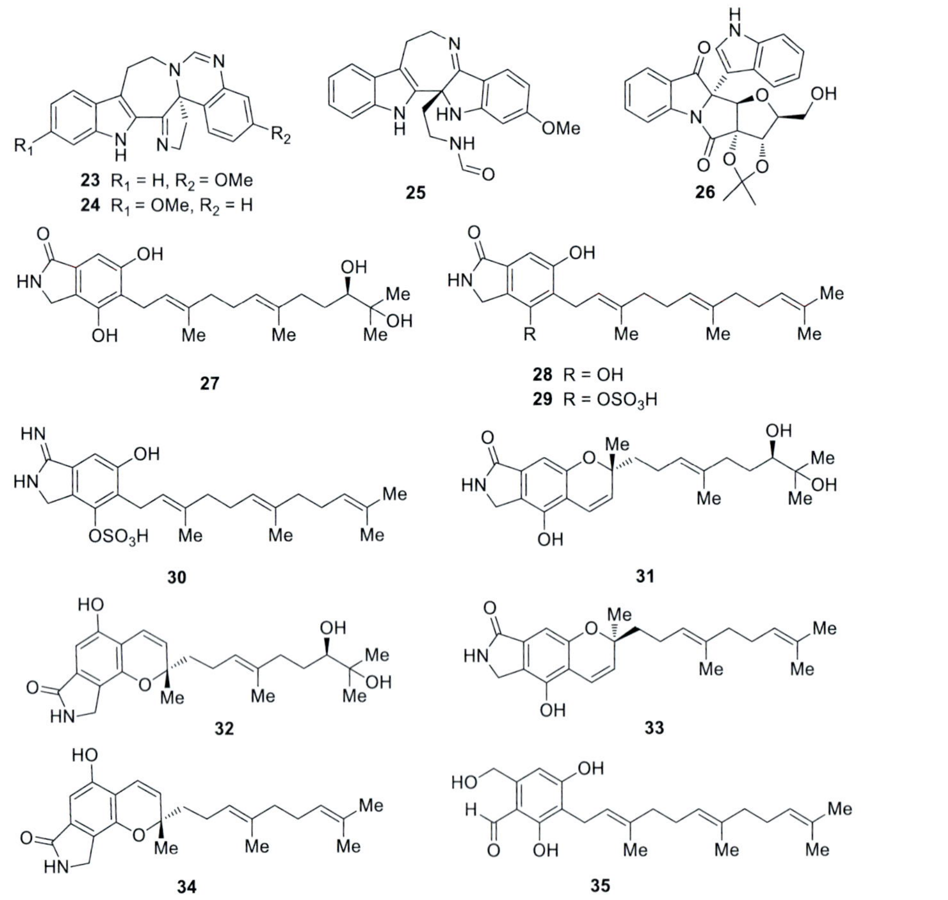

FIGURE 20.5 Anti-HIV indole alkaloids.

Another indole alkaloid, 9α,13α-dihydroxylisopropylidenylisatisine A (**26**), was isolated from the aqueous-alcoholic extract of the leaves of *Isatis indigotica* Fortune ex Lindl and showed moderate anti–HIV-1 activity with EC_{50}, 37.8 μM and SI value of 7.98 (Liu et al., 2007). Few isoindolinone-type alkaloids namely chartarutines A-H (**27–34**) and an analog BR-011 (**35**) isolated from an ethyl acetate extract of sponge-associated fungus, *Stachybotrys chartarum* (Ehrenberg) Hughes, exhibited promising anti–HIV-1 activity. Chartarutines B, G and H (**28, 33** and **34**) showed promising activity (IC_{50}, 4.9–5.58 μM), while other alkaloids showed moderate to poor activity with IC_{50} values ranging between 17.9 μM and 74.0 μM, respectively. However, only chartarutines G and H (**33** and **34**) displayed a considerable selectivity (≥ 18), as the cytotoxic concentration required for the reduction of cell viability (CC_{50}) was found to be > 100 μM. The IC_{50} and CC_{50} values for the positive control (efavirenz) used in that study were 0.65 μM and 40.0 μM, respectively (Li et al., 2014).

20.5.5 Isoquinoline Alkaloids

Norruffscine (**36**) and (-)-8-oxotetrahydropalmatine (**37**) exhibited anti–HIV-1 activity with EC_{50} values of 10.9 μM and 14.1 μM in lymphoblast cells (C8166) and SI values of 45.7 and 18.8, respectively (Figure 20.6). These alkaloids were isolated from a 90% ethanol extract of the aerial parts of

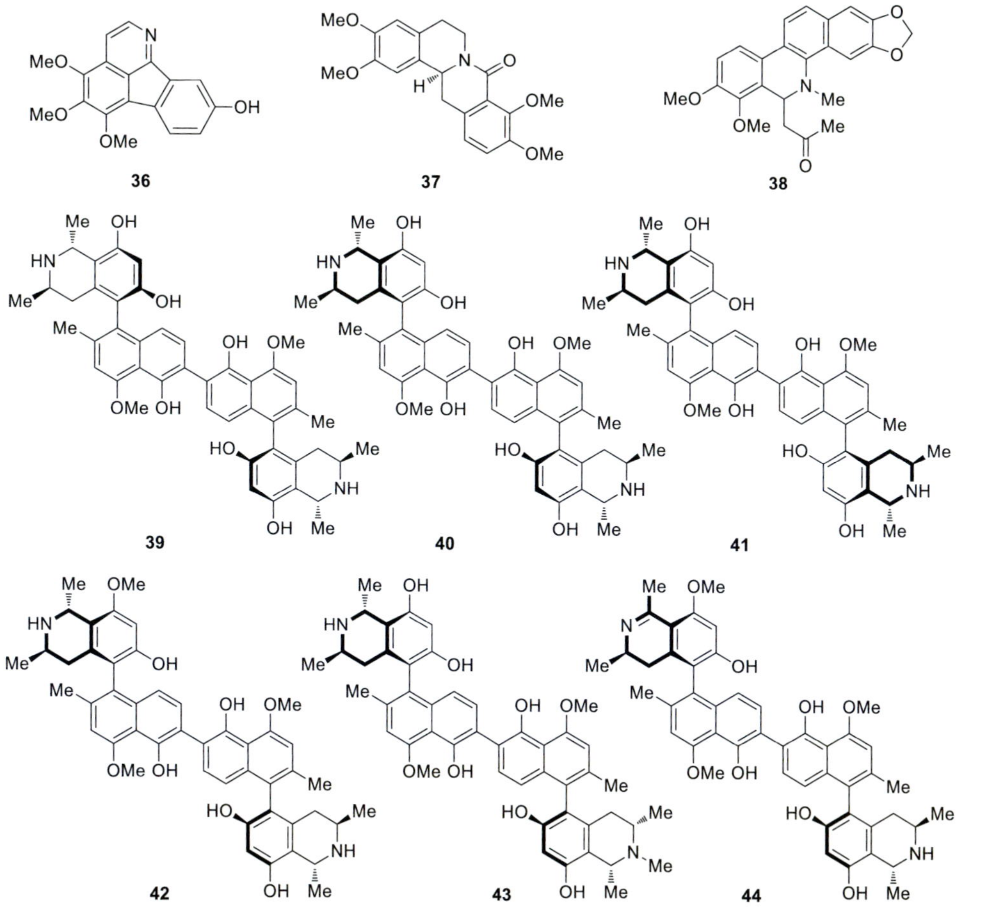

FIGURE 20.6 Anti-HIV isoquinoline alkaloids.

Pericampylus glaucus (Lam.) Merr. The EC_{50} and SI values for the positive control (AZT) used in this study was found to be 0.0108 μM and 471388, respectively (Yan et al., 2008). Another isoquinoline-based (benzo[*c*]phenanthridine) alkaloid, (±)-6-acetonyldihidrochelerythrine (**38**), was isolated from the methanol extract of Formosan *Argemone mexicana* L. (Papaveraceae) and exhibited remarkable anti-HIV activity in H9 lymphocytes with EC_{50} and TI values of 1.77 μg/mL and 14.6, respectively (Chang et al., 2003).

The atropisomeric naphthylisoquinoline alkaloid dimers, michellamines A-C (**39–41**), were isolated from MeOH-CH_2Cl_2 extract of *Ancistrocladus korupensis* D.W.Thomas & Gereau. For comparison of anticytopathic activity, michellamines A-C were tested against HIV-1 (RF strain) and HIV-2 (CBL-20 strain) in CEM-SS human lymphoblastoid target cells. The EC_{50} values for these compounds against HIV-1 were found to be 10 μM, 10 μM and 13 μM, respectively, while those against HIV-2 were 10 μM, 2 μM and 2 μM, respectively. Evidence from this data shows that michellamine B (**40**) is the best active compound among the three (Boyd et al., 1994).

Further confirmation was performed in a separate study of michellamines A and B against a different strain of HIV-2 (NIH-DZ). The result revealed very poor activity of michellamine A compared to B; concentrations of michellamine A were as high as 250 μM, affording only partial protection (20–40%) of MT-2 cells against HIV-2 (NIH-DZ). Michellamines B (**40**) inhibited HIV-induced cell killings and viral replication in a variety of human cell lines, as well as in cultures of human peripheral blood leukocytes and monocytes. It was active against a panel of biologically diverse laboratory (IIIB, RF and LAV) and clinical (214, SKI, TP1, MN, 205, G, MCK and PM16) strains of HIV-1 in CEM-SS cells, as well as the AZT resistant strain G910–6 and the pyridinone-resistant strain A17 of HIV-1 in MT-2 cells. The EC_{50} values against these strains were found to be in the range of 1–13 μM. The EC_{50} values against HIV-2 laboratory strains (ROD, MS, NIH-DZ and CBL-20) in CEM-SS cells were found to be in the range of 2–18 μM (Boyd et al., 1994).

Three other dimeric naphthylisoquinoline alkaloids, michellamines D-F (**42–44**), along michellamine B (**40**), were isolated from the CH_2Cl_2-MeOH and MeOH extracts of the tropical liana *Ancistrocladus korupensis* D.W.Thomas & Gereau. These alkaloids showed *in vitro* anti–HIV-1 activity with EC_{50} values of 3 μM, 4 μM, 2 μM and 6 μM, respectively, in (CEM-SS) host cells against the RF viral strains. The TI values for these alkaloids were found to be 5.7, 4.75, 94 and 20, respectively. These compounds were tested against OC100 strain in the same host cells, and EC_{50} values were found to be 20 μM, 11 μM, 9 μM and 23 μM, respectively. Further tests were conducted on the A17 and G9106 viral strains in MT2 host cells, and EC_{50} values were found from 9–19 μM (Hallock et al., 1997).

20.5.6 Manzamine Alkaloids

Manzamine alkaloids are a unique group of structurally complex natural alkaloids with a *β*-carboline moiety attached to a pentacyclic diamine ring, having eight- and 13-membered rings on a pyrrolo[2,3-*i*]isoquinoline framework and found mainly in marine sponges (Figure 20.7) (Peng et al., 2003).

A variety of manzamine alkaloids namely, (+)-8-hydroxymanzamine A (**45**), 6-deoxymanzamine X (**46**), neo-kauluamine (**47**), manzamine A (**48**), ircinal A (**49**), manzamine F (**50**), manzamine E (**51**), 12,34-oxamanzamine F (**52**), 12,34-oxamanzamine E (**53**), 12,28-oxamanamine A (**54**), manzamine X (**55**) and ircinol A (**56**)—displayed remarkable anti–HIV-1 activity with EC_{50} values from 0.59–22.2 μM. These alkaloids were isolated from the lipophilic extract of an undescribed species of the Indo-Pacific sponge of the genus *Petrosiidae* and a common Indonesian sponge of the genus *Acanthostrongylophora* (Rao et al., 2004, Yousaf et al., 2004).

However, the toxicity and selectivity of the compounds were not mentioned in these reports. The alkaloids manadomanzamine A (**57**), manadomanzamine B (**58**) and xestomanzamine A (**59**) isolated from the acetone extract of an Indonesian sponge *Acanthostrongylophora* sp. (Haplosclerida: Petrosiidae) exhibited anti–HIV-1 activity with EC_{50} values of 7.0 μg/mL, 16.5 μg/mL and 11.2 μg/mL,

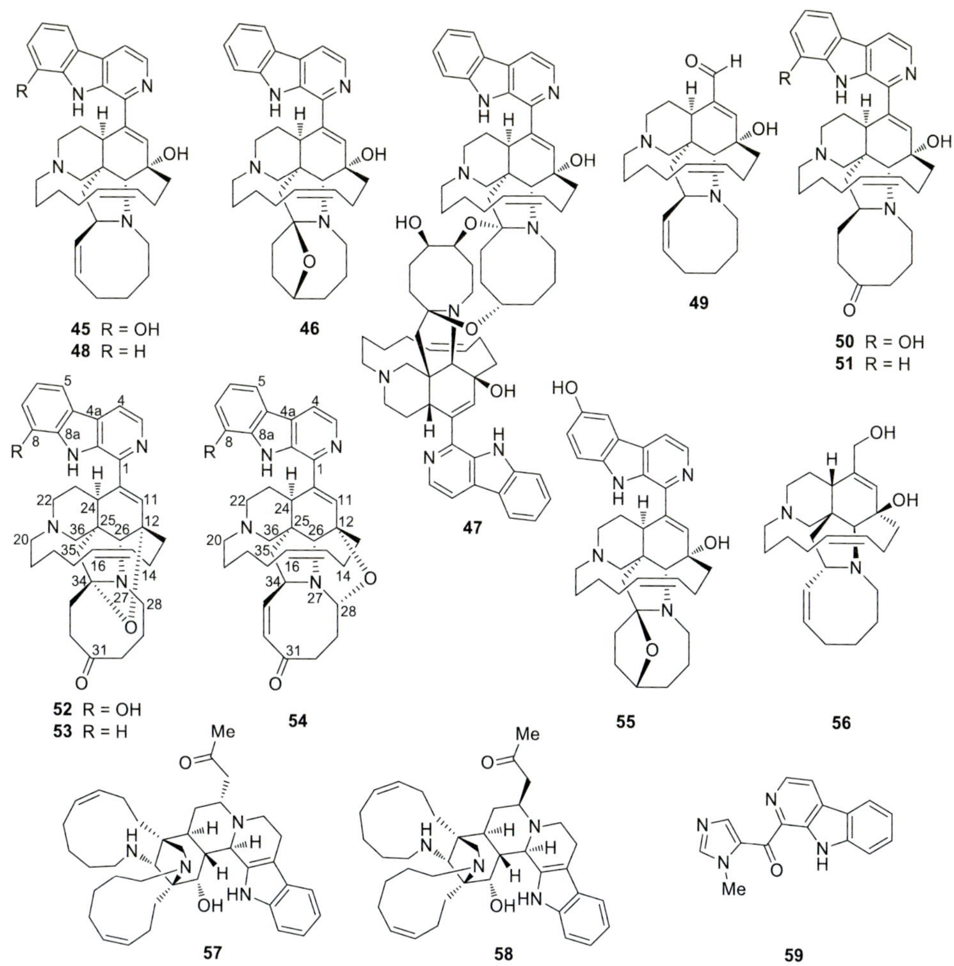

FIGURE 20.7 Anti-HIV manzamine alkaloids.

respectively. Even though the toxicity of these alkaloids was not evaluated, these alkaloids displayed potent cytotoxicity (IC_{50}, 2.5–5.0 μg/mL) against human lung cancer cell lines (A-594) and H-116 colon cancer cells (Peng et al., 2003).

20.5.7 Pyrrole Alkaloids

The amino acid 2-amino-3-(3',4'-dihydroxyphenyl) propionic acid (DOPA) is a key precursor to a broad variety of structurally unique alkaloids in marine invertebrates. A few DOPA-derived pyrrole alkaloids—namely, baculiferins A-O (**60–74**), purpurone (**75**) and ningalin A (**76**)—were isolated from the alcoholic extract of Chinese marine sponge *Iotrochota baculifera* Ridley (Figure 20.8). Alkaloids **60–74** displayed good anti—HIV-1 activity against IIIB strain in human T cells (MT-4) and a HeLa cell clone expressing human CD4 and HIV-LTR-β gal (MAGI or HeLa-CD4-LTR-b-gal cells). The IC_{50} values for these alkaloids against MT-4 cells were found to be in the range of 1.4–8.4 μg/mL, while that against MAGI cells was found to be in the range of < 0.1–4.4 μg/mL. The IC_{50} for purpurone (**75**) was found to be > 25 μg/mL in both the host cells (MT-4 and MAGI), while alkaloid **76** was not tested. These alkaloids act by inhibiting the recombinant gp41 (a trans-membrane protein of HIV-1), viral infectivity factor of HIV-1 (Vif) and a human innate intracellular

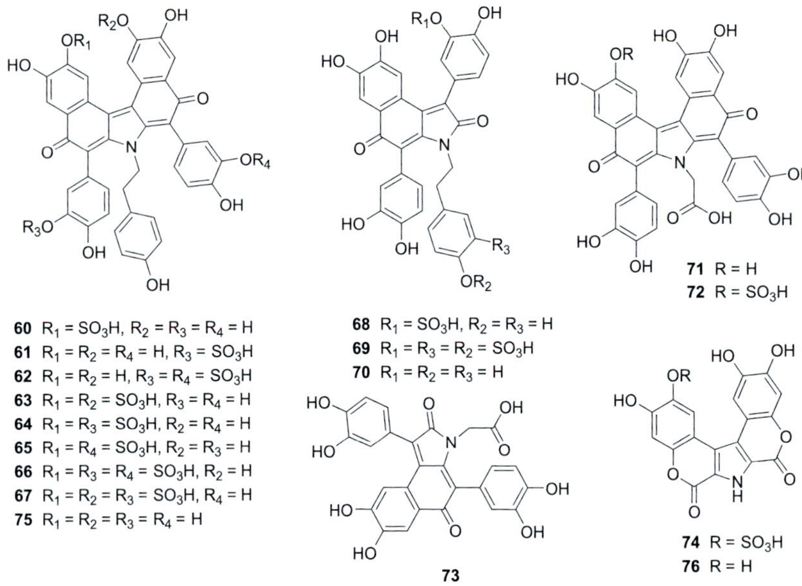

FIGURE 20.8 Anti-HIV pyrrole alkaloids.

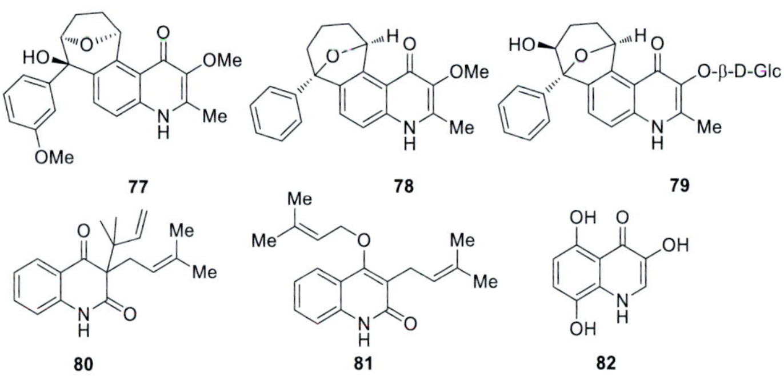

FIGURE 20.9 Anti-HIV quinolone alkaloids.

anti-viral factor (APOBEC3G) as it showed interesting binding affinity to these target proteins (Fan et al., 2010).

20.5.8 Quinolone Alkaloids

Three 4-quinolone alkaloids—waltherione A, C and D (**77–79**)—were isolated from the methanol extract of *Melochia odorata* L. f. (Figure 20.9). Waltherione A and C (**77–78**) showed promising activities in an *in vitro* anti-HIV cytoprotection assay and protected the transformed CEM cells expressing HIV-1 tat and rev genes (CEM-TART) from killing by HIV infection at concentrations of 56.2 and 0.84 μM and inhibition of HIV P24 formation of more than 50% at 1.7 μg/mL and 0.95 μM, respectively. The anti-HIV cytoprotection activity of waltherione C (EC_{50} = 0.84 μM) was significant and outperformed the positive control (AZT) used in that study (EC_{50} = 1.30 μM) (Jadulco et al., 2014).

Two quinolone type alkaloids, buchapine (**80**) and 3-(3-methyl-2-butenyl)-4-[(3-methyl-2-butenyl) oxy]-2(1H)-quinolinone (**81**), exhibited *in vitro* anti–HIV-1 activity in an XTT-tetrazolium assay using human lymphoblastoid (CEM-SS) host cells. These compounds protected the CEM-SS cells from cytopathic effects of HIV-1 with EC_{50} 0.94 μM and 1.64 μM, respectively. The TI values for compounds **80** and **81** were found to be 31 and 16, respectively. Both compounds also showed inhibitory activity in an HIV-1 RT assay with IC_{50} values of 12 μM and 8 μM, respectively. These alkaloids were isolated from the CH_2Cl_2-MeOH extract of *Euodia roxburghiana* (Cham.) Benth. (McCormick et al., 1996).

Another quinolone alkaloid, 3,5,8-trihydroxy-4-quinolone (**82**), was isolated from the EtOAc extract of a sponge of the *Verongia* species. This alkaloid was found to be a potent inhibitor of RNA-directed DNA synthesis of RT of HIV-1 and HIV-2. The inhibition was reversible and non-competitive. Alkaloid **82** bound HIV-1 RTs with high affinity (Ki = 0.46 μM) and inhibited poly$(rA)_n$.oligo$(dT)_{12-18}$-directed DNA synthesis of RT with IC_{50} value of 2.38 μM. However, it was found to be 20-fold less effective when activated DNA was used as substrate (DNA-dependant DNA polymerase activity) for the enzyme (IC_{50}, 43.01 μM) (Loya et al., 1994).

Furthermore, the RNase H function associated with HIV-1 RT was about 100-fold less sensitive to the inhibitor than was the RNA-dependent DNA polymerase activity (IC_{50}, 207 μM). On the other hand, the IC_{50} value for the HIV-2 RT associated RNA-dependent DNA polymerase activity (with poly$[rA]_n$.oligo$[dT]_{12-18}$ as template primer) was found to be 6.44 μM. However, the inhibition of DNA-dependent DNA polymerase of HIV-2 RT with activated DNA was only slightly weaker (IC_{50}, 9.84 μM). As in the case of HIV-1 RT, again **82** was found to be less effective (IC_{50}, 98 μM) in inhibiting RNase H function (Loya et al., 1994).

20.5.9 Sesquiterpene Alkaloids

Several sesquiterpene pyridine alkaloids—tripfordine A (**83**), wilfordine (**84**), wilfortrine (**85**), hypoglaunine A and B (**86** and **87**), triptonine A and B (**88** and **89**), hyponine A and B (**90** and **91**), hyponine E and F (**92** and **93**), forrestine (**94**), cangoronine E-1 (**95**), euonymine (**96**) and neoeuonymine (**97**)—displayed significant anti-HIV activity (Figure 20.10). The EC_{50} values for alkaloids **84–87**, **89**, **90**, **92** and **91**, **94–97** were found to be < 0.1–1.0 μg/mL, which is comparable to the EC_{50} value (0.012 μg/mL) of positive control AZT. The EC_{50} values for alkaloids **88**, **83** and **93** were found to be 2.5 μg/mL, 13.4 μg/mL and 35.2 μg/mL, respectively. The range of TI for these alkaloids was good (11.3–1000+), except for compounds **83** and **93** (1.9 and > 2.8, respectively). The TI value for the AZT was found to be 41,667 μg/mL (Duan et al., 2000; Horiuch et al., 2006).

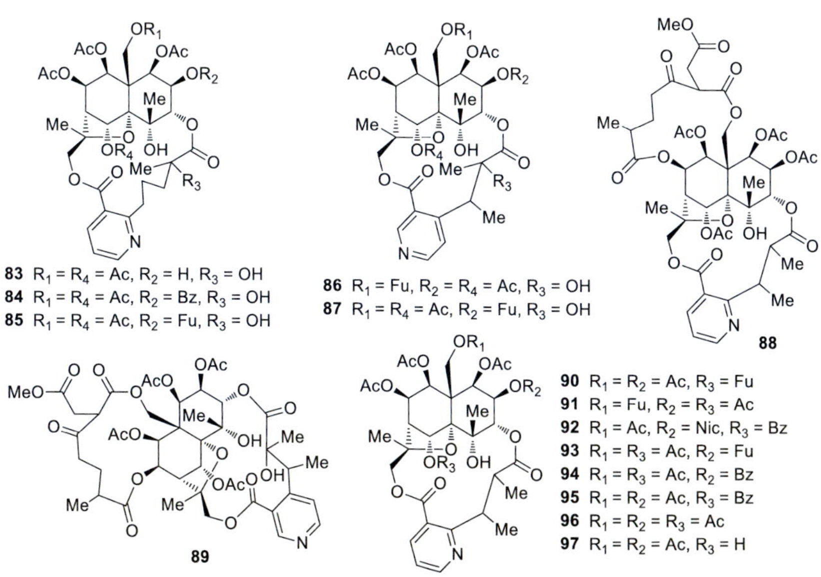

FIGURE 20.10 Anti-HIV sesquiterpene pyridine alkaloids.

20.5.10 Securinega Alkaloids

The Securinega alkaloids are a small family of plant metabolites with fascinating bridged tetracyclic structures and significant biological activity. The name, Securinega alkaloids, was given because of their abundant occurrence in the plant species belonging to the *Securinega* genus in the family Phyllanthaceae. Several *Securinega* alkaloids—including flueggether A (**98**), flueggenines E–I (**99–103**), virosinine A (**104**), fluevirosines E–H (**105–108**), fluevirosinines B-J (**109–117**) and flueggenine D (**118**)—were isolated from the ethanolic extract of *Flueggea virosa* (Roxb. ex Willd.) Royle (Figure 20.11 and Figure 20.12). These alkaloids exhibited weak to moderate *in vitro*

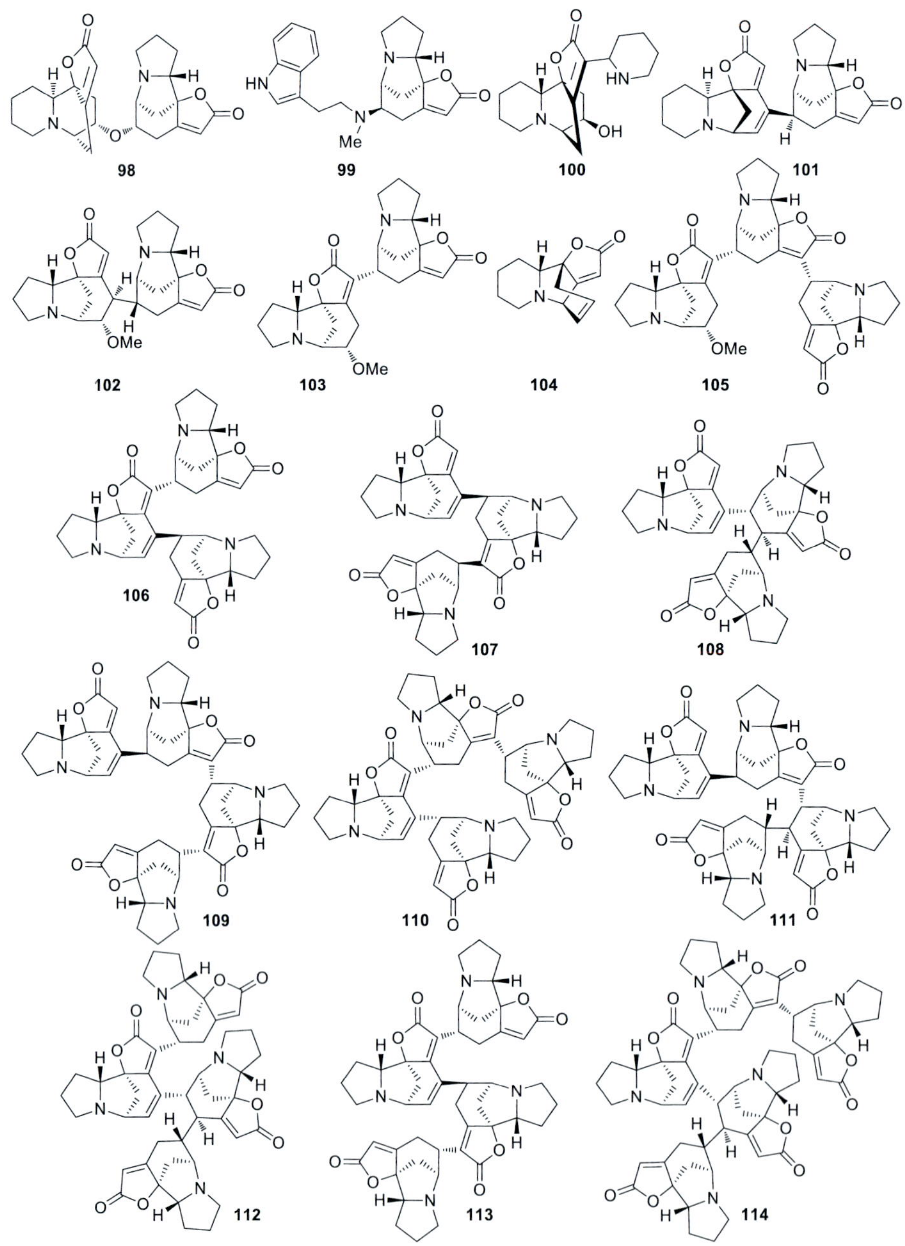

FIGURE 20.11 Anti-HIV Securinega alkaloids.

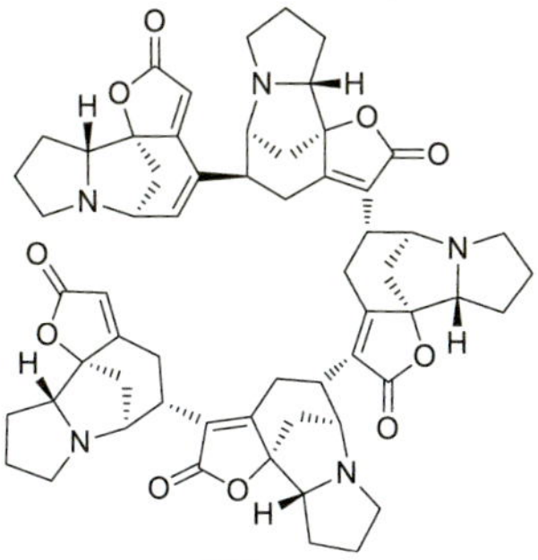

115

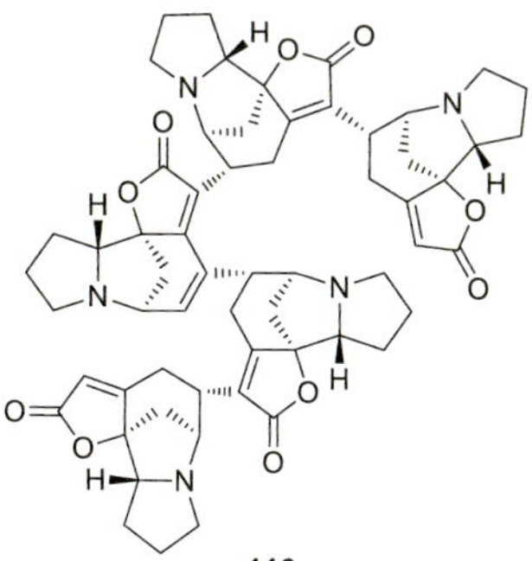

116

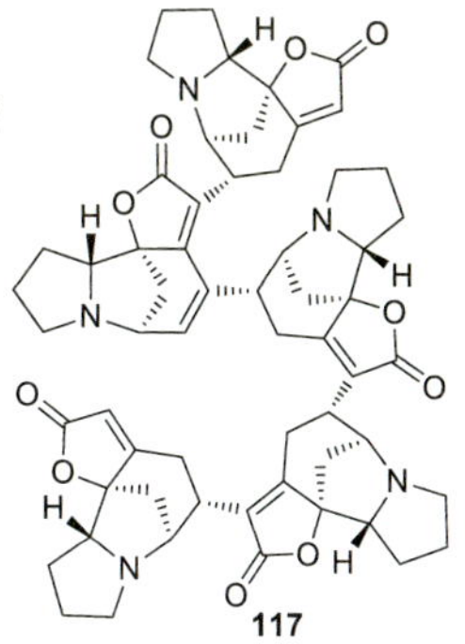

117

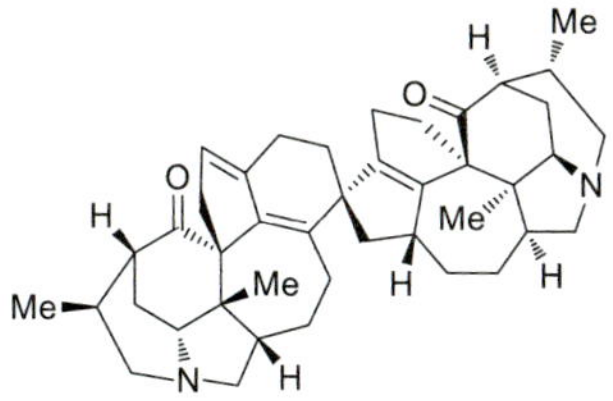

118

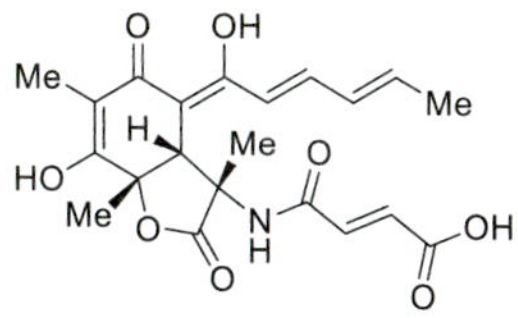

119

FIGURE 20.12 Anti-HIV Securinega, daphniphyllum and sorbicillinoid alkaloids.

anti-HIV activity. The EC_{50} values were found to range from 7.8–122 μM without showing cytotoxicity. The best active compound, flueggenine D. exhibited the EC_{50} value of 7.8 μM (Zhang et al., 2015a, 2015b, 2015c).

Logeracemin A (**119**), the first daphniphyllum alkaloid dimer featuring an unprecedented carbon skeleton with a unique conjugated trispiro[4,5]decane backbone, was isolated from *Daphniphyllum longeracemosum* and exhibited anti-HIV activity on HIV-1 NL 4–3 infected MT-4 cells with an EC_{50} of 4.5 μM and an SI of 6.2 (Xu et al., 2014). A sorbicillinoid alkaloid, sorbicillactone A (**120**), isolated from the saltwater culture of a *Penicillium chrysogenum* strain, isolated from the Mediterranean sponge *Ircinia fasciculate*, showed remarkable anti-HIV activity. Sorbicillactones A protected human T lymphocytes (H9 cells) against the cytopathic effect of HIV-1 and inhibited the expression of viral proteins in the concentration range between 0.3 μg/mL and 3.0 μg/mL. Sorbicillactone A displayed notable activity against the murine leukemic lymphoblasts L5178y tumor cells, with an IC_{50} of 2.2 μg/mL (Bringmann et al., 2005).

20.6 CONCLUSION

HIV is the causative virus for several types of diseases and infections, including cancer. The virus does this by weakening the body's defense system. The most advanced stage of HIV infection is known as AIDS. This syndrome is considered a pandemic, and there are currently no effective treatments for AIDS. For these reasons, AIDS continues to exert a considerable unabated negative effect throughout the world. Surprisingly, the WHO has estimated that around two-thirds of HIV-infected people worldwide are residing in the African region.

Although there is no cure for HIV infections, few antiretroviral drugs (ARVs) have been discovered which can control the virus and help prevent onward transmission to other people. However, these drugs have severe side effects and are very costly; hence, people residing in poor countries or rural areas are unable to access and afford this treatment. This has resulted in the dependence on herbal and traditional medicines for their primary treatment.

Government health agencies and scientists are searching for potentially active, safe and cost-effective medicines from natural resources. As discussed in this chapter, several natural alkaloids and their derivatives have shown more prominent anti-HIV activity than the FDA-approved drugs, though they are not as safe as the positive controls. However, there are several other similar plant species and similar compounds isolated from natural resources, which still need evaluation for their anti-HIV activity. It hoped that scientists will succeed in finding natural medicines for the treatment of HIV-related diseases that are both safe and effective.

REFERENCES

Asres K., Seyoum A., Veeresham C., Bucar F., Gibbons S. 2005. "Naturally derived anti-HIV agents." *Phytotherapy Research* 19: 557–581. doi: 10.1002/ptr.1629.

Bedoya L.M., Alvarez A., Bermejoa M., leza N.G, Beltrana M., Sanchez-Palominoa S., et al. 2008. "Guatemalan plants extracts as virucides against HIV-1 infection." *Phytomedicine* 15: 520–524. doi: 10.1016/j.phymed.2007.10.006.

Beutler J.A., Cardellina II J.H., McMahon J.B., Boyd M.R. 1992. "Anti-HIV and cytotoxic alkaloids from *Buchenavia capitata*." *Journal of Natural Products* 55: 207–213. doi: 10.1021/np50080a008.

BotBin. 2011. "*Tripterygium hypoglaucum*." Obtained from Wikimedia under the Creative Commons Attribution Share-Alike 3.0 license. https://commons.wikimedia.org/wiki/File:Tripterygium_hypoglaucum_BotGardBln07122011D.JPG (Accessed 21 September 2020).

Boyd M.R., Hallock Y.F., Cardellina II J.H., Manfredi K.P., Blunt J.W., McMahon J.B., et al. 1994. "Anti-HIV michellamines from *Ancistrocladus korupensis*." *Journal of Medicinal Chemistry* 37: 1740–1745. doi: 10.1021/jm00038a003.

Bringmann G., Lang G., Gulder T.A.M., Tsuruta H., Muhlbacher J., Maksimenka K., et al. 2005. "The first sorbicillinoid alkaloids, the antileukemic sorbicillactones A and B, from a sponge-derived *Penicillium chrysogenum* strain." *Tetrahedron* 61: 7252–7265. doi: 10.1016/j.tet.2005.05.026.

Chang Y.C., Hsieh P.W., Chang F.R., Wu R.R., Liaw C.C., Lee K.H., et al. 2003. "Two new protopines argemexicaines A and B and the anti-HIV alkaloid 6-acetonyldihydrochelerythrine from formosan *Argemone mexicana*." *Planta Medica* 69: 148–152. doi: 10.1055/s-2003-37710.

Daderot. 2011. "Plant specimen in the Kunming Botanical Garden, Kunming, Yunnan, China." Obtained from Wikimedia under the Public Domain license. https://commons.wikimedia.org/wiki/File:Daphniphyllum_longeracemosa_-_Kunming_Botanical_Garden_-_DSC03188.JPG (Accessed 21 September 2020).

Duan H., Takaishi Y., Imakura Y., Jia Y., Li D., Cosentino L.M., Lee K.H. 2000. "Sesquiterpene alkaloids from *Tripterygium hypoglaucum* and *Tripterygium wilfordii*: A new class of potent anti-HIV agents." *Journal of Natural Products* 63: 357–361. doi: 10.1021/np990281s.

Dupont B. 2015. "S110 Road North of Berg-en-Dal, Kruger NP, South Africa." Obtained from Wikimedia under the Creative Commons Attribution Share-Alike 2.0 license. https://commons.wikimedia.org/wiki/File:Snowberry_Tree_(Flueggea_vrosa)_(16429510476).jpg (Accessed 21 September 2020).

Fan G., Li Z., Shen S., Zeng Y., Yang Y., Xu M., et al. 2010. "Baculiferins A—O, O-sulfated pyrrole alkaloids with anti-HIV-1 activity, from the Chinese marine sponge *Iotrochota baculifera*." *Bioorganic & Medicinal Chemistry* 18: 5466–5474. doi: 10.1016/j.bmc.2010.06.052.

Greta H. 2017. "CD4 cell counts. 2017." www.aidsmap.com/about-hiv/cd4-cell-counts#:~:text=CD4%20cells%20are%20white%20blood,%2Dlymphocytes%2C%20or%20helper%20cells.

Griffiths P. 2016. "*Melochia odorata*." Obtained from Project Noah under the Public Domain license. www.projectnoah.org/spottings/2050126002 (Accessed 21 September 2020).

Hallock Y.F., Manfredi K.P., Dai J.R., Cardellina II J.H., Gulakowski R.J., McMahon J.B., et al. 1997. "Michellamines D-F, new HIV-inhibitory dimeric naphthylisoquinoline alkaloids, and korupensamine e, a new antimalarial monomer, from *Ancistrocladus korupensis*." *Journal of Natural Products* 60: 677–683. doi: 10.1021/np9700679.

Horiuch M., Murakami C., Fukamiya N., Yu D., Chen T.H., Bastow K.F., et al. 2006. "Tripfordines A-C, sesquiterpene pyridine alkaloids from *Tripterygium wilfordii*, and structure Anti-HIV activity relationships of Tripterygium Alkaloids." *Journal of Natural Products* 69: 1271–1274. doi: 10.1021/np060124a.

Houghton P.J., Woldemariam T.Z., Khan A.I., Burke A., Mahmood N. 1994. "Antiviral activity of natural and semi-synthetic chromone alkaloids." *Antiviral Research* 25: 235–244. doi: 10.1016/0166-3542(94)90006-X.

Hsieh P.W., Chang F.R., Lee K.H., Hwang T.L., Chang S.M., Wu Y.C. 2004. "A new anti-HIV alkaloid, drymaritin, and a new C-glycoside flavonoid, diandraflavone, from *Drymaria diandra*." *Journal of Natural Products* 67: 1175–1177. doi: 10.1021/np0400196.

Hua H.M., Peng J., Dunbar D.C., Schinazi R.F., Andrews A.G.C., Cuevas C., et al. 2007. "Batzelladine alkaloids from the caribbean sponge *Monanchora unguifera* and the significant activities against HIV-1 and AIDS opportunistic infectious pathogens." *Tetrahedron* 63: 11179–11188. doi: 10.1016/j.tet.2007.08.005.

Ishida J., Wang H.K., Oyama M., Cosentino M.L., Hu C.Q., Lee K.H. 2001. "Anti-AIDS Agents. 46. anti-HIV activity of harman, an anti-HIV principle from *Symplocos setchuensis*, and its derivatives." *Journal of Natural Products* 64: 958–960. doi: 10.1021/np0101189.

Jadulco R.C., Pond C.D., Wagoner R.M.V., Koch M., Gideon O.G., Matainaho T.K., et al. 2014. "4-Quinolone alkaloids from *Melochia odorata*." *Journal of Natural Products* 77: 183–187. doi: 10.1021/np400847t.

Jiang Y., Wong J.H., Fu M., Ng T.B., Liu Z.K., Wang C.R., et al. 2011. "Isolation of adenosine, iso-sinensetin and dimethylguanosine with antioxidant and HIV-1 protease inhibiting activities from fruiting bodies of *Cordyceps militaris*." *Phytomedicine* 18: 189–193. doi: 10.1016/j.phymed.2010.04.010.

Li Y., Liu D., Cen S., Proksch P., Lin W. 2014. "Isoindolinone-type alkaloids from the sponge-derived fungus *Stachybotrys chartarum*." *Tetrahedron* 70: 7010–7015. doi: 10.1016/j.tet.2014.07.047.

Liu J.F., Jiang Z.Y., Wang R.R., Zheng Y.T., Chen J.J., Zhang X.M., et al. 2007. "Isatisine A, a novel alkaloid with an unprecedented skeleton from leaves of *Isatis indigotica*." *Organic Letters* 9: 4127–4129. doi: 10.1021/ol701540y.

Loya S., Rudi A., Tal R., KashmanY., Loya Y., Hizi A. 1994. "3,5,8-Trihydroxy-4-quinolone, a novel natural inhibitor of the reverse transcriptases of human immunodeficiency viruses type 1 and type 2." *Archives of Biochemistry and Biophysics* 309: 315–322. doi: 10.1006/abbi.1994.1119.

McCormick J.L., McKee T.C., Cardellina II J.H., Boyd M.R. 1996. "HIV inhibitory natural products. 26. quinoline alkaloids from *Euodia roxburghiana*." *Journal of Natural Products* 59: 469–471. doi: 10.1021/np960250m.

Meragelman K.M., McKee T.C., Boyd M.R. 2000 "Siamenol, a new carbazole alkaloid from *Murraya siamensis*." *Journal of Natural Products* 63: 427–428. doi: 10.1021/np990570g.

Mills E., Cooper C., Seely D.I. Kanfer. 2005. "African herbal medicines in the treatment of HIV: Hypoxis and Sutherlandia. An overview of evidence and pharmacology." *Nutrition Journal* 4: 19. doi: 10.1186/1475-2891-4-19.

Peltzer, K., Friend-du Preez, N., Ramlagan, S., Fomundam, H. 2008. "Use of traditional complementary and alternative medicine for HIV patients in KwaZulu-Natal, South Africa." *BMC Public Health* 8: 255. doi: 10.1186/1471-2458-8-255

Peng J., Hu J.F., Kazi A.B., Li Z., Avery M., Peraud O., et al. 2003. "Manadomanzamines A and B: A novel alkaloid ring system with potent activity against Mycobacteria and HIV-1." *Journal of American Chemical Society* 125: 13382–13386. doi: 10.1021/ja030087z.

Plantdrew. 2017. "*Leitneria floridana*." Obtained from Wikimedia under the Creative Commons Attribution Share-Alike 4.0 license. https://commons.wikimedia.org/wiki/File:Leitneria_floridana_at_Missouri_Botanical_Garden.jpg (Accessed 21 September 2020).

Qwert1234 2008. "*Tripterygium regelii*." Obtained from Wikimedia under the Public Domain license. https://commons.wikimedia.org/wiki/File:Tripterygium_regelii_1.JPG (Accessed 21 September 2020).

Rao K.V., Kasanah N., Wahyuono S., Tekwani B.L., Schinazi R.F., Hamann M.T. 2004. "Three new manzamine alkaloids from a common Indonesian sponge and their activity against infectious and tropical parasitic diseases." *Journal of Natural Products* 67: 1314–1318. doi: 10.1021/np0400095.

Tan C.J., Di Y.T., Wang Y.H., Zhang Y., Si Y.K., Zhang Q., et al. 2010. "Three new indole alkaloids from *Trigonostemon lii*." *Organic Letters* 12: 2370–2373. doi: 10.1021/ol100715x.

Valke D. 2012. "Melicope lunu-ankenda (Gaertn.) T.G. Hartley." Obtained from under the Creative Commons Attribution Share-Alike 2.0 license. www.flickr.com/photos/dinesh_valke/6935725823/in/photostream/ (Accessed 21 September 2020).

Vinayaraj 2012. "*Ancistrocladus heyneanus*." Obtained from Wikimedia under the Creative Commons Attribution Share-Alike 3.0 license. https://commons.wikimedia.org/wiki/File:Ancistrocladus_heyneanus_07.JPG (Accessed 22 September 2020).

WHO. 2020. "HIV/AIDS." www.who.int/news-room/fact-sheets/detail/hiv-aids.

Xu J.B., Zhang H., Gan L.S., Han Y.S., Wainberg M.A., Yue J.M. 2014. "Logeracemin A, an anti-HIV daphniphyllum alkaloid dimer with a new carbon skeleton from *Daphniphyllum longeracemosum*." *Journal of American Chemical Society* 136: 7631–7633. doi: 10.1021/ja503995b.

Xu Z., Chang F.R., Wang H.K., Kashiwada Y., McPhail A.T., Bastow K.F., et al. 2000. "Anti-HIV agents 451 and antitumor agents 205.2 two new sesquiterpenes, leitneridanins a and b, and the cytotoxic and anti-HIV principles from *Leitneria floridana*." *Journal of Natural Products* 63: 1712–1715. doi: 10.1021/np000260u.

Yan M.H., Cheng P., Jiang Z.Y., Ma Y.B., Zhang X.M., Zhang F.X., et al. 2008. "Periglaucines A-D, anti-HBV and -HIV-1 alkaloids from *Pericampylus glaucus*." *Journal of Natural Products* 71: 760–763. doi: 10.1021/np070479.

Yousaf M., Hammond N.L., Peng J., Wahyuono S., McIntosh K.A., Charman W.N., et al. 2004. "New manzamine alkaloids from an Indo-Pacific sponge. Pharmacokinetics, oral availability, and the significant activity of several manzamines against HIV-I, AIDS opportunistic infections, and inflammatory diseases." *Journal of Medicinal Chemistry* 47: 3512–3517. doi: 10.1021/jm030475b.

Zhang H., Zhu K.K., Han Y.S., Luo C., Wainberg M.A., Yue J.M. 2015a. "Flueggether A and Virosinine A, anti-HIV alkaloids from *Flueggea virosa*." *Organic Letters* 17: 6274–6277. doi: 10.1021/acs.orglett.5b03320.

Zhang H., Zhu K.K., Han Y.S., Luo C., Wainberg M.A., Yue J.M. 2015b. "New Securinega alkaloids with anti-HIV activity from *Flueggea virosa*." *RSC Advances* 5: 107045–107053. doi: 10.1039/C5RA22191A.

Zhang H., Zhu K.K., Han Y.S., Luo C., Wainberg M.A., Yue J.M. 2015c. "Anti-HIV Securinega alkaloid oligomers from *Flueggea virosa*." *Tetrahedron* 71: 3671–3679. doi: 10.1016/j.tet.2014.10.064.

21 Chemistry and Pharmacology of Naturally Occurring Flavoalkaloids

Rashmi Gaur, Jyoti Gaur and Nikhilesh Kumar

CONTENTS

21.1 INTRODUCTION

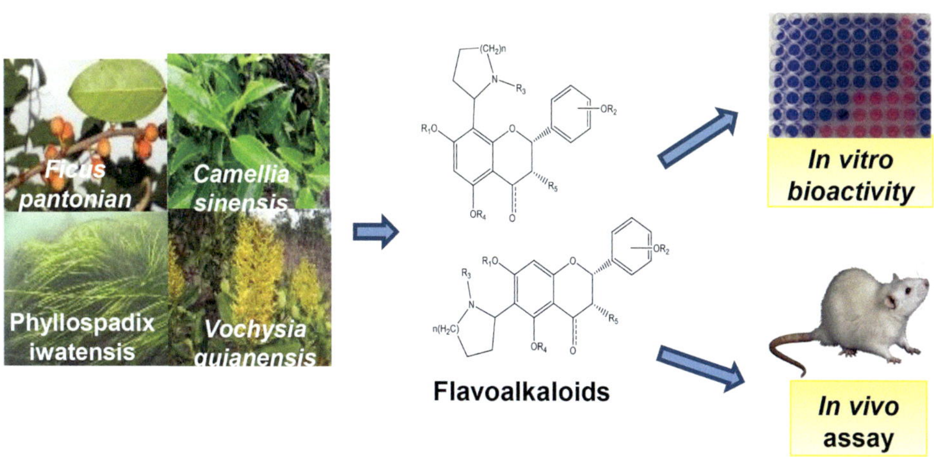

Graphical abstract representing the content of the chapter (AxelBoldt, 2004; Liksakova, 2019; Wikipedie, 2021).

Flavonoids are polyphenolic compounds, which are present in plants and fungi and have various health benefits, especially their antioxidant properties (Pietta, 2000; Veitch and Grayer, 2011). Flavonoids are known to possess the 15-carbon arylchromane structure (Figure 21.1). Flavoalkaloids

DOI: 10.1201/9781003108375-21

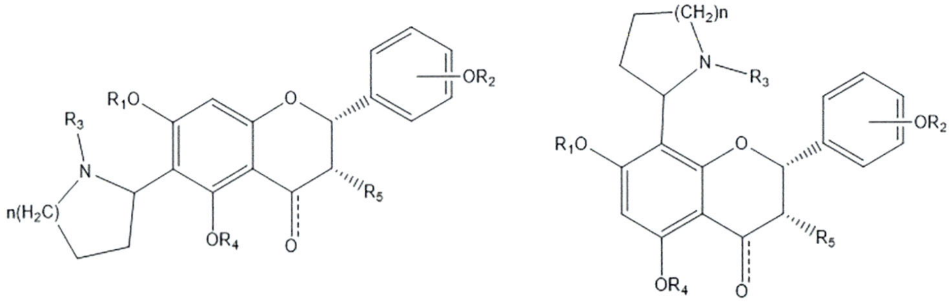

FIGURE 21.1 Basic skeletal structure of flavoalkaloids.

are a special kind of naturally occurring molecules which consist of an alkaloid within the flavonoid structure. The term 'alkaloid' collectively refers to a natural compound, which possesses a basic nitrogen atom. Within flavoalkaloids, the nitrogen is linked at position C (6) or C (8) to the flavonoid structure, which consists of two aromatic rings and an oxygen heterocycle. When compared to flavonoids, flavoalkaloids are fewer in number (Houghton, 1987, 2000; Khadem and Marles, 2012).

The natural products containing a flavonoid skeleton with one or more nitrogen atoms are known as flavoalkaloid, isoflavoalkaloid or neoflavoalkaloid, based upon the skeleton of the flavonoid component. These flavoalkaloids are known for their numerous medicinal properties (Ilkei et al., 2018). The chromone alkaloid's isolation and synthesis have previously been reviewed by Houghton. Later, he reviewed the isolation and bioactivity in 2012. Blair et al. (2017) reviewed the occurrence and synthesis of flavoalkaloids, while Ilkei et al. (2018) reviewed the occurrence and skeletal description of flavoalkaloids. However, no one has covered the biological activity of these flavoalkaloids in depth. Furthermore, after 2018, more flavoalkaloids were identified (Gaur et al., 2020). This chapter will include all the flavoalkaloids that are currently known up until early 2020, describing their isolation and bioactivity.

21.2 EXTRACTION, ISOLATION AND BIOLOGICAL ACTIVITY OF FLAVOALKALOIDS

21.2.1 Flavoalkaloids with Pyrrolidine Ring

The first naturally isolated compounds having both a flavonoid framework and a nitrogen atom were (-)-ficine (**1**) and isoficine (**2**), which were isolated from the fig tree *Ficus pantoniana* King in 1965 (Johns et al., 1965). Both (-)-Ficine (**1**) and isoficine (**2**) are interconvertible via Wesselye Moser rearrangement; the two isomers are in dynamic equilibrium when heated in 70% hydrochloric acid (Johns et al., 1965). The optical rotation of natural (-)-ficine (**1**) is ($[\alpha]_D = -60$), while optical rotation for isoficine (**2**) is unknown (Leete, 1982; Nguyen et al., 2012). (-)-Ficine (**1**) is a potent inhibitor of the cyclin-dependent kinases (CDK) 1 and CDK5 (IC_{50} of 0.04 μM against both isoforms), while the (+)-enantiomer was less active (IC_{50} of 0.39 μM against CDK1 and 0.57 μM against CDK5) (Nguyen et al., 2012). In 1980, a flavoalkaloidsphyllospadine (**3**) was isolated from the Japanese sea-grass *Phyllospadix iwatensis* Makino, which was collected in the Yamagata Prefecture (Figure 21.2) (Takagi et al., 1980).

Later on, the flavoalkaloidvochysine (**4**) was isolated from the fruit of the tropical tree *Vochysia guianensis* Aubl, which occurs in nature as a single racemic diastereomer (Figure 21.2) (Baudouin et al., 1983). In 2006, the epimeric L-proline-flavonoid hybrids prolinalin A (**5**) and B (**6**) were isolated from the shells of the silkworm *Bombyx mori* (Figure 21.2) (Hirayama et al., 2006). The compounds **5** and **6** were subjected to total synthesis, which occurs with most of the pyrrolidine-flavoalkaloids that have been reported. The first total synthesis was performed by Anjaneyulu and Govindachari (1969) and later by Leete (1982). The Mannich reaction of N-methyl-Δ1-pyrrolinium acetate and chrysin resulted in natural products of compounds **1** and **2** in a single step with a very low yield. Nguyen and colleagues, using the Mannich reaction in the presence of pyrroline, afforded a good yield of (±)-isoficine (Nguyen et al., 2011). Using a similar procedure, Baudouin and colleagues synthesized (±)-vochysine (**4**) from a flavan (Baudouin et al., 1983).

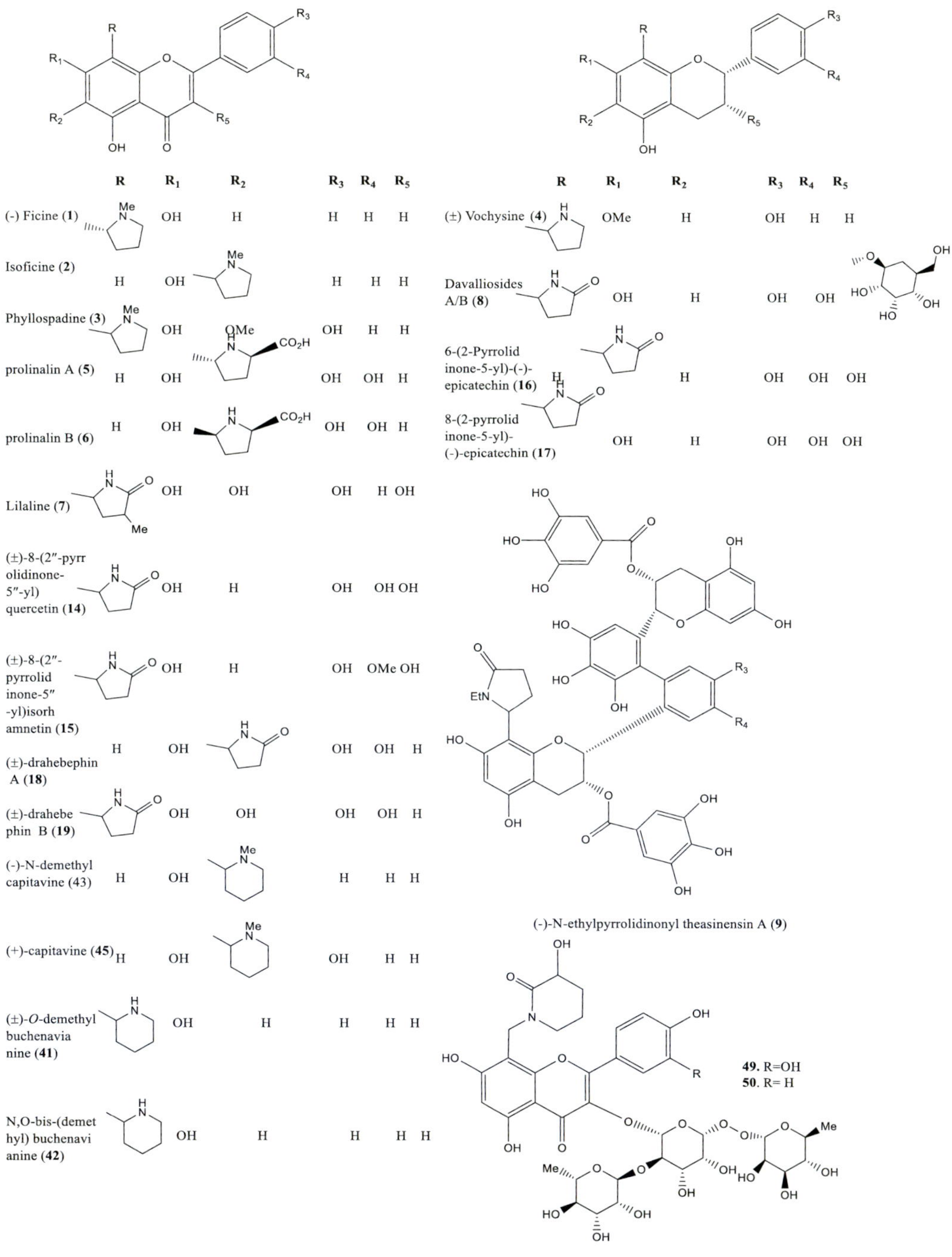

FIGURE 21.2 Structures of naturally occurring flavonoids.

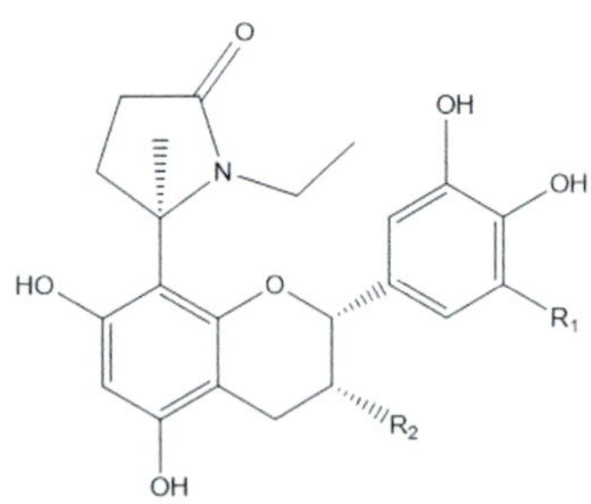

20. R_1=OH, R_2=galloyl (S-EGCG-cThea)
21. R_1=OH, R_2=OH (S-EGC-cThea)
22. R_1=H, R_2=OH (S-EC-cThea)
32. R_1=H, R_2=galloyl
36. R_1=H, R_2=cinnamoyl

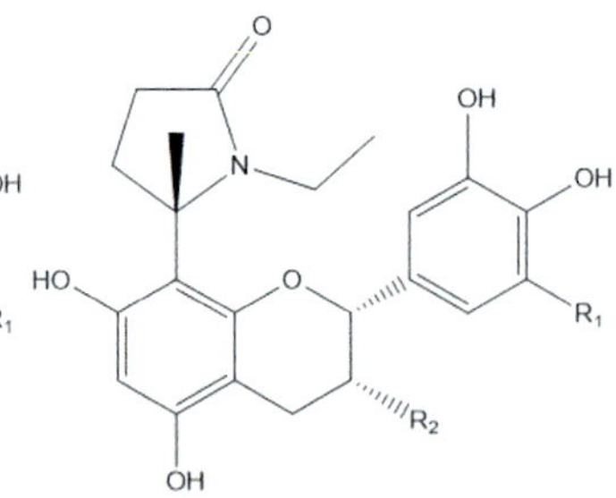

23. R_1=OH, R_2=galloyl (R-EGCG-cThea)
24. R_1=H, R_2=galloyl (R-ECG-cThea)
25. R_1=OH, R_2=OH (S-EGC-cThea)
26. R_1=H, R_2=OH (R-EC-cThea)
29. R_1=OH, R_2=galloyl
35. R_1=H, R_2=cinnamoyl

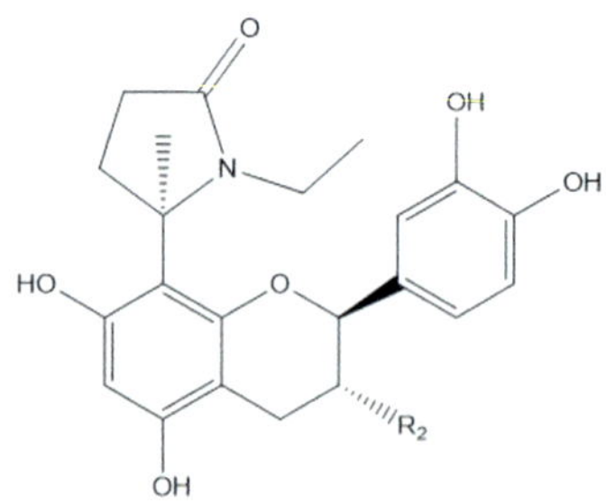

30. R_2 = galloyl

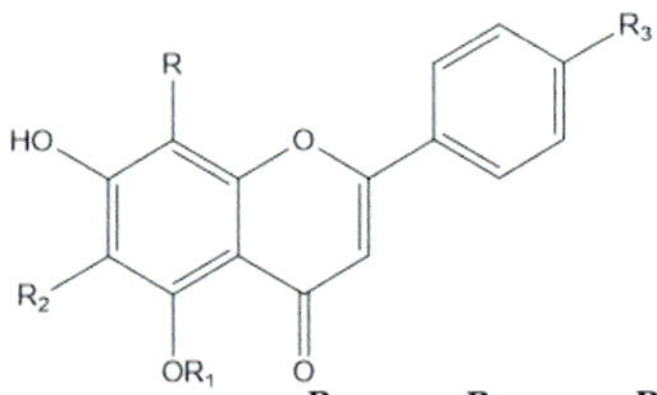

	R1	**R2**	**R3**	**R**
(*S*)-(-)-buchenavianine (**39**)	OMe	H	H	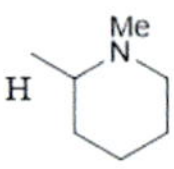
(+)-*N*-demethyl buchenavianine (**40**)	OMe	H	H	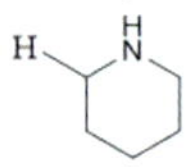
(+)-capitavine (45)	OH	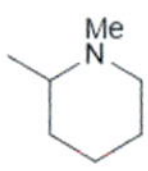	H	H
(-)-4'-hydroxycapitavine (46)	OH	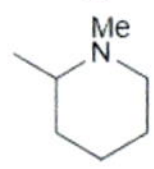	OH	H

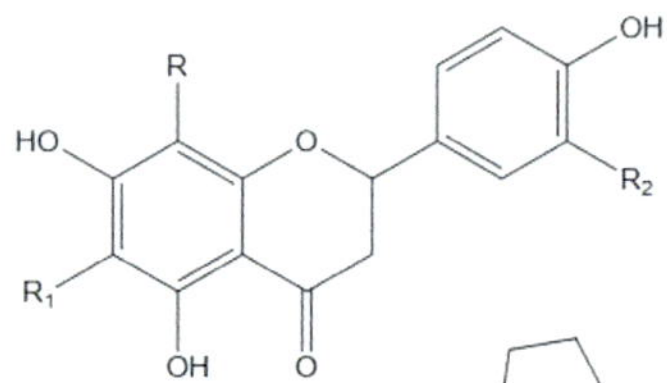

Dracocephin A (**10**); R=R_2=H; R_1= 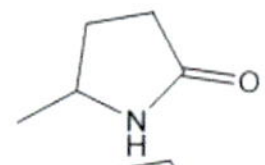

Dracocephin A (**11**); R_1=R_2=H; R= 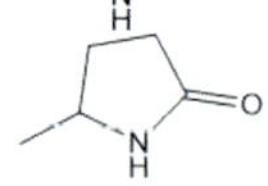

Dracocephin A (**12**); R=H; R_2=OH; R_1= 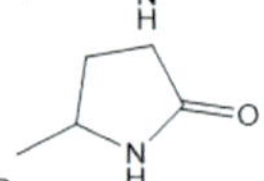

Dracocephin A (**13**); R_1=H; R_2=OH; R_1= 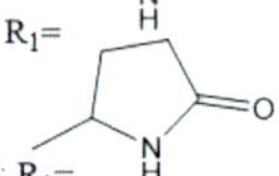

(+)-2,3-dihydro-4'-hydroxycapitavine R=R_2=H; R_1= 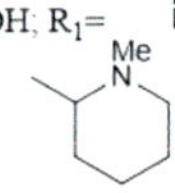

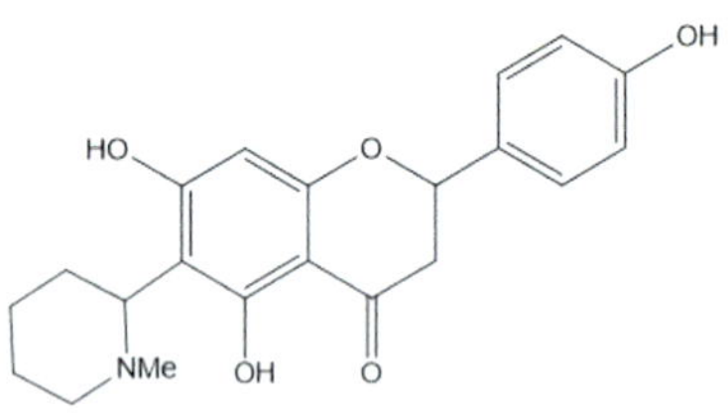

(+)-2,3-dihydro-4'-hydroxycapitavine (47)

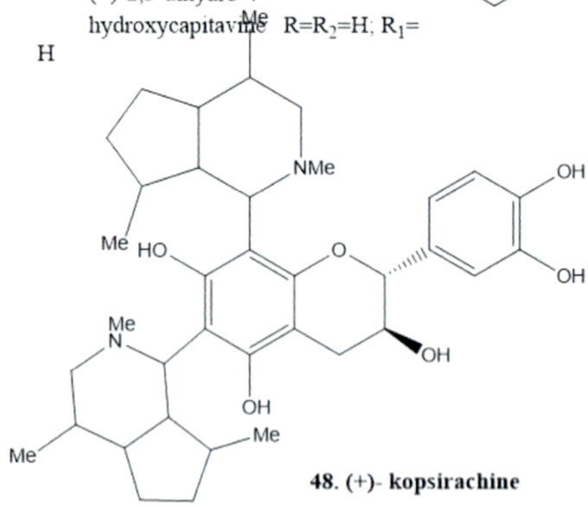

48. (+)- kopsirachine

FIGURE 21.2 (Continued)

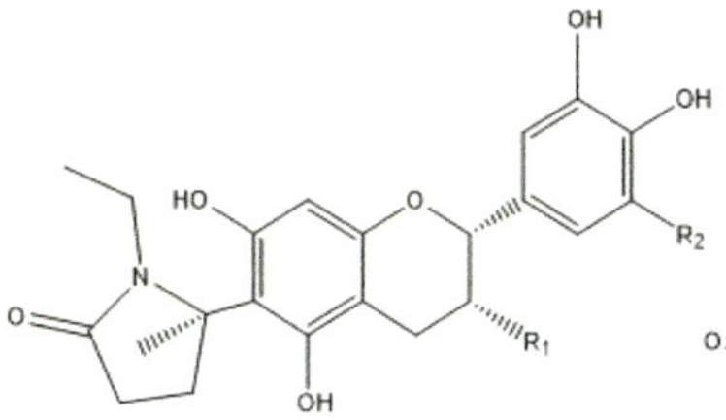

34. R_1=galloyl, R_2=H
27. R_1=galloyl, R_2=OH
37. R_1=cinnamoyl, R_2=H

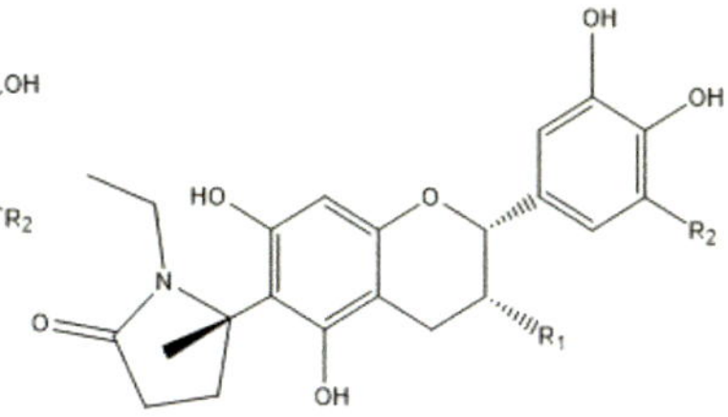

33. R_1=galloyl, R_2=H
28. R_1=galloyl, R_2=OH
38. R_1=cinnamoyl, R_2=H

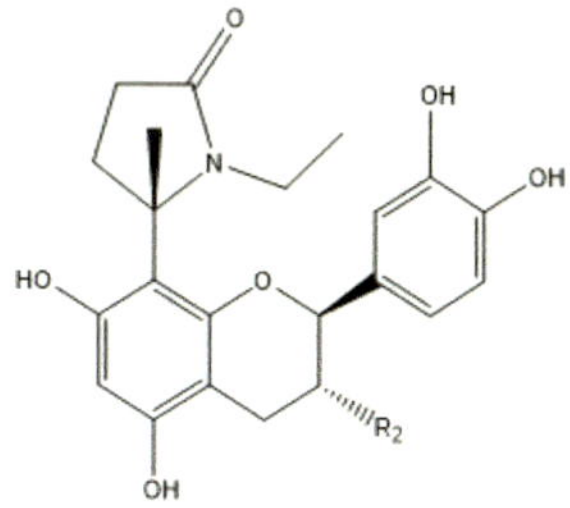

31. R_2 = galloyl

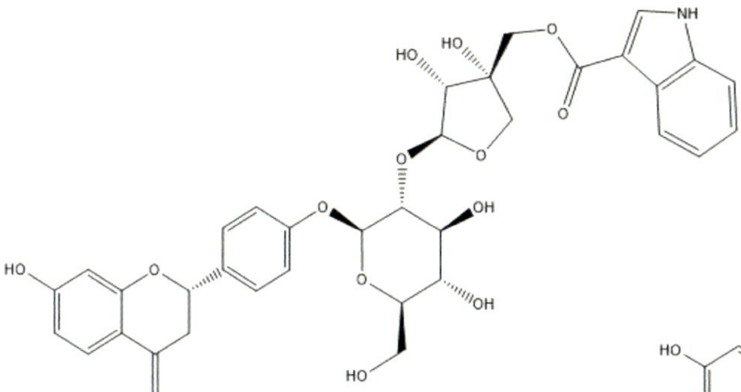

51. (-)-licorice glycoside E

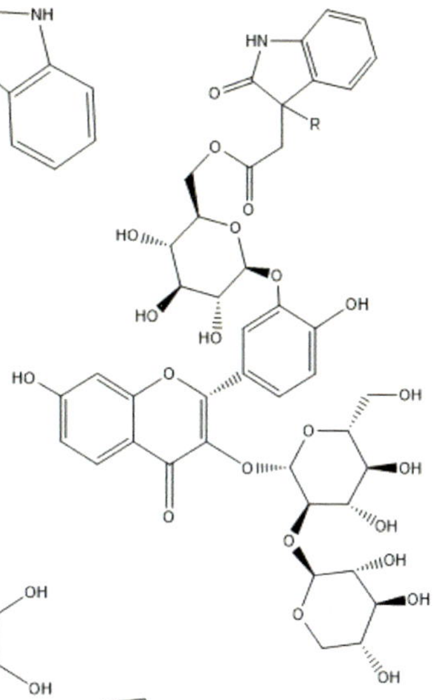

52. R=OH
53. R=H

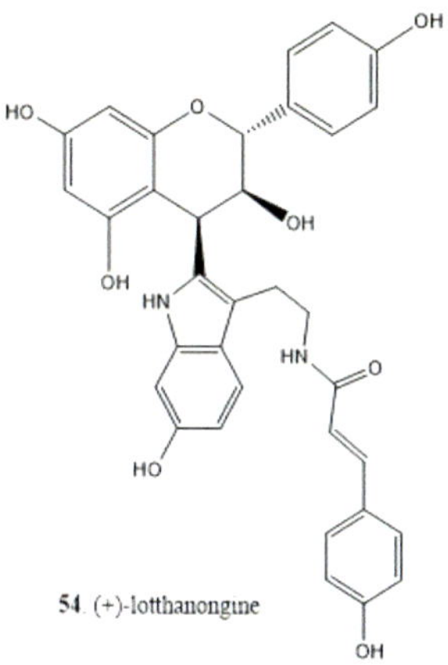

54. (+)-lotthanongine

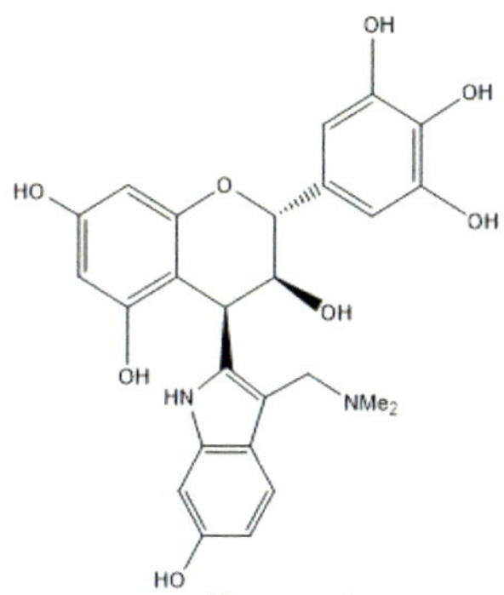

55. yuremamine

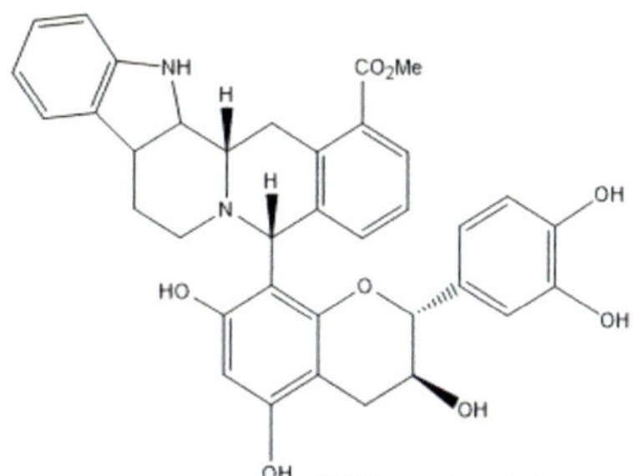

57. (-)-uncariagambirine

FIGURE 21.2 (Continued)

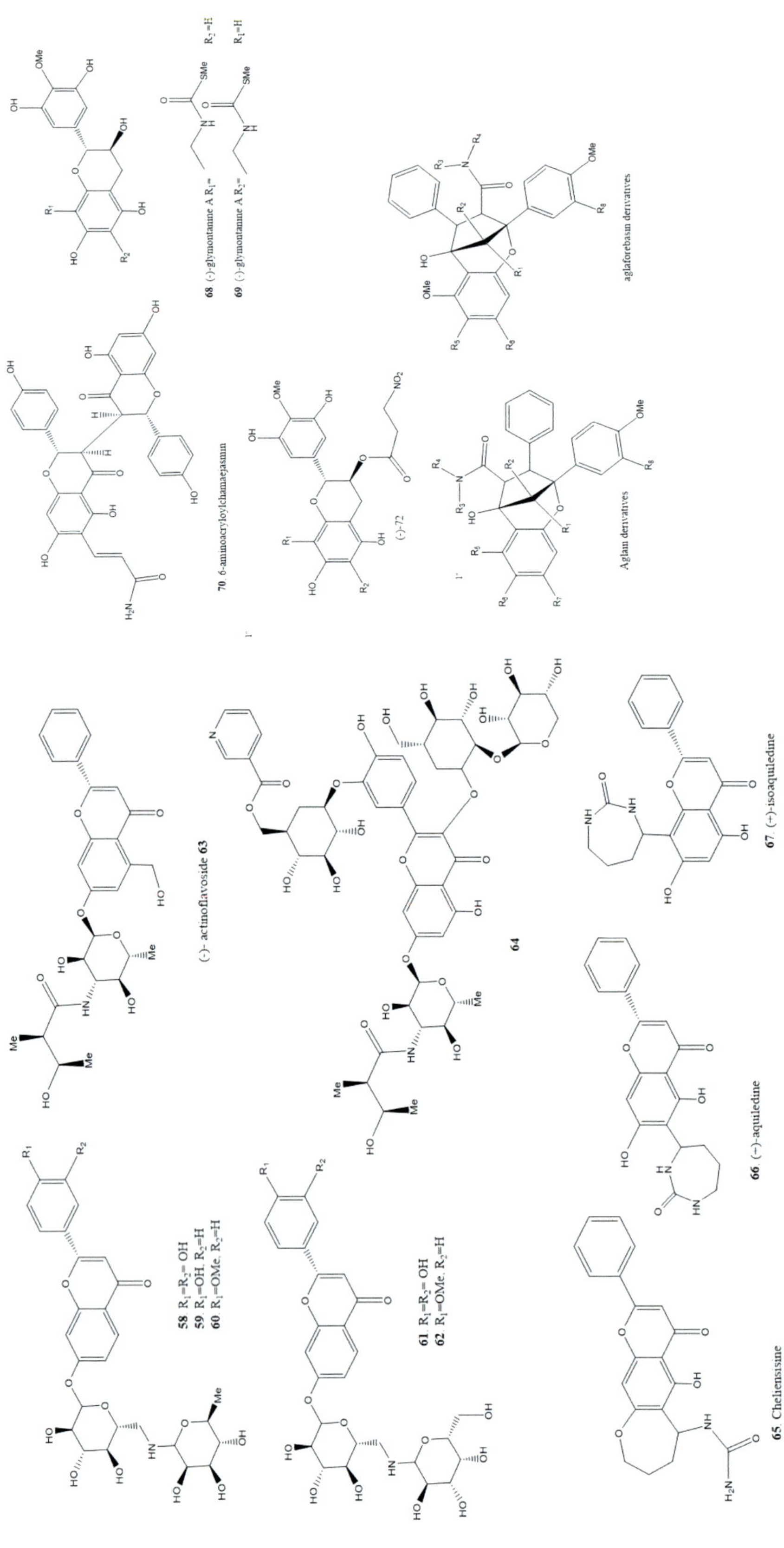

FIGURE 21.2 (Continued)

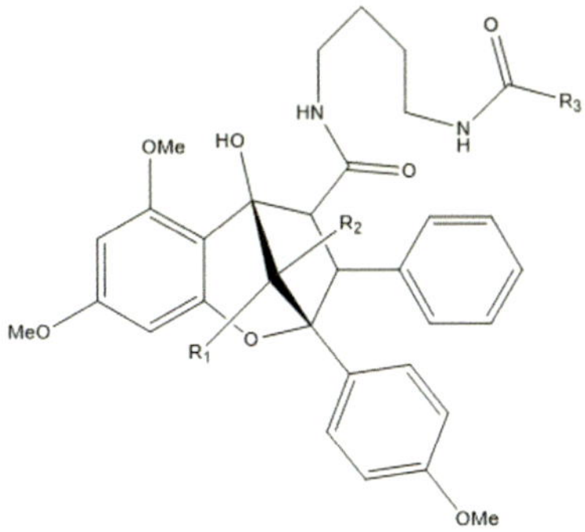

H-3α ; H-4β; R_1=H; R_2=OAc; R3=Ph; (-)-pyramidaglain A **73**
H-3β ; H-4β ; R_1=OAc; R_2=H; R3=Ph; (+)-pyramidaglain B **74**
H-3α ; H-4β ; R_1=*O*-β-glucosyl; R_2=H; R_3= (-) **75**

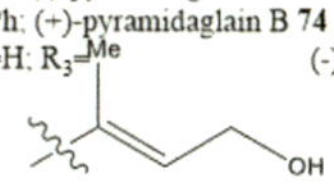

H-3β ; H-4α ; R_1=H; R_2=OH; R_3=Ph; (+)-isofoveoglin **79**
H-3α ; H-4β ; R_1=H; R_2=OH; R_3=Ph; (+)-desacetylpyramidaglain A **82**
H-3α ; H-4β ; R_1=OH; R_2=H; R_3=Ph; (-)-desacetylpyramidaglain C **83**
H-3β ; H-4α ; R_1=H; R_2=OH; R_3=Ph; (+)-desacetylpyramidaglain D **84**

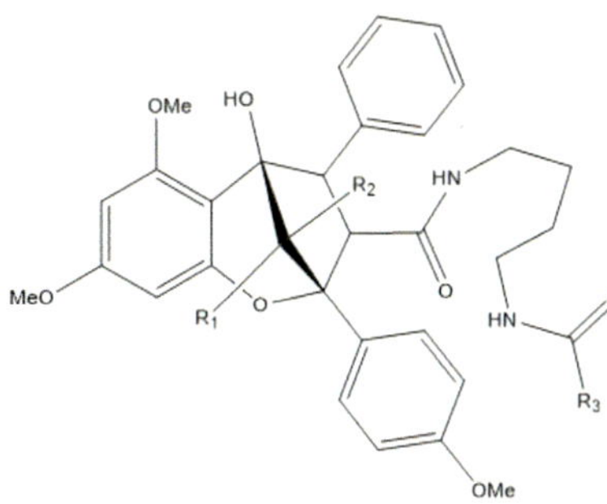

H-3α ; H-4β ; R_1=*O*-β-glucosyl; R_2=H; R_3= (-)**76**

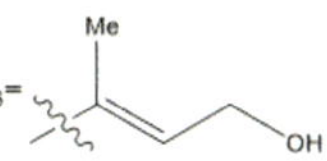

H-3α ; H-4β ; R_1=OH; R_2=H; R_3=Ph; (-)-foveoglin A **77**
H-3α ; H-4β ; R_1=H; R_2=OH; R_3=Ph; (+)-foveoglin B **78**
H-3β ; H-4α ; R_1=OH; R_2=H; R_3= ; (-)-perviridisin A **85**

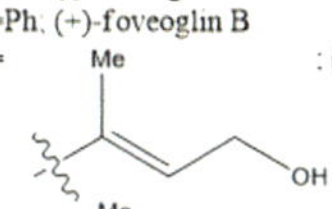

H-3β ; H-4α ; R_1=H; R_2=OH; R_3= ; (+)-perviridisin B **86**

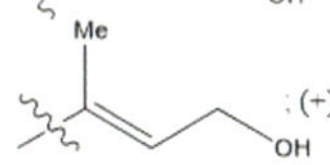

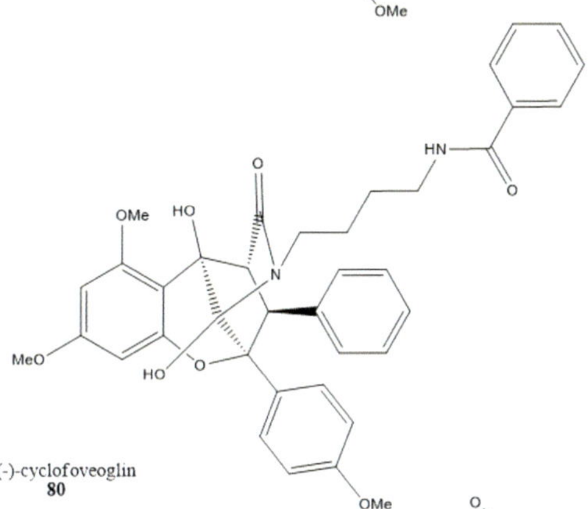

(-)-cyclofoveoglin
80

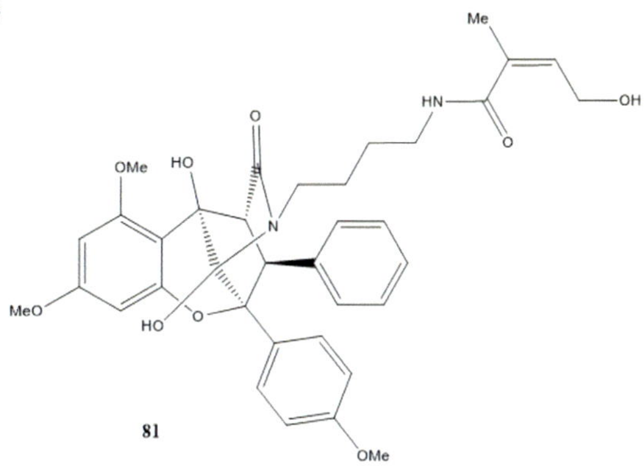

81

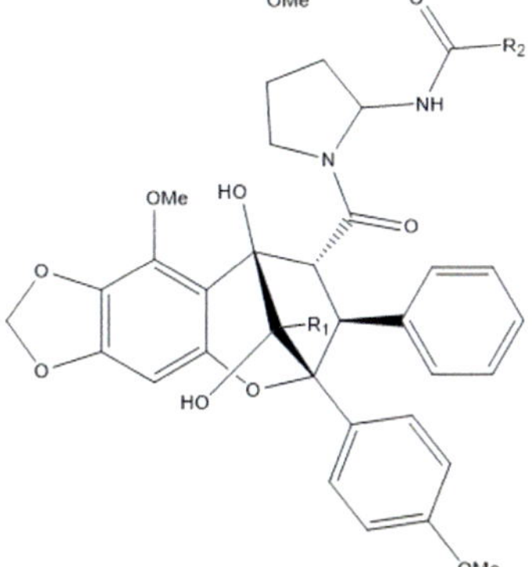

87. R_1=OH; R_2=isobutyl; (-)-edulirin A
88. R_1=OAc; R_2=isobutyl; (-)-edulirin A-10-*O*-acetate
89. R1=OH; R2=isobutenyl; (-)-19,20-dihydroedulirin A

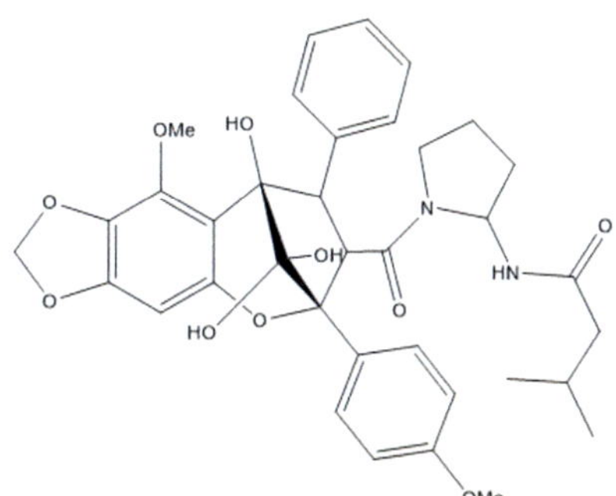

90. H-3α; H-4β; (+)-isoedulirin A
91. H-3β; H-4α; (+)-edulirin B

FIGURE 21.2 (Continued)

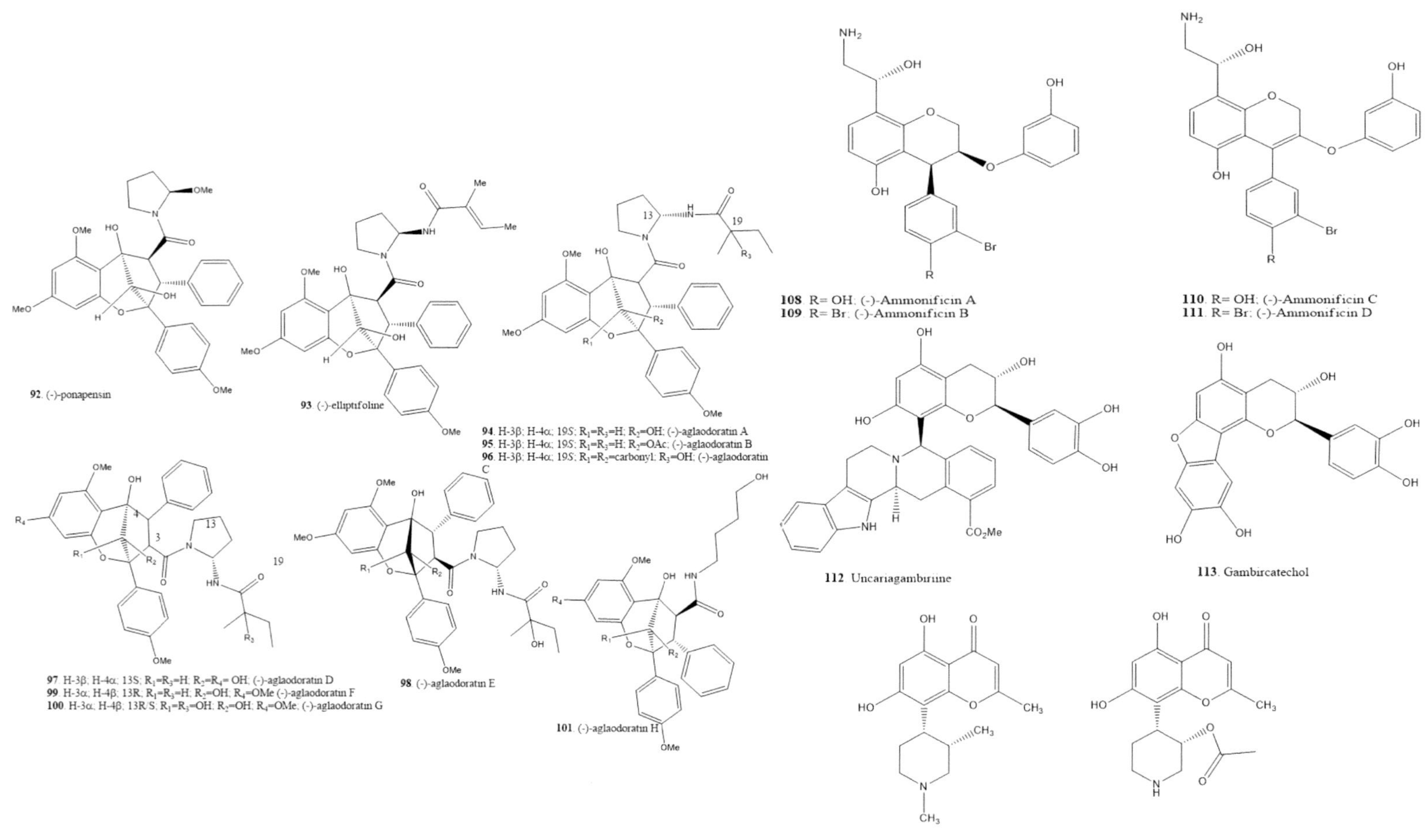

92. (-)-ponapensin

93. (-)-elliptifoline

94. H-3β; H-4α; 19S; R_1=R_3=H; R_2=OH; (-)-aglaodoratin A
95. H-3β; H-4α; 19S; R_1=R_3=H; R_2=OAc; (-)-aglaodoratin B
96. H-3β; H-4α; 19S; R_1=R_2=carbonyl; R_3=OH; (-)-aglaodoratin C

97. H-3β; H-4α; 13S; R_1=R_3=H; R_2=R_4= OH; (-)-aglaodoratin D
99. H-3α; H-4β; 13R; R_1=R_3=H; R_2=OH; R_4=OMe (-)-aglaodoratin F
100. H-3α; H-4β; 13R/S; R_1=R_3=OH; R_2=OH; R_4=OMe; (-)-aglaodoratin G

98. (-)-aglaodoratin E

101. (-)-aglaodoratin H

108. R= OH; (-)-Ammonificin A
109. R= Br; (-)-Ammonificin B

110. R= OH; (-)-Ammonificin C
111. R= Br; (-)-Ammonificin D

112. Uncariagambiriine

113. Gambircatechol

114. Rohitukine

115. N-demethylrohitukine-30-acetate

FIGURE 21.2 (Continued)

21.2.2 Flavoalkaloids with Pyrrolidinone Ring

Pyrrolidinone-flavoalkaloids, (+)-Lilaline (**7**), was identified in 1987 from the flowers of *Lilium candidum* Henry (Madonna lily), collected in Slovakia (Figure 21.2) (Masterova et al., 1987). A few years later, the diastereomeric davalliosides A/B (**8**) were identified in the rhizomes of *Davallia mariesii* Moore, a fern which is used in traditional Korean medicine (Figure 21.2) (Cui et al., 1990). Tanaka et al. isolated (-)-*N*-ethylpyrrolidinonyltheasinensin A, a pseudodimeric molecule (**9**), in 2005 from black tea (Figure 21.2). The biphenyl moiety of **9** was confirmed to possess *R*-configuration by comparing the circular dichroism (CD) spectra to that of a related atropisomer. Tanaka et al. (2005) claimed that **9** was formed during the processing of the tea leaves. Later, the compound **9** was also synthesized via the condensation of the parent flavonoid with *N*-ethyl-5-hydroxy-2-pyrrolidinone (Tanaka et al., 2005).

Ren et al. (2008) isolated dracocephins A-D (**10–13**) from the aerial parts of *Dracocephalum rupestre* Hance collected from the Shangxi Province, China (Figure 21.2). Each of the dracocephins was separated into its enantiomers (16 in total), hence determining their absolute configuration. In 2008, (±)-8-(2″-pyrrolidinone-5″-yl) quercetin (**14**) and (±)-8-(2″-pyrrolidinone-5″-yl) isorhamnetin (**15**) were isolated from the perennial herb *Senecio argunensis* Turcz, which was collected in the Jilin Province, China (Figure 21.2) (Li et al., 2008).

(±)-8-(2″-pyrrolidinone-5″-yl) quercetin (**14**) is present in the aerial parts of the Chinese herb *Emilia sonchifolia* (L) DC, Wight (Shen et al., 2012). Later, in 2009, the flavoalkaloids 6-(2-pyrrolidinone-5-yl)-(-)-epicatechin (**16**) and 8-(2-pyrrolidinone-5-yl)-(-)-epicatechin (**17**) were identified in the roots of *Actinidia arguta* (Siebold & Zucc.) Planch. ex Miq. which was collected in Korea (Figure 21.2) and were isolated as a mixture of diastereomers (16; 5:3, 17; 2:1) (Jang et al., 2009). Compound **17** was known to be present in pericarps of the tree *Litchi chinensis* (Sonn); later it was described as a single undetermined diastereomer and it could be isolated from *Phyllanthus cochinchinensis* Spreng (Ma et al., 2014; Zhao et al., 2014).

The phytomolecules **16** and **17** are known to show advanced glycation end-products activity with IC_{50} values of 36.0 μM and 47.8 μM, respectively (Jang et al., 2009). Compound **17** is known to possess moderate antioxidant activity (Ma et al., 2014). Isolated Drahebephin A (**18**) and B (**19**) from the perennial herb *Dracocephalum heterophyllum* Benth, where both occur naturally as racemates (Figure 21.2) The compounds **18** and **19** were tested for anticancer activity on various cell lines, but they did not show significant results; instead, their parent compound luteolin showed good anticancer activity with IC_{50} = 10 μM, which showed that introduction of pyrrolidinone ring substitution in luteolin decreased the anticancer activity (Wang et al., 2012b).

Dai et al. (2018) reported on 8-*C*-*N*-ethyl-2-pyrrolidinone-substituted flavan-3-ols (EPSFs), along with seven novel compounds (**20–26**) in white teas which were formed due to the long storage of tea (Dai et al., 2018). Later, Li et al. (2018) reported two novel flavoalkaloids, namely (–)-6-(5“ ‘*S*)-*N*-ethyl-2-pyrrolidinone-1 epigallocatechin 2-*O*-gallate (ester-type catechins pyrrolidinone A, etc-pyrrolidinone A (**20**) and (–)-6-(5“ ‘*R*)-*N*-ethyl-2-pyrrolidinone-epigallocatechin-*O*-gallate (etc-pyrrolidinone B (**29**), and four new naturally occurring flavoalkaloids (**20**, **23**, **27**, **28**) from white tea (*Camellia sinensis* (L.) Kuntze) (Li et al., 2018).

The novel flavoalkaloids, along with (–)-epigallocatechin-O-gallate (EGCG), were evaluated for their inhibition against the formation of advanced glycation end products (AGEs) having IC_{50} values ranging between 10.3 μM and 25.3 μM. The isomers **20** and **23** showed IC_{50} values of 10.3 μM (Table 21.1). In the same year, Cheng et al. (2018) reported five new flavoalkaloids, along with one novel compound (–)-8-(5”‘R)-N-ethyl-2-pyrrolidinone-epicatechin-3-O-gallate, etc-pyrrolidinone H (**31**), and some known flavoalkaloids etc-pyrrolidinones A–D (**20, 24, 25, 28, 30, 32, 33, 34**) were isolated from Xi-Gui green tea (Cheng et al., 2018). Xi-Gui tea is one of the ancient cultivated varieties of *Camellia sinensis* var. *assamica* Kuntze. The isolated molecules were evaluated for their protection against high glucose-induced cell senescence on human umbilical vein endothelial cells (HUVECs) and showed significant effects ($p < 0.01$) at both 1.0 μM and 10 μM (Cheng et al., 2018).

TABLE 21.1
Acetylcholinesterase (AChE) Inhibitory Effect of Flavoalkaloids

Compounds	IC_{50}[1] (μM)
EGCG[2]	11.7
20	25.3
23	10.3
28	13.5
35	0.213
36	0.185
37	0.138
38	0.126
Huperzine A	0.085

Notes: **1** Inhibition Concentration; **2** Epigallocatechin gallate

Gaur et al. (2020) reported cinnamoylated flavoalkaloids (**35–38**) from three different Chinese tea varieties viz. Fuding-Dabai, Huangjingui, and Zimudan. The identified flavoalkaloids were evaluated for their acetylcholinesterase (AChE) inhibitory effect (IC_{50} = 0.12–1.02 μM). All the compounds showed an IC_{50} value below 0.5 μM, with the most active compound **38** having IC_{50} of 0.126 μM (Table 21.1).

21.2.3 Piperidine- and Piperidinone-Flavoalkaloids

Ahond et al. (1984), isolated nine flavoalkaloids having a piperidine heterocycle, from different parts of the tropical tree *Buchenavia* sp. (Ahond et al., 1984). (-)-Buchenavianine (**39**) and (+)-N-demethylbuchenavianine (**40**) are present in the leaves of *Buchenavia macrophylla* (Eichl.) (Figure 21.2), while compounds (±)-O-demethylbuchenavianine (**41**), *N*, O-bis-(demethyl) buchenavianine (**42**), (-)-*N*-demethylcapitavine (**43**) and (±)-2,3-dihydrocapitavine (**44**), alongside (-)-buchenavianine (**39**), were isolated from the fruit of *B. macrophylla* (Figure 21.2). The compounds (+)-capitavine (**45**), (-)-4'-hydroxycapitavine (**46**) and (+)-2,3-dihydro-4'-hydroxycapitavine (**47**) were isolated from the seeds of *Buchenavia capitata* (Vahl) Eichl (Figure 21.2) (Ahond et al. 1984).

Phytomolecules **39**, **40**, **43** and **45–47** are optically active with low magnitudes ($[\alpha]_D$ ranging from −11 to +9) aside from **47** ($[\alpha]_D$ + 51). In 1992, Beutler et al. (1992) isolated compounds **39**, **41** and **42** from *Buchenavia capitata*. Among them, **41** was found to be optically active ($[\alpha]_D$ = -30), and the optical rotation of **42** was found to be ($[\alpha]_D$ = −2); on the other hand, the optical rotation of (-)-buchenavianine (**39**) was same as the initial report ($[\alpha]_D$ = -8 vs. −4) (Ahond et al., 1984; Beutler et al., 1992).

The optical rotation of (-)-Buchenavianine was obtained from the Institut de Chimie des Substances Naturelles (ICSN) chemical library. Later, the synthesis of (±)-buchenavianine and (±)-*O*-demethylbuchenavianine was performed and the respective enantiomers were separated with the help of chiral high pressure liquid chromatography (HPLC). The absolute configurations of enantiomers **39** and **41** were determined with the help of X-ray crystallography (Figure 21.2) (Nguyen et al., 2012). An antiviral assay (*S*)-(-)-*O*-demethylbuchenavianine (**41**) showed moderate cytoprotective effects against the antiviral activity of human immunodeficiency virus (HIV) when cultured in human lymphoblastoid cells and is known to inhibit CDK1 (IC_{50} of 0.03 μM) and CDK5 (IC_{50} of 0.05 μM). Likewise, (+)-enantiomer showed moderate activity against CDK1 (IC_{50} of 1.1 μM) and CDK5 (IC_{50} of 0.95 μM) (Leete, 1982).

Homberger and Hesse (1984) and Kam et al. (1999) isolated piperidine-containing flavanol (+)-kopsirachine (**48**) from the leaves of *Kopsia dasyrachis* Ridl (Figure 21.2), while the relative configuration of positions C2 and C3 in the flavonoid unit of **48** was determined (Homberger and Hesse, 1984; Kam et al., 1999). Krasteva et al. (2015) isolated the piperidinoneflavoalkaloids **49** and **50** from the aerial parts of *Astragalus monspessulanus* Bernh. collected from Rodopi Mountain, Bulgaria (Figure 21.2), and the absolute configuration of the stereogenic centers in the sugar moieties of these compounds were determined (Krasteva et al., 2015). The acid-hydrolysis of the phytomolecules **49** and **50** resulted in the corresponding (2S)-2-butyl glycosides (Krasteva et al., 2015).

Luo et al. (2015) isolated kaempferol-3-*O*-methyl-7-*O*-*β*-*D*-glucopyranosyl-8-(1-methyleneproline)-4'-*O*-*β*-*D*-glucopyranoside, a novel proline-containing flavonol glycoside, along with 15 known flavonoids from the aerial part of *Caragana leucophloea* Pojark (Leguminosae) (Luo et al., 2015). In this molecule flavonoid skeleton, 3-*O*-methylkaempferol is connected to a proline moiety via a methylene bridge at position C-8 and two glucoside units at C-7 and C-4' hydroxyl groups. This molecule is the first example of a flavonoid alkaloid having two glycoside moieties (Luo et al., 2015).

The phenolic Mannich reaction used by Wang et al. (2012a) synthesized 6-substituted flavoalkaloids like (±)-*N*-demethylcapitavine (**43**), and (±)-capitavine (**45**) via Mannich reaction using the suitable six-membered cyclic imine or iminium precursor in the procedure (Nguyen et al., 2011; Wang et al., 2012a). Later on, 8-substituted flavoalkaloid derivatives were synthesized including (±)-*N*,*O*-bisdemethylbuchenavianine (**42**) in a 3:1 ratio with (±)-*N*-demethylcapitavine (**43**) (Nguyen et al., 2011).

21.2.4 Flavoalkaloids with an Indole Ring

(-)-licorice glycoside E was the first indole flavoalkaloid identified (**51**), in 1997, from the roots of *Glycyrrhiza uralensis* Fisch, commonly known as Tohoku licorice (Figure 21.2) (Hatano et al., 1998). The absolute configuration of **51** was determined with the help of CD spectra. Hubner et al. (1999) isolated the oxindole-flavoalkaloid (**52**) from the seeds of the horse chestnut *Aesculus hippocastanum* L. (Figure 21.2). The presence of sugar moieties was confirmed by acid-hydrolysis, semi-synthesis of its derivatives and gas chromatography. Kapusta et al. (2007) isolated from the same plant, and an ion consistent with the deoxy analog **53** was analyzed using mass spectrometry (Figure 21.2) (Kapusta et al., 2007).

Kanchanapoom et al. (2002) identified (+)-lotthanongine (**54**) in the roots of *Trigonostemon reidioides* Craib, a shrub found in Southeast Asia which is used in traditional medicines due to its antiseptic and emetic properties and thus could be useful in treating mycetism (Figure 21.2). Vepsalainen et al. (2005) isolated yuremamine from the stem bark of *Mimosa tenuiflora* Willd. The plant is used in the preparation of the psychoactive beverage yurema, which is consumed for medical religious purposes by certain populations of northeastern Brazil. The serotonin agonist N, N-dimethyltryptamine (DMT) is present in the root bark of *M. tenuiflora*; however, when ingested, this compound becomes inactive due to the rapid metabolism by monoamine oxidase (MAO) in the digestive system. No MAO inhibitors have been detected in *M. tenuiflora*; hence, it is interesting for researchers to note the visionary effects yuremamine exerts. Initially, Yuremamine was assigned as apyrrolo [1, 2-a] indole (**55**), but synthetic methods led to assigning yuremamine as flavoalkaloid (**56**) (Figure 21.2) (Calvert et al., 2015).

Yoshikado et al. (2009) isolated (-)-Uncariagambiriine (**57**) from the leaves of *Uncaria gambir* Roxb., which can be collected in Indonesia. This compound contains a dihydrogambirtannine residue, which is linked to the A ring of the flavanol catechin (Figure 21.2) (Merlini et al., 1967). The absolute configuration of compound **57** was assigned based on the CD spectra analysis. In 2005, Hatakeyama and co-workers reported the total synthesis of (±)-lotthanongine (**54**) (Hatakeyama et al., 2005; Kawamoto et al., 1989). Many other groups performed the total synthesis of yuremamine (**55**) (Ohyama et al., 2015; Johansen et al., 2008; Dethe et al., 2013).

21.2.5 Flavoalkaloids Having Aminoglycoside Moiety

In 1984, Estan et al. isolated the aminoglycosylated flavone (**58**) from the leaves of *Myoporum tenuifolium* E. Foster, commonly known as "siempre verde" (Figure 21.2) (Estan, 1984, 1985a). In 1985, four analogs consisting of compounds **59–62** were isolated from the same species (Figure 21.2) (Estan, 1985b). Jiang et al. (1997) isolated an aminoglycoside (-)-actinoflavoside (**63**) from the fermentation broth of the marine-derived bacteria *Streptomyces* sp. CNB-689 (Figure 21.2). (-)-Actinoflavoside (**63**) is known to occur naturally as a 1:1 diastereomeric mixture, and its optical rotation was found to be ($[\alpha]_D$= -110). This compound is a rare example of a bacterial flavonoid (Figure 21.2) (Jiang et al., 1997; Elshahawi et al., 2015). The compound **63** is epimeric at C2 and was confirmed by acid hydrolysis, which provided the aglycone (Jiang et al., 1997). The flavonoid **63** contains a 5-hydroxymethyl group and showed weak antibacterial activity against *Staphylococcus pneumonia* (IC_{50} = 64 µg/ml), *Staphylococcus aureus* (IC_{50} = 64 µg/ml), *Staphylococcus pyrogenes* (IC_{50} = 64 µg/ml) and *Micrococcus buteus* (IC_{50} = 64 µg/ml) when tested against a panel of cell lines. Suzuki and coworkers accomplished the synthesis of the actinoflavoside aglyconeto get the compound in good yield (Suzuki et al., 2004).

21.2.6 Special Flavoalkaloids

Numerous flavoalkaloids cannot be categorized as described previously. One such unusual compound is an acid-containing flavoalkaloid (**64**), which was isolated from the seeds of the horse chestnut *Aesculus hippocastanum* L. in 1999 (Figure 21.2). The presence of sugar moieties was confirmed by acid-hydrolysis, derivatization, and gas chromatography. Gu et al., (2000) isolated the urea-containing flavonoid cheliensisine (**65**) from the Chinese plant *Goniothalamus cheliensis* Hu. (Figure 21.2) (Gu et al., 2000). Chen et al. (2001) isolated two more urea-containing flavoalkaloids, (+)-aquiledine (**66**) and (+)-isoaquiledine (**67**), from *Aquilegia ecalcarata* Maxim. a herb collected in the Si-Chuan Province, China (Figure 21.2) that is used in traditional Chinese medicine; however, no biological activity has been reported (Chen et al., 2001).

Wang et al. (2005) isolated two unusual flavoalkaloids, (-)-Glymontanine A (**68**) and B (**69**), which contained a thiocarbamate group from the twigs and leaves of *Glycosmis montana* Pierre. (Figure 21.2). The absolute configuration of flavoalkaloids **68** and **69** were assigned based on CD spectra. Xu et al. (2012) isolated the pseudodimeric flavoalkaloid 6-aminoacryloylchamaejasmin (**70**) from the roots of *Ormocarpum kirkii* Moore. collected in Tanzania (Figure 21.2). The absolute configuration of **70** was assigned based on CD spectra. Qui et al. (2013) isolated an unusual flavonoid containing a nitroalkyl group (**71**) from the roots of *Indigofera stachyodes*, Lindl. collected in Guizhou Province, China (Figure 21.2). This compound showed significant activity (122% survival rate) against human liver cells from carbon tetrachloride-induced hepatotoxicity.

21.2.7 Aglains and Related Flavoalkaloids

The naturally occurring compounds having the cyclopenta [*b, c*] benzopyran skeleton are regarded as 'aglains' or the regioisomeric 'aglaforbesins.' The genus *Aglaia* belongs to the mahogany family consisting of around 100 species, which occur throughout Southeast Asia and the Pacific (Proksch et al., 2001; Dumontet et al., 1996). Proksch et al. (2001) proposed biosynthetic pathways of the aglains that results with the conjugate addition of a flavonoid to a cinnamamide, having the C-C bond at the flavonoid C2, followed by an aldol cyclization and reduction resulting in \the formation of these natural products (Proksch et al., 2001). A similar pathway involving the inverse approach of the cinnamamide would result in the aglaforbesin derivatives, and interestingly, this annulation can selectively deliver either enantiomer of the aglains/aglaforbesins. The isolation and biology of these flavoalkaloids spanning 1996–2001 were covered previously by Proksch et al. (2001).

(-)-Pyramidaglain A (**73**) and (+)-pyramidaglain B (**74**) were isolated from the leaves of *Abralia andamanica* Hiern, collected in Thailand, and the absolute configuration was not established in either (Puripattanavong et al., 2000). Glycosylated aglain derivative (-)-**75** and aglaforbesin derivative (-)-**76** was isolated in the following year from the leaves of *Amoora dasyclada* Zhengyi collected in the Yunnan Province, China, and their absolute configurations were not stated (Chaidir et al., 2001). Closely related compounds were isolated in 2007 from the leaves of *Acrosticta foveolata* Loew and named (-)-foveoglin A (**77**), (+)-foveoglin B (**78**), (+)-isofoveoglin (**79**), alongside an unusual lactam variant (-)-cyclofoveoglin (**80**) (Figure 21.2) (Salim et al., 2007).

(-)-Foveoglin A (**77**) exhibited cytotoxicity against a panel of cancer cell lines (effective dose [ED_{50}] of 1.4–1.8 μM), while (+)-isofoveoglin (**79**) and (-)-cyclofoveoglin (**80**) were weakly cytotoxic (ED_{50} of 13.5–21 μM), and (+)-foveoglin B (**78**) was inactive. Furthermore, (-)-foveoglin A (**77**) is an active nuclear factor-kappa B (NF-κB) inhibitor (IC_{50} of 0.37 μM). Curiously, the cyclofoveoglin analog (-)-**81** was later isolated from the mangrove *Amoora cucullata*, Roxb., which has the same relative stereochemistry as **80** and is notably isolated from a non-Aglaia species (Figure 21.2) (Ahmed et al., 2010).

The similar $[\alpha]_D$ values reported for **81** and **82** (–51.7 and −63.1, respectively) suggest that they may have the same absolute configuration, which is yet to be conclusively established. Cyclofoveoglin analog (**81**) showed tumor necrosis factor (TNF)-related apoptosis-inducing ligand (TRAIL)-resistance overcoming activity in human gastric adenocarcinoma cells (1.25 μM or 2.5 μM) of which led to 45 or 57% more growth inhibition compared to cells treated with 100 ng/mL of TRAIL. In 2008, (+)-desacetylpyramidaglain A (**82**), (-)-desacetylpyramidaglain C (**83**), and (+)-desacetylpyramidaglain D (**84**) were isolated from the leaves of *Archidendron forbesii*, Bak. collected in Thailand (Figure 21.2) (Joycharat et al., 2008).

Although the authors invoked biosynthetic considerations to propose the absolute stereochemistry of **82–84**, only the relative stereochemistry was conclusive. Each was screened for antibacterial and antiviral activity. (+)-desacetylpyramidaglain D (**84**) showed moderate active against *Mycobacterium tuberculosis* H37Ra (minimum inhibitory concentration [MIC] of 25 μg/mL) and herpes simplex virus type 1 at a non-cytotoxic concentration of 50 μg/mL (Joycharat et al., 2008).

In 2013, (-)-perviridisin A (**85**) and (+)-perviridisin B (**86**) were isolated from the leaves, twigs and fruits of *Aplastodiscus perviridis*, Lutz. collected in Vietnam (Figure 21.2). Both were tested against human colon cancer cells (HT-29), and (+)-perviridisin B (**86**) exhibited significant cytotoxicity (ED_{50} of 0.46 μM) with no significant effect on normal colon cells (CCD-112CoN). Further testing revealed that **86** moderately inhibits NF-κB (ED_{50} of 2.4 μM). The absolute stereochemistry of **85** and **86** is unknown (Pan et al., 2013).

In 2006, a series of aglains possessing a pyrrolidine moiety named (-)-edulirin A (**87**), (-)-edulirin A 10-O-acetate (**88**), (-)-19, 20- dehydroedulirin A (**89**), (+)-isoedulirin A (**90**) and (+)-edulirin B (**91**) were isolated from the bark of *Aglaia edulis* Roxb. Wall. collected in Indonesia (Figure 21.2) (Kim et al., 2006). While the relative configuration of the stereogenic centers in the cyclopenta [b, c] benzopyran skeleton of compounds **87–91** was established, the stereochemistry C-13 could not be determined. All members of the family (**87–91**) were tested against a panel of human cancer cell lines, but displayed no significant activity (Kim et al., 2006).

In the following year, the structurally related (-)-ponapensin (**92**) was isolated from the leaves and stems of *Aglaia ponapensis* Kaneh. (Figure 21.2) (Salim et al., 2007). Enzyme inhibition assays revealed 94 potential NF-κB inhibitors (IC_{50} of 0.064 μM). Ponapensin (**92**) bears a close structural resemblance to elliptifoline (**93**), isolated in 2001 from *Aglaia elliptifolia* Merr. (Wang et al., 2001b). The absolute configurations of (-)-ponapensin (**92**) and (-)-elliptifoline (**93**) were later established following the total syntheses of their (+)-enantiomers by the Porco group. Additionally, Lajkiewicz reassigned the relative configuration of C13 in (-)-elliptifoline (**93**) and proposed insightful considerations on the biosynthesis of enantioenriched aglains through kinetic resolution mechanisms (Lajkiewicz et al., 2012).

Yet another series of aglain and aglaforbesin derivatives were isolated from the leaves of *Aglaia odorata*, Lour. collected in Yunnan Province, China, named aglaodoratins A-H (**94–101**) (Figure 21.2) (An et al., 2015). A combination of X-ray crystallography and analysis of CD spectra enabled the assignment of the absolute configuration in (-)-aglaodoratin A (**94**), and CD spectra were used to assign the configuration of the stereogenic centers in **95–101**; however, the configuration at C19 was not clearly stated for the members of this family. Note that (-)-aglaodoratin G (**100**) is epimeric at C13 and exists as a 1:1 mixture of diastereomers from which the optical rotation was obtained ($[\alpha]_D = -6$). All eight analogs were screened against a panel of cancer cell lines, with aglaodoratin C (**96**) performing well against HT-29 (IC_{50} of 0.097 μM) and MG-63 cells (IC_{50} of 1.2 μM). Aglaodoratin D (**97**) demonstrated selectivity for MG-63 cells (0.75-fold), and (-)-aglaodoratin E (**98**) was cytotoxic to SMMC-7721 cells (IC_{50} of 6.25 μM) (An et al., 2015).

As previously noted, Lajkiewicz and coworkers reported an elegant asymmetric total synthesis of the potent NF-kB inhibitor (+)-ponapensin (**92**) and (+)-elliptifoline (**93**), the first of any member of the aglain family, which has led to the assignment of their absolute configurations and revision of **93**. Using an enantioselective photocycloaddition, the simple flavonoid was rapidly converted to the cyclopenta [*b,c*] benzopyran framework, which was then subjected to stereoselective reduction, furnishing the core of both (+)-ponapensin (**92**) and (+)-elliptifoline (**93**) with high enantioselectivity (Lajkiewicz et al., 2012).

21.3 EXTRACTION, ISOLATION AND BIOLOGICAL ACTIVITY OF ISOFLAVOALKALOIDS

Isoflavoalkaloids were first isolated in 1989, followed by the isolation of pallimamine (**102**) from the whole herb of *Corydalis solida* ssp. Tauricola, collected in Taiwan (Figure 21.2) (Lu ct al., 1989). Pallimamine (**102**) is a racemic mixture, and its relative configuration was determined by X-ray crystallography. Ngamga et al. (2006) isolated drimiopsis A (**103**) and B (**104**) from the plant *Drimiopsis barteri* Bak. collected in Cameroon (Ngamga et al., 2006; Ngamga et al., 2008). Isoflavoalkaloids consist of a heteroaromatic framework having fused isoquinoline and benzopyran units. Li et al. (2008a) isolated (-)-tonkinensine A (**105**) and B (**106**) from the roots of *Sophora tonkinensis* Gagnep., a plant used as traditional Chinese medicine in the treatment of throat infections.

In both the tonkinensines, the alkaloid part is derived from a cytosine unit. The absolute configuration of compounds **105** and **106** was established with the help of X-ray crystallography, CD spectra and their biogenetic origin. When tested against human breast cancer (MDA-MB-231) and cervical carcinoma (HeLa) cell lines, only (-)-tonkinensine B (**106**) showed moderate activity (IC_{50} of 24.3 μM against HeLa and 48.9 μM against MDA-MB-231) (Li et al., 2008a). Dejon et al. (2013) isolated a racemate compound, (-)-11-Azamedicarpin (**107**), from the roots of *Robinia pseudoacacia*, Linn., commonly known as the black locust tree, and is an aza-analog of the pterocarpan medicarpin (Dejon et al., 2013). Compound **107** showed modest activity against human promyelocytic leukemia cells (HL-60) (cell survival of 72% at 40 μM and 27% survival at 200 μM). Oxidation of **107** to its corresponding indole improved cytotoxicity, and the total synthesis of this compound was performed by Dejon and co-workers (2013).

21.4 ISOLATION AND BIOLOGICAL ACTIVITY OF NEOFLAVOALKALOIDS

Andrianasolo et al. (2009), isolated the hydroxyethyl amine-containing neoflavoalkaloids (-)-ammonificin A (**108**) and B (**109**) from the marine hydrothermal derived bacteria *Thermovibrio ammonificans*, collected on the East Pacific Rise at a depth of 2,500 m (Figure 21.2) (Andrianasolo et al., 2009). In 2012, the unsaturated analogs (-)-ammonificin C (**110**) and D (**111**) were isolated by Andrianasolo et al. (2012) from the same species. The absolute configuration of compounds **108** and **109** were determined with the help of CD spectra. Compounds **108–111** were subjected to an apoptosis-induction assay; the ammonificans C (**110**) and D (**111**) induced apoptosis at 2 and 3 μM, respectively (Andrianasolo et al., 2012).

Yoshikado et al. (2009) identified two new polyphenols namely uncariagambiriine (**112**) and gambircatechol (**113**) having nitrogen ring links and isolated from the leaves of *Uncaria gambir* Hunter, Roxb. (family Rubiaceae), a climbing shrub native to Asia cultivated in the Malay Peninsula and Indonesia. The extract of the fresh leaves of *U. gambir* is known to treat diarrhea and dysentery, burns, scurf and lumbago (Yoshikado et al., 2009).

21.5 CONCLUSION

The flavoalkaloids have a unique structure consisting of phenols and a basic nitrogen atom resulting from the convergence of specific biosynthetic pathways. Though these compounds are difficult to extract, they possess various biological properties, which cannot be obtained from flavonoids and alkaloids alone. One such example is flavopiridol (known as alvocidib), a semisynthetic flavoalkaloid that is used currently under clinical investigation as a potential CDK 9 kinase inhibitor for the treatment of various cancers (Senderowicz et al., 1999). Flavopiridol showed promising results in the treatment of chronic lymphocytic leukemia and acute myeloid leukemia, and will undergo stage III clinical trials in due time (Zeidner et al., 2015).

Taking into consideration their unique molecular structure, flavoalkaloids are known to be good synthetic targets via various new similar analogs that can be synthesized. For example, work is currently being done on 8-*C*-*N*-ethyl pyrrolidinone cinnamoylated catechin, which was synthesized and detected in different tea varieties with the help of liquid chromatography-mass spectrometry (LCMS). Blair et al. (2017) synthesized yuremamine, which uncovered the biosynthetic origins of hybrid natural products (Blair et al., 2017).

Lajkiewicz synthesized (+)-ponapensin and (+)-elliptifoline with the discovery of some novel synthetic methodology, which needed to be developed to synthesize these complex flavoalkaloids. This review includes various structurally novel natural products, and it is hoped further research toward their extraction, isolation, synthesis, derivatization and biological evaluation is initiated (Lajkiewicz et al., 2012). In conclusion, plants are a good source of valuable natural products having numerous biological activities (Figure 21.3).

FIGURE 21.3 A, *Camellia sinensis*; B, *Phyllospadix iwatensis*; C, *Vochysia guianensis* (AxelBoldt, 2004; Liksakova, 2019; Wikipedie, 2021).

REFERENCES

Ahmed, F., Toume, K., Sadhu, S. K., Ohtsuki, T., Arai, M. A., Ishibashi, M. 2010. "Constituents of *Amoora cucullata* with TRAIL resistance-overcoming activity." *Organic & Biomolecular Chemistry* 8: 3696–3703.

Ahond, A., Fournet, A., Moretti, C., Philogene, E., Poupat, C., Thoison, O., Potier, P. 1984. "Premiers alcaloïdesvraisisolés de Combretacées: *Buchenavia macrophylla* Eichl. et *Buchenavia capitata*." *Bulletin de la Société Chimique de France*: 41–45.

An, F.-L., Wang, J.-S., Wang, H., Wang, X.-B., Yang, M.-H., Guo, Q.-L., Dai, Y., Luo, J., Kong, L.-Y. 2015. "Cytotoxic flavonol-diamide [3+2] adducts from the leaves of *Aglaia odorata*." *Tetrahedron* 71, 16: 2450–2457.

Andrianasolo, E. H., Haramaty, L., Rosario-Passapera, R., Bidle, K., White, E., Vetriani, C., Falkowski, P., Lutz, R. 2009. "Ammonificins A and B, hydroxyethylamine chroman derivatives from a cultured marine hydrothermal vent bacterium, *Thermovibrio ammonificans*." *Journal of Natural Products* 72: 1216–1219.

Andrianasolo, E. H., Haramaty, L., Rosario-Passapera, R., Vetriani, C.,Falkowski, P., White, E.; Lutz, R. 2012. "Ammonificins C and D, hydroxyethylamine chromene derivatives from a cultured marine hydrothermal vent bacterium, *Thermovibrio ammonificans*." *Marine Drugs* 10: 2300–2311.

Anjaneyulu, B., Govindachari, T. R. 1969. "Sythesis of (±) ficine." *Tetrahedron Let*ter 10: 2847–2848.

AxelBoldt. 2004. "*C. sinensis*." Obtained from Wikimedia under the Public Domain license. https://commons.wikimedia.org/wiki/File:Csinensis.jpg (Accessed 20 June 2021)

Baudouin, G., Tillequin, F., Koch, M., Vuilhorgne, M., Lallemand, J.-Y., Jacquemin, HIsolement. 1983. "Structure et Synthèse de la Vochysine, Pyrrolidinoflavanne de *Vochysiaguianensis*." *Journal of Natural Products* 46: 681–687.

Beutler, J. A., Cardellina, J. H., II, McMahon, J. B., Boyd, M. R., Cragg, G. M. 1992. "Anti-HIV and cytotoxic alkaloids from *Buchenavia capitata*." *Journal of Natural Products* 55: 207–213.

Blair, L. M., Calvert, M. B., Sperry, J. 2017. "Flavoalkaloids-isolation, biological activity, and total synthesis." *Alkaloids: Chemistry and Biology* 77: 85–115. doi: 10.1016/bs.alkal.2016.04.001

Calvert, M. B., Sperry, J. 2015. "Bioinspired total synthesis and structural revision of yuremamine, an alkaloid from the entheogenic plant *Mimosa tenuiflora*." *Journal of the Chemical Society, Chemical Communications* 51: 6202–6205.

Chaidir, Lin W. H., Ebel, R., Edrada, R., Wray, V., Nimtz, M., Sumaryono, W., Proksch, P. 2001. "Rocaglamides, glycosides, and putrescine bisamides from *Aglaia dasyclada*." *Journal of Natural Products* 64: 1216–1220.

Chen, S.-B., Gao, G.-Y., Leung, H.-W., Yeung, H.-W., Yang, J.-S., Xiao, P.-G. 2001. "Aquiledine and isoaquiledine, novel flavonoid alkaloids from *Aquilegia ecalcarata*." *Journal of Natural Products* 64: 85–87.

Cheng, J., Wu, F. H., Wang, P., Ke, J. P., Wan, X. C., Qiu, M. H., Bao, G. H. 2018. "Flavoalkaloids with a pyrrolidinone ring from Chinese ancient cultivated tea Xi-Gui." *Journal of Agricultural and Food Chemistry* 66: 7948–7957.

Cui, C.-B., Tezuka, Y., Kikuchi, T., Nakano, H., Park, J.-H. 1990. "Davallioside A and B, novel flavan-3-ol derivatives with a γ-lactam, from the rhizomes of *Davallia mariesii* moore." *Chemical and Pharmaceutical Bulletin* 38: 2620–2622.

Dai, W., Tan, J., Lu, M., Zhu, Y., Li, P., Peng, Q., Guo, L., Zhang, Y., Xie, D., Hu, Z., Lin, Z. 2018. "Metabolomics investigation reveals that 8-C-N-ethyl-2-pyrrolidinone-substituted flavan-3-ols are potential marker compounds of stored white teas." *Journal of Agricultural and Food Chemistry* 66: 7209–7218.

Dejon, L., Mohammed, H., Du, P., Jacob, C., Speicher, A. 2013. "Synthesis of chromenoindole derivatives from *Robinia pseudoacacia*." *Medicinal Chemistry Communications* 4: 1580–1583.

Dethe, D. H., Boda, R., Das, S. 2013. "$Cu(OTf)_2$ catalysed [6+2] cycloaddition reaction for the synthesis of highly substituted pyrrolo[1,2-*a*]indoles: Rapid construction of the yuremamine core." *Chemical Communications* 49: 3260–3262.

Dumontet, V., Thoison, O., Omobuwajo, O. R., Martin, M.-T., Perromat, G., Chiaroni, A., Riche, C., Païs, M.,Sévenet, T., Hadi, A. H. A. 1996. "New nitrogenous and aromatic derivatives from *Aglaia argentea* and *A. forbesii*." *Tetrahedron* 52: 6931–6942.

Elshahawi, S. I., Shaaban, K. A.,Kharel, M. K., Thorson, J. S. 2015. "A comprehensive review of glycosylated bacterial natural products." *Chemical Society Reviews* 44: 7591–7697.

Estan, M. T. 1984. "Amino-di-glucosido de luteolina en myoporum tenuifolium." *Revista de agroquímica y tecnología de alimentos* 24: 529–532.

Estan, M. T. 1985a. "Di-glicosil-aminas en Myoporum tenuifolium (E. Foster)." *Revista de agroquímica y tecnología de alimentos* 25: 369–372.

Estan, M. T. 1985b. "Diglucósidos de flavonas en myoporum tenuifolium (E. Foster)." *Revista de agroquímica y tecnología de alimentos* 25: 273–278.

Gaur, R., Ke, J. P., Zhang, P., Yang, Z., Bao, G. H. 2020. "Novel cinnamoylated flavoalkaloids identified in tea with acetylcholinesterase inhibition effect." *Journal of Agricultural and Food Chemistry* 68, 10: 3140–3148

Gu, Z.-B., Liang, H.-Q., Chen, H.-S., Xu, Y.-X., Yang, G.-J., Zhang, W.-D. 2000. "A new type of alkaloid from *Goniothalamus cheliensis*." *Acta Botanica Yunnanica* 22: 499–502.

Hatakeyama, K., Ohmori, K., Suzuki, K. 2005. "Synthesis of (±)-lotthanongine, a novel natural product with a flavan-indole hybrid structure." *Synlett*: 1311.

Hatano, T., Takagi, M., Ito, H., Yoshida, T. 1998. "Acylated flavonoid glycosides and accompanying phenolics from licorice." *Phytochemistry* 47: 287–293.

Hirayama, C., Ono, H., Tamura, Y., Nakamura, M. 2006. "C-prolinylquercetins from the yellow cocoon shell of the silkworm, *Bombyx mori*." *Phytochemistry* 67: 579–583.

Homberger, K., Hesse, M. 1984. "Kopsirachine, an unusual alkaloid from the apocynaceae *Kopsia dasyrachis Ridl*." *Helvetica Chimica Acta* 67: 237–248.

Houghton, P. J. 1987. *The Alkaloids: Chemistry and Pharmacology*. 31, Brossi, A., Ed.; Academic Press, New York: 67–100.

Houghton, P. J. 2000. *Studies in Natural Products Chemistry*. Atta-ur-Rahman, Ed.; Elsevier, Amsterdam: 123–155.

Hubner, G., Wray, V., Nahrstedt, A. 1999. "Flavonol oligosaccharides from the seeds of *Aesculus hippocastanum*." *Planta Medica* 65: 636–642.

Ilkei, Viktor, Hazai, L., Antus, S., Bolcskei, H. 2018. "Flavonoid alkaloids: Isolation, bioactivity, and synthesis." *Studies in Natural Products Chemistry* 56. doi: 10.1016/B978-0-444-64058-1.00008-X

Jang, D. S., Lee, G. Y., Lee, Y. M., Kim, Y. S., Sun, H., Kim, D.-H., Kim, J. S. 2009. "Flavan-3-ols having a gamma-lactam from the roots of *Actinidia arguta* inhibit the formation of advanced glycation end products *in vitro*." *Chemical and Pharmaceutical Bulletin* 57: 397–400.

Jiang, Z.-D., Jensen, P. R., Fenical, W. 1997. "Actinoflavoside, a flavonoid-like glycoside produced by a marine bacterium of the genus *Streptomyces*." *Tetrahedron Letter* 38: 5065–5068.

Johansen, M. B., Kerr, M. A. 2008. "Expedient synthesis of pyrrolo[1,2-*a*]indoles: Preparation of the core of yuremamine." *Organic Letters* 10: 3497–3500.

Johns, S. R., Russel, J. H., Heffernan, M. L. 1965. "Ficine, A novel flavonoidal alkaloid from *Ficus pantoniana*." *Tetrahedron Letter* 6: 1987–1991.

Joycharat, N., Greger, H., Hofer, O., Saifah, E. 2008. "Flavaglines and triterpenoids from the leaves of *Aglaia forbesii*." *Phytochemistry* 69: 206–211.

Kam, T.-S., Choo, Y.-M., Chen, W., Yao, J.-X. 1999. "Indole and monoterpene alkaloids from the leaves of *Kopsia dasyrachis*." *Phytochemistry* 52: 959–963.

Kanchanapoom, T., Kasai, R., Chumsri, P., Kraisintu, K., Yamasaki, K. 2002. "Lotthanongine, an unprecedented flavonoidal indole alkaloid from the roots of Thai medicinal plant, *Trigonostemon reidioides*." *Tetrahedron Letter* 43: 2941–2943.

Kapusta, I., Janda, B., Szajwaj, B., Stochmal, A., Piacente, S., Pizza, C., Franceschi, F., Franz, C., Oleszek, W. 2007. "Flavonoids in horse chestnut (*Aesculus hippocastanum*) seeds and powdered waste water byproducts." *Journal of Agricultural and Food Chemistry* 55: 8485–8490.

Kawamoto, H., Nakatsubo, F., Murakami, K. 1989. "Synthesis of a condensed tannin model compound, 4-(2,4,6-trihydroxyphenyl)flavan-3,3',4',5,7-pentaol." *Journal of Wood Chemistry and Technology* 9: 35–52.

Khadem, S.,Marles, R. J. 2012. "Chromone and flavonoid alkaloids: Occurrence and bioactivity." *Molecules* 17: 191–206.

Kim, S., Chin, Y.-W., Su, B.-N., Riswan, S., Kardono, L. B. S., Afriastini, J. J., Chai, H., Farnsworth, N. R., Cordell, G. A., Swanson, S. M. *et al.* 2006. "Cytotoxic flavaglines and bisamides from *Aglaia edulis*." *Journal of Natural Products* 69: 1769–1775.

Krasteva, I., Bratkov, V., Bucar, F., Kunert, O., Kollroser, M., Kondeva-Burdina, M., Ionkova, I. 2015. "Flavoalkaloids and Flavonoids from *Astragalus monspessulanus*." *Journal of Natural Products* 78: 2565–2571.

Lajkiewicz, N. J., Roche, S. P., Gerard, B., Porco, J. A., Jr. 2012. "Enantioselective photocycloaddition of 3-hydroxyflavones: Total syntheses and absolute configuration assignments of (+)-ponapensin and (+)-elliptifoline." *Journal of the American Chemical Society* 134: 13108–13113.

Leete, E. 1982. "A one-step synthesis of ficine and isoficine." *Journal of Natural Products* 45: 605–607.

Li, N., Shao, L., Zhang, C.-F., Zhang, M. 2008a. "Two new flavonoid alkaloids from *Senecio argunensis*." *Journal of Asian Natural Products Research* 10: 1143–1146.

Li, X., Liu, G. J., Zhang, W., Zhou, Y. L., Ling, T. J., Wan, X. C., Bao, G. H. 2018. "Novel flavoalkaloids from white tea with inhibitory activity against the formation of advanced glycation end products." *Journal of Agricultural and Food Chemistry* 66: 4621–4629.

Liksakova, N. 2019. "*Phyllospadix iwatensis* Makino." Obtained from Global Biodiversity Information Facility under the Creative Common Share-Alike Attribution 4.0 license. www.gbif.org/occurrence/2518074535 (Accessed 22 June 2021)

Lu, S.-T., Hwang, J.-F., Wu, T.-S., McPhail, D. R., McPhail, A. T., Lee, K.-H. 1989. "Alkaloids of *Corydalis pallida* var. *sparsimamma* and the structure and stereochemistry of pallimamine." *Phytochemistry* 28: 1245–1249.

Luo, C., Wang, A., Wang, X., Li, J., Liu, H., Wang, M., Wang, L., Lai, D., Zhou, L. 2015. "A new proline-containing flavonol glycoside from *Caragana leucophloea* Pojark." *Journal of Asian Natural Products Research* 29, 19: 1811–1819.

Ma, Q., Xie, H., Li, S., Zhang, R., Zhang, M., Wei, X. 2014. "Flavonoids from the Pericarps of *Litchi chinensis*." *Journal of Agricultural and Food Chemistry* 62: 1073–1078.

Masterova, I., Uhrin, D., Tomko, J. 1987. "Lilaline-A flavonoid alkaloid from *Lilium candidum*." *Phytochemistry* 26: 1844–1845.

Merlini, L., Mondelli, R., Nasini, G., Hesse, M. 1967. "Indole alkaloids from gambir structure of bambirtannine, oxogambirtannine and dihydrogambirtannine." *Tetrahedron* 23: 3129–3145.

Ngamga, D., Bipa, J., Lebatha, P., Hiza, C., Mutanyatta, J., Bezabih, M.-T., Tane, P., Abegaz, B. M. 2008. "Isoquinoline alkaloids and homoisoflavonoids from *Drimiopsis barteri* Bak and *D. burkei* Bak." *Natural Product Communications* 3: 769–777.

Ngamga, D., Tane, P., Bezabih, M.-T., Abegaz, B. M. 2006. "Secondary metabolites from *Drimiopsis baterrii*." *Planta Medica* 72: 1009.

Nguyen, T. B., Lozach, O., Surpateanu, G., Wang, Q., Retailleau, P., Iorga, B. I., Meijer, L., Guéritte, F. 2012. "Synthesis, Biological evaluation, and molecular modeling of natural and unnatural flavonoidal alkaloids, inhibitors of kinases." *Journal of Medicinal Chemistry* 55: 2811–2819.

Nguyen, T. B., Wang, Q., Guéritte, F. 2011. "An efficient one-step synthesis of piperidin-2-yl and pyrrolidin-2-yl flavonoid alkaloids through phenolic mannich reactions." *European Journal of Organic Chemistry* 35: 7076–7079.

Ohyama, T., Uchida, M., Kusama, H., Iwasawa, N. 2015. "Total synthesis of proposed structure of yuremamine and all diastereomers using [3+2]-cycloaddition of platinum-containing azomethine ylide." *Chemistry* 10: 1850–1853.

Pan, L., Acu~na, U. M., Li, J., Jena, N., Ninh, T. N., Pannell, C. M., Chai, H., Fuchs, J. R., Carcache de Blanco, E. J., Soejarto, D. D, et al. 2013. "Bioactive flavaglines and other constituents isolated from *Aglaia perviridis*." *Journal of Natural Products* 76: 394–404.

Pietta, P. G. 2000. "Flavonoids as antioxidants." *Journal of Natural Products* 63: 1035–1042.

Proksch, P., Edrada, R., Ebel, R., Bohnenstengel, F. I., Nugroho, B. W. 2001. "Chemistry and biological activity of rocaglamide derivatives and related compounds in *Aglaia* species (meliaceae)." *Current Organic Chemistry* 5: 923–938.

Puripattanavong, J., Weber, S., Brecht, V., Frahm, A.W. 2000. "Phytochemical investigation of *Aglaia andamanica*." *Planta Medica* 66: 740–745.

Qui, L., Liang, Y., Tang, G.-H., Yuan, C.-M., Zhang, Y., Hao, X.-Y., Hao, X.-J., He, H.-P. 2013. "Two new flavonols, including one flavan dimer, from the roots of *Indigofera stachyodes*." *Phytochemistry Letter* 6: 368–371.

Ren, D.-M., Guo, H.-F., Yu, W.-T., Wang, S.-Q., Ji, M., Lou, H.-X. 2008. "Stereochemistry of flavonoidal alkaloids from *Dracocephalum rupestre*." *Phytochemistry* 69: 1425–1433.

Salim, A. A., Pawlus, A. D., Chai, H.-B., Farnsworth, N. R., Kinghorn, A. D., Carache-Blanco, E. J. 2007. "Ponapensin, a cyclopenta[*bc*]benzopyran with potent NF-κB inhibitory activity from *Aglaia ponapensis*." *Bioorganic & Medicinal Chemistry Letters* 17: 109–112.

Senderowicz, A. M. 1999. "Flavopiridol: The first cyclin-dependent kinase inhibitor in human clinical trials invest." *New Drugs* 17: 313–320.

Shen, S.-M., Shen, L.-G., Lei, Q.-F., Si, J.-Y., Liu, C.-M., Lu, H. Chin. 2012. "Chemical constituents contained in aerial parts of *Emilia sonchifolia*." *Chinese Journal of Natural Medicines* 37: 3249–3251.

Suzuki, K., Tsuruga, T., Hiranuma, K., Yamaura, M. 2004. "Synthesis of 7,8-dihydroxy-5-hydroxymethyl-2-phenyl-chroman-4-one; the aglycon of actinoflavoside." *Synlett*: 116–118.

Takagi, M., Funahashi, S., Ohta, K., Nakabayashi, T. 1980. "Phyllospadine, a new flavonoidal alkaloid from the sea-grass *Phyllosphad ixiwatensis*." *Agricultural and Biological Chemistry* 44: 3019–3020.

Tanaka, T., Watarumi, S., Fujieda, M., Kouno, I. 2005. "New black tea polyphenol having N-ethyl-2-pyrrolidinone moiety derived from tea amino acid theanine: Isolation, characterization and partial synthesis." *Food Chemistry* 93: 81–87.

Veitch, N. C., Grayer, R. 2011. "Flavonoids and their glycosides, including anthocyanins." *Natural Product Reports* 28: 1626–1695.

Vepsalainen, J. J., Auriola, S., Tukiainen, M., Ropponen, N., Callaway, J. C. 2005. "Isolation and characterization of yuremamine, a new phytoindole." *Planta Medica* 71: 1053–1057.

Wang, J., He, H., Shen, Y., Hao, X. 2005. "Sulfur-containing and dimeric flavanols from *Glycosmis Montana*." *Tetrahedron Letter* 46: 169–172.

Wang, L., Wang, S., Yang, S., Guo, X., Lou, H., Ren, D. 2012a. "Phenolic alkaloids from the aerial parts of *Dracocephalum heterophyllum*." *Phytochemistry* 82: 166–171.

Wang, S.-K., Cheng, Y.-J., Duh, C.-Y. 2012b. "Cytotoxic constituents from leaves of *Aglaia elliptifolia*." *Journal of Natural Products* 64: 92–94.

Wikipedie. 2021. "Kýlatec." Obtained from wikipedie under the Creative commons Share-Alike Attribution 3.0 license. https://cs.wikipedia.org/wiki/K%C3%BDlatec (Accessed 20 June 2021).

Xu, Y.-J., Foubert, K., Dhooghe, L., Lemi_ere, F., Maregesi, S., Coleman, C. M., Zou, Y., Ferreira, D., Apers, S., Pieters, L. 2012. "Rapid isolation and identification of minor natural products by LC—MS, LC—SPE—NMR and ECD: Isoflavanones, biflavanones and bisdihydrocoumarins from *Ormocarpum kirkii*." *Phytochemistry* 79: 121–128.

Yoshikado, N., Taniguchi, S., Kasajima, N., Ohashi, F., Doi, K.-I., Shibata, T., Yoshida, T., Hatano, T. 2009. "Uncariagambiriine and gambircatechol: Novel constituents of *Uncaria gambir* leaves." *Heterocycles* 77: 793–800.

Zeidner, J. F., Foster, M. C., Blackford, A. L., Litzow, M. R., Morris, L. E., Strickland, S. A., Lancet, J. E., Bose, P., Levy, M. Y., Tibes, R., et al. 2015."Randomized multicenter phase II study of flavopiridol (alvocidib), cytarabine, and mitoxantrone (FLAM) versus cytarabine/daunorubicin (7+3) in newly diagnosed acute myeloid leukemia." *Haematologica* 100: 1172–1179.

Zhao, J.-Q., Wang, Y.-M., Lv, J.-J., Zhu, H.-T., Wang, D., Yang, C.-R., Xu, M., Zhang, Y.-J. 2014. "New phenolic glycosides from *Phyllanthus cochinchinensis*." *Journal of the Brazilian Chemical Society* 25: 1446–1454.

22 The Role of Plant-Based Natural Compounds in Inflammation

The Effective Inhibitors of Arachidonic Acid Metabolic Pathways

Marcela Dvorakova, Premysl Landa and Lenka Langhansova

CONTENTS

22.1 ARACHIDONIC ACID METABOLIC PATHWAYS

Arachidonic acid (AA) is released from membrane phospholipids by the action of phospholipases A_2 (PLA_2), which include about 30 different enzymes. Particularly, the cytosolic $PLA_2\alpha$ enzyme regulates the AA release for a massive eicosanoid production in response to cytokines, growth

DOI: 10.1201/9781003108375-22

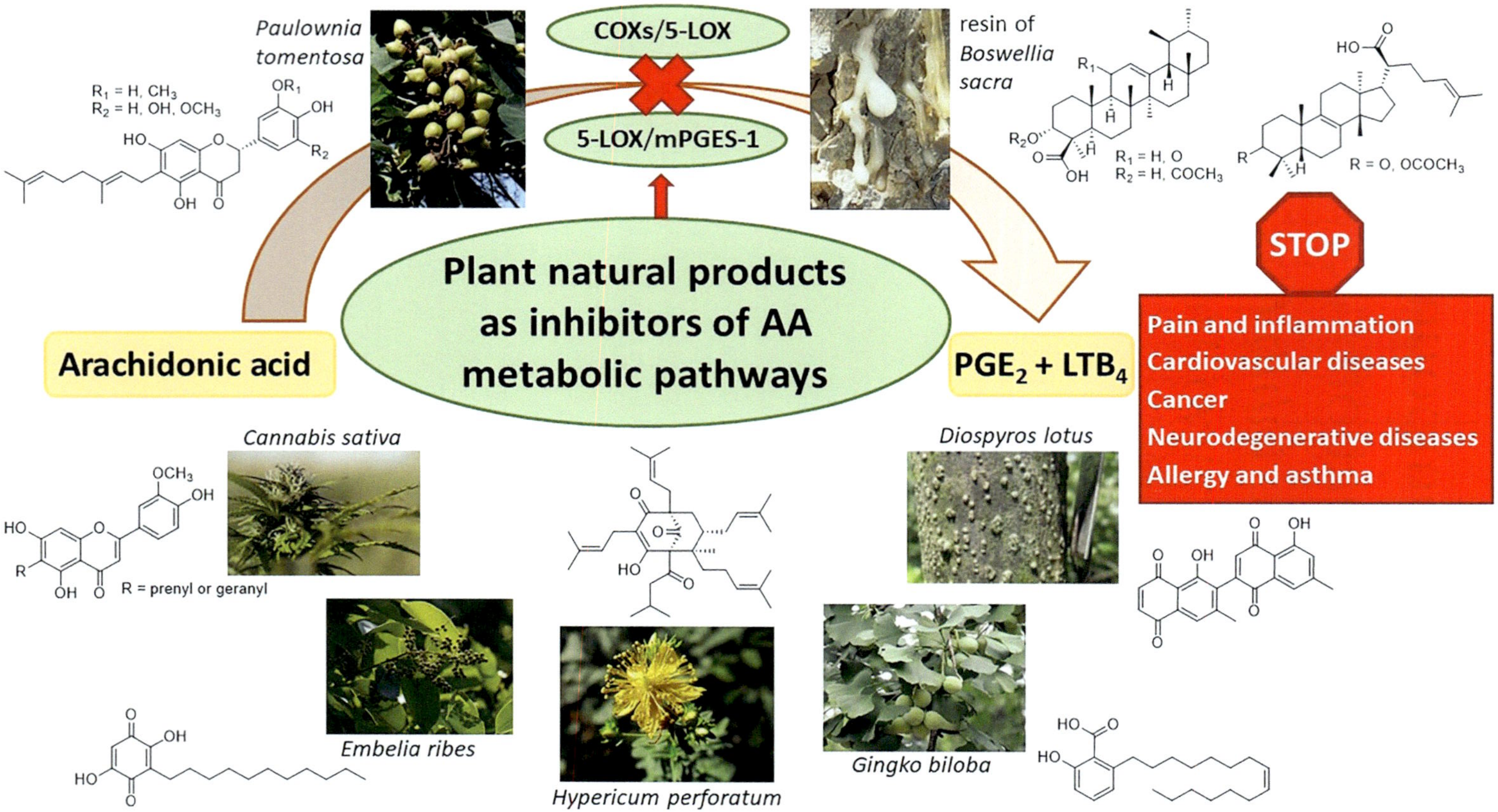

Graphical abstract representing the content of the chapter (AnRo0002, 2015; ©BMK, 2008; KENPEI, 2008; Nova, 2007; Raffaelli, 2012; Skalle-Per Hedenhös, 2012; Valker, 2015).

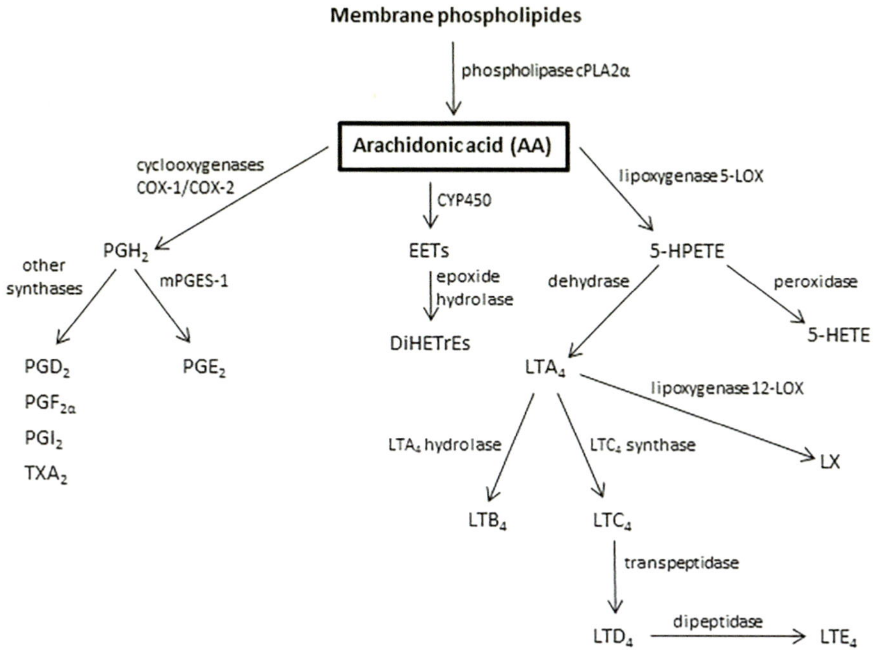

FIGURE 22.1 Metabolism of arachidonic acid into eicosanoids.

factors and hormones, and is activated by intracellular Ca^{2+}-influx and phosphorylation (Koeberle and Werz, 2018). The biosynthesis of eicosanoids occurs in all mammalian tissues and is regulated by an immune system. Eicosanoids are formed from AA via three systems (Figure 22.1): (i) by cyclooxygenases (COXs) and microsomal prostaglandin E synthase-1 (mPGES-1) to prostanoids that include prostaglandins (PGs) and thromboxane; (ii) by lipoxygenases (LOXs) to hydroperoxyeicosatetraenoic acids (HPETEs), leukotrienes (LTs), lipoxins (LXs) and monohydroxyeicosatetraenoic acids (HETEs); and (iii) by cytochrome P450 epoxygenase (CYP450) to epoxyeicosatrienoic acids (EETs). While PGs and LTs trigger inflammation processes, EETs are anti-inflammatory; however, they are rapidly degraded by soluble epoxide hydrolase (sEH) to corresponding diols-dihydroxyeicosatrienoic acids (DiHETrEs), thus losing their anti-inflammatory effect (Garscha et al., 2017).

22.1.1 Conversion of AA by COXs and mPGES-1

Cyclooxygenases (isoforms COX-1 and COX-2) are dimeric membrane-bound proteins. First, COXs convert the released AA to prostaglandin H_2 (PGH_2). Subsequently, tissue-specific PG synthases rapidly convert PGH_2 to bioactive prostanoids: (i) prostaglandins (PGD_2, PGE_2, and $PGF_2\alpha$); (ii) prostacyclin (PGI_2); or (iii) thromboxane (TXA_2) (Ambati and Jachak, 2020). mPGES-1 is situated downstream of COX-2 and promotes an excessive PGE_2 production from COX-2–derived PGH_2 during inflammation.

The COX-1 isoform converts AA mainly to TXA_2, a vasoconstrictor that promotes platelet aggregation. On the other hand, PGE_2 and PGI_2 are formed predominantly by the COX-2 isoform. PGI_2 is a vasodilator and an inhibitor of platelet aggregation. PGE_2 is a major prostanoid linked to inflammation, pain, fever and tumorigenesis (Samuelsson et al., 2007; Tuure et al., 2017; Koeberle and Werz, 2018). During inflammation, the primary prostanoid formation is catalyzed by COX-1, while COX-2 contributes to prostanoid synthesis later on (Curtis et al., 2019; Jiang and Yu, 2020).

22.1.2 Conversion of AA by 5-LOX

Lipoxygenases are non-heme, iron-containing, lipid-peroxidizing enzymes. There are six human lipoxygenases, of which 5-LOX is the one primarily associated with inflammation. 5-LOX is activated by 5-lipoxygenase-activating protein (FLAP) and converts AA to 5-HPETE, which subsequently undergoes dehydration to form non-cysteinyl leukotriene A_4 (LTA_4). LTA_4 is unstable and thus is rapidly converted either in mast cells or eosinophils to cysteinyl-LTs (LTC_4, LTD_4, and LTE_4) or in neutrophils or monocytes to non-cysteine–containing dihydroxy-leukotriene B_4 (LTB_4). The cysteinyl-LTs are formed via successive action of LTC_4 synthase (forms LTC_4), transpeptidase (forms LTD_4) and, finally, dipeptidase (forms LTE_4), while LTB_4 is formed directly by LTA_4 hydrolase (Ghosh et al., 2016; Garscha et al., 2017; Pein et al., 2018; Rahman et al., 2019). The other human LOX enzymes, like 12- and 15-LOX, are expressed in various cell types and organs, and their metabolites exhibit both pro- and anti-inflammatory effects (Singh and Rao, 2019).

22.1.3 Therapeutic Importance of COXs Inhibition

Prostanoids are lipid mediators involved in a broad spectrum of physiological and pathological processes. They play a complex role in the normal function of kidneys, lungs, brain, ovaries, uterus, bones or the cardiovascular system. However, their synthesis is increased in inflamed tissue. Prostaglandins E_2 (PGE_2) and I_2 (PGI_2) are the main pro-inflammatory prostanoids enhancing edema formation and leukocyte infiltration. Moreover, they alter the threshold for pain perception. The application of COX inhibitors called non-steroidal anti-inflammatory drugs (NSAIDs) decreases prostanoid production. NSAIDs are used to relieve pain, fever and inflammation (Dubois et al., 1998; Smith et al., 2000; Smyth et al., 2009). Currently, NSAIDs belong to the most widely used pharmaceuticals consumed in tens of thousands of tons each year, especially as some of them are available over the counter (Warner and Mitchell, 2004).

Since the discovery of COX isoforms, the design of selective COX-2 inhibitors was initiated. The reason was that conventional NSAIDs—such as ibuprofen, naproxen, ketoprofen, indomethacin and diclofenac—inhibit both COX isoforms, and the inhibition of COX-1 leads to the suppression of prostaglandin production in the stomach resulting in an increased risk of ulcers (Smyth et al., 2009; Peskar, 2001).

On the other hand, the inhibitors with high COX-2 selectivity are called "coxibs" (e.g. rofecoxib, celecoxib, valdecoxib, parecoxib, etoricoxib and lumiracoxib) and increase the risk of cardiovascular events due to the changes of the balance in the production of prostanoids influencing vasoconstriction. This side effect resulted in the withdrawal of some coxibs from the market and the formulation of guidelines for the prescription of others (Zarraga and Schwarz, 2007; Curtis et al., 2019).

COX activity may be inhibited via: (i) competitive rapidly reversible inhibition; (ii) competitive tight-binding inhibition; and (iii) covalent modification of the active site. The rapidly reversible inhibition is characterized by a single-step mechanism describing the association of an inhibitor with the enzyme. Ibuprofen is an example of a reversible inhibitor that binds to the native closed conformation of COX. On the other hand, tight-binding inhibition is characterized by a two-step mechanism, in which an initial rapid, reversible binding of the inhibitor to the enzyme is followed by an essentially irreversible time-dependent change into the enzyme-inhibitor complex (flurbiprofen, meclofenamic acid). The only inhibitor known to covalently modify the COXs active site is acetylsalicylic acid, which irreversibly inhibits COXs due to the acetylation of Ser530 (Blobaum and Marnett, 2007).

In addition to the mode of inhibition, COX inhibitors are divided into classes based on their selectivity to the target COX isoform to: (i) non-selective COX inhibitors, which inhibit both COX isoforms with about the same efficiency (e.g. resveratrol); (ii) COX-1 selective inhibitors (e.g. sophoraflavanone G); and (iii) COX-2 selective inhibitors (e.g. chebulagic acid). The selectivity of inhibitors is driven by the difference in the binding site of the two COX isoforms. COX-2 binding site is about 20% larger and contains an additional hydrophilic side pocket. An entrance to this side pocket is hindered in COX-1 by the presence of bulky amino acids Ile523 and Ile434, which in COX-2 are replaced by smaller valines (Cingolani et al., 2017). The selectivity of each inhibitor is given by a selectivity index, which is defined as IC_{50} (COX-2)/IC_{50} (COX-1).

The role of both COX isoforms is associated with other diseases than just inflammation. For example, a 40–50% decreased relative risk of colorectal cancer was observed in persons regularly using NSAIDs (Dubois et al., 1998). Similarly, the risk of lung, breast and prostate cancer was reported to be significantly decreased by NSAIDs administration (Goradel et al., 2019). Recently, the up-regulation of COX-1 was observed in breast, colorectal, skin and ovarian cancer (Pannunzio and Coluccia, 2018; Vitale et al., 2016; Calvello et al., 2017). Furthermore, several studies indicate that NSAIDs may prevent or delay onset of Alzheimer's and Parkinson's disease (Klegeris and McGeer, 2005), and selective COX-2 inhibitors may be effective in the treatment of multiple sclerosis, brain ischemia or epilepsy (Gulla et al., 2020; Jiang and Yu, 2020; Rawat et al., 2019). In addition, COX-1 up-regulation was observed during atherosclerosis or neuroinflammation (Pannunzio and Coluccia, 2018; Vitale et al., 2016). Thus, selective COX-1 inhibitors are considered to have a therapeutic potential in the alleviation of neuroinflammatory diseases in general and in the prevention of cardiovascular diseases (Calvello et al., 2017; Mitchell et al., 2019; O'Banion, 1999).

22.1.4 Therapeutic Importance of mPGES-1 Inhibition

Microsomal prostaglandin E synthase-1 (mPGES-1) is one of three characterized PGE synthases: mPGES-1, microsomal prostaglandin E synthase-2 (mPGES-2) and cytosolic prostaglandin E synthase (cPGES). The latter two are considered to be responsible for physiological PGE_2 formation, whereas the expression of mPGES-1 is induced under inflammatory conditions (Samuelsson et al., 2007). A terminal enzyme in the synthesis of inducible PGE_2, mPGES-1 is therefore a promising drug target to treat inflammation and pain due to a selective attenuation of PGE_2 production under inflammatory conditions; mPGES-1 is a relatively novel therapeutic target, for which inhibitors are under development. Recently, the selective mPGES-1 inhibitor LY3023703 underwent a phase I trial which showed a decreased level of PGE_2 metabolite in urine to a comparable extent as celecoxib (Tuure et al., 2017).

22.1.5 Therapeutic Importance of 5-LOX Inhibition

The production of pro-inflammatory leukotrienes (LTs) from AA is catalyzed by 5-LOX. LTs stimulate processes such as smooth muscle contraction, phagocyte chemotaxis and increased vascular permeability (Gilbert et al., 2011; Radmark and Samuelsson, 2009); thus, LTs are involved in inflammatory diseases such as bronchial asthma, allergic rhinitis, atopic dermatitis, psoriasis, atherosclerosis, rheumatoid arthritis, ischemic renal failure and inflammatory bowel diseases (Jo-Watanabe et al., 2019; Ghosh et al., 2016). Moreover, 5-LOX metabolites are involved in central nervous system diseases (Alzheimer's disease, Parkinson's disease, multiple sclerosis, epilepsy, brain ischemia, etc.), type 2 diabetes and several types of cancer (Moore and Pidgeon, 2017; Ghosh et al., 2016). Currently, the only 5-LOX inhibitor used commercially, Zileuton, is approved for asthma treatment for oral use. However, it has a short half-life and may cause hepatotoxicity (Fiorucci et al., 2001; Bruno et al., 2018; Wood et al., 2020).

5-LOX inhibitors may be divided into classes based on their molecular mechanism of action. These classes include: (i) redox-active inhibitors; (ii) iron-ligand inhibitors; and (iii) non-redox type inhibitors (Werz, 2007). The redox-active inhibitors disrupt the catalytic cycle of the 5-LOX enzyme by keeping the active site iron in the ferrous state (e.g. nordihydroguaiaretic acid). Compounds under this class include flavonoids, coumarins and quinones, which are often distinguished by a lack of selectivity and oral bioavailability (Werz, 2007). Iron-ligand inhibitors are chelating agents that often form hydroxamic acid or hydroxyurea chelates with the active site iron (e.g. Zileuton). Non-redox inhibitors usually compete with AA in binding to the active site or disrupt the binding of AA by blocking the entrance to the active site or by changing enzyme conformation through allosteric inhibition (e.g. boswellic acids) (Werz, 2007).

22.1.6 Dual Inhibitors of AA Metabolic Pathways

Since COXs, mPGES-1 and 5-LOX convert the same substrate (AA), the inhibition of COXs or mPGES-1 may increase AA availability for 5-LOX, resulting in a higher LTs production and thus causing disbalance in eicosanoid production. Therefore, dual COX/5-LOX or mPGES-1/5-LOX inhibition was defined as a promising approach in the development of novel anti-inflammatory drugs. Simultaneous dual COX/5-LOX inhibition may decrease gastric toxicity in comparison with conventional NSAIDs, since LTs are involved in gastric mucosal injuries. Moreover, dual mPGES-1/5-LOX inhibition should avoid the toxicity connected with COX inhibition altogether, as it leads to a selective decrease of inducible PGE_2. Furthermore, since prostanoids and leukotrienes are both involved in the development of some types of cancer, dual inhibition may provide higher anticancer responses. In addition, dual inhibition may be neuroprotective and therefore useful in the prevention of neurodegenerative diseases such as Alzheimer's and Parkinson's diseases (Fiorucci et al., 2001; Leval et al., 2002; Leone et al., 2007).

22.1.7 Tools to Search for the Inhibitors of AA Metabolic Pathways

Various *in vitro* assays are available to test potential inhibitors or plant extracts for their ability to inhibit COXs, mPGES-1 or 5-LOX. Assays are divided into cell-free and cell-based, which may give different results based on the conditions used. In the case of AA metabolism, cell-free assays are defined by direct inhibition of the respective enzyme (COX-1, COX-2, mPGES-1, or 5-LOX), whereas the inhibition in cell-based assays may be obfuscated by the inhibition of the $cPLA_2$ enzyme, which reduces AA supply (Werz, 2007). In the case of 5-LOX, cell-based assays may show an indirect inhibition of 5-LOX enzyme, through the inhibition of 5-lipoxygenase activating protein (FLAP). Enzymatic *in vitro* assays are employed using animal or human recombinant enzymes. Cell-based assays use either isolated cells (e.g. neutrophil granulocytes with 5-LOX activity) or whole blood, which utilizes all the components present within the blood, where cell-cell interactions are preserved (Laufer and Luik, 2010). Furthermore, computational methods are useful tools that enable one to pick potential inhibitors from compound libraries to save the materials needed (Kaserer et al., 2015; Temml et al., 2014).

22.2 PLANTS AS SOURCES OF INHIBITORS OF AA METABOLIC PATHWAYS

There is an enormous number of studies describing plant extracts or their components as potential inhibitors of AA metabolism, and thus suppressors of the activity of COXs, mPGES-1, and 5-LOX. Within this chapter, plant-based compounds that have been reported to efficiently inhibit human COXs, mPGES-1 and 5-LOX ($IC_{50} < 20$ µM) have been grouped according to their structural features (Table 22.1). The inhibition of COXs, mPGES-1 and 5-LOX in cell-free and cell-based assays has been discussed. The active compounds from the last 15 years and those with dual inhibitory activity are predominantly included. The inhibitors of related lipoxygenases, such as 12-LOX, 15-LOX, and soybean or potato LOX has not been documented. Natural compounds of plant origin with selective COX-1, COX-2, mPGES-1 or 5-LOX inhibitory activity, which are not included herein, may be found in the following reviews: COX-1 inhibitors in Vitale et al. (2016), COX-2 inhibitors in Attiq et al. (2018) and Ambati and Jachak (2020), mPGES-1 inhibitors in Khan et al. (2018) and 5-LOX inhibitors in Werz (2007), Rahman et al. (2019) and Orafaie et al. (2020).

TABLE 22.1
Plant Natural Compounds with Potent Dual Inhibitory Activity on Arachidonic Acid Metabolism Pathways

Compounds		5-LOX[1] (CF[2]) (μM)	5-LOX (CB[3]) (μM)	COX-1[4] (μM)	COX-2[5] (μM)	MPGES-1[6] (μM)	References
Flavonoids	Diplacone	0.05	n.d.[7]	1.8	4.2	n.d.	(Hanakova et al., 2017)
	3'-*O*-Methyl-5'-hydroxydiplacone	0.06	n.d.	3.3	10.6	n.d.	(Hanakova et al., 2017)
	Sophoraflavanone G	0.15	n.d.	0.03	n.i.[8]	n.d.	(Chi et al., 2001)
	Xanthohumol	2.1	2.9	16.6	41.5	n.i.[a]	(Forino et al., 2016; Gerhauser et al., 2002)
	Cannflavin A	0.9	1.6–2.4	n.i.[b]	n.i.	1.8	(Werz et al., 2014)
	Cannflavin B	0.8	n.d.	n.d.	n.d.	3.7	(Werz et al., 2014)
	Cudraflavone B	?[c]	n.d.	1.5	2.5	n.d.	(Hosek et al., 2011; Glibo, 2016)
	Luteolin	2.25	0.8–1.8	n.i.	n.i.[d]	n.d.	(Lee et al., 2010; Lee and Kim, 2010; Li et al., 2008; Kutil et al., 2014)
	Quercetin	3.29	0.7–5.1	n.i.[e]	n.i.[f]	n.d.	(Lee et al., 2010; Lee and Kim, 2010; Kutil et al., 2014; Kwon et al., 2011)
Stilbenes	Piceatannol	0.39–1.1	0.24	n.i.	n.i.	n.d.	(Kutil et al., 2014; Boccellino et al., 2019; Lelakova et al., 2019)
	Pinostilbene	4.0-n.i.	n.d.	1.37–1.90	0.25–0.35	n.d.	(Kutil et al., 2015; Lelakova et al., 2019)
	Pterostilbene	1.63–9.32	0,53	5.99–11.7	,6.67–27.04		(Kutil et al., 2015; Lelakova et al., 2019; Boccellino et al., 2019)
	Macasiamenenes A	1.10	n.d.	0.8	4.7	n.d.	(Hosek et al., 2019)
	Macasiamenenes B	0.90	n.d.	0.9	14.7	n.d.	(Hosek et al., 2019)
	2,6-Diprenylresveratrol	0.90	n.d.	0.4	5.7	n.d.	(Hosek et al., 2019)
Phenolic acids	Ginkgolic acid	0.20	2.9–3.8	8.1	n.i.[g]	0.7	(Gerstmeier et al., 2019)
Acylphloroglucinols	Garcinol	0.10	1.9	12.0	n.i.	0.3	(Koeberle, Northoff, and Werz, 2009)
	Hyperforin	0.09	1.5	3.0–12.0	n.i.	1.0	(Albert et al., 2002; Koeberle et al., 2011)
	Arzanol	3.1	2.9–3.1	2.6–17.5	n.i.	0.4	(Bauer et al., 2011; Minassi et al., 2012)
	Acrovestone	2.7	2.3	n.d.	n.d.	1.1	(Svouraki et al., 2017)
	Myrtucommulone	5.0	n.d.	6.0	n.i.	1.0	(Feisst et al., 2005; Koeberle et al., 2009)
Lignans	(+)-de-4''-*O*-Methylmagnolin	5.1	n.d.	35.6	2.3	n.d.	(Coy-Barrera and Cuca-Suarez, 2011)
	Rataniaphenol II	n.i.	n.d.	3.0	1.7	11.0	(Baumgartner et al., 2011)

(Continued)

TABLE 22.1 (Continued)

Compounds		5-LOX[1] (CF[2]) (µM)	5-LOX (CB[3]) (µM)	COX-1[4] (µM)	COX-2[5] (µM)	MPGES-1[6] (µM)	References
Tannins	Chebulagic acid	2.1	n.d.	15.0	0.92	n.d.	(Reddy et al., 2009)
Triterpenes	3-*O*-Acetyl-11-keto-*β*-boswellic acid	2.9–50	1.5–15.0	n.i.[h]	n.i. [h]	3.0	(Werz, 2007; Siemoneit et al., 2009; Siemoneit et al., 2011)
	β-Boswellic acid	23.9	28.1	n.i. [h]	n.i. [h]	5.0	(Siemoneit et al., 2009; Siemoneit et al., 2011; Nhu Ngoc Quynh et al., 2019)
	3-*O*-Acetyl-28-hydroxylupeolic acid	8.3	2.3	n.i. [h]	n.i. [h]	0.9	(Verhoff et al., 2012; Verhoff et al., 2014)
	Timosaponin B-II	1.57	n.d.	n.d.	0.77	n.d.	(Xie et al. 2020)
	Timosaponin B-III	1.82	n.d.	n.d.	0.48	n.d.	(Xie et al., 2020)
Diterpenes	Carnosol	0.3	n.d.	n.i.	n.i.	5.0–10.9	(Maione et al., 2017; Bauer et al., 2012)
Quinones	Embelin	0.07	0.8–2.0	n.i.	n.i.	0.2	(Schaible et al., 2013)
	β-Lapachone	0.24	0.42–8.6	n.d.	n.d.	n.d.	(Rodrigues et al., 2018)
	Primin	1.37	4.0	5.14–18.9	0.09–1.90	n.d.	(Landa et al., 2013; Landa et al., 2012; Sisa et al., 2020)
	Shikonin	n.d.	24.3	1.91	0.33	n.d.	(Landa et al., 2013; Landa et al., 2012)
	Diospyrin	n.d.	n.i.	1.39	0.55	n.d.	(Landa et al., 2012)

Notes: **1** 5-lipoxygenase; **2** results from cell free assays; **3** results from cell-based assays; **4** cyclooxygenase-1; **5** cyclooxygenase-2; **6** microsomal prostaglandine E synthase-1; **7** not determined; **8** no inhibition observed when tested at 50 µM and lower concentrations; **a** 32% residual mPGES-1 enzymatic activity at 10 µM; **b** 35% inhibition at 10 µM; **c** 54% inhibition at 10µ M, unknown IC50 value; **d** lowest reported IC50 value = 59.9 µM; **e** lowest reported IC50 value = 81.1 µM; **f** lowest reported IC50 value = 72.3 µM; **g** no inhibition observed at 10 µM; **h** no inhibition observed below 40 µM

22.2.1 Polyphenols

Polyphenols form the largest group of plant secondary metabolites, which includes flavonoids, stilbenes, phenolic acids and lignans. As early as 3,500 years ago, an extract of willow bark and leaves was utilized for the treatment of fever and inflammation. In the 17th century, salicylic acid was identified as the active compound of this extract (Vane, 2000; Mahesh et al., 2021). Its synthetic derivative, acetylsalicylic acid (aspirin), was introduced to the market in 1869 by Bayer Company. However, it took another century to reveal the mechanistic role that laid behind its activity (Vane, 2000; Mahesh et al., 2021). Vane (1971) found that aspirin and aspirin-like drugs inhibit COX enzymes, which blasted off the development of new generations of COX inhibitors.

On the other hand, the first plant polyphenolic 5-LOX inhibitor, nordihydroguaiaretic acid (NDGA, IC_{50} = 0.8 μM), was not discovered until 1981, when it was isolated from the desert plant *Larrea divaricata* (Sessé et Moc. ex DC.) Coville (Werz, 2007). Since then, polyphenols are widely studied for their anti-inflammatory properties. These studies are supported by the observed lower incidence of chronic inflammatory disorders in humans consuming polyphenol-rich food and beverages (Werz, 2007). The potency of some polyphenols toward COXs and 5-LOX may be attributed to their ability to scavenge reactive oxygen species and suppress lipid peroxidation. On the other hand, even polyphenols lacking antioxidant properties may inhibit these inflammatory enzymes, often due to the resemblance of AA substrate by the presence of prenyl or geranyl side chain (Werz, 2007).

22.2.1.1 Flavonoids

Flavonoids form a large group of plant secondary metabolites, with more than 4,000 representatives. According to a chemical structure (Figure 22.2), they are divided into six classes: flavones, isoflavones, flavonols, flavanols, flavanones and anthocyanins (Mladenka et al., 2010). From these, especially prenylated flavonoids, which possess lipophilic prenyl or geranyl residues, are distinguished by enhanced COXs/mPGES-1/5-LOX inhibitory potency caused by their lipophilic nature, which allows them to penetrate cell membranes more effectively than other flavonoids (Hosek et al., 2011; Werz, 2007).

Recently, Hanakova et al. (2017) isolated a series of prenylated flavanones from the fruits of *Paulownia tomentosa* (Thunb.) (Figure 22.3A), which were studied for their ability to inhibit 5-LOX as well as COX-1 and COX-2 in cell-free assays. Altogether, 19 flavanones and one geranylchromone were isolated. From these, six compounds exhibited inhibitory activity on 5-LOX and five compounds on COXs. The best dual COX/5-LOX inhibitory activity was observed for diplacone, which on 5-LOX exerted superior inhibitory activity (IC_{50} = 0.05 μM) to the reference compound Zileuton (IC_{50} = 0.35 μM), and on COXs comparable inhibitory activity (IC_{50} = 4.2 μM and 1.8 μM for COX-2 and COX-1, respectively) to the reference compound Ibuprofen (IC_{50} = 4.2 μM and 6.3 μM, respectively). In addition, 3'-*O*-methyl-5'-hydroxydiplacone potently inhibited 5-LOX (IC_{50} = 0.06 μM), though its inhibitory activity toward COXs was slightly lower (IC_{50} = 10.6 μM and 3.3 μM for COX-2 and COX-1, respectively) (Hanakova et al., 2017).

Three other compounds, tomentone, 3'-*O*-methyl-5'-methoxydiplacone and paulownione C, exerted selective inhibitory activity on 5-LOX (IC_{50} = 0.35 μM, 0.38 μM and 0.37 μM, respectively), which was comparable to Zileuton. Similarly, two compounds, 3',4'-*O*-dimethyl-5'-hydroxydiplacone and mimulone, exhibited selective COXs inhibition with values comparable to Ibuprofen. From these results, it seems that especially the presence of *ortho*-hydroxyls on B-ring and their methylation pattern together with the presence of unmodified geranyl chain on A-ring are the most important structural features for the measured bioactivity (Hanakova et al., 2017). On the contrary, despite the lack of the vicinal phenolic hydroxyls, the previously reported prenylated flavonoids, kenusanone A and sophoraflavanone G, inhibited 5-LOX (IC_{50} = 0.7 and 0.15 μM, respectively), although with slightly lower potency than diplacone and 3'-*O*-methyl-5'-hydroxydiplacone (Chi et al., 2001). On the other hand, kenusanone A did not inhibit COXs and sophoraflavanone G inhibited only COX-1, though very efficiently (IC_{50} = 0.03 μM) (Chi et al., 2001).

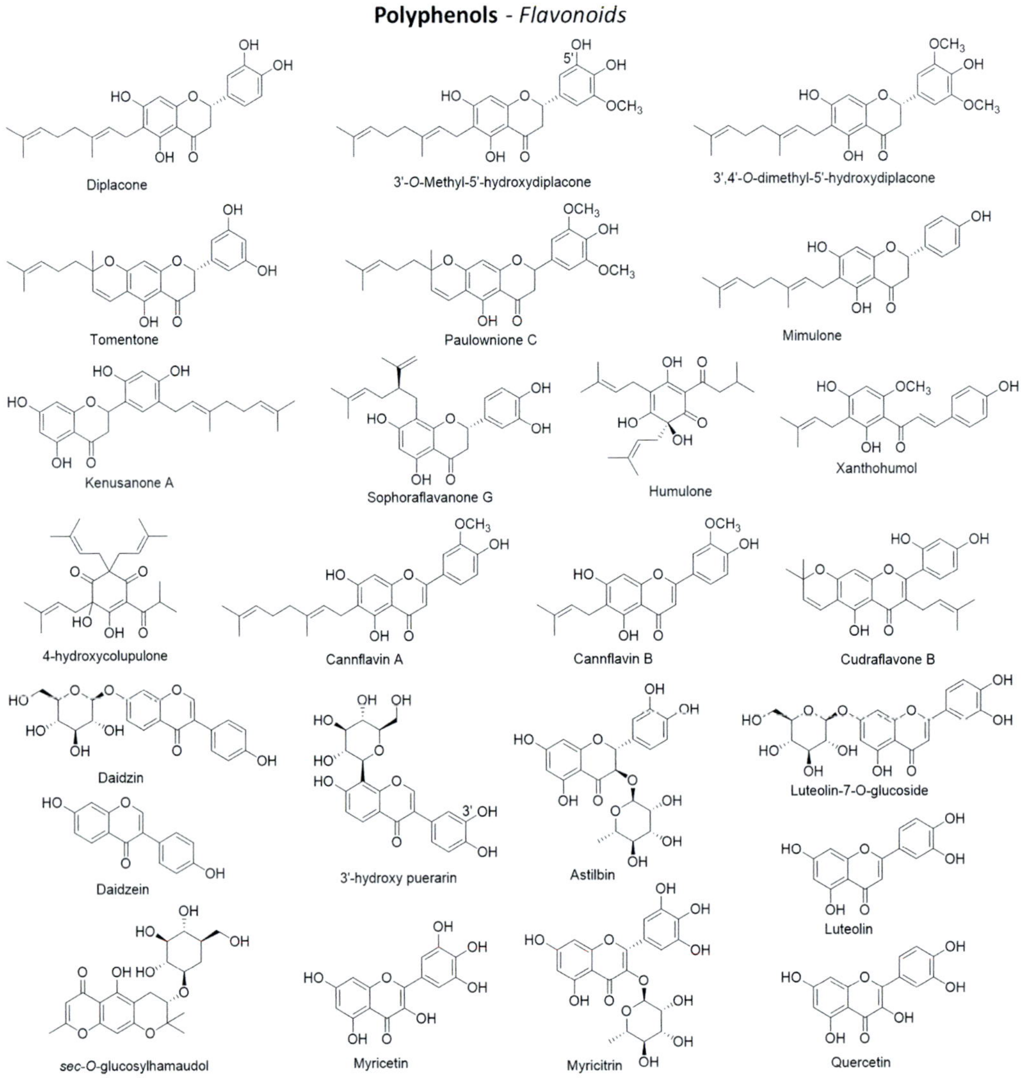

FIGURE 22.2 Structures of selected flavonoids, polyphenols with potent inhibitory activity on arachidonic acid metabolic pathways.

Other prenylated flavonoids were isolated from the inflorescence of the hop plant (*Humulus lupulus* L cultivar "Cascade") (Figure 22.3B) by Forino et al. (2016). From eight isolated compounds, three exerted inhibitory activity toward 5-LOX in the cell-free enzymatic assay (humulone, xanthohumol and 4-hydroxycolupulone, IC_{50} = 2.1–9.3 µM). From these, only one compound, xanthohumol, effectively inhibited 5-LOX using a cell-based assay with an IC_{50} of 2.9 µM (Forino et al., 2016). In another study, humulone was found to strongly inhibit COX-2 enzymatic activity with an IC_{50} = 1.6 µM, while its activity toward COX-1 was very low (IC_{50} = 70 µM) (Tobe, 2009). In addition, xanthohumol weakly inhibited the activity of both COX isoforms with IC_{50} = 16.6 µM and 41.5 µM for COX-1 and COX-2, respectively (Gerhauser et al., 2002). Both xanthohumol and 4-hydroxycolupulone very weakly inhibited mPGES-1, showing residual 32% enzymatic activity when applied at 10 µM concentration (Forino et al., 2016).

Moreover, prenylated flavonoids cannflavins A and B, isolated from industrial *Cannabis sativa* L. (hemp) sprouts (Figure 22.3C), were found to exert dual mPGES-1/5-LOX inhibitory activity.

FIGURE 22.3 Plant sources of flavonoids, potent inhibitors of arachidonic acid metabolic pathways. A, fruits of *Paulownia tomentosa*, a source of diplacone, 3'-*O*-methyl-5'-hydroxydiplacone, 3'-*O*-methyl-5'-methoxydiplacone, 3',4'-*O*-dimethyl-5'-hydroxydiplacone, tomentone, paulownione C, and mimulone; B, inflorescence of *Humulus lupulus*, a source of humulone, xanthohumol, and 4-hydroxycolupulone; C, female inflorescence of *Cannabis sativa*, a source of cannflavins A and B (©BMK, 2008; Skalle-Per Hedenhös, 2012; Stanzilla, 2020).

They potently inhibited 5-LOX in cell-free assay with IC_{50} values of 0.9 μM and 0.8 μM, respectively. In the cell-based assay, only cannflavin A was studied and inhibited 5-LOX with IC_{50} values of 1.6–2.4 μM, depending on the assay conditions. In addition, cannflavins A and B both reversibly inhibited mPGES-1 in cell-free assay with IC_{50} values of 1.8 μM and 3.7 μM, respectively. On the contrary, only weak inhibition of COXs by cannflavin A was observed (Werz et al., 2014).

In *Morus alba* L., a herbal plant with anti-inflammatory properties, another prenylated flavonoid was found in large amounts in the roots. This flavonoid, cudraflavone B, inhibited several pro-inflammatory enzymes, including cyclooxygenases. Its inhibitory activity with IC_{50} values of 1.5 μM and 2.5 μM for COX-1 and COX-2, respectively, was similar to that of indomethacin (IC_{50} = 0.3 μM and 1.9 μM for COX-1 and COX-2, respectively) (Hosek et al., 2011). In addition, cudraflavone B was mentioned as an effective 5-LOX inhibitor (54% inhibition at 10 μM concentration) in a diploma thesis of Glibo (2016); however, no further information was available (Glibo, 2016).

5-LOX inhibitory activity of ethanolic whole plant extracts of four Chinese medicinal plant herbs (*Saposhnikovia divaricata* (Turcz.) Schischk, *Smilax glabra* Roxb., *Pueraria montana* var. *lobata* (Willd) Sanjappa et Pradeep, and *Carthamus tinctorius* L., and their components were studied. All extracts exerted interesting 5-LOX inhibition with IC_{50} values of 0.08–0.66 μg/μL. From their components, the lowest—though moderate—IC_{50} values were demonstrated by daidzin (4.32 μM), followed by daidzein (6.65 μM), 3'-hydroxypuerarin (7.36 μM), sec-*O*-glucosylhamaudol (7.45 μM) and astilbin (8.45 μM). The importance of the presence of the vicinal hydroxyl groups for the activity was clearly shown in puerarin compounds, as the lack of 3'-hydroxy group on puerarin moiety led to the attenuation of the inhibitory activity (IC_{50} = 25.8 μM), while the methylation of 3'-hydroxy group to afford 3'-methoxypuerarin caused complete inactivity (Zhao et al., 2016).

In the case of activity on COXs, daidzein was reported to inhibit LPS-induced COX-2 expression in activated macrophages (Hämäläinen et al., 2011). In addition, astilbin at the final treatment concentration of 50 μM was able to reduce the COX-2 mRNA expression by 80% in LPS-stimulated cells (Huang et al., 2011). On the contrary, inhibitory activity on COXs was not observed for sec-*O*-glucosylhamaudol (Zheng et al., 2005). In addition, COX-1 inhibitory activity of daidzin and daidzein was tested by Applová et al. (2017). While daidzein exerted weak COX-1 inhibition with an IC_{50} of about 30 μM, daidzin was found inactive (Applová et al., 2017).

Luteolin-7-*O*-glucoside from a leaf and branch of *Ailanthus altissima* (Mill.) Swingle was found to inhibit the production of LTC_4 through the inhibition of 5-LOX with an IC_{50} value of 3 μM (Jin et al., 2011). Similarly, luteolin was found to inhibit 5-LOX activity in cell-based assay with an IC_{50} of 0.8 μM (Lee et al., 2010). In addition, the 5-LOX inhibitory activity of luteolin, along with quercetin and myricetin, was studied by Kutil et al. (2014) in a cell-free assay, where these compounds displayed IC_{50} values of 2.25 μM, 3.29 μM, and 4.02 μM, respectively. Their activity was slightly better than that of the reference compound Zileuton (IC_{50} = 4.71 μM) (Kutil et al., 2014). In addition, the inhibitory activity of quercetin on 5-LOX was confirmed in the cell-based assay (IC_{50} = 0.7 μM) (Kwon et al., 2011). However, none of these compounds inhibited COXs in enzymatic assays at concentrations less than 50 μM (Kutil et al., 2014).

Similar results were obtained in other studies. For example, Li et al. (2008) reported luteolin as inactive in COXs inhibition, while its glucoside showed extremely weak inhibition of COX-1 (IC_{50} = 92.3 μM), as well as of COX-2 (IC_{50} = 39.1 μM) (Li et al., 2008). In the study of Lee and Kim (2010), luteolin was found inactive on COX-1, while it inhibited COX-2 with an IC_{50} of 59.9 μM. In the same study, myricetin was found inactive and quercetin showed extremely weak inhibition (IC_{50} = 81.1 and 72.3 μM for COX-1 and COX 2, respectively) (Lee and Kim, 2010). Moreover, Bai and Zhu (2010) observed the stimulatory effect of myricetin and quercetin on COXs in animal studies. Oxyflavone myricitrin (3-*O*-α-L-rhamnopyranoside of myricetin) was reported to inhibit 5-LOX in a cell-free system with an IC_{50} of 7.8 μM, while it did not inhibit mPGES-1 (Winekenstadde et al., 2015). In addition, it was also not able to inhibit COXs (Noreen et al., 1998).

22.2.1.2 Stilbenes

Three major naturally occurring stilbenes—resveratrol, piceatannol and pterostilbene (Figure 22.4)—were assessed for their 5-LOX inhibitory activity against the reference compound, Zileuton. Resveratrol is a compound found in more than 70 plant species and was first identified in *Vitis vinifera* L. (Figure 22.5A). On the contrary, piceatannol was found only in few plants such as *Melaleuca leucadendra* (L.) L., *Cassia garretiana* Craib and *Rheum undulatum* L. A natural resveratrol analog pterostilbene is found in grapes and blueberries, and it was first identified as a dominant compound in *Pterocarpus marsupium* Roxb. (Maurya and Ray, 1984). Piceatannol, as well as pterostilbene, exerted good 5-LOX inhibitory activity in the cell-free enzymatic assay (IC_{50} = 1.1 and 3.9 μM, respectively), as well as in cell-based assay, in which their activity (IC_{50} = 0.24 and 0.53 μM, respectively) surpassed the activity of Zileuton (IC_{50} = 1.1 μM) (Boccellino et al., 2019). On the other hand, an IC_{50} value of resveratrol in the cell-free assay was above 10 μM, while in cell-based assay it reached an IC_{50} of 4.9 μM, which was comparable to the IC_{50} values reported earlier (Werz, 2007).

Following the previously mentioned study, Kutil et al. (2014) reported that resveratrol did not inhibit 5-LOX in a cell-free assay, while piceatannol was identified as a potent 5-LOX inhibitor with an IC_{50} of 0.76 μM (Kutil et al., 2014). Similarly, Lelakova et al. (2019) reported the IC_{50} value for piceatannol to be 0.39 μM (5-LOX, cell-free assay). In this study, the inhibitory activity of other stilbenes on 5-LOX, as well as COXs, was evaluated. Piceatannol was found inactive in inhibiting COXs, while pinostilbene and pterostilbene both inhibited COXs and 5-LOX with IC_{50} values of 1.37 μM, 0.25 μM and 4.01 μM for COX-1, COX-2, and 5-LOX, respectively, for pinostilbene and 5.99 μM, 6.67 μM and 1.63 μM for COX-1, COX-2, and 5-LOX, respectively, for pterostilbene

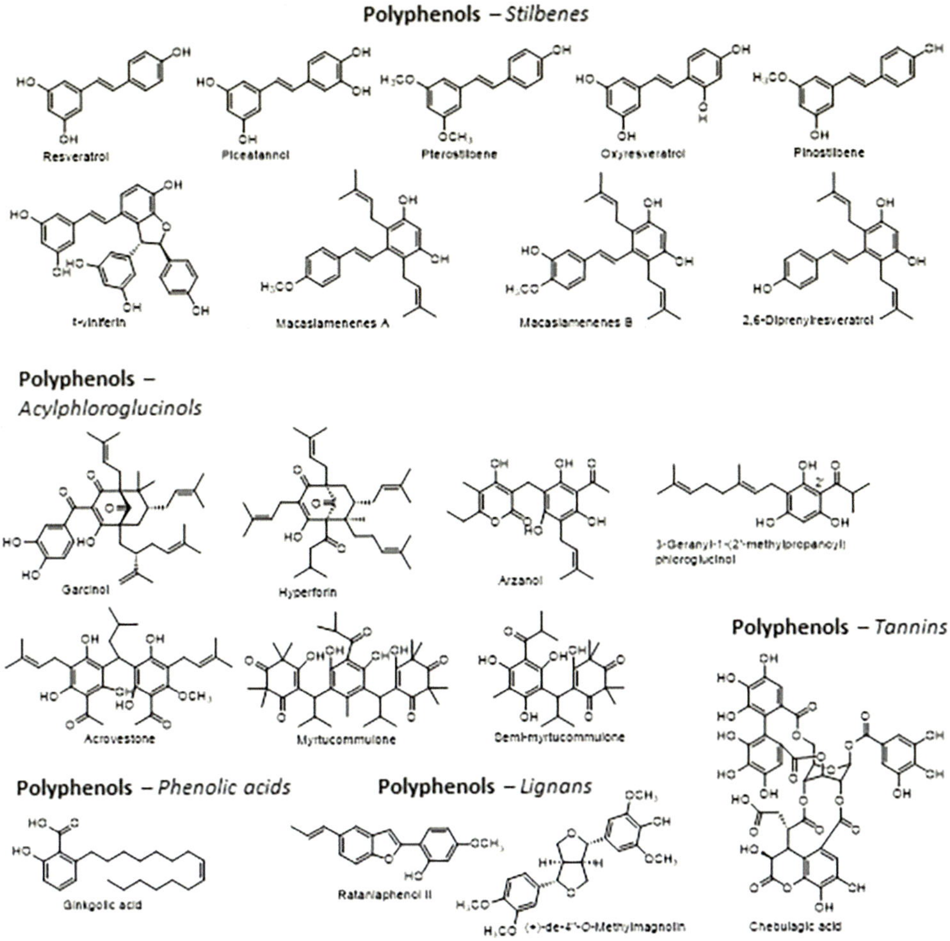

FIGURE 22.4 Structures of other selected polyphenols with potent inhibitory activity on arachidonic acid metabolic pathways.

(Lelakova et al., 2019). Pinostilbene is found in *Gnetum venosum* Spruce ex Benth. and in the bark of *Pinus sibirica* Du Tour (Boralle et al., 1993; Tyukavkina et al., 1972).

In another study on dietary stilbenes, pterostilbene and oxyresveratrol were found to inhibit 5-LOX in cell-free assay with IC_{50} values of 9.32 μM and 18.5 μM, respectively (Kutil et al., 2015). Both these compounds inhibited COX-1 and COX-2 with moderate efficiency with IC_{50} values between 2.79 μM and 27.0 μM. On the other hand, in this study pinostilbene was found to potently inhibit COXs (IC_{50} = 1.90 μM and 0.35 μM for COX-1 and COX-2, respectively); however, it was found inactive in 5-LOX assay below 50 μM, which is in deep contrast with the study of Lelakova et al. (2019). Resveratrol inhibited COXs with IC_{50} values ranging between 0.2 μM and 2.27 μM for COX-1, and between 0.9 μM and 3.40 μM for COX-2, respectively (Kutil et al., 2015; Lelakova et al., 2019). All in all, these results revealed some stilbenes as potent dual COX/5-LOX inhibitors.

On the other hand, in study conducted by Hartung et al. (2019), resveratrol and its dimer, ε-viniferin, were both found to inhibit 5-LOX in cell-free assay with IC_{50} values of 7.3 μM and 0.8 μM, respectively, and in cell-based assay with IC_{50} values of 4.3 μM and 3.1 μM, respectively. In the context of other studies, the evaluated activity of resveratrol in the cell-free assay was surprising, while the IC_{50} value obtained in the cell-based assay was comparable with them (Boccellino et al., 2019; Werz, 2007).

Eight naturally occurring prenylated stilbenoids, isolated from leaves and twigs of *Macaranga siamensis* S.J. Davies, wood and bark of *Artocarpus altilis* (Parkinson ex F.A. Zorn) Fosberg, (Figure 22.5B) and *A. heterophyllus* Lam., were tested for their ability to inhibit 5-LOX and COXs in cell-free assays. From these, three compounds (macasiamenenes A-B, and 2,6-diprenyl-resveratrol) showed dual COX/5-LOX inhibitory activity, which was remarkable on 5-LOX and COX-1, and rather moderate on COX-2. Their IC_{50} values were around 1 μM on 5-LOX and between 0.4 μM and 0.9 μM on COX-1, while on COX-2 they were between 4.7 μM and 14.7 μM (Hosek et al., 2019).

22.2.1.3 Phenolic Acids

Ginkgolic acid (Figure 22.4), isolated from sarcotesta of *Ginkgo biloba* L. (Figure 22.5C), was identified as a virtual hit based on mPGES-1 pharmacophore screening (Gerstmeier et al., 2019). Indeed, it potently inhibited mPGES-1 activity in the cell-free assay in a reversible mode with an IC_{50} of 0.7 μM. Insight into its other potential anti-inflammatory targets revealed its ability to moderately inhibit COX-1 (IC_{50} = 8.1 μM) and thromboxane A_2 synthase (IC_{50} = 5.2 μM). Moreover, it potently inhibited 5-LOX with an IC_{50} of 0.2 μM, thus showing remarkable dual mPGES-1/5-LOX inhibitory activity. Likewise to mPGES-1, the inhibition of 5-LOX was fully reversible. In the cell-based assay, the efficiency to inhibit 5-LOX was slightly attenuated (IC_{50} = 2.9 to 3.8 μM). On the other hand, neither COX-2, nor $cPLA_2$ were inhibited by ginkgolic acid (Gerstmeier et al., 2019).

22.2.1.4 Acylphloroglucinols

A polyprenylated acylphloroglucinol, garcinol (Figure 22.4), from the rind of *Garcinia* sp., which is mainly used as a spice, inhibited 5-LOX, mPGES-1 and COX-1 enzymes. In cell-free assays, garcinol exerted potent sub-micromolar dual mPGES-1/5-LOX inhibitory activity (IC_{50} = 0.1 and 0.3 μM, respectively), accompanied by moderate inhibitory activity toward COX-1 (IC_{50} = 12 μM). In cell-based systems, the IC_{50} value for 5-LOX was slightly higher (1.9 μM) (Koeberle et al., 2009). Similarly, another prenylated acylphloroglucinol, hyperforin, from the herbal plant *Hypericum perforatum* L. (Figure 22.5D), exhibited impressive dual mPGES-1/5-LOX inhibitory activity. Hyperforin potently inhibited 5-LOX in the cell-free assay (IC_{50} = 90 nM), as well as in cell-based assay (IC_{50} = 1.5 μM) (Albert et al., 2002). In the cell-free assay, mPGES-1 enzymatic activity was inhibited with an IC_{50} of 1.0 μM (Koeberle et al., 2011). Moreover, hyperforin inhibited COX-1 with an IC_{50} of 3 μM in both cell-free, as well as cell-based, assays (Albert et al., 2002). However, in a later study by Koeberle et al. (2011), lower IC_{50} value was reported for cell-free assay (12 μM) and no inhibition of COX-1 was observed in cell-based assay up to 30 μM. No inhibition of COX-2 by the aforementioned acylphloroglucinols was observed (Koeberle et al., 2011).

From the fruits of *Hypericum empetrifolium* Willd., two geranylated acylphloroglucinols were isolated. They exerted dual COX/5-LOX inhibitory activity *in vitro* (Crockett et al., 2008). The activity toward 5-LOX was evaluated in a cell-based assay. 3-Geranyl-1-(2'-methylpropanoyl) phloroglucinol inhibited COX-1, COX-2 and 5-LOX with IC_{50} values of 6.0 μM, 29.9 μM and 2.2 μM, respectively. Its methylbutanoyl derivative was less active (Bauer et al., 2011). Unfortunately, their inhibitory activity toward mPGES-1 was not tested.

In addition, from the leaves and flower heads of *Helichrysum italicum* (Roth) G. Don (Figure 22.5E), prenylated phloroglucinol known as arzanol was isolated. Arzanol displayed reversible 5-LOX inhibitory activity with an IC_{50} of 3.1 μM in the cell-free assay (Bauer et al., 2011). In the cell-based assays, arzanol inhibited 5-LOX with IC_{50} values of 2.9 μM and 8.1 μM, depending on the assay conditions. In addition, the inhibition of COX-1 by arzanol in cell-free and cell-based assays was observed with IC_{50} values of 17.5 μM and 2.6 μM, respectively. Arzanol, however, did not inhibit COX-2 in cell-free assay up to 30 μM. On the other hand, arzanol was found to potently inhibit mPGES-1 in a reversible mode in cell-based assay with an IC_{50} of 0.4 μM. Dual mPGES-1/5-LOX inhibitory activity of arzanol was confirmed by Minassi et al. (2012), who studied the inhibitory activity of arzanol and its synthetic derivatives toward mPGES-1 and 5-LOX in cell-based

FIGURE 22.5 Plant sources of other polyphenolic compounds, potent inhibitors of arachidonic acid metabolic pathways. A, fruits of *Vitis vinifera*, a source of resveratrol, pterostilbene and ε-viniferin; B, *Artocarpus altilis*, a source of macasiamenenes A and B; C, fruits of *Ginkgo biloba*, a source of ginkgolic acid; D, flower of *Hypericum perforatum*, a source of hyperforin; E, *Helichrysum italicum*, a source of arzanol (Birding In Spain, 2018; Nova, 2007, 2020; Starr and Starr, 2006; Tonreg, 2012).

assays. They reported the IC_{50} values of arzanol to be 0.4 μM for mPGES-1 and 3.1 μM for 5-LOX (Minassi et al., 2012).

Seven prenylated acetophenone dimers, containing two phloroglucinol units, from the trunk bark of *Acronychia pedunculata* (L.) Miq., was found to act as dual mPGES-1/5-LOX inhibitors. They inhibited mPGES-1 with IC_{50} values ranging from 1.1–4.2 μM in a cell-free assay. Their activity toward 5-LOX was slightly lower with IC_{50} values ranging from 2.0–7.3 μM in cell-free assay and 2.3–7.3 μM in the cell-based assay. The best dual inhibitory activity was observed for acrovestone, which displayed IC_{50} values of 1.1 μM (mPGES-1), 2.7 μM (5-LOX cell-free) and 2.3 μM (5-LOX cell-based) (Svouraki et al., 2017).

Nonprenylated acylphloroglucinols from the leaves of *Myrtus communis* L., myrtucommulone and semi-myrtucommulone (Figure 22.4) was able to inhibit 5-LOX and COX-1 in cell-free assays with IC_{50} values of 5 μM and 8 μM for 5-LOX, and 6 μM and 30 μM for COX-1, respectively. Similarly, the activity of 5-LOX and COX-1 was inhibited in cell-based assays with IC_{50} values of 1.8 μM and 10 μM for 5-LOX, and 17 μM and 29 μM for COX-1, respectively (Feisst et al., 2005). Moreover, myrtucommulone and semi-myrtucommulone both inhibited mPGES-1 in cell-free assay with IC_{50} values of 1 μM and 10 μM, respectively, while they did not inhibit COX-2 up to 30 μM (Koeberle et al., 2009).

22.2.1.5 Lignans

More than 40 different lignans were studied for their ability to inhibit COXs and 5-LOX. They were isolated from the leaves of *Lauraceae* plants—*Pleurothyrium cinereum* van der Werff, *Ocotea macrophylla* Kunth. and *Nectandra amazonum* Nees—or from the roots of *Krameria lappacea* (Dombey) Burdet & B.B. Simpson (Coy-Barrera and Cuca-Suarez, 2011; Coy et al., 2009; Baumgartner et al., 2011). In addition, some of them were tested on mPGES-1 enzyme. All lignans exhibited moderate to low inhibitory activity with the best IC_{50} values observed for rataniaphenol II (Figure 22.4), which were 3.0 μM, 1.7 μM and 11.0 μM for COX-1, COX-2, and mPGES-1, respectively, and for (+)-de-4''-*O*-methylmagnolin, with dual COX-2/5-LOX inhibitory activity, which IC_{50} values were 2.3 μM and 5.1 μM for COX-2 and 5-LOX, respectively (Baumgartner et al., 2011; Coy-Barrera and Cuca-Suarez, 2011).

22.2.1.6 Tannins

Chebulagic acid (Figure 22.4), a tannin isolated from the fruits of *Terminalia chebula* Retz., exerted potent dual COX-2/5-LOX inhibitory activity *in vitro* with IC_{50} values being 0.92 μM for COX-2 and 2.1 μM for 5-LOX. In addition, moderate inhibitory activity toward COX-1 was observed (IC_{50} = 15.0 μM) (Reddy et al., 2009).

22.2.2 Terpenoids

22.2.2.1 Triterpenes

Triterpenes (Figure 22.6) are composed of three terpene units and include steroids, bile acids and saponins (triterpene glycosides). Among anti-inflammatory triterpenes, boswellic acids (BAs) have long been known as non-competitive and non-redox 5-LOX inhibitors (Safayhi et al., 1992). BAs are the main components of an anti-inflammatory remedy, frankincense, a resin obtained from the bark of *Boswellia* species, such as *Boswellia serrata* Roxb.ex Colebr., *B. sacra* Fluck (Figure 22.7A–B), *B. carterii* Birdw, *B. papyrifera* Hochst. or *B. frereana* Birdw. (Verhoff et al., 2014). The main structural features of BAs, important for their 5-LOX activity, include a pentacyclic triterpene ring structure, hydrophilic groups at the C4 position of the A-ring and an 11-keto group (Sharma and Jana, 2020). Especially, 3-*O*-acetyl-11-keto-*β*-boswellic acid (AKBA) has been extensively studied over the years and its reported IC_{50} values for 5-LOX inhibition range between 8 μM and 50 μM in cell-free assays and between 1.5 μM and 15 μM in cell-based assays (Werz, 2007).

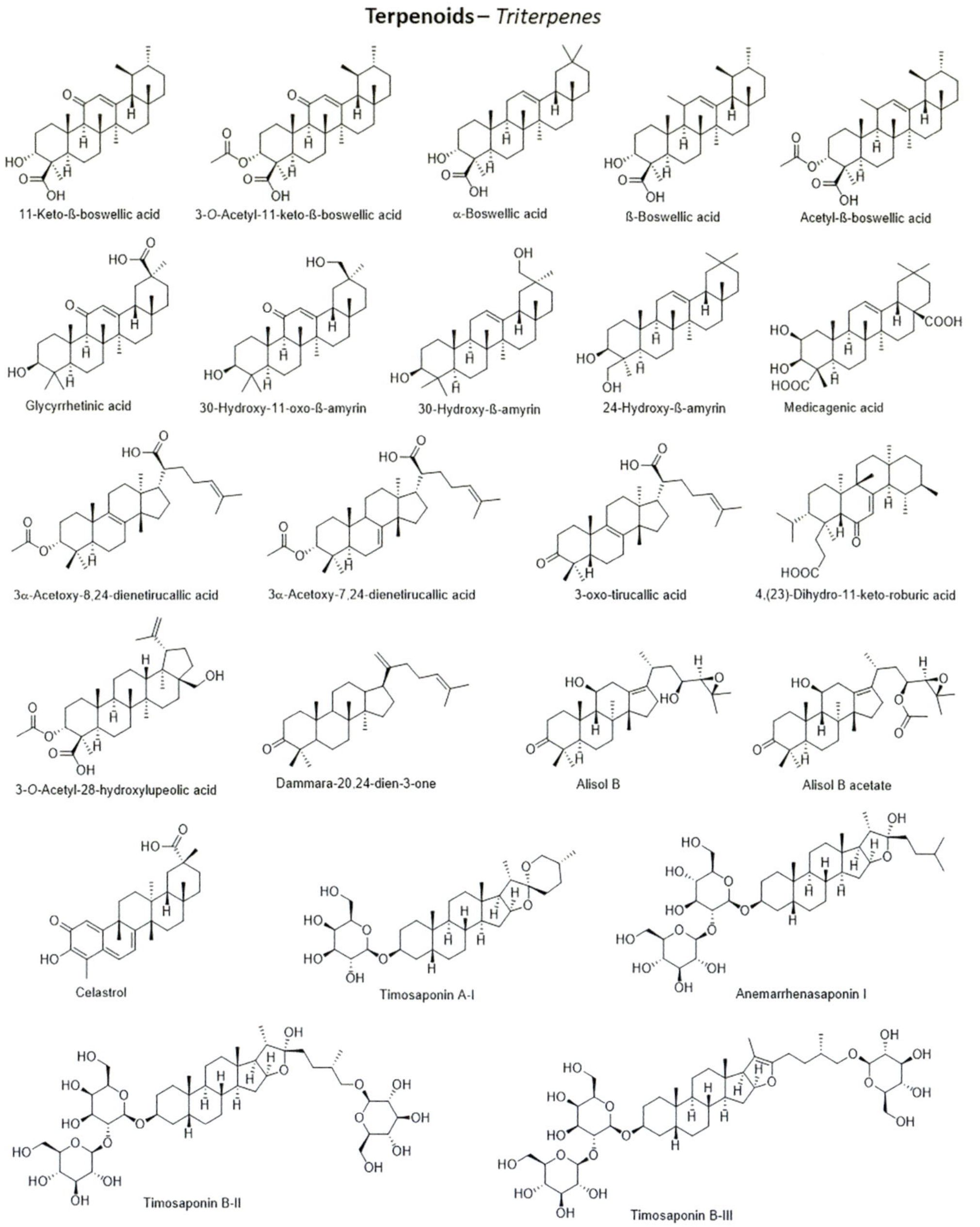

FIGURE 22.6 Structures of selected triterpenes with potent inhibitory activity on arachidonic acid metabolic pathways.

The inhibitory activity of various BAs toward 5-LOX was studied by Siemoneit et al. (2009). In cell-free *in vitro* assay, AKBA showed the highest potency with an IC_{50} of 2.9 μM, followed by 11-keto-*β*-boswellic acid (KBA) with an IC_{50} of 6.3 μM. Rather low inhibitory activity was observed for three other BAs, α-boswellic acid (α-BA, 15.3 μM), *β*-boswellic acid (*β*-BA, 23.9 μM) and acetyl *β*-boswellic acid (A-*β*-BA, 30 μM). However, the activity of these acids in cell-free assays was found to strongly depend on factors regulating 5-LOX catalysis. In cell-based assays,

FIGURE 22.7 Plant sources of triterpenes and quinones, potent inhibitors of arachidonic acid metabolic pathways. A, *Boswellia sacra*, a source of boswellic acids; B, resin of *Boswellia sacra*; C, fruits of *Embelia ribes*, a source of embelin; D, bark of *Diospyros lotus*, a source of diospyrin; E, *Handroanthus impetiginosus*, a source of β-lapachone (Halleypo, 2011; KENPEI, 2008; Raffaelli, 2005, 2012; Valker, 2015).

both AKBA and KBA inhibited 5-LOX in isolated human neutrophils with IC_{50} values of 3.2 μM and 8.8 μM, respectively. From the other acids, only α-BA displayed an IC_{50} value below 30 μM (23.1 μM). When exogenous AA was supplied, IC_{50} value for α-BA decreased (16.0 μM) and β-BA exhibited inhibitory activity with an IC_{50} of 28.1 μM. On the contrary, AKBA and KBA were not able to inhibit 5-LOX in whole blood cells (Siemoneit et al., 2009). This may be caused by their high plasma protein binding. Similarly, their activity in neutrophils was abolished in the presence of albumin. In addition, it was reported that AKBA and KBA suffer from poor bioavailability, and their concentration in plasma is thus negligible (Buchele and Simmet, 2003; Krueger et al., 2008; Krueger et al., 2009; Tausch et al., 2009).

On the other hand, *β*-BAs were reported to reach micromolar plasma levels, suggesting their higher importance in inflammation than that of AKBA and KBA, despite their moderate inhibitory activity in assays (Buchele and Simmet, 2003; Tausch et al., 2009). Moreover, all BAs were found to inhibit mPGES-1 with IC_{50} values between 3 μM and 10 μM in the cell-free assay, while they did not inhibit COXs (Nhu Ngoc Quynh et al., 2019; Siemoneit et al., 2011). In the cell-based assay, their mPGES-1 inhibitory activity was rather low with IC_{50} values ranging from 20–30 μM. In addition, in the whole blood assay, *β*-BA was found to inhibit mPGES-1 with an IC_{50} of 10 μM contrarily to AKBA and KBA, which were inactive (Siemoneit et al., 2011). Considering that *β*-BA is the major component of frankincense and that its plasma levels may reach up to 10 μM, it may be suggested that the pronounced anti-inflammatory activity of frankincense *in vivo* may be caused by the inhibition of mPGES-1 rather than 5-LOX, especially as frankincense exerted therapeutic efficiency in models of rheumatoid arthritis and osteoarthritis, where a role of leukotrienes was excluded (Buchele and Simmet, 2003; Tausch et al., 2009).

From *Boswellia* sp. frankincense. Verhoff et al. (2014) isolated other tetracyclic and pentacyclic triterpene acids and studied their ability to inhibit COXs and mPGES-1. Among these acids belonged tirucallic, lupeolic and roburic acids. From 13 isolated compounds, none of the compounds inhibited COXs; however, seven showed potent inhibition of mPGES-1 with IC_{50} values ranging from 0.4–1.2 μM, and two were moderately active with IC_{50} values of 3 μM and 8.5 μM. The most active compounds were 3α-acetoxy-8,24-dienetirucallic acid and 3α-acetoxy-7,24-dienetirucallic acid (IC_{50} = 0.4 μM). Generally, tirucallic acids were the most potent, while roburic and lupeolic acids were mostly inactive except for 4,(23)-dihydro-11-keto-roburic acid and 3-*O*-acetyl-28-hydroxylupeolic acid, which inhibited mPGES-1 with IC_{50} values of 1.0 μM and 0.9 μM, respectively (Verhoff et al., 2014). Previously, 3-*O*-acetyl-28-hydroxylupeolic acid was found to inhibit 5-LOX in cell-free as well as cell-based assays with IC_{50} values of 8.3 μM and 2.3–3.4 μM, respectively. Other lupeolic acids, those missing 28-hydroxyl, were found to be inactive (Verhoff et al., 2012). In the case of tirucallic acids, inhibition of 5-LOX by some of these compounds (e.g., 3-oxo-tirucallic acid) was reported, while for others, it remains to be elucidated (Koeberle and Werz, 2018; Boden et al., 2001).

A dammarane triterpenoid (dammara-20,24-dien-3-one) was isolated from the leaves of *Chisocheton polyandrous* Merr., a plant used to treat fever or backache. This compound displayed inhibitory activity toward COX-1, COX-2 and 5-LOX with IC_{50} values being 31.9 μM, 3.2 μM, and 24.3 μM, respectively. While the activity on COX-2 was better than that of indomethacin (IC_{50} = 8.9 μM), the activities on COX-1 and 5-LOX were rather low (Chan et al., 2012).

A quinonic pentacyclic triterpenoid celastrol isolated from the roots of *Tripterygium wilfordii* Hook. F. and *Celastrus* sp. was studied for its ability to inhibit 5-LOX and COX-2 enzymatic activity *in vitro*. The obtained IC_{50} values were in a low to moderate range (5 μM for 5-LOX and 20 μM for COX-2). The inhibition of 5-LOX was determined in a cell-based assay, while the inhibition of COX-2 in a cell-free assay (Joshi et al., 2016).

From a rhizome of *Alisma orientale* (Sam.) Juzep., used in Chinese medicine, two triterpenes, alisol B and its acetate, were isolated. These two triterpenes inhibited 5-LOX activity in cell-based assay with IC_{50} values of 3.5 μM and 3.7 μM, respectively (Lee et al., 2012). From a rhizome of another plant, *Anemarrhena asphodeloides* Bunge, seven saponins were isolated and screened for their ability to inhibit COX and 5-LOX enzymes (Xie et al., 2020). In the case of 5-LOX, the compounds exerted inhibitory activity with IC_{50} values between 1.05 μM and 3.29 μM, while the IC_{50} value of Zileuton was 0.23 μM. On COX-2, their IC_{50} values ranged between 0.48 μM and 36.4 μM, with the IC_{50} value of the reference compound celecoxib being 0.06 μM. The best dual COX-2/5-LOX inhibitory activity was observed for timosaponin B-II (IC_{50} = 0.77 μM and 1.57 μM for COX-2 and 5-LOX, respectively) and timosaponin B-III (IC_{50} = 0.48 μM and 1.82 μM for COX-2 and 5-LOX, respectively). In addition, 5-LOX inhibitory activity of timosaponins and related compounds isolated from the Zi-shen pill was studied by the same group (Wang et al., 2020).

Zi-shen pill is a Chinese medicine prescription consisting of three herbs: *Anemarrhena asphodeloides* Bunge rhizome, *Phellodendron chinense* C.K. Schneider bark and *Cinnamonium cassia* (L.) J. Presl bark. The timosaponins, timosaponin A-I, and anemarrhenasaponin I, inhibited 5-LOX with sub-micromolar IC_{50} values of 0.63 μM and 0.72 μM, respectively, while the reference compound Zileuton exhibited an IC_{50} value of 0.13 μM. Both these compounds, timosaponin A-I and anemarrhenasaponin I, were also examined in the study done by Xie et al. (2020) mentioned previously. However, Xie et al. (2020) determined their IC_{50} values to be 3.29 μM and 2.73 μM, respectively. From the seven compounds screened by Xie et al. (2020), these two showed the highest IC_{50} values. Why the values differed between the two studies is not clear, as both studies were performed by the same group of authors.

22.2.2.2 Diterpenes

Two diterpene compounds (Figure 22.8), carnosol (CS) and carnosic acid (CA), are the components of *Salvia* spp. extracts were found to potently inhibit two enzymes of the AA metabolic cascade, 5-LOX and mPGES-1. In cell-free assays, their IC_{50} values were determined to be 0.3 μM and 0.8 μM for 5-LOX, and 10.9 μM and 14.0 μM for mPGES-1, respectively. In neutrophils (cell-based assay), both CS and CA at 3 and 30 μM inhibited the formation of 5-LOX products derived from AA (Maione et al. 2017). Previously, CS and CA were reported to inhibit mPGES-1 with an IC_{50} of 5 μM in a cell-free assay. In the cell-based assay, only CS was found to inhibit mPGES-1 with an IC_{50} of 9.3 μM. Neither CS nor CA inhibited COXs (Bauer et al., 2012).

Recently, Nguyen et al. (2020) isolated five known clerodane diterpenes and subjected them to COX/5-LOX inhibitory assays. All these diterpenes exhibited inhibitory activity in cell-free assays in the nanomolar range for both enzymes (IC_{50} values were between 9.5 nM and 33.4 nM). Such activity would be remarkable; however, the results are obfuscated when compared with the inhibitory activity of chosen reference compounds. The authors reported nanomolar activity of indomethacin on COXs (IC_{50} = 10.4 nM and 12.8 nM on COX-1 and COX-2, respectively), and of diclofenac on 5-LOX (IC_{50} = 23.3 nM) (Nguyen et al., 2020). However, indomethacin is known to inhibit COXs in the micromolar range, as an example may serve IC_{50} values obtained by Chan et al. (2012), IC_{50} = 8.9 μM for COX-2 and 4.9 μM for COX-1 (Chan et al., 2012). Moreover, diclofenac is not 5-LOX inhibitor; thus, we could find only one reported IC_{50} value of 230 μM for its sodium salt (Ku et al., 1986). Thus, the results reported by Nguyen et al. (2020) should be considered with care and may need revision—and for this reason, they were not included in Table 22.1.

22.2.3 Quinones

The profound inhibitory activity of quinones (Figure 22.8) toward 5-LOX may be attributed to their antioxidant potential arising from their quinonic nature, as well as their lipophilicity (Werz 2007). The first reported quinonic 5-LOX inhibitors were ardisiaquinones from the rhizome of *Ardisia japonica* (Blume). They exerted potent 5-LOX inhibitory at a submicromolar level in cell-free assays (IC_{50} = 0.2–1.0 μM) (Fukuyama et al., 1994; Fukuyama et al., 1995). However, their potential to inhibit COX activity was tested only for ardisiaquinone A, which COX-2 inhibitory activity was very low (IC_{50} = 43 μM) (Shahinozzaman et al., 2019).

Similar 5-LOX inhibitory activity was observed for the quinone embelin, obtained from the fruit extract of *Embelia ribes* Burm. F (Figure 22.7C). Embelin was found to remarkably inhibit 5-LOX with an IC_{50} = 70 nM in a cell-free assay, which was superior to the activity of Zileuton (IC_{50} = 760 nM). Embelin acted as a reversible inhibitor of 5-LOX. In cell-based assay, embelin inhibited 5-LOX (IC_{50} = 0.8 μM to 2.0 μM) at a similar level as Zileuton (IC_{50} = 1.2 μM). In addition, embelin inhibited mPGES-1 with an IC_{50} of 0.2 μM in a cell-based assay. On the other hand, embelin failed to inhibit COXs below 10 μM (Schaible et al., 2013).

Twenty-three naturally occurring quinones were tested for their ability to inhibit COXs and 5-LOX. In the case of COXs, shikonin (a major constituent of the root extracts of *Alkanna tinctoria*

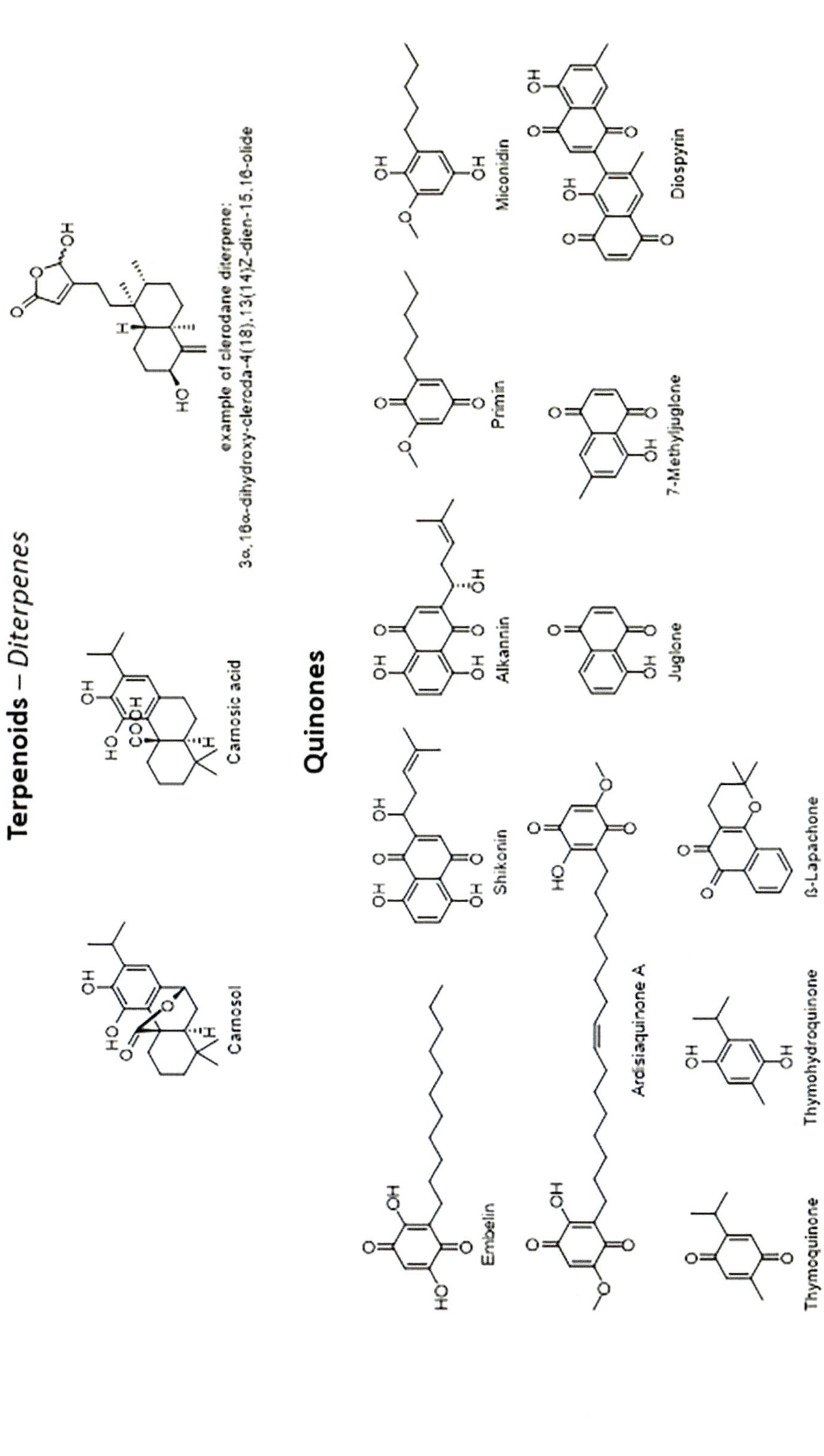

FIGURE 22.8 Structures of selected diterpenes and quinones with potent inhibitory activity on arachidonic acid metabolic pathways.

Tausch., *Lithospermum erythrorhizon* Sieb. Et Zucc or *Arnebia euchroma* (Royle) Johnst.) and primin (found in *Miconia* sp.) exerted sub-micromolar inhibitory activity toward COX-2 (IC_{50} = 0.33 μM and 0.09 μM, respectively) and moderate inhibitory activity toward COX-1 (IC_{50} = 1.91 μM and 5.14 μM, respectively) (Landa et al., 2012). In addition, four other naturally occurring naphthoquinones (juglone, 7-methyljuglone, alkannin and diospyrin), all displayed good inhibitory activities toward COXs with IC_{50} values in the range of 1.39–2.50 μM for COX-1, and 0.55–1.58 μM for COX-2, with diospyrin (found in *Diospyros* sp, Figure 22.7D) giving the best values (Landa et al., 2012). In the case of 5-LOX, the lowest IC_{50} values in the cell-based assay (4.0 μM and 4.1 μM), corresponding to the IC_{50} value of the reference inhibitor Zileuton (IC_{50} = 4.1 μM), were obtained for benzoquinones primin and thymohydroquinone (found in *Nigella sativa* L. seeds), respectively. Thymoquinone, the unreduced form of thymohydroquinone, displayed only low inhibitory activity with IC_{50} = 18.2 μM. In addition, similarly, low inhibitory activity was observed for the naphthoquinone shikonin (IC_{50} = 24.3 μM) (Landa et al., 2013).

Moreover, primin and its hydroquinone form, miconidin (found in *Primula obconica* Hance), were synthesized by Sisa et al. (2020), and their COX and 5-LOX inhibitory activities were studied in cell-free assays. While primin exhibited potent inhibition of all the tested enzymes with IC_{50} values of 1.4 μM, 18.9 μM and 1.9 μM for 5-LOX, COX-1, and COX-2, respectively, miconidin inhibited potently only 5-LOX with an IC_{50} value of 0.94 μM (Sisa et al., 2020). In another study, a natural naphthoquinone, *β*-lapachone, in its hydroquinone form, was able to inhibit 5-LOX in cell-free, as well as cell-based, assays. In a cell-free assay, with the presence of a reducing agent, *β*-lapachone inhibited 5-LOX with an IC_{50} of 0.24 μM. Similarly, in a cell-based assay, IC_{50} value was 0.42 μM in the presence of a reducing agent. Without it, the IC_{50} value increased to 8.6 μM (Rodrigues et al., 2018). *β*-Lapachone was first isolated from the heartwood of *Handroanthus impetiginosus* (Mart. Ex DC.) Mattos. (Figure 22.7E) (Gomes et al., 2021). Its activity on COXs was not studied directly with enzymatic assays; however, it was found to inhibit PGE_2 synthesis in BV2 microglia through the downregulation of COX-2 mRNA with IC_{50} around 5 μM (Moon et al., 2007).

22.3 CONCLUSION

Long-lasting research for an efficient selective inhibitor of a particular enzyme of the AA metabolic cascade did not offer any new natural compound that would successfully go through clinical trials. On the other hand, natural products experience a renaissance as lead structures for the development of multitarget drugs (Koeberle and Werz, 2018). Although the chance that natural products themselves would make it through clinical trials is rather low, due to their complex synthesis, low bioavailability and high lipophilicity, these compounds may serve as starting structural moieties for the synthesis of derivatives with improved efficacy and/or pharmacokinetic and pharmacodynamic profiles (Koeberle and Werz, 2018).

From multitarget drugs, the first attention focused on dual COX-2/5-LOX inhibitors. Many such inhibitors, either isolated or synthesized, were evaluated in preclinical studies and entered clinical trials, but none of them has yet been approved (Koeberle and Werz, 2018). Novel strategy in multitarget therapy focuses on the development of dual mPGES-1/5-LOX inhibitors. In addition, the previously neglected enzyme COX-1 has emerged as a prominent target for the development of anti-inflammatory compounds with potential activity toward diseases other than those treated with conventional NSAIDs (Dvorakova et al., 2021).

From all the compounds discussed in this chapter, the most promising seems those with prenyl or geranyl side chain(s), as they often possess dual mPGES-1/5-LOX inhibitory activity with sub-micromolar IC_{50} values, independent of the core structure (flavonoid/stilbene/phloroglucinol) (Albert et al., 2002; Bauer et al., 2011; Chi et al., 2001; Hanakova et al., 2017; Hosek et al., 2019; Koeberle et al., 2009; Koeberle et al., 2011; Minassi et al., 2012; Svouraki et al., 2017; Werz et al., 2014).

REFERENCES

©BMK. 2008. "*Paulownia tomentosa*." Obtained from Wikimedia under the Creative Commons Attribution-Share Alike 3.0 license. https://commons.wikimedia.org/wiki/File:Paulownia-tomentosa2_BMK.jpg (Accessed 4 May 2021).

Albert, D., I. Zundorf, T. Dingermann, W. E. Muller, D. Steinhilber, and O. Werz. 2002. "Hyperforin is a dual inhibitor of cyclooxygenase-1 and 5-lipoxygenase." *Biochemical Pharmacology* 64, 12: 1767–1775. doi: 10.1016/s0006-2952(02)01387-4.

Ambati, Goutami G, and Sanjay M. Jachak. 2020. "Natural product inhibitors of cyclooxygenase (COX) enzyme: A review on current status and future perspectives." *Current Medicinal Chemistry* 27: 1–28. doi: 10.2174/0929867327666200602131100.

AnRo0002. 2015. "*Hypericum perforatum*." Obtained from Wikimedia under the Creative Commons CC0 1.0 Universal Public Domain Dedication. https://commons.wikimedia.org/wiki/File:20150804Hypericum_perforatum2.jpg (Accessed 4 May 2021)

Applová, Lenka, Jana Karlíčková, Říha Michal, Tomáš Filipský, Kateřina Macáková, Jiřina Spilková, and Přemysl Mladěnka. 2017. "The isoflavonoid tectorigenin has better antiplatelet potential than acetylsalicylic acid." *Phytomedicine* 35: 11–17. doi: 10.1016/j.phymed.2017.08.023.

Attiq, Ali, Juriyati Jalil, Khairana Husain, and Waqas Ahmad. 2018. "Raging the war against inflammation with natural products." *Frontiers in Pharmacology* 9: 976. doi: 10.3389/fphar.2018.00976.

Bai, Hyoung-Woo, and Zhu Bao Ting. 2010. "Myricetin and quercetin are naturally occurring co-substrates of cyclooxygenases *in vivo*." *Prostaglandins, Leukotrienes and Essential Fatty Acids* 82, 1: 45–50. doi: 10.1016/j.plefa.2009.10.006.

Bauer, Julia, Andreas Koeberle, Friederike Dehm, Federica Pollastro, Giovanni Appendino, Hinnak Northoff, Antonietta Rossi, Lidia Sautebin, and Oliver Werz. 2011. "Arzanol, a prenylated heterodimeric phloroglucinyl pyrone, inhibits eicosanoid biosynthesis and exhibits anti-inflammatory efficacy *in vivo*." *Biochemical Pharmacology* 81, 2: 259–268. doi: 10.1016/j.bcp.2010.09.025.

Bauer, Julia, Susanne Kuehnl, Judith M. Rollinger, Olga Scherer, Hinnak Northoff, Hermann Stuppner, Oliver Werz, and Andreas Koeberle. 2012. "Carnosol and carnosic acids from salvia officinalis inhibit microsomal prostaglandin E-2 synthase-1." *Journal of Pharmacology and Experimental Therapeutics* 342, 1: 169–176. doi: 10.1124/jpet.112.193847.

Baumgartner, Lisa, Silvio Sosa, Atanas G. Atanasov, Antje Bodensieck, Nanang Fakhrudin, Julia Bauer, Giorgia Del Favero, Cristina Ponti, Elke H. Heiss, Stefan Schwaiger, Angela Ladurner, Ute Widowitz, Roberto Della Loggia, Judith M. Rollinger, Oliver Werz, Rudolf Bauer, Verena M. Dirsch, Aurelia Tubaro, and Hermann Stuppner. 2011. "Lignan derivatives from krameria lappacea roots inhibit acute inflammation *in Vivo* and Pro-inflammatory mediators *in Vitro*." *Journal of Natural Products* 74, 8: 1779–1786. doi: 10.1021/np200343t.

Birding In Spain. 2018. "*Helichrysum italicum*." Obtained from Wikimedia under the Creative Commons Attribution-Share Alike 4.0 International license. https://commons.wikimedia.org/wiki/File:Curry_plant_Helichrysum_italicum.jpg (Accessed 4 May 2021).

Blobaum, Anna L., and Lawrence J. Marnett. 2007. "Structural and functional basis of cyclooxygenase inhibition." *Journal of Medicinal Chemistry* 50, 7: 1425–1441. doi: 10.1021/jm0613166.

Boccellino, Mariarosaria, Maria Donniacuo, Ferdinando Bruno, Barbara Rinaldi, Lucio Quagliuolo, Marika Ambruosi, Simona Pace, Mario De Rosa, Abdurrahman Olgac, Erden Banoglu, Nicola Alessio, Antonio Massa, Haroon Kahn, Oliver Werz, Antonio Fiorentino, and Rosanna Filosa. 2019. "Protective effect of piceatannol and bioactive stilbene derivatives against hypoxia-induced toxicity in H9c2 cardiomyocytes and structural elucidation as 5-LOX inhibitors." *European Journal of Medicinal Chemistry* 180: 637–647. doi: 10.1016/j.ejmech.2019.07.033.

Boden, S. E., S. Schweizer, T. Bertsche, M. Dufer, G. Drews, and H. Safayhi. 2001. "Stimulation of leukotriene synthesis in intact polymorphonuclear cells by the 5-lipoxygenase inhibitor 3-oxo-tirucallic acid." *Molecular Pharmacology* 60, 2: 267–273. doi: 10.1124/mol.60.2.267.

Boralle, N., H. E. Gottlieb, O. R. Gottlieb, K. Kubitzki, L. M. X. Lopes, M. Yoshida, and M. C. M. Young. 1993. "Oligostilbenoids from Gnetum-venosum." *Phytochemistry* 34, 5: 1403–1407. doi: 10.1016/0031-9422(91)80038-3.

Bruno, Ferdinando, Giuseppe Spaziano, Angela Liparulo, Fiorentina Roviezzo, Seyed Mohammed Nabavi, Antoni Sureda, Rosanna Filosa, and Bruno D'Agostino. 2018. "Recent advances in the search for novel 5-lipoxygenase inhibitors for the treatment of asthma." *European Journal of Medicinal Chemistry* 153: 65–72. doi: 10.1016/j.ejmech.2017.10.020.

Buchele, B., and T. Simmet. 2003. "Analysis of 12 different pentacyclic triterpenic acids from frankincense in human plasma by high-performance liquid chromatography and photodiode array detection." *Journal of Chromatography B-Analytical Technologies in the Biomedical and Life Sciences* 795, 2: 355–362. doi: 10.1016/s1570-0232(03)00555-5.

Calvello, R., D. D. Lofrumento, M. G. Perrone, A. Cianciulli, R. Salvatore, P. Vitale, F. De Nuccio, L. Giannotti, G. Nicolardi, M. A. Panaro, and A. Scilimati. 2017. "Highly selective cyclooxygenase-1 inhibitors P6 and mofezolac counteract inflammatory state both *in vitro* and *in vivo* models of neuroinflammation." *Frontiers in Neurology* 8. doi: 10.3389/fneur.2017.00251.

Chan, Kit Yee, Khalit Mohamad, Audrey J. A. Ooi, Zamrie Imiyabir, and Lip Yong Chung. 2012. "Bioactivity-guided fractionation of the lipoxygenase and cyclooxygenase inhibiting constituents from *Chisocheton polyandrus* Merr." *Fitoterapia* 83, 5: 961–967. doi: 10.1016/j.fitote.2012.04.018.

Chi, Y. S., H. G. Jong, K. H. Son, H. W. Chang, S. S. Kang, and H. P. Kim. 2001. "Effects of naturally occurring prenylated flavonoids on enzymes metabolizing arachidonic acid: Cyclooxygenases and lipoxygenases." *Biochemical Pharmacology* 62, 9: 1185–1191. doi: 10.1016/s0006-2952(01)00773-0.

Cingolani, Gino, Andrea Panella, Maria Grazia Perrone, Paola Vitale, Giuseppe Di Mauro, Cosimo G. Fortuna, Roger S. Armen, Savina Ferorelli, William L. Smith, and Antonio Scilimati. 2017. "Structural basis for selective inhibition of cyclooxygenase-1 (COX-1) by diarylisoxazoles mofezolac and 3-(5-chlorofuran-2-yl)-5-methyl–4-phenylisoxazole (P6)." *European Journal of Medicinal Chemistry* 138: 661–668. doi: 10.1016/j.ejmech.2017.06.045.

Coy, Ericsson David, Luis Enrique Cuca, and Michael Sefkow. 2009. "COX, LOX and platelet aggregation inhibitory properties of Lauraceae neolignans." *Bioorganic & Medicinal Chemistry Letters* 19, 24: 6922–6925. doi: 10.1016/j.bmcl.2009.10.069.

Coy-Barrera, Ericsson D., and Luis E. Cuca-Suarez. 2011. "*In vitro* anti-inflammatory effects of naturally-occurring compounds from two Lauraceae plants." *Anais Da Academia Brasileira De Ciencias* 83, 4: 1397–1402. doi: 10.1590/s0001-37652011005000044.

Crockett, Sara L., Eva-Maria Wenzig, Olaf Kunert, and Rudolf Bauer. 2008. "Anti-inflammatory phloroglucinol derivatives from *Hypericum empetrifolium*." *Phytochemistry Letters* 1, 1: 37–43. doi: 10.1016/j.phytol.2007.12.003.

Curtis, Elizabeth, Nicholas Fuggle, Sarah Shaw, Laura Spooner, Georgia Ntani, Camille Parsons, Nadia Corp, Germain Honvo, Janis Baird, Stefania Maggi, Elaine Dennison, Olivier Bruyere, Jean-Yves Reginster, and Cyrus Cooper. 2019. "Safety of Cyclooxygenase-2 inhibitors in osteoarthritis: Outcomes of a systematic review and meta-analysis." *Drugs & Aging* 36: 25–44. doi: 10.1007/s40266-019-00664-x

Dubois, R. N., S. B. Abramson, L. Crofford, R. A. Gupta, L. S. Simon, L. B. A. Van De Putte, and P. E. Lipsky. 1998. "Cyclooxygenase in biology and disease." *Faseb Journal* 12, 12: 1063–1073. doi: 10.1096/fasebj.12.12.1063

Dvorakova, Marcela, Lenka Langhansova, Veronika Temml, Antonio Pavicic, Tomas Vanek, and Premysl Landa. 2021. "Synthesis, inhibitory activity, and in silico modelling of selective COX-1 inhibitors with a quinazoline core." *ACS Medicinal Chemistry Letters* 12, 4: 610–616. doi: 10.1021/acsmedchemlett.1c00004.

Feisst, C., L. Franke, G. Appendino, and O. Werz. 2005. "Identification of molecular targets of the oligomeric nonprenylated acylphloroglucinols from Myrtus communis and their implication as anti-inflammatory compounds." *Journal of Pharmacology and Experimental Therapeutics* 315, 1: 389–396. doi: 10.1124/jpet.105.090720.

Fiorucci, S., R. Meli, M. Bucci, and G. Cirino. 2001. "Dual inhibitors of cyclooxygenase and 5-lipoxygenase. A new avenue in anti-inflammatory therapy?" *Biochemical Pharmacology* 62, 11: 1433–1438. doi: 10.1016/s0006-2952(01)00747-x.

Forino, Martino, Simona Pace, Giuseppina Chianese, Laura Santagostini, Markus Werner, Christina Weinigel, Silke Rummler, Gelsomina Fico, Oliver Werz, and Orazio Taglialatela-Scafati. 2016. "Humudifucol and bioactive prenylated polyphenols from hops (*Humulus lupulus* cv. 'Cascade')." *Journal of Natural Products* 79, 3: 590–597. doi: 10.1021/acs.jnatprod.5b01052.

Fukuyama, Y., Y. Kiriyama, M. Kodama, H. Iwaki, S. Hosozawa, S. Aki, and K. Matsui. 1994. "Total synthesis of ardisiaquinone-A, a potent 5-lipoxygenase inhibitor, isolated from ardisia-sieboldii and degree of 5-lipoxygenase inhibitory activity of its derivatives." *Chemical & Pharmaceutical Bulletin* 42, 10: 2211–2213.

Fukuyama, Y., Y. Kiriyama, M. Kodama, H. Iwaki, S. Hosozawa, S. Aki, and K. Matsui. 1995. "Naturally-occurring 5-lipoxygenase inhibitors 6. Structures of ardisiaquinone-D, ardisiaquinone-E, and ardisiaquinone-F, from *Ardisia sieboldii*." *Chemical & Pharmaceutical Bulletin* 43, 8: 1391–1394.

Garscha, Ulrike, Erik Romp, Simona Pace, Antonietta Rossi, Veronika Temml, Daniela Schuster, Stefanie Koenig, Jana Gerstmeier, Stefanie Liening, Markus Werner, Heiner Atze, Sandra Wittmann, Christina Weinigel, Silke Rummler, Gerhard K. Scriba, Lidia Sautebin, and Oliver Werz. 2017. "Pharmacological profile and efficiency *in vivo* of diflapolin, the first dual inhibitor of 5-lipoxygenase-activating protein and soluble epoxide hydrolase." *Scientific Reports* 7. doi: 10.1038/s41598-017-09795-w.

Gerhauser, C., A. Alt, E. Heiss, A. Gamal-Eldeen, K. Klimo, J. Knauft, I. Neumann, H. R. Scherf, N. Frank, H. Bartsch, and H. Becker. 2002. "Cancer chemopreventive activity of Xanthohumol, a natural product derived from hop." *Molecular Cancer Therapeutics* 1, 11: 959–969.

Gerstmeier, Jana, Julia Seegers, Finja Witt, Birgit Waltenberger, Veronika Temml, Judith M. Rollinger, Hermann Stuppner, Andreas Koeberle, Daniela Schuster, and Oliver Werz. 2019. "Ginkgolic acid is a multi-target inhibitor of key enzymes in pro-inflammatory lipid mediator biosynthesis." *Frontiers in Pharmacology* 10. doi: 10.3389/fphar.2019.00797.

Ghosh, Arijit, Fang Chen, Abhimanyu Thakur, and Hao Hong. 2016. "Cysteinyl leukotrienes and their receptors: Emerging therapeutic targets in central nervous system disorders." *CNS Neuroscience & Therapeutics* 22: 943–951. doi: 10.1111/cns.12596.

Gilbert, Nathaniel C., Sue G. Bartlett, Maria T. Waight, David B. Neau, William E. Boeglin, Alan R. Brash, and Marcia E. Newcomer. 2011. "The structure of human 5-lipoxygenase." *Science* 331, 6014: 217–219. doi: 10.1126/science.1197203.

Glibo, Ana. 2016. "Anti-inflammatory activity testing using COX (cyclooxygenase) and LOX (lipoxygenase) inhibition." Master's thesis. University of Zagreb. urn:nbn:hr:163:522270. https://repozitorij.pharma.unizg.hr/en/islandora/object/pharma%3A511

Gomes, Camila Luiz, Victor Wanderley Sales de Albuquerque, Camila Gomes de Melo, Rosali Maria Ferreira da Silva, Rodolfo Hideki Vicente Nishimura, Larissa Araujo Rolim, and Pedro Jose Rolim Neto. 2021. "Beta-lapachone: Natural occurrence, physicochemical properties, biological activities, toxicity and synthesis." *Phytochemistry* 186: 112713 doi: 10.1016/j.phytochem.2021.112713.

Goradel, Nasser Hashemi, Masoud Najafi, Eniseh Salehi, Bagher Farhood, and Keywan Mortezaee. 2019. "Cyclooxygenase-2 in cancer: A review." *Journal of Cellular Physiology* 234, 5: 5683–5699. doi: 10.1002/jcp.27411.

Gulla Surendra, Dakshayani Lomada, Anusha Lade, Reddanna Pallu, and Madhava C. Reddy. 2020. "Role of prostaglandins in multiple sclerosis." *Current Pharmaceutical Design* 26: 1–12. doi: 10.2174/13816128 26666200107141328.

Halleypo. 2011. "Ipe rosa tabebuia avellanedae parque flamengo." Obtained from Wikimedia under the Creative Commons Attribution-Share Alike 3.0 Unported license. https://commons.wikimedia.org/wiki/File:Ipe_rosa_tabebuia_avellanedae_parque_flamengo.jpg (Accessed 4 May 2021).

Hämäläinen, Mari, Riina Nieminen, Zaini M. Asmawi, Pia Vuorela, Heikki Vapaatalo, and Eeva Moilanen. 2011. "Effects of flavonoids on prostaglandin E2 production and on COX-2 and mPGES-1 expressions in activated macrophages." *Planta Medica* 77, 13: 1504–1511. doi: 10.1055/s-0030-1270762.

Hanakova, Zuzana, Jan Hosek, Zsofia Kutil, Veronika Temml, Premysl Landa, Tomas Vanek, Daniela Schuster, Stefano Dall'Acqua, Josef Cvacka, Ondrej Polansky, and Karel Smejkal. 2017. "Anti-inflammatory activity of natural geranylated flavonoids: Cyclooxygenase and lipoxygenase inhibitory properties and proteomic analysis." *Journal of Natural Products* 80, 4: 999–1006. doi: 10.1021/acs.jnatprod.6b01011.

Hartung, Nicole M., Jana Fischer, Annika I. Ostermann, Ina Willenberg, Katharina M. Rund, Nils Helge Schebb, and Ulrike Garscha. 2019. "Impact of food polyphenols on oxylipin biosynthesis in human neutrophils." *Biochimica Et Biophysica Acta-Molecular and Cell Biology of Lipids* 10: 1536–1544. doi: 10.1016/j.bbalip.2019.05.002.

Hosek, Jan, Milan Bartos, Stanislav Chudik, Stefano Dall'Acqua, Gabbriella Innocenti, Murat Kartal, Ladislav Kokoska, Peter Kollar, Zsofia Kutil, Premysl Landa, Radek Marek, Veronika Zavalova, Milan Zemlicka, and Karel Smejkal. 2011. "Natural Compound cudraflavone B shows promising anti-inflammatory properties *in vitro*." *Journal of Natural Products* 74, 4: 614–619. doi: 10.1021/np100638h.

Hosek, Jan, Veronika Lelakova, Pavel Bobal, Hana Pizova, Marketa Gazdova, Milan Malanik, Karolina Jakubczyk, Ondrej Vesely, Premysl Landa, Veronika Temml, Daniela Schuster, Vilailak Prachyawarakorn, Phanruethai Pailee, Gang Ren, Filip Zpurny, Michal Oravec, and Karel Smejkal. 2019. "Prenylated stilbenoids affect inflammation by inhibiting the NF-kappa B/AP-1 signaling pathway and cyclooxygenases and lipoxygenase." *Journal of Natural Products* 82, 7: 1839–1848. doi: 10.1021/acs.jnatprod.9b00081.

Huang, Haiqiu, Zhihong Cheng, Haiming Shi, Wenbo Xin, Thomas T. Y. Wang, and Liangli Yu. 2011. "Isolation and characterization of two flavonoids, engeletin and astilbin, from the leaves of *Engelhardia roxburghiana* and their potential anti-inflammatory properties." *Journal of Agricultural and Food Chemistry* 59, 9: 4562–4569. doi: 10.1021/jf2002969.

Jiang, Jianxiong, and Ying Yu. 2020. "Small molecules targeting cyclooxygenase/prostanoid cascade in experimental brain ischemia: Do they translate?" *Medicinal Research Reviews* 41: 828–857. doi: 10.1002/med.21744.

Jin, Meihua, Kun Ho Son, and Hyeun Wook Chang. 2011. "Luteolin-7-O-glucoside suppresses leukotriene C-4 production and degranulation by inhibiting the phosphorylation of mitogen activated protein kinases and phospholipase C gamma 1 in activated mouse bone marrow-derived mast cells." *Biological & Pharmaceutical Bulletin* 34, 7: 1032–1036. doi: 10.1248/bpb.34.1032.

Joshi, Vikram, Shivaprasad H. Venkatesha, Chandrasekaran Ramakrishnan, Ankanahalli N. Nanjaraj Urs, Vilas Hiremath, Kamal D. Moudgil, Devadasan Velmurugan, and Bannikuppe Sannanaik Vishwanath. 2016. "Celastrol modulates inflammation through inhibition of the catalytic activity of mediators of arachidonic acid pathway: Secretory phospholipase A(2) group IIA, 5-lipoxygenase and cyclooxygenase-2." *Pharmacological Research* 113: 265–275. doi: 10.1016/j.phrs.2016.08.035.

Jo-Watanabe, Airi, Toshiaki Okuno, and Takehiko Yokomizo. 2019. "The role of leukotrienes as potential therapeutic targets in allergic disorders." *International Journal of Molecular Sciences* 20, 14. doi: 10.3390/ijms20143580.

Kaserer, Teresa, Veronika Temml, Zsofia Kutil, Tomas Vanek, Premysl Landa, and Daniela Schuster. 2015. "Prospective performance evaluation of selected common virtual screening tools. Case study: Cyclooxygenase (COX) 1 and 2." *European Journal of Medicinal Chemistry* 96: 445–457. doi: 10.1016/j.ejmech.2015.04.017.

KENPEI. 2008. "*Diospyros lotus*." Obtained from Wikimedia under the Creative Commons Attribution-Share Alike 2.1 Japan license. https://commons.wikimedia.org/wiki/File:Diospyros_lotus1.jpg (Accessed 4 May 2021).

Khan, Haroon, Kannan R. R. Rengasamy, Aini Pervaiz, Seyed Mohammad Nabavi, Atanas G. Atanasov, and Mohammad A. Kamal. 2018. "Plant-derived mPGES-1 inhibitors or suppressors: A new emerging trend in the search for small molecules to combat inflammation." *European Journal of Medicinal Chemistry* 135: 2–28. doi: 10.1016/j.ejmech.2017.12.059.

Klegeris Andis, and Patrick L. McGeer. 2005. "Non-steroidal anti-inflammatory drugs (NSAIDs) and other anti-inflammatory agents in the treatment of neurodegenerative disease." *Current Alzheimer Research* 2, 3: 355–365. doi: 10.2174/1567205054367883.

Koeberle, Andreas, Hinnak Northoff, and Oliver Werz. 2009. "Identification of 5-lipoxygenase and microsomal prostaglandin E-2 synthase-1 as functional targets of the anti-inflammatory and anti-carcinogenic garcinol." *Biochemical Pharmacology* 77, 9: 1513–1521. doi: 10.1016/j.bcp.2009.02.005.

Koeberle, Andreas, Federica Pollastro, Hinnak Northoff, and Oliver Werz. 2009. "Myrtucommulone, a natural acylphloroglucinol, inhibits microsomal prostaglandin E-2 synthase-1." *British Journal of Pharmacology* 156, 6: 952–961. doi: 10.1111/j.1476-5381.2009.00070.x.

Koeberle, Andreas, Antonietta Rossi, Julia Bauer, Friederike Dehm, Luisella Verotta, Hinnak Northoff, Lidia Sautebin, and Oliver Werz. 2011. "Hyperforin, an anti-inflammatory constituent from St. John's wort, inhibits microsomal prostaglandin E-2 synthase-1 and suppresses prostaglandin E-2 formation *in vivo*." *Frontiers in Pharmacology* 2. doi: 10.3389/fphar.2011.00007.

Koeberle, Andreas, and Oliver Werz. 2018. "Natural products as inhibitors of prostaglandin E-2 and pro-inflammatory 5-lipoxygenase-derived lipid mediator biosynthesis." *Biotechnology Advances* 36, 6: 1709–1723. doi: 10.1016/j.biotechadv.2018.02.010.

Krueger, Phillip, Rambod Daneshfar, Gunter P. Eckert, Jochen Klein, Dietrich A. Volmer, Ute Bahr, Walter E. Mueller, Michael Karas, Manfred Schubert-Zsilavecz, and Mona Abdel-Tawab. 2008. "Metabolism of boswellic acids in vitro and *in vivo*." *Drug Metabolism and Disposition* 36, 6: 1135–1142. doi: 10.1124/dmd.107.018424.

Krueger, Phillip, Johanna Kanzer, Jessica Hummel, Gert Fricker, Manfred Schubert-Zsilavecz, and Mona Abdel-Tawab. 2009. "Permeation of boswellia extract in the caco-2 model and possible interactions of its constituents KBA and AKBA with OATP1B3 and MRP2." *European Journal of Pharmaceutical Sciences* 36, 2–3: 275–284. doi: 10.1016/j.ejps.2008.10.005.

Ku, Edmond C., Warren Lee, Himanshu V. Kothari, Dietrich W. Scholer. 1986. "Effect of diclofenac sodium on the arachidonic acid cascade." *The American Journal of Medicine* 80, 4: 18–23. doi: 10.1016/0002-9343(86)90074-4.

Kutil, Zsofia, Marie Kvasnicova, Veronika Temml, Daniela Schuster, Petra Marsik, Eloy Fernandez Cusimamani, Ji-Dong Lou, Tomas Vanek, and Premysl Landa. 2015. "Effect of dietary stilbenes on 5-lipoxygenase and cyclooxygenases activities *in vitro*." *International Journal of Food Properties* 18, 7: 1471–1477. doi: 10.1080/10942912.2014.903416.

Kutil, Zsofia, Veronika Temml, David Maghradze, Marie Pribylova, Marcela Dvorakova, Daniela Schuster, Tomas Vanek, and Premysl Landa. 2014. "Impact of wines and wine constituents on cyclooxygenase-1, cyclooxygenase-2, and 5-lipoxygenase catalytic activity." *Mediators of Inflammation* 2014. doi: 10.1155/2014/178931.

Kwon, Oh Song, Jae Sue Choi, Md Nurul Islam, Yeong Shik Kim, and Hyun Pyo Kim. 2011. "Inhibition of 5-lipoxygenase and skin inflammation by the aerial parts of artemisia capillaris and its constituents." *Archives of Pharmacal Research* 34, 9: 1561–1569. doi: 10.1007/s12272-011-0919-0.

Landa, Premysl, Zsofia Kutil, Veronika Temml, Jan Malik, Ladislav Kokoska, Ute Widowitz, Marie Pribylova, Marcela Dvorakova, Petr Marsik, Daniela Schuster, Rudolf Bauer, and Tomas Vanek. 2013. "Inhibition of in vitro leukotriene B-4 biosynthesis in human neutrophil granulocytes and docking studies of natural quinones." *Natural Product Communications* 8, 1: 105–108. doi: 10.1177/1934578X1300800124

Landa, Premysl, Zsofia Kutil, Veronika Temml, Anna Vuorinen, Jan Malik, Marcela Dvorakova, Petr Marsik, Ladislav Kokoska, Marie Pribylova, Daniela Schuster, and Tomas Vanek. 2012. "Redox and non-redox mechanism of *in vitro* cyclooxygenase inhibition by natural quinones." *Planta Medica* 78, 4: 326–333. doi: 10.1055/s-0031-1280430.

Laufer, Stefan, and Sabine Luik. 2010. "Different methods for testing potential cyclooxygenase-1 and cyclooxygenase-2 inhibitors." *Cyclooxygenases: Methods and Protocols* 644: 91–116. doi: 10.1007/978-1-59745-364-6_8.

Lee, Eun Ju, Ju Sun Kim, Hyun Pyo Kim, Je-Hyun Lee, and Sam Sik Kang. 2010. "Phenolic constituents from the flower buds of Lonicera japonica and their 5-lipoxygenase inhibitory activities." *Food Chemistry* 120, 1: 134–139. doi: 10.1016/j.foodchem.2009.09.088.

Lee, Je Hyeong, Oh Song Kwon, Hong-Guang Jin, Eun-Rhan Woo, Yeong Shik Kim, and Hyun Pyo Kim. 2012. "The rhizomes of alisma orientale and alisol derivatives inhibit allergic response and experimental atopic dermatitis." *Biological & Pharmaceutical Bulletin* 35, 9: 1581–1587. doi: 10.1248/bpb.b110689.

Lee, Je-Hyuk, and Gun-Hee Kim. 2010. "Evaluation of antioxidant and inhibitory activities for different subclasses flavonoids on enzymes for rheumatoid arthritis." *Journal of Food Science* 75, 7: H2012–H217. doi: 10.1111/j.1750-3841.2010.01755.x.

Lelakova, Veronika, Karel Šmejkal, Karolina Jakubczyk, Ondřej Veselý, Přemysl Landa, Jiří Václavík, Pavel Bobál, Hana Pížová, Veronika Temml, Theresa Steinacher, Daniela Schuster, Sebastian Granica, Zuzana Hanáková, and Jan Hošek. 2019. "Parallel *in vitro* and *in silico* investigations into anti-inflammatory effects of non-prenylated stilbenoids." *Food Chemistry* 285: 431–440. doi: 10.1016/j.foodchem.2019.01.128.

Leone, S., A. Ottani, and A. Bertolini. 2007. "Dual acting anti-inflammatory drugs." *Current Topics in Medicinal Chemistry* 7, 3: 265–275. doi: 10.2174/156802607779941341.

Leval, Xavier de, Fabien Julemont, Jacques Delarge, Bernard Pirotte, and Jean-Michel Dogne. 2002. "New trends in dual 5-LOX/COX inhibition." *Current Medicinal Chemistry* 9, 9: 941–962. doi: 10.2174/0929867024606713.

Li, Xiang, Zhi Liu, Xin-feng Zhang, Li-juan Wang, Yi-nan Zheng, Chang-chun Yuan, and Guang-zhi Sun. 2008. "Isolation and characterization of phenolic compounds from the leaves of *Salix matsudana*." *Molecules* 13: 1530–1537. doi: 10.3390/molecules13081530.

Mahesh, Gopa, Kotha Anil Kumar, Pallu Reddanna. 2021. "Overview on the discovery and development of anti-inflammatory drugs: Should the focus be on synthesis or degradation of PGE_2?." *Journal of Inflammation Research* 14: 253–263. doi: 10.2147/JIR.S278514.

Maione, Francesco, Vincenza Cantone, Simona Pace, Maria Giovanna Chini, Angela Bisio, Giovanni Romussi, Stefano Pieretti, Oliver Werz, Andreas Koeberle, Nicola Mascolo, and Giuseppe Bifulco. 2017. "Anti-inflammatory and analgesic activity of carnosol and carnosic acid *in vivo* and *in vitro* and *in silico* analysis of their target interactions." *British Journal of Pharmacology* 174, 11: 1497–1508. doi: 10.1111/bph.13545.

Maurya, R., and A. B. Ray. 1984. "Constituents of Pterocarpus marsupium." *Journal of Natural Products* 47, 1: 179–181. doi: 10.1021/np50031a029

Minassi, Alberto, Lavinia Cicione, Andreas Koeberle, Julia Bauer, Stefan Laufer, Oliver Werz, and Giovanni Appendino. 2012. "A multicomponent carba-betti strategy to alkylidene heterodimers-total synthesis and structure-activity relationships of arzanol." *European Journal of Organic Chemistry* 2012, 4: 772–779. doi: 10.1002/ejoc.201101193.

Mitchell, Jane A., Fisnik Shala, Youssef Elghazouli, Timothy D. Warner, Carle Gaston-Massuet, Marilena Crescente, Paul C. Armstrong, Harvey R. Herschman, and Nicholas S. Kirkby. 2019. "Cell-Specific gene deletion reveals the antithrombotic function of COX1 and explains the vascular COX1/prostacyclin paradox." *Circulation Research* 125: 847–854. doi: 10.1161/CIRCRESAHA.119.314927.

Mladenka, Premysl, Libuse Zatloukalova, Tomas Filipsky, and Radomir Hrdina. 2010. "Cardiovascular effects of flavonoids are not caused only by direct antioxidant activity." *Free Radical Biology and Medicine* 49, 6: 963–975. doi: 10.1016/j.freeradbiomed.2010.06.010.

Moon, Dong-Oh, Yung Hyun Choi, Nam-Duk Kim, Yeong-Min Park, and Gi-Young Kim. 2007. "Anti-inflammatory effects of β-lapachone in lipopolysaccharide-stimulated BV2 microglia." *International Immunopharmacology* 7, 4: 506–514. doi: 10.1016/j.intimp.2006.12.006.

Moore, Gillian Y., and Graham P. Pidgeon. 2017. "Cross-talk between cancer cells and the tumour microenvironment: The role of the 5-lipoxygenase pathway." *International Journal of Molecular Sciences* 18, 2. doi: 10.3390/ijms18020236.

Nguyen, Ha Thi, Thien- Y. Vu, Vishala Chandi, Haritha Polimati, and Vinay Bharadwaj Tatipamula. 2020. "Dual COX and 5-LOX inhibition by clerodane diterpenes from seeds of Polyalthia longifolia (Sonn.) Thwaites." *Scientific Reports* 10, 1. doi: 10.1038/s41598-020-72840-8.

Nhu Ngoc Quynh, Vo, Yuhta Nomura, Toshiya Muranaka, and Ery Odette Fukushima. 2019. "Structure-activity relationships of pentacyclic triterpenoids as inhibitors of cyclooxygenase and lipoxygenase enzymes." *Journal of Natural Products* 82, 12: 3311–3320. doi: 10.1021/acs.jnatprod.9b00538.

Noreen, Yiva, Gudelia Serrano, Premila Perera, and L. Bohlin. 1998. "Flavan-3-ols isolated from some medicinal plants inhibiting COX-1 and COX-2 catalysed prostaglandin biosynthesis." *Planta Medica* 64, 6: 520–524. doi: 10.1055/s-2006-957506.

Nova, A. K. 2007. "*Ginkgo biloba*." Obtained from Wikimedia under the Creative Commons Attribution-Share Alike 4.0 International license. https://commons.wikimedia.org/wiki/File:Ginkgo_biloba_Mi%C5%82orz%C4%85b_dwuklapowy_2007-08-11_02.jpg (Accessed 4 May 2021).

Nova, A. K. 2020. "*Hypericum perforatum*." Obtained from Wikimedia under the Creative Commons Attribution-Share Alike 4.0 International license. https://commons.wikimedia.org/wiki/File:Hypericum_perforatum_Dziurawiec_zwyczajny_2020-07-12_05.jpg (Accessed 4 May 2021).

O'Banion, M. Kerry. 1999. "COX-2 and Alzheimer's disease: Potential roles in inflammation and neurodegeneration." *Expert Opinion on Investigational Drugs* 8, 10: 1521–1536. doi: 10.1517/13543784.8.10.1521.

Orafaie, Ala, Mina Mousavian, Hossein Orafai, and Hamid Sadeghian. 2020. "An overview of lipoxygenase inhibitors with approach of *in vivo* studies." *Prostaglandins and Other Lipid Mediators* 148: 106411. doi: 10.1016/j.prostaglandins.2020.106411.

Pannunzio, Alessandra, and Mauro Coluccia. 2018. "Cyclooxygenase-1 (COX-1) and COX-1 inhibitors in cancer: A review of oncology and medicinal chemistry literature." *Pharmaceuticals* 11, 4. doi: 10.3390/ph11040101.

Pein, Helmut, Alexia Ville, Simona Pace, Veronika Temml, Ulrike Garscha, Martin Raasch, Khaled Alsabil, Guillaume Viault, Dinh Chau-Phi, David Guilet, Fabiana Troisi, Konstantin Neukirch, Stefanie Koenig, Rosella Bilancia, Birgit Waltenberger, Hermann Stuppner, Maria Wallert, Stefan Lorkowski, Christina Weinigel, Silke Rummler, Marc Birringer, Fiorentina Roviezzo, Lidia Sautebin, Jean-Jacques Helesbeux, Denis Seraphin, Alexander S. Mosig, Daniela Schuster, Antonietta Rossi, Pascal Richomme, Oliver Werz, and Andreas Koeberle. 2018. "Endogenous metabolites of vitamin E limit inflammation by targeting 5-lipoxygenase." *Nature Communications* 9. doi: 10.1038/s41467-018-06158-5.

Peskar, B. M. 2001. "Role of cyclooxygenase isoforms in gastric mucosal defence." *Journal of Physiology-Paris* 95, 1–6: 3–9. doi: 10.1016/s0928-4257(01)00003-1.

Radmark, Olof, and Bengt Samuelsson. 2009. "5-Lipoxygenase: Mechanisms of regulation." *Journal of Lipid Research* 50: S40–S45. doi: 10.1194/jlr.R800062-JLR200.

Raffaelli, M. 2005. "*Boswellia* Dowkah." Obtained from Wikimedia under the Creative Commons Attribution-Share Alike 3.0 Unported license. https://commons.wikimedia.org/wiki/File:Boswellia-Dowkah-2.JPG (Accessed 4 May 2021).

Raffaelli, M. 2012. "*Boswellia sacra*." Obtained from Wikimedia under the Creative Commons Attribution-Share Alike 3.0 Unported license. https://commons.wikimedia.org/wiki/File:Incenso-2.jpg (Accessed 4 May 2021).

Rahman, Syed Obaidur, Rakesh Kumar Singh, Salman Hussain, Mohd Akhtar, and Abul Kalam Najmi. 2019. "A novel therapeutic potential of cysteinyl leukotrienes and their receptors modulation in the neurological complications associated with Alzheimer's disease." *European Journal of Pharmacology* 842: 208–220. doi: 10.1016/j.ejphar.2018.10.040.

Rawat, Chitra, Samiksha Kukal, Ujjwal Ranjan Dahiya, and Ritushree Kukreti. 2019. "Cyclooxygenase-2 (COX-2) inhibitors: Future therapeutic strategies for epilepsy management." *Journal of Neuroinflammation* 16: 197. doi: 10.1186/s12974-019-1592-3.

Reddy, D. Bharat, T. C. M. Reddy, G. Jyotsna, Satish Sharan, Nalini Priya, V. Lakshmipathi, and Pallu Reddanna. 2009. "Chebulagic acid, a COX-LOX dual inhibitor isolated from the fruits of *Terminalia chebula* Retz., induces apoptosis in COLO-205 cell line." *Journal of Ethnopharmacology* 124, 3: 506–512. doi: 10.1016/j.jep.2009.05.022.

Rodrigues, Tiago, Markus Werner, Jakob Roth, Eduardo H. G. da Cruz, Marta C. Marques, Padma Akkapeddi, Susana A. Lobo, Andreas Koeberle, Francisco Corzana, Eufranio N. da Silva Junior, Oliver Werz, and Goncalo J. L. Bernardes. 2018. "Machine intelligence decrypts -lapachone as an allosteric 5-lipoxygenase inhibitor." *Chemical Science* 9, 34. doi: 10.1039/c8sc02634c.

Safayhi, H., T. Mack, J. Sabieraj, M. I. Anazodo, L. R. Subramanian, and H. P. T. Ammon. 1992. "Boswellic acids—novel, specific, nonredox inhibitors of 5-lipoxygenase." *Journal of Pharmacology and Experimental Therapeutics* 261, 3: 1143–1146.

Samuelsson, Bengt, Ralf Morgenstern, and Petr-Johan Jakobsson. 2007. "Membrane prostaglandin E Synthase-1: A novel therapeutic target." *Pharmacological Reviews* 59: 207–224. doi: 10.1124/pr.59.3.1

Schaible, Anja M., Heidi Traber, Veronika Temml, Stefan M. Noha, Rosanna Filosa, Antonella Peduto, Christina Weinigel, Dagmar Barz, Daniela Schuster, and Oliver Werz. 2013. "Potent inhibition of human 5-lipoxygenase and microsomal prostaglandin E-2 synthase-1 by the anti-carcinogenic and anti-inflammatory agent embelin." *Biochemical Pharmacology* 86, 4: 476–486. doi: 10.1016/j.bcp.2013.04.015.

Shahinozzaman, Md, Takahiro Ishii, Mohammad A. Halim, Md Amzad Hossain, Md Tofazzal Islam, and Shinkichi Tawata. 2019. "Cytotoxic and anti-inflammatory resorcinol and alkylbenzoquinone derivatives from the leaves of *Ardisia sieboldii*." *Zeitschrift fur Naturforschung Section C-A Journal of Biosciences* 74, 11–12: 303–311. doi: 10.1515/znc-2019-0114.

Sharma, Tarun, and Snehasis Jana. 2020. "Boswellic acids as natural anticancer medicine: Precious gift to humankind." *Journal of Herbal Medicine* 20. doi: 10.1016/j.hermed.2019.100313.

Siemoneit, U., A. Koeberle, A. Rossi, F. Dehm, M. Verhoff, S. Reckel, T. J. Maier, J. Jauch, H. Northoff, F. Bernhard, V. Doetsch, L. Sautebin, and O. Werz. 2011. "Inhibition of microsomal prostaglandin E-2 synthase-1 as a molecular basis for the anti-inflammatory actions of boswellic acids from frankincense." *British Journal of Pharmacology* 162, 1: 147–162. doi: 10.1111/j.1476-5381.2010.01020.x.

Siemoneit, Ulf, Carlo Pergola, Bianca Jazzar, Hinnak Northoff, Carsten Skarke, Johann Jauch, and Oliver Werz. 2009. "On the interference of boswellic acids with 5-lipoxygenase: Mechanistic studies in vitro and pharmacological relevance." *European Journal of Pharmacology* 606, 1–3: 246–254. doi: 10.1016/j.ejphar.2009.01.044.

Singh, Nikhlesh K., and Gadiparthi N. Rao. 2019. "Emerging role of 12/15-Lipoxygenase (ALOX15) in human pathologies." *Progress in Lipid Research* 73: 28–45. doi: 10.1016/j.plipres.2018.11.001

Sisa, Miroslav, Marcela Dvorakova, Veronika Temml, Veronika Jarosova, Tomas Vanek, and Premysl Landa. 2020. "Synthesis, inhibitory activity and in silico docking of dual COX/5-LOX inhibitors with quinone and resorcinol core." *European Journal of Medicinal Chemistry* 204. doi: 10.1016/j.ejmech.2020.112620.

Skalle-Per Hedenhös. 2012. "Hampa *Cannabis sativa* L." Obtained from Wikimedia under the Creative Commons Attribution-Share Alike 4.0 International license. https://commons.wikimedia.org/wiki/File:Hampa_Cannabis_sativa_L._(n%C3%A4rbild).jpg (Accessed 4 May 2021).

Smith, W. L., D. L. DeWitt, and R. M. Garavito. 2000. "Cyclooxygenases: Structural, cellular, and molecular biology." *Annual Review of Biochemistry* 69: 145–182. doi: 10.1146/annurev.biochem.69.1.145.

Smyth, Emer M., Tilo Grosser, Miao Wang, Ying Yu, and Garret A. FitzGerald. 2009. "Prostanoids in health and disease." *Journal of Lipid Research* 50: S423–S428. doi: 10.1194/jlr.R800094-JLR200.

Stanzilla. 2020. "*Humulus lupulus*." Obtained from Wikimedia under the Creative Commons Attribution-Share Alike 4.0 International license. https://commons.wikimedia.org/wiki/File:Humulus_lupulus_with_fly.jpg (Accessed 4 May 2021).

Starr, F., and K. Starr. 2006. "*Artocarpus altilis*." Obtained from Wikimedia under the Creative Commons Attribution 3.0 Unported license. https://commons.wikimedia.org/wiki/File:Starr_060703-8345_Artocarpus_altilis.jpg (Accessed 4 May 2021).

Svouraki, Alexandra, Ulrike Garscha, Eirini Kouloura, Simona Pace, Carlo Pergola, Verena Krauth, Antonietta Rossi, Lidia Sautebin, Maria Halabalaki, Oliver Werz, Nicolas Gaboriaud-Kolar, and Alexios-Leandros Skaltsounis. 2017. "Evaluation of dual 5-lipoxygenase/microsomal prostaglandin E2 synthase-1 inhibitory effect of natural and synthetic acronychiatype isoprenylated acetophenones." *Journal of Natural Products* 80, 3: 699–706. doi: 10.1021/acs.jnatprod.6b01008.

Tausch, Lars, Arne Henkel, Ulf Siemoneit, Daniel Poeckel, Nicole Kather, Lutz Franke, Bettina Hofmann, Gisbert Schneider, Carlo Angioni, Gerd Geisslinger, Carsten Skarke, Wolfgang Holtmeier, Tobias Beckhaus, Michael Karas, Johann Jauch, and Oliver Werz. 2009. "Identification of human cathepsin G As a functional target of boswellic acids from the anti-inflammatory remedy frankincense." *Journal of Immunology* 183, 5: 3433–3442. doi: 10.4049/jimmunol.0803574.

Temml, Veronika, Teresa Kaserer, Zsofia Kutil, Premysl Landa, Tomas Vanek, and Daniela Schuster. 2014. "Pharmacophore modeling for COX-1 and-2 inhibitors with LigandScout in comparison to discovery studio." *Future Medicinal Chemistry* 6, 17: 1869–1881. doi: 10.4155/fmc.14.114.

Tobe, Hiroyasu. 2009. "Biological activities of humulone." *Beer in Health and Disease Prevention*: 695–702. doi: 10.1016/B978-0-12-373891-2.00070-5.

Tonreg, R. 2012. "*Vitis vinifera* subsp. *vinifera*." Obtained from Wikimedia under the Creative Commons Attribution 2.0 Generic license. https://commons.wikimedia.org/wiki/File:Vitis_vinifera_subsp._vinifera_(8248156192).jpg (Accessed 4 May 2021).

Tuure, Lauri, Mari Hamalainen, Brendan J. Whittle, and Eeva Moilanen. 2017. "Microsomal prostaglandin E synthase-1 expression in inflammatory conditions is downregulated by dexamethasone: Seminal role of the regulatory phosphatase MKP-1." *Frontiers in Pharmacology* 8: 646. doi: 10.3389/fphar.2017.00646.

Tyukavkina, N. A., A. S. Gromova, V. I. Lutskii, and V. K. Voronov. 1972. "Hydroxystilbenes from the bark of *Pinus sibirica*." *Chemistry of Natural Compounds* 8, 5: 570–572. doi:10.1007/BF00564298.

Valker, D. 2015. "*Embelia ribes*." Obtained from Wikimedia under the Creative Commons Attribution-Share Alike 3.0 Generic license. https://commons.wikimedia.org/wiki/File:Embelia_ribes_(17146786429).jpg (Accessed 4 May 2021).

Vane, J. R. 1971. "Inhibition of prostaglandin synthesis as a mechanism of action for aspirin-like drugs." *Nature New Biology* 231, 25: 232–235. doi: 10.1038/newbio231232a0.

Vane, J. R. 2000. "The fight against rheumatism: From willow bark to COX-1 sparing drugs." *Journal of Physiology and Pharmacology* 51, 4: 573–586.

Verhoff, Moritz, Stefanie Seitz, Hinnak Northoff, Johann Jauch, Anja M. Schaible, and Oliver Werz. 2012. "A novel C(28)-hydroxylated lupeolic acid suppresses the biosynthesis of eicosanoids through inhibition of cytosolic phospholipase A(2)." *Biochemical Pharmacology* 84, 5: 681–691. doi: 10.1016/j.bcp.2012.06.016.

Verhoff, Moritz, Stefanie Seitz, Michael Paul, Stefan M. Noha, Johann Jauch, Daniela Schuster, and Oliver Werz. 2014. "Tetra- and pentacyclic triterpene acids from the ancient anti-inflammatory remedy frankincense as inhibitors of microsomal prostaglandin E-2 synthase-1." *Journal of Natural Products* 77, 6: 1445–1451. doi: 10.1021/np500198g.

Vitale, Paola, Andrea Panella, Antonio Scilimati, and Maria Grazia Perrone. 2016. "COX-1 inhibitors: Beyond structure toward therapy." *Medicinal Research Reviews* 36, 4: 641–671. doi: 10.1002/med.21389.

Wang, Siqi, Jiaxin Huai, Ying Shang, Linlin Xie, Xiaotong Cao, Jun Liao, Teng Zhang, and Ronghua Dai. 2020. "Screening for natural inhibitors of 5-lipoxygenase from Zi-shen pill extract by affinity ultrafiltration coupled with ultra performance liquid chromatography-mass spectrometry." *Journal of Ethnopharmacology* 254. doi: 10.1016/j.jep.2020.112733.

Warner, T. D., and J. A. Mitchell. 2004. "Cyclooxygenases: New forms, new inhibitors, and lessons from the clinic." *Faseb Journal* 18, 7: 790–804. doi: 10.1096/fj.03-0645rev.

Werz, Oliver. 2007. "Inhibition of 5-lipoxygenase product synthesis by natural compounds of plant origin." *Planta Medica* 73, 13: 1331–1357. doi: 10.1055/s-2007-990242.

Werz, Oliver, Julia Seegers, Anja Maria Schaible, Christina Weinigel, Dagmar Barz, Andreas Koeberle, Gianna Allegrone, Federica Pollastro, Lorenzo Zampieri, Gianpaolo Grassi, and Giovanni Appendino. 2014. "Cannflavins from hemp sprouts, a novel cannabinoid-free hemp food product, target microsomal prostaglandin E-2 synthase-1 and 5-lipoxygenase." *PharmaNutrition* 2, 3: 53–60. doi: 10.1016/j.phanu.2014.05.001.

Winekenstadde, D., A. Angelis, B. Waltenberger, S. Schwaiger, J. Tchoumtchoua, S. Konig, O. Werz, N. Aligiannis, A. L. Skaltsounis, and H. Stuppner. 2015. "Phytochemical profile of the aerial parts of Sedum sediforme and anti-inflammatory activity of myricitrin." *Natural Product Communications* 10, 1: 83–88. doi: 10.1177/1934578X1501000122.

Wood, Irene, Andres Trostchansky, and Homero Rubbo. 2020. "Structural considerations on lipoxygenase function, inhibition and crosstalk with nitric oxide pathways." *Biochimie* 178: 170–180. doi: 10.1016/j.biochi.2020.09.021.

Xie, Linlin, David Yue-wei Lee, Ying Shang, Xiaotong Cao, Siqi Wang, Jun Liao, Teng Zhang, and Ronghua Dai. 2020. "Characterization of spirostanol glycosides and furostanol glycosides from anemarrhenae rhizoma as dual targeted inhibitors of 5-lipoxygenase and Cyclooxygenase-2 by employing a combination of affinity ultrafiltration and HPLC/MS." *Phytomedicine* 77. doi: 10.1016/j.phymed.2020.153284.

Zarraga, Ignatius Gerardo E., and Ernst R. Schwarz. 2007. "Coxibs and heart disease—What we have learned and what else we need to know." *Journal of the American College of Cardiology* 49, 1: 1–14. doi: 10.1016/j.jacc.2006.10.003.

Zhao, Anqi, Li Li, Bo Li, Meizhu Zheng, and Rong Tsao. 2016. "Ultrafiltration LC-ESI-MSn screening of 5-lipoxygenase inhibitors from selected Chinese medicinal herbs Saposhnikovia divaricata, Smilax glabra, Pueraria lobata and Carthamus tinctorius." *Journal of Functional Foods* 24: 244–253. doi: 10.1016/j.jff.2016.04.011.

Zheng, Ming Shan, Wen Yi Jin, Kun Ho Son, Kim Hyun Pyo, and Sam Sik Kang. 2005. "The constituents isolated from *Peucedanum japonicum* Thunb. and their cyclooxygenase (COX) inhibitory activity." *Korean Journal of Medicinal Crop Science* 13, 2: 75–79.

23 Botanicals and the Gut Microbiome

Anna-Mari Reid and Namrita Lall

CONTENTS

DOI: 10.1201/9781003108375-23

23.1 INTRODUCTION

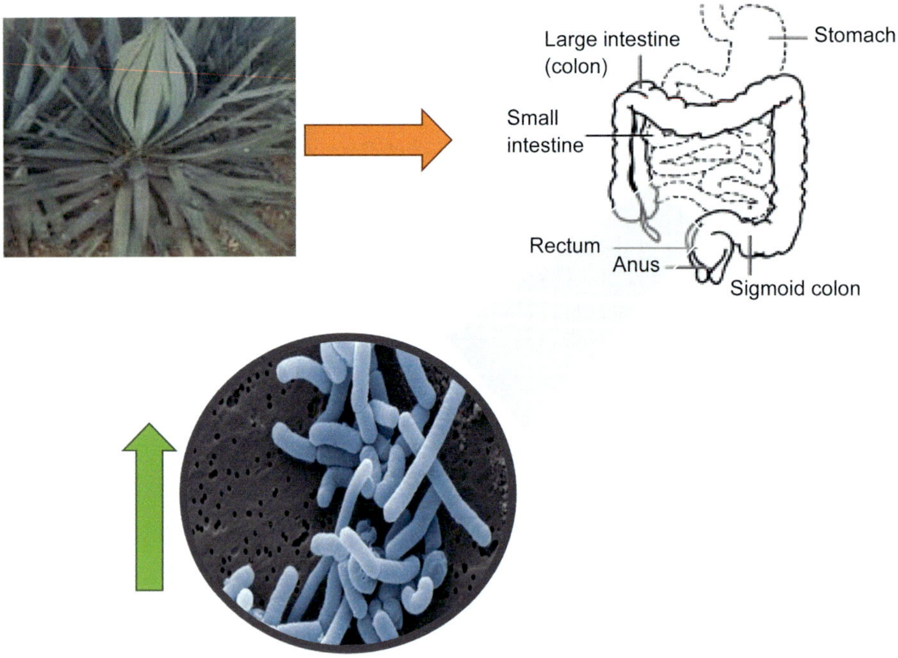

Graphical abstract representing the content of the chapter (Dflock, 2004).

23.1.1 What Is the Gut Microbiome, and What Does It Consist Of?

Micro-organisms can be found on our skin, mouth, nose, ears, intestinal tract and genitals. The microbial gut genome of humans contains more genes than are encoded by the host itself (Bäckhed et al., 2005; Nicholson et al., 2012; Lu et al., 2019). The gut microbiome is known to influence the host's immune system, development and physiology, as well as metabolism (de Vos and de Vos, 2012; Smith, 2015; Lu et al., 2019). Variability exists between the gut microbiome of different individuals and can be influenced by factors such as age, diet and health status (Moschen et al., 2012; O'Toole and Jeffery, 2015; Shreiner et al., 2015). A healthy human gut represents an ecological system that consists of bacteria, fungi and virus domains (Eid et al., 2017). The composition of the microbiome that has been compiled through the studies on 16s rRNA gene sequences from human fecal microbiota compared with 59 other mammalian species shows that the prevalent groups are mostly comprised of Firmicutes (65.7%), Bacteroidetes (16.3%), Proteobacteria (8.8%), Actinobacteria (4,7%), Verrucomicrobia (2.2%), Fusobacteria (0.67%) and others (Possemiers et al., 2011; Habtemariam, 2020).

The gut consists of both harmful and beneficial bacteria. Harmful bacteria that can be found in the gut are mainly *Escherichia coli*, while the beneficial bacteria include *Bifidobacterium adolescentis* and *Lactobacillus acidophilus* (Habtemariam, 2020). As described earlier, each human individual has a specific gut microbiome, as unique as our fingerprints, although we all have main bacterial divisions that are mainly found in the gut and the gastrointestinal tract. Two archaeal species found are *Methanobrevibacter* and *Methanosphaera stadtmanae*, with the former being the most dominant (Possemiers et al., 2011). Each individual's personalized community of microbiota functions in equilibrium with the host to achieve overall homeostasis in relation to the theory of 'evolutionary stable strategy.' In simplified terms, the microbiota will try to grow and overcome the host's immune responses—but not to the extent that it is detrimental to the host's fitness, as this will only lead to limitations within the microbiota organization (Sekirov et al., 2010).

Many important functions have been attributed to the microbiome in the intestinal tract such as extraction of energy, stimulation of the gut immune system, regulation of cells and synthesis of vitamins K and B (Eckburg et al, 2005; Swidsinski et al., 2005; Frank et al., 2007; Vyas and Ranganathan, 2012). Not only are the microbiome linked to these functions, but there are associations with several diseases and disorders such as allergies, inflamed bowels, diabetes type 2, atrophy and obesity (Björkstén et al., 2001; Boulangé et al., 2016; Boursier et al., 2016; Bull et al., 2014; Guarner et al., 2003; Lee et al., 2020).

As we age, the gut microbiota show signs of aging in the phyla. These were identified from the feces of infants containing primarily *Bifidobacterium* and *Clostridium coccoides*, whereas with elderly adults, the major phyla found are Firmicutes and Bacteroidetes (Siegel et al., 2017; Chu et al., 2017; Liu et al., 2019). It is these changes in the gut microbiota that make elderly individuals susceptible to dysbiosis and age-related diseases.

Linkages exist between the gut and some major organs in the host's body, such as the liver and brain. The connection between the gut and the brain includes neural, hormonal and immune signals (Collins et al., 2012). There is a bidirectional communication system whereby the brain has the ability to direct peristalsis, mucin production and immune functionality (Mayer, 2011). It is known that gut microbiota can affect brain functionality and behavior through regulation of tryptophan metabolism, as well as in the production of neurometabolites (Desbonnet et al., 2008; Thomas et al., 2012).

Any compound that is ingested by the host has the ability to affect the composition and the diversity of the gut microbiome. The ingestion of antibiotics, as an antibacterial, has the ability to inhibit or increase certain populations of microbes (Pferschy-Wenzig et al., 2017). A host's stress levels have the ability to affect the integrity of the gut's epithelium, alter the peristalsis, secretions and mucin production, thereby changing the surrounding habitat of the intestinal microbiota and its composition (Collins et al., 2012). Individuals with levels of acute stress can induce certain responses in both the upper and lower gastrointestinal tract by increasing the levels of stress response hormones (Eutamene and Bueno, 2007).

23.2 FUNCTIONS OF THE GUT MICROBIOTA

As the gut and its microbes are recognized as an important feature of the human body, especially when it comes to the ingestion and metabolism of food and drugs, and what their effects will be on the host itself, much research has been conducted on their function. The main functions that have been attributed to the gut microbiota are discussed in what follows.

23.2.1 Metabolism

There are numerous metabolism capabilities within the gut, due to the vast metabolic enzymes present (Lu et al., 2019). The gut can independently metabolize several substrates or co-metabolize together with the host (Tremaroli and Bäckhed, 2012). The gut microbiota mainly synthesizes both essential and nonessential amino acids and various vitamins, and is involved in the biotransformation of the bile (Vyas and Ranganathan, 2012). Other functions include the metabolism of carbohydrates—which are indigestible, such as larger polysaccharides—and some oligopolysaccharides, sugars and alcohols (Cummings et al., 1987). The energy required by the colon is obtained through the metabolism of these carbohydrates (Bull and Plummer, 2014). Regulation of the host's metabolic processes is an important function of short-chain fatty acids (SCFAs). The SCFAs required are synthesized by the gut microbiota (Nicholson et al., 2012; Fernandes et al., 2014).

23.2.2 Development of the Immune System

An individual's immune system development is influenced by their intestinal microbiota and the metabolites they produce. The intestinal epithelial cells have the ability to signal an innate immune response to prevent any threats from pathogens (Vyas and Ranganathan, 2012). The most significant protection provided by the intestinal microbiota is known as the barrier or competitive-exclusion effect. This effect is the ability of the intestinal bacteria to produce antimicrobial compounds such as bacteriocins to compete for nutrients and a site of attachment in the gut lining. In this way, the intestinal bacteria prevent the colonization of pathogenic bacteria in the gut (Guarner et al., 2003). The intestinal bacteria are known for their production of bacteriocins and this is a property distributed throughout most of the gastrointestinal bacteria (Guarner et al., 2003).

Other mechanisms incorporated to control microbial populations are that of bile acids that escape the enterohepatic circulation and finds itself in the colon, whereby the colonic microbiota change the properties and structures of the bile acids. The bile acids, however, possess antimicrobial properties that limit the growth of certain gut microbiota (Rowland et al., 2018). The gut microbiota serve as a source of ligands for the innate immune system. Pattern recognition receptor signaling in this manner that originates from the gut microbiota is deemed as a necessity for a steady state of protection of the intestines. These receptors include toll-like receptors (TLRs) and NOD (nucleotide-binding oligomerization domain)-like receptors (NLRs) that are important for the maturation of the intestinal immune system and protecting against enteric pathogens (Rolhion and Chassaing, 2016).

23.2.3 Quorum Sensing and Gut Microbes

Quorum sensing is the communication pathway between bacterial species that is based on chemical signals and through which their behavior can be determined. The main signaling molecule is primarily *N*-acyl homoserine lactones. Quorum sensing controls a number of features within a bacterial colony, such as their resistance, their motility and how they form biofilms, as well as their level of pathogenicity (McCarthy and O'Gara, 2015). Next-generation antimicrobials have the ability to decrease resistance and target the communication between bacteria known as quorum sensing/signaling. Ingested antibiotic adjuvants can increase the oral bioavailability of the antibiotic by inhibiting the quorum sensing between the pathogenic bacteria and making them less virulent and pathogenic, rendering them more susceptible to antibiotics. The gut microbiota mainly disrupts the communication between the pathogenic bacterial colonies and therefore prevents from taking place the initial stages of infection which involve integration within the host (Clemente et al., 2012).

23.3 GUT MICROBIOME AND EARLY LIFE

The colonization of the human gut starts at birth, and it is colonized during and after delivery (Bull and Plummer, 2014). Babies that are born naturally (vaginally) are exposed to the vaginal microbial population of the mother. Babies that are born through cesarean section have shown to have reduced numbers of microbiota during the first month when compared to babies that were born vaginally (Huurre et al., 2008). Therefore, it is likely that the initial colonization from birth has a significant effect on the gut microbiota in adult life. The gut microbiota of babies undergo several changes as they progress from breastfeeding, formula feeding and introduction to solid foods (Mackie et al., 1999; Bull et al., 2014).

Not only does the diversity of the gut microbiota depend on the initial colonization during birth, but it is dependent on external factors such as the environment the individual is in and the food of which their diet consists. Several internal factors have been listed by Mackie et al. (1999), which include the pH of the intestinal area, interactions with other microbes, the temperature of the environment, peristalsis, bile acid levels, the host's immune responses, drug therapy underwent and the bacterial mucosal receptors (Mackie et al., 1999). An infant's exposure to intestinal bacteria can determine the development of allergies, as it was seen that children with allergies have a different composition of bacteria present in the gut than those who do not develop any allergies. It is hypothesized that the intestinal microbiota can stimulate and train the immune responses toward antigens to tolerate some and overreact to others (Björkstén et al., 2001).

23.4 GUT MICROBIOME AND THE DEVELOPMENT OF CHRONIC AND INFECTIOUS DISEASES SUCH AS OBESITY, IBS AND CANCER

23.4.1 Chronic Liver Diseases

Chronic diseases in the liver include infectious liver diseases and metabolic liver diseases, as well as auto-immune diseases, which can ultimately lead to liver cancer and cirrhosis (Lu et al., 2019). It has been shown that chronic liver diseases are associated with imbalances in the gut microbiota. This is known as dysbiosis, which is the imbalance between the beneficial and pathogenic bacteria in favor of the pathogenic bacteria (Chen et al., 2011; Lu et al 2011; Qin et al., 2014; Michail et al., 2015; Tang et al., 2018; Lu et al., 2019). The current hypothesis is that as gut microbiota are stressed due to various diseases of the host which affects them, this eventually leads to dysbiosis, which then results in an acceleration of liver fibrosis and liver cirrhosis through the increase in inflammation. Data that has supported this theory is explained by alcoholic liver disease. Increased consumption of alcohol is directly intoxicating the liver through the dysbiosis in the form of small and large intestinal bacterial overgrowth or directly through microbial toxicity. The intestinal barriers also undergo direct local injury, which results in translocation of bacteria and the increase in inflammation (Acharya and Bajaj, 2021).

23.4.2 Carcinogenesis

Dysbiosis in the gut is normally associated with a low diversity of microbes—as well as the imbalance between beneficial and pathogenic bacteria, as previously discussed—and this has also been shown to promote tumor formation in the liver, which was found to be reversible with the administration of probiotics (Goedert et al., 2015). The presence of pathogenic bacteria is known to increase inflammation locally and cause damage to DNA and disorders of the immune system (Liu et al., 2019).

It has been proposed that various mechanisms are in play in the manner of how microbiota affects specifically colorectal carcinogenesis. It is the overriding and alteration of the gut microbiota composition that influences the occurrence and progression of colorectal carcinoma. More than colorectal carcinogenesis, dysbiosis of the gut microbiota have also been reported in experimental models to influence the occurrence and progression of extra intestinal cancers such as breast and hepatocellular carcinoma. It has been hypothesized that the mechanisms through which this occurs is through inflammation and metabolic systems (Zitvogel et al., 2015). It is proposed by Liu et al., (2019) that polysaccharides have the potential of promoting bacteria that have a positive outcome on the host's health and minimize cancer risks such as tumors through improvisation of a leaky gut and avoidance of hyper nutrition, and that enforce an immune balance (Liu et al., 2019).

23.4.3 Irritable Bowel Syndrome (IBS)

Irritable bowel syndrome (IBS) refers to a group of bowel disorders that are characterized by abdominal discomfort, pain when passing stools, changes in normal bowel habits and disordered stool passing (Bull and Plummer, 2015). Although the cause of IBS is possibly due to numerous factors such as genetics, infection and immunity, it is the variation in the microbiota of the gut that causes the low-grade intestinal inflammation found associated with the syndrome (Guinane and Cotter, 2013). When dysbiosis of the gut occurs, pathogenic microbes may attach themselves and cause the symptoms seen in IBS (Rinttilä et al., 2011). It was reported that the composition of the gut changes when IBS is present in that there is a twofold increase in the ratio of Firmicutes to Bacteroidetes (Ponnusamy et al., 2011)

23.4.4 Inflammatory Bowel Disease

Inflammatory bowel disease includes Crohn's disease and ulcerative colitis, with the former characterized by affected regions of inflammation in between sections of healthy tissues along the gastrointestinal tract. Including inflammation, this disease is characterized by ulcers found across the wall of the intestines that can result in perforations, which may affect other organs (Nagalingam and Lynch, 2012). Ulcerative colitis is a continuous inflammation that includes the surface layers of the intestinal wall, is mainly found in the colon and originates from the rectum (Baumgart and Sandborn, 2007). Figure 23.1 illustrates the active stage of ulcerative colitis (KGH, 2006). Once again, the abundance of the Firmicutes is greatly reduced in individuals who present with this disease (Bull and Plummer, 2014). Firmicutes are known for their role in SCFAs synthesis such as butyrate and acetate, which are known for their anti-inflammatory properties (Barcenilla et al., 2000).

23.4.5 Obesity and Type 2 Diabetes

In the case of obesity, Firmicutes were found to be more abundant than the Bacteroidetes within a mouse model investigated by Tujrnbaugh et al. (2006), with lean mice having more abundant levels of Bacteroidetes. Despite many other factors such as genetics and the environment, it was found that Type 2 diabetes is influenced by the composition of the microbiota within the intestines (Turnbaugh et al., 2006). It was found that among those individuals who are known to have Type 2 diabetes,

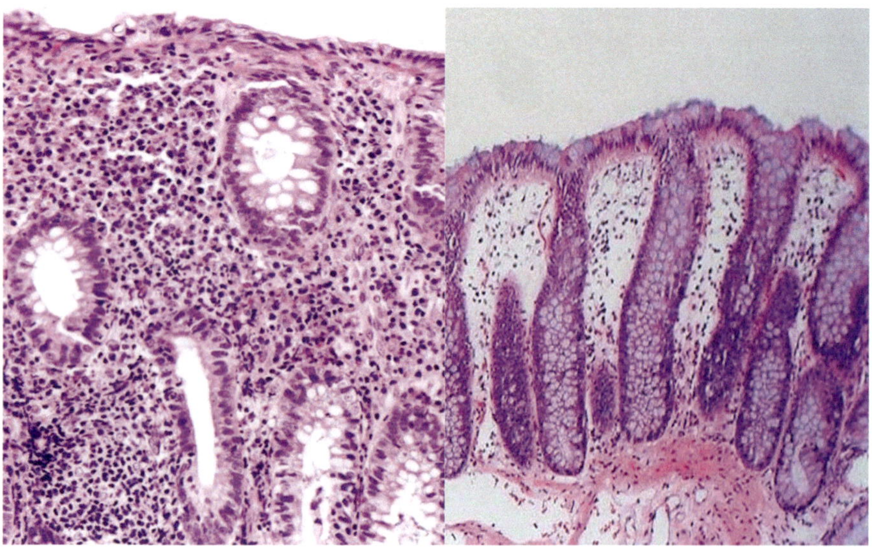

FIGURE 23.1 Active stage of ulcerative colitis (left) versus normal large intestinal tissue (right) (KGH, 2006; Uthman, 2008).

their gut microbiota only slightly differ from the control group, attributed to decline in the levels of butyrate-producing bacteria such as those from the Firmicutes (Qin et al., 2012).

The synthesis of SCFAs is important in regulating the metabolic processes of the host. Higher levels of SCFAs—specifically, the levels of butyrate and propionate—have been linked to obesity (Nicholson et al., 2012; Fernandes et al., 2014; Schwiertz et al., 2010; Payne et al., 2011). There is a tight link between the gut, genetics, environment and immunity described as having starvation and immunosuppression on one side of the scale, and obesity and inflammation on the other side (Wellen and Hotamisligil, 2005).

23.4.6 Allergic Disease Including Ectopic Eczema

There has been an unparalleled increase in the number of individuals infected with Type 1 hypersensitization, which includes atopic eczema, atopic asthma, rhinitis and Type 1 food allergies (Okada et al., 2010). Atopic sensitization is believed to occur in the first two years of life and persists a lifetime (Shen et al., 2013). It is believed that a mutation in a particular gene such as the CARD11 gene, which involves the function of the skin barrier as well as the method of birth (cesarean vs. vaginally), is the cause of atopic eczema. It was shown that infants found to be afflicted with the disease during the first month had very low diversity of gut microbiota present specifically with regard to the Bacteroidetes phylum compared with the infants without atopic eczema, as well as lower levels of Proteobacteria at infants 12 months of age with atopic eczema (Abrahamsson et al., 2012). Proteobacteria cell walls contain lipopolysaccharides, which are known to have the ability to elicit a host response and as seen those with low levels of lipopolysaccharides, i.e. low levels of Proteobacteria in infants, are at a higher risk of developing atopic eczema (Gehring et al., 2001).

23.5 GUT MICROBIOME AND PROPOSED TREATMENTS

There are currently several options available for the treatment of gut microbiota imbalances. As in the case of IBS, the current treatments have not shown any promise, and much more research is required. The current treatment options are described in what follows.

23.5.1 Probiotics

Probiotics have been used for more than 100 years and mainly consist of combinations of *Lactobacillus* and *Bifidobacterium* genera (Bull and Plummer, 2015). In 1907, it was the consumption of these fermented products that contain *Lactobacillus bulgaricus* where the first observations were made with regards to gut microbiota and good health (Bull and Plummer, 2015). According to the World Health Organization (WHO), probiotics are "live organisms that confer a beneficial effect on the health of the host when administered in adequate amounts" (Bull and Plummer, 2015). The expansion of this includes that probiotics must be able to survive traveling through the gut and tolerate exposure to gastric juices and bile. Probiotics must be safe, effective and able to maintain effectiveness and activity for the whole duration of their shelf life (Bull and Plummer, 2015).

Probiotics are currently available in a myriad of ways such as the inclusion in dairy products (milk and cheese), cereals and freeze-dried powders (Bull and Plummer, 2015). There are a few current publications available on the effects of probiotics as well as their ability to improve the intestinal barrier, stimulate the immune system, produce antibacterial effects against other bacteria, cause changes in intestinal motility and reduce visceral pain. Bull and Plummer (2015) summarized all the mechanisms of action of probiotics as follows:

- Intestinal wall integrity and barrier function.
- Modulation of the immune system.
- Host-derived antimicrobial peptide synthesis.

- Epithelial adherence of the bacteria.
- Blocking of communication between pathogens.
- Probiotic antimicrobial factor synthesis.

The major mechanisms of action of probiotics have been summarized in Figure 23.2 (Shoemaker, 2016). Intestinal wall integrity is maintained by the prevention of death of the epithelial cells and by increasing the production of mucin by the host (Mack et al., 2003). The epithelial cells are important in that they prevent the passing of pathogens from the gastrointestinal tract to other parts of the body. The production of certain proteins by the probiotic bacteria means that the bacteria and pathogens are competing for binding sites on the epithelial cells along the wall of the gut (Johnson-Henry et al., 2007). Probiotics have the ability to produce antimicrobial compounds, stimulate the host to produce antimicrobial compounds and lower the intestinal pH levels through the production of SCFAs from the epithelial cells to prevent microbial pathogen growth (Penner et al., 2005; Lebeer et al., 2010; Tejero-Sariñena et al., 2012).

The ability of probiotic bacteria to prevent pathogenic cell-to-cell signaling in the hosts as previously described prevents the establishment of other pathogenic bacteria, especially in the early stages of infection, due to the fact that they need the communication for the establishment and

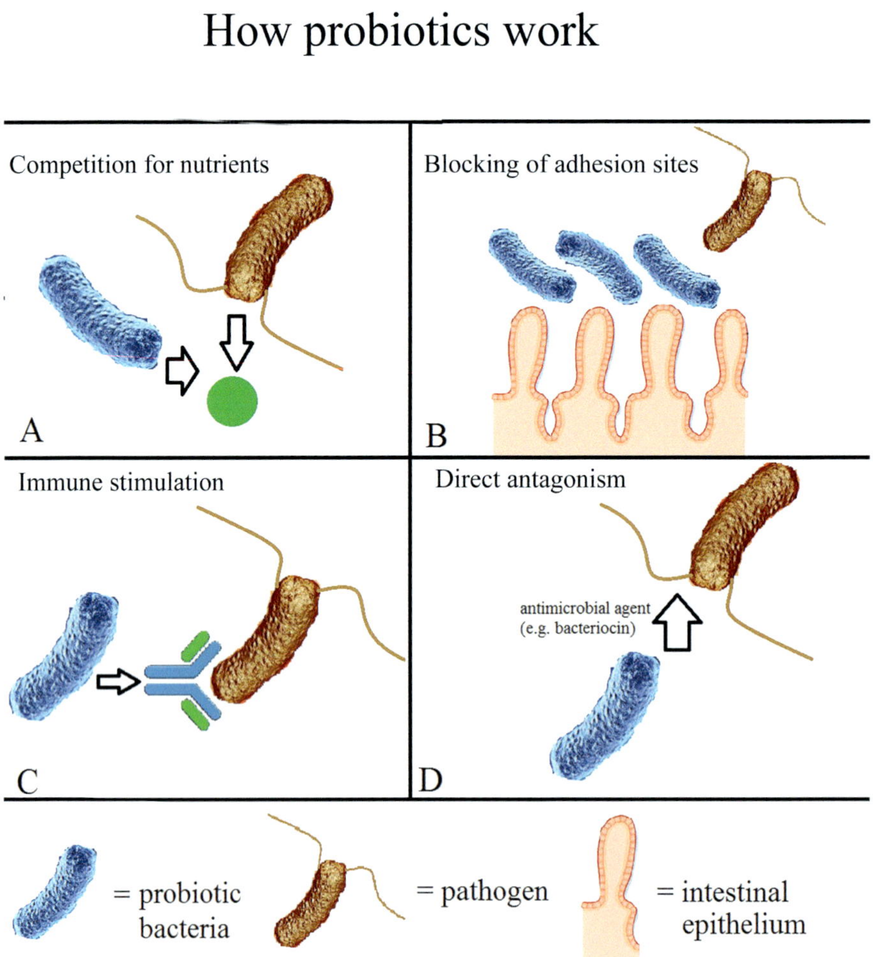

FIGURE 23.2 Major mechanisms of action of probiotics (Shoemaker. 2016).

the maintenance of their infection in the host. The specific factors that may be affected through inhibition of cell-to-cell communication is the decrease in motility, inhibition of the formation of a biofilm and reduction of virulence-specific gene expression (Kendall and Sperandio, 2007). Probiotic bacteria can stimulate the host's immune system to produce anti-inflammatory cytokines and reduce the production of any pro-inflammatory cytokines (Drakes et al., 2004; Kim et al., 2006).

23.5.1.1 Antibiotics and Probiotic Treatment

In many instances when antibiotics are taken by individuals, the gastrointestinal microbiota gets disturbed, which leads to diarrhea. This is a common occurrence when taking aminopenicillins, cephalosporins and clindamycins (Bull and Plummer, 2015). Antibiotic-associated diarrhea is normally associated with a reduction in the integrity of the intestinal wall and mineral and vitamin metabolism (Johnston et al., 2012). It is the dysbiosis of the intestinal microbiota and the subsequent overgrowth of the bacterial pathogens that makes this a common occurrence when taking certain antibiotics. The role of probiotics in the prevention of antibiotic-associated diarrhea is that probiotics have the ability to assist disrupted microbiota, enhancing a response from the host immune system to clear the pathogens from the host (McFarland, 2006).

23.5.1.2 Inflammatory Bowel Disease (IBS) and Probiotic Treatment

IBS is a disease state normally associated with individuals with a certain composition of microbiota compared to healthy individuals. Many trials that have been conducted to prove that the provision of probiotics to IBS patients for the reduction of symptoms show it is a symptomatic improvement that takes place with no further adverse effects (Moayyedi et al., 2010).

23.5.1.3 IBD and Probiotic Treatment

IBD, which includes Crohn's disease and ulcerative colitis, is based on several genetic factors—but not all with these predisposed genetic-based factors will develop the disease, and it is believed that the genes responsible for this disease may be assisted with a certain gut microbiota composition for full development (Bull and Plummer, 2015; Akiho et al., 2015; Lu et al., 2019). Ulcerative colitis is mainly characterized by inflammation and ulcers in the regions of the colon and rectum (Lu et al., 2019). In an animal model of IBD, it was shown that probiotics, especially that of *Bifidobacteria*, can influence the immune response of the host and cause the suppression of mucosal inflammation (O'Mahony et al., 2005). The improvement of IBD in animal models is, however, not directly associated with the same results in humans, and this may be due to the intricacies of the immune system functioning in humans (Sheil et al., 2007). This has led to an overall conclusion that probiotics are not recommended for the treatment of IBD (Prantera and Scribano, 2006).

Necrotizing enterocolitis is a form of IBD found in infancy and causes necrosis of sections of the bowel (Illi et al., 2006). This is a common disorder in infants that present a very low birth weight (Updegrove, 2004). Probiotics can be given to infants for the prevention of this disorder through the promotion of beneficial bacteria to colonize the gut and prevention of any other pathogenic bacteria from colonizing. It improves the mucosal barrier of the gut and causes an enhanced immune system response from the host (Deshpande et al., 2007).

23.5.1.4 Atopic Eczema and Probiotic Treatment

Atopic eczema forms part of the allergic disorders associated with the atopic march (Okada et al., 2010). This is an allergic disorder associated with the composition of the gut microbiota and birthing factors such as birth mode, antibiotic use during early childhood (newborn and infant) and diet (breastfeeding or formula feeding) (Abrahamsson et al., 2012; Williams and Grindlay, 2010). Due to all of this, it has been proposed to feed infants probiotics that can stimulate the immune system early on and create the necessary immune response against allergens (Bull and Plummer, 2015).

23.5.1.5 Obesity, Type 2 Diabetes and Probiotic Treatment

It has been shown by recent studies in humans that there are *Lactobacillus* and *Bifidobacterium* strains that show positive effects on the molecular markers of both obesity and Type 2 diabetes (Aronsson et al., 2010; Lee et al., 2006, Andreasen et al., 2010; Bull and Plummer, 2015).

23.5.2 Prebiotics

The term prebiotics was devised by Roberfroid to describe “a selectively fermented ingredient that allows specific changes, both in the composition of and/or activity in the gastrointestinal microflora that confers benefits upon a host’s well-being and health” (Roberfroid, 2007, pp. 830S–837S). Prebiotics need to be able to travel through the major regions of the gastrointestinal tract so that they may be close enough to their target cells (bacteria) and may not be metabolized by the host cells, but only by gut microbiota that are important for gut health. They must be resistant to gastric acidity, hydrolysis and absorption in the gastrointestinal tract (Roberfroid, 2007).

23.5.2.1 Inulin as a Prebiotic

Inulin is a plant-based, heterogeneous collection of fructose polymers and utilized as storage carbohydrates in both plants and vegetables (Macfarlane, 2010; Niness, 1999; Bull and Plummer, 2015). Inulin is resistant to low levels of pH, hydrolysis by mammalian enzymes and absorption as required. It has been shown to promote the growth of both *Lactobacilli* and *Bifidobacteria* (Roberfroid et al., 1998; Cherbut, 2002; Tako et al., 2008).

23.5.2.2 Obesity, Type 2 Diabetes and Prebiotic Treatment

In both these diseases, the proportion of bacteria that belong to the Firmicutes phylum are found in smaller numbers as compared to lean individuals and those who do not present with Type 2 diabetes; therefore, the levels of *Bifidobacteria* belonging to the Firmicutes phylum are of concern (Wu et al., 2010). In a study involving obese and diabetic mice, when the mice were given inulin, in many mice the population of *Bifidobacteria* increased as compared to mice fed on a standard diet without the addition of a prebiotic (Cani et al., 2007; Cani et al., 2009; Bull and Plummer, 2015). Inulin is said to be functional on a genetic level in that it reduces the expression of several genes associated with adiposity and inflammation (Vyas and Ranganathan, 2012).

23.5.3 Symbiotics

Administering both a probiotic and a prebiotic that work together creates what is known as a symbiotic interaction. This is the increase of beneficial bacteria to increase the levels of anaerobes therefore the production of SCFAs. The function of the prebiotic is to ensure and increase the survival of the probiotics in the mixture by stimulating the activity of the bacteria found in the gut of the host (Rastall and Maitin, 2002).

23.5.4 Fecal Microbial Transplantation

Fecal microbial transplantation is a novel mode of treatment used in particular when dysbiosis has occurred and to re-establish and introduce a community of gastrointestinal bacteria from healthy donors through a transcolonoscopic infusion to an unhealthy recipient (Bull and Plummer, 2015).

23.6 GUT MICROBIOME AND PLANT-BASED MODULATORS THEREOF

23.6.1 *Coptis chinensis* Franch

Coptis chinensis is traditionally used in traditional Chinese Mmedicine and contains berberine, an isoquinoline alkaloid, which has shown an effect with non-alcoholic fatty liver disease (NAFLD), through the increase in the abundance of bacteria associated with metabolic improvements such as those from Bacteroidetes, *Bifidobacterium* and *Lactobacillus* (Cao et al., 2016).

23.6.2 Qushi Huayu Fang

Another Chinese herbal formula, Qushi Huayu Fang, modulates the gut microbiome and alleviates hepatic steatosis and overall decreases the amount of fatty acid in the liver (Yin et al., 2013; Lu et al., 2019). Yin et al. (2013) found structural changes in the gut microbiota in a rat NAFLD model treated with Qushi Huayu Fang. The major effects and hypothesized mechanism of action of this polyherbal was a reduction in opportunistic bacteria such as *Escherichia/Shigella* spp., but also the increase in beneficial species such as *Collinsella.*

23.6.3 Yinchenhao

A decoction known as Yinchenhao, consisting mainly of *Artemisia annua* L., *Gardenia jasminoides* Ellis and *Rheum palmatum* L., is suggested to prevent liver injury, apoptosis of the liver cells, activation of hepatic stellate cells, synthesis of collagen and the promotion of bilirubin metabolism (Sakaida et al., 2003; Yamamoto et al., 2000, Yamamoto et al., 1996, Yamshiki et al., 2000, Huang et al., 2004; Lu et al., 2019). All of these mainly contribute to the Yinchenhao decoction in treating jaundice. Specifically, with regard to the gut microbiome, the study conducted by Liu et al. (2019) summarized that the protective effects seen in the liver injury was specifically related to the production of 3-hydroxybutyric acid through the changes in the availability of Clostridia and Clostridiales.

23.6.4 *Silybum marianum* L. (Gaertn.)

Silymarin is a known hepatoprotective agent and is extracted from the seeds of *Silybum marianum* (Figure 23.3A). Silychristin (Figure 23.3B) is found within silymarin and is associated with both inflammation and immunomodulation. This compound has shown to exhibit hypolipidemic, antioxidant and hepatoprotective activity (Škottová and Krečman, 1998; Krečman et al., 1998; Katiyar, 2005; Lee et al., 2007; Lu et al., 2019). In a study conducted by Shen et al. (2019), it was found that silibinin and silymarin have the ability to improve memory impairment due to Alzheimer's disease through their regulatory effect on the composition of the gut microbiota of APP/PS1 mice (Shen et al., 2019).

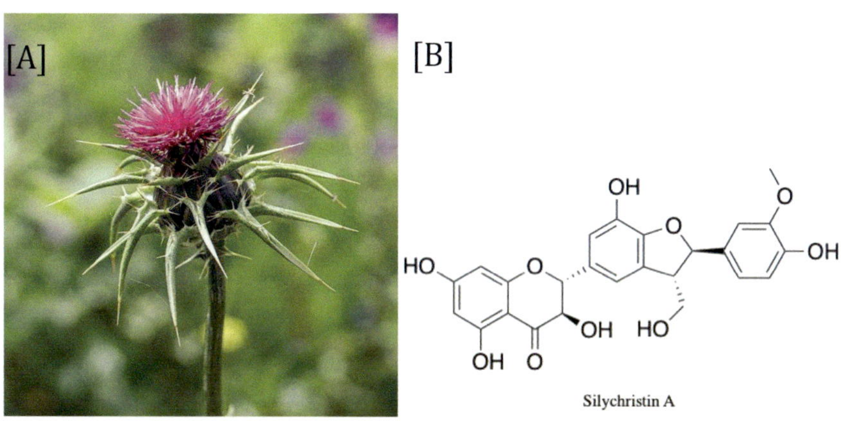

FIGURE 23.3 A, The flower of *Silybum marianum*; B, Silychristin A, which is isolated from the seeds produced by *S. marianum* (Jsawires42, 2020; Zell, 2016).

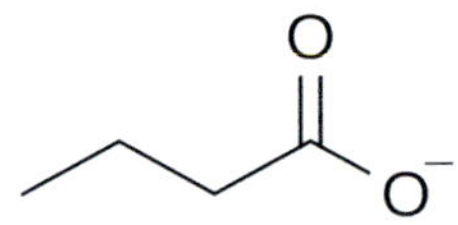

FIGURE 23.4 Butyrate isolated from Bacteroidetes (Aloneinthewild [talk], 2014).

23.6.5 Polysaccharides

Among all the metabolites formed from polysaccharides, the most abundant metabolites formed are SCFAs that include propionate butyrate and acetate (Liu et al., 2019). Butyrate (Figure 23.4) is produced mainly by *Bacteroidetes*, as well as Cluster IV and XIVa, which belong to *Clostridia* (Liu et al., 2019). Many of the other bacterial groups are responsible for the production of acetate in the gut (Ríos-Covian et al., 2016).

Oligosaccharides and polysaccharides increase the phyla of Actinobacteria and Firmicutes, except for the *Staphylococcus* and *Clostridum* species, as they increase SCFAs glucagon-like peptide-1 (GLP-1) and PYY (a peptide hormone), and decrease the triglycerides, very low-density lipoprotein (VLDL) and low-density lipoprotein (LDL), with a mixed trend observed for high-density lipoprotein (HDL) (Morrison and Preston, 2016, Eid et al., 2017). In summary, polysaccharides are therefore responsible for the total amount of SCFAs that are produced, and decreasing the number of polysaccharides will therefore have a profound effect on the amount of SCFAs that are produced (Duncan et al., 2007; Liu et al., 2019). Many anti-tumor properties have been proposed to exist through the consumption of polysaccharides and are summarized in what follows.

23.6.5.1 Protection of the Intestinal Barrier

The main function of the intestinal barrier is to regulate the movement of macromolecules between the host and the environment, and to maintain an equilibrium toward non–self-antigens. As described in the previous sections, a biological layer of bacteria is required to prevent leakage to the liver and prevent the adherence of pathogenic bacteria and production of antimicrobial compounds (Liu et al., 2019). Polysaccharides have the ability to promote the composition of bacteria required to bring forth all these health-promoting functions (Cameron and Sperandio, 2015).

23.6.5.2 Prevention of Hyper Nutrition

Hyper-nutrition is the increase in a body mass index due to an overload of nutrition and is associated with a cancer mortality rate that is much higher (Calle et al., 2003). The mechanism of action of the tumorigenesis through hyper-nutrition takes place through inflammation, accumulation of fat and resistance to insulin with hyperglycemia (Font-Burgada et al., 2016). Polysaccharides have the ability to promote the loss of weight and associate body-mass index and improvement of the dysbiosis that has occurred (Nguyen et al., 2016; Goffredo et al., 2016; Liu et al., 2019). The metabolites formed by polysaccharides, such as acetate and propionate, can be absorbed into the blood and contribute to weight management (Macfarlane and Macfarlane, 2011). Acetate seems to suppress an individual's appetite and can increase the release of leptin from adipose tissues in mice (Xiong et al., 2004). Propionate lowers the synthesis of cholesterol, all indications of the positive outcome of a polysaccharide-enriched diet, and in this case, the potential lowering of a nutrition overload by lowering intake of food (Liu et al., 2019).

23.6.5.3 Immune Balance Reinforcement

Polysaccharides indirectly promote the homeostasis of the immune system by promoting the production of metabolites such as butyrate, which many studies have been shown to induce T-regulation (Treg) cell differentiation and assist in the metabolism of Vitamin B9 (Kunisawa et al., 2012; Yoo and Mazmanian, 2017; Liu et al., 2019). The anti-inflammatory activity found within the gut is regulated by an increase in the levels of SCFAs, which include butyrate, acetate and propionate (Inan et al., 2000; Rodríguez-Cabezas, 2002; Liu et al., 2019).

23.6.6 Resveratrol

Resveratrol, which can be found in fruits such as grapes (*Vitis vinivera* L.), has been indicated in more than one clinical study to partly improve the inflammatory condition found in humans that present with ulcerative colitis (Nunes et al., 2018). Resveratrol has shown to improve glucose homeostasis in obese mice (Sung et al., 2017).

23.6.7 Caffeic Acid

Caffeic acid, which forms part of many human diets and also found in the Southern blue gum (*Eucalyptus globulus* Labill.), has been shown to improve inflammation in a model investigating colitis in mice (Zhang et al., 2016; Lu et al., 2019).

23.6.8 Polyphenols

Polyphenols are mostly unabsorbed by the intestinal tract and are degraded into smaller phenolic compounds and monomers by enzymes in the colon to become physiologically active (Stevens and Maier, 2016; Kumar and Pandey, 2013; Lu et al., 2019). In terms of Type 2 diabetes, several studies have shown that polyphenols have the ability to inhibit activities of the enzymes that digest the carbohydrates and lipids *in vitro* (Nie et al., 2019; Lu et al., 2019). They can modulate transport proteins and, in this way, inhibit the absorption of glucose; they possess antibacterial activity and inhibit the growth of pathogenic bacteria (Lu et al., 2019).

Previous studies have confirmed that dietary components that are rich in polyphenols and plant polysaccharides such as cocoa beverages and green tea potentially have both anti-diabetic and anti-obesity activity (Eid et al., 2017; Lu et al., 2019). It was shown that polysaccharides derived from *Plantago asiatica* L. have the ability to reduce blood glucose levels, insulin, total cholesterol, triglycerides, non-esterified fatty acids and malondialdehyde levels in Type 2 diabetic rats. It was seen that *P. asiatica* can increase the levels of specifically HDL cholesterol and antioxidant activity of enzymes (Nie et al., 2019; Lu et al., 2019).

23.7 GUT MICROBIOME AND FUTURE PROSPECTIVE

The gut microbiota have been shown to be closely related to an individual's health and disease. It is said that the gut microbiota has a higher metabolic rate than the liver with a symbiotic relationship between the host and the gut microbiota (Possemiers et al., 2011; Bibbò et al., 2017; An et al., 2019). When herbal mixtures enter the body, it is normally through oral ingestion where it can then interact with the gut microbiota. It is normally after the ingestion of these herbal mixtures that the gut microbiota composition changes (Creely et al., 2007; An et al., 2019). Both positive and negative interactions between the microbiota and herbal mixtures are possible, and all depends on the final composition as an increase in pathogenic bacteria can leave the host in a diseased state with increased levels of inflammation, whereas positive interaction can have a curative effect, especially if dysbiosis was present and is now balanced.

More research is required to see what effects plant-based treatment has on the effect of gut microbiomes both *in vitro* and ultimately the same effect *in vivo*. As showcased in this chapter, many medicinal plants have shown their potential as treatments for some of the main diseases as a result of gut dysbiosis. Many biodiversity hotspots are present worldwide, and it is in these areas where many different plant families and their biologically active constituents can be found and identified as treatments.

REFERENCES

Abrahamsson, Thomas R., Hedvig E. Jakobsson, Anders F. Andersson, Bengt Björkstén, Lars Engstrand, and Maria C. Jenmalm. 2012 "Low diversity of the gut microbiota in infants with atopic eczema." *Journal of Allergy and Clinical Immunology* 129, 2: 434–440.

Acharya, C., and J.S. Bajaj. 2021 "Chronic liver diseases and the microbiome-Translating our knowledge of gut microbiota to management of chronic liver diseases." *Gastroenterology* 160, 2: 556–572.

Akiho, Hirotada, Azusa Yokoyama, Shuichi Abe, Yuichi Nakazono, Masatoshi Murakami, Yoshihiro Otsuka, Kyoko Fukawa, Mitsuru Esaki, Yusuke Niina, and Haruei Ogino. 2015 "Promising biological therapies for ulcerative colitis: A review of the literature." *World Journal of Gastrointestinal Pathophysiology* 6, 4: 219.

Aloneinthewild (talk). 2014 "Butyrate.svg." Obtained from Wikimedia under the Public domain license. https://commons.wikimedia.org/wiki/File:Butyrate.svg (Accessed 13 May 2021).

An, Xuedong, Qi Bao, Sha Di, Yiru Zhao, Shenghui Zhao, Haiyu Zhang, Fengmei Lian, and Xiaolin Tong. 2019 "The interaction between the gut Microbiota and herbal medicines." *Biomedicine & Pharmacotherapy* 118: 109252. doi: 10.1016/j.biopha.2019.109252

Andreasen, Anne Sofie, Nadja Larsen, Theis Pedersen-Skovsgaard, Ronan M.G. Berg, Kirsten Møller, Kira Dynnes Svendsen, Mogens Jakobsen, and Bente Klarlund Pedersen. 2010 "Effects of Lactobacillus acidophilus NCFM on insulin sensitivity and the systemic inflammatory response in human subjects." *British Journal of Nutrition* 104, 12: 1831–1838.

Aronsson, Linda, Ying Huang, Paolo Parini, Marion Korach-André, Janet Håkansson, Jan-Åke Gustafsson, Sven Pettersson, Velmurugesan Arulampalam, and Joseph Rafter. 2010 "Decreased fat storage by Lactobacillus paracasei is associated with increased levels of angiopoietin-like 4 protein (ANGPTL4)." *PloS One* 5, 9: e13087. doi:10.1371/journal.pone.0013087

Bäckhed, Fredrik, Ruth E. Ley, Justin L. Sonnenburg, Daniel A. Peterson, and Jeffrey I. Gordon. 2005 "Host-bacterial mutualism in the human intestine." *Science* 307, 5717: 1915–1920.

Barcenilla, Adela, Susan E. Pryde, Jennifer C. Martin, Sylvia H. Duncan, Colin S. Stewart, Colin Henderson, and Harry J. Flint. 2000 "Phylogenetic relationships of butyrate-producing bacteria from the human gut." *Applied and Environmental Microbiology* 66, 4: 1654–1661.

Baumgart, Daniel C., and William J. Sandborn. 2007 "Inflammatory bowel disease: Clinical aspects and established and evolving therapies." *The Lancet* 369, 9573: 1641–1657.

Bibbò, Stefano, Maria Pina Dore, Giovanni Mario Pes, Giuseppe Delitala, and Alessandro P. Delitala. 2017 "Is there a role for gut microbiota in type 1 diabetes pathogenesis?" *Annals of medicine* 49, 1: 11–22.

Björkstén, Bengt, Epp Sepp, Kaja Julge, Tiia Voor, and Marika Mikelsaar. 2001 "Allergy development and the intestinal microflora during the first year of life." *Journal of Allergy and Clinical Immunology* 108, 4: 516–520.

Boulangé, Claire L., Ana Luisa Neves, Julien Chilloux, Jeremy K. Nicholson, and Marc-Emmanuel Dumas. 2016 "Impact of the gut microbiota on inflammation, obesity, and metabolic disease." *Genome Medicine* 8, 1: 1–12.

Boursier, Jérôme, Olaf Mueller, Matthieu Barret, Mariana Machado, Lionel Fizanne, Felix Araujo-Perez, Cynthia D. Guy et al. 2016 "The severity of nonalcoholic fatty liver disease is associated with gut dysbiosis and shift in the metabolic function of the gut microbiota." *Hepatology* 63, 3: 764–775.

Bull, Matthew J., and Nigel T. Plummer. 2014 "Part 1: The human gut microbiome in health and disease." *Integrative Medicine: A Clinician's Journal* 13, 6: 17.

Bull, Matthew J., and Nigel T. Plummer. 2015 "Part 2: Treatments for chronic gastrointestinal disease and gut dysbiosis." *Integrative Medicine: A Clinician's Journal* 14, 1: 25.

Calle, Eugenia E., Carmen Rodriguez, Kimberly Walker-Thurmond, and Michael J. Thun. 2003 "Overweight, obesity, and mortality from cancer in a prospectively studied cohort of US adults." *New England Journal of Medicine* 348, 17: 1625–1638.

Cameron, Elizabeth A., and Vanessa Sperandio. 2015 "Frenemies: Signaling and nutritional integration in pathogen-microbiota-host interactions." *Cell Host & Microbe* 18, 3: 275–284.

Cani, Patrice D., Audrey M. Neyrinck, F. Fava, Claude Knauf, Rémy G. Burcelin, Kieran M. Tuohy, G. R. Gibson, and Nathalie M. Delzenne. 2007 "Selective increases of bifidobacteria in gut microflora improve high-fat-diet-induced diabetes in mice through a mechanism associated with endotoxaemia." *Diabetologia* 50, 11: 2374–2383.

Cani, Patrice D., Sam Possemiers, Tom Van de Wiele, Yves Guiot, Amandine Everard, Olivier Rottier, Lucie Geurts et al. 2009 "Changes in gut microbiota control inflammation in obese mice through a mechanism involving GLP-2-driven improvement of gut permeability." *Gut* 58, 8: 1091–1103.

Cao, Yi, Qin Pan, Feng Shen, Guang-Yu Chen, Lei-Ming Xu, and Jian-Gao Fan. 2016 "Modulation of gut microbiota by berberine improves steatohepatitis in high-fat diet-fed BALB/C mice." *Archives of Iranian Medicine* 19, 3: 0–0.

Chen, Yanfei, Fengling Yang, Haifeng Lu, Baohong Wang, Yunbo Chen, Dajiang Lei, Yuezhu Wang, Baoli Zhu, and Lanjuan Li. 2011 "Characterization of fecal microbial communities in patients with liver cirrhosis." *Hepatology* 54, 2: 562–572.

Cherbut, Christine. 2002 "Inulin and oligofructose in the dietary fibre concept." *British Journal of Nutrition* 87, S2: S159–S162. doi: 10.1079/BJN2002532

Chu, Derrick M., Jun Ma, Amanda L. Prince, Kathleen M. Antony, Maxim D. Seferovic, and Kjersti M. Aagaard. 2017 "Maturation of the infant microbiome community structure and function across multiple body sites and in relation to mode of delivery." *Nature Medicine* 23, 3: 314–326.

Clemente, Jose, Luke Ursell, Laura Parfrey, and Rob Knight. 2012 "The impact of the gut microbiota on human health: An integrative view." *Cell* 148, 6: 1258–1270.

Collins, Stephen M., Michael Surette, and Premysl Bercik. 2012 "The interplay between the intestinal microbiota and the brain." *Nature Reviews Microbiology* 10, 11: 735–742.

Creely, Steven James, Philip Gerard McTernan, Christine Maria Kusminski, ff M. Fisher, N.F. Da Silva, M. Khanolkar, M. Evans, A.L. Harte, and S. Kumar. 2007 "Lipopolysaccharide activates an innate immune system response in human adipose tissue in obesity and type 2 diabetes." *American Journal of Physiology-Endocrinology and Metabolism* 292, 3: E740–E747. doi: 10.1152/ajpendo.00302.2006.

Cummings, J.H., E.W. Pomare, W.J. Branch, C.P. Naylor, and G.T. Macfarlane. 1987 "Short chain fatty acids in human large intestine, portal, hepatic and venous blood." *Gut* 28, 10: 1221–1227.

Desbonnet, Lieve, Lillian Garrett, Gerard Clarke, John Bienenstock, and Timothy G. Dinan. 2008 "The probiotic Bifidobacteria infantis: An assessment of potential antidepressant properties in the rat." *Journal of Psychiatric Research* 43, 2: 164–174.

Deshpande, Girish, Shripada Rao, and Sanjay Patole. 2007 "Probiotics for prevention of necrotising enterocolitis in preterm neonates with very low birthweight: A systematic review of randomised controlled trials." *The Lancet* 369, 9573: 1614–1620.

de Vos, Willem M., and Elisabeth AJ de Vos. 2012 "Role of the intestinal microbiome in health and disease: From correlation to causation." *Nutrition reviews* 70, Suppl_1: S45–S56. doi: 10.1111/j.1753-4887.2012.00505.x

Dflock. 2004 "Intestine." Obtained from Wikimedia under the Creative commons Attribution-Share Alike 4.0. https://commons.wikimedia.org/wiki/File:Intestine.png (Accessed 13 May 2021)

Drakes, Maureen, Thomas Blanchard, and Steven Czinn. 2004 "Bacterial probiotic modulation of dendritic cells." *Infection and Immunity* 72, 6: 3299–3309.

Duncan, Sylvia H., Alvaro Belenguer, Grietje Holtrop, Alexandra M. Johnstone, Harry J. Flint, and Gerald E. Lobley. 2007 "Reduced dietary intake of carbohydrates by obese subjects results in decreased concentrations of butyrate and butyrate-producing bacteria in feces." *Applied and Environmental Microbiology* 73, 4: 1073–1078.

Eckburg, Paul B., Elisabeth M. Bik, Charles N. Bernstein, Elizabeth Purdom, Les Dethlefsen, Michael Sargent, Steven R. Gill, Karen E. Nelson, and David A. Relman. 2005 "Diversity of the human intestinal microbial flora." *Science* 308, 5728: 1635–1638.

Eid, Hoda M., Michelle L. Wright, N.V. Anil Kumar, Abdel Qawasmeh, Sherif T.S. Hassan, Andrei Mocan, Seyed M. Nabavi, Luca Rastrelli, Atanas G. Atanasov, and Pierre S. Haddad. 2017 "Significance of microbiota in obesity and metabolic diseases and the modulatory potential by medicinal plant and food ingredients." *Frontiers in Pharmacology* 8: 387.

Eutamene, H., and L. Bueno. 2007. "Role of probiotics in correcting abnormalities of colonic flora induced by stress." *Gut* 56, 11: 1495–1497. doi: 10.1136/gut.2007.124040

Fernandes, J., W. Su, S. Rahat-Rozenbloom, T. M. S. Wolever, and E. M. Comelli. 2014 "Adiposity, gut microbiota and faecal short chain fatty acids are linked in adult humans." *Nutrition & Diabetes* 4, 6: e121–e121. doi: 10.1038/nutd.2014.23.

Font-Burgada, Joan, Beicheng Sun, and Michael Karin. 2016 "Obesity and cancer: The oil that feeds the flame." *Cell Metabolism* 23, 1: 48–62.

Frank, Daniel N., Allison L. St Amand, Robert A. Feldman, Edgar C. Boedeker, Noam Harpaz, and Norman R. Pace. 2007 "Molecular-phylogenetic characterization of microbial community imbalances in human inflammatory bowel diseases." *Proceedings of the National Academy of Sciences* 104, 34: 13780–13785.

Gehring, Ulrike, Gabriele Bolte, Michael Borte, Wolfgang Bischof, Bärbel Fahlbusch, H-Erich Wichmann, Joachim Heinrich, and LISA Study Group. 2001 "Exposure to endotoxin decreases the risk of atopic eczema in infancy: A cohort study." *Journal of Allergy and Clinical Immunology* 108, 5: 847–854.

Goedert, James J., Gieira Jones, Xing Hua, Xia Xu, Guoqin Yu, Roberto Flores, Roni T. Falk et al. 2015 "Investigation of the association between the fecal microbiota and breast cancer in postmenopausal women: A population-based case-control pilot study." *JNCI: Journal of the National Cancer Institute* 107, 8.

Goffredo, Martina, Kendra Mass, Elizabeth J. Parks, David A. Wagner, Emily Ann McClure, Joerg Graf, Mary Savoye, Bridget Pierpont, Gary Cline, and Nicola Santoro. 2016 "Role of gut microbiota and short chain fatty acids in modulating energy harvest and fat partitioning in youth." *The Journal of Clinical Endocrinology & Metabolism* 101, 11: 4367–4376.

Guarner, F., F. Casellas, N. Borruel, M. Antolìn, S. Videla, J. Vilaseca, and R. Malagelada Jr. 2003 "Gut flora in health and disease." *Lancet* 361: 512–519. doi: 10.1016/S0140-6736(03)12489-0.

Guinane, Caitriona M., and Paul D. Cotter. 2013 "Role of the gut microbiota in health and chronic gastrointestinal disease: Understanding a hidden metabolic organ." *Therapeutic Advances in Gastroenterology* 6, 4: 295–308.

Habtemariam, Solomon. 2020 "Berberine pharmacology and the gut Microbiota: A hidden therapeutic link." *Pharmacological Research*: 104722.

Huang, Wendong, Jun Zhang, and David D. Moore. 2004 "A traditional herbal medicine enhances bilirubin clearance by activating the nuclear receptor CAR." *The Journal of clinical investigation* 113, 1: 137–143.

Huurre, Anu, Marko Kalliomäki, Samuli Rautava, Minna Rinne, Seppo Salminen, and Erika Isolauri. 2008 "Mode of delivery—effects on gut microbiota and humoral immunity." *Neonatology* 93, 4: 236–240.

Illi, Sabina, Erika von Mutius, Susanne Lau, Bodo Niggemann, Christoph Grüber, and Ulrich Wahn. 2006 "Perennial allergen sensitisation early in life and chronic asthma in children: A birth cohort study." *The Lancet* 368, 9537: 763–770.

Inan, Mehmet Sait, Reza J. Rasoulpour, Lei Yin, Andrea K. Hubbard, Daniel W. Rosenberg, and Charles Giardina. 2000 "The luminal short-chain fatty acid butyrate modulates NF-κB activity in a human colonic epithelial cell line." *Gastroenterology* 118, 4: 724–734.

Johnson-Henry, Kathene C., Karen E. Hagen, Mahsa Gordonpour, Thomas A. Tompkins, and Philip M. Sherman. 2007 "Surface-layer protein extracts from Lactobacillus helveticus inhibit enterohaemorrhagic Escherichia coli O157: H7 adhesion to epithelial cells." *Cellular Microbiology* 9, 2: 356–367.

Johnston, Bradley C., Stephanie S.Y. Ma, Joshua Z. Goldenberg, Kristian Thorlund, Per O. Vandvik, Mark Loeb, and Gordon H. Guyatt. 2012 "Probiotics for the prevention of Clostridium difficile—associated diarrhea: A systematic review and meta-analysis." *Annals of Internal Medicine* 157, 12: 878–888.

Jsawires42. 2020 "Silychristin A 1996.gif." Obtained from Wikimedia under the Creative commons Attribution-Share Alike 4.0. https://commons.wikimedia.org/wiki/File:Silychristin_A_1996.gif (Accessed 13 May 2021).

Katiyar, Santosh K. 2005 "Silymarin and skin cancer prevention: Anti-inflammatory, antioxidant and immunomodulatory effects." *International Journal of Oncology* 26, 1: 169–176.

Kendall, Melissa M., and Vanessa Sperandio. 2007 "Quorum sensing by enteric pathogens." *Current Opinion in Gastroenterology* 23, 1: 10–15.

KGH. 2006 "Ulcerative colitis." Obtained from Wikimedia under the Creative Commons Attribution-Share Alike 3.0 Unported license. https://commons.wikimedia.org/wiki/File:Ulcerative_colitis_(1)_active.jpg (Accessed 13 May 2021)

Kim, Sung O., Haroon I. Sheikh, Soon-Duck Ha, Andrew Martins, and Gregor Reid. 2006 "G-CSF-mediated inhibition of JNK is a key mechanism for Lactobacillus rhamnosus-induced suppression of TNF production in macrophages." *Cellular Microbiology* 8, 12: 1958–1971.

Krečman, V., N. Škottová, D. Walterová, J. Ulrichová, and V. Šimánek. 1998 "Silymarin inhibits the development of diet-induced hypercholesterolemia in rats." *Planta Medica* 64, 02: 138–142.

Kumar, Shashank, and Abhay K. Pandey. 2013 "Chemistry and biological activities of flavonoids: An overview." *The Scientific World Journal*. doi: 10.1155/2013/162750.

Kunisawa, Jun, Eri Hashimoto, Izumi Ishikawa, and Hiroshi Kiyono. 2012 "A pivotal role of vitamin B9 in the maintenance of regulatory T cells in vitro and in vivo." *PloS One* 7, 2: e32094. doi: 10.1371/journal.pone.0032094.

Lebeer, Sarah, Jos Vanderleyden, and Sigrid CJ De Keersmaecker. 2010 "Host interactions of probiotic bacterial surface molecules: Comparison with commensals and pathogens." *Nature Reviews Microbiology* 8, 3: 171–184.

Lee, Clare J., Cynthia L. Sears, and Nisa Maruthur. 2020 "Gut microbiome and its role in obesity and insulin resistance." *Annals of the New York Academy of Sciences* 1461, 1: 37–52.

Lee, Hui-Young, Jong-Hwan Park, Seung-Hyeok Seok, Min-Won Baek, Dong-Jae Kim, Ki-Eun Lee, Kyung-Soo Paek, Yeonhee Lee, and Jae-Hak Park. 2006 "Human originated bacteria, Lactobacillus rhamnosus

PL60, produce conjugated linoleic acid and show anti-obesity effects in diet-induced obese mice." *Biochimica et Biophysica Acta (BBA)-Molecular and Cell Biology of Lipids* 1761, 7: 736–744.

Lee, Jun Sik, Sang Gap Kim, Hyung Keun Kim, Tae-Hyung Lee, Young-Il Jeong, Chang-Min Lee, Man-Soo Yoon et al. 2007 "Silibinin polarizes Th1/Th2 immune responses through the inhibition of immunostimulatory function of dendritic cells." *Journal of Cellular Physiology* 210, 2: 385–397.

Liu, Liqiao, Mingzhu Li, Mulan Yu, Mingyue Shen, Qun Wang, Yue Yu, and Jianhua Xie. 2019 "Natural polysaccharides exhibit anti-tumor activity by targeting gut microbiota." *International Journal of Biological Macromolecules* 121: 743–751.

Lu, Haifeng, Zhongwen Wu, Wei Xu, Jiezuan Yang, Yunbo Chen, and Lanjuan Li. 2011 "Intestinal microbiota was assessed in cirrhotic patients with hepatitis B virus infection." *Microbial Ecology* 61, 3: 693–703.

Lu, Yan-Meng, Jiao-Jiao Xie, Cong-Gao Peng, Bao-Hong Wang, Kai-Cen Wang, and Lan-Juan Li. 2019 "Enhancing clinical efficacy through the gut microbiota: A new field of traditional Chinese medicine." *Engineering* 5, 1: 40–49.

Macfarlane, George T., and Sandra Macfarlane. 2011 "Fermentation in the human large intestine: Its physiologic consequences and the potential contribution of prebiotics." *Journal of Clinical Gastroenterology* 45: S120-S127.

Macfarlane, Sandra. 2010 "Prebiotics in the gastrointestinal tract." In *Bioactive Foods in Promoting Health*, pp. 145–156. Boston: Academic Press.

Mack, D. R., Siv Ahrné, L. Hyde, S. Wei, and Michael A. Hollingsworth. 2003 "Extracellular MUC3 mucin secretion follows adherence of Lactobacillus strains to intestinal epithelial cells in vitro." *Gut* 52, 6: 827–833.

Mackie, Roderick I., Abdelghani Sghir, and H. Rex Gaskins. 1999 "Developmental microbial ecology of the neonatal gastrointestinal tract." *The American Journal of Clinical Nutrition* 69, 5: 1035s-1045s.

Mayer, Emeran A. 2011 "Gut feelings: The emerging biology of gut—brain communication." *Nature Reviews Neuroscience* 12, 8: 453–466.

McCarthy, R.R., and Fergal O'Gara. 2015 "The impact of phytochemicals present in the diet on microbial signalling in the human gut." *Journal of Functional Foods* 14: 684–691.

McFarland, Lynne V. 2006 "Meta-analysis of probiotics for the prevention of antibiotic associated diarrhea and the treatment of Clostridium difficile disease." *American Journal of Gastroenterology* 101, 4: 812–822.

Michail, Sonia, Malinda Lin, Mark R. Frey, Rob Fanter, Oleg Paliy, Brian Hilbush, and Nicholas V. Reo. 2015 "Altered gut microbial energy and metabolism in children with non-alcoholic fatty liver disease." *FEMS Microbiology Ecology* 91, 2: 1.

Moayyedi, Paul, Alexander Charles Ford, Nicholas J. Talley, Filippo Cremonini, Amy E. Foxx-Orenstein, Lawrence J. Brandt, and Eamonn M.M. Quigley. 2010 "The efficacy of probiotics in the treatment of irritable bowel syndrome: A systematic review." *Gut* 59, 3: 325–332.

Morrison, Douglas J., and Tom Preston. 2016 "Formation of short chain fatty acids by the gut microbiota and their impact on human metabolism." *Gut Microbes* 7, 3: 189–200. doi: 10.1080/19490976.2015.1134082

Moschen, Alexander R., Verena Wieser, and Herbert Tilg. 2012 "Dietary factors: Major regulators of the gut's microbiota." *Gut and Liver* 6, 4: 411.

Nagalingam, Nabeetha A., and Susan V. Lynch. 2012 "Role of the microbiota in inflammatory bowel diseases." *Inflammatory Bowel Diseases* 18, 5: 968–984.

Nguyen, Son G., Jungman Kim, Robin B. Guevarra, Ji-Hoon Lee, Eungpil Kim, Su-il Kim, and Tatsuya Unno. 2016 "Laminarin favorably modulates gut microbiota in mice fed a high-fat diet." *Food & Function* 7, 10: 4193–4201.

Nicholson, Jeremy K., Elaine Holmes, James Kinross, Remy Burcelin, Glenn Gibson, Wei Jia, and Sven Pettersson. 2012 "Host-gut microbiota metabolic interactions." *Science* 336, 6086: 1262–1267.

Nie, Qixing, Jielun Hu, He Gao, Linlin Fan, Haihong Chen, and Shaoping Nie. 2019 "Polysaccharide from Plantago asiatica L. attenuates hyperglycemia, hyperlipidemia and affects colon microbiota in type 2 diabetic rats." *Food Hydrocolloids* 86: 34–42.

Niness, Kathy R. 1999 "Inulin and oligofructose: What are they?." *The Journal of Nutrition* 129, 7: 1402S–1406S.

Nunes, Sandra, Francesca Danesi, Daniele Del Rio, and Paula Silva. 2018 "Resveratrol and inflammatory bowel disease: The evidence so far." *Nutrition Research Reviews* 31, 1: 85–97.

Okada, Hea, C. Kuhn, H. Feillet, and J-F. Bach. 2010 "The 'hygiene hypothesis' for autoimmune and allergic diseases: An update." *Clinical & Experimental Immunology* 160, 1: 1–9.

O'Mahony, Liam, Jane McCarthy, Peter Kelly, George Hurley, Fangyi Luo, Kersang Chen, Gerald C. O'Sullivan et al. 2005 "Lactobacillus and bifidobacterium in irritable bowel syndrome: Symptom responses and relationship to cytokine profiles." *Gastroenterology* 128, 3: 541–551.

O'Toole, Paul W., and Ian B. Jeffery. 2015 "Gut microbiota and aging." *Science* 350, 6265: 1214–1215.
Payne, Amanda N., C. Chassard, M. Zimmermann, P. Müller, S. Stinca, and C. Lacroix. 2011 "The metabolic activity of gut microbiota in obese children is increased compared with normal-weight children and exhibits more exhaustive substrate utilization." *Nutrition & diabetes* 1, 7: e12–e12. doi: 10.1038/nutd.2011.8.
Penner, Robert, Richard N. Fedorak, and Karen L. Madsen. 2005 "Probiotics and nutraceuticals: Non-medicinal treatments of gastrointestinal diseases." *Current Opinion in Pharmacology* 5, 6: 596–603.
Pferschy-Wenzig, Eva-Maria, Kaisa Koskinen, Christine Moissl-Eichinger, and Rudolf Bauer. 2017 "A combined LC-MS metabolomics-and 16S rRNA sequencing platform to assess interactions between herbal medicinal products and human gut bacteria in vitro: A pilot study on Willow Bark extract." *Frontiers in Pharmacology* 8: 893.
Ponnusamy, Kannan, Jung Nam Choi, Jiyoung Kim, Sun-Young Lee, and Choong Hwan Lee. 2011 "Microbial community and metabolomic comparison of irritable bowel syndrome faeces." *Journal of Medical Microbiology* 60, Pt 6: 817. doi: 10.1099/jmm.0.028126-0.
Possemiers, Sam, Selin Bolca, Willy Verstraete, and Arne Heyerick. 2011 "The intestinal microbiome: A separate organ inside the body with the metabolic potential to influence the bioactivity of botanicals." *Fitoterapia* 82, 1: 53–66.
Prantera, Cosimo, and Maria Lia Scribano. 2009 "Antibiotics and probiotics in inflammatory bowel disease: Why, when, and how." *Current Opinion in Gastroenterology* 25, 4: 329–333.
Qin, Junjie, Yingrui Li, Zhiming Cai, Shenghui Li, Jianfeng Zhu, Fan Zhang, Suisha Liang et al. 2012 "A metagenome-wide association study of gut microbiota in type 2 diabetes." *Nature* 490, 7418: 55–60.
Qin, Nan, Fengling Yang, Ang Li, Edi Prifti, Yanfei Chen, Li Shao, Jing Guo et al. 2014 "Alterations of the human gut microbiome in liver cirrhosis." *Nature* 513, 7516: 59–64.
Rastall, Robert A., and Vatsala Maitin. 2002 "Prebiotics and synbiotics: Towards the next generation." *Current Opinion in Biotechnology* 13, 5: 490–496.
Rinttilä, Teemu, Anna Lyra, Lotta Krogius-Kurikka, and Airi Palva. 2011 "Real-time PCR analysis of enteric pathogens from fecal samples of irritable bowel syndrome subjects." *Gut Pathogens* 3, 1: 6.
Ríos-Covián, David, Patricia Ruas-Madiedo, Abelardo Margolles, Miguel Gueimonde, Clara G. De Los Reyes-gavilán, and Nuria Salazar. 2016 "Intestinal short chain fatty acids and their link with diet and human health." *Frontiers in Microbiology* 7: 185.
Roberfroid, Marcel B. 2007 "Prebiotics: The concept revisited." *The Journal of Nutrition* 137, 3: 830S-837S.
Roberfroid, Marcel B., Jan A.E. Van Loo, and Glenn R. Gibson. 1998 "The bifidogenic nature of chicory inulin and its hydrolysis products." *The Journal of Nutrition* 128, 1: 11–19.
Rodríguez-Cabezas, Maria Elena, Julio Galvez, Maria Dolores Lorente, Angel Concha, Desiree Camuesco, Shamira Azzouz, Antonio Osuna, Luis Redondo, and Antonio Zarzuelo. 2002 "Dietary fiber down-regulates colonic tumor necrosis factor α and nitric oxide production in trinitrobenzenesulfonic acid-induced colitic rats." *The Journal of Nutrition* 132, 11: 3263–3271.
Rolhion, N., and B. Chassaing. 2016 "When pathogenic bacteria meet the intestinal microbiota." *Philosophical transactions of the Royal Society of London. Series B, Biological Sciences* 371, 1707: 20150504. doi:10.1098/rstb.2015.0504.
Rowland, I., G. Gibson, A. Heinken, K. Scott, J. Swann, I. Thiele, and K. Tuohy. 2018 "Gut microbiota functions: Metabolism of nutrients and other food components." *European Journal of Nutrition* 57: 1–24.
Sakaida, Isao, Masako Tsuchiya, Kotarou Kawaguchi, Teruaki Kimura, Shuji Terai, and Kiwamu Okita. 2003 "Herbal medicine Inchin-ko-to (TJ-135) prevents liver fibrosis and enzyme-altered lesions in rat liver cirrhosis induced by a choline-deficient L-amino acid-defined diet." *Journal of Hepatology* 38, 6: 762–769.
Schwiertz, Andreas, David Taras, Klaus Schäfer, Silvia Beijer, Nicolaas A. Bos, Christiane Donus, and Philip D. Hardt. 2010 "Microbiota and SCFA in lean and overweight healthy subjects." *Obesity* 18, 1: 190–195.
Sekirov, Inna, Shannon L. Russell, L. Caetano M. Antunes, and B. Brett Finlay. 2010 "Gut microbiota in health and disease." *Physiological Reviews* 90, 3: 859–904.
Sheil, Barbara, Fergus Shanahan, and Liam O'Mahony. 2007 "Probiotic effects on inflammatory bowel disease." *The Journal of Nutrition* 137, 3: 819S-824S.
Shen, Chian-Yin, Ming-Chih Lin, Heng-Kuei Lin, Ching-Heng Lin, Lin-Shien Fu, and Yun-Chin Fu. 2013 "The natural course of eczema from birth to age 7 years and the association with asthma and allergic rhinitis: A population-based birth cohort study." *Allergy & Asthma Proceedings* 34, 1.
Shen, Liang, Lu Liu, Xin-Yu Li, and Hong-Fang Ji. 2019 "Regulation of gut microbiota in Alzheimer's disease mice by silibinin and silymarin and their pharmacological implications." *Applied Microbiology and Biotechnology* 103, 17: 7141–7149.

Shoemaker, R. 2016 "Probiotics. png." Obtained from Wikimedia under the Creative commons Attribution-Share Alike 4.0 license. https://commons.wikimedia.org/wiki/File:Probiotic.png (Accessed 13 May 2021).

Shreiner, Andrew B., John Y. Kao, and Vincent B. Young. 2015 "The gut microbiome in health and in disease." *Current Opinion in Gastroenterology* 31, 1: 69.

Siegel, Rebecca L., Kimberly D. Miller, and Ahmedin Jemal. 2016 "Cancer statistics, 2016." *CA: A Cancer Journal for Clinicians* 66, 1: 7–30.

Škottová, N., and V. Krečman. 1998 "Silymarin as a potential hypocholesterolaemic drug." *Physiol Res* 47, 1: 1–7.

Smith, Peter Andrey. 2015 "The tantalizing links between gut microbes and the brain." *Nature* 526, 7573:312–314.

Stevens, Jan F., and Claudia S. Maier. 2016 "The chemistry of gut microbial metabolism of polyphenols." *Phytochemistry Reviews* 15, 3: 425–444.

Sung, Miranda M., Ty T. Kim, Emmanuel Denou, Carrie-Lynn M. Soltys, Shereen M. Hamza, Nikole J. Byrne, Grant Masson et al. 2017 "Improved glucose homeostasis in obese mice treated with resveratrol is associated with alterations in the gut microbiome." *Diabetes* 66, 2: 418–425.

Swidsinski, Alexander, Petra Schlien, Annelie Pernthaler, Uwe Gottschalk, Eckard Bärlehner, Georg Decker, Sonja Swidsinski et al. 2005 "Bacterial biofilm within diseased pancreatic and biliary tracts." *Gut* 54, 3: 388–395.

Tako, Elad, R.P. Glahn, R.M. Welch, Xingen Lei, K. Yasuda, and D.D. Miller. 2008 "Dietary inulin affects the expression of intestinal enterocyte iron transporters, receptors and storage protein and alters the microbiota in the pig intestine." *British Journal of Nutrition* 99, 3: 472–480.

Tang, Ruqi, Yiran Wei, Yanmei Li, Weihua Chen, Haoyan Chen, Qixia Wang, Fan Yang et al. 2018 "Gut microbial profile is altered in primary biliary cholangitis and partially restored after UDCA therapy." *Gut* 67, 3: 534–541.

Tejero-Sariñena, Sandra, Janine Barlow, Adele Costabile, Glenn R. Gibson, and Ian Rowland. 2012 "*In vitro* evaluation of the antimicrobial activity of a range of probiotics against pathogens: Evidence for the effects of organic acids." *Anaerobe* 18, 5: 530–538.

Thomas, Raymond H., Melissa M. Meeking, Jennifer R. Mepham, Lisa Tichenoff, Fred Possmayer, Suya Liu, and Derrick F. MacFabe. 2012 "The enteric bacterial metabolite propionic acid alters brain and plasma phospholipid molecular species: Further development of a rodent model of autism spectrum disorders." *Journal of Neuroinflammation* 9, 1: 153.

Tremaroli, Valentina, and Fredrik Bäckhed. 2012 "Functional interactions between the gut microbiota and host metabolism." *Nature* 489, 7415: 242–249.

Turnbaugh, Peter J., Ruth E. Ley, Michael A. Mahowald, Vincent Magrini, Elaine R. Mardis, and Jeffrey I. Gordon. 2006 "An obesity-associated gut microbiome with increased capacity for energy harvest." *Nature* 444, 7122: 1027.

Updegrove, Kim. 2004 "Necrotizing enterocolitis: The evidence for use of human milk in prevention and treatment." *Journal of Human Lactation* 20, 3: 335–339.

Uthman, E. 2008 "Normal colonic mucosa." Obtained from Wikimedia under the Creative Commons Attribution-Share Alike 2.0 Generic license. https://commons.wikimedia.org/wiki/File:Normal_colonic_mucosa.jpg (Accessed 13 May 2021)

Vyas, Usha, and Natarajan Ranganathan. 2012 "Probiotics, prebiotics, and synbiotics: Gut and beyond." *Gastroenterology Research and Practice* 2012: 1–16. doi: 10.1155/2012/872716.

Wellen, Kathryn E., and Gökhan S. Hotamisligil. 2005 "Inflammation, stress, and diabetes." *The Journal of Clinical Investigation* 115, 5: 1111–1119.

Williams, H.C., and D.J.C. Grindlay. 2010 "What's new in atopic eczema? An analysis of systematic reviews published in 2007 and 2008. Part 1. Definitions, causes and consequences of eczema." *Clinical and Experimental Dermatology: Clinical Dermatology* 35, 1: 12–15.

Wu, Xiaokang, Chaofeng Ma, Lei Han, Muhammad Nawaz, Fei Gao, Xuyan Zhang, Pengbo Yu et al. 2010 "Molecular characterisation of the faecal microbiota in patients with type II diabetes." *Current Microbiology* 61, 1: 69–78.

Xiong, Yumei, Norimasa Miyamoto, Kenji Shibata, Mark A. Valasek, Toshiyuki Motoike, Rafal M. Kedzierski, and Masashi Yanagisawa. 2004 "Short-chain fatty acids stimulate leptin production in adipocytes through the G protein-coupled receptor GPR41." *Proceedings of the National Academy of Sciences* 101, 4: 1045–1050.

Yamamoto, Masahiro, Naoko Miura, Nobuhiro Ohtake, Sakae Amagaya, Atsushi Ishige, Hiroshi Sasaki, Yasuhiro Komatsu, Kazunori Fukuda, Takashi Ito, and Katsutoshi Terasawa. 2000 "Genipin, a metabolite derived from the herbal medicine Inchin-ko-to, and suppression of Fas-induced lethal liver apoptosis in mice." *Gastroenterology* 118, 2: 380–389.

Yamamoto, Masahiro, Kazuo Ogawa, Makoto Morita, Kazunori Fukuda, and Yasuhiro Komatsu. 1996 "The herbal medicine Inchin-ko-to inhibits liver cell apoptosis induced by transforming growth factor β1." *Hepatology* 23, 3: 552–559.

Yamshiki, Masayoshi, Akihito Mase, Ichiro Arai, Xian-Xi Huang, Tsutomu Nobori, Akira Nishimura, Seigo Sakaguchi, and Kyoichi Inoue. 2000 "Effects of the Japanese herbal medicine 'Inchinko-to'(TJ-135) on concanavalin A-induced hepatitis in mice." *Clinical Science* 99, 5: 421–431.

Yin, Xiaochen, Jinghua Peng, Liping Zhao, Yunpeng Yu, Xu Zhang, Ping Liu, Qin Feng, Yiyang Hu, and Xiaoyan Pang. 2013 "Structural changes of gut microbiota in a rat non-alcoholic fatty liver disease model treated with a Chinese herbal formula." *Systematic and Applied Microbiology* 36, 3: 188–196.

Yoo, Bryan B., and Sarkis K. Mazmanian. 2017 "The enteric network: Interactions between the immune and nervous systems of the gut." *Immunity* 46, 6: 910–926.

Zell, H. 2016 "*Silybum marianum*- Tenerife 01.jpg." Obtained from Wikimedia under the Creative commons Attribution-Share Alike 3.0. https://commons.wikimedia.org/wiki/File:Silybum_marianum_-_Tenerife_01.jpg (Accessed 13 May 2021).

Zhang, Zhan, Xinyue Wu, Shuyuan Cao, Li Wang, Di Wang, Hui Yang, Yiming Feng, Shoulin Wang, and Lei Li. 2016 "Caffeic acid ameliorates colitis in association with increased Akkermansia population in the gut microbiota of mice." *Oncotarget* 7, 22: 31790.

Zitvogel, L., Galluzzi, L., Viaud, S., Vétizou, M., Daillère, R., Merad, M., and Kroemer, G. 2015 "Cancer and the gut microbiota: An unexpected link." *Science Translational Medicine* 7, 271: 271ps1. doi: 10.1126/scitranslmed.3010473.

24 Plant-Derived Compounds as New Therapeutics for Substance Use Disorders

Kevin S. Murnane and Mary Frances Vest

CONTENTS

24.1 EPIDEMIOLOGY OF SUBSTANCE USE DISORDERS (SUDS)

Plant-derived compounds as new therapeutics for substance use disorders

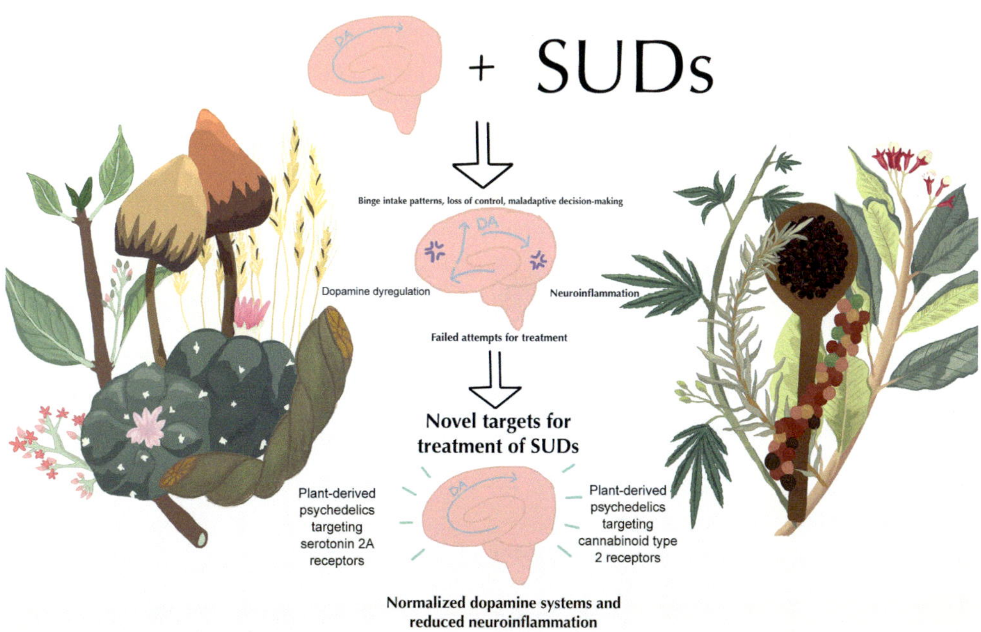

Graphical abstract representing the content of the chapter.

DOI: 10.1201/9781003108375-24

Abused drugs can be broadly categorized into classes of compounds that share chemical properties or pharmacological actions, such as stimulants, cannabinoids, sedatives, opioids, nicotine and alcohol. SUDs are a growing problem in the United States (U.S.) and across the world. Stimulant abuse is an important public health concern, with millions of Americans regularly using classic stimulants such as cocaine and methamphetamine for non-medical purposes. Acute overdoses are a particularly hazardous aspect of stimulant abuse, which can lead to hospitalization and death (Devlin and Henry, 2008). It has been documented that approximately one in three drug-related emergency room episodes involves cocaine abuse (Gaval-Cruz et al., 2008; Substance Abuse and Mental Health Services Administration (SAMHSA, 2010). Epidemiological studies demonstrate that amphetamine-type stimulants are the most widely used illicit drugs worldwide after cannabis, with up to 51 million global users between the ages of 15 and 64 (McKetin et al., 2013; Wearne and Cornish, 2018). In this regard, there is a current resurgence in methamphetamine use due to inexpensive, high-quality products flooding the market (Ben-Yehuda and Siecke, 2018).

According to the National Institute on Drug Abuse (NIDA), methamphetamine overdose deaths have been surging nationwide. U.S. poison control centers report that the prevalence of methamphetamine greatly intensified from 2000–2019 (Chen et al., 2021; Han et al., 2021). Furthermore, it is troubling that methamphetamine use has begun to expand in vulnerable communities such as native peoples and African Americans, who showed the highest death rates and steepest increases in methamphetamine overdoses over this period, respectively (Chen et al., 2021; Han et al., 2021). Moreover, NIDA estimates that substance abuse costs the United States more than $600 billion annually. Despite its immense costs to individuals, families and society, the prevalence and severity of methamphetamine use continue to increase yearly. Developing in parallel to this devastating methamphetamine crisis, as reported by the United Nations Office on Drugs and Crime and others, there has been a recent and substantial increase in the detection of novel psychoactive substances, the majority of which are novel stimulants (United Nations Office on Drugs and Crime, 2018). This will only further exacerbate the medical and public health consequences of stimulant use disorder, as well as the dire need for the development of new medications.

According to the National Institute on Alcohol Abuse and Alcoholism, 17 million adults in the United States have an alcohol use disorder (AUD) (Mokdad et al., 2004; Sacks et al., 2015). The U.S. Centers for Disease Control and Prevention (CDC) reports that excessive drinking is directly responsible for upwards of 80,000 deaths per year making it the third leading cause of preventable death in the United States, followed by tobacco and obesity, and costs $224 billion (Mokdad et al., 2004; Sacks et al., 2015). AUD has a high yearly (13.9%) and lifetime (29.1%) prevalence rate, with roughly 10.5% of U.S. children aged 17 or younger (75 million children) having lived in the past year in a household where at least one parent had an AUD (Grant et al., 2015).

The abuse of other drugs besides stimulants and alcohol is a clear public health crisis. For example, opioid use and abuse are major public health concerns worldwide, which include not only the abuse of illicit opioids (e.g., heroin), but also the misuse of prescription opioids. In parallel to the increase in the use of stimulants and alcohol, there has been an increase in the use of opioids, sedatives and other drugs of abuse, as well as their deleterious consequences. For example, the United Nations Office on Drugs and Crime recently reported there was a greater than 20% increase in drug overdose deaths in the United States, which was largely attributed to an upsurge in deaths associated with pharmaceutical opioids (United Nations Office on Drugs and Crime, 2018).

To give a single example, oxycodone is one of the most widely abused prescription drugs worldwide, with reported global seizures, as well as oxycodone-related abuse and overdose, increasing drastically in recent years (Compton et al., 2016; Kolodny et al., 2015; Manchikanti et al., 2010; Manchikanti and Singh, 2008; United Nations Office on Drugs and Crime, 2018). However, for the sake of focus and brevity, this chapter will largely discuss stimulants and alcohol.

24.2 THE DISEASE OF ADDICTION

Substance abuse is characterized by maladaptive patterns of drug consumption. This drug-taking has been associated with drug-induced feelings of drug 'high' and drug-induced euphoria, as well as elevated levels of the neurotransmitter dopamine in key regions of addiction-relevant neural circuitry, particularly the nucleus accumbens (NAcc). It is believed that the dopaminergic reward pathway, including dopamine cell bodies in the ventral tegmental area (VTA) and projections for active dopamine release into the NAcc, mediates the reinforcing and addictive effects of stimulants, alcohol and other drugs of abuse (Madras et al., 1989; Reith et al., 1986).

For example, the abuse liability of stimulants has been attributed primarily to their actions at the dopamine transporter (DAT), with a significant correlation between DAT occupancy and subjective feelings of drug high and rush in human studies (Ritz et al., 1987; Volkow et al., 1999). Methamphetamine is a dangerous amphetamine derivative that produces significant effects on physical, behavioral, cognitive and psychiatric output. It does so by promoting dopamine signaling through substrate-based release at the DAT (Fleckenstein et al., 2007).

SUDs are characterized by binge intake patterns, loss of control and maladaptive decision making. Prominent models argue that there are key transition periods in the addiction process including drug initiation, bingeing, withdrawal and relapse, which can be modeled in animals using self-administration (SA), appetitive conditioning in the conditioned place preference (CPP) paradigm, withdrawal/extinction and reinstatement procedures (Carroll et al., 2009).

These transitions from abuse to dependence likely involve extended neural systems. While dopaminergic projections from the VTA to NAcc form the core of the reward circuit, this circuit is highly modulated by extended limbic regions, including the amygdala, medial prefrontal cortex (PFC) and hippocampus. In this regard, neuronal projections from the basolateral amygdala to NAcc play an important role in processing reward-related cues and drug craving (Hsu et al., 2020). Likewise, the amygdala may be involved in drug cue reactivity as triggering environments can induce similar electrophysiological activity in the NAcc and amygdala (Hsu et al., 2020; Li et al., 2015; Venniro et al., 2020). The use of plant-derived medications that target the extended reward system and these additional neural substrates may be more promising than approaches that focus exclusively on the core dopamine reward system.

24.2.1 Current Treatments

The current treatments for SUDs primarily consist of pharmacotherapies and behavioral therapies. Pharmacotherapies have conventionally focused on the dopamine 'reward circuit' and its central neurobiological locus in the NAcc. Over several decades, significant efforts have been devoted to developing an effective treatment for stimulant use disorder. These efforts have focused on the development of small-molecule dopamine agonists and antagonists, GABAergic agonists, kappa opioid agonists, mu-opioid partial agonists and serotonergic agonists and antagonists as treatments for cocaine addiction. Despite these efforts, there are no U.S. Food and Drug Administration (FDA)-approved pharmacotherapies to treat stimulant use disorder, representing a critical unmet medical need (Vocci et al., 2005; Vocci and Appel, 2007; Volkow and Li, 2004). Previous medication development efforts have largely focused on antagonist versus substitute agonist therapies, with considerable debate as to the best approach. Antagonist therapy is the more conceptually clear approach, as it relies on stopping cocaine from entering the brain or from reaching its protein targets. However, substance-dependent individuals show unique problems with treatment compliance, and antagonist approaches may exacerbate such problems.

Substitute-agonist therapies mimic key aspects of the abused drug to reduce cravings and withdrawal, and promote abstinence. FDA-approved substitute-agonist therapies include methadone, buprenorphine, varenicline, transdermal and buccal formulations of nicotine. Strengths of this approach include the clinical success of these agents, better compliance, reduced withdrawal and craving, and excellent efficacy profiles in preclinical models. Weaknesses include the risk of toxic drug interactions during relapse and potential diversion. Alternative approaches have variously examined drug vaccines, monoclonal antibodies, dopamine receptor antagonists and agents that induce sickness reactions. Relapse rates are similarly disappointing with such approaches that directly target reward systems.

Behavioral therapies target stress reactivity and behavioral control, and trigger management. Approaches such as contingency management and cognitive behavioral therapy are common and focus on either rewarding individuals for desired behavior or trying to prevent individuals from being in high-risk situations that will cause them to relapse. The use of behavioral therapies alone for some abused substances has been documented to result in relapse rates of 80% within two years (Bart, 2012). Many patients are refractory to either pharmacotherapy or behavioral therapy, perhaps because the reward circuit does not capture the totality of the pathophysiology of SUDs, while behavioral therapies do not alleviate neurobiological maladaptations that can lead to relapse. New plant-derived treatments that can affect neural substrates beyond the reward circuit and can correct maladaptive neurobiology may prove to have significant therapeutic utility, either alone or as adjuncts to conventional therapy.

The current FDA-approved treatments for AUD and/or alcohol dependence are acamprosate, naltrexone and disulfiram. Unfortunately, these pharmacotherapies produce disappointing remission rates. These FDA-approved treatments for AUD can have benefits for patients suffering from AUD; however, a high percentage of individuals remain refractory to these therapies. Meta-analyses of the effects of each of these treatment options have documented significant deficiencies in their clinical efficacy. For example, while disulfiram may reduce the frequency of drinking or induce short-term transient abstinence, it induces an untoward—and potentially severe—sickness reaction (Garbutt et al., 1999; Jorgensen et al., 2011).

While naltrexone can reduce the number of heavy drinking days, its effects on overall abstinence are less robust (Rosner et al., 2010b). Conversely, acamprosate may have some effect on abstinence, but has less influence on heavy drinking (Rosner et al., 2010a). A meta-analysis of the effectiveness of naltrexone and/or acamprosate for the treatment of alcohol dependence showed that the effectiveness of naltrexone compared to placebo was only 12–19% and that of acamprosate was only 7–13% (Kranzler and Van Kirk, 2001). Additionally, disulfiram showed no significant difference between individuals in the treatment and the placebo/control groups across several different parameters, including total abstinence from alcohol, time to first drink, employment, and social stability, thereby indicating the need for new and effective interventions (Fuller et al., 1986). As with stimulants, there is a major unmet need for more effective pharmacotherapies to treat AUD, which was recently reviewed (Oppong-Damoah et al., 2021).

24.3 NEUROPATHOLOGICAL CHANGES ASSOCIATED WITH SUDS

SUDs are characterized by the loss of control over drug consumption and maladaptive decision making, including pursuing drugs of abuse at the cost of important life activities. Sensitivity to, and learning about, behavioral outcomes is derailed, and compulsive drug-taking may result from changes in the corticostriatal structures that regulate outcome-based learning and decision making. Likewise, SUDs and the maintenance of abstinence are heavily affected by stress and stress reactivity (Goeders, 2002; Koob and Moal, 1997). While acute increases in dopamine signaling drive drug-associated euphoria and abuse, there is a wealth of data across rodents, primates and humans that exposure to methamphetamine may cause dopaminergic pathways to become hypofunctional and lead to impaired impulse control, increased drug craving and loss of the capacity to regulate behavior, in a manner that leaves individuals susceptible to relapse (Howell and Murnane, 2011; Hyatt et al., 2020; Knight et al., 2013; Miller et al., 2013; Murnane et al., 2012; Volkow et al., 2001). The neurocognitive decline is one of the most potentially severe consequences of SUDs, as cognitive decline may further predispose the individual to further methamphetamine intake through an impaired capacity to regulate decision making and control the response to stressful triggers.

Exposure to drugs of abuse can produce neurocognitive decline and dysregulation of the brain dopamine systems. Indeed, a drug-induced parkinsonian-like state can be transiently induced by the clinical use of drugs that block dopamine receptors (Christian and Paulson, 1958; Janno et al., 2004; Mamo et al., 1999). Likewise, there is accumulating evidence that non-clinical, high-dose 'binge' use of stimulants such as methamphetamine produce dysregulation of brain dopamine systems, and induce a state that recapitulates the cognitive deficits seen in the early stages of Parkinsonism. In addition to stress reactivity and compulsive drug use, SUDs include deficits in executive decision making and impulse control, as well as increases in depression, psychosis and sleep disturbances. Cognitive deficits in SUDs appear to be related to loss of frontostriatal dopamine circuits and are exemplified by the loss of dopaminergic innervation of the dorsal and ventral striatum and consequential change in basal ganglia outflow, specifically in a basal ganglia-thalamocortical loop that includes the dorsolateral PFC. This is thought to subserve cognitive control over goal-directed behavior and cognitive flexibility (Haber and Calzavara, 2009; Holthoff-Detto et al., 1997; Miller and Cohen, 2001). Although the deterioration of this circuit results in a frontostriatal-like syndrome, evidence suggests other neural circuits may degrade in SUDs. Recent studies have supported the idea of hippocampal degeneration following exposure to stimulants such as methamphetamine that can produce a drug-induced parkinsonian-like state.

There is a wealth of data supporting the contention that oxidative stress and neuroinflammation can impair neurocognitive function and affect the influence of stress reactivity on drug-seeking (Miller et al., 2017). Evidence has emerged that neuroinflammation and oxidative stress may directly mediate the formation of intense conditioning that sustains compulsive drug use (Correia et al., 2020). Cocaine, methamphetamine and other drugs of abuse have been argued to directly activate pattern recognition toll-like receptor 4 (TLR4) on microglia and other phagocytic inflammatory cells, which is a receptor that induces an intense innate inflammatory response. TLR4 receptor stimulation by drugs of abuse is followed by the release of inflammatory cytokines that alter synaptic plasticity in a manner that may lead to maladaptive behavior (Lacagnina et al., 2017).

Research has shown that methamphetamine induces severe vascular damage, including in the neurovascular unit, which may contribute to neurocognitive decline and dopamine dysregulation (Turowski and Kenny, 2015). Chronic stimulant use is associated with increased serum levels of the pro-inflammatory cytokine interleukin 6 (IL-6) and decreased levels of the anti-inflammatory cytokine interleukin 10 (IL-10) (Moreira et al., 2016). The postmortem brains of individuals with stimulant use disorders and non-human primates with prolonged cocaine SA histories exhibit microglia activation associated with dopamine dysregulation or loss of dopamine neurons, suggesting neuroinflammation may intensify dependence, as well as its anhedonic and depressive symptoms (Little et al., 2009; Smith et al., 2019).

There is accumulating evidence that anti-inflammatory and antioxidant drugs may represent a new way forward in medication development efforts for SUDs. Administration of (+)naloxone prevents the development of a cocaine CPP (Northcutt et al., 2015). N-acetylcysteine, an anti-inflammatory agent that has received extensive attention in both psychiatry and neurology, decreases desire and interest toward cocaine, as well as craving and time taken to relapse in human subjects (LaRowe et al., 2007; LaRowe et al., 2013). In animal studies, N-acetylcysteine blocks cocaine-induced locomotor sensitization and prevents increased cocaine intake in a long-access paradigm designed to induce compulsive responding (Madayag et al., 2007). The development of plant-derived medications that address neuroinflammation and protect dopamine systems may have significant benefits for SUDs.

24.3.1 Plant-Derived Medications for SUDs

Plants and plant extracts may provide compounds that can be used to validate novel targets for the treatment of SUDs, as they have provided compounds useful in the development of treatments for other brain disorders such as Alzheimer's disease (Howes and Houghton, 2003). In support of this, studies have examined plant derivatives and phytochemicals such as compounds isolated from *Hypericum perforatum* L. (St. John's wort), isoflavonoids from *Pueraria lobata* (Willd.) Ohwi (kudzu), ibogaine and dietary cannabinoid β-caryophyllene (BCP) in preclinical animal studies with a focus on SUD. Hence, there is a sound basis for the study of phytochemicals and plant extracts in the development of promising targets for the treatment of AUD.

Plant extracts have been shown to reduce ethanol intake by themselves, as well as synergistically with pharmaceuticals. An extract of *H. perforatum* has been shown to reduce ethanol intake in alcohol-preferring rats without affecting food, water or saccharin intake. A subthreshold dose (1 mg/kg) of the opioid receptor antagonist naloxone augmented the suppression of ethanol intake by this extract (Perfumi et al., 2003). Likewise, the phytochemical BCP has been shown to reduce voluntary alcohol intake and attenuate the ethanol-induced CPP in mice, which was recently reviewed (Al Mansouri et al., 2014; Oppong-Damoah et al., 2021).

β-caryophyllene oxide (BCPO) is another naturally occurring sesquiterpene. Both BCP and BCPO are generally regarded as well tolerated and safe. As with the parent BCP, it is found in the essential oils of many plants including black pepper (*Piper nigrum* L.), clove (*Syzygium aromaticum* (L.) Merr. & L.M.Perry), marijuana (*Cannabis sativa* L.), sugar-apple (*Annona squamosa* L.) and rosemary (*Rosmarinus officinalis* L.) (Figure 24.1) (Chavan et al., 2010; Ghelardini et al., 2001;

FIGURE 24.1 Medicinal plants with active agents that can stimulate serotonin 2A (5-HT$_{2A}$) receptors or cannabinoid type 2 (CB2) receptors. 1A and B, *Psychotria viridis* (Shrub that contains the 5-HT$_{2A}$ agonist dimethyltryptamine, which is an ingredient in ayahuasca brew); 2, *Banisteriopsis caapi* (a vine that contains that monoamine oxidase inhibitor that is used in ayahuasca brew to protect dimethyltryptamine from degradation); 3, *Claviceps purpurea* (Rye ergot fungus); 4, *Lophophora williamsii* (peyote cactus that contains the 5-HT$_{2A}$ receptor agonist mescaline); 5, *Psilocybe semilanceata* (one of many fungal species that contain the 5-HT$_{2A}$ receptor agonist psilocybin); 6, *Cannabis sativa* (a heavily cultivated herbaceous flowering plant that contains cannabinoid type 1 and CB2 receptor agonists); 7, *Rosmarinus officinalis* (rosemary contains CB2 receptor agonists); 8A and B, *Piper nigrum* (black pepper contains CB2 receptor agonists); 9A and B, *Syzygium aromaticum* (clove contains CB2 receptor agonists) (Costa, 2016; Guilding, 1832; Hedenhös, 2012; Hir, 2012; McCormick, 2012; Scienceman71, 2013; Shaull, 2021; T137, 2010; Terpsichore, 2004; Tinofrey, 2005; US Drug Enforcement Agency, 2008; Vinayaraj, 2013).

Ormeno et al., 2008). The U.S. FDA and the European Food and Safety Authority have approved both BCP and BCPO as food additives. BCP has been shown to possess antioxidant, anti-inflammatory and analgesic properties, while BCPO has been shown to exhibit antifungal, analgesic and anti-inflammatory properties (Chavan et al., 2010; Klauke et al., 2014; Yang et al., 1999). Furthermore, studies have documented that BCPO decreases ethanol drinking and ethanol CPP. It does so by independently changing the pharmacokinetics of ethanol. Moreover, drug interaction experiments were conducted which demonstrated that BCPO stimulates cannabinoid type 2 (CB2) receptors *in vivo* (Oppong-Damoah et al., 2019a).

24.3.2 Plant-Derived Cannabinoids and Terpenes for the Treatment of SUDs

Cannabinoid systems are one of the most promising targets for plant-derived medicines. The endocannabinoid system (ECS) has been implicated in the abuse-related effects of cannabinoids, nicotine, opioids and alcohol (Maldonado et al., 2006). Evidence for the involvement of the ECS in drug dependence has been derived from randomized controlled trials, as well as animal studies of alcohol-seeking behavior, including voluntary alcohol consumption, operant ethanol SA, ethanol preference and ethanol-induced place conditioning (de Fonseca et al., 1999; Houchi et al., 2005; Hungund et al., 2003; Maldonado et al., 2006; Sloan et al., 2017; Wang et al., 2003). The ECS is composed of two different G protein-coupled receptors (GPCRs), the CB1 and CB2 receptors (Matsuda et al., 1990). These receptors are activated by endogenous endocannabinoids such as 2-arachidonoyl ethanolamine (also known as anandamide) and 2-arachidonoylglycerol (2-AG), as was recently reviewed (Devane et al., 1992; Oppong-Damoah et al., 2021; Sugiura et al., 1995).

The most widely expressed GPCR in the brain is the CB1 receptor, which mediates the psychoactive effects of drugs like delta-9-tetrahydrocannabinol (Δ9-THC), a major psychoactive constituent of the plant *Cannabis sativa*, and is involved in nociception, emotional behavior and reward signaling through the VTA (Ledent et al., 1999; Lichtman and Martin, 1997; Martin et al., 2002; Riegel and Lupica, 2004). In contrast to CB1 receptors, CB2 receptors are predominantly located in peripheral tissues and immune cells, where they play an important role in inflammation and other effects (Anand et al., 2009; Bab et al., 2008; Munro et al., 1993). As with other inflammatory receptors, these CB2 receptors are highly inducible on reactive microglial cells in the brain following injury or inflammation, which may include the stimulation of inflammatory cells through TLR4 receptors. This has been noted as an effect of numerous drugs of abuse (Ashton et al., 2007; Benito et al., 2005).

The ECS, particularly the CB2 receptors, have recently emerged as neuromodulators and therapeutic targets for plant-based medicines for brain disease, including SUDs. For example, BCP protects against dopamine cell death, the generation of reactive oxygen species (ROS) and the suppression of the antioxidant glutathione induced by dopamine neurotoxins (Wang et al., 2018). BCP ameliorates motor dysfunction, protects against DA neuronal loss, and alleviates glial activation by dopamine neurotoxins in a CB2 receptor-dependent manner (Viveros-Paredes et al., 2017). Additionally, BCP attenuates oxidative stress and neuroinflammation and preserves dopaminergic neurons in a CB2 receptor-dependent manner in rats exposed to the dopamine neurotoxin rotenone (Ojha et al., 2016).

There is substantial evidence supporting the procognitive effects of plant-derived compounds that stimulate CB2 receptors. For example, stimulation of CB2 receptors enhances memory consolidation in the passive avoidance task, and improves deficits in both short- and long-term visual recognition memory in rats exposed to ethanol (Ratano et al., 2017; Rivera et al., 2018). Recent reports suggest that BCP reverses cognitive decline in aged mice (Lindsey et al., 2019). The combination of potentially addressing both the reinforcing effects of drugs of abuse and the long-term dysregulations of dopamine systems and associated cognitive impairments in SUDs make plant-derived medications that activate CB2 receptors a compelling target for the discovery of medications for SUDs.

Although ameliorating the abuse liability of drugs of abuse and the cognitive symptoms of SUDs would be an important clinical advance, the greatest impact would come from a plant-derived medication that forestalls or reverses the degeneration of dopamine systems in SUDs. This could modify the course of SUDs. The benefits of CB2 receptor stimulation on these indices have garnered significant support across a broad range of brain injuries (Antonazzo et al., 2019). *In vitro* studies have shown that stimulation of CB2 receptors significantly decreases lipopolysaccharide (a profound innate immune response stimulator) induction of TNF-α and IL-1β release, and more importantly, it reverses neuroinflammation in rats that have been exposed to ethanol (Klegeris et al., 2003; Rivera et al., 2018). Likewise, BCP reduces levels of circulatory proinflammatory cytokines IL-23 and IL-27 in aged mice undergoing normal age-related decline (Lindsey et al., 2019).

In the context of dopamine systems, stimulation of CB2 receptors reduces microglial activation following exposure to a dopamine neurotoxin (Price et al., 2009). CB2 receptor stimulation mitigates dopamine neurotoxin-associated oxidative stress, which occurs when ROS are formed during the endogenous metabolism of dopamine and its metabolites. It can occur with alterations in the expression of antioxidant proteins, including superoxide dismutase, catalase and glutamate-cysteine ligase, which are rate-limiting toward GSH synthesis (Weng et al., 2018). Such oxidative stress is believed to be a critical mediator of cell death and neurodegeneration of dopamine systems (Weng et al., 2018). Both of the plant-derived compounds, cannabidiol and BCP, reverse dopamine neurotoxin-induced degeneration and increase the expression and activity of antioxidant and neuroprotective enzymes (García-Arencibia et al., 2007; Ojha et al., 2016; Wang et al., 2018). These promising results strongly support the development of plant-derived medications for SUDs that target the ECS, particularly the CB2 receptor.

24.3.3 Plant-Derived Psychedelics for the Treatment of SUDs

The term psychedelic has come to be associated with a broad class of drugs with diverse chemical, pharmacological and psychoactive effects. These compounds have also been described as hallucinogens, entheogens, psychotomimetics and other appellations to denote specific concepts related to these drugs. Hallucinations are an obvious and profound drug effect that can be produced by these drugs; however, these effects are not reliably produced at typical doses, thus causing the term hallucinogen to fall out of favor among scientists (Nichols, 2016). Psychedelics have been used by humans in traditional indigenous rituals for thousands of years (Meyer and Quenzer, 2005). Prototypical psychedelics are representatives of a specific subclass which includes psilocybin, a tryptamine found in several genera of mushrooms; lysergic acid diethylamide (LSD), ergotamine originally derived from ergot fungi; and mescaline, a phenethylamine found in peyote and other cacti (Hofmann, 1970; Shulgin and Shulgin, 1991; Stamets, 1996). Mescaline derived from the peyote cactus was incorporated into the indigenous religious ceremonies of the native people of North America. Likewise, a brew of the *Banisteriopsis caapi* (Spruce ex Griseb.) C.V.Morton vine and the *Psychotria viridis* Ruiz & Pav. shrub called ayahuasca was incorporated into the indigenous religious ceremonies of the native people of South America (Carhart-Harris and Goodwin, 2017). Various analogs of these compounds have been synthesized and tested for bioactivity; however, for the focus of this chapter, the discussion will be largely limited to plant-derived psychedelics (Nichols et al., 2015; Shulgin and Shulgin, 1991, 1997; Shulgin et al., 2011).

The fundamental shared pharmacological property of all psychedelics is high affinity and agonist activity at the serotonin 2A (5-HT_{2A}) receptor. Recent studies support the clinical potential of 5-HT_{2A} receptor agonists. Clinical trials at institutions including Johns Hopkins School of Medicine and NYU School of Medicine in the United States, Imperial College in the United Kingdom, and the University of Basel in Switzerland, have examined the safety and efficacy of psilocybin as a treatment for depression, anxiety and substance abuse (Bogenschutz and Johnson, 2016; Carhart-Harris et al., 2018; Griffiths et al., 2016; Johnson et al., 2017; Kraehenmann et al., 2015; Ross, 2012; Ross et al., 2016). FDA-approved Phase II randomized placebo-controlled clinical trials have demonstrated a profound and long-lasting ability to alleviate depression and anxiety compared to active placebo. The effect size was several-fold higher than commonly found with selective serotonin reuptake inhibitor treatment over the same time course (Griffiths et al., 2016; Ross et al., 2016).

In another FDA-approved trial, psilocybin treatment in heavily tobacco-dependent smokers—whose intake were on average 51 years of age, had a mean of six previous lifetime quitting attempts, smoked 19 cigarettes per day and had smoked for 31 years—resulted in approximately 80% abstinence rates at six months and 67% abstinence rates at 12 months (Johnson et al., 2014; Johnson et al., 2017). Another recent clinical study reported significant decreases in alcohol consumption after one to two sessions of psilocybin in ten alcohol-dependent subjects using an open-label design (Bogenschutz et al., 2015). In this six-month study, the percentage of drinking days decreased from ~40% at baseline to ~12% within four weeks, and for the remaining five months. The same research group is now performing a larger double-blind placebo-controlled clinical trial. Moreover, the FDA has assigned breakthrough therapy status for psilocybin for anxiety and depression (Bogenschutz et al., 2015).

Neuroanatomical and neuropharmacological studies have indicated that 5-HT is an important neurotransmitter able to modify key elements of brain reward circuits. The 5-HT neurons originate in the raphe nuclei and project to the VTA, NAcc, amygdala and PFC, brain areas that are critical to drug reward (Haber and Knutson, 2010; Halliday and Tork, 1989); they innervate the cell bodies and terminals of dopamine neurons, often making direct synaptic contact (Beart and McDonald, 1982; Geyer et al., 1976; Herve et al., 1987; Nedergaard et al., 1988; Parent et al., 1981). Functionally, 5-HT provides tonic and phasic control of dopaminergic systems within the limbic pathway. Systemic administration of selective 5-HT transporter inhibitors, which increase extracellular levels of 5-HT, attenuates drug-induced increases in dopamine neurotransmission. 5-HT_{2A} receptors are highly expressed on the cell bodies of VTA dopamine projection neurons and gamma-aminobutyric acid (GABA) medium spiny neurons in the NAcc (Bubar and Cunningham, 2007; Bubar et al., 2005; Ikemoto et al., 2000; Nocjar et al., 2002).

Supporting the notion of a relationship between 5-HT systems and ethanol reward is the finding that a functional polymorphism in the serotonin transporter both results in lower basal levels of extracellular 5-HT, and with symptoms and changes such as early-onset AUD, an increased propensity to consume alcohol, fewer unpleasant effects reported from alcohol consumption and increased craving for alcohol. Recent studies have shown that the synthetic psychedelic 2,5-dimethoxy-4-iodoamphetamine (DOI) reduces ethanol intake and has a preference for ethanol in the two-bottle choice assay without affecting total fluid intake (Oppong-Damoah et al., 2019b). Pre-treatment with the selective 5-HT_{2A} receptor antagonist M100907 demonstrated that DOI is acting through a 5-HT_{2A} receptor-dependent mechanism to reduce alcohol drinking. Further, it was demonstrated that DOI reduces ethanol-induced place conditioning in the CPP paradigm (Oppong-Damoah et al., 2019b).

Alcohol affects both the innate and adaptive immune systems, leading to increased production of proinflammatory cytokines and impaired production of anti-inflammatory cytokines, which can result in alcohol-induced neuroinflammation (Lippai et al., 2013; Szabo and Saha, 2015). It has been proposed that ethanol-related neuroinflammation can contribute to subsequent ethanol-seeking and -taking behavior. Studies have reported that DOI has profound anti-inflammatory properties, which may support the use of psychedelics against SUDs (Nau et al., 2013; Yu et al., 2008). Furthermore, it has been reported that DOI reverses the proinflammatory effects of ethanol, as indexed by increased nitric oxide production within the cell culture model (Oppong-Damoah et al., 2019b).

Similar to both CB2 receptors and other receptors involved in inflammation, 5-HT_{2A} receptors are highly inducible. Across a variety of models, it has been reported that various stressors increase the density of 5-HT_{2A} receptors; however, it is unclear whether this may be a neuroadaptive response that is engaged to mitigate the effects of the stressor (Berton et al., 1998; McKittrick et al., 1995; Metz and Heal, 1986; Ossowska et al., 2001; Takao et al., 1995; Torda et al., 1988). As with the cannabinoids, the capacity of psychedelics to impart both the reward pathway and drug-induced neuroinflammation suggests that plant-derived psychedelics may offer therapeutic promise for SUDs.

A recent study provided a review of human and animal-based studies that have supported the procognitive effects of psychedelics (Murnane, 2019). Neurotrophic factors are involved in the effects of DOI in synaptic plasticity, a cellular basis for learning and memory (Benekareddy et al., 2013). Consistent with a role in synaptic plasticity, a long series of studies has convincingly demonstrated that LSD enhances associative learning (Gimpl et al., 1979; Harvey et al., 1988; Harvey et al., 1982). This is likely mediated by activation of 5-HT$_{2A}$ receptors as ritanserin inhibits associative learning in humans, while 5-HT$_{1A}$ receptor agonists do not exhibit this effect and upregulation of 5-HT$_{2A}$ receptors in the frontal cortex facilitates learning, while this effect is not attenuated by the depletion of 5-HT (Harvey et al., 2004; Hensman et al., 1991; Romano et al., 2006; Welsh et al., 1998).

Administration of LSD directly into the hippocampus promotes learning (Romano et al., 2010). Administration of the novel psychedelic TCB-2 enhances the consolidation of object memory (Zhang et al., 2013). Ayahuasca enhances learned responses, as does the administration of TCB-2 into the basolateral amygdala (Clinard et al., 2015; Favaro et al., 2015). Viral-mediated induced re-expression of 5-HT$_{2A}$ receptors in mice genetically devoid of 5-HT$_{2A}$ receptors rescues learning and synaptic plasticity deficits (Barre et al., 2016). Microdosing of LSD has been reported to improve cognition in humans, and psilocybin microdosing improves attention in low-performing rats (Higgins et al., 2021; Hutten et al., 2020). As with cannabinoids, the fact that plant-based psychedelics may directly affect the dopamine reward system may improve cognition and behavioral flexibility and may have potential disease-modifying anti-inflammatory and antioxidant effects suggest that psychedelics may provide compelling medications for SUDs.

24.4 CONCLUSIONS

SUDs represent a devastating worldwide public health crisis. The reinforcing, behavioral and addictive effects of drugs of abuse have been closely tied to the dopamine 'reward' system. However, medication development efforts targeting this system have not resulted in a robust arsenal of pharmacotherapeutic tools. SUDs include elements of compulsive binge drug intake patterns, loss of control and maladaptive decision making. Plant-derived medications have shown promise as new treatments for SUDs. In particular, plant-derived compounds that target CB2 receptors or 5-HT$_{2A}$ receptors have been shown to modulate behaviors believed to be dependent on the dopamine reward system. They have been shown to have procognitive effects that could address the impacts of SUDs on compulsive behavior, impulse control and maladaptive decision making. Perhaps most importantly, these potential plant-derived medications have neuroprotective effects, as well as anti-inflammatory and antioxidant effects that may stall or reverse the pathophysiological changes associated with SUDs. These plant-based medicines offer promise and hope in the fight against the worldwide and growing devastation of SUDs.

REFERENCES

Al Mansouri S, Ojha S, Al Maamari E, Al Ameri M, Nurulain SM, Bahi A. 2014. "The cannabinoid receptor 2 agonist, beta-caryophyllene, reduced voluntary alcohol intake and attenuated ethanol-induced place preference and sensitivity in mice." *Pharmacology Biochemistry and Behavior* 124: 260–268. doi: 10.1016/j.pbb.2014.06.025

Anand P, Whiteside G, Fowler CJ, Hohmann AG. 2009. "Targeting CB2 receptors and the endocannabinoid system for the treatment of pain." *Brain Research Reviews* 60: 255–266. doi: 10.1016/j.brainresrev.2008.12.003

Antonazzo M, Botta M, Bengoetxea H, Ruiz-Ortega J, Morera-Herreras T. 2019. "Therapeutic potential of cannabinoids as neuroprotective agents for damaged cells conducing to movement disorders." *International Review of Neurobiology* 146: 229–257. doi: 10.1016/bs.irn.2019.06.012

Ashton JC, Rahman RM, Nair SM, Sutherland BA, Glass M, Appleton I. 2007. "Cerebral hypoxia-ischemia and middle cerebral artery occlusion induce expression of the cannabinoid CB2 receptor in the brain." *Neuroscience Letters* 412: 114–117. doi: 10.1016/j.neulet.2006.10.053

Bab I, Ofek O, Tam J, Rehnelt J, Zimmer A. 2008. "Endocannabinoids and the regulation of bone metabolism." *Journal of Neuroendocrinology* 20: 69–74. doi: 10.1111/j.1365-2826.2008.016'75.x

Barre A, Berthoux C, De Bundel D, Valjent E, Bockaert J, Marin P, Becamel C. 2016. "Presynaptic serotonin 2A receptors modulate thalamocortical plasticity and associative learning." *Proceedings of the National Academy of Sciences of the United States of America* 113: E1382–1391. doi: 10.1073/pnas.1525586113

Bart G. 2012. "Maintenance medication for opiate addiction: The foundation of recovery." *Journal of Addictive Diseases* 31: 207–225. doi: 10.1080/10550887.2012.694598

Beart PM, McDonald D. 1982. "Neurochemical studies of the mesolimbic dopaminergic pathway: [3H]spiperone labels two bindings sites in homogenates of the nucleus accumbens of rat brain." *Journal of Neurochemistry* 39: 1452–1460. doi: 10.1111/j.1471-4159.1982.tb12591.x

Benekareddy M, Nair AR, Dias BG, Suri D, Autry AE, Monteggia LM, Vaidya VA. 2013. "Induction of the plasticity-associated immediate early gene Arc by stress and hallucinogens: Role of brain-derived neurotrophic factor." *International Journal of Neuropsychopharmacology* 16: 405–415. doi: 10.1017/S1461145712000168

Benito C, Kim W-K, Chavarría I, Hillard CJ, Mackie K, Tolón RM, Williams K, Romero J. 2005. "A glial endogenous cannabinoid system is upregulated in the brains of macaques with simian immunodeficiency virus-induced encephalitis." *Journal of Neuroscience* 25: 2530–2536. doi: 10.1523/JNEUROSCI.3923-04.2005

Ben-Yehuda O, Siecke N. 2018. "Crystal methamphetamine: A drug and cardiovascular epidemic." *Journal of the American College of Cardiology Heart Failure* 6: 219–221. doi: 10.1016/j.jchf.2018.01.004

Berton O, Aguerre S, Sarrieau A, Mormede P, Chaouloff F. 1998. "Differential effects of social stress on central serotonergic activity and emotional reactivity in Lewis and spontaneously hypertensive rats." *Neuroscience* 82: 147–159. doi: 10.1016/s0306-4522(97)00282-0

Bogenschutz MP, Forcehimes AA, Pommy JA, Wilcox CE, Barbosa PC, Strassman RJ. 2015. "Psilocybin-assisted treatment for alcohol dependence: A proof-of-concept study." *Journal of Psychopharmacology* (Oxford, England) 29: 289–299. doi: 10.1177/0269881114565144

Bogenschutz MP, Johnson MW. 2016. "Classic hallucinogens in the treatment of addictions." *Progress in Neuro-psychopharmacology and Biological Psychiatry* 64: 250–258. doi: 10.1016/j.pnpbp.2015.03.002

Bubar MJ, Cunningham KA. 2007. "Distribution of serotonin 5-HT2C receptors in the ventral tegmental area." *Neuroscience* 146: 286–297. doi: 10.1016/j.neuroscience.2006.12.071

Bubar MJ, Seitz PK, Thomas ML, Cunningham KA. 2005. "Validation of a selective serotonin 5-HT(2C) receptor antibody for utilization in fluorescence immunohistochemistry studies." *Brain Research* 1063: 105–113. doi: 10.1016/j.brainres.2005.09.050

Carhart-Harris RL, Bolstridge M, Day CMJ, Rucker J, Watts R, Erritzoe DE, Kaelen M, Giribaldi B, Bloomfield M, Pilling S, Rickard JA, Forbes B, Feilding A, Taylor D, Curran HV, Nutt DJ. 2018. "Psilocybin with psychological support for treatment-resistant depression: Six-month follow-up." *Psychopharmacology* (Berl) 235: 399–408. doi: 10.1007/s00213-017-4771-x

Carhart-Harris RL, Goodwin GM. 2017. "The therapeutic potential of psychedelic drugs: Past, present, and future." *Neuropsychopharmacology* 42: 2105–2113. doi: 10.1038/npp.2017.84

Carroll ME, Anker JJ, Perry JL. 2009. "Modeling risk factors for nicotine and other drug abuse in the preclinical laboratory." *Drug and Alcohol Dependence* 104, Suppl 1: S70–78. doi: 10.1016/j.drugalcdep.2008.11.011

Chavan MJ, Wakte PS, Shinde DB. 2010. "Analgesic and anti-inflammatory activity of Caryophyllene oxide from *Annona squamosa* L. bark." *Phytomedicine: International Journal of Phytotherapy and Phytopharmacology* 17: 149–151. doi: 10.1016/j.phymed.2009.05.016

Chen T, Spiller HA, Badeti J, Funk AR, Zhu M, Smith GA. 2021. "Methamphetamine exposures reported to United States poison control centers, 2000–2019." *Clinical Toxicology* 59, 8: 705–714. doi: 10.1080/15563650.2020.1861287

Christian C, Paulson G. 1958. "Severe motility disturbance after small doses of prochlorperazine." *New England Journal of Medicine* 259: 828–830. doi: 10.1056/NEJM195810232591708

Clinard CT, Bader LR, Sullivan MA, Cooper MA. 2015. "Activation of 5-HT2a receptors in the basolateral amygdala promotes defeat-induced anxiety and the acquisition of conditioned defeat in Syrian hamsters." *Neuropharmacology* 90: 102–112. doi: 10.1016/j.neuropharm.2014.11.016

Compton WM, Jones CM, Baldwin GT. 2016. "Relationship between nonmedical prescription-opioid use and heroin use." *New England Journal of Medicine* 374: 154–163. doi: 10.1056/NEJMra1508490

Correia C, Romieu P, Olmstead MC, Befort K. 2020. "Can cocaine-induced neuroinflammation explain maladaptive cocaine-associated memories?" *Neuroscience and Biobehavioral Reviews* 111: 69–83. doi: 10.1016/j.neubiorev.2020.01.001

Costa PPPR. 2016. "*Psychotria viridis*." Obtained from Wikimedia under the Creative Commons Attribution-Share Alike 4.0 International license. https://commons.wikimedia.org/wiki/File:PsychotriaviridisFruto DSC74.jpg (Accessed 3 July 2021)

de Fonseca RF, Roberts AJ, Bilbao A, Koob GF, Navarro M. 1999. "Cannabinoid receptor antagonist SR141716A decreases operant ethanol self administration in rats exposed to ethanol-vapor chambers." *Zhongguo yao li xue bao= Acta pharmacologica Sinica* 20: 1109–1114.

Devane WA, Hanus L, Breuer A, Pertwee RG, Stevenson LA, Griffin G, Gibson D, Mandelbaum A, Etinger A, Mechoulam R. 1992. "Isolation and structure of a brain constituent that binds to the cannabinoid receptor." *Science* 258: 1946–1949. doi: 10.1126/science.1470919

Devlin RJ, Henry JA. 2008. "Clinical review: Major consequences of illicit drug consumption." *Critical care (London, England)* 12: 202. doi: 10.1186/cc6166

Favaro VM, Yonamine M, Soares JC, Oliveira MG. 2015. "Effects of long-term ayahuasca administration on memory and anxiety in rats." *Plos One* 10: e0145840. doi: 10.1371/journal.pone.0145840

Fleckenstein AE, Volz TJ, Riddle EL, Gibb JW, Hanson GR. 2007. "New insights into the mechanism of action of amphetamines." *Annual Review of Pharmacology and Toxicology* 47: 681–698. doi: 10.1146/annurev.pharmtox.47.120505.105140

Fuller RK, Branchey L, Brightwell DR, Derman RM, Emrick CD, Iber FL, James KE, Lacoursiere RB, Lee KK, Lowenstam I. 1986. "Disulfiram treatment of alcoholism: A Veterans Administration cooperative study." *Journal of the American Medical Association* 256: 1449–1455.

Garbutt JC, West SL, Carey TS, Lohr KN, Crews FT. 1999. "Pharmacological treatment of alcohol dependence: A review of the evidence." *Journal of the American Medical Association* 281: 1318–1325. doi: 10.1001/jama.281.14.1318

García-Arencibia M, González S, de Lago E, Ramos JA, Mechoulam R, Fernández-Ruiz J. 2007. "Evaluation of the neuroprotective effect of cannabinoids in a rat model of Parkinson's disease: Importance of antioxidant and cannabinoid receptor-independent properties." *Brain Research* 1134: 162–170. doi: 10.1016/j.brainres.2006.11.063

Gaval-Cruz M, Schroeder JP, Liles LC, Javors MA, Weinshenker D. 2008. "Effects of disulfiram and dopamine beta-hydroxylase knockout on cocaine-induced seizures." *Pharmacology Biochemistry and Behavior* 89: 556–562. doi: 10.1016/j.pbb.2008.02.009

Geyer MA, Puerto A, Dawsey WJ, Knapp S, Bullard WP, Mandell AJ. 1976. "Histologic and enzymatic studies of the mesolimbic and mesostriatal serotonergic pathways." *Brain Research* 106: 241–256. doi: 10.1016/0006-8993(76)91023-4

Ghelardini C, Galeotti N, Di Cesare Mannelli L, Mazzanti G, Bartolini A. 2001. "Local anaesthetic activity of beta-caryophyllene." *Farmaco (Societa chimica italiana: 1989)* 56: 387–389. doi: 10.1016/s0014-827x(01)01092-8

Gimpl MP, Gormezano I, Harvey JA. 1979. "Effects of LSD on learning as measured by classical conditioning of the rabbit nictitating membrane response." *The Journal of Pharmacology and Experimental Therapeutics* 208: 330–334.

Goeders NE. 2002. "Stress and cocaine addiction." *The Journal of Pharmacology and Experimental Therapeutics* 301: 785–789. doi: 10.1124/jpet.301.3.785

Grant BF, Goldstein RB, Saha TD, Chou SP, Jung J, Zhang H, Pickering RP, Ruan WJ, Smith SM, Huang B. 2015. "Epidemiology of DSM-5 alcohol use disorder: Results from the National Epidemiologic Survey on Alcohol and Related Conditions III." *JAMA Psychiatry* 72: 757–766. doi: 10.1001/jamapsychiatry.2015.0584

Griffiths RR, Johnson MW, Carducci MA, Umbricht A, Richards WA, Richards BD, Cosimano MP, Klinedinst MA. 2016. "Psilocybin produces substantial and sustained decreases in depression and anxiety in patients with life-threatening cancer: A randomized double-blind trial." *Journal of Psychopharmacology* (Oxford, England) 30: 1181–1197. doi: 10.1177/0269881116675513

Guilding, L. 1832. "*Piper nigrum*." Obtained from Wikimedia under the Public Domain license. https://commons.wikimedia.org/wiki/File:Piper_nigrum_drawing_1832.jpg (Accessed 3 July 2021)

Haber SN, Calzavara R. 2009. "The cortico-basal ganglia integrative network: The role of the thalamus." *Brain Research Bulletin* 78: 69–74. doi: 10.1016/j.brainresbull.2008.09.013

Haber SN, Knutson B. 2010. "The reward circuit: Linking primate anatomy and human imaging." *Neuropsychopharmacology* 35: 4–26. doi: 10.1038/npp.2009.129

Halliday GM, Tork I. 1989. "Serotonin-like immunoreactive cells and fibres in the rat ventromedial mesencephalic tegmentum." *Brain Research Bulletin* 22: 725–735. doi: 10.1016/0361-9230(89)90092-0

Han B, Cotto J, Etz K, Einstein EB, Compton WM, Volkow ND. 2021. "Methamphetamine overdose deaths in the US by sex and race and ethnicity." *JAMA Psychiatry*. doi: 10.1001/jamapsychiatry.2020.4321

Harvey JA, Gormezano I, Cool-Hauser VA. 1982. "Effects of d-lysergic acid diethylamide, d-2-bromolysergic acid diethylamide, dl-2,5-dimethoxy-4-methylamphetamine and d-amphetamine on classical conditioning of the rabbit nictitating membrane response." *The Journal of Pharmacology and Experimental Therapeutics* 221: 289–294.

Harvey JA, Gormezano I, Cool-Hauser VA, Schindler CW. 1988. "Effects of LSD on classical conditioning as a function of CS-UCS interval: Relationship to reflex facilitation." *Pharmacology Biochemistry and Behavior* 30: 433–441. doi: 10.1016/0091-3057(88)90477-7

Harvey JA, Quinn JL, Liu R, Aloyo VJ, Romano AG. 2004. "Selective remodeling of rabbit frontal cortex: Relationship between 5-HT2A receptor density and associative learning." *Psychopharmacology (Berl)* 172: 435–442. doi: 10.1007/s00213-003-1687-4

Hedenhös, S. 2012. "*Cannabis sativa*," Obtained from Wikimedia under the Creative Commons Attribution-Share Alike 4.0 International license. https://commons.wikimedia.org/wiki/File:Hampa_Cannabis_sativa_L.jpg (Accessed 3 July 2021)

Hensman R, Guimaraes FS, Wang M, Deakin JF. 1991. "Effects of ritanserin on aversive classical conditioning in humans." *Psychopharmacology (Berl)* 104: 220–224. doi: 10.1007/BF02244182

Herve D, Pickel VM, Joh TH, Beaudet A. 1987. "Serotonin axon terminals in the ventral tegmental area of the rat: Fine structure and synaptic input to dopaminergic neurons." *Brain research* 435: 71–83. doi: 10.1016/0006-8993(87)91588-5

Higgins GA, Carroll NK, Brown M, MacMillan C, Silenieks LB, Thevarkunnel S, Izhakova J, Magomedova L, DeLannoy I, Sellers EM. 2021. "Low doses of psilocybin and ketamine enhance motivation and attention in poor performing rats: Evidence for an antidepressant property." *Frontiers in Pharmacology* 12: 640241. doi: 10.3389/fphar.2021.640241

Hir D. 2012. "*Lophophora williamsii*." Obtained from Wikimedia under the Creative Commons Attribution-Share Alike 3.0 Unported license. https://commons.wikimedia.org/wiki/File:Lophophora_williamsii_Bl%C3%BCte.JPG (Accessed 3 July 2021)

Hofmann A. 1970. "Notes and documents concerning the discovery of LSD." *Agents Actions* 1: 148–150. doi: 10.1007/BF01986673

Holthoff-Detto VA, Kessler J, Herholz K, Bönner H, Pietrzyk U, Würker M, Ghaemi M, Wienhard K, Wagner R, Heiss WD. 1997. "Functional effects of striatal dysfunction in Parkinson disease." *Archives of Neurology* 54: 145–150. doi: 10.1001/archneur.1997.00550140025008

Houchi H, Babovic D, Pierrefiche O, Ledent C, Daoust M, Naassila M. 2005. "CB 1 receptor knockout mice display reduced ethanol-induced conditioned place preference and increased striatal dopamine D2 receptors." *Neuropsychopharmacology* 30: 339. doi: 10.1038/sj.npp.1300568

Howell LL, Murnane KS. 2011. "Nonhuman primate positron emission tomography neuroimaging in drug abuse research." *The Journal of Pharmacology and Experimental Therapeutics* 337: 324–334. doi: 10.1124/jpet.108.136689

Howes M-JR, Houghton PJ. 2003. "Plants used in Chinese and Indian traditional medicine for improvement of memory and cognitive function." *Pharmacology Biochemistry and Behavior* 75: 513–527. doi: 10.1016/s0091-3057(03)00128-x

Hsu CC, Madsen TE, O'Gorman E, Gourley SL, Rainnie DG. 2020. "Reward-related dynamical coupling between basolateral amygdala and nucleus accumbens." *Brain Structure & Function* 225: 1873–1888. doi: 10.1007/s00429-020-02099-2

Hungund BL, Szakall I, Adam A, Basavarajappa BS, Vadasz C. 2003. "Cannabinoid CB1 receptor knockout mice exhibit markedly reduced voluntary alcohol consumption and lack alcohol-induced dopamine release in the nucleus accumbens." *Journal of Neurochemistry* 84: 698–704. doi: 10.1046/j.1471-4159.2003.01576.x

Hutten N, Mason NL, Dolder PC, Theunissen EL, Holze F, Liechti ME, Feilding A, Ramaekers JG, Kuypers KPC. 2020. "Mood and cognition after administration of low LSD doses in healthy volunteers: A placebo controlled dose-effect finding study." *European Neuropsychopharmacology* 41: 81–91. doi: 10.1016/j.euroneuro.2020.10.002

Hyatt WS, Hirsh CE, Russell LN, Chitre NM, Murnane KS, Rice KC, Fantegrossi WE. 2020. "The synthetic cathinone 3,4-methylenedioxypyrovalerone increases impulsive action in rats." *Behavioural Pharmacology* 31: 309–321. doi: 10.1097/FBP.0000000000000548

Ikemoto K, Nishimura A, Okado N, Mikuni M, Nishi K, Nagatsu I. 2000. "Human midbrain dopamine neurons express serotonin 2A receptor: An immunohistochemical demonstration." *Brain Research* 853: 377–380. doi: 10.1016/s0006-8993(99)02237-4

Janno S, Holi M, Tuisku K, Wahlbeck K. 2004. "Prevalence of neuroleptic-induced movement disorders in chronic schizophrenia inpatients." *American Journal of Psychiatry* 161: 160–163. doi: 10.1176/appi.ajp.161.1.160

Johnson MW, Garcia-Romeu A, Cosimano MP, Griffiths RR. 2014. "Pilot study of the 5-HT2AR agonist psilocybin in the treatment of tobacco addiction." *Journal of psychopharmacology (Oxford, England)* 28: 983–992. doi: 10.1177/0269881114548296

Johnson MW, Garcia-Romeu A, Griffiths RR. 2017. "Long-term follow-up of psilocybin-facilitated smoking cessation." *The American Journal of Drug and Alcohol Abuse* 43: 55–60. doi: 10.3109/00952990.2016.1170135

Jorgensen CH, Pedersen B, Tonnesen H. 2011. "The efficacy of disulfiram for the treatment of alcohol use disorder." *Alcoholism, Clinical and Experimental Research* 35: 1749–1758. doi: 10.1111/j.1530-0277.2011.01523.x

Klauke A-L, Racz I, Pradier B, Markert A, Zimmer A, Gertsch J, Zimmer A. 2014. "The cannabinoid CB2 receptor-selective phytocannabinoid beta-caryophyllene exerts analgesic effects in mouse models of inflammatory and neuropathic pain." *European Neuropsychopharmacology* 24: 608–620. doi: 10.1016/j.euroneuro.2013.10.008

Klegeris A, Bissonnette CJ, McGeer PL. 2003. "Reduction of human monocytic cell neurotoxicity and cytokine secretion by ligands of the cannabinoid-type CB2 receptor." *British Journal of Pharmacology* 139: 775–786. doi: 10.1038/sj.bjp.0705304

Knight EJ, Min HK, Hwang SC, Marsh MP, Paek S, Kim I, Felmlee JP, Abulseoud OA, Bennet KE, Frye MA, Lee KH. 2013. "Nucleus accumbens deep brain stimulation results in insula and prefrontal activation: A large animal FMRI study." *Plos One* 8: e56640. doi: 10.1371/journal.pone.0056640

Kolodny A, Courtwright DT, Hwang CS, Kreiner P, Eadie JL, Clark TW, Alexander GC. 2015. "The prescription opioid and heroin crisis: A public health approach to an epidemic of addiction." *Annual Review of Public Health* 36: 559–574. doi: 10.1146/annurev-publhealth-031914-122957

Koob GF, Moal ML. 1997. "Drug abuse: Hedonic homeostatic dysregulation." *Science* 278: 52–58. doi: 10.1126/science.278.5335.52

Kraehenmann R, Preller KH, Scheidegger M, Pokorny T, Bosch OG, Seifritz E, Vollenweider FX. 2015. "Psilocybin-induced decrease in amygdala reactivity correlates with enhanced positive mood in healthy volunteers." *Biological Psychiatry* 78: 572–581. doi: 10.1016/j.biopsych.2014.04.010

Kranzler HR, Van Kirk J. 2001. "Efficacy of naltrexone and acamprosate for alcoholism treatment: A meta-analysis." *Alcoholism: Clinical and Experimental Research* 25: 1335–1341.

Lacagnina MJ, Rivera PD, Bilbo SD. 2017. "Glial and neuroimmune mechanisms as critical modulators of drug use and abuse." *Neuropsychopharmacology* 42: 156–177. doi: 10.1038/npp.2016.121

LaRowe SD, Kalivas PW, Nicholas JS, Randall PK, Mardikian PN, Malcolm RJ. 2013. "A double-blind placebo-controlled trial of N-acetylcysteine in the treatment of cocaine dependence." *The American Journal on Addictions/American Academy of Psychiatrists in Alcoholism and Addictions* 22: 443–452. doi: 10.1111/j.1521-0391.2013.12034.x

LaRowe SD, Myrick H, Hedden S, Mardikian P, Saladin M, McRae A, Brady K, Kalivas PW, Malcolm R. 2007. "Is cocaine desire reduced by N-acetylcysteine?" *The American Journal of Psychiatry* 164: 1115–1117. doi: 10.1176/ajp.2007.164.7.1115

Ledent C, Valverde O, Cossu G, Petitet F, Aubert J-F, Beslot F, Böhme GA, Imperato A, Pedrazzini T, Roques BP. 1999. "Unresponsiveness to cannabinoids and reduced addictive effects of opiates in CB1 receptor knockout mice." *Science* 283: 401–404. doi: 10.1126/science.283.5400.401

Li X, Zeric T, Kambhampati S, Bossert JM, Shaham Y. 2015. "The central amygdala nucleus is critical for incubation of methamphetamine craving." *Neuropsychopharmacology* 40: 1297–1306. doi: 10.1038/npp.2014.320

Lichtman AH, Martin BR. 1997. "The selective cannabinoid antagonist SR 141716A blocks cannabinoid-induced antinociception in rats." *Pharmacology Biochemistry and Behavior* 57: 7–12. doi: 10.1016/s0091-3057(96)00121-9

Lindsey LP, Daphney CM, Aboagyewaah O-D, Uchakin PN, Abney SE, Uchakina ON, Khusial RD, Akil A, Murnane KS. 2019. "The cannabinoid receptor 2 agonist, β-caryophyllene, improves working memory and reduces circulating levels of specific proinflammatory cytokines in aged male mice." *Behavioural Brain Research*: 112012. doi: 10.1016/j.bbr.2019.112012

Lippai D, Bala S, Petrasek J, Csak T, Levin I, Kurt-Jones EA, Szabo G. 2013. "Alcohol-induced IL-1β in the brain is mediated by NLRP3/ASC inflammasome activation that amplifies neuroinflammation." *Journal of Leukocyte Biology* 94: 171–182. doi: 10.1189/jlb.1212659

Little KY, Ramssen E, Welchko R, Volberg V, Roland CJ, Cassin B. 2009. "Decreased brain dopamine cell numbers in human cocaine users." *Psychiatry Research* 168: 173–180. doi: 10.1016/j.psychres.2008.10.034

Madayag A, Lobner D, Kau KS, Mantsch JR, Abdulhameed O, Hearing M, Grier MD, Baker DA. 2007. "Repeated N-acetylcysteine administration alters plasticity-dependent effects of cocaine." *Journal of Neuroscience* 27: 13968–13976. doi: 10.1523/JNEUROSCI.2808-07.2007

Madras BK, Fahey MA, Bergman J, Canfield DR, Spealman RD. 1989. "Effects of cocaine and related drugs in nonhuman primates. I. [3H]cocaine binding sites in caudate-putamen." *The Journal of Pharmacology and Experimental Therapeutics* 251: 131–141.

Maldonado R, Valverde O, Berrendero F. 2006. "Involvement of the endocannabinoid system in drug addiction." *Trends in Neurosciences* 29: 225–232. doi: 10.1016/j.tins.2006.01.008

Mamo DC, Sweet RA, Keshavan MS. 1999. "Managing antipsychotic-induced Parkinsonism." *Drug Safety* 20: 269–275. doi: 10.2165/00002018-199920030-00006

Manchikanti L, Fellows B, Ailinani H, Pampati V. 2010. "Therapeutic use, abuse, and nonmedical use of opioids: A ten-year perspective." *Pain Physician* 13: 401–435.

Manchikanti L, Singh A. 2008. "Therapeutic opioids: A ten-year perspective on the complexities and complications of the escalating use, abuse, and nonmedical use of opioids." *Pain Physician* 11: S63–88.

Martin M, Ledent C, Parmentier M, Maldonado R, Valverde O. 2002. "Involvement of CB1 cannabinoid receptors in emotional behaviour." *Psychopharmacology (Berl)* 159: 379–387. doi: 10.1007/s00213-001-0946-5

Matsuda LA, Lolait SJ, Brownstein MJ, Young AC, Bonner TI. 1990. "Structure of a cannabinoid receptor and functional expression of the cloned cDNA." *Nature* 346: 561–564. doi: 10.1038/346561a0

McCormick. 2012. "*Syzygium aromaticum*." Obtained from Wikimedia under the Public Domain license. https://commons.wikimedia.org/wiki/File:Cloves-penang-zanzibar.png (Accessed 3 July 2021)

McKetin R, Lubman DI, Baker AL, Dawe S, Ali RL. 2013. "Dose-related psychotic symptoms in chronic methamphetamine users: Evidence from a prospective longitudinal study." *JAMA Psychiatry* 70: 319–324. doi: 10.1001/jamapsychiatry.2013.283

McKittrick CR, Blanchard DC, Blanchard RJ, McEwen BS, Sakai RR. 1995. "Serotonin receptor binding in a colony model of chronic social stress." *Biological Psychiatry* 37: 383–393. doi: 10.1016/0006-3223(94)00152-s

Metz A, Heal DJ. 1986. "In mice repeated administration of electroconvulsive shock or desmethylimipramine produces rapid alterations in 5-HT2-mediated head-twitch responses and cortical 5-HT2 receptor number." *European Journal of Pharmacology* 126: 159–162. doi: 10.1016/0014-2999(86)90754-5

Meyer JS, Quenzer LF. 2005. *Psychopharmacology: Drugs, the brain, and behavior.* Sinauer Associates, Sunderland, MA

Miller AH, Haroon E, Felger JC. 2017. "Therapeutic implications of brain-immune interactions: Treatment in translation." *Neuropsychopharmacology* 42: 334–359. doi: 10.1038/npp.2016.167

Miller AH, Haroon E, Raison CL, Felger JC. 2013. "Cytokine targets in the brain: Impact on neurotransmitters and neurocircuits." *Depression and Anxiety* 30: 297–306. doi: 10.1002/da.22084

Miller EK, Cohen JD. 2001. "An integrative theory of prefrontal cortex function." *Annual Review of Neuroscience* 24: 167–202. doi: 10.1146/annurev.neuro.24.1.167

Mokdad AH, Marks JS, Stroup DF, Gerberding JL. 2004. "Actual causes of death in the United States, 2000." *JAMA* 291: 1238–1245. doi: 10.1001/jama.291.10.1238

Moreira FP, Medeiros JR, Lhullier AC, Souza LD, Jansen K, Portela LV, Lara DR, da Silva RA, Wiener CD, Oses JP. 2016. "Cocaine abuse and effects in the serum levels of cytokines IL-6 and IL-10." *Drug and Alcohol Dependence* 158: 181–185. doi: 10.1016/j.drugalcdep.2015.11.024

Munro S, Thomas KL, Abu-Shaar M. 1993. "Molecular characterization of a peripheral receptor for cannabinoids." *Nature* 365: 61. doi: 10.1038/365061a0

Murnane KS. 2019. "Serotonin 2A receptors are a stress response system: Implications for post-traumatic stress disorder." *Behavioural Pharmacology*. doi: 10.1097/FBP.0000000000000459

Murnane KS, Perrine SA, Finton BJ, Galloway MP, Howell LL, Fantegrossi WE. 2012. "Effects of exposure to amphetamine derivatives on passive avoidance performance and the central levels of monoamines and their metabolites in mice: Correlations between behavior and neurochemistry." *Psychopharmacology (Berl)* 220: 495–508. doi: 10.1007/s00213-011-2504-0

Nau F, Jr., Yu B, Martin D, Nichols CD. 2013. "Serotonin 5-HT2A receptor activation blocks TNF-alpha mediated inflammation *in vivo*." *Plos One* 8: e75426. doi: 10.1371/journal.pone.0075426

Nedergaard S, Bolam JP, Greenfield SA. 1988. "Facilitation of a dendritic calcium conductance by 5-hydroxytryptamine in the substantia nigra." *Nature* 333: 174–177. doi: 10.1038/333174a0

Nichols DE. 2016. "Psychedelics." *Pharmacological Reviews* 68: 264–355. doi: 10.1124/pr.115.011478

Nichols DE, Sassano MF, Halberstadt AL, Klein LM, Brandt SD, Elliott SP, Fiedler WJ. 2015. "N-benzyl-5-methoxytryptamines as potent serotonin 5-HT2 receptor family agonists and comparison with a series of phenethylamine analogues." *ACS Chemical Neuroscience* 6: 1165–1175. doi: 10.1021/cn500292d

Nocjar C, Roth BL, Pehek EA. 2002. "Localization of 5-HT(2A) receptors on dopamine cells in subnuclei of the midbrain A10 cell group." *Neuroscience* 111: 163–176. doi: 10.1016/s0306-4522(01)00593-0

Northcutt AL, Hutchinson MR, Wang X, Baratta MV, Hiranita T, Cochran TA, Pomrenze MB, Galer EL, Kopajtic TA, Li CM, Amat J, Larson G, Cooper DC, Huang Y, O'Neill CE, Yin H, Zahniser NR, Katz JL, Rice KC, Maier SF, Bachtell RK, Watkins LR. 2015. "DAT isn't all that: Cocaine reward and reinforcement require Toll-like receptor 4 signaling." *Molecular Psychiatry* 20: 1525–1537. doi: 10.1038/mp.2014.177

Ojha S, Javed H, Azimullah S, Haque ME. 2016. "β-Caryophyllene, a phytocannabinoid attenuates oxidative stress, neuroinflammation, glial activation, and salvages dopaminergic neurons in a rat model of Parkinson disease." *Molecular and Cellular Biochemistry* 418: 59–70. doi: 10.1007/s11010-016-2733-y

Oppong-Damoah A, Blough BE, Makriyannis A, Murnane KS. 2019a. "The sesquiterpene beta-caryophyllene oxide attenuates ethanol drinking and place conditioning in mice." *Heliyon* 5: e01915. doi: 10.1016/j.heliyon.2019.e01915

Oppong-Damoah A, Curry KE, Blough BE, Rice KC, Murnane KS. 2019b. "Effects of the synthetic psychedelic 2,5-dimethoxy-4-iodoamphetamine (DOI) on ethanol consumption and place conditioning in male mice." *Psychopharmacology (Berl)* 236: 3567–3578. doi: 10.1007/s00213-019-05328-7

Oppong-Damoah A, Gannon BM, Murnane KS. 2021. "The endocannabinoid system and alcohol dependence: Will cannabinoid receptor 2 agonism be more fruitful than cannabinoid receptor 1 antagonism?" *CNS & Neurological Disorders Drug Targets*. doi: 10.2174/1871527320666210211115007

Ormeno E, Baldy V, Ballini C, Fernandez C. 2008. "Production and diversity of volatile terpenes from plants on calcareous and siliceous soils: Effect of soil nutrients." *Journal of Chemical Ecology* 34: 1219–1229. doi: 10.1007/s10886-008-9515-2

Ossowska G, Nowa G, Kata R, Klenk-Majewska B, Danilczuk Z, Zebrowska-Lupina I. 2001. "Brain monoamine receptors in a chronic unpredictable stress model in rats." *Journal of Neural Transmission* 108: 311–319. doi: 10.1007/s007020170077

Parent A, Descarries L, Beaudet A. 1981. "Organization of ascending serotonin systems in the adult rat brain. A radioautographic study after intraventricular administration of [3H]5-hydroxytryptamine." *Neuroscience* 6: 115–138. doi: 10.1016/0306-4522(81)90050-6

Perfumi M, Santoni M, Cippitelli A, Ciccocioppo R, Froldi R, Massi M. 2003. "Hypericum perforatum CO2 extract and opioid receptor antagonists act synergistically to reduce ethanol intake in alcohol-preferring rats." *Alcoholism, Clinical and Experimental Research* 27: 1554–1562. doi: 10.1097/01.ALC.0000092062.60924.56

Price DA, Martinez AA, Seillier A, Koek W, Acosta Y, Fernandez E, Strong R, Lutz B, Marsicano G, Roberts JL, Giuffrida A. 2009. "WIN55,212–2, a cannabinoid receptor agonist, protects against nigrostriatal cell loss in the 1-methyl-4-phenyl-1,2,3,6-tetrahydropyridine mouse model of Parkinson's disease." *The European Journal of Neuroscience* 29: 2177–2186. doi: 10.1111/j.1460-9568.2009.06764.x

Ratano P, Palmery M, Trezza V, Campolongo P. 2017. "Cannabinoid modulation of memory consolidation in rats: Beyond the role of cannabinoid receptor subtype 1." *Frontiers in Pharmacology* 8: 200. doi: 10.3389/fphar.2017.00200

Reith MEA, Meisler BE, Sershen H, Lajtha A. 1986. "Structural requirements for cocaine congeners to interact with dopamine and serotonin uptake sites in mouse brain and to induce stereotyped behavior." *Biochemical Pharmacology* 35: 1123–1129.

Riegel AC, Lupica CR. 2004. "Independent presynaptic and postsynaptic mechanisms regulate endocannabinoid signaling at multiple synapses in the ventral tegmental area." *Journal of Neuroscience* 24: 11070–11078.

Ritz MC, Lamb RJ, Goldberg SR, Kuhar MJ. 1987. "Cocaine receptors on dopamine transporters are related to self-administration of cocaine." *Science* 237: 1219–1223.

Rivera P, Fernandez-Arjona MDM, Silva-Pena D, Blanco E, Vargas A, Lopez-Avalos MD, Grondona JM, Serrano A, Pavon FJ, Rodriguez de Fonseca F, Suarez J. 2018. "Pharmacological blockade of fatty acid amide hydrolase (FAAH) by URB597 improves memory and changes the phenotype of hippocampal microglia despite ethanol exposure." *Biochemical Pharmacology* 157: 244–257.

Romano AG, Quinn JL, Li L, Dave KD, Schindler EA, Aloyo VJ, Harvey JA. 2010. "Intrahippocampal LSD accelerates learning and desensitizes the 5-HT(2A) receptor in the rabbit." *Psychopharmacology (Berl)* 212: 441–448.

Romano AG, Quinn JL, Liu R, Dave KD, Schwab D, Alexander G, Aloyo VJ, Harvey JA. 2006. "Effect of serotonin depletion on 5-HT2A-mediated learning in the rabbit: Evidence for constitutive activity of the 5-HT2A receptor *in vivo*." *Psychopharmacology (Berl)* 184: 173–181.

Rosner S, Hackl-Herrwerth A, Leucht S, Lehert P, Vecchi S, Soyka M. 2010a. "Acamprosate for alcohol dependence." *The Cochrane Database of Systematic Reviews*: Cd004332.

Rosner S, Hackl-Herrwerth A, Leucht S, Vecchi S, Srisurapanont M, Soyka M. 2010b. "Opioid antagonists for alcohol dependence." *The Cochrane Database of Systematic Reviews*: Cd001867.

Ross S. 2012. "Serotonergic hallucinogens and emerging targets for addiction pharmacotherapies." *Psychiatric Clinics of North America* 35: 357–374.

Ross S, Bossis A, Guss J, Agin-Liebes G, Malone T, Cohen B, Mennenga SE, Belser A, Kalliontzi K, Babb J, Su Z, Corby P, Schmidt BL. 2016. "Rapid and sustained symptom reduction following psilocybin treatment for anxiety and depression in patients with life-threatening cancer: A randomized controlled trial." *Journal of Psychopharmacology (Oxford, England)* 30: 1165–1180.

Sacks JJ, Gonzales KR, Bouchery EE, Tomedi LE, Brewer RD. 2015. "2010 national and state costs of excessive alcohol consumption." *American Journal of Preventive Medicine* 49: e73–e79.

Scienceman71.2013. "*Psilocybe semilanceata*." Obtained from Wikimedia under the Creative Commons Attribution-Share Alike 4.0 International license. https://commons.wikimedia.org/wiki/File:IMAG0172.jpg (Accessed 3 July 2021)

Shaull, L. 2021. "Ergot on wheat." Obtained from Wikimedia under the Creative Commons Attribution-Share Alike 2.0 Generic license. https://commons.wikimedia.org/wiki/File:Ergot_on_wheat_at_the_Dassel_History_Center_and_Ergot_Museum.jpg (Accessed 3 July 2021)

Shulgin A, Shulgin A. 1991. *PIHKAL: A chemical love story*. Transform Press, Berkeley, CA, USA

Shulgin A, Shulgin A. 1997. *TIHKAL: The continuation*. Transform Press, Berkeley, CA, USA, p. 804

Shulgin AT, Manning T, Daley PF. 2011. *The Shulgin index*. Transform Press 1, Berkeley, CA, USA

Sloan ME, Gowin JL, Ramchandani VA, Hurd YL, Le Foll B. 2017. "The endocannabinoid system as a target for addiction treatment: Trials and tribulations." *Neuropharmacology* 124: 73–83.

Smith HR, Beveridge TJR, Nader SH, Nader MA, Porrino LJ. 2019. "Regional elevations in microglial activation and cerebral glucose utilization in frontal white matter tracts of rhesus monkeys following prolonged cocaine self-administration." *Brain Structure & Function* 224: 1417–1428.

Stamets P. 1996. *Psilocybin mushrooms of the world: An identification guide*. Ten Speed Press, Berkeley, CA, USA

Substance Abuse and Mental Health Services Administration (SAMHSA) USPHS. 2010. *Results from the 2006 national survey on drug use and health: National findings NSDUH series H-32*. Office of Applied Studies, Rockville, MD

Sugiura T, Kondo S, Sukagawa A, Nakane S, Shinoda A, Itoh K, Yamashita A, Waku K. 1995. "2-Arachidonoylgylcerol: A possible endogenous cannabinoid receptor ligand in brain." *Biochemical and Biophysical Research Communications* 215: 89–97.

Szabo G, Saha B. 2015. "Alcohol's effect on host defense." *Alcohol Research: Current Reviews* 37: 159–170.

T137. 2010. "*Rosmarinus officinalis*." Obtained from Wikimedia under the Creative Commons Attribution-Share Alike 3.0 Unported license. https://commons.wikimedia.org/wiki/File:Rosmarinus_officinalis_(1).jpg (Accessed 3 July 2021)

Takao K, Nagatani T, Kitamura Y, Kawasaki K, Hayakawa H, Yamawaki S. 1995. "Chronic forced swim stress of rats increases frontal cortical 5-HT2 receptors and the wet-dog shakes they mediate, but not frontal cortical beta-adrenoceptors." *European Journal of Pharmacology* 294: 721–726.

Terpsichore. 2004. "*Banisteriopsis caapi*." Obtained from Wikimedia under the Creative Commons Attribution-Share Alike 3.0 Unported, 2.5 Generic, 2.0 Generic and 1.0 Generic license. https://commons.wikimedia.org/wiki/File:Banisteriopsis_caapi_vine.JPG (Accessed 3 July 2021)

Tinofrey. 2005. "*Syzygium aromaticum*." Obtained from Wikimedia under the Public Domain license. https://commons.wikimedia.org/wiki/File:Syzygium_aromaticum_on_tree.jpg (Accessed 3 July 2021)

Torda T, Culman J, Cechova E, Murgas K. 1988. "3-H-ketanserin (serotonin type 2) binding in the rat frontal cortex: Effect of immobilization stress." *Endocrinologia experimentalis* 22: 99–105.

Turowski P, Kenny BA. 2015. "The blood-brain barrier and methamphetamine: Open sesame?" *Frontiers in Neuroscience* 9: 156.

United Nations Office on Drugs and Crime. 2018. "Analysis of Drug Markets: Opiates, cocaine, cannabis, synthetic drugs." *World Drug Report*.

US Drug Enforcement Agency, 2008. "*Psychotria viridis* dried leaves." Obtained from Wikimedia under the Public Domain license. https://commons.wikimedia.org/wiki/File:Psychotria-viridis-dried-leaves.jpg (Accessed 3 July 2021)

Venniro M, Russell TI, Ramsey LA, Richie CT, Lesscher HMB, Giovanetti SM, Messing RO, Shaham Y. 2020. "Abstinence-dependent dissociable central amygdala microcircuits control drug craving." *Proceedings of the National Academy of Sciences of the United States of America* 117: 8126–8134.

Vinayaraj. 2013. "*Piper nigrum*." Obtained from Wikimedia under the Creative Commons Attribution-Share Alike 3.0 Unported license. https://commons.wikimedia.org/wiki/File:Piper_nigrum_03.JPG (Accessed 3 July 2021)

Viveros-Paredes JM, González-Castañeda RE, Gertsch J, Chaparro-Huerta V, López-Roa RI, Vázquez-Valls E, Beas-Zarate C, Camins-Espuny A, Flores-Soto ME. 2017. "Neuroprotective effects of β-caryophyllene against dopaminergic neuron injury in a murine model of Parkinson's disease induced by MPTP." *Pharmaceuticals* 10.

Vocci FJ, Acri J, Elkashef A. 2005. "Medication development for addictive disorders: The state of the science." *American Journal of Psychiatry* 162: 1432–1440.

Vocci FJ, Appel NM. 2007. "Approaches to the development of medications for the treatment of methamphetamine dependence." *Addiction* 102, Suppl 1: 96–106.

Volkow ND, Chang L, Wang GJ, Fowler JS, Franceschi D, Sedler M, Gatley SJ, Miller E, Hitzemann R, Ding YS, Logan J. 2001. "Loss of dopamine transporters in methamphetamine abusers recovers with protracted abstinence." *Journal of Neuroscience* 21: 9414–9418.

Volkow ND, Li TK. 2004. "Drug addiction: The neurobiology of behaviour gone awry." *Nature Reviews Neuroscience* 5: 963–970.

Volkow ND, Wang GJ, Fowler JS, Gatley SJ, Logan J, Ding YS, Dewey SL, Hitzemann R, Gifford AN, Pappas NR. 1999. "Blockade of striatal dopamine transporters by intravenous methylphenidate is not sufficient to induce self-reports of "high"." *The Journal of Pharmacology and Experimental Therapeutics* 288: 14–20.

Wang G, Ma W, Du J. 2018. "β-Caryophyllene (BCP) ameliorates MPP+ induced cytotoxicity." *Biomedicine & Pharmacotherapy* 103: 1086–1091.

Wang L, Liu J, Harvey-White J, Zimmer A, Kunos G. 2003. "Endocannabinoid signaling via cannabinoid receptor 1 is involved in ethanol preference and its age-dependent decline in mice." *Proceedings of the National Academy of Sciences* 100: 1393–1398.

Wearne TA, Cornish JL. 2018. "A comparison of methamphetamine-induced psychosis and schizophrenia: A review of positive, negative, and cognitive symptomatology." *Front Psychiatry* 9: 491.

Welsh SE, Kachelries WJ, Romano AG, Simansky KJ, Harvey JA. 1998. "Effects of LSD, ritanserin, 8-OH-DPAT, and lisuride on classical conditioning in the rabbit." *Pharmacology Biochemistry and Behavior* 59: 469–475.

Weng M, Xie X, Liu C, Lim KL, Zhang CW, Li L. 2018. "The sources of reactive oxygen species and its possible role in the pathogenesis of Parkinson's disease." *Parkinson's Disease* 2018: 9163040.

Yang D, Michel L, Chaumont JP, Millet-Clerc J. 1999. "Use of caryophyllene oxide as an antifungal agent in an in vitro experimental model of onychomycosis." *Mycopathologia* 148: 79–82.

Yu B, Becnel J, Zerfaoui M, Rohatgi R, Boulares AH, Nichols CD. 2008. "Serotonin 5-hydroxytryptamine(2A) receptor activation suppresses tumor necrosis factor-alpha-induced inflammation with extraordinary potency." *The Journal of Pharmacology and Experimental Therapeutics* 327: 316–323.

Zhang G, Asgeirsdottir HN, Cohen SJ, Munchow AH, Barrera MP, Stackman RW, Jr. 2013. "Stimulation of serotonin 2A receptors facilitates consolidation and extinction of fear memory in C57BL/6J mice." *Neuropharmacology* 64: 403–413.

25 Agro-Technology of Important Medicinal Plants

From Farm to Pharma

Nikita Patel, Swetal Patel, Abdullahi A. AbdulRahaman and Ramar Krishnamurthy

CONTENTS

25.1 INTRODUCTION

25.1.1 Medicinal Plants in India and the Global Scenario

Every nation has its unique traditional medicinal wisdom. However, due to overriding current technological advances, many countries across the globe have lost their glory. But the Indian traditional system of medicine (ISM)—Ayurveda, literally meaning the science of life—has not only retained its traditional know-how, but also popularized its practice across the world. Due to health consciousness and its necessary awareness, people adopt ISMs like Ayurveda, Siddha, Unani, homeopathy and other traditional medicinal systems worldwide. The demand for medicinal plants and herbs has been rapidly increasing over the last decade and is still growing. It is estimated that more than 80% of the world's population currently relies on traditional plant-based medicines for their primary

DOI: 10.1201/9781003108375-25

Graphical abstract representing the content of the chapter.

healthcare needs (World Health Organization, 2003). The global trade demand for plant-based products is estimated to be at 62–120 billion USD, and it is expected to go up to above 7 trillion USD by the year 2050 (NMPB, 2015).

India and China are the two major medicinal plants producers, occupying more than 40% of the global biodiversity. India possesses one of the world's richest medicinal plant heritages, with about 8,000 plant species being used by rural communities. Medicinal plants of India are utilized by about 60% of the world's population. India ranks second among major exporters: 70% of crude extracts and 30% of finished products. In India, 15 agro-climatic zones have 17,000–18,000 species of flowering plants; out of that, 6,000–7,000 species have medicinal uses in folk and Indian documented systems of medicines (FRLHT report, 2007; NMPB, 2015). It is estimated that about 960 plant species are involved in the business trade, of which 178 species are consumed in excess levels of over 100 metric tons. A large proportion of the Indian population depends on medicinal plants as a significant resource for livelihood and health security. The Indian domestic trade industry has business to the tune of Rs. 80–90 billion and exports products worth about Rs. 10 billion per annum (Pathak et al., 2005).

Medicinal and aromatic plants are economically important as they provide raw materials for pharmaceutical, perfumery, flavor, soap and cosmetic industries worldwide. Due to India's standard of living, increased population and several constraints in the long-term use of modern medicine, there has been a continuous demand for traditional medicinal plants and their products during recent years. Even though India earns considerable foreign exchange revenue from medicinal plants, it remains challenging to keep up with long-term supply, due to over-exploitation from natural resources. India is faced with tough competition from other developing countries for quality medicinal plant raw materials and competitive pricing. There is an urgent need to encourage Indian farmers/growers to develop organized cultivation practices of medicinal plant species, which are in high demand. With all of this in mind, the following details are discussed about selected critical medicinal plants, while sharing our experiences with farmers to promote and popularize the cultivation of medicinal plants, which in turn supports sustainable availability of quality medicinal plant raw materials—and thereby prevents overexploitation of natural resources (Pathak et al., 2005).

25.1.2 Why Medicinal Plants Require Special Cultivation

- To ensure quality and purity.
- Consistency of inactive ingredients and chemical constituents.
- Better yield and sustainable availability of quality raw materials to the user industry.
- Systematic cultivation results in many medicinal crops with an increased content of active constitutes and oils levels, e.g. in ginger, turmeric, licorice, etc.
- Organized and systematic harvesting periods, drying and post-harvesting operations help to improve the quality of crude drugs and value addition in many medicinal plants.
- Systematic plantation and the cultivation of medicinal plants tree species help conserve many plant species from overexploitation in the wild source.

25.1.3 Selected Medicinal Plants and their Cultivation Practices

25.1.3.1 *Centella asiatica* Linn./*Hydrocotyle asiatica* L. (Mandukaparni)

It belongs to the Apiaceae family. It is a small perennial creeping herb with slender jointed stems (Figure 25.1A) and a tropical plant found in damp, moist and marshy areas, preferably under partial shade conditions. It is native to India, China, Indonesia, Madagascar, Africa and through tropical Asia, ascending in Abyssinia to an elevation of 6,000 feet. It also occurs in the Americas from South Carolina to Valdivia in Chile, and in the West Indies, the Pacific, New Zealand and Australian islands (Patel et al., 2020).

FIGURE 25.1 Cultivation practices of medicinal plants. A, *Centella asiatica*, plant; B, *C. asiatica*, cultivation under shade net; C, *C. asiatica*, crop ready to harvesting of leaves for drying; D, *Cassia angustifolia*, cultivation field; E, *Gymnema sylvestre*, cultivation field; F, *G. sylvestre*, leaves harvested and shade dried; G, *Mucuna pruriens*, itching/non-itching trichome lines harvesting of pods and seeds; H, *M. pruriens*, cultivation field; I, *M. pruriens*, at flowering stage (Krishnamurthy et al., 2005c, 2006).

Hydrocotyle is known as Mandukaparni in Sanskrit, Khulakhudi in Hindi and is also known as 'Brahmi.' It has long been used in medicine by the natives of Java and the Coromandel Coast. Boilean, a French physician of Mauritius, pointed out in 1852 its virtues in the treatment of leprosy, and this plant was later tried mainly for leprosy in the hospitals of Madras by Hunter in 1855. Since then, the Mandukaparni has been included in the Pharmacopoeia of India (Gohil et al., 2010).

The plant is propagated through stem cutting, comprising a rooted node. It can be cultivated under mango orchards and artificial 50–75% shade-net conditions in commercial production. The dense plantation is maintained at a plant spacing of 15 × 15 cm (Figure 25.1B–C), and the same crop is continued for two years. The crop gives about 3 metric tons of dry herb yield/hectare in six harvestings in two years, and is recommended as intercropping in mango orchards (Krishnamurthy et al., 2006). Furthermore, soil applications of nitrogen, phosphorus and potassium (NPK)-based bio-fertilizer were found to improve yield and recorded 2,400 kg/hectare (Krishnamurthy et al., 2003a). The plant is used as a nerve and hair tonic, for skin disease, antiulcer, and used in chutneys, as a vegetable, pickles and as refreshing soft drinks in India (Patel et al., 2020).

The major chemical constituents of the plant are asiaticosides and triterpenoids, which are comprised of about 1.0% in the leaves and stems. Apart from asiaticosides, these plants possess asiatic acid and madecassoside, which are of great pharmacological importance (Prakash et al., 2017). The aerial part of this plant possesses centellin, asiaticin and centellicin, and their structures were determined by two-dimensional nuclear magnetic resonance spectrometry (NMR) techniques (Siddiqui et al., 2007). Mandukaparni is vegetative propagating plants through its rooted nodes; for commercial cultivation, about 1.0 lakhs of plants are required per hectare for a dense plantation. Therefore, producing such a large quantity of planting material under nursery conditions becomes difficult due to space constraints. Micropropagation techniques through plant tissue culture could be adopted in research and on a commercial scale by multiplying Brahmi genotypes with high triterpenoids content (above 1.0%) and having characteristics of higher biomass yield. In a previous study, *C. asiatica* accessions from India were surveyed for higher triterpenoid content (asiaticosides) and a genotype was found with 1.351% total triterpenoid concentration, named cv. Zandu Brahmi (Krishnamurthy et al., 2006).

Moreover, the triterpenoid content of the leaf of Zandu Brahmi is high (above 1.00%) at 180 days after planting. Therefore, the first harvesting of leaves is done at six months of the crop age for higher active ingredients content. The National Medicinal Plant Board located in New Delhi, India, reported that the market demand for Mandukaparni (*C. asiatica*) is 500–1,000 MT per annum (NMPB, 2015). It was reported that this is one of the 46 wasteland species used in high trading (FRLHT, 2007).

25.1.3.2 Senna (*Cassia angustifolia* [Mill.]-Alexandrian Senna; *C. actifolia* [Delile]-Thirunelveli Senna)

The two species of Senna leaves, popularly known as Alexandrian and Thirunelveli Senna (*Cassia angustifolia*), are used in traditional medicine. It is also known as Sonamukhi among locals in India. They belong to the family Caesalpiniaceae. The plant grows in Yemen and Hadramaut in Southern Arabia, and can be found on the Somali Coast, in Sind and Punjab and some parts of India (Balasankar et al., 2013). It is considered one of the most important sources of organic laxatives. The wild, uncultivated plant of Arabia is distributed worldwide. However, the so-called Bombay Senna comes from a cultivated source that occupies the majority of the global market. The crop was propagated initially from Arabian seeds and is popularly known as 'Thirunelveli Senna,' which is in the drug market (Balaji et al., 2016).

The elaborate work of Carl Martins indicated that the knowledge of Senna could not be traced back earlier than the time of the Elder Serapion in the ninth and tenth centuries. Issac Judaeus, who wrote during A.D. 850–900, mentioned that, as a native of Egypt, the best Senna was brought from Mecca. The Senna (known as Sinen or Senen) was enumerated among the commodities liable to duty at Acre in Palestine during the 12th Century. In the year 1542, in France, a pound of Senna was valued as an official tariff at 15 sols as equivalent to the price of pepper or ginger (Dymock et al., 1893). *Cassia obovate* Coll. was the first species known to botanists and was cultivated for medicinal purposes in Italy during the 16th century. Indian companies export Senna leaves and pods worth over Rs. 239 million. The Senna crops are generally grown through seeds. Therefore, developing planting materials by other methods such as through tissue culture is not required except in the multiplication of high Sennosides yielding accession (more than 2.50% in leaves) (Chadha, 2001).

The crop is sensitive to heavy rainfall, and prefers warm and dry weather during the growing season. It grows well on red loams to coarse gravelly soils and alluvial soils (Figure 25.1D). This crop's advantages are that it is best situated for sandy loam, lateritic soils of low fertility with a higher pH range of 7.0–8.5. The crop is raised through seeds; its first harvesting is done at 90–120 days, and the subsequent harvesting is done after 40 days. It is estimated that the crop produces 600–1,000 kg of dry leaf yield and 400 kg of dry pod yield/hectare. The leaves and young pods possess Sennosides (2–3%), which are believed to have laxative properties. It has been recommended as a laxative or purgative since ancient times in the Indian system of medicines; it is considered a safe and effective drug for habitual constipation, and is available in modern medical stores (Ramchander and Middha, 2017). It is estimated that 5,000–10,000 kg of Senna raw materials are used annually in global trade (Sastry et al., 2010).

25.1.3.3 *Gymnema sylvestre* R.Br. (Gudmar/Madhunashini)

Madhunashini is commonly known as 'Gudmar' in Hindi. This important medicinal climbing plant belongs to the family Asclepiadaceae and is acclaimed for its anti-diabetic properties. It grows in tropical conditions in the Deccan Penninsula, Uttar Pradesh, Madhya Pradesh, Tamil Nadu, Kerala, Punjab and some parts of Asia and Malaysia. This plant grows abundantly in the forests of Karnataka, Tamil Nadu and Bihar. The National Medicinal Plants Board at New Delhi, India, projected that 500–1,000 MT of Gudmar is used annually in trading (NMPB, 2015). This plant is considered and reported as one of the 70th medicinal plant species in high trade sourced from the tropical forest (FRLHT, 2007).

There are two types of varieties available, and they are classified according to leaf size (Farooqi et al., 2001):

- Small leaves type: Its leaves are oval, measuring 1.0–3.5 cm in length and 1.5–2.5 cm in width, and are very soft. They are found in dry regions.
- Broad and pubescent type: Its leaves are oval, measuring 3.0–6.0 cm in length and 3.5–5.0 cm in width. The leaves are dark green compared to the tiny leaves type, and all are pubescent.

The plant grows well in red sandy loam medium-deep soil and dry soils. It is sensitive to waterlogging soil conditions. Plants can be reproduced by stem cutting under nursery conditions, as their seeds show poor germination. The vegetative propagation of this plant by stem cutting takes around two months for rooting and therefore, tissue culture-based micropropagation method is desirable for raising commercial-scale planting materials for more extensive area cultivation. The leaves of *Gymnema sylvestre* (R. Br.) possess important primary and secondary metabolites such as triterpene saponins, phytin, resins, d-Quercitol, tartaric acid, formic acid, butyric acid, lupeol, ß-amyrin–related glycosides and stigmasterol (Srivastava et al., 2019). In previous studies, five elite accessions of *Gymnema sylvestre* were collected from various regions of the Western Ghats and Konkan, India, for higher foliage (leaf) and gymnemic acid yield at 12 months and 18 months crop ages under cultivation. This study recommended that accession ZFGS-23 could be utilized for commercial cultivation and production of gymnemic acid (Krishnamurthy et al., 2015).

In commercial cultivation, the plantation is done at 2 × 2 m plant spacing (Figure 25.1E), and leaf harvesting is performed at six months. The crop produces about 2 tons of dry leaves per hectare/year in 3–4 coppicing leaves (Figure 25.1F). It was found that the biofertilizers like *Azospirillum* and phosphate solubilizing bacteria (PSB) improved the leaf freshness and dry matter content of *G. sylvestre* (Patel et al., 2016). The leaves contain 2.0–3.5% gymnemic acid (Krishnamurthy et al., 2016). It is believed that the anti-diabetic properties of the plant are attributed to the presence of a mixture of triterpenes and saponins in the leaves. In addition to this, the plant is extensively used in almost all the Indian systems of medicine as a remedy for rheumatism, cough, ulcers and pain in the eyes. In some parts of India, the root of this plant is recommended as a remedy for snake bites (Vijaya et al., 2013).

25.1.3.4 *Andrographis paniculata* (Burm.f.) Wall. ex Nees (Kalmegh)

Kalmegh is an annual tropical herb belonging to the Acanthaceae family, 1–2 feet in height, commonly found throughout India and growing under the shade of trees (Figure 25.3G). It is found in Sri Lanka and Java, and it has been introduced into the West Indies. The plant, popularly known as 'king of bitters,' is a bitter annual herb and the whole plant is medicinally used. Kalmegh is known in Bengali (Indian language) as Maha-tita, which translates into 'king of bitter,' and it has been included in the Pharmacopoeia of India. It possesses various phytochemicals with important pharmacological properties, which are used in the treatment of fever, common cold, liver disorders, etc. (Gabrielian et al., 2002; Pandey et al., 2019).

It grows well in a variety of soils, i.e. from sandy to loam soils. *Andrographis paniculata* is propagated either by seeds (Figure 25.3H) or stem cuttings. In commercial cultivation, seeds are used for raising seedlings in nurseries and transplanted in a dense plant population (222,222 plants/ha) at a spacing of 30 × 15 cm apart (Figure 25.3I). The whole plant is harvested at 90 days and the second harvesting (known as ratoon crop) is performed 60 days after the first harvest. In our previous study, the crop was found to respond with improved plant height, number of branches, number of leaves and herbage yield under biofertilizer application, i.e. *Azospirillum* and PSB (Patel, 2016). The crop gives an average yield of 2.0–2.5 tons dry herb yield (whole shoot system) per hectare. After harvesting plants, the drying of plants under shade conditions is highly recommended (Farooqi et al., 2001).

The whole herb is the source of several diterpenoids, of which the bitter water-soluble lactone Andrographolide is concentrated up to 2.5% in leaves and 0.5–1.0% in the stem. The whole plant is used as a liver stimulant (hepatoprotective), hepatitis and anti-typhoid. In some parts of India, it is also used for appetizers and dysentery (Jayakumar et al., 2013).

25.1.3.5 *Mucuna pruriens* (L) DC. (Velvet Bean)

Velvet bean is commonly known as cowhage and atmagupta/kewach in Hindi. It is an annual and tropical climber belonging to the family Leguminosae. It is reported to be found throughout tropical conditions of Indian regions and in the Andaman and Nicobar Islands. The pods (Figure 25.1G) of this plant bear trichomes/hairs (on the outer surface of pods), which causes itching upon contact. The seeds of *Mucuna pruriens* have shown considerable variation in their morphology and chemical composition including 3,4-dihydroxyphenylalanine (L-Dopa) at 2–6%, and yield associated characteristics (Krishnamurthy et al., 2002, 2003b, 2005b). Different parts of *Mucuna pruriens* are used in traditional and modern medicine. They possess critical bioactive compounds which are used in the treatment of bone fractures, snake bites, cough, syphilis and wounds. Apart from critical bioactive compounds, *Mucuna pruriens* possess antimicrobial, anti-inflammatory, anti-Parkinson and aphrodisiac properties (Divya et al., 2017). Collections of *Mucuna pruriens* assembled from different regions of India showed that there is a direct correlation between seed yield and other seed characteristics (Krishnamurthy et al., 2005b).

It can withstand drought and grow in sandy to clayey-loam soils, provided that there is adequate drainage. It produces copious foliage growth and is generally grown as a fodder crop in India (CSIR, New Delhi, 1962). The crop duration is 9–10 months, and it is planted at a spacing of 1.0 × 1.0 m apart (Figure 25.1H–I). In commercial cultivation, the crop is raised through direct sowing of seeds and the plants require support (stakes) to climb and achieve a maximum seed yield of 1.0–1.5 ton/ha. The proximate analysis of seeds of cultivated crops recorded 25% protein, 58% carbohydrates, 5% fat, 385 calorie value (K.Cal/100 gm), Calcium (Ca) 353 mg/100 gm, Phosphorous (P) 254 mg/100 gm, Iron (Fe) 117 mg/kg, Potassium (K) 10 gm/kg and Sodium (Na) 261 mg/kg

(Krishnamurthy et al., 2003b). Nitrogen (N)- and Phosphorous (P)-based biofertilizer application improved the seed yield (1.4–1.5 tons/ha) of *Mucuna pruriens*, though it did not increase L-Dopa content, which remained constant at 3.33–3.66% (Krishnamurthy et al., 2003b).

In Ayurveda, the seeds of velvet bean are used as a tonic and for aphrodisiac properties. It is a part of one of the ingredients in many Ayurvedic products. Seed powder has been tried in Parkinson's disease (Kampavata) and for depression, with positive results (Dymock et al., 1980; Mahajani et al., 1996; Manayam, 1995; Manayam and Parikh, 2000). In the Indian system of medicines, the seeds of *M. pruriens* have been recommended as a nervine tonic since ancient times, since the findings of these plant seeds containing L-Dopa, which is an anti-Parkinson's drug (Manyam et al., 2004). Notably, this crop has drawn significant attention from farmers/growers and plant-based industries for commercializing this plant.

25.1.3.6 *Asparagus racemosus* Willd (Shatavari)

Shatavari belongs to the family Liliaceae, and its roots (tuber) are used in Ayurveda, Siddha and homeopathic medicines. It is a tropical perennial plant found in India, Asia, Africa and Australia. The plant prefers to grow in red loamy soils and sandy, loamy soils with adequate drainage. Since it is a shallow-rooted (tuberous) crop, it best suits shallow and rocky soil conditions where soil depth is up to 30 cm with a pH range of 7.0–8. It grows in dry soil conditions (50–100 cm annual rainfall) and up to 1,400 m in elevation (Krishnamurthy et al., 2005a).

Asparagus racemosus is vegetative propagated through root suckers with rhizomatous discs for commercial cultivation. The plant produces seeds in the months of February–May. In commercial plantations, the development of planting materials through root suckers or seeds requires long periods (2–3 months). Therefore, raising plants through tissue culture technology (TCT) is desirable and could be more economical. In dense plantations, the plant spacing is maintained at 1.0 × 0.5 m (Figure 25.2A), and the crop is ready to harvest its tubers after 24 months. The maximum tuber yield of 3 tons dry/ha (Figure 25.2B) could be achieved with the crop cultivated under systematic

FIGURE 25.2 Cultivation practices of medicinal plants. A, *Asparagus racemosus*, cultivation field; B, *A. racemosus*, fresh and dry tubers after removing outer skin; C, *Withania somnifera*, cultivation field; D, *W. somnifera*, root harvesting and drying; E, *Plumbago zeylanica*, (i) uprooting, (ii) root harvesting and drying; F, *P. zeylanica*, intercropping with *Phyllanthus emblica* (Amla); G, *Acorus calamus*, cultivation under shade net condition; H, *A. calamus*, (i) rhizome, (ii) crop in open cultivation field; I, Pristiparni, Dashmool plants at harvesting stage of 12 months, (i) *Uraria picta*, (ii) *Pseudarthria viscida* (Krishnamurthy et al., 2005a).

cultivation practice, irrigation and manuring after 24 months (Krishnamurthy et al., 2005a). It is estimated by the National Medicinal Plants Board, New Delhi that the annual market demand of Shatavari dry roots in India ranges from 2,000–5,000 MT (NMPB, 2015).

Shatavari roots, which contain galactagogue, are the primary source of medicinal use as it stimulates the secretion of breast milk. In the ISM, it is used as an aphrodisiac and immune support tonic. It can be used as potential energy to support weakened body conditions during prolonged/chronic illness. The tuber contains main chemical constituents, namely shatavarin and saponins. The flowers and mature fruits of Shatavari contain quercetin, hyperoside and rutin, whereas the leaves contain quercetin and diosgenin (Negi et al., 2010).

25.1.3.7 *Withania sonmifera* Dunal (Ashwagandha)

Withania sonmifera Dunal belongs to the family Solanaceae. Ashwagandha is an important medicinal plant found in the dry regions of tropical and subtropical areas of India and other countries, i.e. in the Indian states of Rajasthan, Madhya Pradesh and Uttar Pradesh, and in South Africa, Egypt, Pakistan and Afghanistan. Its roots are used in various preparations of Ayurvedic and Unani systems of Indian medicines. Ashwagandha requires a relatively dry season for optimal growth; its roots are fully developed at six months and thereby ready for harvesting. The crop grows successfully in sandy loam or light red soils, having a pH of 7.5–8.0 (Farooqi et al., 2001).

In commercial cultivation, direct planting of seeds along with monsoon rains is preferred with a dense population of 111,111 plants/ha at a desirable plant spacing of 30 × 30 cm (Figure 25.2C). Generally, plant-based industries prefer to buy only cultivated *Withania sonmifera* roots rather than the wild root collections, due to the poor quality of root traits of wild accessions. The crop yields 600 kg of dry root and 50 kg of seed/ha (Figure 25.2D). The significant alkaloids found in the plant are Withanolides in the roots and Somniferin in the leaves. In the ISM, the Ashwagandha roots are recommended as a tonic, an aphrodisiac, as a rejuvenator and for sexual weakness. The Indian market demand for Ashwagandha root is estimated as above 5,000 MT/annum. Industries are prepared to buy superior quality root materials from cultivated sources. Ashwagandha was rated as one of the highest commercially traded source materials sourced from cultivation. This plant is one among 36 medicinal plant species listed in their report, since Ashwagandha is recommended to be cultivated through seeds and less significant in tissue culture technology for commercial cultivation, except for purposes of research and plant breeding (FRLHT, 2007; Patel and Krishnamurthy, 2013).

25.1.3.8 *Plumbago zeylanica* (L.) (Chitrak)

In Ayurvedic and other Indian medicine systems, the roots of the white-flowered *Plumbago zeylanica* are used in different formulations. The plant is found in the wild in the Western Peninsula, Bengal and some of Madhya Pradesh and Chhattisgarh forests of India. It is a tropical plant that performs well under partial shade conditions of tree forest coverage. It is a perennial shrub that grows up to a height of 1.5–2.0 m. During the rainy season, the plant grows actively, and in winter, growth stops after flowering and fruiting. However, in summer, the plant appears almost leafless. The new shoots develop after drying old shoots within the shoot system (Patel, 2016).

The plant is best suited for well-drained, sandy loam and clayey loam soils, having higher organic content. It does not withstand waterlogging or sandy or gravelly soil with lower water holding capacity. The plant is propagated through stem cuttings of 10–15 cm long, having 2–3 nodes. However, this type of vegetative propagation takes time (two months for complete rooting) and requires a larger area of nursery setup for developing commercial-scale planting materials (around 80,000 plants are needed per hectare). Therefore, it is preferred to have tissue culture micropropagation technology for the production of an economically viable number of plants for commercial cultivation. In commercial plantations, the optimum plant spacing of 50 × 25 cm (80,000 plants/ha) is desirable for achieving maximum root yield and generating maximal income. The crop yields aerated root of 1,700 kg/ha after 12 months (Figure 25.2E) (Patel, 2016).

Indian industries buy Chitrak roots for their various drug formulations and herbal products. There are herbal products available in the Indian market prepared from their roots. Intercropping of Chitrak with other tree crops like *Phyllanthus emblica* (Amala) is recommended for commercial cultivation for obtaining a remunerative yield and economic viability (Figure 25.2F). In our earlier study, we found that application of *Azospirillum* and PSB to the soil improved plant height and root yield, and thereby, the crop is ready to harvest after 12 months (Patel, 2016; Patel et al., 2016). Furthermore, silver nanoparticles synthesized from the root extract of *Plumbago zeylanica* showed a toxic effect on *Bacillus cereus*, *Staphylococcus aureus*, *Serratia marcescens* and *Escherichia coli*, and therefore is recommended as an antimicrobial agent (Patel and Krishnamurthy, 2015).

The root of Chitrak contains the alkaloid Plumbagin, which is considered to stimulate the central nervous system and secretion of sweat, urine and bile. The root is believed to be a potent abortifacient when taken orally during the early stage of pregnancy. The root is an appetizer and is used in skin diseases. The root extracted in oil is effective in rheumatism, joint pain and paralysis. It was reported that Chitrak is one of the highly traded commercial source materials from wastelands out of 46 medicinal plants listed. The annual demand for Chitrak root is estimated at 2,000–5,000 MT. (FRLHT, 2007).

25.1.3.9 *Acorus calamus* Linn. (Sweet Flag)

The *Acorus calamus*, commonly known as Vacha, belongs to the family Araceae and is a perennial plant propagating through the rhizome (root). The plant is found in Uttaranchal, Himachal Pradesh, Jammu, throughout India and in China and Arabian countries. The plant grows naturally along streams and river boundaries and in low-lying areas. In our earlier studies, chemical profiling of 20 accessions of *Acorus calamus* was collected from different geographical regions to shortlist low ß-asarone content diploid lines (Kasture and Krishnamurthy, 2016; Kasture et al., 2016). Because the diploid plants are known to lack ß-asarone, the triploid plants contain a small portion of ß-asarone (5%) in their oil, and tetraploid plants are high in ß-asarone (90–96%) (Pattanaik et al., 2013). It is propagated through its rhizomes under nursery conditions (Figure 25.2G) and transplanted infields at a density of 30 × 30 cm plant spacing (Figure 25.2 Hii) (Krishnamurthy and Kasture, 2015).

Since it is a vegetative propagating plant through rhizome (Figure 25.2Hi), the laboratory (TCT) is desirable only in research and large-scale propagating of the most elite planting materials and superior accessions. In commercial cultivation, the crop requires frequent irrigations with a recommended dose of fertilizers and other field management practices for maximum yield and rhizome quality. The rhizomes mature and are ready to harvest at 12 months under regular cultivation practice and field management. The Vacha crop recorded 16,470–27,310 kg/ha fresh rhizome yield at 12 months under the well-planned adaptation of cultivation practices. The Indian domestic consumption of *Acorus calamus* root is estimated as high as 500–1,000 MT dry rhizome annually (Tiwari et al., 2012; Krishnamurthy and Kasture, 2015).

The calamus rhizome has demand in plant-based industries for the preparation of medicines and flavoring liquors. The oil extracted from the rhizome and leaves showed antimicrobial activity (Kasture et al., 2015). It is used as an appetizer and for epilepsy. It is commonly used in insecticides and deterrents against flies, mosquitoes and bedbugs, and perfumery and toiletry industries (Rajput et al., 2014).

25.1.3.10 *Desmodium gangeticum* (L.) DC) (Shalparni)

Shalparni is a tall, under shrub bearing broad alternative unifoliate leaves and white-purple colored flowers that grows 1.0–1.2 m tall in monsoon rain forests in the dry scrub up to 900 m elevation of Indian tropical regions. It belongs to the family of Leguminosae (Kirubha et al., 2011). In Ayurveda, the roots of ten medicinal plants are used together (five herbs and five tree species) as 'Dashmool,' considered highly efficacious in managing several common ailments. *Desmodium gangeticum* is one among ten Dashmool medicinal plants. Among the classical formulations, 'Dashmoolarishta' is the most popular and is used as an essential ingredient in manufacturing over 109 drug formulations (Chopra et al., 1999). Due to the great demand for raw materials in the industry, the roots of some of these species are in short supply. This has led industries to use associated species of Dashmool plants and their whole plants (viz. Panchang). Salparni is among ten plants which grow all over plains in Indian regions (Manivel et al., 2019).

The plant is propagated through seeds and developed seedlings under nursery conditions and planted in fields in a dense population from 49,000–55,000 plants per hectare for commercial cultivation. In our earlier study, 35 accessions of *Desmodium gangeticum* were collected and evaluated for two growing cycles (years), and identified were the two promising lines (named ZFDG 1 and ZFDG 21) based on the analysis of their growth, yield and chemical constituents in the root and shoot samples at 12 months harvest age, determining that they were superior in quality compared to the wild-growing materials used in the industry. Its cultivation practices have been developed for superior accession ZFDG 1 (Figure 25.3C). It produces 3–4 tons of roots (dry) per hectare (Figure 25.3D–F) in a growing season of 12 months crop cycle under pilot-scale trial experiments. This work was carried out on *Desmodium gangeticum* as a regular crop in the high rainfall tracts of South Gujarat, India (Pathak et al., 2005).

The crop can be profitable under cultivation over low fertile marginal lands in high-rainfall tracts. The shoot system (dry) can be sold to industry and considered as additional income to cultivators/farmers. Therefore, Shalparni cultivation can be introduced in partial shades of tree species orchards in regions experiencing short and mild winter seasons. Shalparni is used for the treatment of various diseases such as piles, asthma, bronchitis and typhoid. It contains essential primary and secondary metabolites such as flavonoids and terpenes, which act as potent wound healing agents (Deshpande and Sanjivani, 2014).

FIGURE 25.3 Cultivation practices of medicinal plants. A, *Uraria picta*, cultivation field; B, *Pseudarthria viscida*, cultivation field; C, *Desmodium gangeticum*, superior line used for cultivation; D, *D. gangeticum*, after uprooting of cultivated plant; E, *D. gangeticum*, root harvesting and drying; F, *D. gangeticum*, cultivation field; G, *Andographis paniculata*, young plant; H, *A. paniculata*, seed production field; I, *A. paniculata*, cultivation field.

25.1.3.11 *Uraria picta* (Jacq.) DC. & *Pseudarthria viscida* (L.) Wight and Arnott- (Pristiparni)

In Dhashmoolarishta, both *Uraria picta* and *Pseudarthria viscida* are known as Pristiparni. In Ayurveda, the roots of both species are recommended in Dashmool drug preparations. However, in both western and northern parts of India, the industries use *U. picta* or *U. lagopoides* as Pristiparni; whereas, in southern parts of India, the industries buy raw materials of *P. viscida* as Pristiparni for their drug formulations. Our earlier study found that *P. viscida* exhibited potential antimicrobial activities (Kunal and Krishnamurthy, 2014). According to Hari and colleagues, the aerial part of *U. picta* possesses various phytoconstituents such as alkaloids, cardiac glycosides, flavonoids, saponins, steroids, tannins and terpenoids with numerous biological properties that were validated through HPTLC analysis (Hari et al., 2016). GC-MS chromatogram of the methanol extract of *Pseudarthria viscida* showed 18 peaks indicating the presence of 3-O-methyl-d-glucose (61.33%), n-hexadecanoic acid (12.66%), oleic acid (7.93%) and 9,12 octadecanoic acid (Z,Z) (4.88%) (Muthukrishnan and Rajan, 2012). *U. picta* recorded 54 g/plant whole plant yield (shoot and root dry) and it is estimated from our earlier experience that in hectare cultivation, about 3 tons dry whole plant yield of *P. viscida* and *U. picta* could be achieved (Figure 25.2Ii and Figure 25.2ii, Figure 25.3A–B) (Kunal, 2019). Furthermore, in other studies, the methanolic extract of *P. viscida* and *U. picta* exhibited anti-inflammatory activity in carrageenan-induced paw edema in albino rats (Kunal and Krishnamurthy, 2018).

Since all the Pristiparni species produce small root quantities even after one year of the plant age under cultivation, there remains an acute shortage of natural cultivated root raw materials in the market, and hence, drug manufacturers are forced to use 'Panchang' (all plant parts mixed) for their products. Furthermore, *P. viscida* is a red-listed plant species. Therefore there is a necessity for systematic and regular cultivation of this species. In commercial cultivation of *P. viscida*, around 55,555 plants per hectare at a spacing of 30 × 60 cm are required as planting material for raising dense plant population under regular commercial cultivation. Generating such large amounts of uniform planting material through seeds is relatively time consuming and non-viable economically. Hence, it is advisable to adopt plant tissue culture–based micropropagation methods to produce bulk planting material to meet this red-listed plant species (Kunal, 2019).

25.2 SCOPE OF ORGANIC FERTILIZERS AND NATURAL PESTICIDES FOR THE CULTIVATION OF MEDICINAL PLANTS

Pesticides are defined as chemical compounds that are used to kill pests and insects that affect plants' growth. Pesticides or chemical fertilizers are potentially toxic to the health of both humans and animals. Chemical fertilizers and pesticides cannot be used in the cultivation of medicinal plants, as their roots and other parts are used directly in the preparation of drug formulations. We cannot nullify the toxic effect of those chemicals quickly (Naguib, 2011). To overcome the problems faced by using chemical fertilizers/insecticides, natural plant-based compounds of plants such as *Azadirachta indica* (A.Juss.), *Calotropis procera* (Aiton) W.T.Aiton) and *Allium sativum* (L.) can be used as they possess azadiractin, limonides and tri-sulfide, respectively, as their active constituents, which inhibit various pests and insects (Satej and Yogesh, 2014). Moreover, organic fertilizers show beneficial effects in terms of the growth and yield of medicinal plants. Soil organic matter is important for the proper growth and development of plants, and it can be maintained through organic farming. There is a massive demand for natural active constituents from medicinal plants in the herbal market for human health. The quality and quantity of these compounds extracted from organically grown plants are superior as compared to inorganic fertilizers. Thus, developing more sustainable farming practices on a large scale is of utmost importance (Badalingappanavar et al., 2018).

25.3 CONCLUSION

In the present chapter, we have focused on important medicinal plants which should be propagated in large quantities to meet the demand of global plant-based medicinal industries, with an emphasis on their organic cultivation.

REFERENCES

Badalingappanavar, Rekha, M. Hanumanthappa, H. K. Veeranna, Shashikala Kolakar, and Gajendra Khidrapure. 2018. "Organic fertilizer management in cultivation of medicinal and aromatic crops: A review." *Journal of Pharmacognosy and Phytochemistry*: 126–129.

Balaji, P., N. Lily Miracline Fernando, A. Vidhyavathi, and M. Santhasheela. 2016. "Production and marketing of Senna in Thoothukudi district of Tamil Nadu: Problems and prospects." *International Research Journal of Agricultural Economics and Statistics* 7, 1: 29–33. doi: 10.15740/HAS/IRJAES/7.1/29-33

Balasankar, D., K. Vanilarasu, P. Selva Preetha, S. M. U. Rajeswari, and B. Debjit. 2013. "Senna—A medical miracle plant." *Journal of Medicinal Plants Studies* 1, 3: 41–47.

Chadha, K. L. 2001."Handbook of horticulture." In: *Handbook of horticulture,* ICAR, New Delhi: 622–1031

Chopra, R. N., S. L. Nayar, and I. C. Chopra. 1999. "Glossary of Indian medicinal plants." National Institute of Science and Communications, CSIR, New Delhi.

CSIR, New Delhi, 1962."The wealth of India: Raw materials." *Council of Scientific and Industrial Research, New Delhi* 6: 639–644.

Deshpande, Harshal A., and Sanjivani R. Bhalsing. 2014. "A review of phytochemical profile of *desmodiumgangeticum* (L.) DC: A valued endangered medicinal plant." *International Journal of Pharma Research and Health Sciences* 1, 4: 1–12

Divya, B. J., B. Suman, M. Venkataswamy, and K. Thyaga Raju. 2017. "The traditional uses and pharmacological activities of *Mucuna pruriens* (l) Dc: A comprehensive review." *Indo American Journal of Pharmaceutical Research* 7, 1: 7516–7525.

Dymock, W., C. J. H. Warden, and D. Hooper. 1893. "A history of the principal drugs of vegetable origin, met with in British India." *Pharmacographia Indica* 3, 6: 1889–1893.

Dymock, W., C. J. H. Warden, and D. Hooper. 1980. "*Mucuna*." *Pharmacographia India* 1: 447–450.

Farooqi, A. A., M. M. Khan, and M. Vasundhara. 2001. *Production technology of medical and aromatic crops*. Natural Remedies Pvt Ltd., Bangalore: 7–8

FRLHT (Foundation for Revitalisation of Local Health Traditions). 2007. "Sustainable harvest of medicinal plants in India." WWF, TRAFFIC and FRLHT, Bangalore. www.wwf.de/fileadmin/fmwwf/pdfneu/HG_Sustainable_Harvest_of_Medicinal_Plants_in_India.pdf (Accessed 5th June 2020)

Gabrielian, E. S., A. K. Shukarian, G. I. Goukasova, G. L. Chandanian, Alexander George Panossian, G. Wikman, and Hildebert Wagner. 2002. "A double blind, placebo-controlled study of *Andrographis paniculata* fixed combination Kan Jang in the treatment of acute upper respiratory tract infections including sinusitis." *Phytomedicine* 9, 7: 589–597. doi: 10.1078/094471102321616391

Gohil, K. J., J. A. Patel, and A. K. Gajjar. 2010. "Pharmacological review on *Centella asiatica*: A potential herbal cure-all". *Indian Journal of Pharmaceutical Sciences* 72, 5: 546–556. doi: 10.4103/0250-474X.78519

Hari, Om Saxena, Brij Mohanb, Arun Kakkarb, Ganesha, and Satosh Kumar Choubeya. 2016. "Phytochemical screening and hptlc finger print analysis of aerial parts of *Uraria picta* desv. A Dashmool species" *Journal of Pharma. Research* 5, 5: 87–93.

Jayakumar, Thanasekaran, Cheng-Ying Hsieh, Jie-Jen Lee, and Joen-Rong Sheu. 2013. "Experimental and clinical pharmacology of *Andrographis paniculata*and its major bioactive phytoconstituent andrographolide." *Evidence-Based Complementary and Alternative Medicine*: 1–16. doi: 10.1155/2013/846740.

Kasture, Avani, and Ramar Krishnamurthy. 2016. "Evaluation of the chemical variability and phytochemical analysis among Indian Germplasm of *Acorus calamus* linn using GC-MS and FTIR." *European Journal of Medicinal Plants*: 1–11. doi: 10.9734/EJMP/2016/25774

Kasture, Avani, Ramar Krishnamurthy, and Katagi Rajkumar. 2016. "Genetic variation in the endangered Indian sweet flag (*Acorus calamus* L.) Estimated using ISSR and RAPD markers." *Journal of Applied Research on Medicinal and Aromatic Plants* 3, 3: 112–119. doi: 10.1016/j.jarmap.2016.03.001

Kasture, Avani, Shuchi Patel, Jigna Chauhan, and R. Krishnamurthy. 2015. "*In vitro* antimicrobial effect of essential oil from leaf and rhizome of various accessions of *Acorus calamus* Linn., and its phytochemical screening." *European Journal of Medicinal Plants*: 1–13. doi: 10.9734/EJMP/2015/17318

Kirubha, T. Shri Vijaya, M. Jegadeesan, and S. Kavimani. 2011. "Studies on *Desmodium gangeticum*: A review." *Journal of Chemical and Pharmaceutical Research* 3, 6: 850–855.

Krishnamurthy, R., D. AdedayoAnimasaun, R. T. Patel, and R. S. Ingalhalli. 2016. "Phytochemical constituents and hypoglycemic effect of gymnemic acid extracts from big and small leaf varieties of *Gymnema sylvestre* R. Br." *Indonesian Journal of Pharmacy* 27, 2: 59. doi: 10.14499/indonesianjpharm27iss2pp59

Krishnamurthy, R., M. S. Chandorkar, B. G. Kalzunkar, M. R. Palsuledesai, J. M. Pathak, and Rajendra Gupta. 2003a. "Studies on use of bio-fertilizers on Kewach (*Mucuna pruriens*) and Brahmi (*Centella asiatica*) for yield and content." In: *Organic production of medicinal aromatic and dye yielding plants* Published by FAO, India: 1–11.

Krishnamurthy, R., M. S. Chandorkar, B. G. Kalzunkar, M. R. Palsuledesai, J. M. Pathak and Rajendra Gupta. 2005a. "Studies in for improving tuber yield in Shatavari (*Asparagus racemosus*) as irrigated and rainfed crop." *Journal of Medicinal and Aromatic plants* 26, 1: 740–745.

Krishnamurthy, R., M. S. Chandorkar, B. G. Kalzunkar, M. R. Palsuledesai, J. M. Pathak and Rajendra Gupta. 2005b. "Diversity evaluation in velvet bean (*Mucuna pruriens*) germplasm for seed yield and associated agronomical traits." *Journal of Medicinal and Aromatic Plants* 27, 1: 291–296.

Krishnamurthy, R., M. S. Chandorkar, B. G. Kalzunkar, J. M. Pathak, and Rajendra Gupta. 2005c. "Genetic variability and correlation studies in yield associated traits of *Mucuna pruriens* (L.) DC. germplasm." *Journal of Non Timber Forest Products* 12, 3: 135–143.

Krishnamurthy, R., M. S. Chandorkar, E. G. Kalzunkar, J. M. Pathak, and Rajendra Gupta. 2006. "Studies on agronomic practices for growing *Centella asiatica* (L.) Urban in high rainfall localities under open and partial shade of mango orchards." *Indian Journal of Horticulture* 63, 1: 76–80.

Krishnamurthy, R., M. S. Chandorkar, M. R. Palsuledesai, B. G. Kalzunkar, J. M. Pathak, and Rajendra Gupta. 2003b. "Standardization of cultivation and harvesting stage of velvet bean (*Mucuna pruriens* var *utilis*) for optimum yield and quality." *Indian Journal of Agricultural Science* 73, 11: 585–589.

Krishnamurthy, R., M. S. Chandorkar, M. R. Palsuledesai, J. M. Pathak, and Rajendra Gupta. 2002. "Breeding in velvet bean (*Mucuna pruriens*) for improvement in seed yield and quality traits." *Indian Journal of Agricultural Science* 72, 12: 709–715.

Krishnamurthy, R., M. S. Chandorkar, J. M. Pathak, Animasaun David Adedayo, and Gupta R. 2015. "Selection of elite lines from accessions of *Gymnema sylvestre* (Gudmar) based on characterization of foliage and gymnemic acid yield." *International Journal of Medicinal Plants Photon*, 108: 596–605

Krishnamurthy, R., and Avani Kasture. 2015. "Impact of planting space, irrigation pattern and different concentration of fertilizers on the yield and yield related parameters of sweet flag (*Acorus calamus* Linn) in South Gujarat region of India." *International Journal of Agriculture Photon*, 126: 405–411.

Kunal, H. Naik. 2019. "PhD thesis on plant growth, yield and phyto-pharmacological evaluation of *Pseudarthria viscida* (L.) wight and arn. and *Uraria picta* (Jacq.) DC. as anti-inflammatory and wound healing agents submitted to UkaTarsadia University, Bardoli, and Gujarat, India." Under the guidance of Prof. Dr. R. Krishnamurthy: 1–145

Kunal, H. Naik, and R. Krishnamurthy. 2014. "Inhibitory activity of methanolic extract of bark of *Pseudharthria viscida* against microorganisms." *International Conference on Current Status, Opportunities and Challenges in Medicinal Plants and Natural Product Research* September: 24–26.

Kunal, H. Naik, and R. Krishnamurthy. 2018. "Anti-inflammatory activity of methanolic extracts of *Pseudarthria viscida* and *Uraria picta* against carrageenan induced paw edema in albino rat." *Open Access Journal of Medicinal and Aromatic Plants* 9, 1: 1–11.

Mahajani, S. S., V. J. Doshi, K. M. Parikh, and B. V. Manyam. 1996. "Bioavailability of l-DOPA from HP-200—a formulation of seed powder of *Mucuna pruriens* (Bak): A pharmacokinetic and pharmacodynamic study." *Phytotherapy Research* 10, 3: 254–256.

Manayam, B. V. 1995. "HP-200–1995. A alternative medicine treatment for Parkinson's' disease." *Journal of Alternative and Complementary Medicine* 1: 249–255.

Manayam, B. V., and K. M. Parikh. 2000. "Anti-Parkinson activity of *Mucuna pruriens* seeds." *International Journal of Neuropsychopharmacology* 3: 52.

Manivel, P., R. N. Reddy, Raju Saravanan, V. Thondaiman, and R. S. Ganvit. 2019. "Exploration and collection of genetic resources of Salparni (*Desmodium gangeticum* L.) in India." *International Journal of Minor Fruits, Medicinal and Aromatic Plants* 5, 1: 21–28.

Manyam, B. V., Muralikrishnan Dhanasekaran, and Theodore A. Hare. 2004. "Neuroprotective effects of the anti-Parkinson drug *Mucuna pruriens*." *Phytotherapy Research: An International Journal Devoted to Pharmacological and Toxicological Evaluation of Natural Product Derivatives* 18, 9: 706–712.

Muthukrishnan, Suriyavathana, and Rajan Thinakaran. 2012. "Chemical investigation of *Pseudarthria viscida* root by GC-MS analysis." *Pharmacognosy Communications* 2, 3: 26–29. doi: 10.5530/pc.2012.3.6

Naguib, Nabila Yehya Mohamed. 2011. "Organic vs chemical fertilization of medicinal plants: A Concise Review of researches." *Advances in Environmental Biology* 5, 2: 394–400.

National Medicinal Plants Board. 2015. "Ministry of AYUSH." *Government of India*. https://nmpb.nic.in/cultivation-status

Negi, J. S., P. Singh, G. P. Joshi, M. S. Rawat, and V. K. Bisht. 2010. "Chemical constituents of Asparagus." *Pharmacognosy Reviews* 4, 8: 215–220. doi: 10.4103/0973-7847.70921

Pandey, Jyoti, Vimal K. Saini, and Wasim Raja. 2019. "Evaluation of phytochemical analysis of *Andrographis paniculata* leaf and stem extract." *World Journal of Pharmaceutical and Life Sciences* 5, 2: 188–190

Patel, Hetal. 2016. "PhD Thesis on Effect of Biofertilizers on yield and quality of Medicinal plants submitted to Veer Narmad South Gujarat University, Surat, India." Under the guidance of Prof. Dr. R. Krishnamurthy: 96–243

Patel, Hetal D., and Ramar Krishnamurthy. 2015. "Antimicrobial efficiency of biologically synthesized nanoparticles using root extract of *Plumbago zeylanica* as biofertilizer application." *International Journal of Bioassays* 4, 11: 4473–4475.

Patel, Hetal D., Ramar Krishnamurthy, and Musibau A. Azeez. 2016. "Effect of biofertilizer on growth, yield and bioactive component of *Plumbago zeylanica* (Lead Wort)." *Journal of Agricultural Science* 8, 5: 142–155. doi:10.5539/jas.v8n5p141

Patel, Nikita, Mohini Rana, A. Rahaman, and Ramar Krishnamurthy. 2020. "Medicinal plant *Centella asiatica* (Mandukaparni): From farm to pharma." *Trends in Pharmaceutical Research and Development* 4, 1: 10–22. doi: 10.9734/bpi/tprd/v4.

Patel, Pooja, and R. Krishnamurthy. 2013. "Feasibility studies on *in vitro* mass-scale propagation of Indian Ashwagandha (*Withania somnifera*) cultivars for commercial purpose." *Journal of Pharmacognosy and Phytochemistry* 2, 2: 168–174

Pathak, J. M., R. Krishnamurthy, M. S. Chandorkar, Vijay D. Gulkari, and Rajendra Gupta. 2005. "Identification of high yielding genotypes of Dashmool Shalparni (*Desmodium gangeticum*) drug plant and its cultivation under high density planting." *Indian Journal of Horticulture* 62, 4: 378–384.

Pattanaik, Jina, Yogesh Kumar, and Ravi Shankar Khatri. 2013. "*Acorus calamus* Linn. A herbal tonic for central nervous system." *Journal of Scientific and Innovative Research* 2, 5: 950–954.

Prakash, Ved, Jaiswal Nishita, and Mrinal Srivastava. 2017. "A review on medicinal properties of *Centella asiatica*." *Asian Journal of Pharmaceutical and Clinical Research* 10, 10: 69–77. doi: 10.22159/ajpcr.2017.v10i10.20760

Rajput, Sandeep B., Madan B. Tonge, and S. Mohan Karuppayil. 2014. "An overview on traditional uses and pharmacological profile of *Acorus calamus* Linn. (Sweet flag) and other *Acorus* species." *Phytomedicine* 21, 3: 268–276. doi: 10.1016/j.phymed.2013.09.020

Ramchander, Jalwal P., and Anil Middha. 2017. "Recent advances on Senna as a laxative: A comprehensive review." *Journal of Pharmacognosy and Phytochemistry* 6, 2: 349–353.

Sastry, K. P., B. R. Rajeswara Rao, C. P. Singh, and Kamla Singh. 2010. "Development of A rural industry based on the cultivation and processing of *cassia senna*." *Rural Development in India: Scope for Industrial Development*: 189–196.

Satej Banne and Yogesh Girigaon. 2014. "Scope of natural pesticides in cultivation of medicinal plants." *International Journal of Scientific Research* 3, 10: 368–370

Siddiqui, B. S., H. Aslam, S. T. Ali, S. Khan, and S. Begum. 2007."Chemical constituents of *Centella asiatica*." *Journal of Asian Natural Products Research* 9, 4: 407–414. doi: 10.1080/10286020600782454

Srivastava, Ashutosh, Rajan Bhatt, and Rinki Khobra. 2019. "*Gymnema sylvestre*. Medicinal properties and biological action." *Medicinal Plants in India: Importance and Cultivation* Chapter: 13, 231–248.

Tiwari, R. K. S., K. Das, D. Pandey, R. B. Tiwari, and Jola Dubey. 2012. "Rhizome yield of sweet flag (*Acorus calamus* L.) as influenced by planting season, harvest time, and spacing." *International Journal of Agronomy* 2012: 1–8. doi: 10.1155/2012/731375

Vijaya, P., R. Ranjani, M. R. Rao, and G. Sudarsanam. 2013. "Identification of antidote medicinal plants against snake venom-a field survey." *International Journal of Medicine and Pharmaceutical Sciences* 3, 5: 21–32.

World Health Organization. 2003. *WHO Guidelines on Good Agricultural and Collection Practices [GACP] for Medicinal Plants*." World Health Organization, 1–80. https://apps.who.int/iris/handle/10665/42783.

26 Aquatic Phytotherapy

Prospects, Challenges and Bibliometric Analysis of Global Research Output on Medicinal Aquatic Plants from 2011–2020

Muhali O. Jimoh and Learnmore Kambizi

CONTENTS

26.1 INTRODUCTION

Aquatic plants spend at least a part of their life cycle in marshy or swampy wetlands, freshwater, marine and estuarine ecosystems for their growth, development, regeneration and reproduction (Aasim et al., 2018; FAO, 2012). As primary producers, they exert important ecological functions in the supply of oxygen and vital nutrients to the aquatic ecosystem, this owing to some cell ultrastructures that are capable of performing metabolic functions, surviving turbulence in the water and making them adaptable to the flooded environment (Campbell et al., 2014). Aquatic plants can be free-floating, submerged or deeply rooted in riverbank sediments, which provide anchorage for their establishment (Lall, 2020).

Being the producers of the food web in aquatic environments, most organisms in aquatic ecosystems depend on them for survival. Different parts of aquatic plants such as roots, stems and leaves play significant geomorphological and biological roles ranging from manipulating the soil geochemistry, slowing water flow, absorbing light, trapping sediments and bioaccumulating pollutants

DOI: 10.1201/9781003108375-26

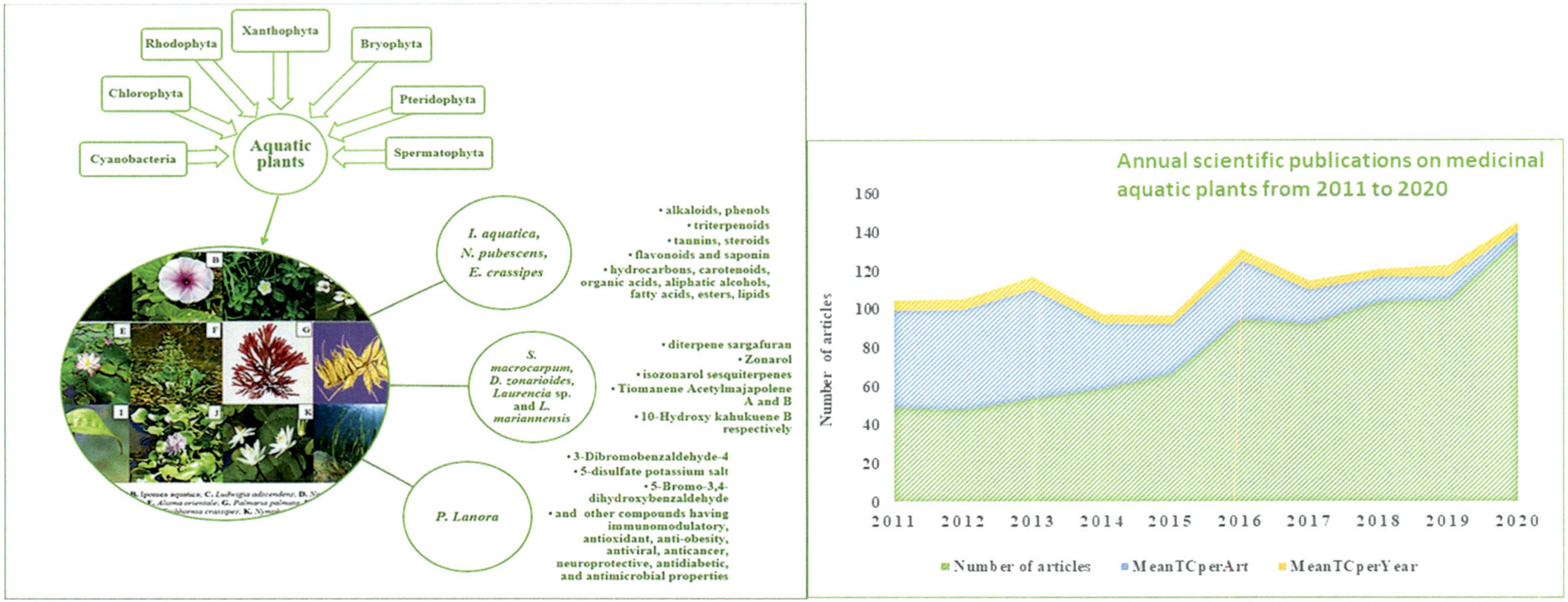

Graphical abstract representing the content of the chapter (Crouan and Crouan, 1852; Garg, 2009; Gloor, 2016; Hendrik, 2011; Hernandez, 2010; Mata, 2020; Prenn, 2013; Subhash, 2015; Vengolis, 2013; Venter, 2017; Zezhong, 2013; Ziarnek, 2012).

in the tissues (Brix, 1997; Gurnell, 2014). The relative abundance of aquatic plants is predicted by herbivory and other factors like turbulence, which control nutrients and energy flow in the ecosystem (Wood et al., 2017). It was recently established that the magnitude of herbivory is significantly influenced by the plant-herbivore relationships across a wide range of species. The limited information on factors that regulate the consumption of vascular plants in aquatic environments has mired our critical understanding of the dynamics and effective management of aquatic ecosystems (Stillman et al., 2015; Wood et al., 2017).

The survival of aquatic plants in deep water or areas saturated with water is achieved through the different hydraulic adaptations to a wide range of habitats, with their shoots and roots attuned to flooded environments (Gurnell, 2014). Likewise, the prevalence of seedling mortality in hydrophytes may be connected with constant erosion triggered by flow throbs during summer, reorganization of the riverbed by winter snows and periodic desiccation occasioned by low flows during summer (Gurnell, 2014; O'Hare et al., 2018). In addition, the demand for oxygen in the aquatic ecosystem exceeds its delivery, resulting in a limited supply, which hinders biogeochemical activities in aquatic ecosystems, despite wetlands receiving substantial quantities of minerals and organic matters from immediate terrestrial catchments (Schlesinger and Bernhardt, 2020).

Like their terrestrial counterparts, submerged plant species are depots of phytonutrients used for various cosmeceutical and pharmaceutical purposes (Aasim et al., 2019; Lall, 2020). The dearth of information on the classification and pharmacological uses of most aquatic plants is due to restricted access to riverine areas, patchy natural surroundings of many aquatic habitats, adverse recruitment conditions and potential risks associated with their medicinal uses (Ceschin et al., 2021; FAO, 2012). Populations of the identified plants in aquatic ecosystems are very low compared to available species. Most literature on submersed plants lists only a few of the species, leaving behind close relatives of those species with scanty taxonomic data. The rising demand for natural products has made it imperative to explore plants found in various ranges of aquatic habitats having the potentials to serve as food or pharmaceutical precursors to complement the existing phytotherapeutic options.

26.2 AQUATIC PLANTS: DIVERSITY AND DENSITIES IN NATURE

Traditionally, assessments of diversity, composition and abundance of aquatic plants were quite problematic. However, these challenges arising from temporal and spatial coverage were overcome by the advent of improved technologies, such as telescopic poles and kites, and near-infrared–sensitive digital single-lens reflex (DSLR) cameras. These technologies have been applied in mapping the spatial distribution and depth of submerged species. Remotely piloted aircraft systems have been used to take high-resolution aerial images of rivers and lakes and to map, identify and estimate the diversity of non-submerged species (O'Hare et al., 2018; Visser et al., 2015).

In aquatic environments, plant diversity is unidirectional due to downstream accumulation of genetic diversity initiated by hydrochoric dispersal of seeds in the downstream direction, leaving upstream populations genetically deprived or facing the threat of extinction (Honnay et al., 2010; Markwith and Scanlon, 2007). Aquatic autotrophs are diverse. They may either grow above the water surface, permanently submerged or periodically floating on water. Chambers et al. (2008, pp. 9–26) reported that aquatic plants can be grouped into seven representative plant divisions: "Cyanobacteria, Chlorophyta, Rhodophyta, Xanthophyta, Bryophyta, Pteridophyta and Spermatophyta." A total of 2,614 species from 412 genera representing about 1% of the entire population of vascular plants worldwide had a terrestrial origin, and nearly 39% of these aquatic genera are widespread in a particular biogeographic region (Chambers et al., 2008).

The earth's surface is divided into eight major ecozones, which are the Afrotropics, Antarctica, Australasia, Nearctic, Neotropics, Oceania, Indo-Malay and Palaearctic realms (BBC Nature, 2014; Chambers et al., 2008). Each biogeographic realm is geographically isolated, with distinct environmental pressures leading to species diversification through natural selection. Available literature establishes that the Neotropics has the richest population of aquatic plants, while the Antarctic ecozone has been severally reported to have the lowest species populations (Chambers et al., 2008; Murphy et al., 2019). On a broader scale, most of the earth's macrophyte diversity is constituted by small-ranged species, and on a global scale, the diversity and distribution of aquatic macrophytes are driven by many factors. The factors include biological characteristics of the plants, anthropogenic factors, geographical changes in biotic pressures and other factors mediated by latitudinal geo-climatic conditions (García-Girón et al., 2021; Getnet et al., 2021; Murphy et al., 2019).

26.2.1 Adaptive Strategies for Survival

Aquatic plants adopt different survival strategies for their environment. These include heterophylly, whereby striking variability is observed in aerial and submerged leaves. The former are stomata bearing, cutinized and thicker, while the latter have thin leaves lacking stomata and cuticle (Wells and Pigliucci, 2000). The development of fast-growing aerenchyma cells enhances the longitudinal expansion of roots into flooded areas and facilitates efficient diffusion of oxygen to the roots from the shoots (Yamauchi et al., 2013). Cortical aerenchyma has been implicated as an adaptive trait for optimizing phosphorus acquisition and remobilization from the root cortex, and to reduce the metabolic costs of root respiration during soil exploration (Bouranis et al., 2006; Evans, 2004; Postma and Lynch, 2011). Apart from enhancing oxygen transport, aerenchyma tissues act as barriers that prevent the excessive loss of oxygen around the root tips of macrophytes and facilitate the extracellular removal of facultative gases from the roots and soils (Evans, 2004).

26.2.2 Ecological Functions of Aquatic Plants

Aquatic plants play a crucial role in environmental remediation as they mediate the uptake and release of nutrients (Lu et al., 2018). The process of phytoremediation is embedded in the ability of aquatic plants to photosynthetically produce, transform and mobilize molecules to rhizospheric organisms, resulting in oxidation, uptake of bio-nutrients and contaminants suspended in the rhizosphere in a species-specific manner (Jimoh and Jimoh, 2021; Stottmeister et al., 2003). Conventional waste stabilization and removal in water treatment involve different mechanisms, such as physical sedimentation and the use of a trickling filter and activated sludges. However, recent findings show that aquatic plants such as *Myriophyllum spicatum* L., *Eichhornia crassipes* Mart., *Hydrilla verticillate* (L.f.) Royle and *Nelumbo nucifera* Gaertn. are good candidates for regulating bacterial metabolism in water treatment systems (Kanabkaew and Puetpaiboon, 2004; Lu et al., 2018).

Earlier, Ferrat et al., (2003) reviewed the use of seagrasses as biomarkers of environmental quality and as diagnostic tools in ecotoxicological studies using detectable responses in the synthesis and accumulation of secondary metabolites, photosynthetic activity, enzymatic processes involved in nitrogen and phosphorus assimilation during light and dark reactions, hydrolysis of organic phosphate monoesters to inorganic phosphate, carbohydrate synthesis and mobilization, and other oxidative mechanisms. The identified bioindicators of metallic pollutants were *Amphibolis Antarctica* (Labill.) Asch., *Cymodocea nodosa* (Ucrea.) Asch., *Cymodocea rotundata* Asch. & Schweinf., *Enhalus acoroides* (L.f) Royles, *Halodule uninervis* (Forssk.) Boiss., *Halophila ovalis* (R.Br) Hook.f., *Halophila stipulacea* (Forssk.) Asch., *Heterozostera tasmanica* (Martens ex Ascherson), *Posidonia australis* Hook.f., *Syringodium isoetifolium* (Asch.) Dandy, *Thalassia hemprichii* (Ehrenb. ex Solms) Asch., *Thalassodendron ciliatum* (Forssk.) Hartog, *Zostera marina* L., *Thalassia testudinum* Banks ex Konig. and *Zostera muelleri* Irmisch ex Asch. (Ferrat et al., 2003). In addition to those just listed, *Posidonia oceanica* (L.) Delile was regarded as a 'classical biomarker species' due to its high sensitivity to changes in environmental quality (Ferrat et al., 2002).

Macrophytes are essential for assessing the ecological status of aquatic environments. Their presence or absence may serve as a measure of ecosystem functioning, or evaluate the effects of abiotic processes on plant succession and biodiversity in the ecosystem (Bornette and Puijalon, 2011). Aquatic plants can also be used to biomark geological and geochemical parameters, substrate characteristics, anchorage strength of sediments and the magnitude of disturbances arising from abiotic stresses. Thus, abiotic stresses imposed by elevated temperature, light, carbon dioxide concentration and hydrodynamic waves can be biomarked (Boeger, 1992; Carter and Rybicki, 1990; Koch, 2001; Pilon and Santamaría, 2001; Puijalon et al., 2008; Schutten et al., 2005).

26.3 COMPOUNDS ISOLATED FROM AQUATIC PLANTS AND THEIR POTENTIAL THERAPEUTIC FUNCTIONS

Aquatic plants are carbon sinks and reservoirs of novel metabolites that are useful in the development of functional food and natural products (Bornette and Puijalon, 2011). They are rich depots of natural minerals, soluble proteins, dietary fibers, vitamins, carotenoids, polyunsaturated fatty acids, antioxidant phytonutrients and other functional ingredients with potential applications in disease management, cosmetics and nutraceutical formulations (Lall, 2020; Pooja et al., 2020). Recent bioassay-guided separation techniques have made possible the understanding, elucidation, characterization and isolation of underexploited novel compounds of high pharmacological significance from aquatic plants (Lall, 2020; Rinehart et al., 1990). Several bioactive compounds of diverse chemical composition isolated from aquatic plants and have been reported to be of significant interest to modern medicine and chemistry (Figure 26.1). Ordinarily, organic precursors of these metabolites are not limited to alkaloids, hydrocarbons, carotenoids, organic acids, aliphatic alcohols, fatty acids, esters and lipids, but also their chemical relatives from triterpenes, tannins, steroids, and glycoside groups (Bandaranayake, 2002).

Some novel antimicrobial metabolites have been derived from seaweeds (Zerrifi et al., 2018). These include Diterpene sargafuran isolated from *Sargassum macrocarpum* C. Agardh.; Peyssonoic acid A and B isolated from *Peyssonnelia* sp. Decaisne; 10-Hydroxy kahukuene B isolated from *Laurencia mariannensis* Yamada; 3-Dibromobenzaldehyde-4, 5-disulfate potassium salt and 5-Bromo-3,4-dihydroxybenzaldehyde isolated from *Polysiphonia Lanora* J.V. Lamouroux; Tiomanene Acetylmajapolene (A and B) from *Laurencia* sp. J.V. Lamouroux; and Zonarol and isozonarol sesquiterpenes obtained from *Dictyopteris zonarioides* Farlow (Zerrifi et al., 2018). Earlier, Vasu et al., (2009) had screened three aquatic plants species (*Ipomoea aquatica* Forssk., *Nymphaea pubescens* Willd. and *Eichhornia crassipes* Mart.) and reported the presence of alkaloids, phenols, tannins, steroids, triterpenoids, flavonoids and saponin, and other biomolecules critical to the therapeutic properties of the plants (Vasu et al., 2009). Recently, Pooja et al., (2020) published some novel bioactive compounds derived from marine plants with immunomodulatory, antioxidant, anti-obesity, antiviral, anticancer, neuroprotective, antidiabetic, and antimicrobial properties (Figure 26.1) (Pooja et al., 2020). These compounds may also serve as important components of industrial raw chemicals, functional food biomolecules, medicinal precursors and a potential resource for biofuel production.

The antioxidant capacity of a seaweed, *Palmaria palmata* (L.) F.Weber & D.Mohr (dulse)—which is traditionally eaten as garnish or snacks—was attributed to reducing soluble compounds characterized by phenolic functional groups (Yuan et al., 2005). Although the identity of the antioxidant molecules was not elucidated, it was reported that crude extracts from the seaweed quenched reactive oxygen species, inhibited the synthesis of thiobarbituric acid and conjugated diene production in a linoleic acid emulsion. About 1.0 mg of *P. palmata* was also observed to be equivalent to 10.3 μg gallic acid in terms of total polyphenol content, and having reducing efficacy as 9.68 μg L-ascorbic acid (Altemimi et al., 2017; Yuan et al., 2005). Preliminary screening of aqueous and ethanolic extracts of three macrophytes, namely *Echinochloa stagnina* (Retz.) P.Beauv., *Nymphaea lotus* L. and *Pistia stratiotes* L., harvested from different water bodies in the southern part of Manzalah Lake, El-Serw village, northeastern Egypt, showed some sterols and tannins in *E. stagnina*, resins and sterols in *P. stratiotes* and tannins, resins, sterols and flavonoids in *N. lotus*. The plant extracts were also reported to have exhibited inhibitory effects on fungal and bacterial isolates (Daboor and Haroon, 2012).

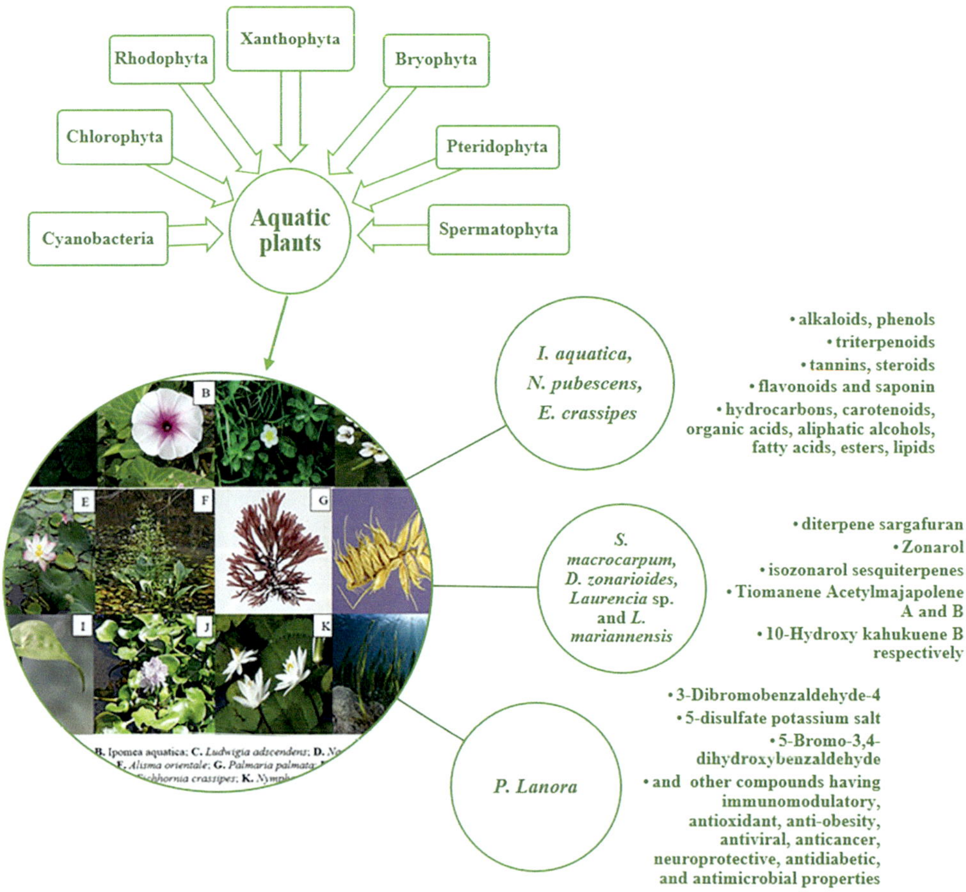

FIGURE 26.1 A synopsis of some compounds isolated from aquatic plants (Crouan & Crouan, 1852; Garg, 2009; Gloor, 2016; Hendrik, 2011; Hernandez, 2010; Mata, 2020; Prenn, 2013; Subhash, 2015; Vengolis, 2013; Venter, 2017; Zezhong, 2013; Ziarnek, 2012).

Several researchers identified *Cymodocea serrulata* (R.Brown) as a potent natural drug due to its cytotoxic effects against pathogenic microbes and cancer cells (Chanthini et al., 2015; Kumar et al., 2008; Palaniappan et al., 2015; Ramalingam et al., 2013). The seagrass has been widely recommended as a source of antioxidants in food formulation, as a natural remedy for microbial infections, and as a stabilizing biomolecule suitable as an organic preservative and valuable bio-nutrient useful in the management of cervical and lung cancers (Chanthini et al., 2015; Kumar et al., 2008; Palaniappan et al., 2015; Ramalingam et al., 2013).

Pharmacological evaluations of some freshwater macrophytes indicated that they are potential sources of antioxidants and anticancer natural products. Crude extracts of leaves and fruits of *Monochoria hastata* (L.) Solms, *Ipomea aquatica* Forssk., *Nasturtium officinale* W.T.Aiton, *Nelumbo nucifera* Gaertn., and particularly *Ludwigia adscendens* (L.) H.Hara, *Colocasia esculenta* L. Schott. and *Oryza sativa* L., had higher alpha-glucosidase inhibitory effects than anti-diabetic acarbose and their decoction had different applications in traditional medicine (Chai et al., 2015; Ooh et al., 2014). Aqueous extracts of different parts of *M. hastata* and *L. adscendens* exhibited lipoxygenase inhibitory activity (Ooh et al., 2014), while seeds of *Alisma orientale* are a natural source of 2,4,5-Trihydroxycinnamic acid, a bioactive component of rooibos tea (Zhao et al., 2012). These among other pharmacological reports suggest that there are untapped potentials in aquatic plants in drug discovery, development and production.

26.4 BIBLIOMETRIC ANALYSIS OF RESEARCH OUTPUTS ON THERAPEUTIC MACROPHYTES FROM 2011–2020

Bibliometrics is a technique used to monitor trends of research outputs on a particular topic, assess authors' contributions and citation classics, and measure the scientific impacts of publications on the academic community and society (Iftikhar et al., 2019). The bibliometric analysis uses detailed technology to extract data from published research findings in books, journals, conference proceedings and other outlets via which research findings are communicated to society (Diem and Wolter, 2013; Okaiyeto and Oguntibeju, 2021). The methods of bibliometrics can be used to develop thesauri, assess academic outputs, evaluate the pattern of readership of a journal and its relevance to the scientific community, and identify the impacts of a research field or group to predict areas of research with knowledge gaps (Linnenluecke et al., 2020; Rousseau, 2014; Smith et al., 2021).

The core objective of this analysis was to harmonize global research outputs on the medicinal applications of aquatic plants in the last decade, considering that the period had witnessed a surge in demand for plant-based products for food and medical applications (Jimoh et al., 2020; Srivastava et al., 2019). Although herbal medicine has been in practice since ancient times, its optimal application was restricted by different biological and chemical barriers such as inappropriate dosage, high toxicity, hydrophobicity, insolubility, low bioavailability of bioactive substances and other factors that limit the delivery of herbal drugs to targeted areas (Kurhekar, 2021; Sajid et al., 2019; Srivastava et al., 2019). This bibliometric study seeks to review existing literature on therapeutic uses of aquatic plants to explore notable scientific outputs and researchers in the field and identify key areas craving for attention to meet the rising demand for natural products and conserve biodiversity.

26.4.1 Material and Methods

26.4.1.1 Data Extraction

Data were sourced from the Web of Science database which has been adjudged globally as a leading multidisciplinary academic repository covering over 100 million records from 33,000 top-rated journals (Clarivate Analytics, 2017; Orimoloye and Ololade, 2021). Different keywords like "aquatic plant medicine," "aquatic plant therapy," "seaweed therapy," "medicinal seaweeds," "medicinal hydrophytes," "therapeutic hydrophytes," "medicinal macrophytes," "therapeutic macrophytes," "medicinal marine plants" and "therapeutic marine plants" that are closely related to the topic under review were adopted with the Boolean operator "OR" in between key terms. To prevent false results arising from the accumulation of unrelated datasets from sources that are at variance with target results, the "title" search was used as reported by Okaiyeto and Oguntibeju (2021).

As listed in Table 26.1, a sum of 809 articles was retrieved from the Web of Science after excluding documents from unrelated sources or early access publications from 2011–2020. The documents were downloaded in plain text file format from Web of Science and exported to *Biblioshyny* in R-studio (version 4.0.2) to analyze extracted bibliometric indices according to Aria and Cuccurullo (2017) and Tenen (2017). The following bibliometric indices were used to evaluate the existing and future research trends on aquatic phytotherapy, annual scientific production, most productive authors, most productive journals/source, most productive countries, collaboration network, bibliographic coupling and thematic map. Citation analysis was used to evaluate the relative productivity of authors or their publications as cited by other researchers in the field using parameters like bibliographic coupling and co-citation obtained from the publication-citation matrix (Smith et al., 2021).

26.4.2 Results and Discussion

26.4.2.1 Annual Scientific Production

In the period under review, the annual scientific publication had a growth rate of 12.01%. This may imply a steady growth in research on medicinal macrophytes. The smallest number of articles (48)

TABLE 26.1
Web of Science Data Summary

Descriptions	Results
Main information about data	
Period	2011–2020
Sources (journals, books, etc.)	427
Documents	809
Average years from publication	4.54
Average citations per document	24.06
Average citations per year per doc	3.934
References	52129
Document types	
Article	570
Article; proceedings paper	8
Editorial material	2
Meeting abstract	2
Review	226
Review; book chapter	1
Document contents	
Keywords plus (ID)	3366
Author's keywords (DE)	2734
Authors	
Authors	3794
Author appearances	4296
Authors of single-authored documents	32
Authors of multi-authored documents	3762
Authors' collaboration	
Single-authored documents	33
Documents per author	0.213
Authors per document	4.69
Co-authors per document	5.31
Collaboration Index	4.85

was published in 2012, while in 2020, the highest number of publication outputs (136) representing 16.4% of total publications recorded in ten years (Figure 26.2). The mean total citation per article was highest in 2013 (56.19), while the smallest value (4.51) was recorded in 2020. A surge in publication trend was observed in 2020, when research interest in medicinal macrophytes grew by 30% compared to the previous year (Table 26.1). This result corroborates growing interest in natural products as reported by Jimoh et al. (2019) and Kambizi et al. (2017), especially in aquatic plants as more macrophytes were investigated for medicinal uses (Jimoh et al., 2019; Kambizi et al., 2017; Lall, 2020).

26.4.2.2 Most Productive Authors

A total number of 3,794 authors contributed to research on medicinal macrophytes in the 10-year period studied. Not less than 99.2% of the total documents (809) retrieved from the Web of Science database were multi-authored (Table 26.1). This translates to high collaboration networks among authors (Figure 26.3), with similar interests in aquatic phytotherapy. Further bibliometric indices

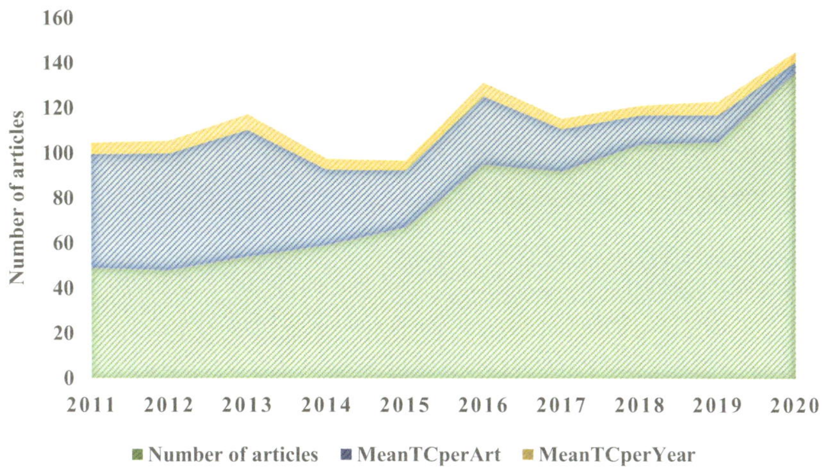

FIGURE 26.2 Annual scientific publications on medicinal aquatic plants from 2011–2020 in the Web of Science database. MeanTCperArt = Mean total citation per article; MeanTCperYear = Mean total citation per year.

TABLE 26.2
Top 25 Authors on Medicinal Macrophytes and Their Bibliometric Impact

Elements	Articles	H_index[1]	G_Index	M_Index	TC[2]	NP[3]	Py_Start[4]
Aravindan N	6	4	4	0.571	44	4	2015
Nunes B	6	4	6	0.8	48	6	2017
Teixeira VL	6	4	6	0.571	42	6	2015
Andrade PB	5	5	5	0.5	207	5	2012
Aravindan S	5	3	3	0.429	38	3	2015
Barbosa JM	5	4	4	0.364	163	4	2011
Ermakova SP	5	3	4	0.273	49	4	2011
Freitas ALP	5	5	5	0.556	139	5	2013
Kumar A	5	4	5	0.4	182	5	2012
Lee JH	5	3	5	0.273	96	5	2011
Lopes G	5	5	5	0.5	207	5	2012
Medeiros JVR	5	5	5	0.556	139	5	2013
Valentao P	5	5	5	0.5	207	5	2012
Wang J	5	5	5	0.5	172	5	2012
Barros CD	4	3	4	0.429	25	4	2015
Brooks BW	4	4	4	0.667	118	4	2016
Fuly AL	4	4	4	0.571	25	4	2015
Herman TS	4	2	2	0.286	26	2	2015
Hyun JW	4	3	4	0.375	42	4	2014
Liu J	4	3	4	0.429	88	4	2015
Mishra A	4	4	4	0.4	106	4	2012
Noseda MD	4	2	3	0.286	16	3	2015
Okoh AI	4	3	4	0.5	79	4	2016
Ramos AA	4	3	4	0.6	48	4	2017
Rocha E	4	3	4	0.6	48	4	2017

Notes: **1** Hirsch index; **2** Total citations; **3** Number of publications; **4** Publication year

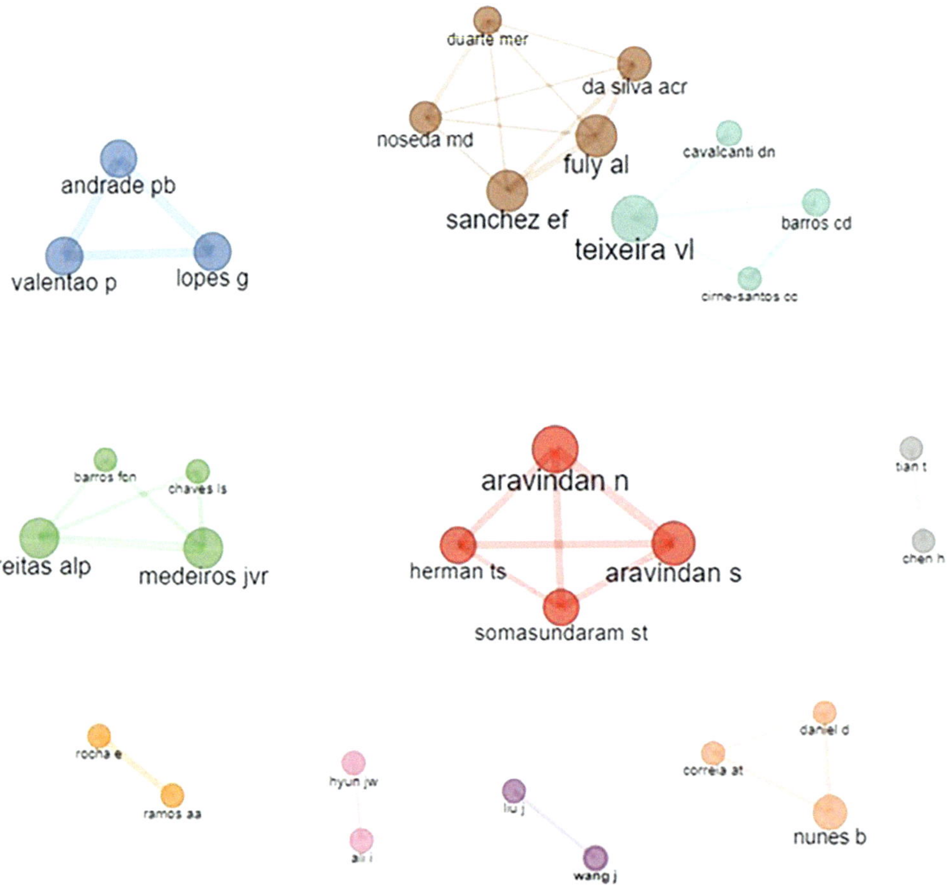

FIGURE 26.3 Collaborative networks between numerous authors.

indicate that the trio of Aravindan N, Nunes B and Teixeira VL contributed the highest number of publications on medicinal macrophytes in these ten years (Table 26.2). Other bibliometric parameters for the topmost 25 authors are listed in Table 26.2.

26.4.2.3 Most Frequently Used Words

Figure 26.4 presents the top 25 most frequently used words in the 809 documents obtained from Web of Science. The word '*in vitro*' was the most patronized word, having 104 counts concerning medicinal macrophytes. This may suggest that several assays were conducted *in vitro* to validate ethnobotanical reports on therapeutic uses of aquatic plants. Besides, other words frequently used by researchers in the field are listed in the chart in order of mention.

26.4.2.4 Collaboration Networks

As presented in Figure 26.5 and Figure 26.6, data from this study show that there is a huge gap in collaboration between authors, institutions and countries. This creates a window for the exchange of human and material resources among authors and institutions worldwide to explore opportunities in the study, cultivation and exchange of more aquatic plant species for medicinal exploits.

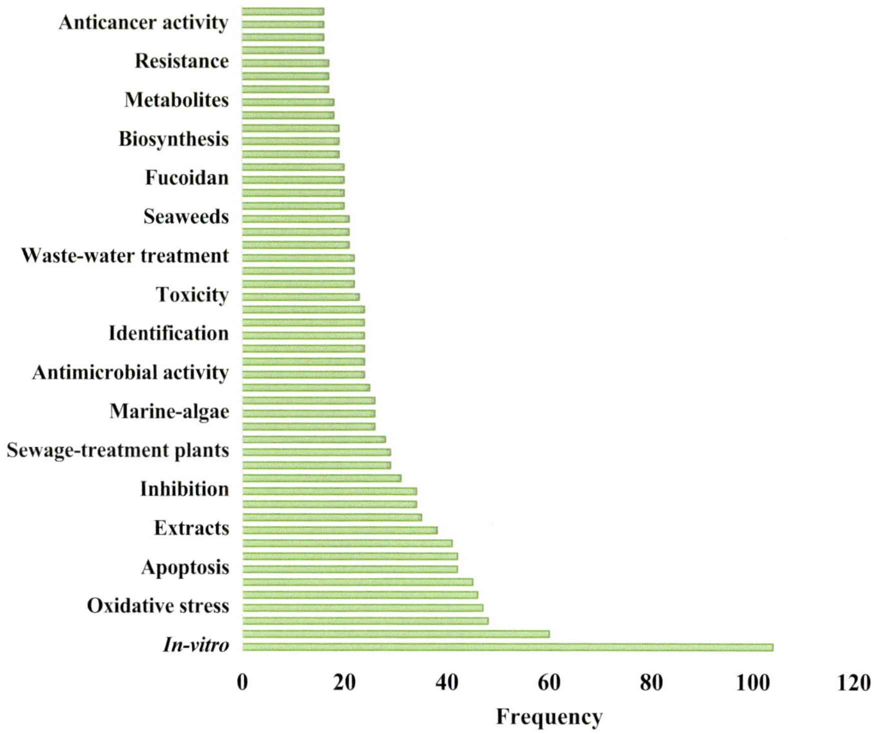

FIGURE 26.4 The 25 most frequently used words in aquatic phytotherapy research.

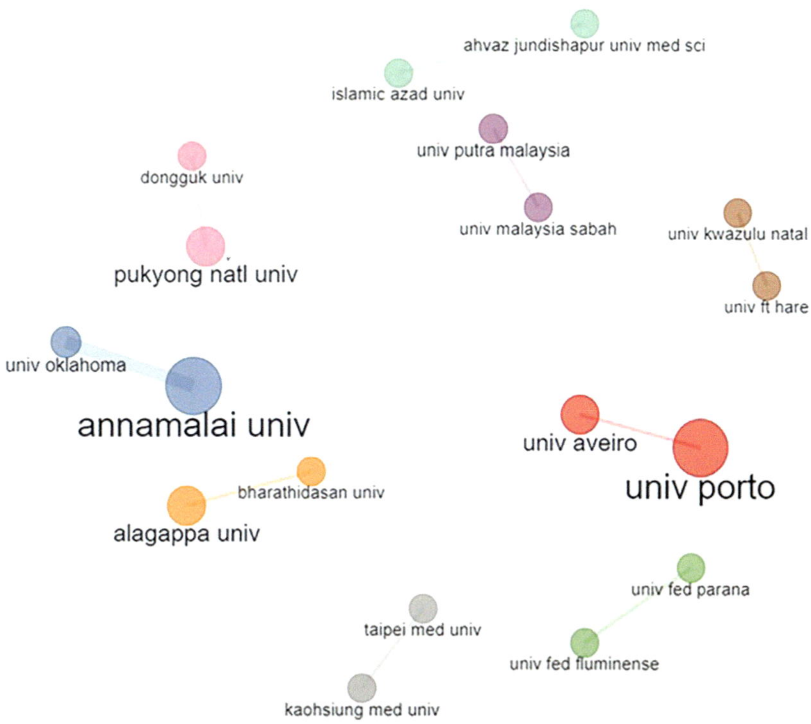

FIGURE 26.5 Collaborations between different institutions worldwide.

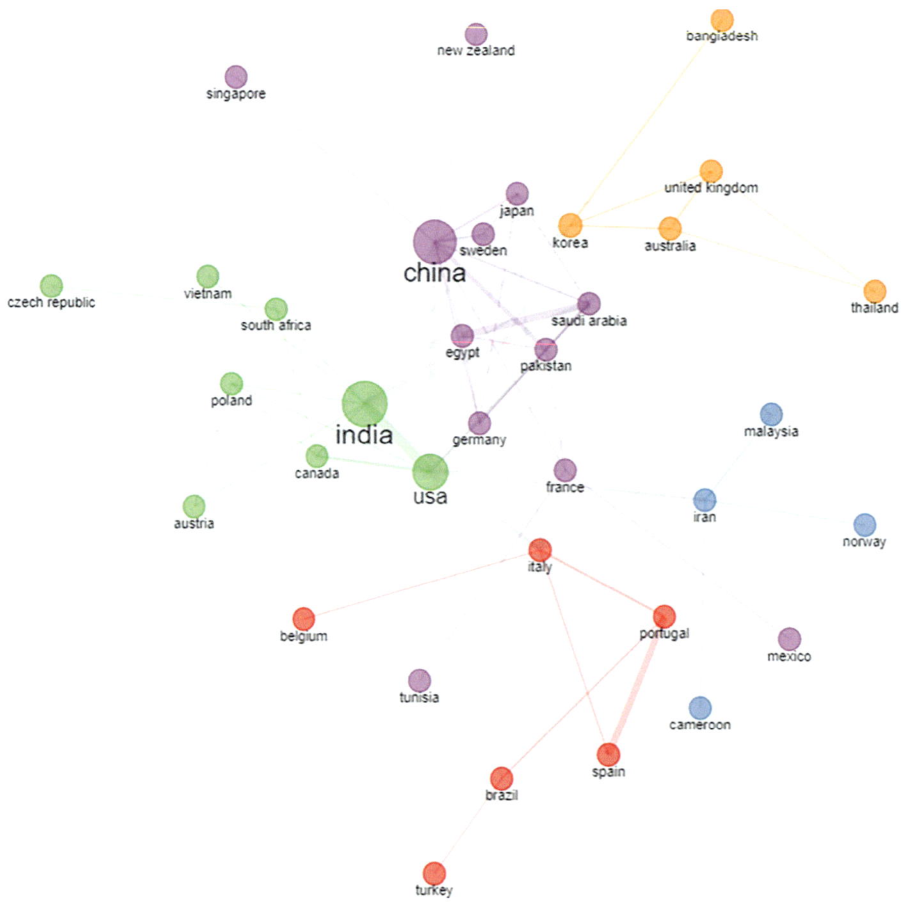

FIGURE 26.6 Networking between countries.

26.5 CONCLUSION

The global distribution and diversity of macrophytes are dictated by latitude. Although there is a steady growth in research related to medicinal uses of macrophytes in the period under review, bibliometric indices used in this study indicate that a huge gap exists in collaboration among researchers and exchange of materials and human resources. Researchers and food industries should focus on beneficial uses of macrophytes as food supplements to safeguard society and create more leafy options for households.

REFERENCES

Aasim, M., Bakhsh, A., Sameeullah, M., Karataş, M., Khawar, K.M., 2018. "Aquatic plants as human food." In: *Global Perspectives on Underutilized Crops*. Springer International Publishing: 165–187. doi: 10.1007/978-3-319-77776-4_6

Aasim, M., Khawar, K.M., Ahmed, S.I., Karatas, M., 2019. "Multiple uses of some important aquatic and semiaquatic medicinal plants." In: *Plant and Human Health, Phytochemistry and Molecular Aspects*. Springer Nature, Switzerland 2: 541–577.

Altemimi, A., Lakhssassi, N., Baharlouei, A., Watson, D.G., Lightfoot, D.A., 2017. "Phytochemicals: Extraction, isolation, and identification of bioactive compounds from plant extracts." *Plants* 6. doi: 10.3390/plants6040042

Aria, M., Cuccurullo, C., 2017. "Bibliometrix: An R-tool for comprehensive science mapping analysis." *Journal of Informetrics* 11: 959–975. doi: 10.1016/j.joi.2017.08.007

Bandaranayake, W.M., 2002. "Bioactivities, bioactive compounds and chemical constituents of mangrove plants." *Wetland Ecology Management* 10: 421–452. doi: 10.1023/A:1021397624349

BBC Nature, 2014. "Ecozones." https://web.archive.org/web/20180710161951/www.bbc.co.uk/nature/ecozones (Accessed 28 April 2021).

Boeger, R.T., 1992. "The influence of substratum and water velocity on growth of *Ranunculus aquatilis* L. (Ranunculaceae)." *Aquatic Botany* 42: 351–359. doi: 10.1016/0304-3770(92)90054-M

Bornette, G., Puijalon, S., 2011. "Response of aquatic plants to abiotic factors: A review." *Aquatic Science* 73: 1–14. doi: 10.1007/s00027-010-0162-7

Bouranis, D.L., Chorianopoulou, S.N., Kollias, C., Maniou, P., Protonotarios, V.E., Siyiannis, V.F., Hawkesford, M.J., 2006. "Dynamics of aerenchyma distribution in the cortex of sulfate-deprived adventitious roots of maize." *Annals of Botany* 97, 695–704. doi: 10.1093/aob/mcl024

Brix, H., 1997. "Do macrophytes play a role in constructed treatment wetlands?" *Water Science and Technology* 35: 11–17. doi: 10.2166/wst.1997.0154

Campbell, C.J., Johns, C. V., Nielsen, D.L., 2014. "The value of plant functional groups in demonstrating and communicating vegetation responses to environmental flows." *Freshwater Biology* 59: 858–869. doi: 10.1111/fwb.12309

Carter, V., Rybicki, N.B., 1990. "Light attenuation and submersed macrophyte distribution in the tidal Potomac River and estuary." *Estuaries* 13: 441–452. doi: 10.2307/1351788

Ceschin, S., Bellini, A., Scalici, M., 2021. "Aquatic plants and ecotoxicological assessment in freshwater ecosystems: A review." *Environmental Science and Pollution Research* 28: 4975–4988. doi: 10.1007/s11356-020-11496-3

Chai, T.T., Ooh, K.F., Quah, Y., Wong, F.C., 2015. "Edible freshwater macrophytes: A source of anticancer and antioxidative natural products—a mini-review." *Phytochemistry Review* 14: 443–457. doi: 10.1007/s11101-015-9399-z

Chambers, P.A., Lacoul, P., Murphy, K.J., Thomaz, S.M., 2008. "Global diversity of aquatic macrophytes in freshwater." *Hydrobiologia*. doi: 10.1007/s10750-007-9154-6

Chanthini, A.B., Balasubramani, G., Ramkumar, R., Sowmiya, R., Balakumaran, M.D., Kalaichelvan, P.T., Perumal, P., 2015. "Structural characterization, antioxidant and in vitro cytotoxic properties of seagrass, *Cymodocea serrulata* (R.Br.) Asch. & Magnus mediated silver nanoparticles." *Journal of Photochemistry and Photobiology B: Biology* 153: 145–152. doi: 10.1016/j.jphotobiol.2015.09.014

Clarivate Analytics, 2017. "Web of science fact book." Web of Science Fact Book.

Crouan, P., Crouan, H., 1852. "*Palmaria palmate*." Obtained from Wikimedia under the Public domain license. https://commons.wikimedia.org/wiki/File:Palmaria_palmata_2_Crouan.jpg (Accessed 30 June 2021)

Daboor, S.M., Haroon, A.M., 2012. "In vitro: Antimicrobial potential and phytochemical screening of some Egyptian aquatic plants." *Egyptian Journal of Aquatic Research* 38: 233–239. doi: 10.1016/j.ejar.2013.01.005

Diem, A., Wolter, S.C., 2013. "The use of bibliometrics to measure research performance in education sciences." *Research in Higher Education* 54: 86–114. doi: 10.1007/s11162-012-9264-5

Evans, D.E., 2004. "Aerenchyma formation." *New Phytologist*. doi: 10.1046/j.1469-8137.2003.00907.x

FAO, 2012. "Aquatic plants their uses and risks." In: Wersal, R., Madsen, J. (Eds.), *International Plant Protection Convention*. FAO, Viale delle Terme di Caracalla, 00153 Rome, Italy: 94.

Ferrat, L., Bingert, A., Roméo, M., Gnassia-Barelli, M., Pergent-Martini, C., 2002. "Mercury uptake and enzymatic response of *Posidonia oceanica* after an experimental exposure to organic and inorganic forms." *Environmental Toxicology and Chemistry* 21: 2365–2371. doi: 10.1002/etc.5620211115

Ferrat, L., Pergent-Martini, C., Roméo, M., 2003. "Assessment of the use of biomarkers in aquatic plants for the evaluation of environmental quality: Application to seagrasses." *Aquatic Toxicology* 65: 187–204. doi: 10.1016/S0166-445X(03)00133-4

García-Girón, J., Heino, J., Iversen, L.L., Helm, A., Alahuhta, J., 2021. "Rarity in freshwater vascular plants across Europe and North America: Patterns, mechanisms and future scenarios." *Science of the Total Environment* 786: 147–491. doi: 10.1016/j.scitotenv.2021.147491

Garg, J.M., 2009. "*Ipomoea aquatica*." Obtained from Wikimedia under the Creative Commons Attribution 3.0 Unported license. https://commons.wikimedia.org/wiki/File:Ipomoea_aquatica_(Marsh_Glory)_flower_W_IMG_0405.jpg (Accessed 30 June 2021)

Getnet, H., Mengistou, S., Warkineh, B., 2021. "Diversity of macrophytes in relation to environmental conditions in wetlands along the lower part of the Gilgel Abay River catchment in Ethiopia. African." *Journal of Aquatic Science*. doi: 10.2989/16085914.2020.1842721

Gloor, C., 2016. "*Enhalus acoroides*." Obtained from Wikimedia under the Creative Commons Attribution 2.0 Generic license. https://commons.wikimedia.org/wiki/File:Enhalus_acoroides_(30774220094).jpg (Accessed 30 June 2021)

Gurnell, A., 2014. "Plants as river system engineers. Earth Surf. Process." *Landforms* 39: 4–25. doi: 10.1002/esp.3397

Hendrik, W., 2011. "*Monochoria hastata*." Obtained from Wikimedia under the Creative Commons Attribution 3.0 Unported license. https://commons.wikimedia.org/wiki/File:Monochoria_hastata,_plant.JPG (Accessed 30 June 2021)

Hernandez, J., 2010. "*Echinochloa stagnina*." Obtained from Wikimedia under the Public Domain license. https://commons.wikimedia.org/wiki/File:Echinochloa_stagnina_seeds_1.jpg (Accessed 30 June 2021)

Honnay, O., Jacquemyn, H., Nackaerts, K., Breyne, P., Van Looy, K., 2010. "Patterns of population genetic diversity in riparian and aquatic plant species along rivers." *Journal of Biogeography* 37: 1730–1739. doi: 10.1111/j.1365-2699.2010.02331.x

Iftikhar, P.M., Ali, F., Faisaluddin, M., Khayyat, A., De Gouvia De Sa, M., Rao, T., 2019. "A bibliometric analysis of the top 30 most-cited articles in gestational diabetes mellitus literature (1946–2019)." *Cureus* 11: 4131. doi: 10.7759/cureus.4131

Jimoh, M.A., Jimoh, M.O., 2021. "Economic consequences of plant biodiversity loss." In: Aliero, A.A., Agboola, D.A., Vwioko, E.D. (Eds.), *Plants and the Ecosystems*. FUK Press, Federal University of Kashere, Gombe State, Nigeria: 397–411 Publishe.

Jimoh, M.O., Afolayan, A.J., Lewu, F.B., 2019. "Antioxidant and phytochemical activities of *Amaranthus caudatus* L. harvested from different soils at various growth stages." *Scientific Reports* 9: 12965. doi: 10.1038/s41598-019-49276-w

Jimoh, M.O., Afolayan, A.J., Lewu, F.B., 2020. "Nutrients and antinutrient constituents of *Amaranthus caudatus* L. Cultivated on different soils." *Saudi Journal of Biological Sciences* 27: 3570–3580. doi: 10.1016/j.sjbs.2020.07.029

Kambizi, L., Bakare-Odunola, M.T., Oladiji, A.T., Kola-Mustapha, A.T., Amusa, T.O., Atolani, O., Njinga, N.S., Quadri, A.L., 2017. "Proteinease inhibition, membrane stabilization, antioxidant and phytochemical evaluations of leaves, seeds and calyces of four selected edible medicinal plants." *Cogent Chemistry* 3: 1–15. doi: 10.1080/23312009.2017.1314064

Kanabkaew, T., Puetpaiboon, U., 2004. "Aquatic plants for domestic wastewater treatment: Lotus (*Nelumbo nucifera*) and Hydrilla (*Hydrilla verticillata*) systems." *Songklanakarin Journal of Science and Technology* 26: 749–756.

Koch, E.W., 2001. "Beyond light: Physical, geological, and geochemical parameters as possible submersed aquatic vegetation habitat requirements." *Estuaries*. doi: 10.2307/1352808

Kumar, C.S., Sarada, D.V.L., Thomas, P.G., Rengasamy, R., 2008. "Antibacterial activity of three South Indian seagrasses, *Cymodocea serrulata*, *Halophila ovalis* and *Zostera capensis*." *World Journal of Microbiology and Biotechnology* 24: 1989–1992. doi: 10.1007/s11274-008-9695-5

Kurhekar, J.V., 2021. "Ancient and modern practices in phytomedicine." In: *Preparation of Phytopharmaceuticals for the Management of Disorders*. Elsevier: 55–75. doi: 10.1016/b978-0-12-820284-5.00019-8

Lall, N. (Ed.), 2020. *Aquatic Plants: Pharmaceutical and Cosmetic Applications*. CRC Press, Taylor & Francis Group.

Linnenluecke, M.K., Marrone, M., Singh, A.K., 2020. "Conducting systematic literature reviews and bibliometric analyses." *Australian Journal of Management* 45: 175–194. doi: 10.1177/0312896219877678

Lu, B., Xu, Z., Li, J., Chai, X., 2018. "Removal of water nutrients by different aquatic plant species: An alternative way to remediate polluted rural rivers." *Ecological Engineering* 110: 18–26. doi: 10.1016/j.ecoleng.2017.09.016

Markwith, S.H., Scanlon, M.J., 2007. "Multiscale analysis of *Hymenocallis coronaria* (Amaryllidaceae) genetic diversity, genetic structure, and gene movement under the influence of unidirectional stream flow." *American Journal of Botany* 94: 151–160. doi: 10.3732/ajb.94.2.151

Mata, J.C.F. 2020. "*Eichhornia crassipes*." Obtained from Wikimedia under the Creative Commons Attribution-Share Alike 4.0 International license. https://commons.wikimedia.org/wiki/File:Eichhornia_crassipes_(or_Pontederia_crassipes)_-_Family_Pontederiaceae.jpg (Accessed 30 June 2021)

Murphy, K., Efremov, A., Davidson, T.A., Molina-Navarro, E., Fidanza, K., Crivelari Betiol, T.C., Chambers, P., Tapia Grimaldo, J., Varandas Martins, S., Springuel, I., Kennedy, M., Mormul, R.P., Dibble, E., Hofstra, D., Lukács, B.A., Gebler, D., Baastrup-Spohr, L., Urrutia-Estrada, J., 2019. "World distribution, diversity and endemism of aquatic macrophytes." *Aquatic Botany* 158: 103127. doi: 10.1016/j.aquabot.2019.06.006

O'Hare, M.T., Aguiar, F.C., Asaeda, T., Bakker, E.S., Chambers, P.A., Clayton, J.S., Elger, A., Ferreira, T.M., Gross, E.M., Gunn, I.D.M., Gurnell, A.M., Hellsten, S., Hofstra, D.E., Li, W., Mohr, S., Puijalon, S.,

Szoszkiewicz, K., Willby, N.J., Wood, K.A., 2018. "Plants in aquatic ecosystems: Current trends and future directions." *Hydrobiologia*. doi: 10.1007/s10750-017-3190-7

Okaiyeto, K., Oguntibeju, O.O., 2021. "Trends in diabetes research outputs in South Africa over 30 years from 2010 to 2019: A bibliometric analysis." *Saudi Journal of Biological Sciences* 28: 2914–2924. doi: 10.1016/j.sjbs.2021.02.025

Ooh, K.F., Ong, H.C., Wong, F.C., Sit, N.W., Chai, T.T., 2014. "High performance liquid chromatography profiling of health-promoting phytochemicals and evaluation of antioxidant, anti-lipoxygenase, iron chelating and anti-glucosidase activities of wetland macrophytes." *Pharmacognosy Magazine* 10: S443—S455. doi: 10.4103/0973-1296.139767

Orimoloye, I.R., Ololade, O.O., 2021. "Global trends assessment of environmental health degradation studies from 1990 to 2018." *Environment, Development and Sustainability* 23: 3251–3264. doi: 10.1007/s10668-020-00716-y

Palaniappan, P., Sathishkumar, G., Sankar, R., 2015. "Fabrication of nano-silver particles using *Cymodocea serrulata* and its cytotoxicity effect against human lung cancer A549 cells line." *Spectrochimica Acta Part A: Molecular Spectroscopy* 138: 885–890. doi: 10.1016/j.saa.2014.10.072

Pilon, J., Santamaría, J., 2001."Seasonal acclimation in the photosynthetic and respiratory temperature responses of three submerged freshwater macrophyte species." *New Phytologist* 151: 659–670. doi: 10.1046/j.0028-646x.2001.00212.x

Pooja, K., Rani, S., Rana, V., Pal, G.K., 2020. "Aquatic plants as a natural source of antimicrobial and functional ingredients." In: *Functional and Preservative Properties of Phytochemicals*. Elsevier: 93–118. doi: 10.1016/b978-0-12-818593-3.00003-8

Postma, J.A., Lynch, J.P., 2011. "Theoretical evidence for the functional benefit of root cortical aerenchyma in soils with low phosphorus availability." *Annals of Botany* 107: 829–841. doi: 10.1093/aob/mcq199

Prenn. 2013. "*Nelumbo nucifera*." Obtained from Wikimedia under the Creative Commons Attribution 3.0 Unported license. https://commons.wikimedia.org/wiki/File:Nelumbo_nucifera_01.JPG (Accessed 30 June 2021)

Puijalon, S., Léna, J.P., Rivière, N., Champagne, J.Y., Rostan, J.C., Bornette, G., 2008. "Phenotypic plasticity in response to mechanical stress: Hydrodynamic performance and fitness of four aquatic plant species." *New Phytology* 177: 907–917. doi: 10.1111/j.1469-8137.2007.02314.x

Ramalingam, N., Ramakrishnan, K., Krishnan, C., Sankar, S., 2013. "Phytochemical analysis, HPTLC finger printing, *in vitro* antioxidant and cytotoxic activity of *Cymodocea serrulata*." *Pharmacognosy Journal* 5: 238–241. doi: 10.1016/j.phcgj.2013.09.006

Rinehart, K.L., Holt, T.G., Fregeau, N.L., Keifer, P.A., Robert Wilson, G., Thomas, J.P., Sakai, R., Thompson, A.G., Stroh, J.G., Shield, L.S., Seigler, D.S., Li, L.H., Martin, D.G., Grimmelikhuijzen, C.J.P., Gäde, G., 1990. "Bioactive compounds from aquatic and terrestrial sources." *Journal of Natural Products* 53: 771–792. doi: 10.1021/np50070a001

Rousseau, R., 2014. "Library science: Forgotten founder of bibliometrics." *Nature*. doi: 10.1038/510218e

Sajid, M., Cameotra, S.S., Ahmad Khan, M.S., Ahmad, I., 2019. "Nanoparticle-based delivery of phytomedicines: Challenges and opportunities." In: *New Look to Phytomedicine: Advancements in Herbal Products as Novel Drug Leads*. Elsevier: 597–623. doi: 10.1016/B978-0-12-814619-4.00024-0

Schlesinger, W.H., Bernhardt, E.S., 2020. "Wetland ecosystems." *Biogeochemistry*: 249–291. doi: 10.1016/b978-0-12-814608-8.00007-4

Schutten, J., Dainty, J., Davy, A.J., 2005. "Root anchorage and its significance for submerged plants in shallow lakes." *Journal of Ecology* 93: 556–571. doi: 10.1111/j.1365-2745.2005.00980.x

Smith, H.H., Idris, O.A., Maboeta, M.S., 2021. "Global trends of green pesticide research from 1994 to 2019: A bibliometric analysis." *Journal of Toxicology*. doi: 10.1155/2021/6637516

Srivastava, A., Srivastava, P., Pandey, A., Khanna, V.K., Pant, A.B., 2019. "Phytomedicine: A Potential Alternative Medicine in Controlling Neurological Disorders." In: *New Look to Phytomedicine: Advancements in Herbal Products as Novel Drug Leads*. Elsevier: 625–655. doi: 10.1016/B978-0-12-814619-4.00025-2

Stillman, R.A., Wood, K.A., Gilkerson, W., Elkinton, E., Black, J.M., Ward, D.H., Petrie, M., 2015. "Predicting effects of environmental change on a migratory herbivore." *Ecosphere* 6: art114. doi: 10.1890/es14-00455.1

Stottmeister, U., Wießner, A., Kuschk, P., Kappelmeyer, U., Kästner, M., Bederski, O., Müller, R.A., Moormann, H., 2003. "Effects of plants and microorganisms in constructed wetlands for wastewater treatment." *Biotechnology Advances* 22: 93–117. doi: 10.1016/j.biotechadv.2003.08.010

Subhash, M., 2015. "*Nymphaea lotus*." Obtained from Wikimedia under the Creative Commons CC0 1.0 Universal Public Domain Dedication license. https://commons.wikimedia.org/wiki/File:Nymphaea_Lotus.jpg (Accessed 30 June 2021)

Tenen, D., 2017. *Plain Text: The Poetics of Computation*. Stanford University Press, Indiana.

Vasu, K., Goud, J.V., Suryam, A., Charya, M.A.S., 2009. "Biomolecular and phytochemical analyses of three aquatic angiosperms." *African Journal of Microbiology Research* 3: 418–421.

Vengolis, 2013. "*Ludwigia adscendens*." Obtained from Wikimedia under the Creative Commons Attribution 3.0 Unported license. https://commons.wikimedia.org/wiki/File:Ludwigia_adscendens_08495.jpg (Accessed 30 June 2021)

Venter, P., 2017. "*Nasturtium officinale*." Obtained from Wikimedia under the Creative Commons Attribution-Share Alike 4.0 International license. https://commons.wikimedia.org/wiki/File:Nasturtium_officinale00.jpg (Accessed 30 June 2021)

Visser, F., Buis, K., Verschoren, V., Meire, P., 2015. "Depth estimation of submerged aquatic vegetation in clear water streams using low-altitude optical remote sensing." *Sensors* 15: 25287–25312. doi: 10.3390/s151025287

Wells, C.L., Pigliucci, M., 2000. "Adaptive phenotypic plasticity: The case of heterophylly in aquatic plants." *Perspectives in Plant Ecology, Evolution and Systematics* 3: 1–18. doi: 10.1078/1433-8319-00001

Wood, K.A., O'Hare, M.T., McDonald, C., Searle, K.R., Daunt, F., Stillman, R.A., 2017. "Herbivore regulation of plant abundance in aquatic ecosystems." *Biological Reviews* 92: 1128–1141. doi: 10.1111/brv.12272

Yamauchi, T., Shimamura, S., Nakazono, M., Mochizuki, T., 2013. "Aerenchyma formation in crop species: A review." *Field Crops Research* 152: 8–16. doi: 10.1016/j.fcr.2012.12.008

Yuan, Y.V., Bone, D.E., Carrington, M.F., 2005. "Antioxidant activity of dulse (*Palmaria palmata*) extract evaluated *in vitro*." *Food Chemistry* 91: 485–494. doi: 10.1016/j.foodchem.2004.04.039

Zerrifi, S.E.A., Khalloufi, F. El, Oudra, B., Vasconcelos, V., 2018. "Seaweed bioactive compounds against pathogens and microalgae: Potential uses on pharmacology and harmful algae bloom control." *Marine Drugs* 16: 1–22. doi: 10.3390/md16020055

Zezhong, T., 2013. "*Nymphaea pubescens*." Obtained from Wikimedia under the Creative Commons Attribution 2.0 Generic license. https://commons.wikimedia.org/wiki/File:Nymphaea_pubescens_(9149720913).jpg (Accessed 30 June 2021)

Zhao, M., Chen, J.Y., Xu, L.J., Goedecke, T., Zhang, X.Q., Duan, J.A., Che, C.T., 2012. "cis-aconitic anhydride ethyl ester and phenolic compounds from the seeds of *Alisma orientale*." *Natural Product Communications* 7: 785–787. doi: 10.1177/1934578x1200700624

Ziarnek, K., 2012. "*Alisma*." Obtained from Wikimedia under the Creative Commons Attribution-Share Alike 4.0 International license. https://commons.wikimedia.org/wiki/File:Alisma_plantago-aquatica_kz04.jpg (Accessed 30 June 2021)

Index

B

C

I

N

O

P

Q

R

S

U

V

W

Y